AF831492

Advances in Intelligent Systems and Computing

1453

The series "Advances in Intelligent Systems and Computing" contains publications on theory, applications, and design methods of Intelligent Systems and Intelligent Computing. Virtually all disciplines such as engineering, natural sciences, computer and information science, ICT, economics, business, e-commerce, environment, healthcare, life science are covered. The list of topics spans all the areas of modern intelligent systems and computing such as: computational intelligence, soft computing including neural networks, fuzzy systems, evolutionary computing and the fusion of these paradigms, social intelligence, ambient intelligence, computational neuroscience, artificial life, virtual worlds and society, cognitive science and systems, Perception and Vision, DNA and immune based systems, self-organizing and adaptive systems, e-Learning and teaching, human-centered and human-centric computing, recommender systems, intelligent control, robotics and mechatronics including human-machine teaming, knowledge-based paradigms, learning paradigms, machine ethics, intelligent data analysis, knowledge management, intelligent agents, intelligent decision making and support, intelligent network security, trust management, interactive entertainment, Web intelligence and multimedia.

The publications within "Advances in Intelligent Systems and Computing" are primarily proceedings of important conferences, symposia and congresses. They cover significant recent developments in the field, both of a foundational and applicable character. An important characteristic feature of the series is the short publication time and world-wide distribution. This permits a rapid and broad dissemination of research results.

Indexed by DBLP, INSPEC, WTI Frankfurt eG, zbMATH, Japanese Science and Technology Agency (JST).

All books published in the series are submitted for consideration in Web of Science.

For proposals from Asia please contact Aninda Bose (aninda.bose@springer.com).

Nitin Naik · Paul Jenkins · Paul Grace ·
Longzhi Yang · Shaligram Prajapat
Editors

Advances in Computational Intelligence Systems

Contributions Presented at the 22nd UK Workshop on Computational Intelligence (UKCI 2023), September 6–8, 2023 Birmingham, UK

Springer

Editors
Nitin Naik
Aston University
Birmingham, UK

Paul Jenkins
Cardiff Metropolitan University
Cardiff, UK

Paul Grace
Aston University
Birmingham, UK

Longzhi Yang
Northumbria University
Newcastle upon Tyne, UK

Shaligram Prajapat
Devi Ahilya University
Indore, India

ISSN 2194-5357 ISSN 2194-5365 (electronic)
Advances in Intelligent Systems and Computing
ISBN 978-3-031-47507-8 ISBN 978-3-031-47508-5 (eBook)
https://doi.org/10.1007/978-3-031-47508-5

This Springer imprint is published by the registered company Springer Nature Switzerland AG
The registered company address is: Gewerbestrasse 11, 6330 Cham, Switzerland

Contents

Machine Learning

Federated Learning

An Introduction to Federated Learning: Working, Types, Benefits and Limitations

Dishita Naik[1] and Nitin Naik[2](✉)

[1] Birmingham City University, Birmingham, UK
dishita.naik@mail.bcu.ac.uk
[2] School of Computer Science and Digital Technologies, Aston University, Birmingham, UK
n.naik1@aston.ac.uk

Abstract. Machine learning has been constantly evolving and revolutionizing every aspect of our lives. There is ongoing research to enhance and modify machine learning models where scientists and researchers are finding ways to improve the effectiveness and adaptability of models with the changing technology moulding to user requirements for real life application. The main challenges in this endeavour of enhancing machine learning models are obtaining quality data, selecting an appropriate model, and ensuring the data privacy. Federated learning has been developed to address the aforementioned challenges, which is an effective way to train machine learning models in a collaborative manner by using the local data from a large number of devices without directly exchanging their raw data whilst simultaneously delivering on model performance. Federated learning is not just a type of machine learning, it is an amalgamation of several technologies and techniques. To fully understand its concepts a comprehensive study is required. This paper aims to simplify the fundamentals of federated learning in order to provide a better understanding of it. It explains federated learning in a step-by-step manner covering its comprehensive definition, detailed working, different types, benefits and limitations.

Keywords: Distributed machine learning (DML) · Federated learning (FL) · Federated machine learning (FML) · Centralized federated learning · Decentralized federated learning · Cross-device federated learning · Cross-silo federated learning · Horizontal federated learning · Vertical federated learning

1 Introduction

Machine learning (ML) is a field of artificial intelligence (AI) that allows systems to learn from data, identify patterns, and make logical decisions and predictions with little to no human intervention. The effectiveness of a machine learning model is dependent on two major aspects: the type and quality of the

N. Naik et al. (Eds.): UKCI 2023, AISC 1453, pp. 3–17, 2024.
https://doi.org/10.1007/978-3-031-47508-5_1

data as well as the choice of the model itself [5]. The better the machine learning model, the more accurately it can make decisions and predictions. ML has been progressively implemented in a distributed manner to harness real-time, high-quality data generated by a large number of end-user devices. This gradual evolution has changed the several features of machine learning such as moving centralized data to distributed data, shifting the computational load to end-user devices, significantly improving the model in a collaborative manner, and applying strict data privacy. A distributed machine learning model that is comprised of the aforementioned features is known as federated learning (FL) or federated machine learning (FML).

In FL or FML, a collaborative learning is performed by a large number of nodes utilising their local data, while the central server simply acts as an aggregator or coordinator for accumulating all the learning performed by individual nodes. The server normally sends the current machine learning model to all the nodes, where each node implements the sent model, trains this model on their local data, and sends the updated model (i.e., updated parameters or weights) to the central server. The central server then aggregates all the received model updates in order to produce an improved and consolidated global model, which is then sent to all the nodes again. This is an iterative process that enhances the machine learning model through collaborative learning, while keeping the training data locally without exchanging it with the central server. Therefore, in FL, the data does not move to the model, but it is the model that moves to the data. As this is an evolving area of machine learning, it is important to simplify and understand the fundamentals of federated learning. Therefore, this paper explains federated learning in a step-by-step manner covering its comprehensive definition, detailed working, different types, benefits and limitations.

The rest of the paper is structured as follows: Sect. 2 explains federated learning and its comprehensive definition. Section 3 presents the step-by-step working of federated learning. Section 4 describes the different types of federated learning. Section 5 elucidates several benefits and limitations of federated learning. Section 6 concludes the paper and suggests some future work.

2 What Is Federated Learning (FL)?

Federated machine learning is a type of distributed machine learning that further decentralizes learning operations using the local data on each participating node. Generally, in this type of machine learning, a large number of nodes collaboratively train a machine learning model under the orchestration of a central server, while keeping the raw data decentralized without being moved to a centralized location. The term federated learning was introduced by McMahan et al. to describe this type of collaborative learning [3]. Federated learning is also known as federated machine learning, collaborative learning or decentralized learning.

In federated learning, the central server acts as an assistant or aggregator that coordinates all the nodes to work together instead of controlling all the operations unlike traditional DML. It sends the current machine learning model to all the nodes, where each node implements the sent model, trains this model on their local data, and sends the updated local model (i.e., updated parameters or weights) to the central server. The central server then aggregates all the received model updates in order to produce the improved and consolidated global model, which is then sent to all the nodes again. This is an iterative process that enhances the machine learning model through collaborative learning, while keeping the training data locally without exchanging it with the central server.

In this type of machine learning, it is not that the data moves to the model, but the model moves to the data, hence the model is sent and trained locally on a large number nodes with their local data. Federated machine learning is mainly a combination of distributed computing, machine learning and privacy-preserving techniques. That provides enhanced models comprising of privacy-by-default, with several benefits such as lower latency, communication overhead, and power consumption [4]. It is well suited in scenarios where the on-device data is more relevant than the data that exists on the central location [2].

It not only optimises the machine learning process by utilising distributed resources efficiently, but also ensures the privacy of the decentralized raw data without revealing its sensitive information to the central server. This strong privacy guarantee makes federated machine learning a popular choice in several application areas where data breaches and information theft are common and serious threats. This also ensures that the data in each node adheres to data privacy policies and protects any data leaks or breaches. Additionally, another unique feature of federated machine learning is that it utilises the unbalanced and non-Independent and Identically Distributed (non-IID) data. Here, the unbalanced data means that the amount of data at each node can be very different depending on the usage and environment. The non-Independent and Identically Distributed (IID) data means that the type of data at each node can be very different depending on the usage and environment.

3 Working of Federated Learning

There are different categories of federated learning and their working may be slightly different based on their design and implementation. However, the working of a generic federated learning model can be explained in the following steps: model selection, model distribution, model training, model aggregation, and model iteration.

3.1 Step 1 - Model Selection

A baseline machine learning model is selected at the central server to train in a federated manner as shown in Fig. 1. The selected machine learning model is either pre-trained on the central server or not trained at all.

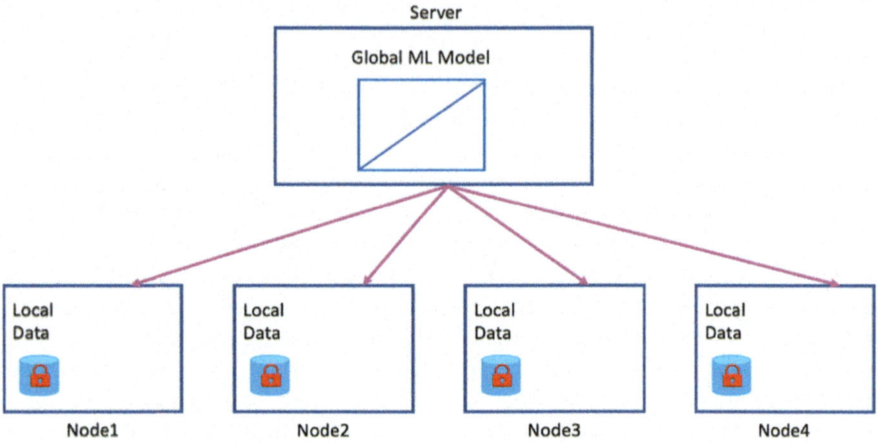

Fig. 1. Step 1 - Model selection at central server

3.2 Step 2 - Model Distribution

The copies of the baseline machine learning model are distributed to all the selected nodes (e.g., computers, smartphones, IoT devices or local servers) to be trained locally as shown in Fig. 2. The selection criteria for nodes could be different for different learning networks, such as sufficient data, processing power, processing memory, and battery life.

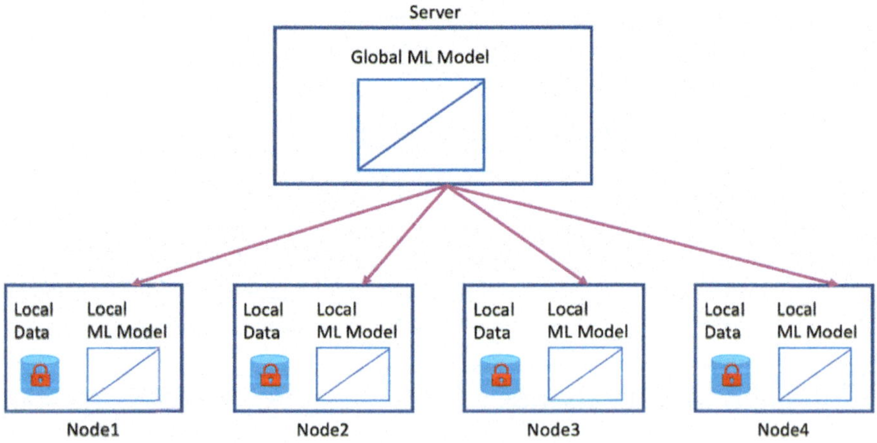

Fig. 2. Step 2 - Model distribution to selected nodes

3.3 Step 3 - Model Training

Each selected node then trains the model using the local data or its subset which they generate on-site as shown in Fig. 3. The local data is collected on different nodes from diverse sources; therefore, it broadens the training scope of the machine learning model. The central server does not have direct access to the local training data of any node, nor is it able to process it at its end. It brings the model to the data, rather than bringing the data to the model, therefore, it enables multiple nodes to train a model while satisfying the legal data restrictions. As a result of this changing learning style, this approach helps to address the fundamental problems of privacy, ownership, and locality of data.

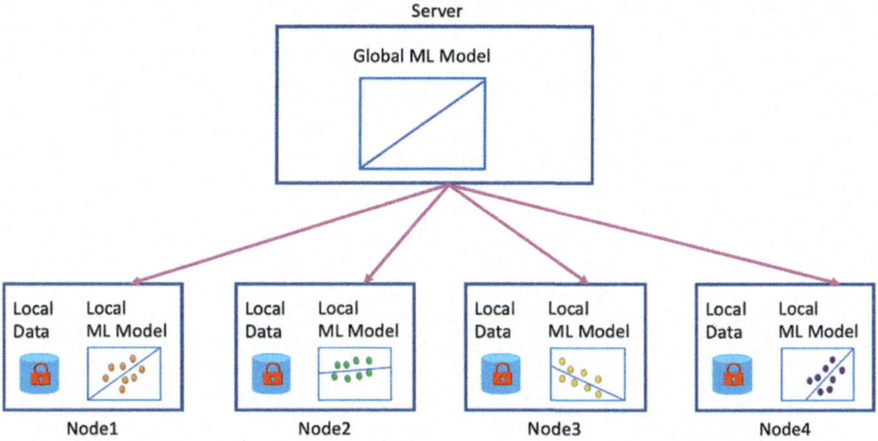

Fig. 3. Step 3 - Model training on selected nodes using local data

3.4 Step 4 - Model Aggregation

The model updates (e.g., model parameters or weights) from the locally trained models on the nodes are uploaded to the central server, where these model updates are aggregated using the secure aggregation technique in order to produce a federated global model as shown in Fig. 4. Note all the training data remains on the nodes, and the locally trained model summarizes their changes as a small focused update, and only this model update is sent to the central server, thus maintaining the privacy of data. Here the straggler nodes might be dropped once a sufficient number of nodes have reported their model updates in the current iteration. Several aggregation techniques are available to perform the aggregation task, for example, one technique to combine the model updates is to take the average of each coefficient, weighting by the amount of training data available on the corresponding node.

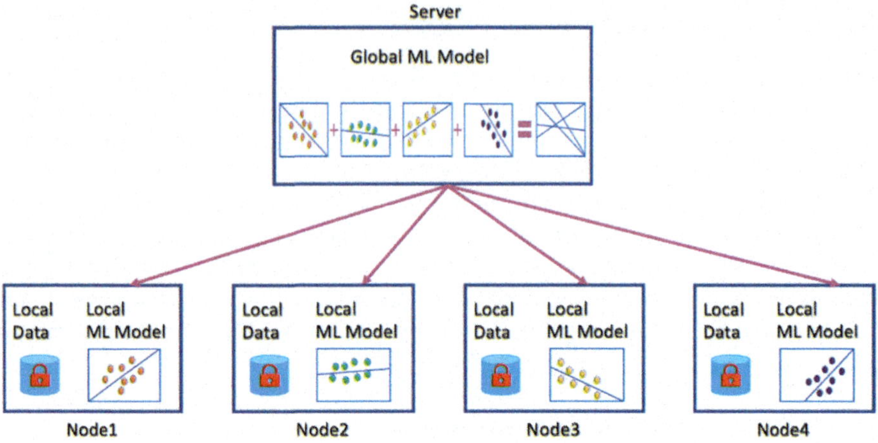

Fig. 4. Step 4 - Model aggregation at central server

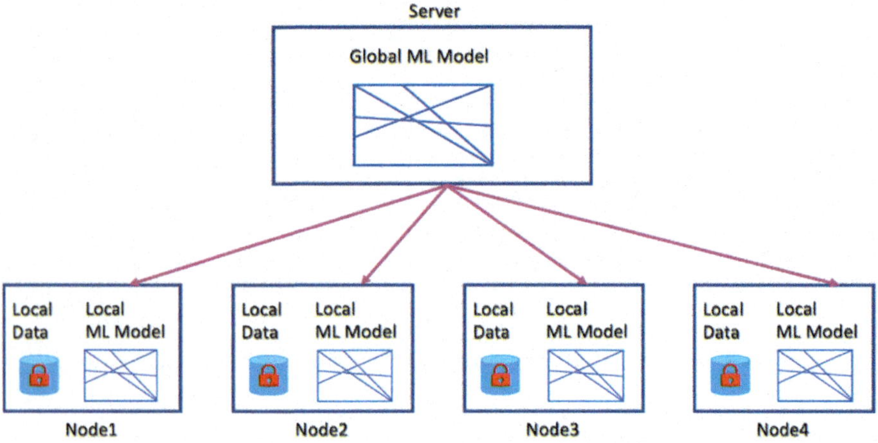

Fig. 5. Step 5 - Model training iteration

3.5 Step 5 - Model Iteration

The aggregated global model is sent again to all the selected nodes for the next iteration as shown in Fig. 5; and the learning process is repeated if required. It is important to note that in each iteration, the nodes can acquire new training data, and some nodes may leave, and others may join, which may help to further generalize the model. With every iteration, the model is further refined and improved based on changing nodes and data while maintaining the data privacy. Over time, the models on individual nodes become personalized and provide a better user experience.

4 Types of Federated Learning

Federated learning can be classified into several different ways depending on various factors such as system architecture, federation scope and data partition [1]. Here it is classified into three types: federated learning based on system architecture, federated learning based on federation scope, and federated learning based on data partition.

4.1 Federated Learning Based on System Architecture

Federated learning can be implemented based on two different types of system architecture: centralized architecture and decentralized architecture [1]. The difference between these two types of architecture is based on the type of nodes and their roles in the learning process.

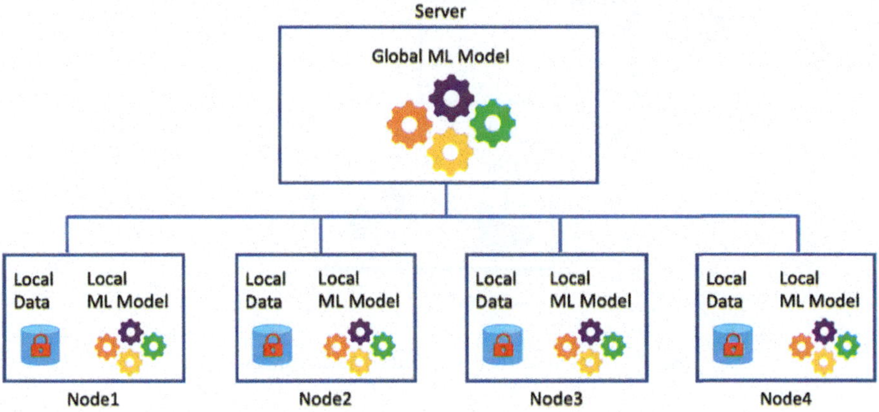

Fig. 6. Centralized federated machine learning

Centralized Federated Machine Learning In centralized federated learning, the learning process is orchestrated by a central server that holds a global ML Model as shown in Fig. 6. All the other nodes perform model training on their local data, then updates of the local model are sent to the central server. The central server aggregates all the local updates in order to produce an improved global model. Here, only the local model update is sent to the central server, however, no individual model updates are stored in the central location, whilst the local training data remains preserved on the local nodes. The communication between the central server and all the other local nodes can be synchronous or asynchronous.

The central server is crucial for the learning process, it should be powerful as well as reliable. However, this central server can pose a bottleneck problem as it is a single point of failure due to several reasons such as network failures, hardware failures, software problems, which can affect the collaborative learning process. Another recurring problem with this architecture is the traffic congestion due to high load or unexpected demand on the central server, when too many nodes communicate with the same server.

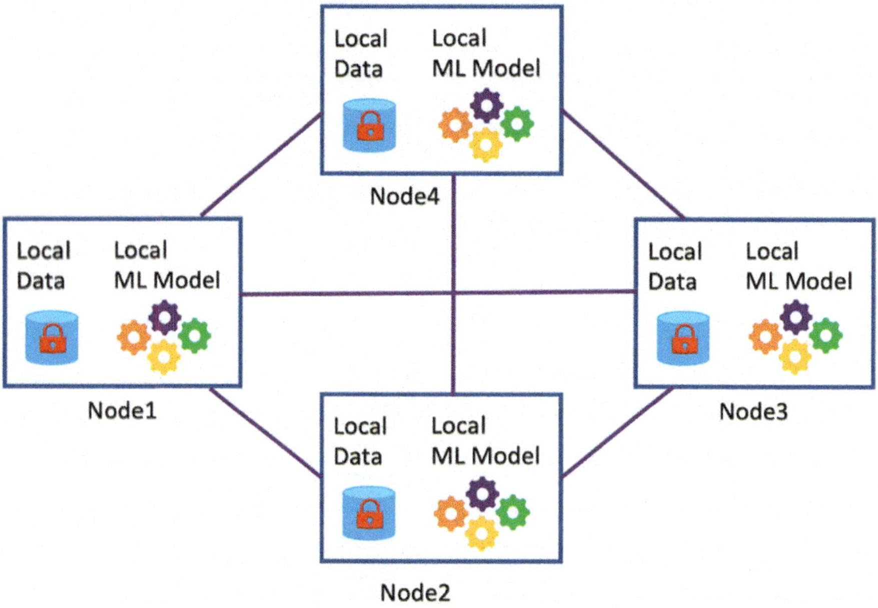

Fig. 7. Decentralized federated machine learning

Decentralized Federated Machine Learning In decentralized federated learning, the learning process is not orchestrated by a central server, instead all the nodes coordinate with each other in order to perform the learning process and update the global model without requiring a dedicated server as shown in Fig. 7. Here, each node performs a model training on their local data, updates their local model and exchanges the update with their neighbours in the network in order to produce the improved global model. The model training process and the accuracy of the model is dependent on the network topology and the global model update method.

The decentralized federated learning removes dependency on the central server, and replaces communication with the server by peer-to-peer communication between individual nodes, which prevents the possibility of a single point of failure. However, the design of the decentralized architecture is complex and challenging, and incurs significant communication overhead due to a large number of nodes involved in the learning process. Additionally, despite the decentralized architecture, sometimes, a central authority may still be in charge of setting up the learning task.

4.2 Federated Learning Based on Federation Scope

Federated learning can be implemented based on two different scopes of federation: cross-device and cross-silo [1]. The difference between these two scopes of federation is based on the type and number of users or devices and the amount of data involved in the learning process.

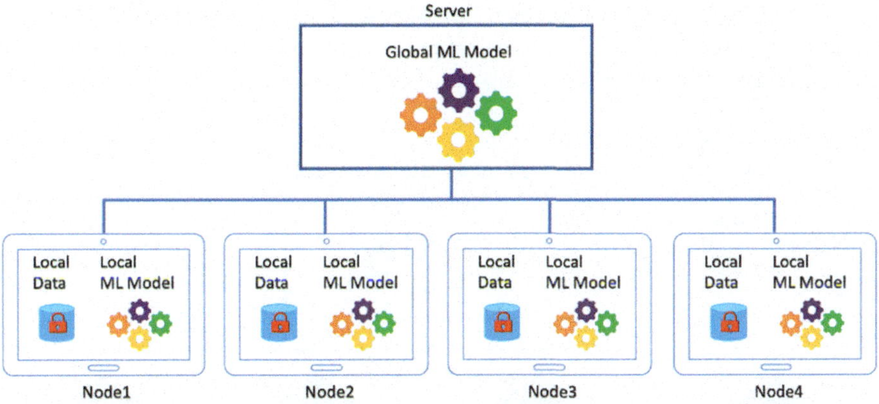

Fig. 8. Cross-device federated learning

Cross-Device Federated Learning In a cross-device federated learning, the learning process is based on the local data of the large number of end-user devices in the network as shown in Fig. 8. These end-user devices are generally, computers, smartphones, and IoT devices, which act as the source of data to train the model locally as shown in Fig. 8. Generally, the model learning process is similar to centralized federated learning which is orchestrated by a central server. All the end-user devices perform training of the local model on their local data, update the local model and send it to the central server. The central server aggregates all the local updates received from the end-user devices in order to produce the improved global model.

This type of federated setup requires millions of devices to provide effective training required for the global model due to the several limitations of the end-user devices such as offline devices and insufficient data. One of the best examples of this type of federated learning is the Google's GBoard, that is the next-word prediction model which uses this cross-device federated setup.

Cross-Silo Federated Learning In a cross-silo federated learning, the learning process is based on the local data of the selected organisations which form the learning network as shown in Fig. 9. A silo is an isolated data storage place for an organisation which contains raw data with the restricted access within that organisation as shown in Fig. 9. Consequently, this data is not readily available for usage or further processing to the outside network. These silos are utilised collaboratively and act as a source of data by which the model can be trained locally. Wherein organizations have a common goal and incentive to train a model based on their data without sharing it directly and keeping their raw data separate in their silos.

In this type of learning process, the number of nodes is normally small but possess significant data and computational power. It is a more flexible design but challenging to implement and distribute computational resources across organisations effectively under the constraint of the privacy framework.

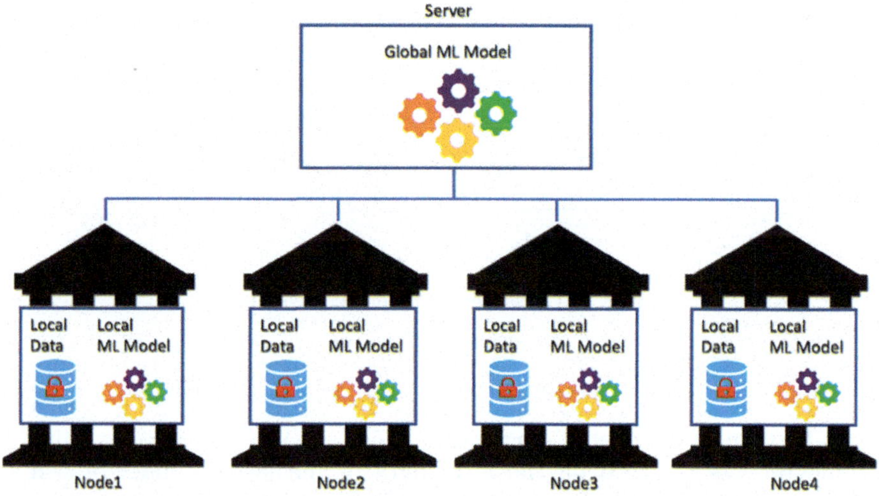

Fig. 9. Cross-silo federated learning

4.3 Federated Learning Based on Data Partition

Federated learning can be implemented based on two different types of data partitioning: horizontal and vertical [6]. The difference between these two types of data partitioning is based on the feature space and sample space involved in the learning process. Feature space refers to the collection of features that are used to characterise the data in the learning network. Whereas sample space refers to the collection data samples provided by users (i.e., end-user devices or silos) in the learning network.

Horizontal Federated Learning Horizontal federated learning uses the data with the same feature space but different sample spaces across all nodes in the learning network to collaboratively train a global model as shown in Fig. 10. It primarily deals with nodes having a homogenous set of data meaning that the use of the same features across all nodes in the learning network as shown in Fig. 10; therefore, this type of federated learning is also known as homogenous federated learning.

This type of federated learning is commonly used for the cross-device setting, where different nodes improve the model performance on the task related to the same features. Therefore, nodes can train the local models using their local data with the same model architecture. Finally, the global model can simply be updated by averaging all the local model updates.

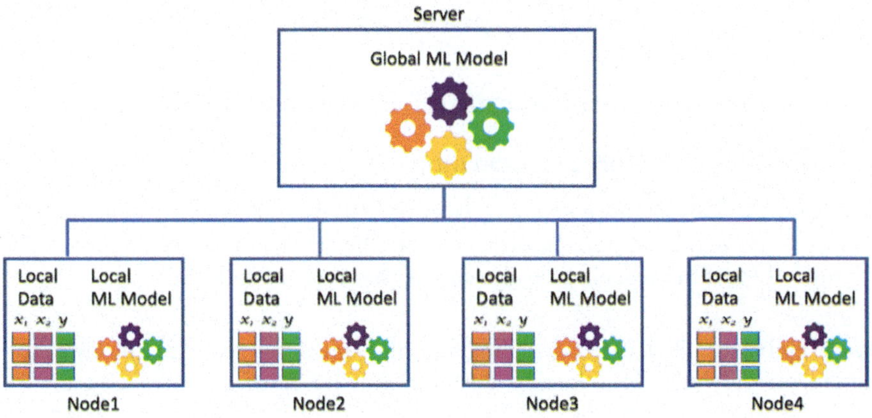

Fig. 10. Horizontal federated learning

Vertical Federated Learning Vertical federated learning uses the data with different feature spaces but with the same sample space across all the nodes in the learning network to collaboratively train a global model as shown in Fig. 11. It primarily deals with nodes having a heterogeneous set of data meaning that the use of different features across all nodes in the network as shown in Fig. 11; therefore, this type of federated learning is also known as heterogeneous federated learning.

This type of federated learning is commonly used for the cross-silo setting, where different nodes improve the model performance on the task related to the same sample space. Therefore, nodes can train the local models using their local data with the specific model architecture depending on the specific requirements. Finally, the improved global model can be obtained by combing all the local model updates.

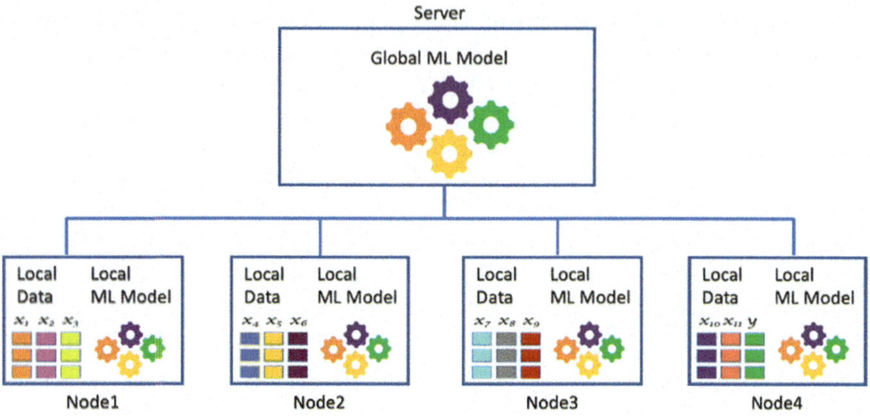

Fig. 11. Vertical federated learning

5 Benefits and Limitations of Federated Learning

5.1 Benefits of Federated Learning

Federated learning is more secured form of distributed machine learning which leverages decentralized data in order to provide several benefits over traditional centralized machine learning, and some of them are as follows:

Data Privacy Federated learning enables machine learning models to be trained on personal data locally without moving it to a centralized location, and it leverages secure aggregation of local model updates to maintain users' updates privately. The transmitted model updates consist minimal information to improve the accuracy of a machine learning model. The updates themselves can be brief, and will never contain more information than the raw training data. Therefore, the server cannot determine the value or source of the individual model updates that the users provide. This minimises the risk of personal data leakage and reduces the possibility of privacy related attacks such as inference and data attribution attacks.

Data Security In federated learning, nodes only share encrypted model updates with the central server without sending their data to the server, ensuring the security of model updates and data. Moreover, secure aggregation techniques are used to aggregate local model updates in order to improve the global model which may require the decryption of only aggregated results. This protects the data from unauthorized access and ensures the confidentiality and integrity of data.

Data Sovereignty In federated learning, the device owner who owns the data have the full sovereign control of their data and personal information, and it is not controlled by any other party in the learning network. The model owner can only train their model on the user data but cannot own or control it. The data owner is able to access, update, share, hide, or delete their data and personal information without necessarily notifying the model owner. This data sovereignty is crucial in several application areas, such as medical, financial and government organisations.

Data Diversity In federated learning, the training data is significantly diverse due to the variations in participating users' attributes such as age, ethnicity, gender and nationality. Additionally, the data comes in different formats and languages due to different devices, locations, and organizations. The model training on this diverse data from a variety of sources can enhance the model in such a way that can generalise new data, handle variations, and reduce bias.

Scalability In federated learning, the data is decentralized and maintained on multiple nodes, and model training is performed on the local data on these nodes simultaneously. Therefore, depending on the training requirements, the number of training nodes can be increased in order to improve the scalability of the training process.

5.2 Limitations of Federated Learning

Federated learning offers several benefits as discussed earlier; however, it has some limitations as well which are as follows:

Communication Overheads Federated learning normally involves millions of nodes in one learning network, where nodes iteratively send model updates or small messages as part of the distributed learning process rather than sending the data over the network. The communication overheads for bringing the models to the device should be moderately low for the success of federated learning, otherwise it may impact the federated learning process negatively. However, the communication may be affected due to several factors such as low bandwidth, lack of resources, or geographical location. This can be resolved by minimising the size of transmitted model updates in each iteration or reducing the number of iterations.

Privacy Concerns In federated learning, the local data never leaves the node and only model updates are sent to the central server; however, sharing model updates can also potentially reveal sensitive information, either to the central server or a third party. For example, an internal agent may be an adversary participating in the training process who can influence the model updates, or an external agent can only observe the learning and update process but is still able to make inferences that can compromise the data privacy. Consequently, federated learning is still vulnerable to many attacks such as information leaks, backdoor attacks, model poisoning attacks and inference attacks.

Several privacy techniques are employed in FL to solve issues regarding privacy, and some of the common privacy-preserving techniques are: differential privacy, homomorphic encryption, secure multiparty computation. However, these techniques usually provide privacy at the cost of reduced model performance or system efficiency, and balancing these trade-offs is a considerable challenge in implementing federated learning systems. Therefore, such privacy techniques have to be computationally economical, communication-efficient, and tolerant to dropped devices.

Systems/Nodes Heterogeneity Federated learning normally involves millions of heterogeneous systems/nodes with differing computational capabilities, storage, and communication capabilities. Additionally, a small number of nodes may be active at any particular time of learning, and a large number of nodes may not be available or reliable due to connectivity or energy constraints. Therefore, providing an effective and unbiased model training in FL is a significant challenge. Such heterogeneities in nodes can be handled using various techniques such as asynchronous communication, active device sampling, and fault tolerance.

Data/Statistical Heterogeneity Federated learning normally involves millions of heterogeneous nodes with differing types and size of data, i.e., the unbalanced and non-Independent and Identically Distributed (non-IID) data, which is in contrast with the assumption of IID data in traditional machine learning. This may increase the complexity of the learning process with respect to data structuring, modelling, analysis and inferencing. Additionally, the non-identical distribution of the data at each node may introduce a bias that may produce lower accuracy compared to a centralized dataset. Such heterogeneities in data can be handled using various heterogeneity-aware optimization techniques.

6 Conclusion

This paper presented the fundamentals of federated learning in order to provide a better understanding of it. It explained the comprehensive definition of federated learning, its working steps, different types, benefits and limitations. The working of a generic federated learning model was explained in the following steps: model selection, model distribution, model training, model aggregation, and model iteration. The paper described different types of FL: centralized FL, decentralized FL, cross-device FL, cross-silo FL, horizontal FL and vertical FL. In future, it would be worthwhile to conduct a practical analysis of different types federated machine learning and privacy techniques.

References

1. Kairouz, P., McMahan, H.B., Avent, B., Bellet, A., Bennis, M., Bhagoji, A.N., Bonawitz, K., Charles, Z., Cormode, G., Cummings, R., et al.: Advances and open problems in federated learning. Found. Trends Mach. Learn. **14**(1–2), 1–210 (2021)
2. Liu, J., Huang, J., Zhou, Y., Li, X., Ji, S., Xiong, H., Dou, D.: From distributed machine learning to federated learning: a survey. Knowl. Inf. Syst. **64**(4), 885–917 (2022)
3. McMahan, B., Moore, E., Ramage, D., Hampson, S., Arcas, B.A.: Communication-efficient learning of deep networks from decentralized data. In: Artificial Intelligence and Statistics, pp. 1273–1282. PMLR (2017)
4. McMahan, B., Ramage, D.: Federated learning: collaborative machine learning without centralized training data (2017). https://ai.googleblog.com/2017/04/federated-learning-collaborative.html
5. Naik, D., Naik, N.: The changing landscape of machine learning: a comparative analysis of centralized machine learning, distributed machine learning and federated machine learning. In: UK Workshop on Computational Intelligence (UKCI). Springer (2023)
6. Verbraeken, J., Wolting, M., Katzy, J., Kloppenburg, J., Verbelen, T., Rellermeyer, J.S.: A survey on distributed machine learning. ACM Comput. Surv. (CSUR) **53**(2), 1–33 (2020)

The Changing Landscape of Machine Learning: A Comparative Analysis of Centralized Machine Learning, Distributed Machine Learning and Federated Machine Learning

Dishita Naik[1(✉)] and Nitin Naik[2]

[1] Birmingham City University, Birmingham, UK
dishita.naik@mail.bcu.ac.uk
[2] School of Computer Science and Digital Technologies, Aston University, Birmingham, UK
n.naik1@aston.ac.uk

Abstract. The landscape of machine learning is changing rapidly due to the ever-evolving nature of data and devices. The large centralized data is replaced by the distributed data and a central server is replaced with a large number of geographically distributed, loosely connected devices, such as smartphones, laptops, and other IoT devices. Therefore, the centralized machine learning (CML) which involves centralized data training on a central server is no longer an effective solution when the data is inherently distributed or too big to process on a central server, or data privacy is paramount; and the quest for a suitable machine learning to resolve these issues led to the evolution of distributed machine learning (DML). For large-scale learning tasks, DML has evolved to effectively handle enormous data within big data and distributed computing environment. Resolving most limitations faced by CML with the implementation of parallel learning on a large number of nodes to optimise time, learning resources and performance. However, DML may not necessarily ensure strict data privacy leading to further development and innovation of federated machine learning (FML) which is a type of DML that further decentralizes learning operations using local data on each participating node incorporating data privacy adherence. This paper analyses the transformation journey of machine learning whilst explaining its evolution from centralized, distributed to federated machine learning. Examining these three variants of machine learning exemplifies their coherent and comparative analysis. Which helps grasp a better understanding of each machine learning type as well as presenting the reason for the changing landscape. Additionally, the paper will address each type of machine learning alongside their different types, strengths and limitations.

Keywords: Centralized machine learning (CML) · Distributed machine learning (DML) · Data-parallel distributed machine learning ·

N. Naik et al. (Eds.): UKCI 2023, AISC 1453, pp. 18–28, 2024.
https://doi.org/10.1007/978-3-031-47508-5_2

Model-parallel distributed machine learning · Federated machine learning · Federated learning · FML · FL · Centralized federated machine learning · Decentralized federated machine learning

1 Introduction

Machine learning (ML) is a data analysis technique that automates the building of analytical models using data to make decisions and predictions to continuously improve the accuracy of models and outputs. Incessant technological advancements have progressed machine learning to accommodate the changing computing environments. Two important changes in computing which transformed the machine learning landscape are distributed computing environment and privacy of data. The distributed computing environment has replaced the large centralized data with the distributed data, and a central server is replaced with a large number of geographically distributed, loosely connected devices, such as smartphones, laptops, and other IoT devices [2]. The strict privacy requirement for data has enabled the use of the decentralized data on the end-user device without revealing their sensitive information.

Centralized machine learning (CML) or traditional machine learning involves centralized data training on a central server, where all data is collected into one centralized location and the entire model is trained on the central server. However, CML is no longer an effective solution when the data is inherently distributed or too big to process on a central server, or data privacy is paramount [8]. This limitations of CML led to the evolution of distributed machine learning (DML).

DML has evolved to handle large-scale learning on the enormous data in big data and distributed computing environment. It has resolved most of the limitations of CML by performing parallel learning on a large number of nodes to optimise the time, learning resources and performance. However, DML may not necessarily ensures strict data privacy. This limitation of DML led to the evolution of federated machine learning (FML).

Federated machine learning is a type of DML that further decentralizes learning operations using the local data on each participating node. Generally, in this type of machine learning, a large number of nodes collaboratively train a machine learning model under the orchestration of a central server, while keeping the raw data decentralized without being moved to a central server and ensuring strict data privacy.

This paper will analyse this transformation journey of machine learning and explain the evolution of centralized, distributed and federated machine learning. It will examine these three variants of machine learning to provide their coherent and comparative analysis, whilst illustrating each type of machine learning and its contribution to the changing landscape in depth. Additionally, this paper will discuss types, strengths and limitations of each type of machine learning.

The rest of the paper is structured as follows: Sect. 2 explains centralized machine learning (CML) and its limitations. Section 3 presents distributed machine learning (DML), its types and limitations. Section 4 describes federated machine learning (FML), its types and limitations. Section 5 provides the comparative analysis of CML, DML and FML. Section 6 presents conclusion and future work.

2 Centralized Machine Learning (CML)

Centralized machine learning or traditional machine learning involves centralized data training on a central server. In this type of machine learning, the complete data is collected into one centralized location and the entire model is trained on a central server as shown in Fig. 1. Generally, the data is collected from multiple sources into one centralized location, and this central location can be a data warehouse, data lake, or the combination of both, i.e., lakehouse. The role of the central server is to combine all the data from all the nodes, extract the features from it, and train a machine learning model using the centralized data. Subsequently, this developed centralized model can be used on the central server itself or distributed across all other nodes in the network as shown in Fig. 1. Since the complete learning task is performed on the central server, therefore, it requires powerful processors and significant memory for the learning process. However, other nodes only perform inference or prediction tasks using the trained model and do not require similar power or significant resources as compared to the central server.

The centralized machine learning has several advantages, such as simplicity, consistency, affordability and efficiency. Unlike, distributed and federated machine learning, it does not require distributed communication and coordination among nodes for training the machine learning model, which saves significant communication overheads. However, centralized machine learning also has several limitations, such as limited scalability, adaptability, privacy, and robustness [3]. Additionally, it relies on a centralized data and server, therefore, it is prone to a single point of failure (SPOF) and may violate the laws on user privacy and data confidentiality.

3 Distributed Machine Learning (DML)

Distributed machine learning has evolved to handle large-scale learning on the enormous data in big data and distributed computing environment. When working with big data, a vast amount of data is used to train the machine learning model daily, which may exponentially increase the training time and resources, and affect the overall performance of the model. Therefore, distributed machine learning systems can be built and training can be parallelized on a large number of nodes, allowing for optimised conditions. Distributed machine learning is mainly a combination of distributed computing and machine learning. It makes use of several processing nodes to overcome the limitations of centralized machine learning. In this type of machine learning, a huge task is divided into subtasks and executed on several nodes in parallel in order to optimise the time, resources

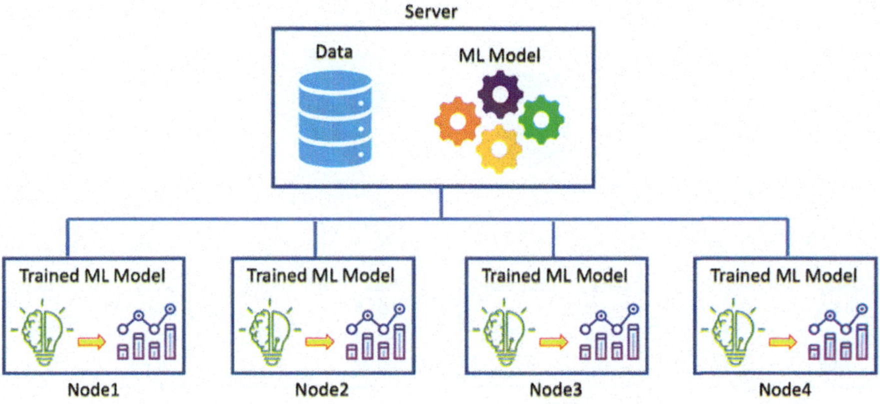

Fig. 1. Centralized machine learning

and performance. In distributed system, different nodes can be assigned different roles, generally, most nodes are employed as training or worker nodes, and one or more nodes can be used as a server. Generally, the central server (or parameter server) is responsible for most of the distributed management operations such as partitioning the model or/and the training data; scheduling of worker nodes; allocating subtasks to worker nodes; and aggregating the subtasks.

All worker nodes perform their allocated subtask in parallel and, when all nodes have completed their subtasks, the server (or parameter server) aggregates all the subtasks together and generates a complete task. All the worker nodes can complete their subtasks independently, however, if required, a worker node can also communicate or share necessary data with other worker nodes for completing a subtask due to their interdependence. Normally, the server can have access to the entire dataset in order to create its partitions irrespective of how the data is collected and managed. The distributed computing process requires significant communications between the server and worker nodes. Consequently, communication overheads may affect the performance of the distributed machine learning depending on the available networking resources. The training of the machine learning model can be performed either synchronously, where a node waits for all other nodes in the network to complete their task for that particular iteration, or asynchronously, where a node does not require to wait for any other node in the network to complete their task for that particular iteration. Distributed machine learning can be designed and implemented in two different ways either parallelizing the data or model; however, these two methods can also be applied simultaneously [6,8].

3.1 Data-Parallel Distributed Machine Learning

In data-parallel distributed machine learning, the data is partitioned into multiple smaller parts depending on the number of worker nodes, and subsequently,

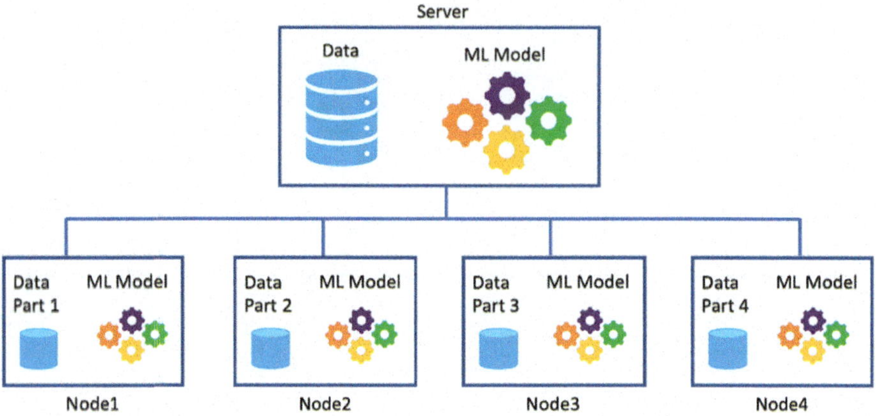

Fig. 2. Data-parallel distributed machine learning

the same model is applied to all the data partitions on the worker nodes as shown in Fig. 2. The same model is available to all the worker nodes either through centralization or through replication process, and each worker node operates on its own subset of data in order to produce the consistent output. However, it is crucial that each worker node has to have the capacity to support and run the model that it is being trained on it.

Each node can independently compute the outputs and errors, and update its model which is sent to the central server or across all other worker nodes to update their corresponding models depending on the learning requirement. Therefore, all the worker nodes require synchronization of their model parameters, or gradients after each iteration in order to ensure the consistent training of the model. Data-parallel DML is widely used and relatively easy to implement as compared to model-parallel DML.

3.2 Model-Parallel Distributed Machine Learning

In model-parallel distributed machine learning, the model is partitioned into multiple smaller parts, therefore, each part of the model can operate on different worker nodes with the same data as shown in Fig. 3. The model is divided either horizontally or vertically into different parts that can run concurrently on different worker nodes where each worker node runs the part of the model on the same data.

Consequently, the global model is the aggregate of all the parts of the model trained on all the individual worker nodes. The model-parallel DML cannot automatically be applied to every machine learning algorithm, because the model parameters cannot normally be split up. In model parallelism, worker nodes may require to synchronize the shared parameters in order to obtain the consistent global model. Model-parallel DML is used in rare situations when the model is

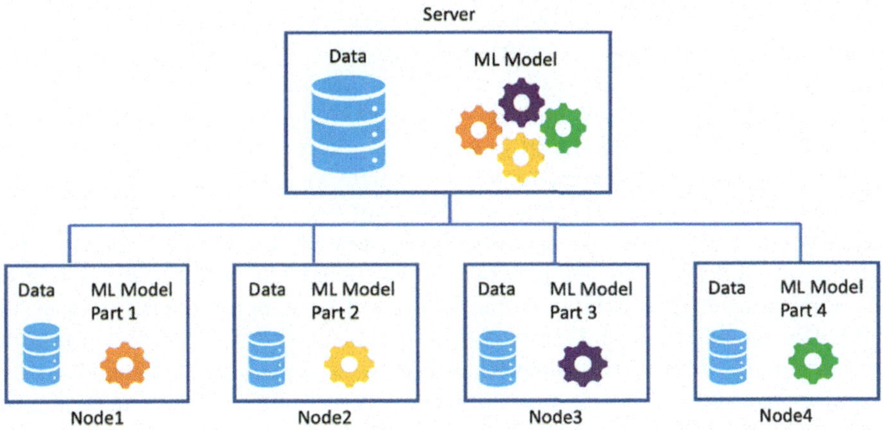

Fig. 3. Model-parallel distributed machine learning

too large and computationally complex for a single worker node. Also, model-parallel DML is relatively difficult to implement as compared to data-parallel DML.

4 Federated Machine Learning (FML)/Federated Learning (FL)

Federated machine learning is a type of distributed machine learning that further decentralizes learning operations using the local data on each participating node [7]. Generally, in this type of machine learning, a large number of nodes collaboratively train a machine learning model under the orchestration of a central server, while keeping the raw data decentralized without being moved to a centralized location [7]. The term federated learning was introduced by McMahan et al. to describe this type of collaborative learning [4]. Federated machine learning is also known as federated learning, collaborative learning or decentralized learning.

In federated learning, the central server acts as an assistant or aggregator that coordinates all the nodes to work together instead of controlling all the operations unlike traditional DML. It sends the current machine learning model to all the nodes, where each node implements the sent model, trains this model on their local data, and sends the updated local model (i.e., updated parameters or weights) to the central server. The central server then aggregates all the received model updates in order to produce the improved and consolidated global model, which is then sent to all the nodes again. This is an iterative process that enhances the machine learning model through collaborative learning, while keeping the training data locally without exchanging it with the central server.

In this type of machine learning, it is not that the data moves to the model, but the model moves to the data, hence the model is sent and trained locally on a large number nodes with their local data. Federated machine learning is mainly a combination of distributed computing, machine learning and privacy-preserving techniques. That provides enhanced models comprising of privacy-by-default, with several benefits such as lower latency, communication overhead, and power consumption [5]. It is well suited in scenarios where the on-device data is more relevant than the data that exists on the central location [3].

It not only optimises the machine learning process by utilising distributed resources efficiently, but also ensures the privacy of the decentralized raw data without revealing its sensitive information to the central server. This strong privacy guarantee makes federated machine learning a popular choice in several application areas where data breaches and information theft are common and serious threats. This also ensures that the data in each node adheres to data privacy policies and protects any data leaks or breaches. Additionally, another unique feature of federated machine learning is that it utilises the unbalanced and non-Independent and Identically Distributed (non-IID) data. Here, the unbalanced data means that the amount of data at each node can be very different depending on the usage and environment. The non-Independent and Identically Distributed (IID) data means that the type of data at each node can be very different depending on the usage and environment.

There are several ways to classify federated machine learning, however, here it is classified into two different types based on the system architecture: centralized architecture and decentralized architecture [1]. The difference between these two architectures is based on the types of nodes and their roles in the learning process.

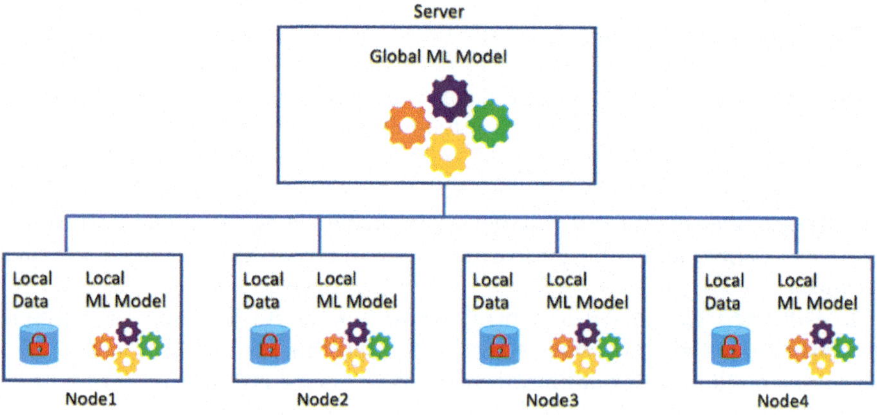

Fig. 4. Centralized federated machine learning

4.1 Centralized Federated Machine Learning

In centralized federated machine learning, the learning process is orchestrated by a central server that holds a global ML Model as shown in Fig. 4. All the other nodes perform model training on their local data, then updates the local model and sends it to the central server. The central server aggregates all the local updates in order to produce the improved global model. Here, only the local model update is sent to the central server, however, no individual model updates are stored in the central location, whilst the local training data remains preserved on the local nodes. The communication between the central server and all the other local nodes can be synchronous or asynchronous.

The central server is crucial for learning process, it should be powerful as well as reliable. However, this central server can pose a bottleneck problem as it is a single point of failure due to several reasons such as network failures, hardware failures, software problems, which can affect the collaborative learning process. Another recurring problem with this architecture is the traffic congestion due to high load or unexpected demand on the central server, when too many nodes communicate with the same server.

4.2 Decentralized Federated Machine Learning

In decentralized federated machine learning, the learning process is not orchestrated by a central server, instead all the nodes coordinate with each other in order to perform the learning process and update the global model without requiring a dedicated server as shown in Fig. 5. Here, each node performs a model training on their local data, updates their local model and exchanges the update with their neighbours in the network in order to produce the improved global model. The model training process and the accuracy of the model is dependent on the network topology and the global model update method.

The decentralized federated machine learning removes dependency on the central server, and replaces communication with the server by peer-to-peer communication between individual nodes, which prevents the possibility of a single point of failure. However, the design of the decentralized architecture is complex and challenging, and incurs significant communication overhead due to a large number of nodes involved in the learning process. Additionally, despite the decentralized architecture, sometimes, a central authority may still be in charge of setting up the learning task.

5 Comparative Analysis of Centralized Machine Learning, Distributed Machine Learning and Federated Machine Learning

This section aims to present a comparative analysis of three distinct types of machine learning CML, DML and FML evaluating factors such as data distribution, resource requirements, ML model training, communication overhead, fault tolerance, data privacy and scalability which mirrors the content depicted in the Table 1.

Table 1. Comparative analysis of centralized machine learning, distributed machine learning and federated machine learning

Criteria	Centralized machine learning	Distributed machine learning	Federated machine learning
Data distribution	Data is aggregated and stored on a centralized location.	Data is distributed across multiple nodes.	Data is distributed across multiple nodes.
Computing resources	Centralized server is employed, therefore, the computational load is managed by it, which requires substantial computing resources.	Distributed resources employed across several nodes, therefore, the computational load is distributed on these nodes, which reduces the resource requirements on each node; however, the central server still requires significant resources, as it deals with the data.	Distributed resources employed across several nodes, therefore, the computational load is distributed on these nodes, which reduces the resource requirements on each node; additionally, the central server also requires less computing resources, as it only deals with model updates and aggregation.
ML model training	ML Model is trained on a centralized server.	ML Model is trained on multiple nodes.	ML Model is trained on multiple nodes.
Communication overhead	Communication overhead is relatively low, as the model learning process is performed on the central server.	Communication overhead is the highest, as the data needs to be transmitted between nodes and the central server during the model learning process.	Communication overhead is relatively lower as compared to DML, because only model updates are shared between nodes and the central server, but not the data.
Fault tolerance	It relies on a central server and data repository, therefore, it is prone to a single point of failure, which can affect the entire learning process.	The data processing and model learning is spread across multiple nodes. Therefore, if some nodes fail, the learning process can still continue on the remaining nodes as long as sufficient number of participating nodes are available.	The data processing and model learning is spread across multiple nodes. Therefore, if some nodes fail, the learning process can still continue on the remaining nodes as long as sufficient number of participating nodes are available.
Data privacy	Centralized data contains all the sensitive information from multiple sources, which potentially increases the risk of data privacy, and may violate the laws on user privacy and data confidentiality.	Data distribution can provide some data privacy; however, it might not be enough to address strict privacy and regulatory requirements due to data movement and processing across different nodes.	Decentralised data and the appropriate privacy framework provide the privacy by default and are compliant with data protection regulations, where the data never leaves the individual nodes and only encrypted model updates are sent to the central server for aggregation.
Scalability	Limited scalability, as data processing and model learning is performed on the central server which requires significant resources and limits the scalability	Higher scalability than CML, as data processing and model learning is performed on multiple nodes which can be easily scaled	Highly scalable than other two categories, as each node can process its local data and train model on it, and the model aggregation is performed on the central server, which can be easily scaled

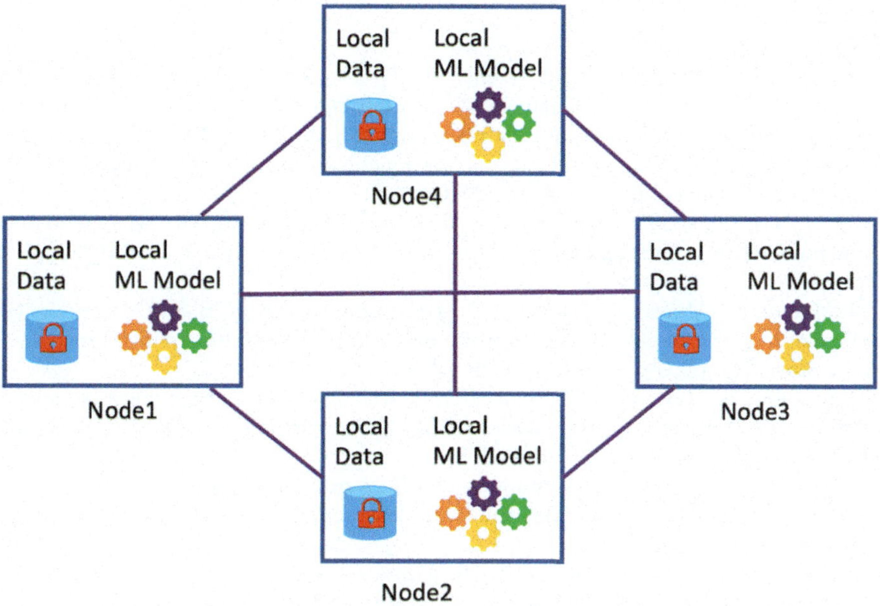

Fig. 5. Decentralized federated machine learning

6 Conclusion

This paper presented the transformation journey of machine learning and explained the evolution of centralized, distributed and federated machine learning. It examined these three variants of machine learning to provide their coherent and comparative analysis, where it illustrated each type of machine learning with its contribution to the changing landscape in depth. The paper discussed types, strengths and limitations of each type of machine learning. It explained two types of distributed machine learning, data-parallel and model-parallel; and two types of federated machine learning, centralized and decentralized. Finally, it presented a comparative analysis of these three types of machine learning based on several important criteria such as: data distribution, computing resources, ML model training, communication overhead, fault tolerance, data privacy, and scalability. In future, it would be worthwhile to conduct a practical analysis of these three types of machine learning.

References

1. Kairouz, P., McMahan, H.B., Avent, B., Bellet, A., Bennis, M., Bhagoji, A.N., Bonawitz, K., Charles, Z., Cormode, G., Cummings, R., et al.: Advances and open problems in federated learning. Found. Trends Mach. Learn. **14**(1–2), 1–210 (2021)
2. Kamp, M.: Black-box parallelization for machine learning. Ph.D. thesis, Universitäts-und Landesbibliothek Bonn (2019)
3. Liu, J., Huang, J., Zhou, Y., Li, X., Ji, S., Xiong, H., Dou, D.: From distributed machine learning to federated learning: a survey. Knowl. Inf. Syst. **64**(4), 885–917 (2022)
4. McMahan, B., Moore, E., Ramage, D., Hampson, S., Arcas, B.A.: Communication-efficient learning of deep networks from decentralized data. In: Artificial Intelligence and Statistics, pp. 1273–1282. PMLR (2017)
5. McMahan, B., Ramage, D.: Federated learning: collaborative machine learning without centralized training data. https://ai.googleblog.com/2017/04/federated-learning-collaborative.html (2017)
6. Microsoft.com: Distributed training with Azure Machine Learning. https://learn.microsoft.com/en-us/azure/machine-learning/concept-distributed-training?view=azureml-api-2 (2023)
7. Naik, D., Naik, N.: An introduction to federated learning: working, types, benefits and limitations. In: UK Workshop on Computational Intelligence (UKCI). Springer (2023)
8. Verbraeken, J., Wolting, M., Katzy, J., Kloppenburg, J., Verbelen, T., Rellermeyer, J.S.: A survey on distributed machine learning. ACM Comput. Surv. (CSUR) **53**(2), 1–33 (2020)

Arabic Sentiment Analysis with Federated Deep Learning

Mohammed Al-refai, Ahmad Alzu'bi(✉), Naba Bani Yaseen, and Taymaa Obeidat

Faculty of Computer and Information Technology, Jordan University of Science and Technology, Irbid 22110, Jordan
agalzubi@just.edu.jo

Abstract. The application of deep learning techniques in federated learning environments has shown remarkable performance across various domains. This has enabled the development of large-scale systems that enhance responsiveness, reduce processing costs and complexity, and maintain data privacy. In this research paper, we propose a federated deep learning model specifically designed for Arabic sentiment analysis using a Twitter-based benchmark dataset. Our approach leverages the effectiveness of fine-tuning the BERT model as a global learning model to extract discriminating embeddings from Arabic tweets. Through the efficient federated environment, we successfully learn the text patterns and train a classifier with the ability to accurately categorize tweets as positive, negative, or neutral. Despite the inherent complexity of Arabic language processing, extensive experiments were conducted to evaluate the performance of the federated approach in Arabic sentiment analysis. The results demonstrated significant advantages over centralized learning, particularly in terms of training time. Furthermore, our proposed model achieved a weighted average accuracy of 90% across various training and aggregation setups.

Keywords: Arabic sentiment analysis · Federated learning · Deep learning · Edge computing

1 Introduction

The proliferation of Internet-connected devices and the rapid growth of social media has led to an unprecedented generation of vast amounts of data across various fields and domains. However, traditional cloud computing approaches face significant challenges in effectively handling this massive volume of data. These challenges include high response times, security threats, and bandwidth limitations, which hinder their efficiency.

Edge computing refers to a distributed architecture that aims to bring computation and data storage closer to the sources of data [1–3]. This approach involves deploying smaller-scale computing resources at the network edge,

N. Naik et al. (Eds.): UKCI 2023, AISC 1453, pp. 29–38, 2024.
https://doi.org/10.1007/978-3-031-47508-5_3

which offers several benefits for different applications including the Internet of Things (IoT), mobile computing, autonomous vehicles, and augmented reality [2]. By moving computation closer to the data sources, edge computing enables faster response times, reduces network latency, minimizes bandwidth usage, and enhances privacy and security measures.

Social media sentiment analysis is a prevalent application that involves identifying and extracting subjective information, opinions, attitudes, and emotions from textual or multimedia data. Its significance has grown across various domains, including social media monitoring, customer feedback analysis, political campaigning, and brand reputation management. However, performing sentiment analysis on vast and diverse data streams poses considerable challenges in terms of scalability, efficiency, and accuracy [4]. Traditional cloud-based sentiment analysis approaches might encounter issues like high network latency, limited bandwidth, and privacy concerns, particularly in situations where real-time or near-real-time analysis is necessary [2].

To address these challenges, research efforts have put forward a range of edge computing frameworks and methodologies for sentiment analysis [3]. For instance, certain investigations have delved into employing edge devices equipped with GPUs or FPGAs to expedite sentiment analysis algorithms [5], some works have explored the amalgamation of edge and cloud resources for distributed sentiment analysis [6], whereas others have introduced machine learning algorithms tailored for edge devices, enabling sentiment analysis with the federated deep learning technique [7]. Federated learning is a decentralized strategy that enables several edges (devices) to be trained cooperatively without sharing their raw data, where user data privacy is preserved. The secrecy of individual data is ensured by only exchanging model updates or aggregated data rather than sending local data to a central server.

In this paper, we develop a federated deep learning system for the task of sentiment analysis of tweets written in the Arabic language. More specifically, we used BERT which has been pre-trained on very big and varied datasets to assist transfer learning on both Modern Standard Arabic (MSA) and Arabic dialects [8]. Our proposed methodology involves fine-tuning the BERT model using the Twitter-based benchmark dataset (ASAD) [9]. Unlike traditional centralized approaches, our approach distributes the dataset across multiple client devices, enabling local training without exposing sensitive data. The central server aggregates updates from all devices. Given the complexity of the Arabic language and the resource-intensive preprocessing steps involved, we aim to investigate the effectiveness of federated deep learning compared to the conventional centralized deep learning approach for analyzing Arabic tweets. To the best of our knowledge, this is the first study to apply the federated deep learning approach to Arabic sentiment analysis.

The rest of the paper is organized as follows. Section 2 reviews the related works, Sect. 3 illustrates the proposed methodology, Sect. 4 discusses the experimental results, and Sect. 5 concludes this study.

2 Related Work

Edge computing and deep learning are dynamic technologies that are continuously advancing and synergizing with each other in numerous ways. Researchers have put forward various approaches to facilitate the integration of these technologies. Ran et al. [10] developed a measurement-driven framework, DeepDecision, to make smart decisions under variable network conditions. This framework allows deep learning to be executed locally or remotely in the cloud or edge. Li et al. [11] introduced a deep learning model for IoT into the edge computing environment to enhance network performance and ensure user privacy.

Many algorithms have been presented to enable resource-constrained edge devices to perform NLP tasks. Corcoran et al. [12] discussed the potential of edge computing for NLP tasks, highlighting the challenges in implementing edge computing, such as data compression and model optimization. Talagala et al. [13] addressed the computational complexity with the optimization of edge devices. Basu et al. [14] proposed an energy-efficient approach to NLP on resource-constrained devices and to minimise energy consumption by selecting the most energy-efficient processing configurations. Liu et al. [15] also contributed to the field with a comprehensive survey on edge computing for NLP tasks, covering various aspects, such as architecture, deployment, and optimization. More recent works [16–18] also proposed a real-time Sentiment Analysis approaches optimized for edge devices and NLP tasks.

Ruiz-Millán et al. [8] explored the use of federated learning for customized sentiment analysis on Twitter with BERT fine-tuning. They delved into customized sentiment analysis on Twitter by fine-tuning BERT models. Their primary focus was on leveraging federated learning to train BERT models directly on users' devices, ensuring the privacy of their data. Singh et al. [19] proposed a method for analyzing tweets related to vaccinations using federated learning while ensuring user privacy and data security. In order to solve private language processing problems on edge devices without the usage of resource-hungry language models or privacy-violating client data collection, Nagy et al. [20] proposed a privacy-preserving federated learning model evaluated on the social movies dataset. Zhou et al. [21] proposed the Distilled One-Shot Federated Learning (DOSFL) to reduce communication costs in federated learning processing environments.

Prior research has given limited attention to the application of federated learning specifically in the realm of Arabic sentiment analysis. Our work is distinguished by investigating the performance of the federated learning paradigm in analyzing Arabic tweets, considering both accuracy and training speed. We employed a deep learning model to learn deep embeddings, incorporating an efficient transfer learning procedure tailored to Arabic sentiment analysis.

3 Methodology

3.1 The Generic Pipeline

In this study, our primary focus was on the integration of federated deep learning with sentiment analysis specifically for the Arabic language. The pipeline of the proposed methodology is depicted in Fig. 1. The dataset is stored locally on edge devices, also referred to as clients, for processing. Each edge device runs the global deep learning model, configured by the central server, to perform the training phase on its local data and then sends the results back to the server. The server aggregates these results and updates the model weights accordingly. The federated paradigm manages all communications between the edge devices and the central server, as well as the aggregation of the results. By utilising federated learning, instead of collecting and centralising all the data in a single processing node, the training process is decentralized and distributed across multiple edge devices or servers, each equipped with a local partition of the training dataset.

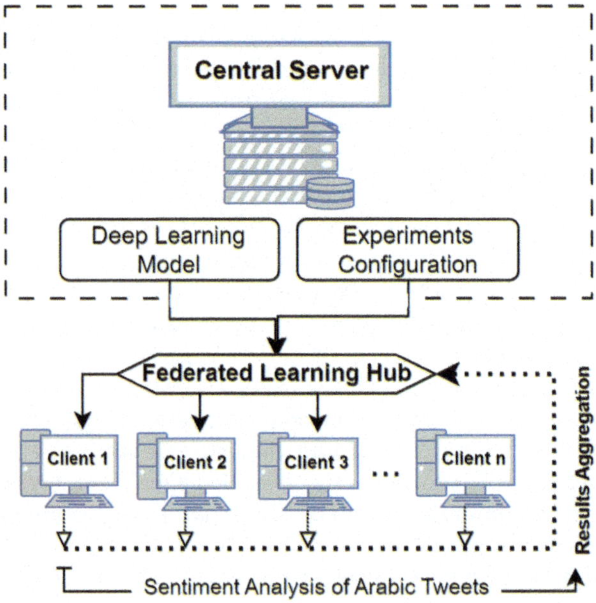

Fig. 1. The generic pipeline of the proposed deep federated learning.

3.2 Dataset Preparation

We used the Twitter-based benchmark dataset for Arabic Sentiment Analysis (ASAD) [9], which is a public dataset intended to speed up research in Arabic natural language processing (NLP) in general and Arabic sentiment classification in particular. ASAD is a high-quality annotated dataset including 100K tweets, with three-class sentiment tweet labels: 15,282 positive, 15,349 negative, and 69,369 neutral tweets. The tweets were collected between May 2012 and April 2020. They are written in a variety of Arabic dialects, including Egyptian, Modern Standard Arabic, Khaleeji, and Hijazi.

The dataset contains three fields for each tweet: tweet-id, sentiment polarity, and tweet content. Tweet-id is an identifier for each tweet. The sentiment polarity for each tweet can be positive, negative, or neutral. The tweet content shows the original tweet that was collected by the API, where the content includes a mix of Arabic text, emojis, punctuation marks, numbers, and URLs. Before using the BERT model to classify the entity of a token, the processing step is needed to make the input data suitable for making sentiment analysis. Therefore, the following preprocessing steps are applied:

- URL Removal: Remove all URLs that are present in the input text and replace them with empty strings.
- Numeric Removal: Remove any standalone number from the input text and replace it with an empty string.
- Arabic Text Normalization: Remove diacritics, remove elongation marks, and replace them with some specific Arabic characters.
- Punctuation Removal: Remove punctuation marks from the input text and replace them with spaces.
- Emoji Handling: Split emojis from the input text then join only the substrings back together with a space.
- Stemming: Apply Arabic word stemming then join the stemmed words back into a string.

3.3 The Implementation of Deep Federated Learning

The clients are the edge devices involved in the federated learning system. The execution setup for every client device is prepared by installing the required dependencies, including interpreters, e.g. Python, and ML libraries, e.g., TensorFlow and PyTorch. Each client device will work on a partition of the data considered as its own local tweets corpus.

The pre-trained UBC-NLP/MARBERT model, a specialized model with the Arabic Language, is used in federated learning. Initially, this model is prepared and set with the basic parameters in the central server. Then, the model is sent to the client devices to ensure consistency. Thereafter, each client trains its local model and adjusts its parameters using its local partition of the dataset. After each round of local training, the federated learning hub receives the models' updates from the clients and aggregates the results using a weighted averaging technique to generate an update for the global model at the central server. The central server sends the updated global model to the clients, allowing them to incorporate the federated learning process's pooled knowledge.

Figure 2 shows the fine-tuned BERT transformer model developed in this study. As can be observed, cleaning and preprocessing the input Arabic tweet text is first applied, e.g., diacritics removal and stemming. Then, the preprocessed Arabic text is tokenised. Thereafter, the tokenised Arabic text is converted into the input format expected by BERT. This involves adding special tokens such as [CLS] (for classification) at the beginning and [SEP] (for separation) between sentences or segments. The tokenised Arabic text is then passed through the BERT model to obtain token embeddings, as demonstrated in Fig. 2. BERT generates contextualised word embeddings by considering the surrounding words in a tweet. BERT embeddings represent contextualised word representations in a high-dimensional vector space. These embeddings capture the semantic and syntactic information of the input text. These BERT embeddings are provided into classification layers (e.g., a fully connected neural network) of the fine-tuned BERT model that transforms the embeddings into text vectorisation, i.e., a fixed-length vector representation for the entire text, then maps the vector to sentiment labels, i.e., positive, negative, neutral.

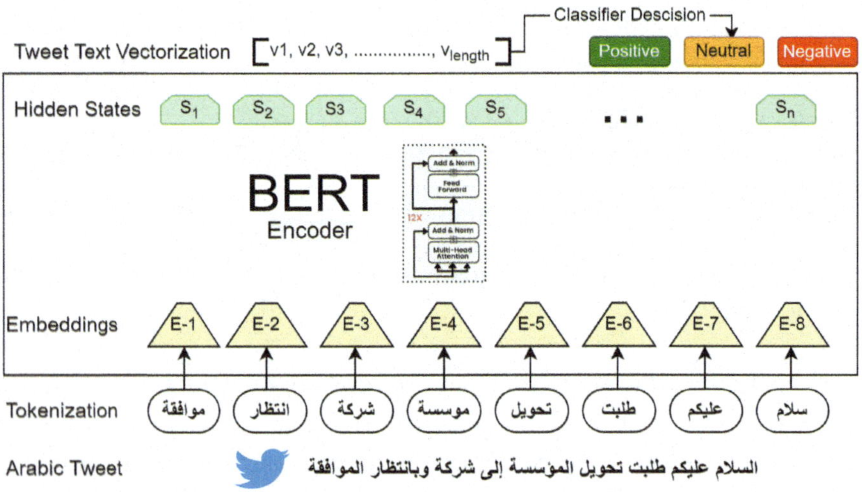

Fig. 2. The constructed BERT model for Arabic tweets analysis.

4 Experiments and Results

4.1 Experiments Configuration

All experiments were conducted by PyTorch [23] in a cloud-based processing environment. PyTorch is an open-source machine learning framework that is widely used for deep learning tasks, including federated deep learning. In this work's setup, the client-server architecture, model sharing, distributed training, and results aggregation have been implemented.

The evaluation metric used in this study is the weighted average of accuracy aggregated from each client's local node. Accuracy serves as a measure of the model's capacity to identify correlations and patterns among variables, and a higher accuracy reflects a stronger ability to generalize to unseen data, thereby yielding improved predictions and insights. To calculate the weighted average accuracy metric, the accuracy achieved by each client is multiplied by the number of tweet data samples used locally. Subsequently, the summation of all accuracy values is averaged across the clients' data.

The Federated Average (FedAvg) [22] optimizer follows a process where each client independently performs a gradient descent step on the current model using its local data. Subsequently, the server calculates a weighted average of the resulting models. This approach allows for additional computation on each client by iterating the local update multiple times prior to the averaging step. Table 1 summarises the hyperparameters configured for the global BERT encoder that run on the local client host.

Table 1. The hyperparameters of the global BERT deep model.

Parameter	Value
Optimizer	FedAvg
Learning rate	0.001
Momentum	0.9
Epochs	10
Loss function	CrossEntropy

4.2 The Performance Results

The performance evaluation of the federated deep learning model was conducted on the ASAD dataset using the default train and test data splits. The training set of tweets was utilized for fine-tuning the BERT encoder, extracting embeddings, and acquiring the essential discriminating descriptors. The accuracy metric provides insights into the model's generalization ability to identify Arabic tweets and classify them as positive, negative, or neutral. Through extensive experiments, the proposed federated deep learning model was thoroughly evaluated in terms of accuracy and training speed. The results were then compared between the federated and centralized learning approaches, as presented in Table 2. Given the number of local clients or edges involved in communication, it is noticeable that the accuracy attained in both federated and centralized learning modes exhibits a relatively similar performance, with a slight advantage for the centralized mode, showing a slight increment of 3%.

In terms of training speed, the federated learning (FL) deep model demonstrated a significant reduction in training time, taking only 15 min on average.

This marks a notable 75% decrease compared to the centralized deep learning (DL) model, which required 59 min for training. The FL model's execution remained stable when using 2–10 clients in the setup. However, as the number of clients increased, the training time began to increase noticeably. This was attributed to the communication overhead experienced by the server when handling larger client populations. The increased execution rounds, learned parameters, and updated weights contributed to slower training speed in the decentralized environment. Despite this, the average accuracy achieved by the FL model remained relatively stable across all experimental setups, reaching an accuracy of 90%(±1).

Table 2. The performance results of the FL and centralised DL models.

Setup	Clients	Train time (min)	Weighted avg. accuracy
Central DL	0	59	0.93
D-FL	2	16	0.91
D-FL	5	13	0.89
D-FL	7	12	0.91
D-FL	10	16	0.90
D-FL	12	23	0.89
D-FL	15	45	0.91
D-FL	18	70	0.90

5 Conclusion

In this research paper, we introduced a federated learning system for Arabic Sentiment Analysis utilizing a fine-tuned BERT encoder. The deep learning model, along with its execution configuration, serves as a global model shared by a central server and multiple edge devices acting as data clients. This approach enables the preservation of sensitive client data by performing the Arabic tweet analysis locally on each device while aggregating the training updates on the central server. The experimental results and findings provided evidence for the effectiveness of the proposed federated learning deep model in maintaining data privacy, reducing training time, and achieving accurate tweet classification. As initial findings, there is potential for further improvement by exploring additional aggregation approaches on the server side, which would enhance scalability when dealing with a larger number of edge devices or clients. Another potential direction for future investigation entails overcoming the constraint of the server having to wait for all diverse clients to finish their local work. While this condition guarantees synchronization for updating and creating a new global model, it also results in the optimization round's duration being influenced by the slowest update among the diverse clients.

References

1. Cao, K., Liu, Y., Meng, G., Sun, Q.: An overview on edge computing research. IEEE Access **8**, 85714–85728 (2020)
2. Shi, W., Cao, J., Zhang, Q., Li, Y., Xu, L.: Edge computing: vision and challenges. IEEE Internet Things J. **3**(5), 637–646 (2016)
3. Bonomi, F., Milito, R., Zhu, J., Addepalli, S.: Fog computing and its role in the internet of things. In: Proceedings of the First Edition of the MCC Workshop on Mobile Cloud Computing, pp. 13–16 (2012)
4. Farooq, M., Jan, M.S., Liu, Y.: A survey on edge computing for sentiment analysis: recent advances and future trends. IEEE Access, **9**, 45317–45337 (2021)
5. Huang, Y., Shen, L., Li, P., Li, J., Wu, J.: GPU-enabled real-time sentiment analysis at the network edge. Futur. Gener. Comput. Syst. **97**, 249–259 (2019)
6. Hasan, M.A., Mohammad, R.T., Rahim, M.S.: Distributed sentiment analysis at the edge and cloud for social media big data. J. Ambient. Intell. Humaniz. Comput. **10**(1), 45–56 (2019)
7. Sun, X., Chen, Z., Yu, Y., Yang, X.: Towards efficient edge-based sentiment analysis with federated learning. Inf. Sci. **573**, 180–194 (2021)
8. Ruiz-Millán, J.A., Martínez-Cámara, E., Victoria Luzón, M., Herrera, F.: Personalised federated learning with BERT fine tuning. Case study on twitter sentiment analysis. In: Advances in Deep Learning, Artificial Intelligence and Robotics: Proceedings of the 2nd International Conference on Deep Learning, Artificial Intelligence and Robotics (ICDLAIR), pp. 193–202. Springer International Publishing (2022)
9. Alharbi, B., Alamro, H., Alshehri, M., Khayyat, Z., Kalkatawi, M., Jaber, I.I., Zhang, X.: ASAD: a twitter-based benchmark Arabic sentiment analysis dataset. arXiv:2011.00578 (2020)
10. Ran, X., Chen, H., Zhu, X., Liu, Z., Chen, J.: Deepdecision: a mobile deep learning framework for edge video analytics. In: IEEE INFOCOM 2018-IEEE Conference on Computer Communications, pp. 1421–1429. IEEE (2018)
11. Li, H., Ota, K., Dong, M.: Learning IoT in edge: deep learning for the internet of things with edge computing. IEEE Netw. **32**(1), 96–101 (2018)
12. Corcoran, C.M., Benavides, C., Cecchi, G.: Natural language processing: opportunities and challenges for patients, providers, and hospital systems. Psychiatr. Ann. **49**(5), 202–208 (2019)
13. Talagala, N., Sundararaman, S., Sridhar, V., Arteaga, D., Luo, Q., Subramanian, S., Roselli, D.: ECO: harmonizing edge and cloud with ML/DL Orchestration. In: USENIX Workshop on Hot Topics in Edge Computing (HotEdge 18) (2018)
14. Basu, P., Roy, T.S., Naidu, R., Muftuoglu, Z.: Privacy enabled financial text classification using differential privacy and federated learning. arXiv:2110.01643 (2021)
15. Liu, M., Ho, S., Wang, M., Gao, L., Jin, Y., Zhang, H.: Federated learning meets natural language processing: a survey. arXiv:2107.12603 (2021)
16. Naranjo, P.G.V., Pooranian, Z., Shojafar, M., Conti, M., Buyya, R.: FOCAN: a fog-supported smart city network architecture for management of applications in the internet of everything environments. J. Parallel Distrib. Comput. **132**, 274–283 (2019)
17. Deshpande, A., Sultan, M.A., Ferritto, A., Kalyan, A., Narasimhan, K., Sil, A.: SPARTAN: sparse hierarchical memory for parameter-efficient transformers. arXiv:2211.16634 (2022)

18. Ding, Q., Zhu, R., Liu, H., Ma, M.: An overview of machine learning-based energy-efficient routing algorithms in wireless sensor networks. Electronics **10**(13), 1539 (2021)
19. Singh, M., Madhulika, Bansal, S.: A proposed federated learning model for vaccination tweets. In: International Conference on Computational Intelligence in Pattern Recognition, pp. 383–392. Springer Nature Singapore, Singapore (2022)
20. Nagy, B., Hegedűs, I., Sándor, N., Egedi, B., Mehmood, H., Saravanan, K., Loki, G., Kiss, Á.: Privacy-preserving federated learning and its application to natural language processing. Knowl. Based Syst. **268**, 110475 (2023)
21. Zhou, Y., Pu, G., Ma, X., Li, X., Wu, D.: Distilled one-shot federated learning. arXiv:2009.07999 (2020)
22. McMahan, B., Moore, E., Ramage, D., Hampson, S., Arcas, B.A.: Communication-efficient learning of deep networks from decentralized data. In Artificial Intelligence and Statistics, pp. 1273–1282. PMLR (2017)
23. Paszke, A., Gross, S., Massa, F., Lerer, A., Bradbury, J., Chanan, G., Chintala, S.: Pytorch: an imperative style, high-performance deep learning library. Adv. Neural Inf. Process. Syst. **32** (2019)

Neural Networks/Deep Learning

Towards Reinforcement Learning for Non-stationary Environments

Sebastian Gregory Dal Toé, Bernard Tiddeman, and Neil Mac Parthaláin(✉)

Department of Computer Science, Aberystwyth University, Aberystwyth SY23 3RE, Ceredigion, Wales, UK
ncm@aber.ac.uk

Abstract. In the Reinforcement Learning paradigm, environments that change between (or during) episodes are known as non-stationary. This property poses a challenge for traditional Reinforcement Learning approaches due to the fact that such methods typically rely on information that has been learned in previous episodes. However, in a non-stationary task, such information is usually an obstacle to learning that can lead to worse-than-random performance because it misleads the agent. Although an active area of research, most existing RL approaches also suffer from poor sample efficiency and struggle when the size of the state space increases. This work introduces a novel Reinforcement Learning approach that performs well in non-stationary environments, irrespective of the size of the state space. The approach proposed is termed Idea-based Reinforcement Learning, a symbolic method that can be applied to any problem that is a fully observable Markov Decision Process. IbRL is proven to perform statistically significantly above chance-level in all experiments and performs better than PPO2 in non-stationary problems where the state space is large.

Keywords: Reinforcement Learning · Non-stationary environments · Variational Autoencoder · Directed Exploration

1 Background

The field of Reinforcement Learning (RL) has seen many recent successes for complex problems such as *Chess*, *Shogi*, *Go*, *Atari* and *Starcraft II* [1–5]. The environments for such problems tend to have large state-spaces and require a computationally intensive training phase in order to converge to an optimal policy [3, 4]. In [3], *AlphaZero* took 9, 12 and 34 h of training for the games *Chess*, *Shogi* and *Go* respectively. This is equivalent to 44, 24 and 21 million self-play training episodes to achieve superhuman performance in each respective game. For *Starcraft II* [4], 12 agents were trained in parallel with 32 dedicated TPUs each, for a duration of 44 days.

In all of these environments, the underlying dynamics model learned by the agent is stationary; e.g. in *Chess,* the rules and how each piece is allowed to move does not change during or between episodes. It is this property that allows an agent to exploit information it has learned over time from previous episodes. However, there is also a set

N. Naik et al. (Eds.): UKCI 2023, AISC 1453, pp. 41–52, 2024.
https://doi.org/10.1007/978-3-031-47508-5_4

of problems for which the environment dynamics change (either completely or partially), between each episode. This is known as a non-stationary environment or problem. The aforementioned approaches fail in such environments because the agents require a huge amount of training time in order to derive an optimal policy.

One of the primary reasons for such an intensive training phase in these approaches is due to the exploration strategy - they either perform non-linear function approximation e.g., DQN [1], a form of Monte-Carlo Tree Search (MCTS) [2, 3], or a combination of both. These methods rely on the random sampling [6] of the environment in order to better approximate the value of unexplored states or actions [7]. In every example environment mentioned previously (as is typical of RL), the length of an episode (or number of actions) is significantly smaller than the size of the state space. It is therefore highly unlikely that an agent can sufficiently sample the state-action pairs for non-linear function approximation or MCTS to be effective within a single episode. Considering that *Chess* is a relatively constrained environment when compared with e.g., a real-world robotic navigation problem, it is revealing that the agent still took 44 million episodes to converge on this stationary problem. Though this work will only consider a contrived example problem, the non-stationary property arises in a number of real-world tasks such as autonomous control systems where the most appropriate current action depends on several external factors. This also includes use cases with an intrinsic temporal aspect such as trading and advertising where the agent must react quickly to changes in the environment without needing to retrain on this new data.

This paper proposes a novel approach termed Idea-based Reinforcement Learning (IbRL) proven to perform statistically significantly above chance-level in all experiments and performs better than state-of-the-art algorithms in non-stationary problems with large state spaces. The remainder of this paper is structured as follows. Section 2 defines the non-stationary environment and details the motivation for the proposed novel approach and associated algorithm. Section 3 presents the results of a series of experimental evaluations and comparisons with existing RL methods. Finally, the paper is concluded with discussion of the unique attributes of the approach and potential avenues for further investigation and future work.

2 Motivation

This section first provides an example non-stationary problem, then outlines the requirements and rationale for an algorithm that addresses this task.

2.1 Non-stationary Environments

In order to offer a frame of reference and comparison for the novel approach with existing ones, it is necessary to define a non-stationary environment. Consider an $n \times n$ grid where each cell can contain exactly one token or is empty. There are red, green and blue tokens. The agent interacts with this environment by altering the contents of a cell. An agent can: place a token in an empty cell, change the colour of an occupied cell or remove the token from a cell, leaving it empty. A global score exists for the entire grid based on some simple rules. These hidden rules (defined in Sect. 3.1) are unknown to the agent.

Thus far, this environment poses two problems. An *optimisation* problem where the agent must act in a way that maximises the score, however, in order to do so, the agent must discover the hidden rules, making it an *exploration* problem. In this environment, the hidden rules may change between each episode, meaning that the agent has limited time (a single episode) in order to explore the dynamic landscape and exploit it before having to restart.

The proposed approach termed Idea-based Reinforcement Learning (IbRL), can be divided into three primary phases of development: 1) A model that converts an input observation into a generic symbol using a learned structure. 2) A classifier learner that can be used to predict the reward of unknown symbols (prospective actions) in the latent space, based on facts (explored actions) recorded during the current episode. 3) A pre-trained model used as a 'common sense prior' to further augment the action selection. The following subsections discuss these phases and the motivation for development of the approach in further detail (see Fig. 1).

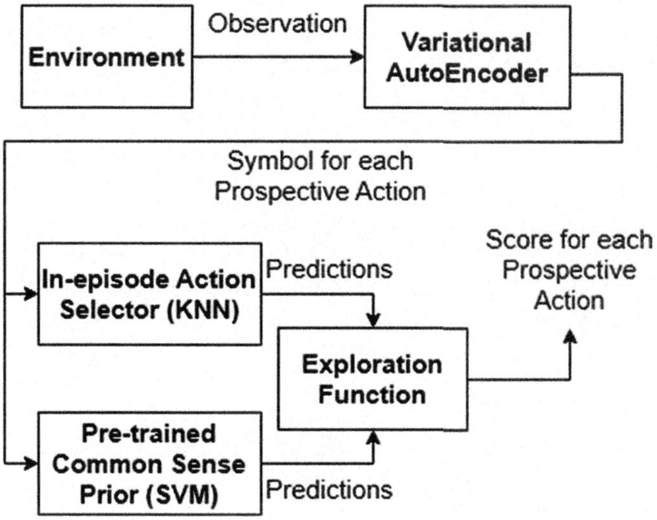

Fig. 1. Overview of the proposed approach

2.2 Symbolic Representation

As outlined in Sect. 1, a solution involving an undirected exploration strategy is not suitable for this problem, due to the reliance on a large number of random samples. In order to implement a directed exploration strategy, the approach must instead rely on domain knowledge or a heuristic. To avoid encoding environment-specific elements into the algorithm whilst taking advantage of domain knowledge, a symbolic method can be used to learn this information automatically.

An example of symbolic planning in the context of Reinforcement Learning is detailed in [8], where the authors use a 'neural back end' to map the high-dimensional raw inputs of a game environment to a lower-dimensional conceptual representation. The resulting conceptual state-space can then be used to perform symbolic reasoning and planning anonymously. What the authors term a 'neural back end' is in fact an autoencoder (a form of unsupervised ANN). The upshot of the work in [8] is that an autoencoder offers a method of converting an arbitrary observation into an anonymous symbol that can be manipulated, used in calculation and compared with other symbols, without ever referring to environment specific concepts such as tokens, cells or neighbours.

The use of this architecture ensures that the algorithm can reason effectively in any environment. However, it is known that standard autoencoders can suffer from an inconsistent latent space structure and that variational autoencoders (VAE) are a way to address this issue [9–11]. A VAE will also provide better generalisation over observations that are largely the same. Changes to the dynamics model that affect the observation dimensionality would require that the VAE is retrained with the appropriate architecture. However, to mitigate this, the input to the VAE is in fact, not the entire state observation. Instead, only the components of the observation that are directly relevant to the transition that has occurred. Using the earlier example environment in Sect. 2, to demonstrate, when a token is placed/changed/removed, the reward is derived only from that token and those directly adjacent to it. The remainder of the grid is irrelevant to the transition and can therefore be ignored. The advantage of focusing on the salient components of the observation in this way are two-fold: firstly, it addresses the issue of needing to change the architecture of the VAE and retrain should the observation shape change. Secondly, it alleviates the computational overhead of generating symbols as well as ensuring that the cost is constant over increasing grid sizes, because the first layer of the VAE is invariably 9 neurons. Figure 2 shows the salient features extracted from a typical state transition and the resultant symbol that represents them.

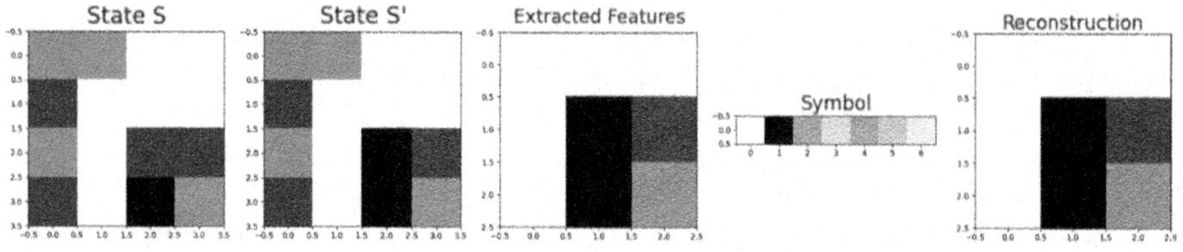

Fig. 2. Example state transition, extracted salient features, transition symbol and reconstruction using a VAE.

2.3 Episodic Action Selection

The following section describes how the latent space can be used to implement action selection. The VAE described in Sect. 3.1, provides an encoding for representing actions (state-transitions) that can be performed within the environment. From this, a directed exploration strategy can be implemented to maximise desired properties by performing an assessment within the latent space.

When using VAEs, inputs that are similar should map to symbols that are similar and are therefore 'close' in the latent space. Properties such as novelty or similarity are therefore trivial to derive by calculating the mean Euclidean distance between every symbol in the collection of known facts (explored actions) and the symbol associated with the prospective action that is currently being evaluated:

$$d(a,b) = \sqrt{\sum_{i=0}^{n} (a_i - b_i)^2} \tag{1}$$

where n is the number of dimensions in the latent space and d is the distance between a known fact a and the prospective point b. Equation 1 is repeated for each known fact and the mean is taken. In the case of novelty, it should be intuitive that choosing an action that maximises $\overline{d}$ means that the prospective symbol resides in a less-explored area of the latent space.

However, action selection based on this simple property alone would be naive since choosing an action that is dissimilar to the previously sampled actions offers no guarantee of reward. It is therefore important to consider the commonalities between each of the known facts, for which the k-Nearest Neighbour (kNN) algorithm is employed. This classifier is used in-episode, where the data is the collection of known facts that have been discovered during the trial.

2.4 Pre-trained Action Selection

In this section, the process of augmenting the selection phase by taking advantage of the latent space structure is described. In order to identify actions that are typically associated with a certain reward, the task of exploring the latent space to find some predictable structure, must be tackled. The motivation here is that if a model can confidently identify regions of the latent space that are generally associated with a certain reward (positive or negative), this can be used as a 'common sense prior' [8] to improve the action selection. The agent can then identify certain situations that are 'non-rewarding' and guide the exploration away from such actions.

Figure 3 shows that the latent space for this environment does indeed have a coherent structure. Most notably, there is a prominent cluster in which positive samples do not exist. It is therefore apparent that a model could learn which regions of the latent space are strongly associated with a certain reward.

In order to implement this, a Support Vector Machine (SVM) [12] was pre-trained on random state transitions from 50 episodes, each with a duration of 100 actions. The outputs of this model and the kNN described previously are combined in an exploration function. During many preliminary experiments, the following exploration function was found to yield the best performance.

$$W(s) = quality + prior \tag{2}$$

where s is the symbol corresponding to the prospective state transition, quality is the kNN output and prior is the SVM output.

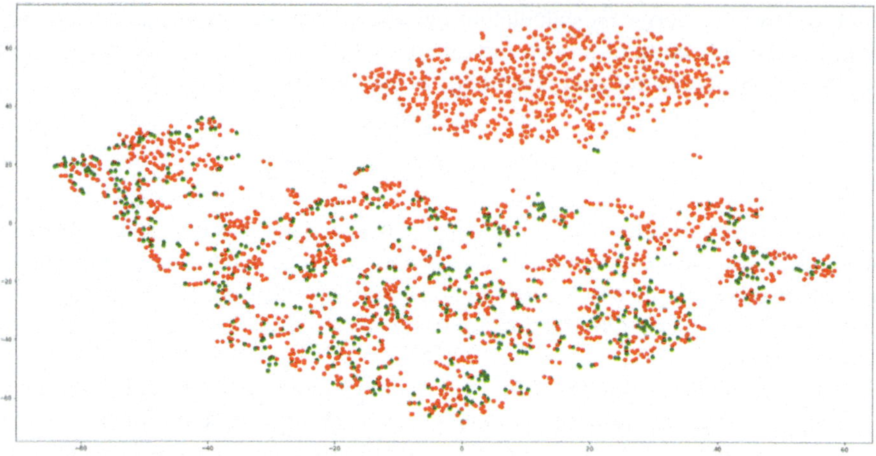

Fig. 3. A visualisation of 3000 symbols sampled from random actions in a single episode, mapped to 2D using Principal Component Analysis.

2.5 Proposed Algorithm

Algorithm 1

```
function select_action(observation)
    max_w = -1
    best_action = null
    for action in legal_actions do
        prospective_symbol = VAE_encode(action)
        distances = []
        for i in 0..length(known_facts) do
            distances[i] = euclidean(
                prospective_symbol,
                known_facts[i].symbol)
        end for
        nearest_k = []
        for i in 0..k do
            fact_index = argmin(distances)
            nearest_k[i] = known_facts[fact_index].reward
        end for
        quality = sum(nearest_k)   # given that reward is {0,1}
        prior = svm_predict(prospective_symbol)
        w = quality + prior
        if w > max_w then
            max_w = w
            best_action = action
        end if
    end for
    return best_action
end function
```

As an aside, the term 'Idea-based Reinforcement Learning' was chosen because the algorithm models the behaviour observed in human players tasked with the problem described in Sects. 2.1 and 3.1. Players would formulate ideas despite knowing that their information was incomplete and approximations carry some level of error.

3 Experimental Evaluation

This section provides an overview of the specific problem used in the series of experiments, then presents the collected results and discusses these findings.

3.1 Experimental Setup

Section 2.1 described the non-stationary environment and refers to 'hidden rules' that are used to calculate the global score. The hidden rules used in the following evaluation are as follows: Each token of a particular colour has a preferred number of 'neighbours'; chosen randomly at the start of an episode. Tokens that are directly adjacent on the grid in any of the cardinal directions can be considered neighbours.

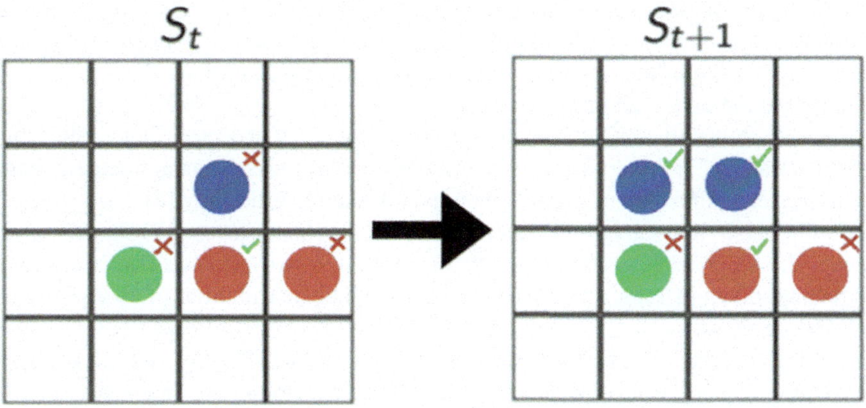

Fig. 4. Example state transition where red, green and blue require 3, 0 and 2 neighbours respectively.

In state S_t of Fig. 4, one of the red tokens is satisfied because there are three neighbouring tokens adjacent to it. The other tokens are not satisfied because their current number of neighbours does not match the required value. In state S_{t+1}, an extra blue token is added. Both blue tokens on the grid now have two neighbours which matches the predefined value. The score in state S_t and S_{t+1} are 1 and 3 respectively. Alternatively, one can consider the chosen action as being worth +2.

Finally, to frame this as a non-stationary problem, the preferred number of neighbours for each colour will change with every episode. It is therefore not possible to learn a good policy by simply sampling over many episodes, because a state-action pair that was rewarding in one episode may not be in the next.

The code for this environment is compliant with the OpenAI gym API [13] for easy comparison with algorithms included in the Stable Baselines API [14]. Each experiment was run with the following parameters: 4x4 grid, 1000 test episodes and 50 actions per episode. For IbRL, the VAE was pre-trained for 5e6 epochs on random observations and the SVM was pre-trained on 5000 randomly sampled symbol-reward pairs (50 episodes, 100 actions each). For the kNN algorithm, parameter $k = 3$. All parameters related to training were found to be optimal using parameter-search and manual experimentation.

3.2 Model Comparison

The following section compares the proposed algorithm and several existing methods on the test environment described in Sects. 2.1 and 3.1. In Fig. 5, IbRL is compared with a traditional algorithm DQN [1], a more recent algorithm 'ACER' [15], a state-of-the-art algorithm PPO2 [16] and a fully random agent to act as a baseline.

The first experiment compares each algorithm on two performance statistics. Average reward over the 1000 test episodes, which indicates the typical performance over a large sample size. The other evaluation method is '% of the 1000 episodes with a positive gain', where 'gain' is the difference between the score of the grid on initialisation compared to the grid at the end of an episode.

Figure 5 shows that DQN fails in this environment, as expected. DQN relies on the ϵ-greedy strategy [7] (random exploration) which as discussed, is an undirected strategy, unsuitable for such an environment given its low sample efficiency [6].

Figure 5 also shows that IbRL is performing above chance level. In order to determine whether this improvement is statistically significant, a two-sample t-test is conducted. The results shown in Table 1 compare the performance distributions of all algorithms to the Random agent.

In Table 1, IbRL achieves a P-value significant at the 1% level and the positive T-value shows that its performance is higher than the random agent. In the case of DQN, the P-value is extremely significant, but the T-value shows its performance is well below chance. Whilst Fig. 5 shows that ACER can somewhat function in this environment, the resultant P-value reveals that its performance is not significantly above chance.

The dominating algorithm in this experiment is PPO2. One of the notable qualities of Proximal Policy Optimisation is the gain in sample efficiency compared to other algorithms. Given a 4x4 grid as in all previous experiments, the state space is small enough for PPO2 to successfully track the non-stationary dynamics of the environment. Whilst PPO2 is the least susceptible algorithm to this sample efficiency problem, it is not immune. The following section will test whether this property holds in large state spaces.

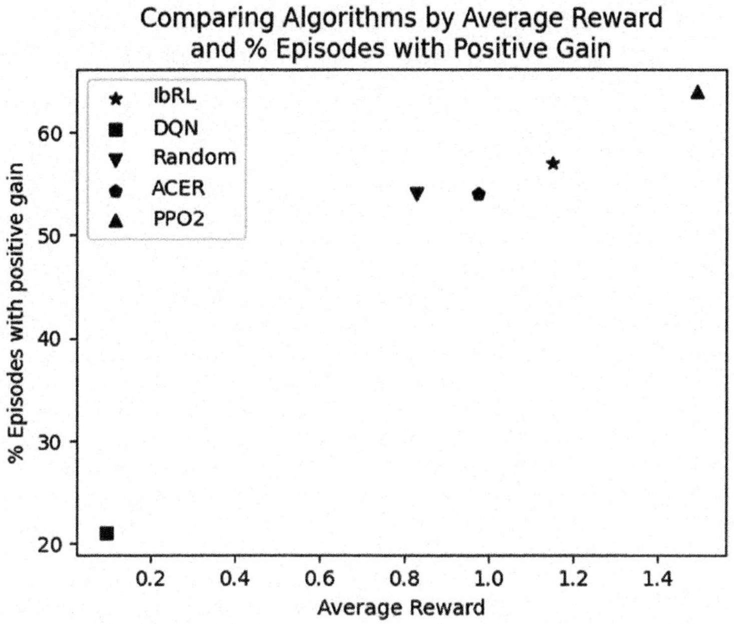

Fig. 5. A scatter plot showing the average reward and % episodes with positive gain for each model.

Table 1. Two-sample *t*-test of various algorithms when compared to a fully random agent.

Model	T-value	P-value
IbRL	3.187	0.0014
DQN	−10.063	2.8e-23
ACER	1.471	0.1413
PPO2	6.365	2.4e-10

3.3 Sample Efficiency

This section investigates the limits of PPO2's sample efficiency by increasing the size of the grid and therefore, the size of the state space. Note that all unmentioned parameters are left the same as the previous experiment. Figure 6 shows the results of PPO2, IbRL and the random agent on a 12×12 grid.

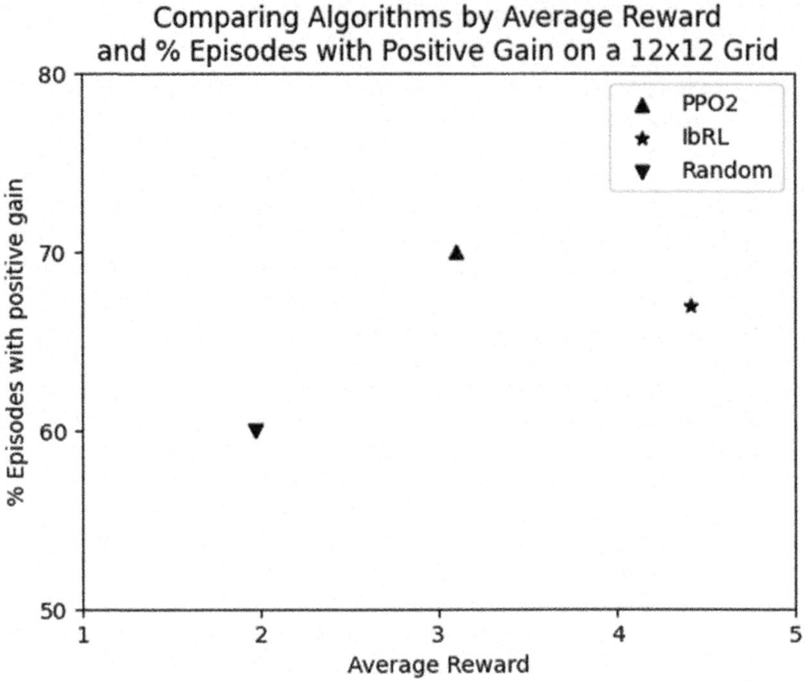

Fig. 6. Random, PPO2 and IbRL compared on a 12×12 grid.

From Fig. 6, it appears that PPO2 sees diminishing returns as the grid is increased in size and that IbRL's performance appears uninhibited. To confirm this, a two-sample *t*-test between their performance distributions is performed and results in a P-value of 2e−06. This is very strong evidence at the 0.1% level that IbRL is performing better than PPO2 on the 12×12 grid.

Since every subprocess involved in IbRL is working within the latent space and the latent space is invariably 9-dimensional despite the change in grid size, the size of the state space has no bearing on the algorithm's performance.

Below, Fig. 7 shows that as the grid size continues to increase, the P-value of PPO2 as compared to the random agent becomes less and less significant (larger value). This demonstrates that the algorithm inevitably reaches the limit of its sample efficiency and is reduced to chance level performance.

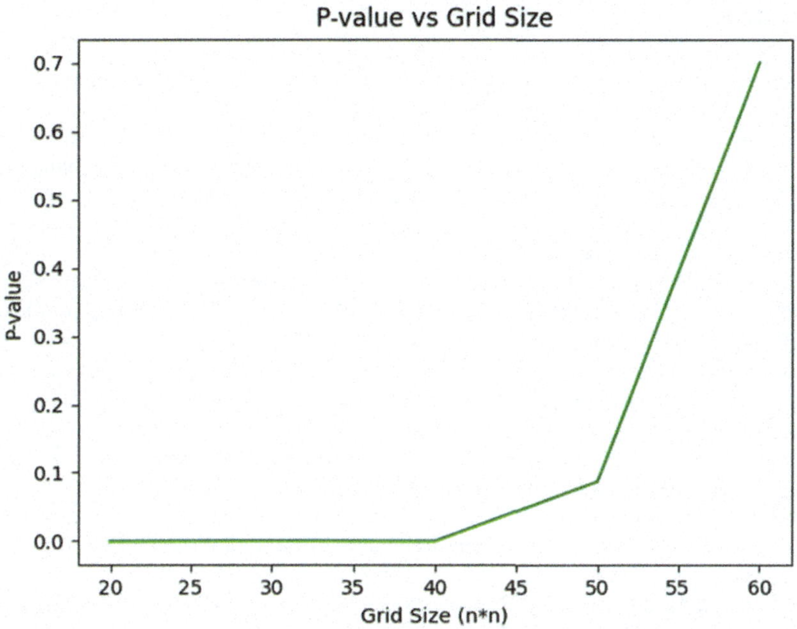

Fig. 7. P-value produced in two-sample *t*-test between PPO2 and the random agent on various grid sizes.

4 Conclusion

This paper has presented a novel approach to Reinforcement Learning for non-stationary environments. The inability to sample over many episodes as is common in canonical Reinforcement Learning motivated the choice of a symbolic approach as opposed to the more traditional connectionist and policy-based methods. The proposed algorithm IbRL has been shown to be statistically significant in all experiments and performs well regardless of the state space size, outperforming approaches such as PPO2, which fail on large state spaces when tasked with non-stationary problems. There are several avenues for further investigation which could help to improve the performance of the approach even further. For instance, the algorithm evaluates every single available action and greedily selects the one that optimises the exploration function. This is a naive approach and causes issues for computation time as the action space increases, though the theoretical performance guarantees of the algorithm are maintained regardless. The feature extraction method implemented for the VAE input observation is also rather rudimentary and would in future be replaced by some autonomous method that is not task specific. Other aspects include new evaluation metrics; improved quality metric (replacing KNN); methods of selecting a subset of the prospective actions to evaluate rather than computing every possible action; true planning rather than greedy action selection; scheduling between exploration and exploitation; weighting and decaying terms in the exploration function; and investigating the use of clustering algorithms within the VAE latent space.

References

1. Mnih, V., Kavukcuoglu, K., Silver, D. et al.: Playing atari with deep reinforcement learning (19 Dec 2013). arXiv:1312.5602. https://doi.org/10.48550/arXiv.1312.5602
2. Silver, D., Huang, A., Maddison, C., et al.: Mastering the game of Go with deep neural networks and tree search. Nature **529**, 484–489 (2016). https://doi.org/10.1038/nature16961
3. Silver, D., Hubert, T., Schrittwieser, J., et al.: Mastering chess and shogi by self-play with a general reinforcement learning algorithm (5 Dec 2017). arXiv:1712.01815. https://doi.org/10.48550/arXiv.1712.01815
4. Vinyals, O., Babushkin, I., Czarnecki, W., et al.: Grandmaster level in StarCraft II using multi-agent reinforcement learning. Nature **575**(7782), 350–354 (2019). https://doi.org/10.1038/s41586-019-1724-z
5. Schrittwieser, J., Antonoglou, I., Hubert, T., et al.: Mastering atari, go, chess and shogi by planning with a learned model. Nature **588**(7839), 604–609 (2020). https://doi.org/10.1038/s41586-020-03051-4
6. Thrun, S.: Efficient Exploration in Reinforcement Learning. Carnegie Mellon University (1992). https://citeseerx.ist.psu.edu/viewdoc/summary?doi=10.1.1.45.2894
7. Watkins, C.: Learning from Delayed Rewards. PhD thesis, University of Cambridge, Cambridge, England (1989). https://www.academia.edu/3294050/Learning_from_delayed_rewards?from=cover_page
8. Garnelo, M., Arulkumaran, K., Shanahan, M.: Towards deep symbolic reinforcement learning (18 Sep 2016). arXiv:1609.05518. https://doi.org/10.48550/arXiv.1609.05518
9. Asai, M., Kajino, H., Fukunaga, A., et al.: Classical planning in deep latent space (30 Jun 2021). arXiv:2107.00110. https://doi.org/10.48550/arXiv.2107.00110
10. Doersch, C.: Tutorial on variational autoencoders (19 Jun 2016). arXiv:1606.05908. https://doi.org/10.48550/arXiv.1606.05908
11. Asperti, A., Trentin, M.: Balancing reconstruction error and Kullback-Leibler divergence in variational autoencoders. IEEE Access **8**, 199440–199448 (2020). https://doi.org/10.48550/arXiv.2002.07514
12. Cortes, C., Vapnik, V.: Support-vector networks. Mach. Learn. **20**(3), 273–297 (1995). https://doi.org/10.1007/BF00994018
13. OpenAI, Gym Documentation (2022). https://www.gymlibrary.dev/content/api/
14. Hill, A., Raffin, A., Ernestus, M., et al.: Stable Baselines (2018). https://stable-baselines.readthedocs.io/en/master/
15. Wang, Z., Bapst, V., Heess, N., et al.: Sample efficient actor-critic with experience replay (3 Nov 2016). arXiv:1611.01224. https://doi.org/10.48550/arXiv.1611.01224
16. Schulman, J., Wolski, F., Dhariwal, P., et al.: Proximal policy optimization algorithms (20 Jul 2017). arXiv:1707.06347. https://doi.org/10.48550/arXiv.1707.06347

Predictive World Models for Social Navigation

Goodluck Oguzie(✉), Aniko Ekart, and Luis J. Manso

Aston University, B4 7ET Birmingham, UK
190212683@aston.ac.uk
https://cs.aston.ac.uk/arp

Abstract. As robots begin to coexist with humans, the need for efficient and safe social robot navigation becomes increasingly pressing. In this paper we investigate how world models can enhance the effectiveness of reinforcement learning in social navigation tasks. We introduce three approaches that leverage predictive world models, which are then benchmarked against state-of-the-art algorithms. For a comprehensive and reliable evaluation, we employed multiple metrics during the training and testing phases. The key novelty of our approach consists in the integration and evaluation of predictive world models within the context of social navigation, as well as in the models themselves. Based on a diverse set of performance metrics, the experimental results provide evidence that predictive world models help improve reinforcement learning techniques for social navigation.

1 Introduction

With the increased sharing of space between humans and robots, the need for effective robot Social Navigation (SocNav) has become paramount [19]. Most state-of-the-art approaches for SocNav depend on hand-crafted algorithms that are difficult to scale to consider additional variables [7], the most common variables being the goal position, free space, and the 2D poses of humans and robots [7].

Reinforcement Learning (RL) provides a framework to overcome the reliance on hand-crafted algorithms, but current RL algorithms often exhibit prolonged convergence times, requiring extensive interactions with the environment before they can learn a near-optimal policy. Despite the significant success of RL in numerous tasks [2], more research is needed before RL-based SocNav can be successfully applied in complex real-world scenarios [7].

RL approaches using world models capable of predicting future states of the environment have outperformed more traditional approaches in multiple RL environments [4,8,15]. In this paper, we propose three methods that integrate predictive world models into an RL algorithm for SocNav tasks. Our methods leverage a world model similar to the one proposed by Ha and Schmidhuber [8], combining a Variational Autoencoder (VAE) [16] and a Long-Short Term Memory network (LSTM) [13]. The first method, termed *2StepAhead*, builds on top

N. Naik et al. (Eds.): UKCI 2023, AISC 1453, pp. 53–64, 2024.
https://doi.org/10.1007/978-3-031-47508-5_5

of Ha and Schmidhuber [8], but makes the predictions two steps ahead (assuming that the same action is taken twice) and uses Dueling DQN [28] instead of Covariance Matrix Adaptation (CMA) [12]. Our second method, *MASPM*, also expands upon that of Ha and Schmidhuber [8] by considering multiple actions while performing single-step predictions. The third method, *2StepAhead-MASPM*, combines the ideas of both prior approaches by performing two steps ahead predictions and considering multiple actions.

2 Related Work

Reinforcement Learning is a learning paradigm where an agent learns to interact near-optimally with its environment to maximise a given reward, operating within a Markov Decision Process framework [11,26]. In the domain of robotics, RL has been leveraged to teach robots complex manipulation tasks [1], and in gaming, it has been employed to develop agents that can play games proficiently [18].

Despite its wide-ranging successes, RL has well-known limitations. Adaptability to novel environments poses a significant challenge [21]. Moreover, RL often requires large volumes of data for training, making it computationally expensive [17]. Specifically in the domain of social navigation, these issues become even more pronounced due to the rich and complex nature of social dynamics [20,23,30]. The complex interactions that happen in social settings are difficult to model and predict, making RL agents' learning of optimal policies even more challenging [5].

In response to these challenges, world models have emerged as a promising solution. For instance, MuZero, an RL-based method using world models, has demonstrated its efficacy in learning ATARI game rules using observed image data and action sequences, even with limited computational resources [10,22]. AlphaGo, another RL-based method using world models, outperformed human experts in the game of Go in 2016 [24].

Unlike traditional RL-based predictions that rely on the current environment state represented by a state-action pair, predictive world models can consider both past and present states to anticipate future ones [8]. Arguably, we can interpret that the inclusion of these models into RL algorithms incorporates into the algorithms the prior that predicting future states is useful. This methodology has been applied successfully in environments such as CarRacing [4] and Doom [15], outperforming traditional RL [8]. Furthermore, world models introduce a predictive component to the RL dynamics, enabling the agent to anticipate future actions. This can lead to faster learning and potentially improved results in fewer episodes [27].

Additionally, world models augment the MDP framework by shifting from reliance solely on current observations and actions to a broader decision optimisation perspective. This allows agents to generate more informed policies based on a predictive understanding of the environment [6]. Notably, Dreamer, a model-based RL agent, has demonstrated the capability of combining world models and policy learning to achieve state-of-the-art performance in various tasks [9].

This study delves deeper into these predictive world models, specifically within RL-based social navigation which leads to our research question: "Can world models help us improve RL-based social navigation?"

3 Methodology

In this paper we explore the use of predictive world models to improve RL-based SocNav using the three aforementioned proposed methods –2StepAhead, MASPM, and 2StepAhead-MASPM; in this section we describe the three approaches and provide experimental details.

Our experiments are conducted in SocNavEnv [14], a configurable environment specifically designed for social navigation scenarios. This environment has the capacity to incorporate a wide range of entities such as humans (static or moving), plants, tables, and laptop computers. For our experiments, SocNavEnv was configured to work with a discrete action space of four actions (stop, move forward, rotate left, and rotate right), three moving humans, and a social navigation reward function [3]. The goal of the agent in SocNavGym is to train the agent to navigate towards the target while (1) avoiding collisions with surrounding entities and (2) minimising the discomfort caused to the humans. A screenshot of SocNavEnv is shown in Fig. 1.

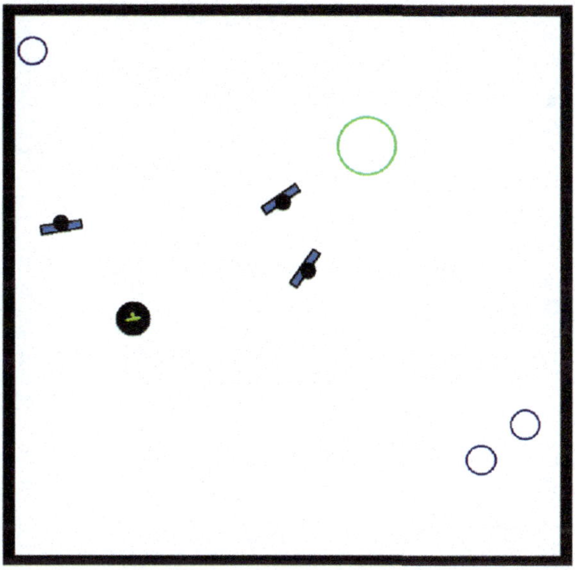

Fig. 1. Screenshot of SocNavEnv, the environment used for the experiments [14]. Blue squares represent humans, blue circles indicate humans' goals (which are unknown to the robot), green circles represent the robot's goals, and black-green circles represent robot agents.

Although we are aware that in real-life settings the number of individuals involved is frequently greater than three, we found that including three humans was sufficient for the experiments to be challenging for the RL algorithm used as a baseline. Dueling DQN was chosen as the baseline, because it is a well known algorithm that performs generally well even when dealing with high-dimensional state spaces and it is suitable for discrete action spaces [28]. Dueling DQN is an evolution of DQN where the final layer of the network is split into two distinct pathways: one computes the state-value function and the other estimates the advantage function for each discrete action [28]. This design allows Dueling DQN to better distinguish between the impact of different actions, thus optimising learning outcomes. To our knowledge, there is no reason to believe that the methods would not be applicable to other RL algorithms.

Our proposed methods build upon the architecture developed by Ha and Schmidhuber [8] (see Fig. 2), where a VAE, parameterised by ϕ, compresses the observation (s) into a latent state (z) (of sizes 23 and 16, respectively), as shown by the relationship $z = \text{VAE}(s; \phi)$. The role of the VAE is to improve the efficiency and performance by compressing important information within this reduced dimensionality. Following this, the LSTM, parameterised by ψ, utilises z and the chosen action (a) to predict the next latent state (z') and hidden state (h') following

$$(z', h') = \text{LSTM}(z, a; \psi).$$

These predicted states are then input into the Dueling DQN, forming the foundation for the predictive world models in our methods.

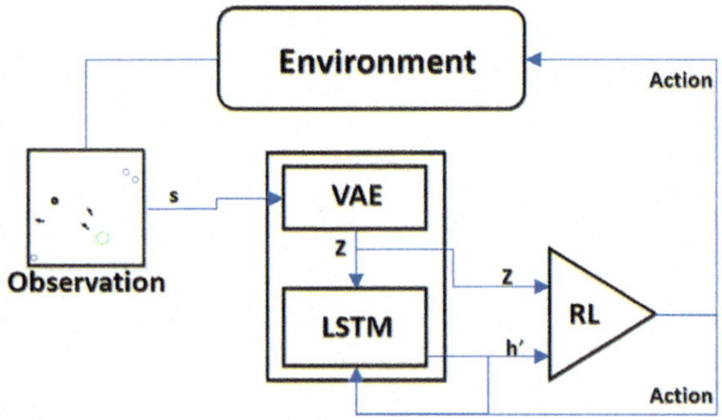

Fig. 2. Predictive world model as proposed by Ha and Schmidhuber [8].

3.1 Two Step Ahead Predictive World Model: 2StepAhead

2StepAhead extends the vanilla approach of Ha and Schmidhuber [8] by predicting the hidden state and the latent state two steps ahead. The number of steps that the model is predicting ahead was empirically determined out of 2, 4, 8, and 16 steps. Although this number arguably depends on the environment, predicting more than 2 steps ahead did not improve the results in our SocNavGym setup and made training slower.

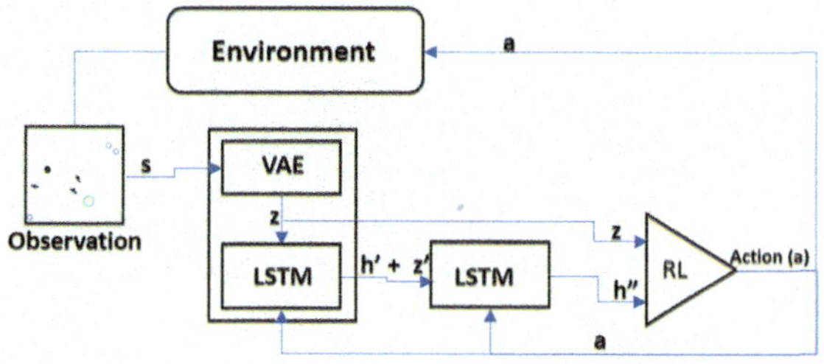

Fig. 3. In 2StepAhead, the same LSTM is used recursively to predict two steps ahead.

As depicted in Fig. 3, our model predicts two steps ahead for the hidden state (h'') and the latent state (z'') by using the predicted next and hidden states (z', h') and the current action (a):

$$(z'', h'') = \text{LSTM}(z', h', a; \psi).$$

Subsequently, the environment's current latent state (z) and the two steps ahead hidden state (h'') are fed into the Dueling DQN to choose the next action (a^*):

$$a^* = \text{Dueling DQN}(z|h''; \xi),$$

where ξ represents the parameters of our Dueling DQN. By predicting the latent state of the environment two steps ahead, we hope to provide to the RL algorithm richer information regarding the future state in case the robot keeps taking the current action, potentially improving performance and robustness in a dynamic environment.

3.2 Multi Action State Predictive Model: MASPM

This model provides the Dueling DQN with a comprehensive view of future state possibilities, encompassing all four available actions, potentially enabling more informed decision-making and thereby improving the model robustness and performance (see Fig. 4).

The latent state (z) along with the action serve as inputs for an LSTM, which predicts the next state and hidden state based on the given action. We denote the action index by i, ranging from 0 to 3 indicating that we only have four possible actions to consider. Thus, for each action i, the latent state and action are input to the LSTM model to predict the subsequent state and hidden state:

$$(z_i', h_i') = \text{LSTM}(z|a_i; \psi),$$

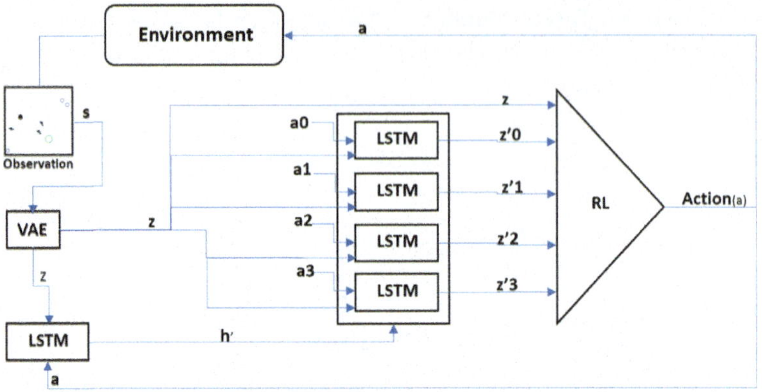

Fig. 4. In MASPM, the LSTM is not used recursively, but it is provided with the four possible actions and all the resulting data are fed into the RL algorithm.

where z is the current latent state, a_i is the i-th action (provided to the network as a one-hot encoding), and ψ represents the LSTM parameters. The four next predicted states, z'_1, z'_2, z'_3, z'_4, together with the current latent state z then serve as inputs for the Dueling DQN to estimate the best action a^*:

$$a^* = \text{Dueling DQN}(z|z'_1|z'_2|z'_3|z'_4; \xi),$$

where ξ represents the Dueling DQN parameters. MASPM provides a broadened perspective of future states across multiple actions, offering the Dueling DQN a richer foundation for decision-making.

3.3 Combining 2StepAhead and MASPM: 2StepAhead-MASPM

The 2StepAhead-MASPM is a combination of MASPM and the 2StepAhead method and aims to combine their advantages. This model provides a two-step-ahead prediction for each potential action. The two-step-ahead prediction horizon facilitates the Dueling DQN algorithm with a more refined decision-making capability. It achieves this by leveraging the current latent state and the predicted two-step-ahead state for each possible action to determine its subsequent action.

Figure 5 illustrates the architecture of the proposed 2StepAhead-MASPM. The latent state (z), coupled with the related action, is fed into the LSTM. The LSTM uses these inputs to predict the next state and the hidden state conditioned on the input action. The action index i can range from 0 to 3, representing the four possible actions. For each action i, the LSTM model processes the latent state and action as input and predicts the corresponding next state and hidden state. The model repeats this process, using the same action and the previously predicted latent state for the second prediction.

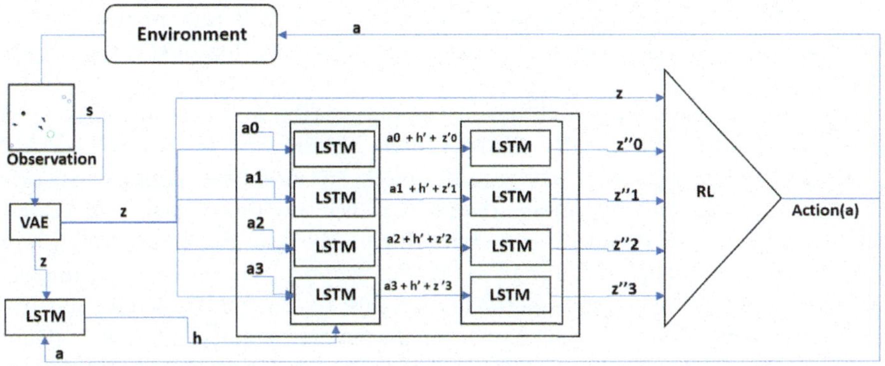

Fig. 5. 2StepAhead-MASPM combines the advantages of 2StepAhead and MASPM. It predicts two steps ahead and considers all actions instead of just the current action.

Given a latent state z and an action a_i at a time t, the LSTM predicts the next state z_{t+1} and hidden state h_{t+1}. The process is repeated using the new latent state z_{t+1} and the same action a_i to predict the next latent state z_{t+2} and hidden state h_{t+2}:

$$(s_{t+1}, h_{t+1}) = LSTM(s_t, a_i, h_t; \phi)$$
$$(s_{t+2}, h_{t+2}) = LSTM(s_{t+1}, a_i, h_{t+1}; \phi)$$

We hypothesise that combining two steps ahead predictions with a coverage of all actions can improve Dueling DQN's decision-making. The next section benchmarks the three proposed methods against the selected baselines to evaluate whether the use of Predictive World Models is beneficial in the context of SocNav.

4 Experimental Results

All the developed models are based on the Dueling DQN reinforcement learning algorithm and are trained within the SocNavEnv environment [14]. To ascertain the influence of predictive world models on RL-based social navigation, Dueling DQN is also used as a baseline. The hyperparameters of Dueling DQN, particularly the size of the hidden layers, are critical in determining the agent's learning capabilities [25]. Therefore, we evaluated two Dueling DQN MLP model architectures -one with two hidden layers of size 128 each, and another with layers of sizes 512 and 128, respectively. After 200,000 episodes –the number of episodes required for all experiments to converge in this paper– the model with hidden layers of size 512 and 128 achieved a slightly higher expected cumulative reward for the vanilla Dueling DQN. Therefore, we selected this architecture for the rest of the Dueling DQN-based agents. Subsequently, we integrated predictive world models into the RL framework according to the three proposed methods in Sec. 3. We evaluated the proposed methods using different metrics [7], each uniquely designed with predictive capabilities, in the context of social navigation tasks.

The novelty of our approach lies in the integration and evaluation of predictive world models –specifically, 2StepAhead, MASP, and 2StepAhead-MASP– within the context of social navigation, which has not been explored in previous work, as well as in the models themselves. For a comprehensive and reliable evaluation, we employed multiple metrics during the training and testing phases.

Using only a single metric can limit the scope of the evaluation and may not fully capture the model's performance due to the multi-faceted nature of social navigation tasks. Metrics such as discomfort counts, human collisions, and personal space compliance are as important as the traditionally employed metrics in RL such as reward or convergence time. Therefore, we use this broad range of metrics to ensure a holistic analysis that comprehensively reflects the performance in a human-robot interactive environment. Furthermore, our comparative analysis extends beyond our baseline Dueling DQN models. For the testing phase, we also include comparisons with other established models in the domain, like the RVO2 and social force model, to provide a broader context for the performance of our models. These benchmarks were chosen due to their widespread use in social navigation tasks.

4.1 Training Phase Metric Evaluation

The training phase is focused on the cumulative reward, training time, and episodes to convergence. The results from this phase showed significant improvements in our proposed models over the baseline Dueling DQN models. The 2StepAhead model was particularly efficient, solving the task in about 3200 episodes, as depicted in Fig. 6. The 2StepAhead-MASPM model outperformed all the other models, achieving the highest average cumulative reward of 0.67.

4.2 Testing Phase Metric Evaluation

In the testing phase, we used a broad range of metrics related to human-robot interactions, navigation efficiency, and overall performance, and measured those metrics for 500 episodes per algorithm. In Figs. 7 and 8 the histograms of the following metrics are shown:

- Human discomfort: Average human discomfort caused to humans, as described in [3].
- Distance travelled: Distance travelled by the agent, per episode (in meters).
- Simulation time: Calculated as the number of steps multiplied by the step time (in seconds).
- Human collisions: Whether the robot collides in a trajectory or not (binary metric).
- Max steps: Whether the agent reaches the maximum number of steps in a particular episode (binary metric).

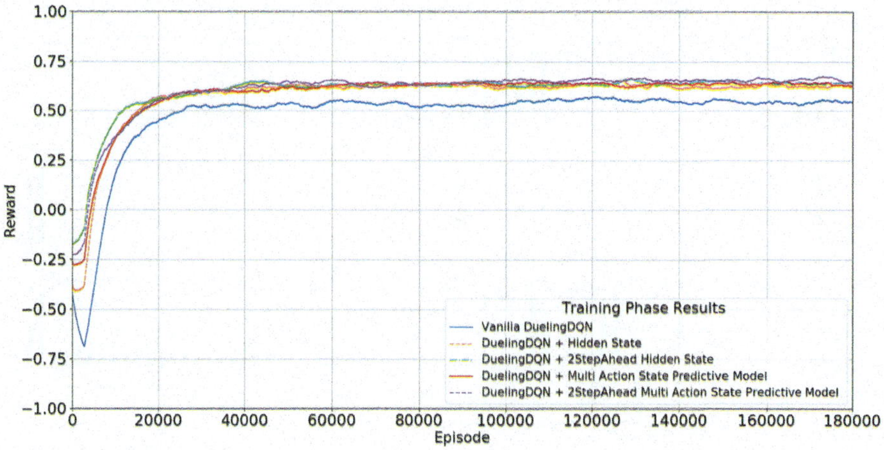

Fig. 6. Smoothed cumulative reward during training.

- Reward: The cumulative reward per episode (scalar).
- Successful run: Whether the agent reaches the goal or not in an episode (binary metric).
- Idle time: Steps where the robot moves less than 0.05m (in seconds).
- Personal space compliance rate: Ratio of the time where robot is further away than 0.5 m from any human divided by the total time (scalar).

The 2StepAhead-MASPM achieved higher average cumulative reward than the baseline models. Success rate, human collision, and cumulative reward were also improved with our 2StepAhead-MASPM model. Our model performed well overall, achieving the second-best in minimal idleness and ranking third for personal space compliance, simulation time, and distance travelled, respectively. However, it is important to remember that optimising one aspect of social norms may have unintended consequences on others. For example, while reducing the time to reach the goal by finding the shortest path may be desirable, this could compromise human personal space. Therefore, the ideal solution is not to maximise one specific metric but to strike a balance across all metrics.

While our 2StepAhead-MASPM model might not have achieved the highest score in all individual metrics, it excelled in achieving well-rounded results over most metrics used, respecting the Pareto nondomination criterion [29], i.e., no other method performed better across all metrics. It improved critical aspects of social norms such as avoiding collisions with humans and maintaining a high success rate, all without excessively compromising personal space compliance. Moving forward, our aim is to continue refining our models to obtain an even better balance across the multiple dimensions involved, thereby further improving performance in complex, multi-faceted tasks such as social navigation.

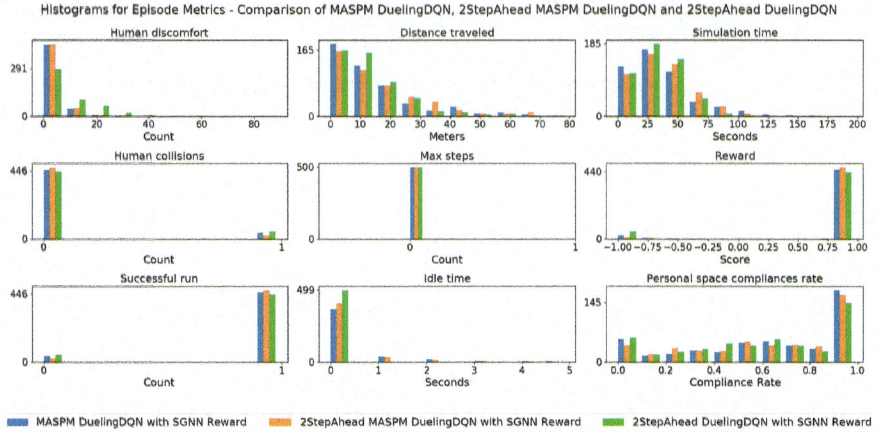

Fig. 7. Histograms of the metrics used for comparison, applied to the three proposed models.

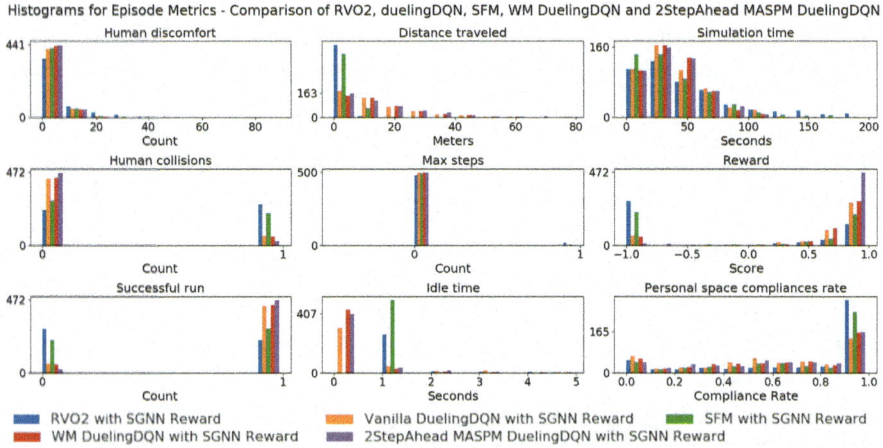

Fig. 8. Histograms of the metrics used for comparison, applied to RVO2, Dueling DQN, SFM, WM Dueling DQN, and 2StepAhead-MASPM Dueling DQN.

5 Conclusions and Future Work

The experimental results confirm the value of integrating world models in RL-based social navigation. We present a novel contribution –the 2StepAhead-MASPM predictive model integrated into the Dueling DQN framework– which demonstrated superior performance over the baseline models across various metrics, particularly in terms of success rate, cumulative reward and human collision. However, our study also revealed areas where improvement can be made, most notably in terms of maintaining personal space –an essential aspect in social navigation. This insight highlights the importance of the Pareto nondomination criterion [29] in dealing with such multi-faceted tasks.

As future work, we are planning to experiment on more complex navigation environments and continuous action spaces. By introducing a range of different obstacles such as tables, chairs, and laptops, and varying the number of humans present in the environment, we aim to simulate more realistic and dynamic scenarios. With these, we want to further test the limits of predictive world models and refine the performance of our models. Ultimately, our goal is to develop an RL agent that not only navigates efficiently through complex social environments but also maintains respect for personal boundaries and pedestrians' comfort.

References

1. Andrychowicz, O.A.M., Baker, B., Chociej, M., Józefowicz, R., McGrew, B., Pachocki, J., Petron, A., Plappert, M., Powell, G., Ray, A., Schneider, J., Sidor, S., Tobin, J., Welinder, P., Weng, L., Zaremba, W.: Learning dexterous in-hand manipulation. Int. J. Robot. Res. **39**(1), 3–20 (2020). https://doi.org/10.1177/0278364919887447
2. Arulkumaran, K., Deisenroth, M.P., Brundage, M., Bharath, A.A.: Deep reinforcement learning: a brief survey. IEEE Signal Process. Mag. **34**(6), 26–38 (2017)
3. Bachiller, P., Rodriguez-Criado, D., Jorvekar, R.R., Bustos, P., Faria, D.R., Manso, L.J.: A graph neural network to model disruption in human-aware robot navigation. Multimed. Tools Appl. 1–19 (2021). https://doi.org/10.1007/s11042-021-11113-6
4. Brockman, G., Cheung, V., Pettersson, L., Schneider, J., Schulman, J., Tang, J., Zaremba, W.: Openai gym. arXiv:1606.01540 (2016)
5. Chen, Y.F., Everett, M., Liu, M., How, J.P.: Socially aware motion planning with deep reinforcement learning. In: 2017 IEEE/RSJ International Conference on Intelligent Robots and Systems (IROS), pp. 1343–1350. IEEE (2017)
6. Chua, K., Calandra, R., McAllister, R., Levine, S.: Deep reinforcement learning in a handful of trials using probabilistic dynamics models. Adv. Neural Inf. Process. Syst. **31** (2018)
7. Francis, A., Perez-D'Arpino, C., Li, C., Xia, F., Alahi, A., Alami, R., Bera, A., Biswas, A., Biswas, J., Chandra, R., et al.: Principles and guidelines for evaluating social robot navigation algorithms. arXiv:2306.16740 (2023)
8. Ha, D., Schmidhuber, J.: World models. arXiv:1803.10122 (2018)
9. Hafner, D., Lillicrap, T., Ba, J., Norouzi, M.: Dream to control: learning behaviors by latent imagination. arXiv:1912.01603 (2019)
10. Hafner, D., Lillicrap, T., Norouzi, M., Ba, J.: Mastering Atari with discrete world models. arXiv:2010.02193 (2020)
11. Han, X.: A mathematical introduction to reinforcement learning. Semantic Scholar pp. 1–4 (2018)
12. Hansen, N.: The CMA evolution strategy: a comparing review. In: Towards a New Evolutionary Computation: Advances in the Estimation of Distribution Algorithms, pp. 75–102 (2006)
13. Hochreiter, S., Schmidhuber, J.: Long short-term memory. Neural Comput. **9**(8), 1735–1780 (1997)
14. Kapoor, A., Swamy, S., Manso, L., Bachiller, P.: Socnavgym: a reinforcement learning gym for social navigation. arXiv:2304.14102 (2023)

15. Kempka, M., Wydmuch, M., Runc, G., Toczek, J., Jaśkowski, W.: Vizdoom: a doom-based AI research platform for visual reinforcement learning. In: 2016 IEEE Conference on Computational Intelligence and Games (CIG), pp. 1–8. IEEE (2016)
16. Kingma, D.P., Welling, M.: Auto-encoding variational Bayes. arXiv:1312.6114 (2013)
17. Makoviychuk, V., Wawrzyniak, L., Guo, Y., Lu, M., Storey, K., Macklin, M., Hoeller, D., Rudin, N., Allshire, A., Handa, A., et al.: Isaac gym: high performance GPU-based physics simulation for robot learning. arXiv:2108.10470 (2021)
18. Matsuo, Y., LeCun, Y., Sahani, M., Precup, D., Silver, D., Sugiyama, M., Uchibe, E., Morimoto, J.: Deep learning, reinforcement learning, and world models. Neural Netw. **152**, 267–275 (2022). https://doi.org/10.1016/j.neunet.2022.03.037
19. Mavrogiannis, C., Baldini, F., Wang, A., Zhao, D., Trautman, P., Steinfeld, A., Oh, J.: Core challenges of social robot navigation: a survey. ACM Trans. Human-Robot Interact. **12**(3), 1–39 (2023)
20. Rao, K., Harris, C., Irpan, A., Levine, S., Ibarz, J., Khansari, M.: RL-CycleGan: reinforcement learning aware simulation-to-real. In: Proceedings of the IEEE Computer Society Conference on Computer Vision and Pattern Recognition, pp. 11154–11163 (2020). https://doi.org/10.1109/CVPR42600.2020.01117
21. Rusu, A.A., Večerík, M., Rothörl, T., Heess, N., Pascanu, R., Hadsell, R.: Sim-to-real robot learning from pixels with progressive nets. In: Conference on Robot Learning, pp. 262–270. PMLR (2017)
22. Schrittwieser, J., Antonoglou, I., Hubert, T., Simonyan, K., Sifre, L., Schmitt, S., Guez, A., Lockhart, E., Hassabis, D., Graepel, T., Lillicrap, T., Silver, D.: Mastering Atari, Go, chess and shogi by planning with a learned model. Nature **588**(7839), 604–609 (2020). https://doi.org/10.1038/s41586-020-03051-4
23. Siekmann, J., Green, K., Warila, J., Fern, A., Hurst, J.: Blind Bipedal stair traversal via sim-to-real reinforcement learning. Robot. Sci. Syst. (2021). https://doi.org/10.15607/RSS.2021.XVII.061
24. Silver, D., Huang, A., Maddison, C.J., Guez, A., Sifre, L., Van Den Driessche, G., Schrittwieser, J., Antonoglou, I., Panneershelvam, V., Lanctot, M., et al.: Mastering the game of go with deep neural networks and tree search. Nature **529**(7587), 484–489 (2016)
25. Stathakis, D.: How many hidden layers and nodes? Int. J. Remote Sens. **30**(8), 2133–2147 (2009)
26. Sutton, R.S., Barto, A.G.: Reinforcement learning: an introduction. Robotica **17**(2), 229–235 (1999)
27. Wang, X., Wang, S., Liang, X., Zhao, D., Huang, J., Xu, X., Dai, B., Miao, Q.: Deep reinforcement learning: a survey. IEEE Trans. Neural Netw. Learn. Syst. 1–15 (2022). https://doi.org/10.1109/TNNLS.2022.3207346
28. Wang, Z., Schaul, T., Hessel, M., Hasselt, H., Lanctot, M., Freitas, N.: Dueling network architectures for deep reinforcement learning. In: International Conference on Machine Learning, pp. 1995–2003. PMLR (2016)
29. Yu, P.L.: Cone convexity, cone extreme points, and nondominated solutions in decision problems with multiobjectives. J. Optim. Theory Appl. **14**, 319–377 (1974)
30. Yu, T., Kumar, A., Rafailov, R., Rajeswaran, A., Levine, S., Finn, C.: COMBO: conservative offline model-based policy optimization. Adv. Neural Inf. Process. Syst. (NeurIPS) **35**, 28954–28967 (2021)

FireNet-Micro: Compact Fire Detection Model with High Recall

Simi Issac Marakkaparambil[1], Reshma Rameshkumar[1], Manju Punnanilkunnathil Dinesh[1], Asra Aslam[2], and Mohammad Samar Ansari[1(✉)]

[1] University of Chester, Chester, UK
{2229464,2229948,2225186,m.ansari}@chester.ac.uk, mdsamar@gmail.com
[2] Faculty of Medicine and Health, University of Leeds, Leeds, UK
a.aslam2@leeds.ac.uk

Abstract. Fire occurrences and threats in everyday life incur substantial costs on ecological, economic, and even social levels. It is crucial to equip establishments with fire prevention systems due to the notable increase in fire incidents. Numerous studies have been conducted to develop efficient and optimal fire detection models in order to prevent such mishaps. Initially, thermal/chemical methods were used, but later, image processing techniques were also employed to identify fire occurrences. Recent approaches have capitalized on the advancements in deep learning models for computer vision. However, most deep learning models face a trade-off between detection speed and performance (accuracy/recall/precision) to maintain a reasonable inference time (for real-time applications) and parameter count. In this paper, we present a bespoke and highly lightweight convolutional neural network specifically designed for fire detection. This model can be integrated into real-time fire monitoring equipment and potentially applied in future methods suhc as CCTV surveillance cameras, traffic lights, and unmanned aerial vehicles (drones) for fire monitoring in futuristic smart city scenarios. Despite having significantly fewer trainable parameters, our customized model, FireNet-Micro, outperforms existing low-parameter-count models in fire detection. When evaluated on the FireNet dataset, FireNet-Micro, with only 171,234 parameters, achieved an impressive overall accuracy of 96.78%. In comparison, FireNet-v2 attained 94.95% accuracy with 318,460 parameters (which is almost double the parameter count of the proposed FireNet-Micro).

1 Introduction

Fire hazards pose a significant threat to human life and property, making fire detection a crucial factor in preventing potentially fatal fire incidents. In the year ending March 2020, the overall economic and social impact of fires in England amounted to £12.0 billion. Out of this total, £3.2 billion represents the additional costs incurred after the fires, known as marginal costs. The remaining £8.8 billion

N. Naik et al. (Eds.): UKCI 2023, AISC 1453, pp. 65–78, 2024.
https://doi.org/10.1007/978-3-031-47508-5_6

corresponds to the proactive measures implemented to prevent fires or minimize their damage and effects, referred to as anticipation costs [1].

Traditional fire detection methods, such as thermal, photometric, and chemical detectors, have limitations as they require a substantial amount of smoke or fire to activate and are prone to false triggering [2,3]. Moreover, these methods are not suitable for outdoor detection in places like forests, streets, playgrounds, and industries. To address these shortcomings, there is a shift towards utilizing hand-designed detectors that leverage advancements in digital camera and computer vision technology, along with intelligent video approaches. Initially, researchers focused on extracting color and shape characteristics of smoke to differentiate it from fire in images and videos using hand-designed features. However, with technological advancements, image processing, computer vision, and artificial intelligence (AI) techniques have been introduced to overcome previous challenges. These cutting-edge fire detectors offer enhanced resilience, speed, and reliability. Unlike traditional methods, they do not rely on smoke accumulation, allowing them to operate effectively in outdoor settings with fewer false triggers and quicker response times. This improvement can be mainly attributed to the reduced inference times of the models used in these detectors. With the widespread presence of cameras and closed-circuit television (CCTV) systems in various locations such as streets, highways, businesses, shopping centers, parks, and buildings, the integration of visual-based fire detectors with surveillance setups has become feasible. This integration takes advantage of the interconnected networking infrastructure and the Internet of Things (IoT) to visualize objects and activities through cameras and CCTVs, enabling efficient monitoring.

One of the notable advantages of visual-based fire detectors is the elimination of hardware and materials used in thermal/chemical fire detectors. Instead, these detectors rely on software constructs, specifically AI models, to analyze real-time images or video frames for fire detection. Modern fire detectors leverage deep learning (DL) models, which eliminate the need for manual feature extraction and enable automatic feature extraction directly from images or video frames in real-time [4]. This approach offers several benefits, including improved accuracy, reduced false triggering rates, enhanced robustness, and increased reliability compared to traditional methods.

Contribution and Novelty: In this research paper, our focus is on an improved version of FireNet[5] and FireNet-*v2* [6]. This enhanced version exhibits a significant reduction in trainable parameters and a notable better Recall (sensitivity) as compared to its predecessors – which is particularly relevant in fire detection scenarios where the detection of true positive cases is crucial, and a high Recall helps ensure that the model is capturing as many positive instances (fire) as possible. What sets this lightweight model apart is its bespoke design for fire detection purposes. We proceed to demonstrate that, despite having substantially fewer parameters, the proposed model outperforms its predecessors, making it a formidable contender among modern integrated fire detectors driven by deep learning based computer vision methods.

The paper is structured as follows: Sect. 2 provides a concise yet relevant discussion on previous research regarding manually created features and AI methods used for fire detection. Section 3 presents the detailed architecture of the proposed DL model, FireNet-Micro, along with the rationale behind choosing it over FireNet and FireNet-*v2*. Section 4 includes a comprehensive analysis of the results obtained from the experiments, along with a comparison to other published works in the field. Lastly, Sect. 5 presents concluding remarks summarizing the findings and highlighting the key contributions of the study.

2 Related Works

Extensive research has been conducted to develop effective fire detection systems in order to mitigate fire risks. Modern fire detection systems commonly employ ion-based, infrared, or optical sensors, which require proximity to the fire source or location for activation. However, these sensors may not be suitable for certain settings such as markets, schools, and open areas.

As an alternative to these sensors, vision-based sensors have emerged as a preferred substitute, offering several advantages over traditional sensors. These advantages include lower costs, faster response times, and broader surveillance coverage [7]. Nevertheless, vision-based sensors also come with significant drawbacks, including their reliance on scene complexity, varying lighting conditions, and variable image quality.

In the early stages, researchers focused on the motion and color aspects of flame detection, developing custom algorithms specifically designed for fire detection [8–10]. These algorithms, known as manually-engineered fire detection algorithms, are computationally efficient and can be implemented on resource-constrained embedded hardware like the Raspberry Pi, achieving reasonable performance in terms of frame rates. However, they have a drawback of requiring human extraction of features from raw fire-scene images. This flaw results in time-consuming and often ineffective manual feature engineering, particularly when dealing with large datasets containing numerous images.

In view of this, manually crafted vision-based fire detection systems are being substituted by deep learning (DL)-based techniques that have shown superior performance across various parameters compared to less accurate alternatives with a higher false triggering rate. DL approaches have the advantage of automatically extracting meaningful features from the input data, making the overall process more efficient, less dependent on human operators, and significantly advancing the state-of-the-art in image classification and object recognition [4,11].

Numerous DL algorithms have been proposed in the technical literature for fire detection [11–17]. However, these research models are not suitable for practical fire detection applications in the field, where low-cost, resource-constrained hardware is typically available. Consequently, there has been a significant amount of academic work focusing on developing lightweight DL models specifically designed for edge devices with limited resources [18]. Examples of such lightweight DL models can be found in areas like vehicle and drone trajectory prediction [19,20], machine learning supported disease detection [21], image forgery detection [22], and various other real-life applications [18]. Notable models for fire detection in diverse situations include, but are not limited to, [23–30].

Although these works are innovative and effective, their reported results are based on different datasets. In this paper, we will only consider works that have presented their test results on the FireNet dataset, which has been utilized in our study, for the purpose of comparing the performance of the proposed model.

3 Proposed Model

The remarkable success of Convolutional Neural Networks (CNNs) in various computer vision tasks has led researchers to explore their application in fire detection from images and videos. CNNs consist of three main types of layers. The convolutional layer plays a crucial role by applying convolutions to the input data. Non-linear activation functions like ReLU are commonly employed after the convolutional layer to introduce non-linearity and capture complex interactions between input and output. Pooling layers such as max pooling or average pooling are used to reduce the spatial dimensions of the data, enhancing computational efficiency and reducing sensitivity to minor spatial variations. By combining these layers, CNNs can capture important details while discarding less critical information. The final component of a CNN typically consists of one or more fully-connected layers, sometimes with dropout layers in between, enabling the network to learn increasingly complex and abstract features from raw pixel input. CNNs are highly effective for various computer vision tasks due to their hierarchical nature.

As mentioned earlier, several CNN variants have been employed to improve the accuracy of fire detection and reduce false alerts, inspired by advancements in edge computing capabilities and deep feature extraction [12–16,31–33]. However, these CNN models have a heavy architecture that makes them challenging to deploy on low-cost hardware. To address this, we propose a lightweight neural network called FireNet-Micro, which maintains fire detection performance while significantly reducing the number of parameters compared to its predecessors, namely FireNet [5] and FireNet-*v2* [6].

3.1 Motivation for FireNet-Micro

The selection of an appropriate DL model has always been crucial for diverse applications where deep learning is leveraged. In such scenarios, having a model that is both swift and accurate is indispensable, as even minor delays can lead to significant consequences in terms of human lives and financial resources. Initially developed, FireNet aimed to surpass other available CNN models of its time. Its advantage resided in its relative simplicity compared to earlier deep learning-based fire detection techniques, which typically involved extensive Convolutional Neural Networks (CNN) capable of real-time fire detection at a minimum frame rate of 24 frames per second or higher. FireNet effectively operated in real-time applications for continuous fire detection and was also well-suited for deployment on resource-constrained embedded and mobile devices. It could run on affordable single-board computing platforms like the Raspberry Pi, achieving a frame rate surpassing 24 frames per second. Subsequently, FireNet-*v2* was introduced as an improved iteration of FireNet, offering enhanced performance metrics and a reduced number of parameters.

FireNet-Micro was motivated by the desire to reduce both the computational requirements and the number of trainable parameters compared to its predecessors, while still achieving improved accuracy. In this paper, we will provide a detailed explanation of how FireNet-Micro surpasses the original FireNet and the enhanced FireNet-*v2*, despite having a reduced parameter count. Specifically, the modifications made in our work from the previous versions can be summarized as follows:

- In contrast to the previous iterations, FireNet-Micro utilizes only 2 fully connected (Dense) layers, leading to a substantial decrease in the number of trainable parameters. The most noteworthy advantage of this approach is the remarkably low parameter count of the network, which amounts to only 171,234 parameters.
- The activation function of choice employed throughout the network is 'ReLU', except for the last Dense layer, which utilizes a SoftMax activation. The SoftMax activation is necessary for the final layer as the desired output type involves two classes[1].
- Considering the significance of Recall in scenarios where the detection of true positive cases is crucial, such as in the fire detection use-case (other examples include medical diagnosis, fraud detection), and considering that Precision and Recall oftentimes exhibit a 'trade-off', the FireNet-Micro model was trained specifically to improve the Recall – and achieving a Sensitivity of 97.47%.

3.2 FireNet-Micro Architecture

In the FireNet-Micro model, we utilized the initial (input) layer with image dimensions of $64 \times 64 \times 3$, serving as the first convolutional layer. The intermediate layers consist of a sequential model comprising convolutional layers, average pooling layers, and dropout layers, with the activation function being ReLU (rectified linear unit). All three convolutional layers are accompanied by Maxpooling layers. The first layer employs 16 filters, the second layer uses 32 filters, and the third layer employs 64 filters. The dropout values for these layers can be observed in Fig. 1. The kernel size is fixed at 3×3 for all convolutional

[1] Selecting SoftMax over Sigmoid for binary classification does not result in any performance degradation since both functions essentially serve the same purpose in this context.

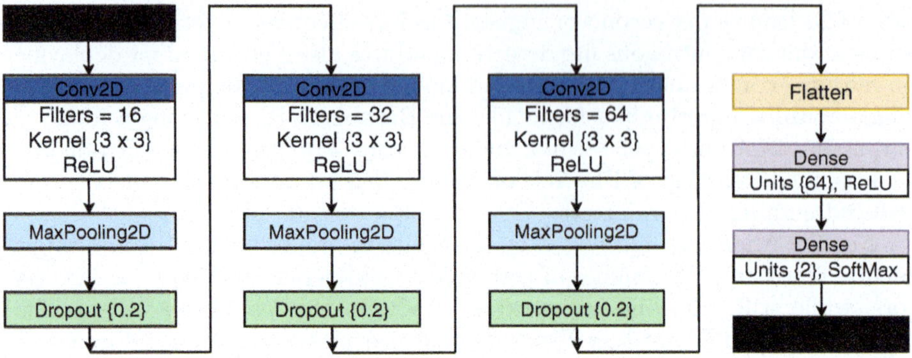

Fig. 1. Architecture of the proposed FireNet-Micro

layers. Following the convolutional layers, there is a Flatten layer. This Flatten layer serves the purpose of reshaping the tensor at the output of the third convolutional layer into a one-dimensional vector. This allows for the transition from the convolutional layers, which process spatial information, to the fully connected layers, which require one-dimensional input. The fully-connected part of the network architecture contains a Dense layer with 64 neuron, employing ReLU as the activation function. The final fully-connected Dense layer, which has 2 neurons, utilizes the 'Softmax' activation function and serves as the prediction layer, producing either 'Fire' or 'Non-Fire' signals (only one of these signals will be dominant at any given time). The detailed model architecture, consisting of only 9 layers (excluding Dropout), is depicted in Fig. 1. Overall, the model contains a minimal number of 171,234 parameters, and is only 2.1 MB on disk.

4 Results

This section presents comprehensive information regarding the dataset employed, the conducted tests, and the resulting outcomes.

4.1 Dataset

Despite the remarkable progress of deep learning (DL) models in computer vision applications over the past decade, their performance is heavily dependent on the quality and quantity of available training data. Unfortunately, there is a scarcity of high-quality datasets specifically tailored for DL models in the field of fire detection. In this study, the dataset initially created for FireNet [5] was utilized to train and test the FireNet-Micro model. This dataset consists of 46 movies depicting fire scenarios and 16 videos where fire is absent (non-fire).[2] It is important to note that this dataset is openly accessible for further research. A total of 1124 images depicting fire and 1301 images without fire (non-fire images)

[2] https://github.com/arpit-jadon/FireNet-LightWeight-Network-for-Fire-Detection.

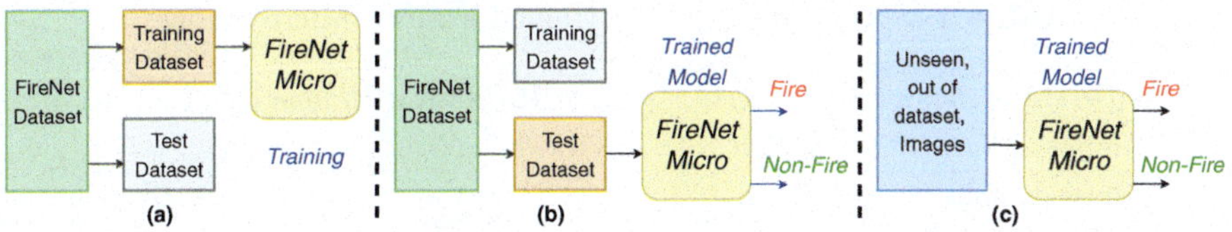

Fig. 2. (a) Training and (b) testing phases of the proposed FireNet-Micro lightweight deep learning on the FireNet dataset; (c) testing the trained FireNet-Micro model on unseen images from the World Wide Web

were employed in this study. Although the training samples from the FireNet dataset may not be numerous, they exhibit a wide range of realistic images, thereby posing a greater challenge for training the model. This assertion will be substantiated in a subsequent section, where it will be demonstrated that the accuracy of the FireNet-Micro model remains uncompromised even when confronted with unseen samples obtained from the World Wide Web.

4.2 Training Details

Figure 2 illustrates the training and testing phases of our approach. The hyper-parameters and activation functions employed in FireNet-Micro are summarized in Table 1. Since the model produces outputs in a 'one-hot' encoding format, where there are two outputs and only one is activated for each test image (representing fire or non-fire), we chose the Softmax activation for the final layer instead of Sigmoid. This choice ensures that the model decreases the estimated probabilities of the other classes while increasing the estimated probability of a specific class. By using Softmax activation, the assigned probabilities for the two output classes sum up to unity, leading to a clearer classification of fire and non-fire scenarios.

For the training process, we utilized a batch size of 32 and employed the ADAM optimizer, which incorporates adaptive learning rates based on estimates of first and second moments of gradients. ADAM combines the advantages of root mean square propagation and momentum-based methods. Its adaptive learning rate and momentum contribute to faster convergence, particularly when dealing with sparse gradients or noisy data. We trained the model for 50 epochs using the ADAM optimizer.

The training and testing of the model for FireNet-Micro was performed in the Google Colaboratory environment with the following resource specifications: Google Compute Engine backend running TensorFlow/Keras 2.12.0 on Python 3.10.12 utilizing 12 GB of RAM.

Table 1. Specifications of the trained FireNet-Micro model

(a) Details of parameters

Parameter	Value
Validation split	0.3
Number of parameters	171,234
Batch size	32
Epochs	50
Optimizer	Adam
Loss	Sparse categorical Crossentropy

(b) Activation functions

Layer	Activation
Conv2D Layer 1	ReLU
Conv2D Layer 1	ReLU
Conv2D Layer 1	ReLU
Fully-connected layer	ReLU
Output dense layer	Softmax

4.3 Model Performance

The FireNet-Micro model underwent training and testing using the FireNet dataset. The model's predictions were categorized into four classes: False Positives (FP), True Positives (TP), False Negatives (FN), and True Negatives (TN). These classifications are visualized in the form of a Confusion matrix in Fig. 3, which demonstrates the strong performance of FireNet-Micro on the dataset, as evidenced by the low percentages of FN and FP. Table 2 provides a comprehensive comparison between the proposed FireNet-Micro model and its predecessors, clearly demonstrating that FireNet-Micro surpasses the others in terms of Accuracy, Recall, and F1-Score metrics, while maintaining a lower parameter count. Additionally, Fig. 4 in the first two rows, showcases several examples of correctly classified images by FireNet-Micro, encompassing both 'fire' and 'no-fire' scenarios, and the last row contains examples of images that FireNet-Micro failed to classify accurately.

Table 2. Performance comparison of FireNet-Micro with FireNet-*v2* and FireNet

Models	FireNet-Micro	FireNet-***v2***	FireNet
Metrics	(%)	(%)	(%)
Accuracy	**96.78**	94.95	93.91
Recall	**97.47**	93.25	94
Precision	97.80	**99.28**	97
F1-Score	**97.63**	96.17	95
No of parameters	**171,234**	318,460	646,818

In the following portion of this section, we present a comparison between the proposed FireNet-Micro model and existing state-of-the-art (SoA) fire identification systems that demonstrate high performance. As mentioned earlier, the key distinguishing feature of the proposed model lies in its notably shallow

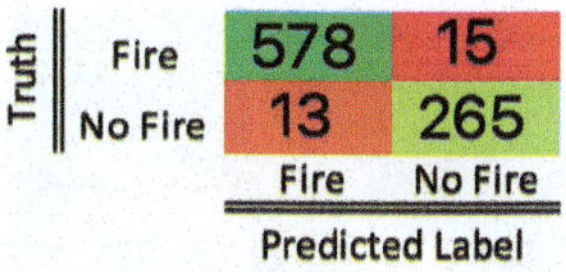

Fig. 3. Confusion Matrix for the FireNet-Micro model over the FireNet dataset

Table 3. Comparison of testing accuracy of FireNet-Micro model with other existing works on FireNet dataset

Work	Accuracy (%)	Number of parameters
Saponara [34]	93.60	**23,482**
FireNet [5]	93.91	646,818
FireNet-*v2* [6]	94.95	318,460
Ayala [35]	96.33	956,226
Saponara [36]	96.58	≈50 Million
FireNet-Micro	**96.78**	171,234

architecture, resulting in a significantly reduced number of trainable parameters (171,234). It is important to acknowledge that there are advanced fire detection technologies available in the literature that offer enhanced performance. However, their practical implementation and commercial utilization are hindered by their heavyweight architectures, characterized by large parameter counts and/or long inference times. Therefore, we consider the comparison of the proposed model with other lightweight models to be highly relevant in this context.

This part of this section presents the quantitative comparisons between the proposed approach and other existing methods. Table 3 provides a comparison of accuracy and the number of parameters among different techniques. These results are based on the inferences made by the trained model using the Test subset from the FireNet dataset, which consists of 778 images. Additionally, Table 3 includes relevant recent publications on fire detection that utilize the same dataset, to the best of our knowledge [34–37]. The performance comparison of FireNet-Micro with other available approaches, which report test results on the FireNet dataset, is visualized in Fig. 5. It is evident from Fig. 5 that FireNet-Micro achieves higher accuracy while maintaining a significantly lower parameter count. It should be noted that for [36], the published work does not provide the exact number of parameters, and the value presented in Table 3 is an estimated count considering the use of YOLOv2 in the fire detection model of [36].

Finally, to evaluate the classification performance of the proposed model on new and unseen images, a set of 200 random images was collected from the world wide web.[3] This set consisted of 100 images with fire and 100 images without fire. The trained FireNet-Micro model was utilized to make predictions on these

[3] https://github.com/Asra-Aslam/UnseenNet.

Fig. 4. The first two rows depict samples of images which were correctly classified as fire and non-fire images. The third row presents samples of incorrectly classified images by the FireNet-Micro model from test subset of the FireNet dataset.

images. The tests yielded the following performance metrics: Precision: 89.90%, Recall: 98%, F1-Score: 93.77%, Accuracy: 93.50%. An important observation was that the Recall metric achieved was quite high, indicating that the model successfully identified almost all instances of fire, missing only 2 (out of 100). Moreover, upon further investigation, it was discovered that the lower-than-expected accuracy on the 200 Internet images was due to several 'NoFire' images that closely resembled 'Fire' images. Some examples of such images are illustrated in Fig. 6. The model misclassified these similar-looking images as 'Fire' instead of 'NoFire.'

5 Conclusion

In this study, a lightweight deep learning model with a remarkably low number of parameters was introduced for the classification of images into fire and non-fire categories. The proposed model, called FireNet-Micro, is an enhanced version of the previous models FireNet and FireNet-*v2*, and it boasts a significantly reduced number of trainable parameters compared to its predecessors. To evaluate its performance, FireNet-Micro was tested on the FireNet dataset,

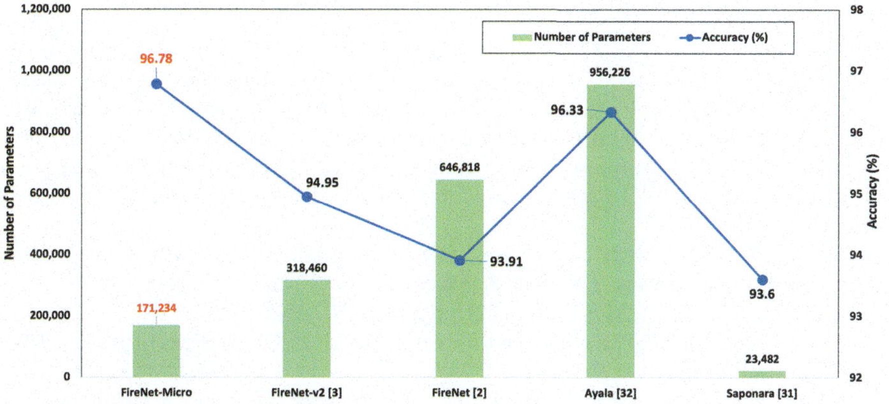

Fig. 5. Comparison of accuracy and number of parameters of FireNet-Micro with available counterparts over FireNet dataset.

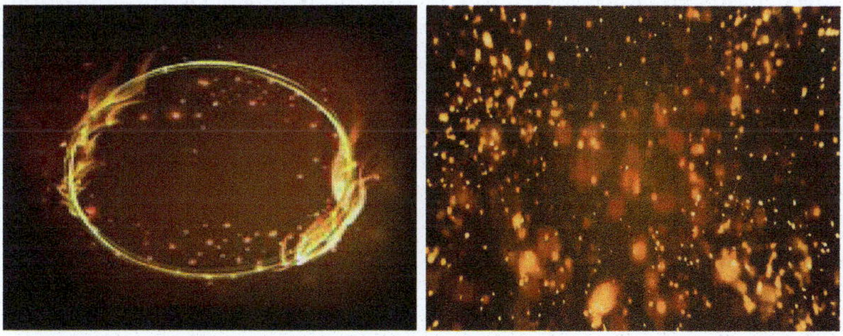

Fig. 6. The FireNet-Micro model's overall classification accuracy was impacted by the presence of certain 'NoFire' images from the Internet that closely resembled 'Fire' images. This figure contains a few examples of such images that contributed to this effect.

and its inference accuracy was found to be superior to existing low-parameter-count models. For instance, FireNet-Micro achieved an accuracy of 96.78% using only 171,234 parameters, outperforming FireNet, which achieved an accuracy of 96.53% with 646,818 parameters, and FireNet-*v2*, which achieved an accuracy of 94.95% with 318,460 parameters. Another significant feature was the high recall of 97.47% which implies that the proposed model missed only a very small proportion of the actual fire scenarios.

References

1. Pickering, J., Beall, W., Phillips, W.: Economic and social cost of fire. https://www.gov.uk/government/publications/economic-and-social-cost-of-fire/economic-and-social-cost-of-fire. Last Accessed 10 July 2023
2. da Penha, O.S., Nakamura, E.F.: Fusing light and temperature data for fire detection. In: The IEEE Symposium on Computers and Communications, pp. 107–112. IEEE (2010)
3. Chen, S.J., Hovde, D.C., Peterson, K.A., Marshall, A.W.: Fire detection using smoke and gas sensors. Fire Saf. J. **42**(8), 507–515 (2007)
4. Zaidi, S.S.A., Ansari, M.S., Aslam, A., Kanwal, N., Asghar, M., Lee, B.: A survey of modern deep learning based object detection models. Digit. Signal Process. 103514 (2022)
5. Jadon, A., Omama, M., Varshney, A., Ansari, M.S., Sharma, R.: Firenet: a specialized lightweight fire & smoke detection model for real-time IoT applications. arXiv:1905.11922 (2019)
6. Shees, A., Ansari, M.S., Varshney, A., Asghar, M.N., Kanwal, N.: Firenet-v2: improved lightweight fire detection model for real-time IoT applications. Procedia Comput. Sci. **218**, 2233–2242 (2023)
7. Dimitropoulos, K., Barmpoutis, P., Grammalidis, N.: Spatio-temporal flame modeling and dynamic texture analysis for automatic video-based fire detection. IEEE Trans. Circuits Syst. Video Technol. **25**, 339–351 (2015)
8. Chen, T.H., Wu, P.H., Chiou, Y.C.: An early fire-detection method based on image processing. In: 2004 International Conference on Image Processing, vol. 3, pp. 1707–1710 (2004)
9. Çelik, T., Özkaramanlı, H., Demirel, H.: Fire and smoke detection without sensors: image processing based approach, pp. 1794–1798 (2007)
10. Rafiee, A., Dianat, R., Jamshidi, M., Tavakoli, R., Abbaspour, S.: Fire and smoke detection using wavelet analysis and disorder characteristics. In: 2011 3rd International Conference on Computer Research and Development, vol. 3, pp. 262–265. IEEE (2011)
11. Krizhevsky, A., Sutskever, I., Hinton, G.E.: Imagenet classification with deep convolutional neural networks. Adv. Neural Inf. Process. Syst. **25** (2012)
12. Muhammad, K., Ahmad, J., Baik, S.W.: Early fire detection using convolutional neural networks during surveillance for effective disaster management. Neurocomputing **288**, 30–42 (2018)
13. Muhammad, K., Ahmad, J., Lv, Z., Bellavista, P., Yang, P., Baik, S.W.: Efficient deep CNN-based fire detection and localization in video surveillance applications. IEEE Trans. Syst., Man, Cybern. Syst. **49**(7), 1419–1434 (2018)
14. Muhammad, K., Ahmad, J., Mehmood, I., Rho, S., Baik, S.W.: Convolutional neural networks based fire detection in surveillance videos. IEEE Access **6**, 18174–18183 (2018)
15. Muhammad, K., Khan, S., Elhoseny, M., Ahmed, S.H., Baik, S.W.: Efficient fire detection for uncertain surveillance environment. IEEE Trans. Ind. Inf. (2019)
16. Zhang, Q., Xu, J., Xu, L., Guo, H.: Deep convolutional neural networks for forest fire detection (2016)
17. Sharma, J., Granmo, O.C., Goodwin, M., Fidje, J.T.: Deep convolutional neural networks for fire detection in images. In: International Conference on Engineering Applications of Neural Networks, pp. 183–193 (2017)

18. Wang, C.H., Huang, K.Y., Yao, Y., Chen, J.C., Shuai, H.H., Cheng, W.H.: Lightweight deep learning: an overview. IEEE Consum. Electron. Mag. (2022)
19. Katariya, V., Baharani, M., Morris, N., Shoghli, O., Tabkhi, H.: Deeptrack: Lightweight deep learning for vehicle trajectory prediction in highways. IEEE Trans. Intell. Transp. Syst. (2022)
20. Alsamhi, S.H., Almalki, F., Ma, O., Ansari, M.S., Lee, B.: Predictive estimation of optimal signal strength from drones over IoT frameworks in smart cities. IEEE Trans. Mobile Comput. (2021)
21. Tiwari, S., Jain, A.: A lightweight capsule network architecture for detection of COVID-19 from lung CT scans. Int. J. Imaging Syst. Technol. **32**(2), 419–434 (2022)
22. Abbas, M.N., Ansari, M.S., Asghar, M.N., Kanwal, N., O'Neill, T., Lee, B.: Lightweight deep learning model for detection of copy-move image forgery with post-processed attacks. In: 2021 IEEE 19th World Symposium on Applied Machine Intelligence and Informatics (SAMI), pp. 000125–000130. IEEE (2021)
23. Xing, Zhong, Y., Zhong, L., X.: An encoder-decoder network based FCN architecture for semantic segmentation. Wirel. Commun. Mob. Comput. (2020)
24. J., Z., Zhu, H., Wang, P., X., L.: ATT squeeze U-Net: a lightweight network for forest fire detection and recognition. IEEE Access (2021)
25. Akhloufi, M.A., Tokime, R.B., Elassady, H.: Wildland fires detection and segmentation using deep learning. In: Pattern Recognition and Tracking XXIX. The International Society for Optical Engineering is Proceedings of SPIE 2018, pp. 10649, 106490B
26. Bochkov, V., Kataeva, L.Y.: wUUNet: advanced fully convolutional neural network for multiclass fire segmentation. Symmetry (2021)
27. Xu, R., Lin, H., Lu, K., Cao, L., Liu, Y.: A forest fire detection system based on ensemble learning. Forests **12**(2), 217 (2021)
28. Jocher, G., Chaurasia, A., Stoken, A., Borovec, J., NanoCode012, Kwon, Y., TaoXie, Fang, J., imyhxy, Michael, K., Lorna, V, A., Montes, D., Nadar, J., Laughing, tkianai, yxNONG, Skalski, P., Wang, Z., Hogan, A., Fati, C., Mammana, L., AlexWang1900, Patel, D., Yiwei, D., You, F., Hajek, J., Diaconu, L., Minh, M.T.: ultralytics/yolov5: v6.1 - TensorRT, TensorFlow Edge TPU and OpenVINO Export and Inference (2022). https://doi.org/10.5281/zenodo.6222936
29. Tan, M., Pang, R., Le, Q.V.: Efficientdet: scalable and efficient object detection. In: Proceedings of the IEEE/CVF Conference on Computer Vision and Pattern Recognition, pp. 10781–10790 (2020)
30. Tan, M., Le, Q.: Efficientnet: rethinking model scaling for convolutional neural networks 6105–6114 (2019)
31. Simonyan, K., Zisserman, A.: Very deep convolutional networks for large-scale image recognition. arXiv:1409.1556 (2014)
32. He, K., Zhang, X., Ren, S., Sun, J.: Deep residual learning for image recognition. In: Proceedings of the IEEE Conference on Computer Vision and Pattern Recognition, pp. 770–778 (2016)
33. Foggia, P., Saggese, A., Vento, M.: Real-time fire detection for video-surveillance applications using a combination of experts based on color, shape, and motion. IEEE Trans. Circuits Syst. Video Technol. **25**(9), 1545–1556 (2015)
34. Saponara, S., Elhanashi, A., Gagliardi, A.: Exploiting R-CNN for video smoke/fire sensing in antifire surveillance indoor and outdoor systems for smart cities. In: 2020 IEEE International Conference on Smart Computing (SMARTCOMP), pp. 392–397. IEEE (2020)

35. Ayala, A., Lima, E., Fernandes, B., Bezerra, B.L., Cruz, F.: Lightweight and efficient octave convolutional neural network for fire recognition 1–6 (2019)
36. Saponara, S., Elhanashi, A., Gagliardi, A.: Real-time video fire/smoke detection based on CNN in antifire surveillance systems. J. Real-Time Image Proc. **18**(3), 889–900 (2021)
37. Altowaijri, A.H., Alfaifi, M.S., Alshawi, T.A., Alshebeil, S.A.: A privacy-preserving IoT-based fire detector. IEEE Access 99 (2020)

Reward-Guided Individualised Communication for Deep Reinforcement Learning in Multi-Agent Systems

Yi-Yu Lin and Xiao-Jun Zeng(✉)

Department of Computer Science, The University of Manchester, Manchester, UK
yi-yu.lin@postgrad.manchester.ac.uk, x.zeng@manchester.ac.uk

Abstract. Broadcasting communication poses a fundamental challenge in Multi-Agent Deep Reinforcement Learning, prompting the emergence of decentralised communication as a promising paradigm. However, the current approach in decentralised communication exhibits drawbacks, including message overhead and asynchronous network updating. More critically, the method predominantly relies on divergence as a metric for communication network updating, which fails to align with the goal of performance maximisation in Reinforcement Learning (RL). To address these limitations, this paper introduces Reward-Guided Individualised Communication (RGIC), a method that integrates rewards into the communication network. By adhering to RL principles, RGIC facilitates purposeful one-to-one interactions and enhances overall performance. The optimised learning process of RGIC leads to accelerated convergence, enhanced efficiency, and reduced computational requirements. Extensive experimentation validates the efficacy of RGIC, establishing its suitability for real-world multi-agent scenarios that demand real-time decision-making and reward-driven actions.

Keywords: Multi-agent system · Multi-agent deep reinforcement learning · Multi-agent communication

1 Introduction

Artificial agents, equipped with autonomous sequential decision-making capabilities, have experienced an escalating prevalence across diverse domains, attributing to the emergence of challenges that necessitate their utilisation. In the past, agents relied on supervised learning, which limited their ability to react to different situations. Yet, advancements in machine learning have paved the way for progress in Reinforcement Learning (RL), aiming to enhance agents' ability to develop knowledge [22]. Within the realm of RL, Single-Agent RL (SARL) has been extensively studied by applying Markov Decision Processes (MDP) [19]. Building upon this foundation, researchers have ventured into the field of Multi-Agent Systems (MAS). This leads to the development of Multi-Agent RL

N. Naik et al. (Eds.): UKCI 2023, AISC 1453, pp. 79–94, 2024.
https://doi.org/10.1007/978-3-031-47508-5_7

(MARL), which extends the principles of MDP to encompass agent-agent interaction [25].

Despite the effectiveness of SARL, the integration of RL in MAS presents several problems: non-stationarity, credit assignment, partial observability, and the curse of dimensionality [6]. MAS inherently exhibits non-stationarity due to the interdependencies among agents. This characteristic poses a major obstacle for RL algorithms, which are designed for stationary environments [22]. The intricate interactions among agents also give rise to the challenge of credit assignment, making it difficult to determine the impact of each agent's actions on the overall system [5]. Additionally, the assumption of full observability in RL is unrealistic in real-world MAS applications [1]. Another significant concern for RL in MAS is the curse of dimensionality, where the state and action spaces expand exponentially as the number of agents increases. This exponential growth introduces a formidable scalability barrier [4].

Deep RL (DRL), which combines deep learning and RL, along with Decentralised Partially Observable MDP (Dec-POMDP) have emerged as a solution to overcome the limitations of MARL [2]. DRL leverages the strength of deep Neural Networks (NN) to learn abstract representations from raw data, enabling agents to generalise to new scenarios without extensive prior knowledge [13]. On the other hand, Dec-POMDP, which accounts for partial observability, provides a framework for studying complex cooperation problems in MAS [1]. The combination of DRL and Dec-POMDP has led to the development of Multi-Agent DRL (MADRL) [16]. However, the inherent partial observability in MAS hampers decision-making and coordination in MAS, necessitating the utilisation of communication mechanism into MADRL [26].

Recent research on MADRL with communication has directed its attention towards broadcasting, which involves one agent sharing its observations with all others [17]. Nonetheless, this mechanism is associated with several limitations. Primarily, broadcasting encounters challenges in scaling up to large-scale MAS due to increased overhead [26]. In addition, it raises concerns over security and privacy as it may inadvertently expose sensitive information. Moreover, broadcasting can lead to sub-optimal behaviours as agents indiscriminately disseminate and gather information without considering its relevance [3].

In contrast to undifferentiated communication, humans exhibit a natural inclination to selectively engage with individuals based on prior knowledge. This tendency has drawn attention to the one-to-one request-reply communication paradigm. Motivated by these considerations, the research intends to address the research problem of developing an effective MADRL method with Communication that minimises the amount of communication required while maximising performance outcomes. In response, this paper proposes Reward-Guided Individualised Communication (RGIC), which leverages rewards to stimulate fruitful decentralised targeted communication. The main contribution of this work lies in the provision of a framework that bolsters the effectiveness and efficiency of communication, thereby enhancing performance outcomes in MAS. To the best of our knowledge, RGIC represents a pioneering endeavour in this field.

The subsequent sections of this paper are structured as follows. Section 2 provides a comprehensive review of relevant MADRL methods with Communication. Section 3 outlines the proposed methodology, including the proposed approach, details on the testing environments, the establishment of a baseline model, and the learning algorithm. Then, Sect. 4 presents a comparative analysis of the proposed method against the established benchmark. Finally, Sect. 5 concludes the paper, summarises the findings, and discusses potential avenues for future research.

2 Related Work

The domain of multi-agent communication has witnessed remarkable progress in recent years, with notable contributions from various researchers. Among them, Foerster et al. play a pioneering role by introducing Reinforced Inter-Agent Learning (RIAL) and Differentiable Inter-Agent Learning (DIAL) [9]. RIAL enables agents to autonomously learn network parameters while considering other agents as part of the environment, whereas DIAL provides a unified learning framework for communication and environmental policies [9]. Both methods, however, have certain limitations in terms of discrete messages and evaluation primarily in fully cooperative environments [26].

In the pursuit of enhanced coordination and information exchange among agents, Sukhbaatar et al. put forward CommNet, an approach that utilises continuous vectors as communication messages to streamline training procedures and improve information throughput [21]. However, CommNet encounters challenges associated with information loss during message generation and credit assignment [26]. Expanding upon CommNet, Hoshen presents Vertex Attention Interaction Networks (VAIN), which incorporates a graph NN (GNN) and an attention mechanism to capture intricate relationships and leverage structural patterns [10]. Nonetheless, VAIN causes additional computational burden and is primarily focused on supervised prediction [7]. Furthermore, Singh et al. extend the versatility of the CommNet framework with Individualised Controlled Continuous Communication Model (IC3Net), enabling individual agents to have independent control and rewards [20]. Despite addressing the credit assignment issue, IC3Net faces obstacles related to information loss and an increased computational cost due to its gating mechanism [15].

Embracing the concept of selective communication, Jiang and Lu propose Attentional Communication Model (ATOC), which incorporates an attention unit to determine optimal timing for communication with agents within the observable field [11]. ATOC forms dynamic communication groups; however, the predefined nature of these groups can be time-consuming and labour-intensive [15]. In a similar vein, Das et al. introduce Targeted Multi-Agent Communication (TarMAC) which employs a signature-based soft attention mechanism to facilitate one-to-one communication while highlighting relevant information [7]. Although aiming to enable one-to-one communication, the underlying mechanism of TarMAC retains properties of a conventional broadcasting approach [8].

Prior multi-agent communication approaches have relied on broadcasting paradigms, leading to redundant information and potential obstacles in the learning process. Moreover, these approaches often assume unlimited bandwidth, which is unrealistic in practical scenarios. To address these limitations, Ding et al. propose Individually Inferred Communication (I2C) as a solution. I2C emphasises targeted one-to-one communication by leveraging prior knowledge and local observations to determine appropriate communication targets [8]. It employs causal inference to establish belief and incorporates correlation regularisation to refine an agent's policy [8]. Despite its promise, I2C encounters substantial message size, delayed updates of the communication network, and limitations associated with using KL divergence as a metric for network updating [24].

Notwithstanding the significant progress in multi-agent communication, current approaches suffer from several weaknesses and limitations, such as the presence of redundant information, reliance on unrealistic assumptions, information loss, credit assignment challenges, and increased communication costs. To bridge the identified gaps, this paper introduces RGIC, which endeavours to surpass the constraints of current methods and elevate the effectiveness and efficiency of multi-agent communication.

3 Methodology and Process

3.1 Proposed Method

Existing methods in multi-agent communication have made significant advancements but are not without their limitations. Many of these approaches rely on broadcasting communication, which proves to be ineffective and impractical in real-world scenarios. The presence of redundant information in such methods can hinder the learning process and potentially lead to sub-optimal performance. While I2C has been put forward as a solution that emphasises one-to-one communication, it still faces certain challenges. In light of these drawbacks, this research proposes a reward-driven MADRL with Communication method, called Reward-Guided Individualised Communication (RGIC), that minimises unnecessary communication while simultaneously maximising performance.

RGIC presents a fundamental change in the communication network by incorporating rewards. This approach empowers the communication network to acquire the capability of learning and prioritising interactions that directly contribute to the attainment of higher rewards. By emphasising the correlation between communication actions and rewards, RGIC fosters the communication network to give precedence to purposeful and meaningful communication, as shown in Fig. 1. As a result, the objectives of the communication network align more closely with the overarching objective of maximising performance in the MAS.

3.2 Environment Setting

Two testing environments are employed in this study: Cooperative Navigation (CN) and Predator Prey (PP) scenarios. To ensure the robustness and reliability

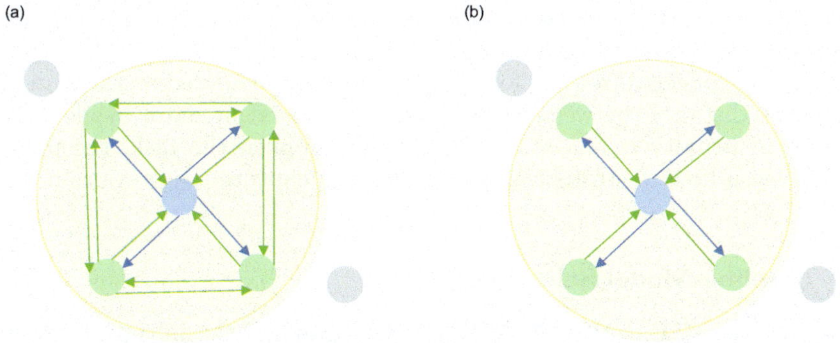

Fig. 1. RGIC introduces a distinct communication mechanism that differs from existing approaches. Unlike method a), which enables communication between all agents, RGIC restricts communication to agents within the observable field. Agents do not communicate with entities outside the boundary, referred to as 'outsiders' (represented by grey circles). Additionally, RGIC implements a specific communication pattern where a 'student' agent (green circle) seeking knowledge initiates communication with a 'teacher' agent (blue circle) possessing the desired knowledge. This communication pattern allows the 'student' agent to acquire valuable knowledge that subsequently leads to more advantageous actions.

of the obtained results, each method is executed five times. The mean performance of each method is depicted by a solid line in the result figures. In order to capture the performance variations comprehensively, the shaded area in the figures represents the range between the minimum and maximum values obtained from the five training runs. This approach facilitates a comprehensive analysis and comparison of the performance across different methods, taking into account both the average performance and the observed variability across multiple runs.

1. Cooperative Navigation: A group of $N = 7$ agents is tasked with occupying a set of $L = 7$ landmarks. Each agent has access to only partial observations of the environment. The observation includes the positional information of the three nearest agents and the three nearest landmarks, relative to the agent's location. Moreover, their communication is restricted to other agents within their observable range. The collective reward is computed as the sum of the negative distances between each landmark and its closest agent. In case of collisions, a penalty of $r_{collide} = -1$ is applied. The primary objective of the agents is to ensure coverage of all the landmarks [14]. To achieve this, agents must take into account the motives of others to determine the most suitable landmark to occupy while simultaneously avoiding collisions with other agents.
2. Predator Prey: A group of $N = 7$ predators (agents) are assigned the objective of capturing $M = 3$ preys. Each predator's observations include the relative positions of the three nearest predators and the three nearest preys. The preys are not stationary; instead, they have designated activity areas. Further, the preys exhibit higher mobility compared to the predators and demonstrate

evasive behaviour, actively avoiding the closest predators. This behavioural characteristic implies that a single predator alone is incapable of capturing prey. The collective reward for the predators is calculated as the sum of the negative distances between each prey and its closest predator, while collisions result in penalty $r_{collide} = -1$ [14]. To accomplish the task, the predators must learn how to strategically surround and capture the preys as a team.

3.3 Baseline Model Selection

Comparison of Existing MADRL Methods As the initial phase of this study, a baseline is established through a comparative analysis of existing methods in multi-agent communication. This analysis aims to provide a comprehensive understanding of the strengths and limitations exhibited by IC3Net [20], TarMAC [7], and I2C [8], which employ diverse communication strategies and learning mechanisms. The evaluation encompasses critical aspects such as convergence speed, learning stability, and overall effectiveness in achieving desired outcomes. The analysis serves as a valuable reference for evaluating the impact and effectiveness of the proposed approach. Additionally, *the objective is to identify the most effective method among the existing approaches, which can serve as a foundation for the development of the proposed method.* Given the significance of the addressed issues, leveraging the most effective established method facilitates the advancement of a more sophisticated algorithm with improved efficacy. To ensure experimental integrity, two sets of hyperparameters were employed for each experiment. The first set adhered to the recommended hyperparameters from the original papers of the respective methods, while the second set adopted the hyperparameters proposed in the I2C paper. The results of the comparative analysis are presented in Table 1.

Table 1. Comparative performance analysis of existing MADRL methods

Methods	Environments
	CN (Fig. 2) & PP (Fig. 3)
IC3Net	It seems to have constraints on exploration to uncover more optimal actions, as indicated by the stagnant learning trajectories. The agents lack foresight and may initially discover high-reward actions that are ultimately unsustainable.
TarMAC	It appears to exhibit some instability in the learning trajectory. This behaviour indicates that it may struggle to consistently converge to optimal solutions.
I2C	It experiences a significant decrease in performance at around iteration 3.2×10^4, which coincides with the start of its updating phase. This observation suggests that during the initial update, its agents engage in environment exploration and action experimentation, which may inadvertently lead to a temporary decrease in performance.

Summary: Across all hyperparameter settings, I2C consistently achieves the highest performance, followed by TarMAC and IC3Net. When employing the hyperparameters from the I2C paper, both IC3Net and TarMAC show inferior performance compared to using their own.

Evaluation of I2C's Targeted Communication In Sect. 3.3, the study established that I2C demonstrates superior performance compared to TarMAC and IC3Net. Building upon this finding, the investigation proceeds to examine the specific communication capabilities of I2C in comparison to two alternative strategies: communicating with all agents within the field of view (FC) and not communicating at all (NC). Table 2 presents a comparative analysis of the performance of NC and FC in relation to I2C.

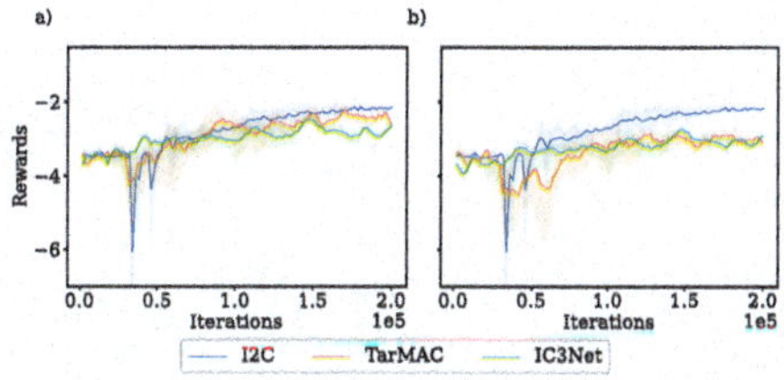

Fig. 2. Comparative performance analysis of existing MADRL Methods in CN. a) Hyperparameters from the respective paper; b) Hyperparameters in the I2C paper.

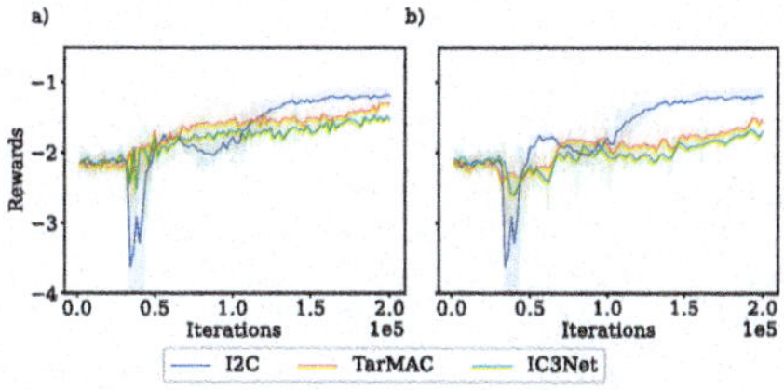

Fig. 3. Comparative performance analysis of existing MADRL Methods in PP. a) Hyperparameters from the respective paper; b) Hyperparameters in the I2C paper.

Summary: The targeted communication strategy employed by I2C demonstrates its effectiveness in attaining preeminent performance. This approach facilitates accelerated convergence and improved performance. Importantly, the targeted communication achieves these outcomes by effectively sharing and extracting relevant information while minimising the amount of necessary information.

Table 2. Comparative performance analysis of NC and FC in relation to I2C's targeted communication

Variants	Environments	
	CN (Fig. 4)	PP (Fig. 5)
NC	It shows signs of early convergence to a sub-optimal policy, possibly due to limited access to comprehensive knowledge. Additionally, it displays a considerably higher standard deviation.	The tendency of early convergence observed in CN persists.
FC	It demonstrates slightly inferior performance compared to I2C, suggesting that even full communication among observable agents may introduce redundant information that impedes the learning process [8]. Also, it demonstrates a greater degree of variability in the learning processes.	Communication plays a critical role in PP to facilitate agent coordination, so it achieves comparable outcomes as I2C.

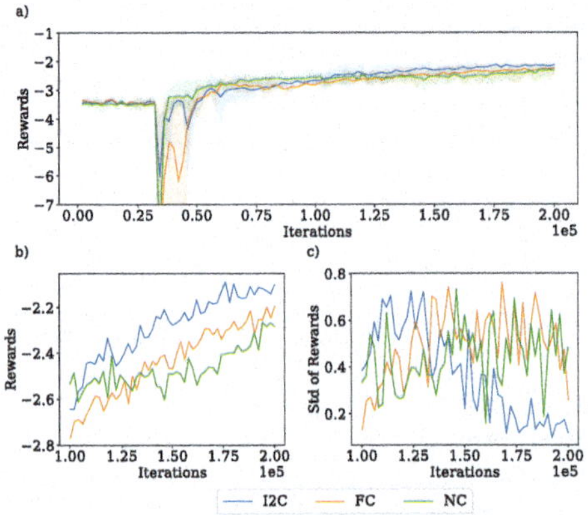

Fig. 4. Comparative performance analysis of I2C's targeted communication versus NC and FC in CN. a) Overall learning trajectory; b) Mean rewards after 1×10^5 iterations; c) Standard deviation of the mean rewards after 1×10^5 iterations.

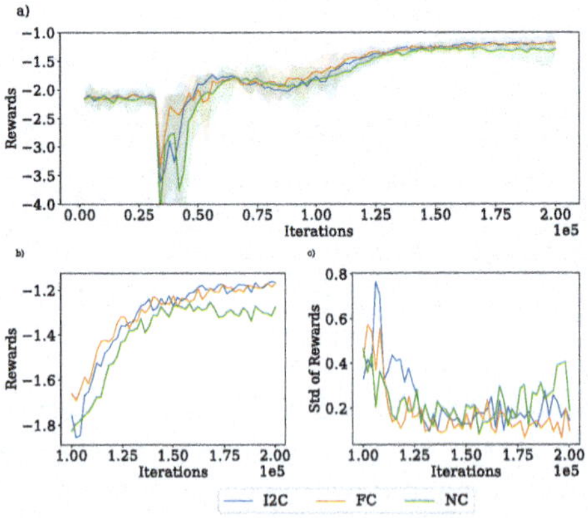

Fig. 5. Comparative performance analysis of I2C's targeted communication versus NC and FC in PP. a) Overall learning trajectory; b) Mean rewards after 1×10^5 iterations; c) Standard deviation of the mean rewards after 1×10^5 iterations.

Limitations of I2C The preceding sections have established the efficacy of the I2C method when compared to other existing approaches. Consequently, further investigations were conducted to scrutinise the implementation details of the I2C algorithm. These investigations revealed that while I2C demonstrates effectiveness, it is not devoid of limitations and shortcomings.

1. Burden of Message Size: The transmission of complete observations as the communicated message between agents can lead to large message sizes, introducing communication overhead and increasing computational and communication costs. This can limit scalability and pose challenges in terms of memory usage and communication bandwidth.
2. Delayed Update of Communication Network: The delayed update timing of the communication network compared to the policy and centralised Critic networks raises concerns about its contribution to the overall performance. By the time the communication network is updated, the other networks may have already converged to a near-optimal solution. This could potentially limit the effectiveness of the communication network in further improving the system's performance.
3. Limitations of KL Divergence as an Update Indicator: Relying on KL divergence as an indicator for updating the communication network in I2C may not capture the true performance of the communication strategy in maximising rewards. While it aims to match the distribution of the communication policy, KL divergence does not directly consider the impact on overall reward maximisation. This approach may lead to inferior communication policies that prioritise distribution matching over maximising cumulative rewards.

3.4 RGIC

Based on the identified limitations of I2C, the research progresses by proposing RGIC that addresses these challenges. RGIC builds upon the structure of I2C and specifically targets the communication network for improvement.

Firstly, to tackle the issue of large message size and communication overhead, RGIC utilises the agent's coordinate as the communication message. By minimising the message content, the burden of transmitting extensive observations across agents is alleviated.

Secondly, RGIC emphasises synchronising the update of the communication network with the policy and centralised Critic networks. By aligning the update timing and frequency of these networks, it ensures that all networks contribute to the learning process simultaneously. This synchronisation allows for a more coordinated learning process, where different networks complement each other in achieving improved performance.

Thirdly, RGIC introduces a fundamental change in the loss function of the communication network by incorporating rewards from the buffer. Unlike the use of KL divergence in I2C, RGIC directly integrates the concept of performance optimisation. This explicit guidance towards actions that yield higher cumulative rewards aligns better with the goal of RL. Further, the method investigates two types of rewards: instant rewards obtained when agents communicate with the target (direct reward *Dir*) and average rewards obtained when communicating with the target (average reward *Avg*).

Moreover, RGIC explores a diverse range of NNs in addition to the fully connected NN (FCNN) used in I2C. These include convolutional (CNN) [12], Inception (INCEPT) [23], and RNN [18]. Each architecture is chosen based on its specific strengths in learning complex patterns and extracting relevant features from the input data. By investigating the effectiveness of these NNs and comparing their performance, the proposed method aims to identify the most suitable architecture and reward type for improving the communication network.

Learning Algorithm RGIC follows the training process that involves the use of a centralised joint action-value function $\hat{Q}_i(o,a)$, parameterised by continuous policies π_{θ_i} with regards to θ_i. It takes as input the joint actions a and observations o of all agents from the buffer and guides the optimisation of the policy. To update the centralised Critic, the following equation is employed.

$$\mathcal{L}(\theta_i) = \mathbb{E}[\,(\hat{Q}_i(o,a) - r_i - \gamma \hat{Q}'_i(o',a')|_{a'\sim\pi_{\theta'}(o)})^2] \tag{1}$$

where $\pi_{\theta'}$ refers to target policies with delayed parameters θ'_i and a' are sampled from $\pi_{\theta'}(o')$.

The regularised policy's gradient parameterised by π_{θ_i} can be derived as follows, with η being the coefficient for correlation regularisation.

$$\begin{aligned}\nabla_{\pi_{\theta_i}} J(\pi_{\theta_i}) &= \mathbb{E}[\mathbb{E}[\,\nabla_{\pi_{\theta_i}} y \hat{Q}_i(o,a_i,a_{-i})] \,-\eta \nabla_{\pi_{\theta_i}} \rho]\,, \\ y &= \log \pi_i(a_i|c_i,o_i), \\ \rho &= D_{KL}(P(a_i|a_{-i},o)\parallel \pi_i(a_i|c_i,o_i))\end{aligned} \tag{2}$$

In addition, the gradient of the message encoder, parameterised by e_{θ_i}, can be expressed as:

$$\nabla_{e_{\theta_i}} J(e_{\theta_i}) = \mathbb{E}[\mathbb{E}[\,\nabla_{e_{\theta_i}} e_i(c_i|m_i)\nabla_{c_i} y\hat{Q}_i(o,a_i,a_{-i}) - \eta \nabla_{e_{\theta_i}} e_i(c_i|m_i)\nabla_{c_i}\rho] \tag{3}$$

Lastly, the prior network, parameterised by b_{θ_i}, serves as a binary classifier and is trained using the below loss function. The labels l_i^j for the training samples are generated based on the normalised rewards ζ_i^j obtained from the buffer. Specifically, $l_i^j = 1$ if $\zeta_i^j > 0.5$, and 0 otherwise. For clarity, x is defined as $x = b_i(o_i, d_j)$. This loss function incorporates reward weighting, thus amplifying the learning from communicated targets associated with higher rewards.

$$\mathcal{L}(b_{\theta_i}) = \mathbb{E}[(-(1-l_i^j)\log(1-\frac{1}{1+e^{-x}}) - l_i^j \log(\frac{1}{1+e^{-x}})) \times \frac{1}{\zeta_i^j}] \tag{4}$$

The pseudocode outlining the workflow of the proposed method for N agents is presented in Algorithm 1.

Algorithm 1: RGIC for N agents

for episode $ep = 1$ to ep_{max} **do**
  Initialise agents and action exploration process
  for step $t = 1$ to $step_{max}$ **do**
    for agent $i = 1$ to N **do**
      Receive initial observation o_i
      Identify agents within the field of view
      Using the prior network b_i
      for observable agent $j = 1$ to N_{obs} **do**
        if $b_i(o_i, d_j)$ **then**
          Send a request to agent j
          Receive a message m^j from agent j
      $m_i \leftarrow$ all the received messages
      Generate encoded message c_i using the message encoder $e_i(m_i)$
      Execute action based on $\pi_i(a_i|c_i, o_i)$
      Obtain reward r_i and new observation o_i'
      Store (o_i, o_i', a_i, r_i) in the reply buffer B
      $o_i \leftarrow o_i'$
      if $t >= D_{len}$ and not $t\%100 = 0$ **then**
        $(o^j, o'^j, a^j, r^j) \leftarrow$ a random minibatch of S samples from D
        Update the centralised Critic $\hat{Q}_i(o, a)$ using Eq. 1
        Update the actor π_i using Eq. 2
        Update the message encoder e_i using Eq. 3
        Update prior network b_i using Eq. 4

4 Experimental Results

The research incorporates eight distinct algorithms, comprising four types of NN architectures (FCNN, CNN, INCEPT, and RNN), each paired with two types of rewards (*Dir* and *Avg*). In this section, a comparative analysis is conducted between I2C and a specific NN architecture, coupled with direct reward and average reward. To maintain fairness in the comparison, the proposed method adopts the same hyperparameters as outlined in the I2C paper [8]. The experimental findings of RGICs are presented in Table 3. The column titled 'RGICs' provides an analysis of the outcomes observed across all proposed methods within a specific testing environment. Subsequently, the column, 'Best Performing RGICs', compares the top-performing method for each NN against I2C.

Summary: RGICs consistently demonstrate superior performance over I2C across various NN architectures and reward types. They exhibit faster convergence rates, suggesting their efficacy in learning optimal policies. Moreover, the utilisation of average reward has been consistently observed to yield improved performance compared to direct reward. This can be attributed to the stability and reliability provided by the average reward metric, which mitigates the effects of noise and significant reward fluctuations.

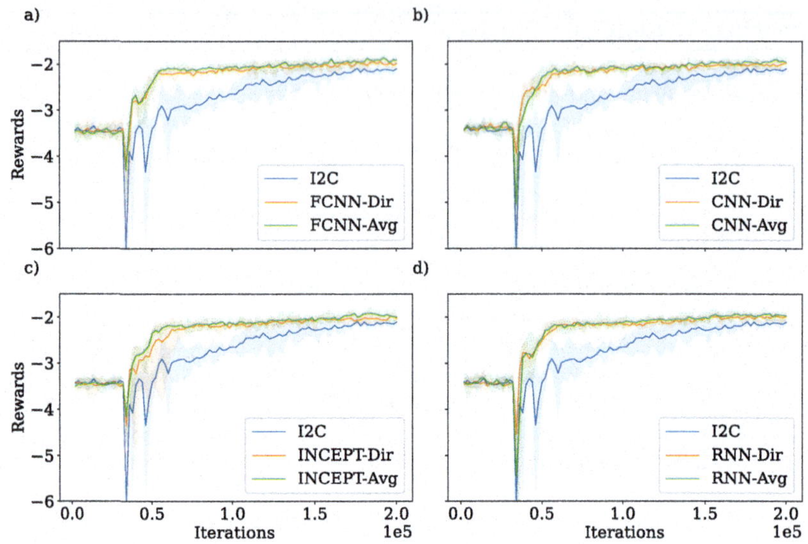

Fig. 6. Comparative performance analysis of RGICs versus I2C in CN. a) FCNN; b) CNN; c) INCEPT; d) RNN.

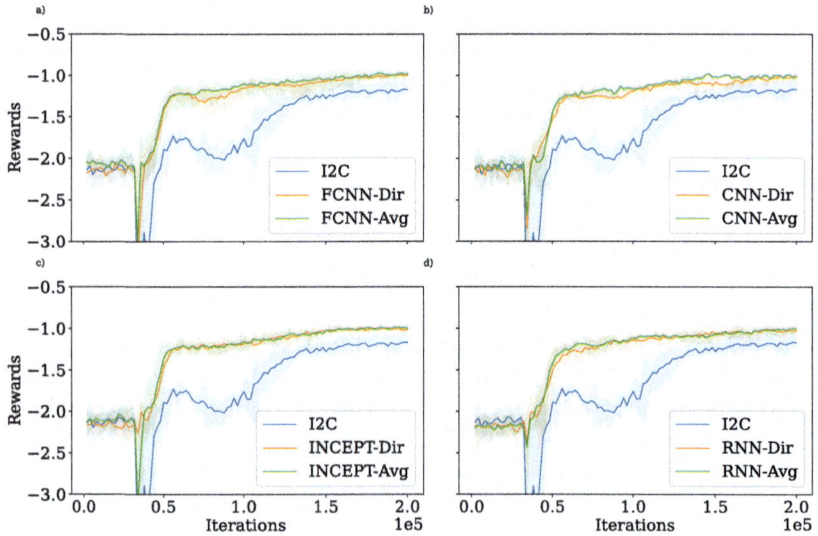

Fig. 7. Comparative performance analysis of RGICs versus I2C in PP. a) FCNN; b) CNN; c) INCEPT; d) RNN.

Among the proposed methods, *FCNN-Avg* stands out as the most effective approach, considering its performance in both CN and PP. Notably, *FCNN-Avg* achieves the highest reward in both CN and PP scenarios, despite exhibiting a slightly higher standard deviation of mean reward in PP than other proposed

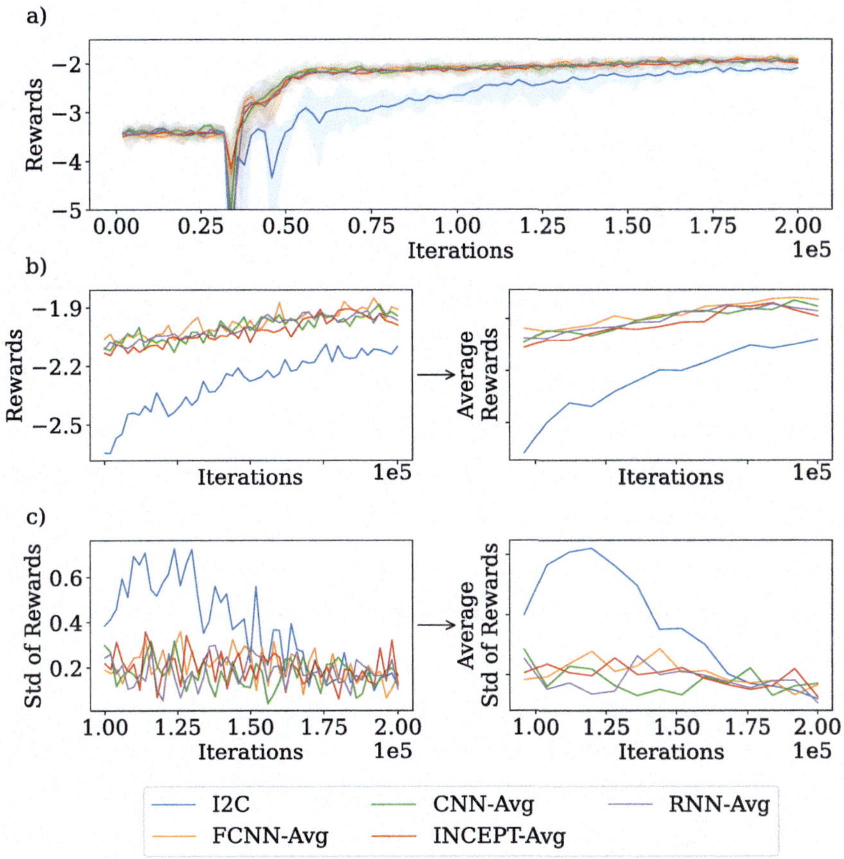

Fig. 8. Comparative performance analysis of best RGICs versus I2C in CN. a) Overall learning trajectories; b) left: Mean rewards obtained after 1×10^5 iterations, right: Averaged mean rewards using a moving window of 1×10^4 iterations; c) left: Standard deviation of mean rewards obtained after 1×10^5 iterations, right: Averaged standard deviation of mean rewards using a moving window of 1×10^4 iterations.

methods. Nonetheless, this standard deviation remains lower than that of I2C, indicating *FCNN-Avg*'s overall robustness and consistency.

5 Conclusion

This paper proposes RGIC as a solution to overcome the limitations of existing multi-agent communication methods. RGIC incorporates rewards into the communication network to enable purposeful one-to-one communication and maximise performance in MAS. By minimising communication overhead, synchronising network updates, and integrating rewards into the loss function, RGIC

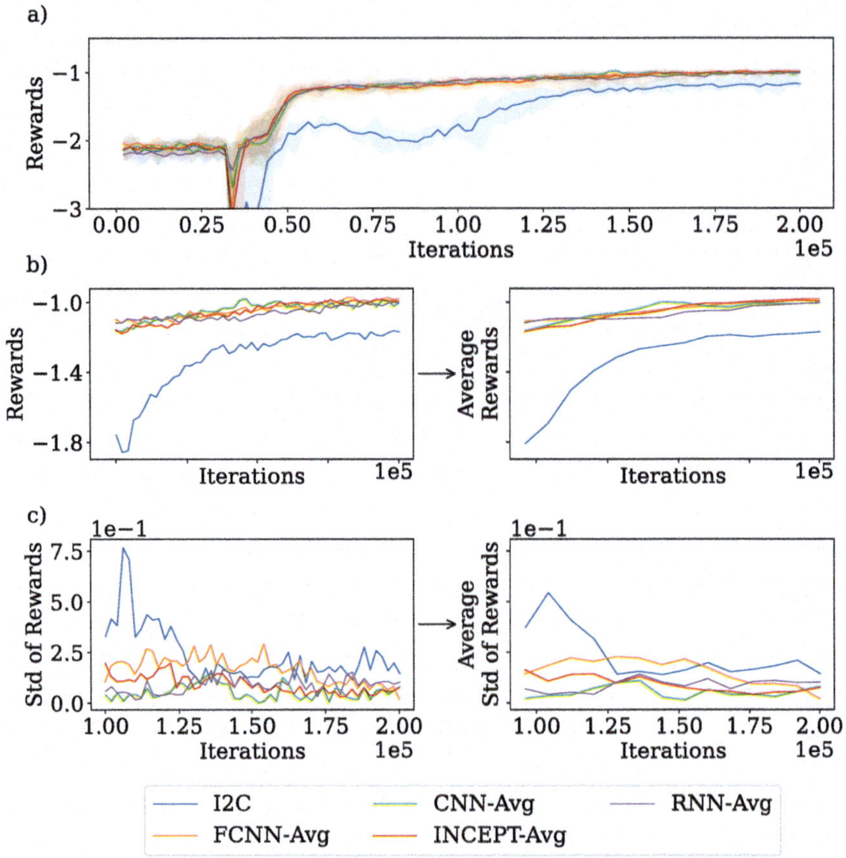

Fig. 9. Comparative performance analysis of best RGICs versus I2C in PP. a) Overall learning trajectories; b) left: Mean rewards obtained after 1×10^5 iterations, right: Averaged mean rewards using a moving window of 1×10^4 iterations; c) left: Standard deviation of mean rewards obtained after 1×10^5 iterations, right: Averaged standard deviation of mean rewards using a moving window of 1×10^4 iterations.

improves the learning process. Extensive experimentation demonstrates the effectiveness of RGIC, surpassing existing methods in different scenarios. The empirical results indicate faster convergence, enhanced efficiency, and reduced computational time. Further, the utilisation of average rewards provides a reliable signal, capturing the agent's cumulative performance and aligning with RL principles. These accelerated and reward-guided capabilities make RGIC more suitable for real-world MAS applications that require real-time decision-making and reward-oriented actions. Overall, RGIC represents an advancement in multi-agent communication, promoting effectiveness and efficiency.

Future research directions could include testing RGIC in more complex scenarios, such as 3D navigation, to assess its performance in challenging and real-

Table 3. Comparative performance analysis of RGICs versus I2C

Methods	Environments	
	CN	PP
RGICs	They outperform I2C across various types of NN and reward types (Fig. 6). Additionally, irrespective of the specific type of NN employed, it is observed that utilising average reward for updating the communication network yields superior results.	They achieve better performance and the utilisation of average reward leads to higher rewards, as shown in Fig. 7. However, the discrepancy between using average and direct reward for each NN is relatively smaller than in CN, particularly for INCEPT and RNN, as illustrated in Fig. 7 c) and d) respectively.
Best Performing RGICs	From Fig. 8 b), it is evident that the rewards obtained by *FCNN-Avg* surpass those of the others. On the other hand, Fig. 8 c) demonstrates that the proposed approaches exhibit similar standard deviation of rewards, indicating consistent performance. Yet, I2C exhibits larger variation in rewards, particularly before 1.6×10^5 iterations.	*FCNN-Avg* and *INCEPT-Avg* demonstrate similar performance, with *FCNN-Avg* slightly surpassing towards the end, as shown in Fig. 9 b). However, *FCNN-Avg* generally exhibits higher standard deviation in rewards compared to other proposed methods, although it achieves the least at the end, as indicated in Fig. 9 c).

istic environments. Additionally, studying the dynamic messaging strategy of agents within RGIC could be explored, allowing agents to adaptively determine the appropriate message type based on the current environmental state and the behaviour of other agents. By investigating these areas, the RGIC framework can be further optimised and tailored to meet the demands of real-world scenarios in MAS.

References

1. Amato, C., Chowdhary, G., Geramifard, A., Üre, N.K., Kochenderfer, M.J.: Decentralized control of partially observable Markov decision processes. In: 52nd IEEE Conference on Decision and Control, pp. 2398–2405. IEEE (2013)
2. Bernstein, D.S., Givan, R., Immerman, N., Zilberstein, S.: The complexity of decentralized control of Markov decision processes. Math. Oper. Res. **27**(4), 819–840 (2002)
3. Búrdalo, L., Terrasa, A., Julián, V., García-Fornes, A.: The information flow problem in multi-agent systems. Eng. Appl. Artif. Intell. **70**, 130–141 (2018)
4. Busoniu, L., Babuska, R., De Schutter, B.: A comprehensive survey of multiagent reinforcement learning. IEEE Trans. Syst., Man, Cybern. Part C (Appl. Rev.) **38**(2), 156–172 (2008)
5. Buşoniu, L., Babuška, R., De Schutter, B.: Multi-agent reinforcement learning: an overview. In: Innovations in Multi-agent Systems and Applications-1, pp. 183–221 (2010)
6. Canese, L., Cardarilli, G.C., Di Nunzio, L., Fazzolari, R., Giardino, D., Re, M., Spanò, S.: Multi-agent reinforcement learning: a review of challenges and applications. Appl. Sci. **11**(11), 4948 (2021)
7. Das, A., Gervet, T., Romoff, J., Batra, D., Parikh, D., Rabbat, M., Pineau, J.: Tarmac: targeted multi-agent communication. In: International Conference on Machine Learning, pp. 1538–1546. PMLR (2019)
8. Ding, Z., Huang, T., Lu, Z.: Learning individually inferred communication for multi-agent cooperation. Adv. Neural. Inf. Process. Syst. **33**, 22069–22079 (2020)

9. Foerster, J., Assael, I.A., De Freitas, N., Whiteson, S.: Learning to communicate with deep multi-agent reinforcement learning. Adv. Neural Inf. Process. Syst. **29** (2016)
10. Hoshen, Y.: Vain: attentional multi-agent predictive modeling. Adv. Neural Inf. Process. Syst. **30** (2017)
11. Jiang, J., Lu, Z.: Learning attentional communication for multi-agent cooperation. Adv. Neural Inf. Process. Syst. **31** (2018)
12. Lecun, Y., Bengio, Y.: Convolutional Networks for Images, Speech and Time Series, pp. 255–258. The MIT Press (1995)
13. Li, Y.: Deep reinforcement learning: an overview. arXiv:1701.07274 (2017)
14. Lowe, R., Wu, Y.I., Tamar, A., Harb, J., Pieter Abbeel, O., Mordatch, I.: Multi-agent actor-critic for mixed cooperative-competitive environments. Adv. Neural Inf. Process. Syst. **30** (2017)
15. Niu, Y., Paleja, R.R., Gombolay, M.C.: Multi-agent graph-attention communication and teaming. In: AAMAS, pp. 964–973 (2021)
16. Omidshafiei, S., Pazis, J., Amato, C., How, J.P., Vian, J.: Deep decentralized multi-task multi-agent reinforcement learning under partial observability. In: International Conference on Machine Learning, pp. 2681–2690. PMLR (2017)
17. Oroojlooy, A., Hajinezhad, D.: A review of cooperative multi-agent deep reinforcement learning. Appl. Intell. 1–46 (2022)
18. Rumelhart, D.E., Hinton, G.E., Williams, R.J.: Learning internal representations by error propagation. Technical report, California Univ San Diego La Jolla Inst for Cognitive Science (1985)
19. Shoham, Y., Powers, R., Grenager, T.: Multi-agent reinforcement learning: a critical survey. Technical report, Citeseer (2003)
20. Singh, A., Jain, T., Sukhbaatar, S.: Learning when to communicate at scale in multiagent cooperative and competitive tasks. arXiv:1812.09755 (2018)
21. Sukhbaatar, S., Fergus, R., et al.: Learning multiagent communication with backpropagation. Adv. Neural Inf. Process. Syst. **29** (2016)
22. Sutton, R.S., Barto, A.G.: Reinforcement Learning: An Introduction. MIT Press (2018)
23. Szegedy, C., Liu, W., Jia, Y., Sermanet, P., Reed, S., Anguelov, D., Erhan, D., Vanhoucke, V., Rabinovich, A.: Going deeper with convolutions. In: Proceedings of the IEEE Conference on Computer Vision and Pattern Recognition (CVPR) (2015)
24. Wang, Y., Zhong, F., Xu, J., Wang, Y.: Tom2c: target-oriented multi-agent communication and cooperation with theory of mind. arXiv:2111.09189 (2021)
25. Zhang, K., Yang, Z., Başar, T.: Multi-agent reinforcement learning: a selective overview of theories and algorithms. In: Handbook of Reinforcement Learning and Control, pp. 321–384 (2021)
26. Zhu, C., Dastani, M., Wang, S.: A survey of multi-agent reinforcement learning with communication. arXiv:2203.08975 (2022)

An Evaluation of Handwriting Digit Recognition Using Multilayer SAM Spiking Neural Network

Minoru Motoki[1(✉)], Heitaro Hirooka[1], Youta Murakami[2], Ryuji Waseda[3], and Terumitsu Nishimuta[3]

[1] National Institute of Technology (KOSEN), Kumamoto College, Suya, Koshi, Kumamoto 2659-2, Japan
motoki@kumamoto-nct.ac.jp

[2] ISB Corporation, 5-1-11 Ohsaki, Shinagawa-ku, Tokyo, Japan

[3] Maviss Design Co., Ltd., 5-1-1 Minamikumamoto, Cyuou-ku, Kumamoto-City, Kumamoto, Japan

Abstract. This paper describes evaluation results of the hand writing digit recognition MNIST using the on-chip trainable multilayer SAM spiking neural network(SAM-SNN). The SAM spiking neuron model is the model that added one more parameter to the LIF neuron model and has a higher resolution ability than the LIF neuron model for implementing onto digital circuitry such as FPGAs. So far, we have proposed a supervised training algorithm for the SAM-SNN and implemented it into FPGAs. In this research, we evaluated the image recognition performance of the SAM-SNN by using the MNIST dataset. As the result, the SAM-SNN performed 99.29% for 60000 training data and 94.52% for 10000 test data. Moreover, the SAM-SNN that used the data made by CNN of ANN performed 99.72% for training data and 97.54% for test data. Decreasing the training rate η made the training progress faster.

Keywords: SAM spiking neural network · On-chip trainable · FPGA · MNIST

1 Introduction

Recently, research of the Spiking Neural Network (SNN), called neuromorphic computing, is expected as next generation that it is compared with the other Artificial Neural Networks (ANN) as for implementation of low-powered hardware. Because the SNN has an advantage of affinity with digital circuitry that is derived from biological plausibility than ANN has. We have been focusing on the SAM neuron model that is added a parameter to the Leaky Integrated and Fire (LIF) neuron model as an SNN neuron model. The SAM neuron model is proposed by Shigematsu et al. in 1996 [1], and the model has the advantage that it maintains a more accurate correspondence between the continuous and discrete representations, avoiding a reduction in the frequency of output spikes even if the sampling interval is wide (The model is very similar to ALIF introduced by Bellec et al. [2]).

N. Naik et al. (Eds.): UKCI 2023, AISC 1453, pp. 95–102, 2024.
https://doi.org/10.1007/978-3-031-47508-5_8

We derived a supervised training algorithm for the SAM neuron model and the multilayer SAM spiking neural network as the steepest descent method (a gradient method) [3]. This training algorithm allows a multiplier-less structure when we implement it into a digital circuit, therefore the SAM neuron network can be implemented into FPGAs as a relatively smaller scale circuit, say as an on-device trainable embedded AI hardware. The intellectual property (IP) that the proposed algorithm is implemented into FPGAs, already has been registered as SAMACT® in Japan [4, 5]. This paper reports an evaluation of the performance of the multilayer SAM-SNN applied to MNIST handwriting digit recognition as the one of image recognition tasks.

Recently, the number of SNN research is increasing. In those research that uses MNIST image processing, such as ANN of ReLU neuron was trained by BP and converted to IF neuron and make fast by the weight normalization [6], devising neuron parameter to achieve a balance between computing cost and biological plausibility [7], and back-propagating based on synfire-gate and achieving on-chip by Intel's Loihi [8].

2 SAM Spiking Neural Network

This section describes the SAM neuron model [1], the supervised algorithm proposed by Motoki et al. [3], and the advantage of the SAM neuron model [9] in order to readers understanding. The main contribution of this paper is an evaluation of the performance of SAM-SNN using the MNIST handwriting dataset as described from Chap. 3.

2.1 SAM Neuron Model

The Spike Accumulation and Modulation (SAM) neuron model is one of the IF type spiking neuron models. At a discrete time t, if the j-th neuron receives a spike $X_j(t) \in \boldsymbol{B} = \{0, 1\}$ from the i-th neuron, then the inner potential $U_j(t) \in \boldsymbol{R}$ of the j-th neuron is calculated by the sum of product between the link weights $W_{ji} \in \boldsymbol{R}$ and input spikes $X_i(t)$, plus the addition of the weighted previous inner potential $aV_j(t-1)$ (where a is a decay parameter). Thus,

$$U_j(t) = \sum_{i=1}^{n_1} W_{ji} X_i(t) + aV_j(t-1) \tag{1}$$

The j-th neuron output $X_j(t) \in \boldsymbol{B}$ is obtained by applying the activation function $g()$ to $U_j(t)$, where $g()$ is a step function such that if $U_j(t)$ is less than θ the output is 0, corresponding to no spike; if $U_j(t)$ is greater than θ the output is 1 and a spike is generated. θ is the threshold, i.e.,

$$X_j(t) = g\big(U_j(t) - \theta\big), \tag{2}$$

$$u = U_j(t) - \theta, \tag{3}$$

$$g(u) = \begin{cases} 0(u < 0) \\ 1(u \geq 0). \end{cases} \tag{4}$$

Moreover, the inner potential is decreased by an amount p upon activation, such that (see Fig. 1)

$$V_j(t) = U_j(t) - pX_j(t). \tag{5}$$

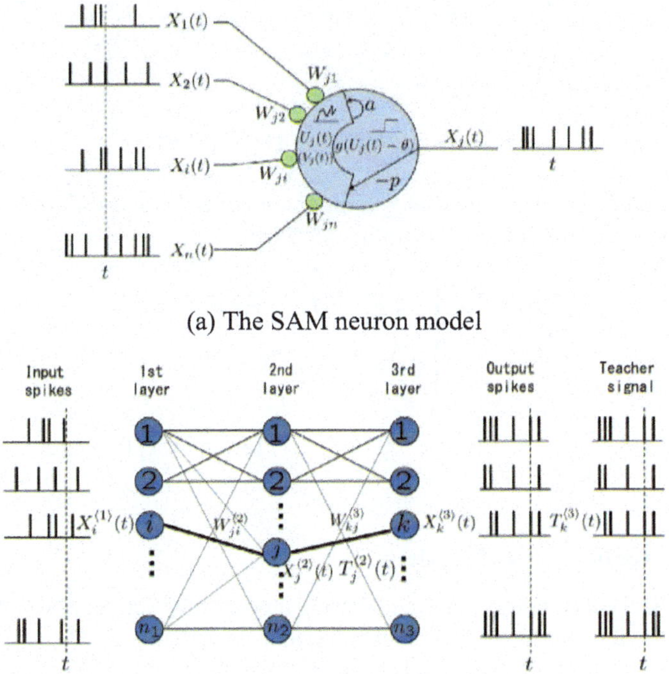

(a) The SAM neuron model

(b) Multilayer SAM spiking neural network.

Fig. 1. The SAM neuron model and the multilayer SAM spiking neural network [3].

2.2 Supervised Training Algorithm for Multilayer SAM-SNN

The training approach for the multilayer SAM neural network is first proposed by Motoki *et al.*[3]. We define the objective function E (loss) as

$$E \stackrel{\text{def}}{=} \frac{1}{N} \sum_{P=1}^{N} \sum_{t=1}^{t_N} \frac{1}{2} \sum_{k=1}^{n_3} \left(X_{P,k}^{\langle 3 \rangle}(t) - T_{P,K}^{\langle 3 \rangle}(t) \right)^2, \tag{6}$$

where $T_{P,K}^{\langle 3 \rangle}(t) \in \boldsymbol{B}$ is the teacher signal of k-th output neuron for a pattern P at a time t. We derived the training algorithm by calculating the gradient for link weights W of E and updating the training parameters (link weights and thresholds). To simplify we considered the number of pattern $P = 1$ in the followings.

For the neurons of the output layer, we can state

$$\frac{\partial E}{\partial W_{kj}^{\langle 3\rangle}} = \sum_{t=1}^{t_N}\left(X_k^{\langle 3\rangle}(t) - T_k^{\langle 3\rangle}(t)\right)\frac{\partial X_k^{\langle 3\rangle}(t)}{\partial W_{kj}^{\langle 3\rangle}}$$

As the result, by approximating, generally, for the hidden layer,

$$\frac{\partial E(t)}{\partial W_{ji}^{\langle 2\rangle}(t)} = \left(X_j^{\langle 2\rangle}(t) - T_j^{\langle 2\rangle}(t)\right)g\prime\left(u_j^{\langle 2\rangle}(t)\right)H_{ji}^{\langle 2\rangle}(t) \tag{7}$$

here,

$$H_{ji}^{\langle 2\rangle}(t) = aH_{ji}^{\langle 2\rangle}(t-1) + X_i^{\langle 1\rangle}(t), \tag{8}$$

and $T_j^{\langle 2\rangle}(t) \in \boldsymbol{B}$ is Teacher signal of j-th hidden neuron, by approximating backpropagation algorithm, (we call this algorithm SAMSBP),

$$T_j^{\langle 2\rangle}(t) = \begin{cases} 1\left(X_j^{\langle 2\rangle}(t) - \frac{\partial E(t)}{\partial X_j^{\langle 2\rangle}(t)}\right) \geq 0.5 \\ 0\left(X_j^{\langle 2\rangle}(t) - \frac{\partial E(t)}{\partial X_j^{\langle 2\rangle}(t)}\right) < 0.5. \end{cases} \tag{9}$$

$$\frac{\partial E(t)}{\partial X_j^{\langle 2\rangle}(t)} = \sum_{k=1}^{n_3}\left(X_k^{\langle 3\rangle}(t) - T_k^{\langle 3\rangle}(t)\right)g\prime\left(u_k^{\langle 3\rangle}(t)\right)W_{kj}^{\langle 3\rangle}(t). \tag{10}$$

2.3 Advantage of the SAM Model

The SAM model is essentially similar to the LIF model, however, the SAM model is more suitable for efficient hardware implementation than the LIF model. The LIF model is one of the simplest models for implementing spiking neurons in FPGAs. Compared to the digitalized LIF model, the SAM model can express output spike frequency (density) with better resolution, even when the discrete time (sampling time) width is wide. Therefore, the SAM model has an important advantage compared to the discrete LIF model when implemented on an FPGA. Figure 2 shows the variation in output spike frequency (F-I curve) for both the LIF and SAM models as shown by Motoki et al. [9]. This figure shows that in the discrete LIF model, even though the input current is increased, there is poor resolution in the output spike frequency. In contrast, in the SAM model, the output spike frequency is expressed quite finely in the same input current range.

3 MNIST Simulation

3.1 SAM-SNN for MNIST Digit Recognition

In this research, we evaluated the SAM-SNN performance in two types of network structures. One is 'with filter', and the other is 'without filter' (Fig. 3 (a) and (b)). 'Filter' stands for feature extraction, namely convolution, max pooling, and binarization, similar to CNN. In the 'with filter' network, we used 12 kernels shown in Fig. 4. These kernels achieve to detect straight and bent lines of the MNIST data.

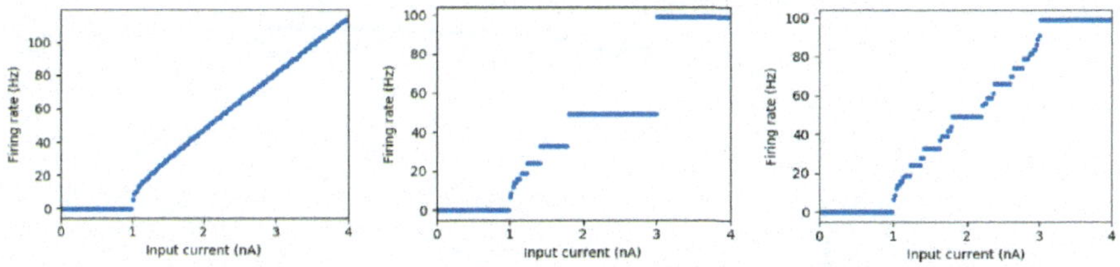

(a) LIF(t_s = 0.1ms) (b) LIF(t_s = 10ms) (c) SAM(t_s = 10ms)

Fig. 2. F-I curve of the LIF model and the SAM model that express $\tau\frac{dU(t)}{dt} = -(U(t) - U_{rest}) + RI(t)$, $\tau = 0.03(\mathrm{s})$, $R = 1.0$, $U_{rset} = 0(mV)$, $\theta = p = 1.0(mV)$. Even though the sampling interval $t_s = 10\mathrm{ms}$ (wider), the SAM model (c) is possible to express the firing rate quite finely than the LIF model (b) [9].

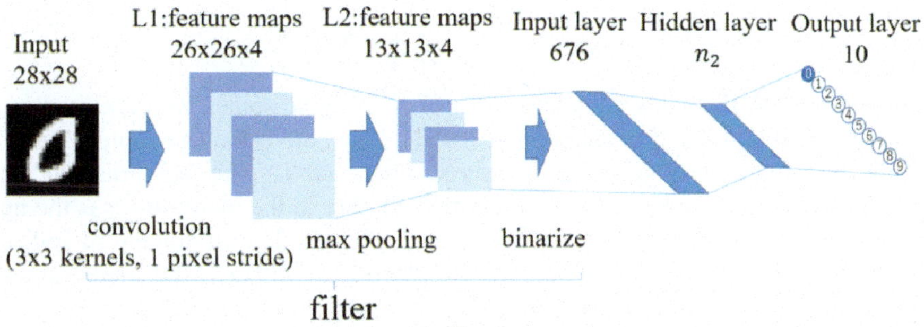

(a) 'with filter' network

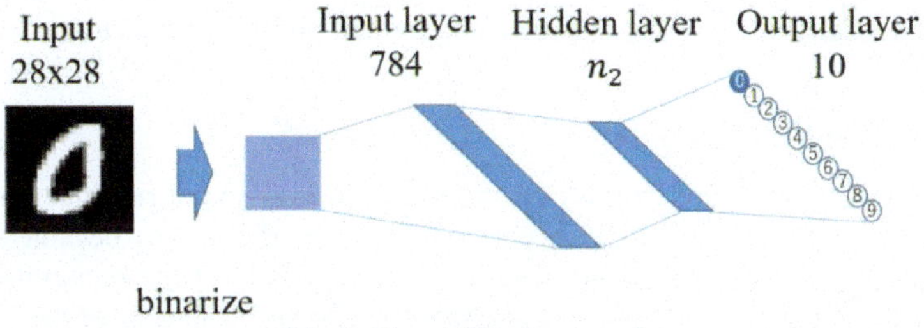

(b) 'without filter' network

Fig. 3. Network structures of the multilayer SAM spiking neural network for MNIST.

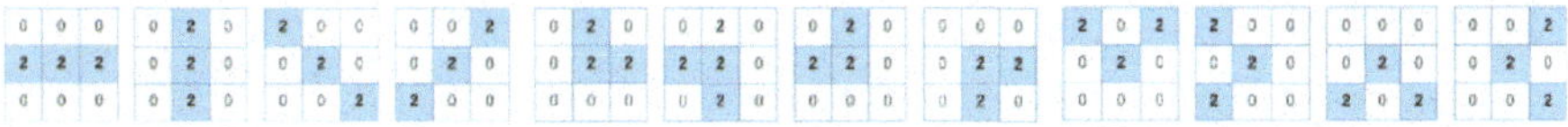

Fig. 4. The 12 kernels used in 'with filter' network shown in Fig. 3 (a).

After inputting MNIST data, the input value was normalized to [0, 1], after that, the data is passed by 'with filter' or 'without filter'. When the binarized data is inputted

Fig. 5. Accuracy curves

to the SAM-SNN, the value is encoded to spikes. Periodic spikes are generated as the encode method. In the output side of SAM-SNN, 'one hot vector' is adopted. Here,

periodic spikes are 1 spike per 1 discrete time, and 1 spike per 2 discrete times (we defined this discrete time as t_C). Identified number in the output layer is the number that corresponds to the output neuron that outputs the largest amount of spikes. If there are more than two number outputs, the identified number is the smaller number corresponding to the output neuron.

The simulation was carried out with varying hyper-parameters such as the number of hidden neurons n_2, training rate η/ι, the shape of the derivative of the step function $g\prime(u)$, and t_C. We simulated the nets by using python programming language.

3.2 Simulation Results

In the training process, the accuracy curves were not smooth but rattling. However, by decreasing the training coefficient η per some epochs, the curves became somewhat smooth, and the training was faster (Fig. 5). A comparison of the recognition performances of the MNIST dataset is shown in Table 1. The performances of the SAM-SNN could not reach state-of-the-art results. However, it was a comparable result as well, therefore it has a possibility to achieve higher performance with parameter tunings.

4 Discussion

We analyzed the loss function surfaces with varying two link-weight parameters after 100 epoch training as shown in Fig. 6. First, we checked the loss values every 0.5 steps around the trained w1, w2 (Fig. 6(a)), the loss surfaces looked smooth. However, we also checked the loss values every 0.05 steps around the trained w1, w2 (Fig. 6(b)), the loss surfaces looked rough. Therefore, this loss surfaces were rough in practice, looking into the details. We could recognize that the SAM-SNN forms a very complex function surfaces in the MNIST task and that this is one of the characteristics of the SAM-SNN.

Table 1. Performance comparison of the MNIST literature in SNN

Publication/Model	Learning Algorithm	Network structure	Training/Test Accuracy(%)
[6] Diehi, F.U, et al. (2015)	Spiking ConvNet	28x28-12x5-2s-64x5-2s-10o	99.19 / 99.14
[10] Neftci et al. (2014)	Constructive divergence	784-500-40	/ 91.9
[11] Zheng et al. (2018)	SGD-MWD-STDP	784-300-10	/ 97.2
[8] Renner et al. (2021)	on-chip sBP	400-400-10	/ 96.2
[7] Fang et al. (2021)	Spike-based BP		/ 99.72
SAM-SNN (without filter)	**SAMSBP**	**784-400-10**	**99.29 / 94.52**
SAM-SNN (with filter)	**SAMSBP**	**784-(12F)-2028-80-10**	**96.09 / 93.95**
SAM-SNN (CNN-ANN pre-trained data) *1	**SAMSBP**	**CNN-256-40-10**	**99.72 / 97.54**

*1 The SAM-SNN that used the data trained (made) by CNN of ANN

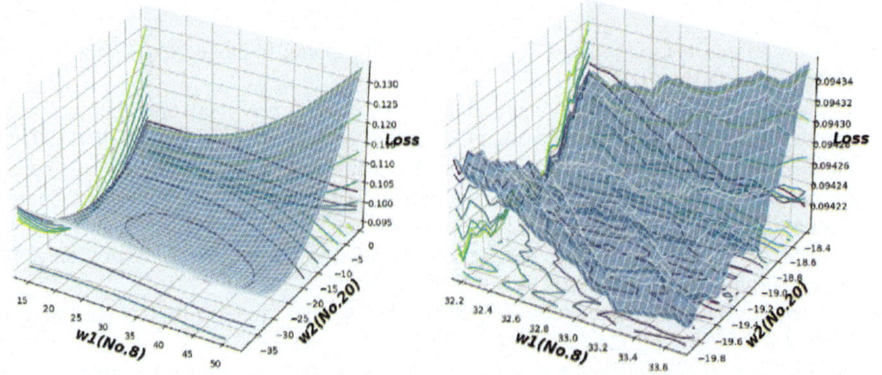

(a) Every 0.5 steps around the trained w1, w2 (b) Every 0.05 steps around the trained w1,w2

Fig. 6. Loss surfaces varying two link-weights w1, w2 of the hidden layer after 100 epoch training in a 784-40-10 net trained by training 60000 data.

That is to say, it is considered that the decreasing η technic was efficient for speed-up the training.

Even though the accuracies of SAM-SNN were relatively high, we are predicting that the numeric expressing resolution (fixed point expression) is the key parameter for the hardware implementation. Probably, from 32 to 64 bits will be required, and there will be a trade-off between the resolution of numeric expression and the circuit resource cost.

5 Conclusions and Future Works

This research made it clear that the on-chip trainable SAM-SNN has comparable performances to the other SNN's performances for the MNIST task. The SAM-SNN is possible to stack as deeper layers, and we would like to explore some other possibilities such as applying reinforcement learning as we have already presented [12].

Acknowledgements. A part of this work was supported by JSPS KAKENHI Grant Number JP19K12176.

References

1. Shigematsu, Y., Matsumoto, G.: Article title. Journal **2**(5), 99–110 (2016)
2. Bellec, G., Scherr, F., et al.: A solution to the learning dilemma for recurrent networks of spiking neurons. Nat. Commun. **11**(1), 1 (2020)
3. Motoki, M., Shintani, H., Matsuo, K., McGinnity, T.M.: Function approximation using multilayer SAM spiking neural network. In: Proceedings of IEEE 8th International Innovative Computing Technology (INTECH2018), pp. 65–70 (2018)
4. Motoki, M., Waseda, R., Nishimuta, T.: An FPGA implementation of on-chip trainable multilayer SAM spiking neural network. In: Proceedings of the 9th IIAE, ICIAE, pp. 144–148 (2021)
5. Maviss Design Co. Ltd.: SAMACT, No.2021-111474 applied in Japan, No.6508894 registered trademark in Japan (2022)
6. Diehi, F.U., et.al.: Fast-classifying, high-accuracy spiking need networks thorough weight and threshold balancing. In: 2015 Joint Conference on Neural Networks (IJCNN) (2015)
7. Fang, W., et. al.: Incorporating learnable membrane time constant to enhance learning of spiking neural networks. In: Proceedings of the IEEE/CVF International Conference on Computer Vision (ICCV), pp. 2641–2651 (2021)
8. Renner, A., Sheldon, F., et al.: The Backpropagation Algorithm Implemented on Spiking Neuromorphic Hardware (2021). arXiv:2106.07030
9. Motoki, M.: Basic Characteristics of SAM Spiking Neuron Model with Rate Coding. IEICE Technical Report, NC2021-63, pp.88–93 (2022). (in Japanese)
10. Neftci, E., et al.: Event-driven constructive divergence for spiking neuromorphic systems. Front. Neurosci., 10 (2014)
11. Zheng, N., Mazunder, P.: Online supervised learning for hardware-based multilayer spiking neural networks through the modulation of weight-dependent spike-time-dependent plasticity. IEEE Trans. Neural Netw. Learn. Syst. **29**(9), 4287–4302 (2018)
12. Motoki, M., Oshiro, Y., Waseda, R., Nishimuta, T.: Actor-critic reinforcement learning using on-chip trainable multilayer SAM spiking neural network. In: Proceeding of 4th International Symposium of Neuromorphic AI Hardware, p. 47, P2-15 (2022)

Exploring the Linguistic Model to Operate on Architectural Façade Illumination Design

Yanting Liu[1] and Fangyi Li[2(✉)]

[1] The University of Tokyo, Bunkyo. 113-8641, Tokyo, Japan
[2] School of Artificial Intelligence, Beijing Normal University, Beijing, China
lify@bnu.edu.cn

Abstract. Designers utilize the compositions to express their feelings and thoughts. However, it is difficult to have corresponding command of design theory for people in different fields. For architects, it is not common to both have the experience in construction design and illumination design. In this research, we aim to be able to effectively express design through verbal manipulation, consider the relationship between architectural façade lighting design parameters and words, pursue the possibility of becoming a tool for design support and manipulate the façade lighting by language. After creating an abstract 3D lighting model, we have explored a design system for linguistic manipulation using an autoencoder and a neural network and have examined the practicability of the model.

Keywords: Facade · Linguistic operation · Design tool · Autoencoder · Neural network

1 Introduction

Architects convey their consciousness and emotions to the world by creating the shapes of buildings. However, in the pitch-black environment of the night, it becomes difficult to discern the shape and material characteristics of the building. In particular, there is a great need for light-up designs for buildings that can serve as landmarks, such as shopping centers, office buildings, and station squares. If the city shines at night, it will be easier for people to gather. Therefore, architectural outdoor lighting is an important part of the design process. However, architects and lighting designers are often not the same person. Therefore, it is not surprising that there may be discrepancies in the concept that is conveyed. When designing lighting, we have to consider various aspects such as the position of the light, the intensity of the light, the combination of light and shadow, the presentation of colors, and energy conservation. One of them is to express in lighting design with sensible words.

For example, in the lighting design for the Yaesu exit of Tokyo Station [1], they first drafted the overall policy, decided on the lighting concept, and based on that concept, designed the lighting for each part. In the actual design, the design and specific operations are determined by the experience and intuition of the designers. However, if the sensibilities of certain lighting designers are different, or if clients who have no

N. Naik et al. (Eds.): UKCI 2023, AISC 1453, pp. 103–113, 2024.
https://doi.org/10.1007/978-3-031-47508-5_9

design experience want to express their own sensibilities, it is difficult to rely solely on experience and intuition.

Numerous studies are also being conducted in the field of architecture, such as simulation-based design support tools and evaluation methods for multiple lighting elements. Liang et al. [2] have constructed the evaluation system by using group search function of nondominated genetic algorithm with different evaluation functions. M H A Edytia et al. [3] have used simulation software to construct the daylight lamination evaluation system. Nakayama et al. [5] conducted an impression evaluation experiment using the SD method in order to quantitatively grasp the impression of architectural exterior colors by adjective pairs, followed by training with a neural network, and developed a building exterior color selection method. The study of Masuyama et al. [6] analyzed the relationship between the bright and dark parts of a building illuminated by appropriate design and the luminance distribution of the three-dimensional effect of building elements.

In the field of architectural planning, Aoki et al. [4] aimed to elucidate the roles and functions of language in planning and design, sought an associative image space between words and lighting patterns, and reproduced the situation in which design proposals were modified using language expressions. A model was constructed and its manipulation experiments were repeated. However, due to the calculation of the function can be limited, it is considered to have improvement by using new calculation methods.

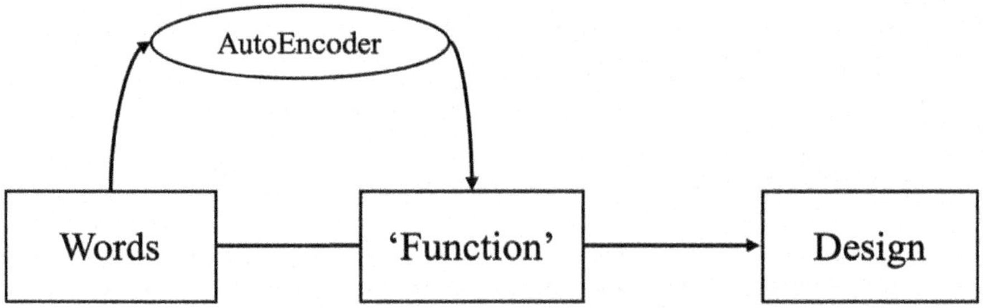

Fig. 1. The schematic diagram of the research theory.

In this study, we aim to realize linguistic operations in the field of lighting design by referring to the methods using autoencoder to extract the 'function' that connecting language and design (Fig. 1). After creating an abstract 3D lighting model, we explore a design system for linguistic manipulation using an autoencoder and a neural network, which could extract the characteristic of language better than previous calculation methods.

2 Methods

2.1 Research Framework

In order to explore architectural façade lighting design models by linguistic manipulation, it is necessary to construct lighting design models and manipulation systems.

When creating a design model, there are two variables: a design object and a design parameter. Regarding the selection of the design target, lighting design has different effects even if the parameters are the same, depending on factors such as the shape of the building and surface materials. Since it is difficult to construct a façade model that can be applied to all buildings, this research focuses on exploring the possibility of manipulating verbal design, and creating a 3D architectural facade model using a specific building as a reference. Set L lighting design parameters D_I (I = 1,...,L). Vary the parameters to create N lighting patterns F_j (j = 1,...,N).

To create an operating system, it is necessary to explore the relationship between language and lighting design. Evaluating design through words means that design and words are related through a certain mechanism. In order to analyze this mechanism mathematically, this study considers applying an autoencoder and a neural network. Select M words w_i (i = 1,...,M), use lighting pattern F_j as evaluation data by word w_i, and compress the evaluation data into K dimensions using an autoencoder, and extract the features of the data. A neural network is used to obtain a design function representing the relationship between the compressed data and design parameters.

2.2 Autoencoder

An Autoencoder [7] is a neural network that trains itself by encoding and then decoding input data so as to output data that reproduces the input data. The number of dimensions of data can be reduced by using an autoencoder. An activation function is a nonlinear function or identity function applied after linear transformation in a neural network. Here, if a nonlinear function is used as the activation function, nonlinear mapping can be achieved, so that the principal component of the data (data better representing the original data, that is, the code) can be obtained. As shown in Fig. 2, the autoencoder model in this research consists of an input layer, an intermediate layer, and an output layer with the same number of units as the input layer. Repeat learning and update the weights each time to reduce the error.

Here, the input data is the evaluation data p, the hidden layer x passed through the autoencoder is a code that expresses the data in a lower dimension, K-dimensional, and the weight a is a vector whose components are the features of the network. This dimension K could be considered as a parameter that we could discuss in the calculations.

The displayed equation is the calculation from the input layer to the hidden layer.

$$x_k = f\left(\sum_{i=1}^{M} a_{ki}p_i + \theta_k\right)(i = 1, \ldots, \mathrm{M})$$

And written in a matrix form.

$$\mathbf{x} = f(\mathbf{ap} + \boldsymbol{\theta})\ (\boldsymbol{\theta}\text{ is the threshold})$$

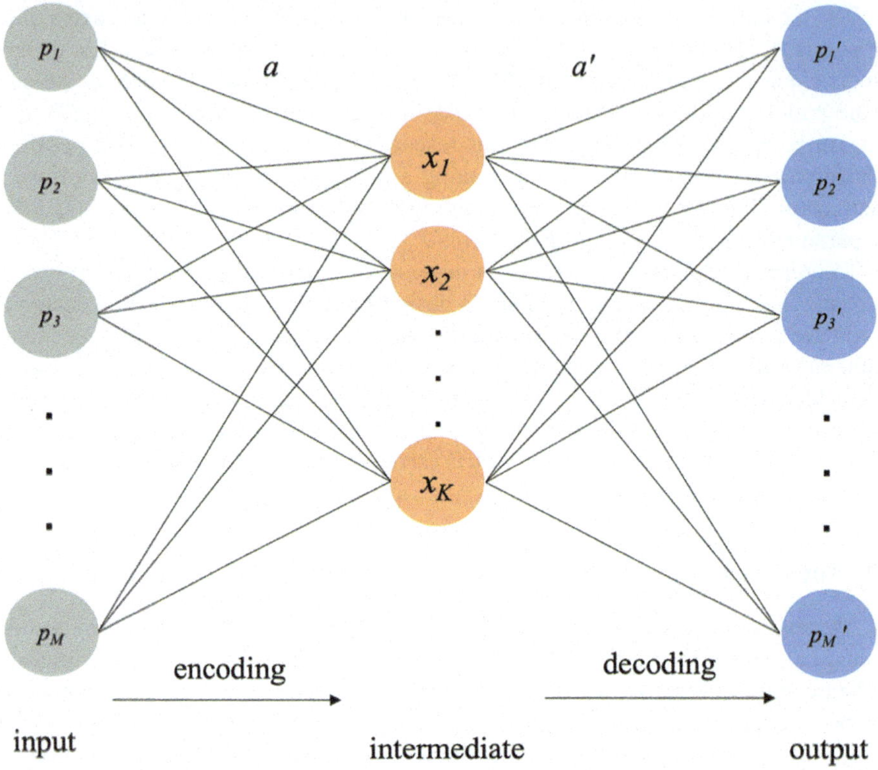

Fig. 2. The framework of Autoencoder model. Input layer p: evaluation data. Intermediate layer x: a code that represents the data in lower dimensions. Weights a, a': weights with data features.

$$\mathbf{p} = \begin{pmatrix} p_1 \\ p_2 \\ \cdot \\ \cdot \\ \cdot \\ p_M \end{pmatrix}, \mathbf{x} = \begin{pmatrix} x_1 \\ x_2 \\ \cdot \\ \cdot \\ \cdot \\ x_K \end{pmatrix}, \mathbf{a} = \begin{pmatrix} a_{11} & \cdots & a_{1\mathrm{M}} \\ \vdots & \ddots & \vdots \\ a_{\mathrm{K}1} & \cdots & a_{\mathrm{KM}} \end{pmatrix},$$

$$\theta = \begin{pmatrix} \theta_1 \\ \theta_2 \\ \cdot \\ \cdot \\ \cdot \\ \theta_K \end{pmatrix}, \mathbf{a}\prime = \begin{pmatrix} a'_{11} & \cdots & a'_{1\mathrm{K}} \\ \vdots & \ddots & \vdots \\ a'_{\mathrm{M}1} & \cdots & a'_{\mathrm{MK}} \end{pmatrix}, \boldsymbol{\theta}' = \begin{pmatrix} \theta'_1 \\ \theta'_2 \\ \cdot \\ \cdot \\ \cdot \\ \theta'_K \end{pmatrix}$$

The computation from the hidden layer to the output layer is

$$p_i' = f\left(\sum_{k=1}^{K} a_{ik}' x_k + \theta_i'\right)(k = 1, \ldots, \mathrm{K}),$$

and written in a matrix form is showed as

$$\mathbf{p}' = f(\mathbf{a}'\mathbf{x} + \boldsymbol{\theta}')(\boldsymbol{\theta}' \text{ is the threshold})$$

f is the activation function, which in this study the sigmoid function was adopted.
***\prime

$$f(x) = \frac{1}{1 + \exp(-x)}$$

The error function is

$$\begin{aligned}\mathbf{E} &= \sum_{i=1}^{N} (\mathrm{p_i} - p_i')^2 \\ &= \sum_{i=1}^{N} \mathrm{p_i} - f(a' f(ap_i + \theta) + \theta')^2\end{aligned}$$

To reduce the error, we use stochastic gradient descent here. The weight a that has been learned and has features is

$$a^{(t+1)} = a^{(t)} - \epsilon \nabla E_i (t \text{ is the number of learning times})$$

to update more precisely, such as ϵ is a positive minute constant representing the degree of one update. ∇E_i is the gradient of the error function.

2.3 Exploring Design Functions and Linguistic Operations

From the method described above, the sign of the data and the weight of the data are obtained from the evaluation data. There is a correspondence relationship between one evaluation data and one design pattern. That is, It is necessary to find the correspondence with the sign of the evaluation data

$$\mathbf{x} = (x_1, x_2, \ldots, x_K)$$

and design parameters.

$$\mathbf{y} = (y_1, y_2, \ldots, y_L)$$

The function from this sign to the design parameter is like

$$\mathbf{y} = \varnothing(\mathbf{x})$$

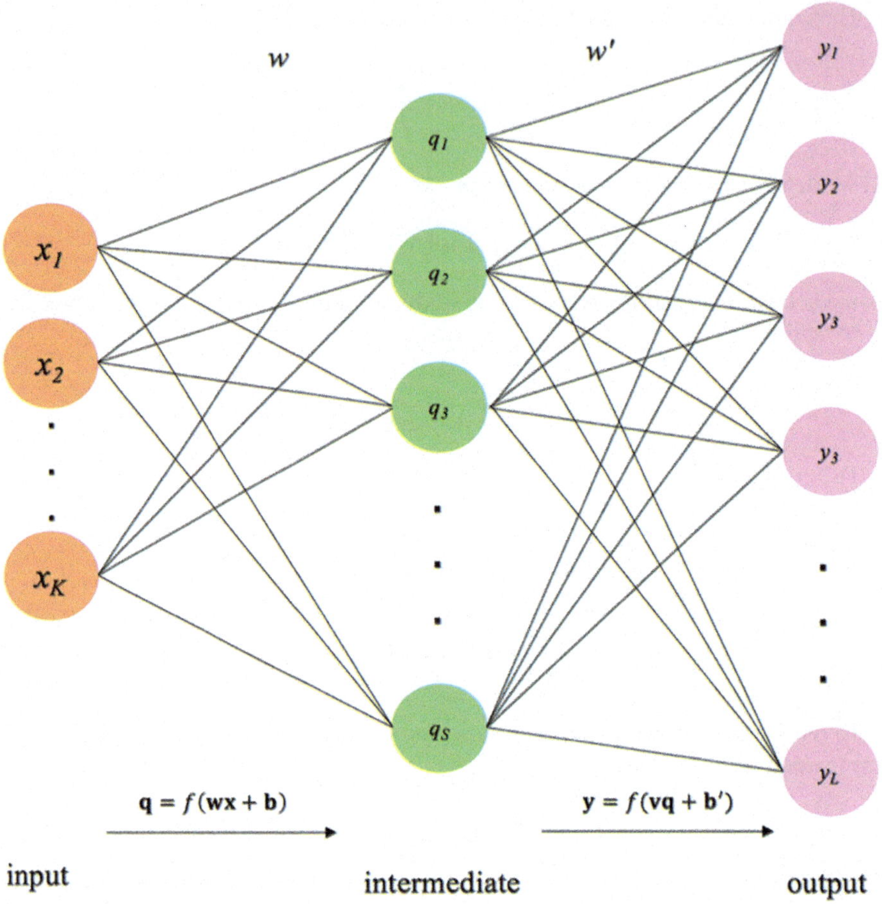

Fig. 3. The framework of neural network model. Sigmoid function: $f(x) = \frac{1}{1+exp(-x)}$. Weights: w, w'. Threshold: b, b'. Input layer: sign of evaluation data. Output Layer: Design Parameters.

However, such functions are generally nonlinear and difficult to obtain analytically. Here, we examined whether a neural network can be applied.

A neural network is a mathematical model of a human neuron (nerve cell), and it is possible to generate a prediction system that improves cognition through learning. Here, we used a hierarchical neural network with three layers, an input layer, an intermediate layer, and an output layer, as shown in Fig. 3.

3 layers where the first input layer is $x_1, x_2, \ldots, x_K$, the second intermediate layer is $q_1, q_2, \ldots, q_S$, and the third output layer is $y_1, y_2, \ldots, y_L$. A neural network model was used. Learning with this neural network is repeated to establish an approximation, and a design function is obtained.

After obtaining design functions, we have discussed the linguistic operations. When a K-dimensional weight $\mathbf{a}_i$ is added to a K-dimensional code x that encodes the evaluation data using an autoencoder, the data decoded into the code x expresses the features of the weight $\mathbf{a}_i$ more strongly than the original evaluation data. This code modification operation is expressed as follows.

$$\mathbf{x}' = \mathbf{x} + d\mathbf{a}_i (d \text{ is the degree of correction})$$

The modified design parameters by the weights are the design functions

$$\mathbf{y}' = \varnothing(\mathbf{x}')$$

can be obtained.

3 Design Model Execution

3.1 Collecting Evaluation Data

Since it is difficult to construct a façade model that can be applied to every building, this research tried using an existing building as a reference. Especially in the daytime, the building has a weak impression, but it was found that the night lighting gave a new feeling. A flat building does not give a strong impression when viewed in the daytime, so the flatness is considered as one of the characteristics of the building. In this study, the Shiseido building in Ginza was used as a reference for experiments.

When actually designing lighting, we have to consider various factors. Regarding the setting of design parameters, Ishida et al. [8] investigated the psychological determinants of the feeling of brightness in a space. It has been found that when humans perceive illumination, the intensity of the light source affects their perception. In this research, we first experimented with a combination of position, color, angle, and light intensity as parameters, but because the data was too large, we set only the light intensity. A spotlight was used as the light source.

Table 1 summarizes the experimental situation. From the previous research on the evaluation of words on the facades of many illuminated buildings, it was found that the choice of words is important. Nakayama et al. [5] collected adjectives from existing literature and examined a color selection method using the SD method. In this research, from the review evaluation literature of works that won the Lighting Design Award of the Illuminating Engineering Institute of Japan and the Lighting Design Award of Japan, we examined the following: We have selected 16 words such as "Integral, Calm, Comfortable, Graceful, Soft, Enjoyable, Light, Beautiful, Vibrant, Rich, Harmonious, Soothing, Simple, Dazzling, Serene, Three-Dimensional".

As for the light source positions, since the original Shiseido building is not lit up, 23 spotlights were placed under the windows, in the areas where the roof, floor, and walls are connected and uneven. It is set as shown in Fig. 4. The intensity of brightness can be changed from 0 to 300. The default units of V-Ray were used. An example of a rendered model is shown in Fig. 4.

Furthermore, by changing the intensity of the light, 40 patterns were created as shown in Fig. 5, and when looking at the 40 lighting pattern renderings created by humans using these words, the words that I felt were "1" and didn't feel them were evaluate as "0". We have done the investigation and survey with 20 people who are all architecture students and obtained the evaluation data.

Table 1. Experiment status of the linguistic evaluation

Items	Numbers
Subjects	4
Subjects specialty	Architecture
Words	16 (Integral, Calm, Comfortable, Graceful, Soft, Enjoyable, Light, Beautiful, Vibrant, Rich, Harmonious, Soothing, Simple, Dazzling, Serene, Three-Dimensional)
Patterns	40
Light sources Numbers of the survey	23 20

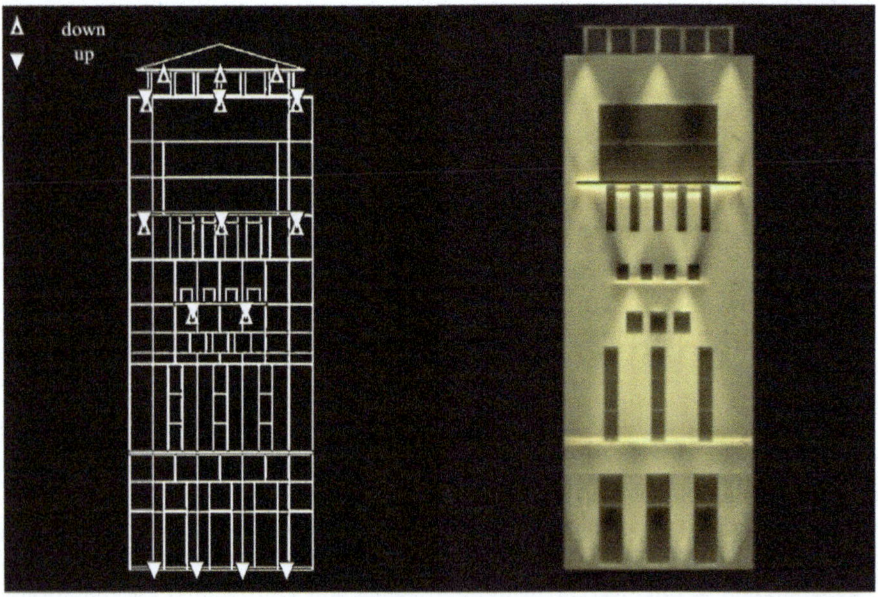

Fig. 4. The distribution map of the light source and one of the examples of the rendering.

3.2 Learning Results

The K-dimensional codes and weights encoding the evaluation data as described above were obtained through an autoencoder. We do not know what the best dimension could be, so here, using an autoencoder, we tried from 16 to 8 dimensions for calculations, and adopted 10 dimensions with the smallest error. The error results we tried are shown in Table 2. Figure 6(a) shows the change in error after learning 3000 times. The horizontal axis of the figure represents the number of times of learning, and the vertical axis represents the error value.

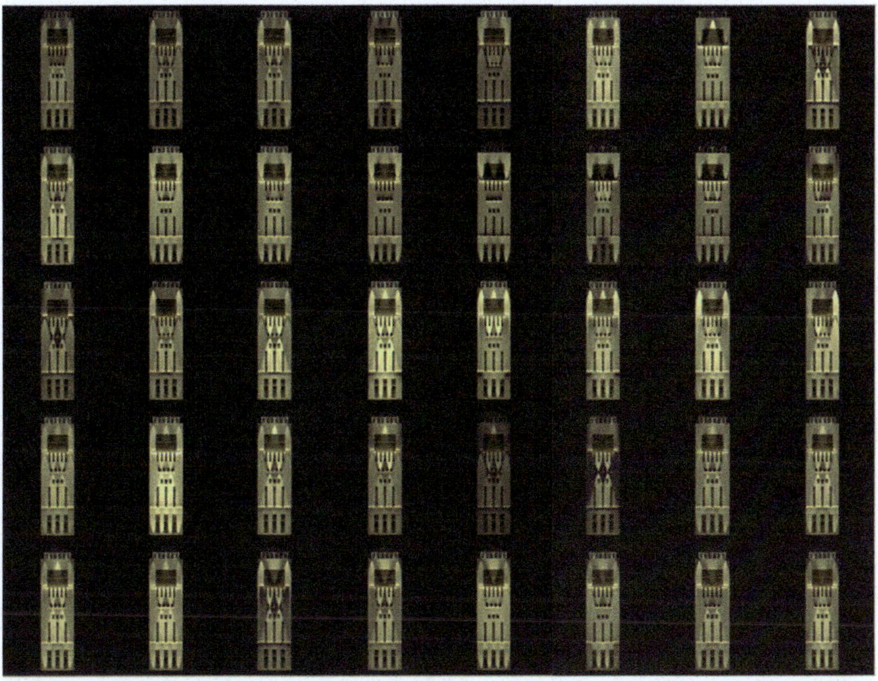

Fig. 5. The rendering patterns that has been used in the experiment.

Table 2. The error learned by numbers of dimensions

dimentions	16	15	14	13	12	11	10	9	8
error	0.191	0.184	0.195	0.192	0.185	0.178	0.176	0.183	0.202

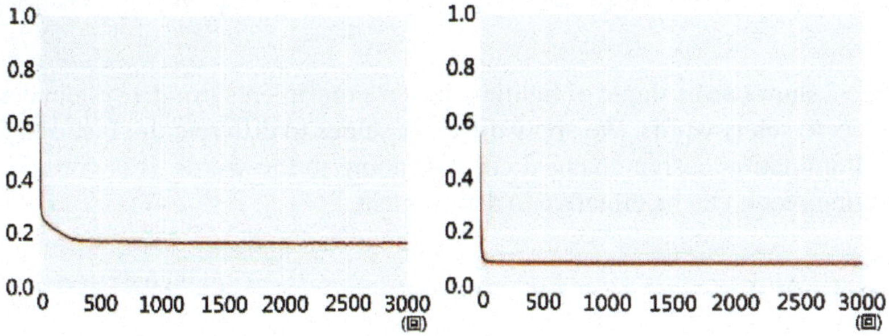

(a) (b)

Fig. 6. The changing of the calculating errors. (a) The changing of errors that from the autoencoder. (b) The changing of errors from the neural network.

Through a three-layer neural network, the sign of the evaluation data and the design function of the design parameters were obtained. A 16-dimensional intermediate layer was adopted to reduce the error. Figure 6(b) shows the change in error after learning 3000 times. The final error is 0.079, and we consider that the learning is sufficient.

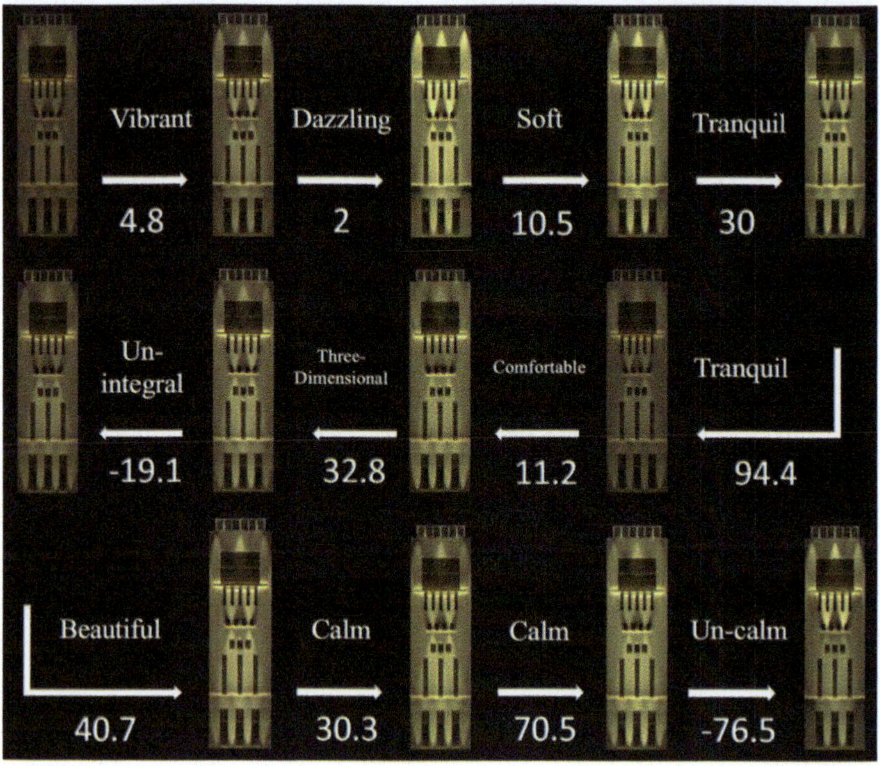

Fig. 7. An example of changing rendering results by changing linguistic parameters.

Figure 7 shows an example of lighting design experiments using a system developed with sliders to select words. We apply different values to different linguistic commands and the illumination design changed corresponding to the words. It is considered that the changing result can be effective to some extent.

4 Conclusion

In this paper, we constructed an architectural façade lighting design model with linguistic manipulation and repeated the manipulation experiment.

Through this research, we have constructed a lighting design model for architectural façades, and used autoencoders and neural networks to clarify the mathematical mechanisms that relate words and patterns. We have created lighting patterns, obtains evaluation data verbally while looking at the patterns, explored the connection between the words and design parameters through an autoencoder and a neural network, and made it possible to verbally correct the lighting design of the building façade simulation model. The specific relationship between the code and its weight obtained from the autoencoder and the lighting pattern and words may be a subject for future work.

References

1. Tomita, Y.: New sights of light-up in Tokyo III: Tokyo station Yaesu area development project(<Feature>Illumination until now and in future∼: Tokyo branch). J. Luminating Eng. Inst. Jpn. **98**(7), 293–295 (2014)
2. Liang, D., Jing, F.: Evaluation of thermal insulation performance of building exterior wall based on multiobjective optimization algorithm. Mob. Inf. Syst. **2672894**, 8 (2022)
3. Edytia, M.H.A., Meutia, E., Sahputra, Z., Billah, M.A., Shafwa, P.: Optimization the use of artificial lighting at architectural design studios in architecture study program of Universitas Syiah Kuala. In: IOP Conference Series: Earth and Environmental Science, vol. 738(1) (2021)
4. Aoki, Y., Inage, M.: A computational model of linguistic instructions on architectural form with adjectives. J. Arch. Plan. **551**, 143–147 (2001)
5. Sato, M., Nakayama, K.: Building color selection system using neural network systems. J. Arch. Plan. **510**, 6–15 (1998)
6. Masuyama, M., Nakamura, Y.: Luminance variation and sense of perspective for façade components of illuminated buildings. J. Arch. Environ. **622**, 9–16 (2007)
7. Okatani, T.: Deep Learning. Kodansha, Japan (2015)
8. Ishida, T., Ogiuchi, Y.: Psychological determinants of brightness of a space-Perceived strength of light source and amount of light in space. J. Illum. Eng. Inst. Jpn. **84**(8), 473–479 (2000)

Towards Accurate Rainfall Volume Prediction: An Initial Approach with Deep Learning, Advanced Feature Selection, Parameter Optimisation, and Ensemble Techniques for Time-Series Forecasting

Bamikole Olaleye Akinsehinde(✉), Changjing Shang, and Qiang Shen

Faculty of Business and Physical Sciences, Department of Computer Science, Aberystwyth University, Aberystwyth, Ceredigion, Wales, UK
{baa32,cns,qqs}@aber.ac.uk

Abstract. Accurate rainfall forecasting is crucial in sectors such as agriculture, transportation, and disaster prevention. This study introduces an initial approach that combines deep forecasting techniques, advanced feature selection, parameter optimisation, and ensemble method to enhance the accuracy of rainfall volume prediction. The proposed methodology is evaluated using a historical weather dataset from Bath, United Kingdom, spanning from January 1, 2000, to April 21, 2020. To address challenges related to generalisation, uncertainty, reliability, and inappropriate predictors, a hybrid mechanism is created by combining various LSTM (Long Short-Term Memory) and GRU (Gated Recurrent Unit) models with a Fuzzy Inference System. The resulting ensemble system comprises five individual hybrid models. Through baseline experiments and comparisons with benchmarks, the effectiveness of the methodology is demonstrated, revealing significant performance improvements over previous studies, across a range of performance indices. Overall, the proposed ensemble approach exhibits better generalisation compared to benchmarks. This research has the potential to revolutionise rainfall volume predictions by leveraging deep learning, advanced feature selection, parameter optimisation and ensemble techniques, overcoming many limitations of the existing approaches.

Keywords: Rainfall Prediction · Weather Forecasting · Deep Learning · Ensemble Techniques · Fuzzy Rough Feature Selection · Optimisation Techniques · Hybrid Method

1 Introduction

Rainfall forecasting is essential for agriculture, disaster preparedness, and water resource management. However, existing models encounter challenges due to the complexity, uncertainty, and dynamism of weather systems [1, 5, 29, 33]. Issues such as underfitting, overfitting, inappropriate predictor features, and lack of reliability in accounting for

N. Naik et al. (Eds.): UKCI 2023, AISC 1453, pp. 114–132, 2024.
https://doi.org/10.1007/978-3-031-47508-5_10

uncertainty hinder these models [3, 7, 27]. Overcoming these challenges is crucial for developing improved models that enhance the understanding and prediction of rainfall events, impacting natural and human processes. To address these challenges, a desirable approach would be a hybrid system, as reported in [35, 36], that combines various algorithms and techniques to enhance accuracy and reliability.

Popular deep learning algorithms, such as Long Short-Term Memory (LSTM) [3, 9, 12, 20, 27, 30], and Gated Recurrent Unit (GRU) [2], show promise in capturing temporal dependencies in rainfall data [4, 27]. They are commonly used for sequence modelling and prediction tasks, including time series analysis. By utilising appropriate feature selection methods [6, 13, 14, 16, 17], parameter optimisation, and Fuzzy Inference Systems (FIS) [31, 32], the accuracy of rainfall forecasting can be improved. Ensemble methods can indeed help reduce combined model errors [8, 10, 23, 24].

This study aims to enhance rainfall forecasting accuracy and reliability through the integration of LSTM, GRU, and FIS, in conjunction with advanced feature selection (AFS), parameter optimisation (PO), and ensemble technique (ET). In particular, by exploiting Fuzzy Rough Feature Selection (FRFS) and RandomisedSearchCV for hyperparameter optimisation, the resulting hybrid system can be expected to significantly improve prediction performance.

As an initial attempt to implement the above aim, the following research questions are addressed, with the corresponding solution mechanisms proposed:

1. How does the hybrid approach of LSTM/GRU combined with FIS enhance rainfall forecasting accuracy, as compared to the standalone Bidirectional-LSTM model [3]?
2. Can an AFS technique using Fuzzy Rough Feature Selection based on Fuzzy C-Means Clustering and Rough Membership (FRFS-FCMRM) effectively address underfitting, overfitting, and inappropriate predictive features by selecting only a small number (say, three) of features (excluding the target variable) from a much larger set (of 46 features), compared to the manual approach of selecting (11) features guided by the computation of the Correlation Matrix through Pearson correlation coefficient of a high dimensional datasets [3]?
3. How can rainfall prediction uncertainty be quantified, considering variations in intensity and frequency throughout the year [1, 11, 33]?
4. What are the main advantages and limitations of combining LSTMs [3], GRUs [2], and FIS [21] in a hybrid system [2] for rainfall forecasting, and how does RandomisedSearchCV [2, 3] and FIS optimise performance?

With reference to the earlier study [3], this study utilises the historical weather data (HWD) from Bath City, United Kingdom (UK), obtained through a subscription to the History Bulk download provided by OpenWeather Limited, United Kingdom[1]. The HWD from 1st January 2000 to 21st April 2020, is used to train the nine setup models for this study. The performance of the resulting hybrid and proposed models is compared against that of baseline and benchmark [3] models. This study contributes to advancements in rainfall forecasting accuracy, addressing challenges faced by existing models [1, 3, 4, 25, 27, 30].

---

[1] Data Source, http://openweathermap.org.

The proposed ensemble system (Hybrid LSTM-GRU-FIS-RandomisedSearchCV) offers a novel approach that overcomes the major limitations of benchmark [3] models. It improves generalisation, reduces uncertainties, ensures reliable feature selection, optimises model parameters, and achieves superior performance in rainfall volume predictions. The summary of the main contributions of this study are:

- Generalisation improvement: By combining LSTM, GRU, and FIS techniques, the ensemble system enhances generalisation for accurate predictions across diverse scenarios and datasets.
- Uncertainty reduction: The ensemble model reduces uncertainties and helps mitigate biases by integrating predictions from multiple hybrid models, thereby combining the strengths of individual models and providing a comprehensive understanding of rainfall patterns. It achieves accounting for fluctuations in intensity and frequency across different season, leveraging the power of deep learning and FIS.
- Reliable feature selection: The core mechanism, FRFS-FCMRM, utilises Fuzzy C-Means Clustering and Rough Membership values to select the most relevant features, addressing the issue of inappropriate predictors in the benchmark models.
- Parameter optimisation: RandomisedSearchCV optimises parameters for each hybrid model, ensuring optimal performance in predicting rainfall volume.
- Performance improvements: The ensemble system exhibits significant enhancements over benchmarks, achieving superior accuracy and reliability in predicting rainfall volume.

The rest of this paper is organised as follows. Section 2 reviews the closely relevant work to the present research. Section 3 details the proposed methodological approach. Section 4 discusses the setup predictive models, Sect. 5 reports on the initial results of the experimental investigation regarding the performance of the proposed approach. Section 6 concludes this work and points out directions for further developments.

2 Related Work

This section provides an overview of the existing literature focused on enhancing the accuracy and reliability of rainfall forecasting. The reviewed literature highlights the advancements made in rainfall forecasting methodologies (RFM), emphasising the importance of incorporating different data sources, AFS, deep learning models, ensemble methods, and hybrid machine learning (ML) techniques. By examining the relevant work, this review delves into the key aspects associated with accurate predictions, offering valuable insights into recent advancements in improving RFM [1, 3, 4, 25, 27].

A comprehensive analysis of relevant techniques in this area has emphasised the importance of incorporating different data sources [2] and predictors [3] to improve accuracy in rainfall forecasting while reducing uncertainties [5] in weather station data. AFS techniques, such as FRFS [6, 28], can reduce model complexity and improve performance.

Artificial neural networks (ANN) like LSTM and GRU have been examined for capturing temporal dependencies in rainfall data [3, 20, 34]. In a recent benchmark study [3], various models were evaluated for hourly rainfall forecasting using HWD datasets from five UK cities. The LSTM-based models, including Stacked-LSTM and Bidirectional-LSTM, outperformed other models, including Extreme Gradient Boosting (XGBoost) and classical ML approaches (CMLA). These results indicate that LSTM-based models can achieve better performance in rainfall volume prediction. Hence, LSTM architectures are adapted as the hybrid model components in this original study.

The deployment of the Complete Ensemble Empirical Mode Decomposition with Adaptive Noise (CEEMDAN)-Convolutional Neural Network (CNN)-GRU model, which involves deep learning techniques, for time series forecasting of soil moisture [2], has provided valuable insights for rainfall volume prediction. By integrating diverse data sources such as satellite-derived data, climate indices, and ground-based variables, the hybrid CEEMDAN-CNN-GRU system outperforms other models in terms of statistical metrics and infographics, demonstrating the potential of hybrid models for accurate rainfall volume prediction. The use of ensemble methods and hyperparameter tuning further improves the system's performance.

The simple averaging ensemble approach [26] has proven successful in various domains, such as wastewater treatment plant prediction using artificial intelligence models. Similarly, the ensemble method that implements assigning weights to ensemble members helps effectively reduce uncertainties in rainfall-runoff simulations and flood risk predictions [8]. Drawing inspiration from the use of ensemble methods in operational space weather forecasting [24], where they have enhanced models and forecasts. Combining the efficiency of ensemble methods with hybrid models is promising in improving the accuracy and reliability of rainfall prediction.

Integration of design choices in ML algorithms enhances model performance for time series problems, as seen in studies involving Neuro-Fuzzy Inference Systems [31] and hybrid fuzzy intelligent agent-based systems [36]. Additionally, hybrid ML techniques, such as the Neural Fuzzy Inference System-Based Weather Prediction Model (NFIS-WPM) [21], have shown improved outcomes in weather forecasting. These models combine fuzzy rule-based neural networks with neural fuzzy inference systems, achieving improved accuracy in precipitation predictions compared to conventional ANN. Furthermore, the Generalised Dynamic Fuzzy Neural Network (GDFNN) model has been introduced for short-term wind speed forecasting, overcoming overfitting issues through optimisation with the brainstorm optimisation algorithm [22]. These insights serve as a foundation for the proposed approach in this study.

Considering the various design choices and approaches as reviewed in this section, the LSTM architecture, including unidirectional LSTM (LSTM), Bidirectional LSTM (BiLSTM), stacked LSTM (StLSTM), and multi-layer LSTM (MtLSTM), along with Gated Recurrent Unit (GRU) and FIS, are being adapted in this original study. Additionally, techniques such as RandomisedSearchCV and ensemble methods are incorporated to perform accurate and reliable rainfall forecasting, as detailed below.

3 Methodology

This section describes an initial approach working towards accurate and reliable rainfall volume forecasting through two major computational processes: the baseline and the hybrid. The methodology can be summarised as the architecture depicted in Fig. 1.

The following subsections discuss the component steps involved, including data collection, data preprocessing, setup of baseline models, and the later section discusses the configuration of various hybrid models. Additionally, the hybrid models are integrated using ensemble techniques.

3.1 Architecture

Figure 1 provides an overview of the architectural design for the approach undertaken in this research. It incorporates techniques such as deep forecasting, feature selection, parameter optimisation, and model ensemble to improve rainfall predictions performance.

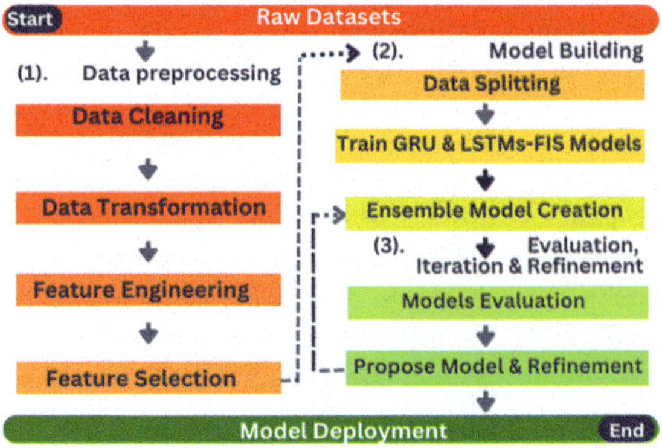

Fig. 1. Flowchart of the proposed framework.

3.2 Data Collection and Pre-processing

The raw historical weather dataset (HWD) from Bath city in the UK is utilised in this study. It contains various weather-related measurements recorded at regular intervals, including temperature, humidity, pressure, wind speed, precipitation, and other meteorological features. Table 1 shows all the features in the dataset, along with the description of the weather element.

The following sub-sections describe several data preprocessing steps undertaken to clean and prepare the dataset for subsequent analysis and verification of the proposed approach.

Table 1. All features in Bath dataset, with description and percentage of missing values.

S/N	Variable Name	Description of Variable	Missing Value	S/N	Variable Name	Description of Variable	Missing Value
01	dt	Time of data calculation	0.00%	15	grnd_level	Earth surface	100%
02	dt_iso	Date and time in UTC	0.0%	16	humidity	Percentage of humidity	0.0%
03	timezone	UTC shift (sec)	0.0%	17	wind_speed	Wind Speed	0.0%
04	city_name	City name	0.0%	18	wind_deg	Wind direction	0.0%
05	lat	Latitude	0.0%	19	wind_gust	Brief increase in wind speed	94.0%
06	lon	Longitude	0.0%	20	rain_1h	Hourly rainfall volume	83.4%
07	temp	Temperature	0.0%	21	rain_3h	3-h rainfall volume	100%
08	visibility	Average visibility (metres)	87.5%	22	snow_1h	Hourly Snow volume	99.7%
09	dew_point	Droplet formation temperature	0.0%	23	snow_3h	3-h rainfall volume	100%
10	feels_like	Weather perception	0.0%	24	clouds_all	Percentage of cloud cover	0.0%
11	temp_min	Minimum temperature	0.0%	25	weather_id	Weather condition id	0.0%
12	temp_max	Maximum temperature	0.0%	26	weather_main	Weather parameter group	0.0%
13	pressure	Atmospheric pressure	0.0%	27	weather_description	Group weather state	0.0%
14	sea_level	Level of sea surface	100%	28	weather_id	Weather condition id	0.0%

3.2.1 Exploratory Data Analysis

Exploratory Data Analysis (EDA) is conducted to gain insights into the dataset, using summary statistics and visualisations such as box plots, histograms, scatter plots, and correlation analysis to understand the data structure and identify any issues. The line graph depicted in Figs. 2, 3, 4 and 5 offer insights into patterns of the target (forecast) variable and the three associated most important features. The visualisation reveals variations in a number of aspects, including: (1) rainfall intensity and frequency, (2) atmospheric pressure, (3) temperature, and (4) degree of wind throughout the year. These observations indicate a complex system which demands accurate and reliable rainfall prediction models that can capture the sophisticated patterns of such nature.

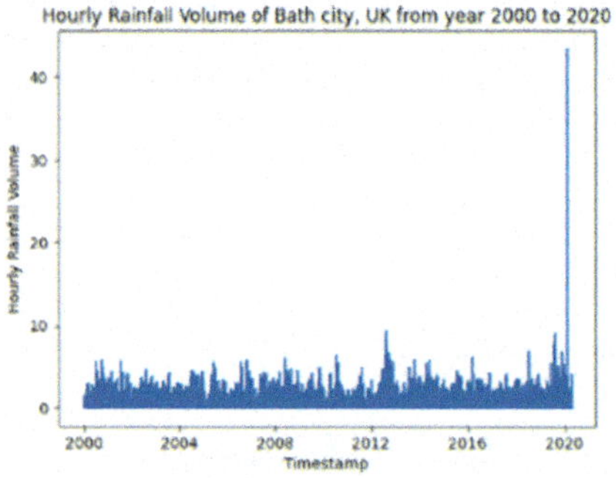

Fig. 2. Hourly Rainfall Volume.

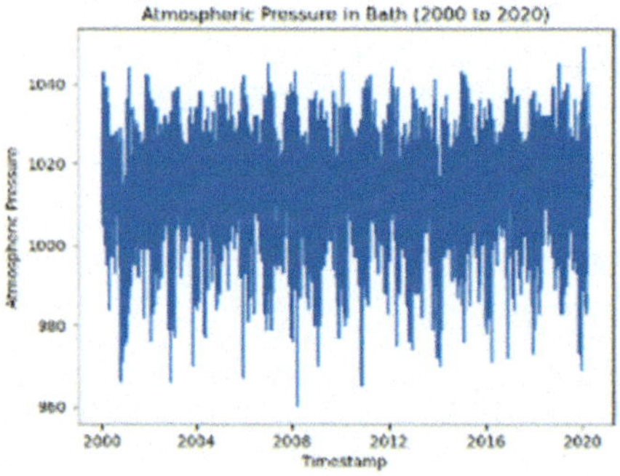

Fig. 3. Atmospheric Pressure.

3.2.2 Data Cleaning and Missing Value Treatment

The HWD dataset initially contains missing values and null entries. Figure 6 presents the bar chart with the missing numbers depicted.

Data cleaning is an essential step that involves addressing missing values and eliminating irrelevant features in this initial approach. The following steps are taken for data cleaning and missing value treatment:

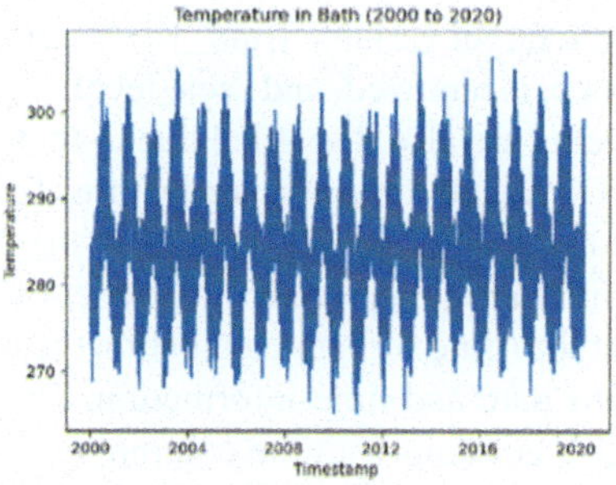

Fig. 4. Temperature in Bath.

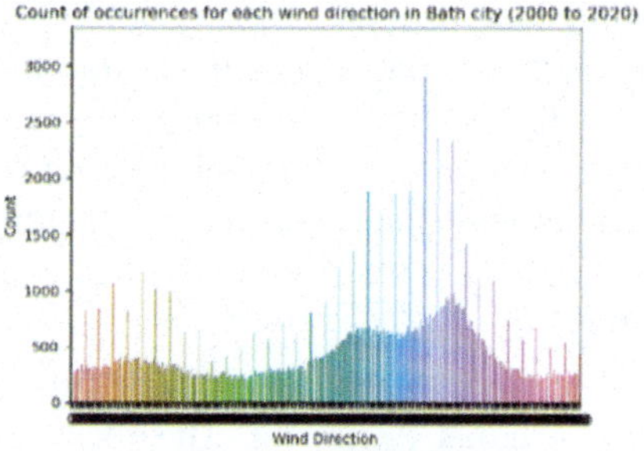

Fig. 5. Degree of Wind direction.

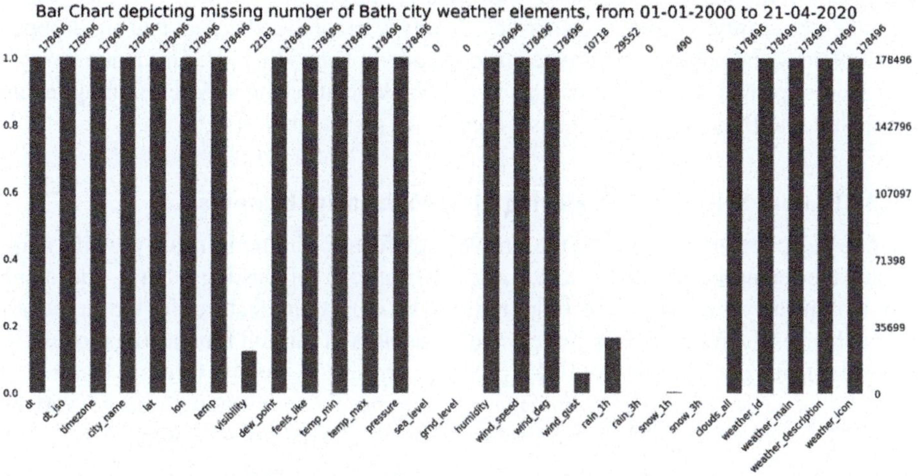

Fig. 6. Missing Data: Bar Chart of Bath - Missing and Nullness of Features (2000–2020).

(1) **Null variables**: Null features (whose values are largely missing) are identified and dropped from the dataset. Of such is the "visibility" variable being removed due to its having 87.57% missing values, as statistical analysis of such variables reveals

little or no impact on the target variable (rain_1h). Similarly, "wind_gust" has 94% missing values and hence, is removed, and "sea_level," "grnd_level," "rain_3h," and "snow_3h" measurements are also discarded due to their 100% missing values.

(2) **Irrelevant Features**: Obviously unnecessary features for the present objectives are excluded. These include the feature "city_name", as the dataset is already separated by city during collection (and herein, only the subset of data taken for Bath city is utilised). The "dt" feature (time of data calculation) is removed since the "dt_iso" feature already contains date and time information. The " + 0000 UTC" string is removed from the dt_iso column, with the column converted into datetime format. The dataFrames are filtered to include HWD from 2000-01-01 00:00:00 to 2020-04-21 00:00:00, to facilitate a comparative analysis with an earlier study [3]. The "timezone" feature is also excluded to avoid restricting the learning process of the models to highly specific time patterns.

(3) **Missing Values:** To handle missing values in specific features such as "rain_1h," "snow_1h," and "snow_3h," where the amount of the missing ones is limited, the missing values are imputed with zeros. Imputing these features with zeros assumes that if a value is absent for any hour, it indicates no precipitation during that period.

3.2.3 Feature Extraction

Feature extraction is applied here to the pre-processed dataset, expanding the variable description (from the original 27 to 46). The focus is on the categorical feature "weather_main," summing information on different weather parameters. It is grouped using "weather_id" and encoded as 0 or 1 to indicate the absence or presence of a specific weather condition, respectively. The original features "weather_main," "weather_description," and "weather_id" therefore become redundant and are dropped, with (19) new and raw features created through one-hot encoding [3]. This feature extraction process provides valuable insights into variable relationships. After running feature extraction, the Bath city HWD dataset consists of 46 features.

3.2.4 Feature Selection – Removing Multicollinearity Features

Feature selection helps address the problem of feature multicollinearity while maintaining the semantics of those selected ones. To improve forecasting models, addressing multicollinearity is crucial. Selecting a reduced set of relevant features for the target variable enhances the reliability, accuracy, and robustness of rainfall forecasting models. As a popular feature selection tool, FRFS provides a means to identify relevant features for rainfall prediction, ensuring the final feature set to be most relevant to the target variable (rain_1h) for an explainable, generalised, and accurate prediction system.

Feature Selection for Baseline model 1 through FRFS-CMM: The first baseline model (BM1) applies a conjunctive approach of FRFS and the Correlation Matrix Method (CMM) [3]. Figure 7 displays the correlation value of variables in the Bath city dataset. The heatmap assists in conducting correlation analysis, helping to identify variable relationships to address multicollinearity issues.

To implement FRFS and automate the selection of important predictive features, the initial step involves identifying variables with correlations equal to or greater than the set threshold (i.e., 0.7 for the present work), following established practices in the literature. By defining the correlation threshold as 0.7, those (39) predictive features (below the threshold) are obtained from the initial (46) ones. Variables such as "dew_point," "feels_like," "temp_min," "temp_max," "snow_1h," and "Snow_id_601" are identified as highly correlated features based on the threshold and subsequently dropped. The selected independent variable for BM1 is split into a training, validation, and test set as inputs for BM1 with the target variable "rain_1h". Table 2 shows the resulting list of independent selected variables for BM1 and other models.

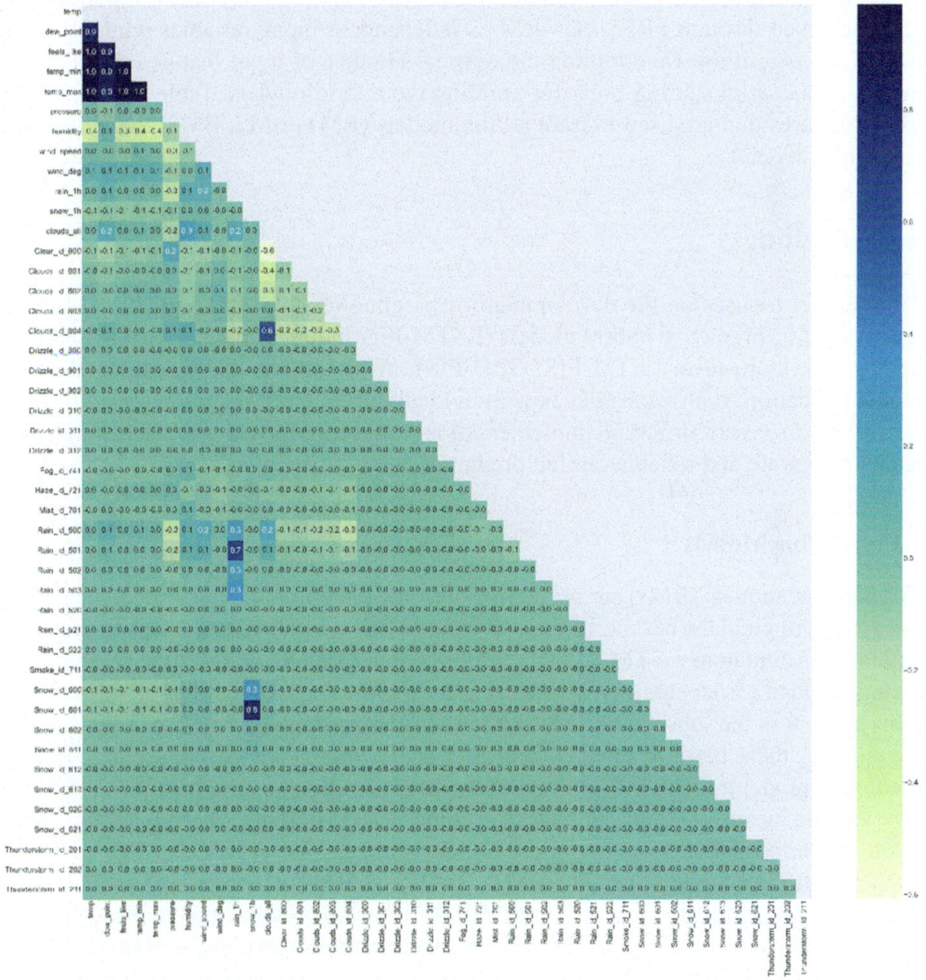

Fig. 7. Heatmap for variables in Bath city dataset (01/01/00–21/04/20).

Feature Selection for Baseline models 2 and 3 through FRFS-FCMRM: The core feature selection mechanism for the proposed rainfall forecasting system is FRFS-FCMRM, which utilises Fuzzy C-Means Clustering and Rough Membership values to determine feature importance. The top five features are selected through this approach and undergo preprocessing (splitting, standardisation) for training and evaluation. The second baseline model (BM2) validates FRFS in a multi-dimensional dataset using these five important features. BM2 broaden the understanding of the importance of using the optimal minimum of the most important variables, as opposed to utilising a larger set of features (39 features) as done by BM1. Because BM2 exhibits similar metric performance with BM1, while offering the benefits of model simplicity, interpretability, reduced overfitting and computational efficiency, it paves the way for discovery the optimal number of features for Baseline model 3 (BM3). BM3 employs the three most important features achieved through FRFS-FCMRM as independent input variables while utilising the LSTM algorithm. Determining the optimal number of input features through BM models is a crucial starting point for building the hybrid models. Table 2 provides the list of features that are used to train all the models (BM1, BM2, BM3, the hybrid and ensemble models).

4 Modelling

This section focuses on the development of baseline models (BM1, BM2, BM3) and that of the five predictive hybrid models (LSTM-FIS, bidirectional LSTM-FIS, Stacked LSTM-FIS, multi-layer LSTM-FIS, GRU-FIS), involving parameter optimisation and model validation. It also specifies how individual hybrid models are integrated through ensembles to create different implementations of the proposed approach, aiming to achieve accurate and reliable rainfall prediction.

4.1 Baseline Models

The baseline models (BMs) are set up to determine the optimal number of selected features that will yield the best performance in predicting rainfall volume. This approach of employing standalone LSTM variants (unidirectional LSTM) follows the conventional method used in earlier studies [3]. The decision to use a minimal set of three selected features as input in the subsequent construction of the hybrid model with the target variable is guided by these baseline models, aiming to achieve an enhanced and accurate rainfall prediction model through model integration. Additionally, the baseline models serve as a systematic means of comparing the performances of different algorithm combinations in the hybrid model.

Three LSTM models, namely BM1, BM2, and BM3, are built as baseline models. They only differ in terms of input features. For the present implementation, BM1 utilises 39 features as independent variables selected through FRFS-CMM. BM2 uses five selected independent variables, along with the target variable (rain_1h). BM3 utilises the top three most important features, selected through the same FRFS-FCMRM method. The features employed by each model are listed on Table 2.

The baseline models (BM1, BM2 and BM3) employ a unidirectional LSTM with a batch size of 64, 150 epochs, and a patience of 100. The primary purpose of the baseline models is to determine the dimensionality (or the number of independent variables)

Table 2. List of feature sets used to train all models.

Type of Model	Baseline Model 1 (BM1)	Baseline Model 2 (BM2)	Baseline Model 3 (BM3)	The five Hybrid Models	Ensemble (Proposed) Model
Implemented Technique	Conjunctive application of FRFS and CMM (FRFS-CMM)	FRFS-FCMRM	FRFS-FCMRM	FRFS-FCMRM	FRFS-FCMRM
Selected Variables	temp, pressure, humidity, wind_speed, wind deg, snow 1h, clouds_all, Clear_id_800, Clouds_id_801, Clouds_id_802, Clouds_id_803, Clouds_id_804, Drizzle_id_300, Drizzle_id_301, Drizzle_id_302, Drizzle_id_310, Drizzle_id_311, Drizzle_id_312, Fog_id_741, Haze_id_721, Mist_id_701, Rain_id_500, Rain_id_501, Rain_id_502, Rain_id_503, Rain_id_520, Rain_id_521, Rain_id_522, Smoke_id_711, Snow_id_600, Snow_id_602, Snow_id_611, Snow_id_612, Snow_id_613, Snow_id_620, Snow_id_621, Thunderstorm_id_201, Thunderstorm_id_202, Thunderstorm_id_211	pressure, temp, wind_deg, humidity, and clouds_all	pressure, temp, wind_deg	pressure, temp, wind_deg	pressure, temp, wind_deg

of hybrid models, aiming to enhance generalisation, prevent overfitting, and reduce complexity in the model architecture.

4.2 Five Predictive Hybrid Models

Through adapting the standalone LSTM-based models [3], including GRU [2] and FIS [21, 22, 36], five predictive hybrid models for rainfall volume prediction are built through the combination of FIS with LSTM variants, namely, unidirectional LSTM, bidirectional LSTM, stacked LSTM, multi-layer LSTM (respectively referred to as LSTM, BiLSTM, StLSTM, MtLSTM, hereafter) and GRU. This leads to the hybrid models of LSTM-FIS, BiLSTM-FIS, StLSTM-FIS, MtLSTM-FIS, and GRU-FIS.

Note that GRU is a modified version of LSTM and is also a type of recurrent neural network (RNN). As with LSTM, GRU is designed to handle sequential data and address the vanishing gradient problem that can occur in traditional RNNs. The five-hybrid models are all designed with an input shape of (1, 3). This implies that each sample in the data is represented by a single time step with three features, as used with BM3. These models utilise a layer consisting of 30 units to control complexity in a similar way to the BMs. Mean squared error (MSE) is used as the loss function, and the 'adam' optimiser [3] is exploited.

The training process involves iterations over the dataset for a specified number of epochs, which can be chosen from a grid of values like 10, 20, and 50. Batch size (32, 64, and 72), i.e., the number of samples processed before updating the model, is also adjustable. RandomisedSearchCV is used for training with the meta-parameter monitor = 'loss' and patience = 30. These parameters will of course affect the architecture, optimisation, and training of the hybrid (LSTM-FIS and GRU-FIS) models.

4.3 Integrated Hybrid Model (Ensemble Model)

The success of utilising simple-averaging ensemble approach [26] in various problem domains provides valuable insights for its applications to addressing the challenges of accurate and reliable rainfall volume prediction. In recognition of the above, this study combines the strengths of different hybrid models (based on LSTM and GRU) to help reduce their individual weaknesses, variance, and bias, leading to more reliable rainfall forecasts.

The proposed ensemble system is created using the RandomForestRegressor algorithm and trained on the same weather data acquired from Bath city as used for developing individual predictive hybrid models (namely, LSTM-FIS, BiLSTM-FIS, StLSTM-FIS, MtLSTM-FIS, and GRU-FIS) being integrated as an ensemble model (Hybrid LSTM-GRU-FIS-RandomisedSearchCV) to make predictions on the test data. The individual predictions of the hybrid models are integrated using the simple averaging ensemble approach. This leads to novel contributions to the literature, including the combination of LSTM variants with FIS and that of GRU with FIS, for predicting rainfall volume or other weather conditions, as well as the integration of hybrid models through an ensemble method to enhance rainfall or weather (element) forecasting.

5 Experimental Investigation

To verify the potential of the proposed approach, initial experimental evaluations are carried out in this section.

5.1 Performance Criteria

Following the existing literature [3, 9, 27, 30], the following criteria are utilised to evaluate the performance of all models in this comparative experimental study:

1. Loss: Measuring the discrepancies between predicted and actual outputs, indicating the extent of incorrect predictions.
2. RMSE (Root Mean Squared Error): Quantifying the discrepancies between predicted and actual values, emphasising significant errors.
3. RMSLE (Root Mean Squared Logarithmic Error): Determining the accuracy of predictions on a logarithmic scale, suitable for variables with a wide value range or outliers.
4. MAE (Mean Absolute Error): Measuring the average absolute divergence between predicted and actual values, providing an estimation of typical prediction inaccuracy.

The above performance metrics provide insights into the accuracy and reliability of the predictions made by different models, with lower values indicating better performance.

5.2 Results and Observations

5.2.1 Results and Performance of Compared Models

The performance outcomes (as per Loss, RMSE, RMSLE and MAE) for all compared models are presented in Table 3 and Fig. 8. Each predicted value is also compared with the actual rainfall volume using the proposed model in Table 4. In addition, the performance of different models is compared against the benchmark models [3].

Table 4 displays observed and predicted rainfall volumes along with prediction accuracy calculated over hourly measurements, using the proposed integrated hybrid LSTM-GRU-FIS-RSCV (ensemble) system. Index_1 indicates the index number in the preprocessed dataset, while Index_2 represents the index number for non-zero rainfall volume (rain_1h). The first twelve timestamps of the predicted values are shown for discussion.

5.2.2 Performance Comparison

The ensemble model is shown to achieve the best performance in terms of RMSE and Loss over the tests, surpassing expected accuracy levels. The hybrid models, particularly the GRU-FIS model, exhibit robustness to uncertain data and provide more accurate and reliable predictions. The proposed system, of integrated hybrid LSTM-GRU-FIS-RandomisedSearchCV (ensemble), performs the best with a test Loss of −0.061956, RMSE of 0.325547, RMSLE of 0.1974, and MAE of 0.162241.

Table 3. Performance of all different models

Type of Experiment	Model Name	Input Feature (No)	Loss Performance	Test Loss	RMSE Performance	Test RMSE	Test RMSLE	Test MAE
Baseline	BM1	39	Rated 2	0.1087	Rated 2	0.3297	0.1944	0.1579
Baseline	BM2	5	Rated 3	0.1096	Rated 3	0.3310	0.1946	0.1582
Baseline	BM3	3	Rated 1	0.1032	Rated 1	0.3212	0.1911	0.1569
Main	LSTM-FIS	3	Rated 4	0.1204	Rated 5	0.3470	0.1981	0.1605
Main	BiLSTM-FIS	3	Rated 3	0.1137	Rated 3	0.3371	0.1981	0.1591
Main	StLSTM-FIS	3	Rated 6	0.1403	Rated 6	0.4125	0.2947	0.3415
Main	MtLSTM-FIS	3	Rated 5	0.1204	Rated 4	0.3431	0.1963	0.1644
Main	GRU-FIS	3	Rated 2	0.1108	Rated 2	0.3329	0.1944	0.1558
Main	Ensemble	3	Rated 1	−0.062	Rated 1	0.3256	0.1974	0.1622

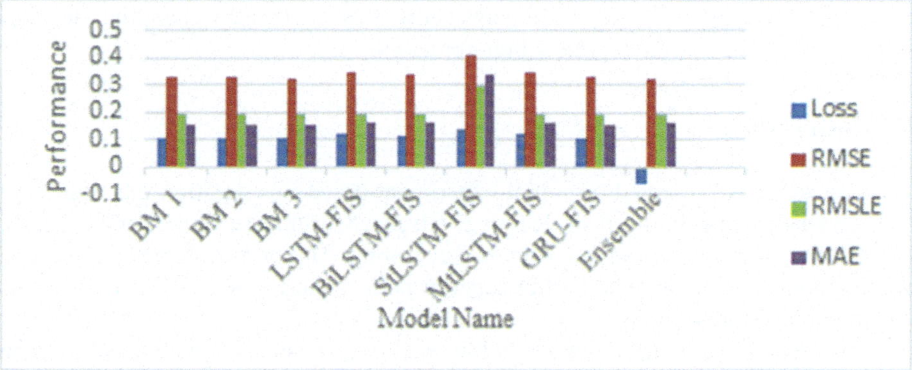

Fig. 8. Model performance

Table 4. Observed and predicted rainfall volumes including percentage accuracy at each timestamp using ensemble (integrated Hybrid LSTM-GRU-FIS-RandomisedSearchCV) model.

S/N	Index_1	Index_2	Observed	Predicted	% Accuracy of Prediction
1	302768	1	0.10	0.097019	97.019464
2	190052	14	0.12	0.096395	80.328969
3	251794	21	0.14	0.097647	69.747988
4	253323	36	1.45	0.097111	6.697309
5	330959	41	0.18	0.097096	53.941992
6	194209	45	0.50	0.097317	19.463494
7	270717	51	0.55	0.097125	17.659052
8	248653	52	1.19	0.097113	8.160774
9	240122	57	0.12	0.097727	81.439006
10	310880	84	0.51	0.098724	19.357673
11	299199	86	0.16	0.096817	60.510384
12	362647	101	0.11	0.096200	87.454387

Comparing the performance of the resulting system to that attainable in the previous study [3], the proposed approach and hybrid models outperform the benchmark ensemble prediction, achieving lower RMSE values. The benchmark models [3] in the existing work encounter challenges in adapting to abrupt changes and suffer from generalisation limitations, thereby often resulting in inaccurate predictions. In sharp contrast, the proposed approach enables better generalisation with superior accuracy.

The experimental results have highlighted the effectiveness of both the ensemble model and the hybrid LSTM-FIS and GRU-FIS models in predicting rainfall volume, outperforming alternative approaches in terms of accuracy and reliability. The integration of LSTM/GRU with FIS facilitates the capturing of long-term dependencies and temporal patterns. In particular, the use of FIS helps address the challenge of handling uncertain and imprecise data, enhancing model interpretability.

6 Conclusion

This research has presented an initial experiment-based investigation that showcases the potential of integrating deep learning, advanced feature selection, and ensemble techniques to strengthen rainfall volume forecasting. The proposed methodology has been demonstrated to be able to offer superior performance in terms of generalisation, interpretability and accuracy.

This original study has employed a low-quality raw dataset, which may have contributed to differences between predicted and observed values. Missing values in the weather dataset need to be addressed in future work. Further improvement of the integrated system's accuracy and reliability can be expected, by incorporating additional data sources such as satellite and remote sensing data, along with standardised datasets, and exploring a wider range of model hyperparameters. This remains active research. Continual efforts to refine and expand the integrated model through additional data sources and parameter optimisation will contribute to even more precise rainfall volume prediction, benefiting various industries that rely on accurate predictions.

Declaration This work is free from any financial interest, competing interest, or personal interest that could influence its outcome.

Acknowledgements. The authors would like to acknowledge Barrera-Animas et al., 2021, for their benchmark study. The first author is grateful to Aberystwyth University for offering the PhD scholarship in support of this research.

References

1. Aguasca-Colomo, R., Castellanos-Nieves, D., Méndez, M.: Comparative analysis of rainfall prediction models using machine learning in islands with complex orography: Tenerife Island. Appl. Sci. **9**, 4931 (2019)
2. Ahmed, A.A.M., et al.: Deep learning forecasts of soil moisture: convolutional neural network and gated recurrent unit models coupled with satellite-derived MODIS, observations and synoptic-scale climate index data. Remote Sens. **13**, 55 (2021)
3. Barrera-Animas, A.Y., Oyedele, L.O., Bilal, M., Akinosho, T.D., Delgado, J.M.D., Akanbi, L.A.: Rainfall prediction: a comparative analysis of modern machine learning algorithms for time-series forecasting. Mach. Learn. Appl., 100204 (2021)
4. Basha, C.Z., Bhavana, N., Bhavya, P., Sowmya, V.: Rainfall prediction using machine learning & deep learning techniques. In: IEEE Xplore (2020)
5. Buizza, R.: Chaos and weather prediction. In: ECMWF (2002)
6. Cornelis, C., Jensen, R., Shen, Q.: Hybrid fuzzy-rough rule induction, and feature selection. In: Aberystwyth Research Portal (2009)
7. Deman, V.M.H., Koppa, A., Waegeman, W., MacLeod, D.A., Bliss Singer, M., Miralles, D.G.: Seasonal prediction of Horn of Africa long rains using machine learning: the pitfalls of preselecting correlated predictors. Front. Water **4**, 1053020 (2022)

8. Doycheva, K., Horn, G., Koch, C., Schumann, A., König, M.: Assessment, and weighting of meteorological ensemble forecast members based on supervised machine learning with application to runoff simulations and flood warning. AEI **33**, 427–439 (2017)
9. Gauch, M., Kratzert, F., Klotz, D., Nearing, G., Lin, J., Hochreiter, S.: Rainfall–runoff prediction at multiple timescales with a single Long Short-Term Memory network. Hydrol. Earth Syst. Sci.. Earth Syst. Sci. **25**, 2045–2062 (2021)
10. Gneiting, T., Raftery, A.E.: Atmospheric science: weather forecasting with ensemble methods. Science **310**, 248–249 (2005)
11. GOV.UK; UK Department for Business, Energy and Industrial Strategy—Average Annual Rainfall in the United Kingdom (UK) from 2001 to 2022. (Statista 2023)
12. Hu, C., Wu, Q., Li, H., Jian, S., Li, N., Lou, Z.: Deep learning with a long ShortTerm memory networks approach for rainfall-runoff simulation. Water **10**, 1543 (2018)
13. Jensen, R., Mac Parthaláin, N.: Towards scalable fuzzy–rough feature selection. Inf. Sci. **323**, 1–15 (2015)
14. Jensen, R., Shen, Q.: Computational Intelligence and Feature Selection (2008)
15. Jensen, R., Shen, Q.: New approaches to fuzzy-rough feature selection. IEEE Trans. Fuzzy Syst. **17**, 824–838 (2009)
16. Ji, W., et al.: Fuzzy rough sets and fuzzy rough neural networks for feature selection: a review. WIREs Data Min. Knowl. Discov. **11**(3), e1402 (2021)
17. Li, F., Shang, C., Li, Y., Shen, Q.: Feature Ranking-Guided Fuzzy Rule Interpolation. IEEE Press (2017)
18. Li, F., Shang, C., Li, Y., Yang, J., Shen, Q.: Fuzzy rule based interpolative reasoning supported by attribute ranking. IEEE Trans. Fuzzy Syst. **26**, 2758–2773 (2018)
19. Li, F., Shang, C., Li, Y., Yang, J., Shen, Q.: Approximate reasoning with fuzzy rule interpolation: background and recent advances. **54**, 4543–4590 (2021). SpringerLink
20. Lindemann, B., Müller, T., Vietz, H., Jazdi, N., Weyrich, M.: A survey on long short-term memory networks for time series prediction. Procedia CIRP **99**, 650–655 (2021)
21. Lu, J., Xue, S., Zhang, X., Zhang, S., Lu, W.: Neural fuzzy inference system—based weather prediction model and its precipitation predicting experiment. Atmosphere **5**, 788–805 (2014)
22. Ma, X., Jin, Y., Dong, Q.: A generalized dynamic fuzzy neural network based on singular spectrum analysis optimized by brainstorm optimization for short-term wind speed forecasting. Appl. Soft Comput.Comput. **54**, 296–312 (2017)
23. Maqsood, I., Khan, M., Abraham, A.: An ensemble of neural networks for weather forecasting. Neural Comput. Appl. **13**, 112–122 (2004)
24. Murray, S.A.: The importance of ensemble techniques for operational space weather forecasting. Space Weather **16**, 777–783 (2018)
25. Nayak, D.R., Mahapatra, A., Mishra, P.: A survey on rainfall prediction using artificial neural network. Int. J. Comput. Appl.Comput. Appl. **72**, 32–40 (2013)
26. Nourani, V., Elkiran, G., Abba, S.I.: Wastewater treatment plant performance analysis using artificial intelligence—an ensemble approach. Water Sci. Technol. **78**, 2064–2076 (2018)
27. Poornima, S., Pushpalatha, M.: Prediction of rainfall using intensified LSTM based recurrent neural network with weighted linear units. Atmosphere **10**, 668 (2019)
28. Qian, Y., Wang, Q., Cheng, H., Liang, J., Dang, C.: Fuzzy-rough feature selection accelerator. Fuzzy Sets Syst. (2015)
29. SciJinks: How reliable are weather forecasts? | NOAA SciJinks—all about weather. In: Scijinks.gov. (2016)
30. Siami-Namini, S., Tavakoli, N., Namin, A.S.: The performance of LSTM and BiLSTM in forecasting time series. In: 2019 IEEE International Conference on Big Data (2019)
31. Sun, Z.-L., Au, K.-F., Choi, T.-M.: A neuro-fuzzy inference system through integration of fuzzy logic and extreme learning machines. IEEE Trans. Syst. Man Cybern. Part B (Cybernetics) **37**, 1321–1331 (2007)

32. Takagi, H., Hayashi, I.: NN-driven fuzzy reasoning. Int. J. Approx. Reason. **5**, 191–212 (1991)
33. Ukhurebor, K.E., Abiodun, I.C.: Variation in annual rainfall data of forty years (1978–2017) for south-south, Nigeria. J. Appl. Sci. Environ. Manag.Manag. **22**, 511 (2018)
34. Ukhurebor, K.E., Azi, S.O., Aigbe, U.O., Onyancha, R.B., Emegha, J.O.: Analyzing the uncertainties between reanalysis meteorological data and ground measured meteorological data. Measurement **165**, 108110 (2020)
35. Wahyuni, I., Mahmudy, W.F., Iriany, A.: Rainfall prediction using hybrid adaptive neuro fuzzy inference system (ANFIS) and genetic algorithm. J. Telecommun. Electron. Comput. Eng. (JTEC 2017)
36. Zarandi, M.H.F., Hadavandi, E., Turksen, I.B.: A hybrid fuzzy intelligent agent-based system for stock price prediction. Int. J. Intell. Syst.Intell. Syst. **27**, 947–969 (2012)

Hierarchies of Power: Identifying Expertise in Anonymous Online Interactions

Amal Htait(✉), Lucia Busso, and Tim Grant

Aston University, Birmingham B4 7ET, England
{a.htait,l.busso,t.d.grant}@aston.ac.uk

Abstract. This paper sets the stage for our primary objective, which is to identify and examine various forms of claimed expertise in anonymous online interactions. By building upon the findings and incorporating the proposed enhancements, we aim to gain a deeper understanding of the nature and implications of different expertise claims within the context of power hierarchies. A combination of various machine learning techniques is employed in this work, including classical methods, deep learning models, and transformer-based approaches to create classification models, while using three datasets collected by specialists and annotated by linguistics experts. The first experiments' results in binary classification, indicating whether a given post reflects expertise or not, are particularly promising, especially when utilising transformer-based approaches. The second set of experiments, focusing on the classification of different types of expertise, produced a diverse range of results with the less favourable results primarily caused by an imbalance in labelling between different classes.

Keywords: Hierarchies of power · Personal expertise · Community expertise · Broad topic expertise · Machine learning · Classification · BERT

1 Introduction

Power is undeniably one of the most complex phenomena that has been explored in the domain of social sciences [3]. And in an effort to provide a simple definition, Farfán and Holzscheiter [3] describe it as "*the ability of an individual to pursue their own interest even against the resistance of another person*". Within the context of discourse, this perspective would be reflected in how individuals employ language exchange and communication as a means to assert dominance and power over others, which created a connection between linguistics and claim of power, and made the topic of *Language and Power* the subject of much study in social sciences [2,9]. These studies frequently explore how differences in social or institutional power can be reflected at the linguistic level (e.g., between teacher and student). A set of low-level linguistic features –such as polite forms (e.g.,

N. Naik et al. (Eds.): UKCI 2023, AISC 1453, pp. 133–139, 2024.
https://doi.org/10.1007/978-3-031-47508-5_11

'sir', 'please')– were initially identified to be connected to authority or power granting. However, they were discovered to be incompetent as direct predictors of true power dynamics in anonymous online interactions due to their multi-functional nature, which adds complexity to the interpretation of results [14]. Furthermore, it has been observed that individuals draw upon various resources to assert and demonstrate power, such as the claim of various forms of expertise. In addition, multiple hierarchies of power are detected simultaneously in online interactions [14].

The purpose of our project is to identify people in positions of power and authority in anonymous online fora –more specifically in online criminal networks– solely through their answers to requests for advice posed by original posters. For that purpose, we considered the expertise claim embedded within the answers, which aimed to influence the actions of the original poster, as an indication of power achieved without resorting to direct face-threatening assertions. In this paper, we introduce our primary objective to identify different forms of expertise with the intention of laying the groundwork for future investigations on assessing their significance within a range of power hierarchies.

Throughout this paper, we explore the intriguing question: *Can machine learning effectively detect expertise in an anonymous online forum by solely analysing conversation text?* To answer it, we worked with three distinct datasets annotated by linguistics experts [13]. First, for ethical reasons (researchers well-being), we used a dataset collected from a benign open web parenting discussion forum, with mainly adoption topics (Forum 1). Then, we worked on two datasets forensically intriguing and potentially distressing in nature: a white nationalist discussion forum, characterised by discussions encompassing potentially illegal activities and extreme ideologies (Forum 2), and a dark web child sexual exploitation and abuse discussion forum where offenders openly discussed their criminal activities (Forum 3).[1]

In the field of online expertise identification, the predominant focus has traditionally been on the users' profile and their network ranking (e.g., followers count, activity levels) [7,15,16]. The consideration of syntax and semantics features[2] in identifying expertise has been relatively neglected, with only a limited number of studies examining the syntax features significance in this context. For example, Horne et al. [8] conducted experiments to evaluate the predictive capability of various features, including syntax features, by employing several machine learning algorithms such as decision trees, support vector machines (SVM), and logistic regression. The experiments covered diverse expertise topics such as Science, Business, Fitness, and more. In the findings, Horne et al. presented results demonstrating the superiority of other features (e.g., Network feature) compared to syntax in terms of their predictive power. However, in scenarios involving anonymous forums where access to other features such as network and profile information is unavailable, we encountered the need to explore

[1] All members of the team with direct access to the data were offered psycho-education and consultations with a supervising counsellor.

[2] Syntax refers to grammar, Semantics refers to meaning.

more advanced technologies to effectively identify expertise solely based on syntax and semantics features present in the discussions. In this paper, to develop an efficient expertise classifier, we experiment with a combination of various machine learning techniques, including classical methods, deep learning models, and transformer-based approaches.

2 Dataset

The datasets used are collected by specialists,[3] and annotated by linguistics experts from fora that ensure anonymity offering no clear differences in power and status between users[4]: Forum 1 (open web parenting discussion forum), Forum 2 (white nationalist discussion forum), and Forum 3 (dark web child sexual exploitation and abuse). In a preliminary observation of the datasets, three types of expertise were identified (example provided from Forum 1): (1) Claim of **personal expertise** on a given topic (e.g., "*When we received the phone call for our lo my husband just knew.*"). (2) Claim of veteran status or accredited expertise, which we interpret as a **community expertise** (e.g., "*how I would cope with going from being a career woman to a SAHM*"). (3) Claims of expertise on a **broad topic** (e.g., "*I'd keep a running list of questions and allow a good long time to talk when you get to see the Medical Advisor*"). Consequently, based on this observation, the collected data is annotated at two levels. Firstly, a binary annotation is applied to each post, indicating whether it reflects expertise or not. Secondly, for the posts annotated as reflecting expertise, they are further classified into one of three specific types of expertise (Personal Expertise, Community Expertise, and Broad Topic Expertise). Table 1 provides an overview of the collection statistics, highlighting the fact that a post can belong to multiple types of expertise.

Table 1. Collection statistics: the total number of posts in each expertise category

Dataset	Binary classification		Expertise		
	Expert	Non-Expert	Pers-Exp	Community-Exp	Broad Topic-Exp
Forum 1	211	141	180	13	82
Forum 2	251	73	117	16	204
Forum 3	198	182	138	53	87

3 Experiments and Preliminary Results

In our experiments, a diverse range of machine learning technologies is employed. Classical machine learning algorithms (with default parameters), such as decision trees [12], Logistic regression [10] and support vector machines (SVM) [6],

[3] Specialist company scraped the fora while excluding any media files and illegal data.
[4] Ethical Approval via Aston University AIFL Ethics subcommittee approval reference AIFL-REC-21-018.

are utilised as baseline. Deep learning models (with epochs = 10), such as LSTM (Long Short-Term Memory) [4] and bi-LSTM (bi-directional Long Short-Term Memory) [5], are employed for their capacity to capture context and dependencies in text. Additionally, transformer-based models (with train_batch = 32, tokens_length = 128 and epochs = 10), such as BERT (Bidirectional Encoder Representations from Transformers) [1] and RoBERTa (Robustly Optimized BERT Pretraining Approach) [11], are used to benefit from their contextual understanding and language representation capabilities. By harnessing the strengths of these various machine learning technologies, we aim to enhance the accuracy of our expertise classification system.

For evaluation purposes, two widely used metrics are employed: AUC-ROC[5] (Area Under the Receiver Operating Characteristic Curve) and F1-measure. AUC-ROC assesses the performance of the classification model by measuring the trade-off between true positive rate and false positive rate. It provides a comprehensive evaluation of the model's ability to distinguish between different classes. On the other hand, F1-measure combines precision[6] and recall,[7] offering a balanced assessment of the model's accuracy.

The performance of the models created using the different machine learning technologies, employed on the three datasets, is presented in Table 2 (binary classification), and Table 3 (multiple types of expertise classification). In all of our experiments, we partitioned the datasets into training and testing sets, with an 80% and 20% split, respectively. To address any class imbalance within the training section, we applied the oversampling technique.

The results in Table 2, for binary classification, exhibit promising outcomes, with the transformer-based models (BERT and RoBERTa) demonstrating high effectiveness in accurately categorising posts as reflecting expertise or not. In contrast, the results presented in Table 3, which focused on classifying multiple types of expertise, exhibit a mix of good and poor performance. The variability in results highlights the complexity of distinguishing between different types of expertise and also the influence of imbalanced classes, since certain expertise categories were underrepresented compared to others. For example, only 13 posts were labelled as *Community Expertise* in the dataset of Forum 1, comprising 211 posts (see Table 1). This severe class imbalance clearly affected the models' performance, as shown in Table 3. In comparison, the balanced labelling of expertise category *Personal Expertise* in the dataset from Forum 1 resulted in favourable outcomes.

4 Conclusion and Future Work

In this paper, we presented our work in recognising claims of expertise in anonymous online fora. A combination of various machine learning techniques was

[5] sklearn default parameters were used (https://scikit-learn.org/stable/modules/generated/sklearn.metrics.roc_auc_score.html).

[6] Precision: The quality of a positive prediction made by the model.

[7] Recall: The percentage of data samples that a machine learning model correctly identifies as belonging to a class of interest.

Table 2. Experiments' results on the three datasets - Binary classification

Method	Forum 1		Forum 2		Forum 3	
	F1-measure	AUC-ROC	f1-measure	AUC-ROC	f1-measure	AUC-ROC
Decision Tree	0.79	0.71	0.65	0.54	0.5	0.54
Logistic Reg	0.78	0.71	0.75	0.64	0.51	0.56
SVM	0.77	0.62	0.76	0.55	0.64	0.63
LSTM	0.8	0.69	0.81	0.64	0.65	0.65
bi-LSTM	0.76	0.64	0.87	0.64	0.69	0.67
BERT	0.79	0.75	0.81	0.66	0.74	0.74
RoBERTa	0.8	0.78	0.80	0.65	0.76	0.74

Table 3. Experiments' results on the three datasets - Multiple types of expertise classification

Expertise	Method	Forum 1		Forum 2		Forum 3	
		f1-measure	AUC-ROC	f1-measure	AUC-ROC	f1-measure	AUC-ROC
Personal Exp.	Decision Tree	0.7	0.64	0.55	0.7	0.63	0.71
	Logistic Reg	0.63	0.64	0.34	0.55	0.65	0.74
	SVM	0.76	0.71	0.45	0.60	0.61	0.69
	LSTM	0.75	0.72	0.26	0.5	0.47	0.64
	bi-LSTM	0.78	0.64	0.4	0.64	0.49	0.53
	BERT	0.79	0.7	0.65	0.74	0.75	0.81
	RoBERTa	0.81	0.74	0.47	0.63	0.63	0.71
Community Exp.	Decision Tree	0.1	0.48	0.2	0.8	0.84	0.88
	Logistic Reg	0.1	0.5	0.4	0.9	0.95	0.95
	SVM	0.1	0.5	1.0	1.0	0.89	0.9
	LSTM	0.1	0.48	0.67	0.99	0.34	0.63
	bi-LSTM	0.1	0.5	0.34	0.96	0.5	0.68
	BERT	0.1	0.5	0.1	0.5	0.78	0.84
	RoBERTa	0.1	0.5	0.1	0.5	0.7	0.82
Broad Topic Exp.	Decision Tree	0.4	0.56	0.53	0.54	0.51	0.66
	Logistic Reg	0.3	0.56	0.63	0.59	0.67	0.76
	SVM	0.4	0.63	0.51	0.58	0.58	0.7
	LSTM	0.4	0.63	0.76	0.74	0.5	0.6
	bi-LSTM	0.24	0.5	0.78	0.71	0.52	0.64
	BERT	0.15	0.46	0.76	0.67	0.68	0.76
	RoBERTa	0.34	0.6	0.67	0.57	0.56	0.69

employed, using three anonymous datasets. The transformer-based models have demonstrated their efficiency in accurately categorising instances as reflecting expertise or not. Additionally, transformer-based models have showcased effectiveness in classifying multiple types of expertise in the majority of cases. The class distribution in the datasets clearly affected the models' performance, where the balanced labelling of an expertise category in a dataset resulted in encouraging outcomes.

To address our research question, we can confidently state based on the promising experiments results, that machine learning technologies –particularly transformer-based models– facilitate expertise detection in anonymous online fora by solely analysing conversation text. These results provide a solid foundation for future investigations into assessing the significance of these expertise claims within various power hierarchies. And to address the limitations of our work, the next step will involve improving the weak points by undertaking the following measures: collecting a larger volume of data, extracting a wider range of expertise types, and focusing on balancing the dataset. The insights gained from our work, coupled with the proposed improvements, will enable us to conduct a more comprehensive analysis, providing valuable insights into the dynamics of expertise and its influence within different power hierarchies.

References

1. Devlin, J., Chang, M.W., Lee, K., Toutanova, K.: Bert: pre-training of deep bidirectional transformers for language understanding. arXiv:1810.04805 (2018)
2. Fairclough, A.: Martin Luther King, Jr. and the war in Vietnam. Phylon (1960-) **45**(1), 19–39 (1984)
3. Farfán, J.A.F., Holzscheiter, A.: The power of discourse and the discourse of power. In: The SAGE Handbook of Sociolinguistics, pp. 139–152 (2010)
4. Graves, A.: Long short-term memory. In: Supervised Sequence Labelling with Recurrent Neural Networks, pp. 37–45 (2012)
5. Graves, A., Jaitly, N., Mohamed, A.R.: Hybrid speech recognition with bidirectional LSTM. In: Automatic Speech Recognition and Understanding Workshop (2013)
6. Hearst, M.A., Dumais, S.T., Osuna, E., Platt, J., Scholkopf, B.: Support vector machines. IEEE Intell. Syst. Appl. **13**(4), 18–28 (1998)
7. Horne, B., Nevo, D., Freitas, J., Ji, H., Adali, S.: Expertise in social networks: how do experts differ from other users? In: Proceedings of the International AAAI Conference on Web and Social Media, vol. 10, pp. 583–586 (2016)
8. Horne, B.D., Nevo, D., Adalı, S.: Recognizing experts on social media: a heuristics-based approach. ACM SIGMIS Database: Data Base Adv. Inf. Syst. **50**(3), 66–84 (2019)
9. Kacewicz, E., Pennebaker, J.W., Davis, M., Jeon, M., Graesser, A.C.: Pronoun use reflects standings in social hierarchies. J. Lang. Soc. Psychol. **33**(2), 125–143 (2014)
10. LaValley, M.P.: Logistic regression. Circulation **117**(18), 2395–2399 (2008)
11. Liu, Y., Ott, M., Goyal, N., Du, J., Joshi, M., Chen, D., Levy, O., Lewis, M., Zettlemoyer, L., Stoyanov, V.: Roberta: a robustly optimized Bert pretraining approach. arXiv:1907.11692 (2019)
12. Mitchell, T.M., et al.: Machine Learning, vol. 1. McGraw-Hill, New York (2007)
13. Newsome, H., Grant, T.: Developing a resource model of power and authority in anonymous online criminal interactions. Linguagem e Direito (in press)
14. Siegel, J.A., Saukko, P.J.: Encyclopedia of Forensic Sciences. Academic Press (2012)

15. Vydiswaran, V.V., Reddy, M.: Identifying peer experts in online health forums. BMC Med. Inform. Decis. Mak. **19**, 41–49 (2019)
16. Zhang, J., Ackerman, M.S., Adamic, L.: Expertise networks in online communities: structure and algorithms. In: Proceedings of the 16th International Conference on World Wide Web, pp. 221–230 (2007)

Noise Profiling for ANNs: A Bio-inspired Approach

Sanjay Dutta[1(✉)], Jay Burk[2], Roger Santer[2], Reyer Zwiggelaar[1], and Tossapon Boongoen[1]

[1] Department of Computer Science, Aberystwyth University, Aberystwyth SY23 3DB, UK
{sad64,rrz,tob45}@aber.ac.uk

[2] Department of Life Sciences, Aberystwyth University, Aberystwyth SY23 3FG, UK
{jab169,rds5}@aber.ac.uk

Abstract. Artificial neural networks (ANNs) are potent computational models, which are capable of completing a range of perception-related tasks. However, sometimes it is difficult for them to learn from complex data. Therefore, it is preferable to introduce noise into the input or hidden layers of the ANN during model training in order to get around this problem. As a result, it can enhance the adaptability of the model. This paper is an approach to noise profiling for ANNs that draws inspiration from the biological workings of insect sensory systems. By using specialized sense organs, insects are evolved to deal with noisy environments. The using of both Gaussian and Chaotic noises have various statistical characteristics and both have remarkable effects on ANNs. Gaussian noise is smooth and continuous, which works as a regularizer for artificial neural networks. On the other hand, Chaotic noise is irregular and also unpredictable and that works as a stimulus. Both the application of noises was compared to the baseline ANN on real data sets. The assessment of the accuracy and robustness of ANN performance under various types and amounts of noise was done. It was demonstrated that the noise profiling approach outperforms the baseline approach. It also examined the impact of Gaussian and Chaotic noise on the internal dynamics and representations of ANNs, providing some intriguing new information on how noise can affect ANN functionality and behaviour. In this research, two datasets were used: Animal and Shaded. The results demonstrated that bio-inspired noise profiling techniques can offer a straightforward yet efficient means of improving ANN performance for insect perception issues as well as diminish the overfitting of the model.

Keywords: Artificial neural networks · Noise profiling · Overfitting · Gaussian noise · Chaotic noise

1 Introduction

Artificial Neural Networks (ANNs) are becoming significant computational models capable of handling challenging tasks in a variety of domains [1–4]. Occasionally, it is seen that training ANNs with some datasets can be a little complicated

N. Naik et al. (Eds.): UKCI 2023, AISC 1453, pp. 140–153, 2024.
https://doi.org/10.1007/978-3-031-47508-5_12

as they find it difficult to generalise from data, which restricts their capacity for learning due to some factors such as high degree of variability, noise, intricate patterns, high dimensionality, overfitting, lack of representative training data and the need for appropriate network architectures and hyperparameters. The idea of adding noise to ANNs while they are being trained has received a lot of interest recently as a means of resolving this problem [5,6]. As a regularisation strategy, adding noise to ANNs aims to increase their capacity for adaptation and generalisation. In this regard, ANNs are encouraged to create more reliable and adaptable representations of the underlying patterns in the data when they are trained with noisy inputs or disturbances in their hidden layers. Instead of relying on overly particular and sensitive characteristics, ANNs are pushed by controlled fluctuations through the noise to learn more generalised features that are applicable to a wider variety of inputs.

Another key benefit of injecting noise into ANNs is the reduction of the overfitting phenomenon [7,8]. Overfitting occurs when a model becomes overly specialised to the training set, which impairs the model's ability to generalise to unseen data. In order to prevent the model from memorising the training data as well as encouraging the identification of more durable patterns, noise is added to the training process. As a result, it exposes the model to a wider variety of input fluctuations. This regularisation technique reduces overfitting and increases the model's capacity to generalise to unobserved data [9].

The efficacy of noise injection in increasing ANNs' learning capacity also has been shown in various applications including image classification, speech recognition and natural language processing. This research has observed improvements in the model's ability to generalise to new cases such as handling noisy or incomplete inputs and reduce the impacts of overfitting by adding controlled perturbations to the training data or network parameters [10]. Despite substantial progress in using noise to enhance ANN training, the applicability of this idea to the particular area of insect perception is still largely unexplored. Insects possess exceptional sensory systems that enable them to communicate and navigate their surroundings. Despite disturbances, these sensory systems have evolved to function in loud and unreliable environments, allowing insects to perceive accurately [11]. Drawing inspiration from the biological mechanisms underlying insect perception and investigating the potential benefits of including noise in ANNs for perception-related tasks, especially those relevant to insect-like perception, becomes fascinating. The research aims to examine the use of noise profiling techniques inspired by insect sensory systems to enhance the performance and resilience of ANNs in handling perception issues. It attempts to bridge the gap between ANN noise injection techniques and the field of insect perception. Additionally, these ideas and methodologies could improve the adaptability and efficiency of ANNs in addressing perception-related tasks by examining the impact of noise on the learning dynamics and generalization capacities of ANNs in the context of insect perception. This research investigated two different kinds of noise: Gaussian [5,12] and Chaotic noise [13,14]. Gaussian and Chaotic noise types were selected due to their distinctive qualities and potential advantages for model training. The training data was subjected to random fluctuations by introducing Gaussian noise, mimicking the inherent uncertainties and variability present in natural situations. Conversely, with the addition of Chaotic noise, the learning process became more complex and nonlinear, enabling the model to identify subtle patterns and connections in the data. Gaussian noise provided a straightforward yet effective method of introducing unpredictability, thereby reducing overfitting and improving the network's ability to generalize to new input. On the other hand, Chaotic noise introduced a more complex and nonlinear perturbation, which may enable the model to capture more subtle patterns and enhance its ability to discriminate between them.

The structure of this article is as follows: The research issue is introduced and the project's goals are laid out in Sect. 1 of this article. In Sect. 2, the background and related work is presented, with a focus on earlier works that investigated noise-based methods for enhancing ANN training. The approach and suggested bio-inspired noise profiling strategies are described in Sect. 3. The findings are provided and a thorough discussion of them is conducted in Sect. 4. Finally, in Sect. 5, the work is concluded, summarizing the major contributions and exploring potential future research trajectories.

2 Background and Related Work

The largest and most diversified group of living organisms is insects, which are essential to many ecosystems. Insect visual perception is the ability of insects to take in and process visual information from their surroundings. Insects possess photoreceptors that enable them to differentiate colors based on the specific wavelengths of light they receive [11,15]. One area of interest is how insects perceive their environment in the field of bio-inspired noise profiling for improving model performance. Insects have remarkable sensory systems that help them communicate and navigate their environment. Although, it is found that the design of ANNs or the pertinent visual modeling tasks have considered whether noise exists in the original data-collection system. However, there is still some gaps present and it is an appealing opportunity for investigation and innovation.

It is essential to collect datasets relevant to the visual perception of insects in order to study the issue at hand and it can be modeled as a classification problem, where the goal is to categorize visual stimuli based on the data perceived by insects. The data collection encompasses a range of visual perception abilities in insects, including their ability to detect objects, recognize patterns, and differentiate between diverse visual stimuli. These datasets form the foundation for developing and evaluating the performance of ANNs.

The idea suggested in the research [6] used hidden multipliers and noise injection to enhance the network structure of multilayer perceptrons (MLPs). They proposed two noise-injected pruning techniques, ANI-MLPHM and MNI-MLPHM, which could be used to select the optimal number of hidden units for MLPs and streamline network construction. The techniques assigned a multiplier function to each hidden unit in an MLP. The multiplier function acted as a 'gate' on the corresponding hidden unit during training, controlling the learnable parameters. A new penalty term was introduced to drive the multiplier functions of MLPs towards either one or zero, effectively opening or closing the gates. This innovative method aimed to improve the effectiveness of hidden unit pruning in MLPs. The researchers rigorously demonstrated the asymptotic convergence of the proposed algorithms and supported their theoretical claims with simulation results. The study showcased the pruning capabilities of the two proposed strategies and highlighted the method's advantages in terms of pruning ability and generalization compared to existing approaches. The suggested method was applied to various UCI datasets to demonstrate its effectiveness.

In the paper [5], the authors discussed the utilization of noise addition during neural network training to enhance the network's resilience against data noise. They suggested a method for selecting the optimal network or network subset for a specific out-of-sample pattern from a collection of networks trained with varying degrees of noise. The researchers proposed a procedure for choosing the best network or network subset for a specific out-of-sample pattern, using a set of networks trained with different levels of noise. Their study focused on a model inverse problem, which served as a simplified analogy to magnetotelluric sounding inverse problems. Inverse problems presented significant challenges in fields such as geophysics, spectroscopy, and tomography. The proposed approach involved the use of ANNs trained with different levels of added noise for each pattern. The ANN outputs were then employed as input data for solving the direct problem, and the results from different sets of networks were compared to the target pattern within the observable space. The set of networks that exhibited the minimum residual determined the response of the entire ANN system for the given pattern. The influence of a parameter depended on its 'position' which was determined by the coefficients and the conditional 'distance' between inputs and outputs. The study adopted a widely-used nature-inspired dependence where the influence was inversely proportional to the square of the distance.

In the study [12], the backpropagation method was modified by incorporating white Gaussian noise into the weighted sum entity of the backpropagation to enhance convergence. The authors tested the suggested technique on several standard benchmark problems, including the iris dataset and 2-bit parity. When applied to the 2-bit parity and Iris datasets, it was observed that the modified backpropagation method required fewer epochs to converge compared to the standard backpropagation approach.

In the paper [10], a novel method was proposed for tuning the noise level of each neuron in ANNs during training. The method computed the pathwise stochastic gradient estimate based on the standard deviation of the Gaussian noise applied to each neuron in the ANN. As a byproduct of the backpropagation process for estimating the gradient with respect to synaptic weights in the ANN, the method also obtained the gradient estimate for the noise levels. When evaluated on various computer vision datasets using both black box and white box attacks, the suggested approach demonstrated significant improvements in the robustness of several well-known ANN architectures.

In order to enhance the robustness of ANNs, a novel noise injection-based training method was suggested in the research [16]. To estimate the gradient with respect to both synaptic weights and noise levels during stochastic gradient descent training, the authors introduced a likelihood ratio technique. Additionally, they developed an approximation for the standard noise injection-based training method, aiming to improve computational performance and conserve memory. The effectiveness of the proposed method was evaluated using the MNIST and Fashion-MNIST datasets, as well as applied to spiking neural networks. The experimental results demonstrated that the suggested method achieved superior performance compared to the traditional gradient-based training method in terms of adversarial robustness, and it also exhibited comparable performance in terms of original accuracy.

The study [17] examined the impact of noise on artificial neural network learning. The authors presented an approach that involved training with noise to improve classification, generalization, and memory performance by introducing structure into noisy training data. They also showed that, in certain cases, the training-with-noise method aligned with the unlearning rule. Furthermore, the authors proposed a sampling strategy to identify the most effective noisy data for both training with noise and unlearning. The research explored the training process parameters that facilitated precise classification and generalization.

The research [18] investigated the regularisation effect of Gaussian noise injections (GNIs) in neural networks. The authors formulated an explicit regularizer for GNIs and illustrated how it imposed penalties on functions with high-frequency Fourier domain components, particularly in layers near the network's output. They further provided analytical and empirical evidence demonstrating how this regularisation led to well-calibrated classifiers with significant classification margins. In summary, the study examined the regularisation impact of GNIs on neural networks and provided insights into how it could improve classifier performance.

The research [19] investigated the effects of introducing noise to neural networks during a decentralized training process. The authors showcased that while noise injection did not enhance linear models, it considerably elevated the performance of non-linear neural networks and facilitated the generalization of a local model that closely resembled the serial baseline. The article highlighted that introducing noise in a distributed setting holds promise as a means to enhance locally trained models.

The challenges associated with training a neural network using the gradient descent learning algorithm and persistent weight noise were discussed in the paper [20]. The paper introduced the performance measure J(w) for a perfect neural network with noisy weight vectors, which represented the desired performance. However, in the presence of persistent multiplicative weight noise, the learning objective became a scalar function L(w), which was not equivalent to J(w). The research compared the characteristics of the intended models to those obtained using the gradient descent learning algorithm and provided simulation results on a simple regression problem and the MNIST handwritten digit recognition dataset to support its claims.

The paper [21] discussed the application of the quasi-Newton method-based BP neural network for aerodynamic modelling. The authors focused on the inclusion of a penalty term and noise data in the models to enhance their generalization capacity. They conducted tests using a dataset of numerical aerodynamic data for an aircraft and compared the coefficient of determination for regression in different scenarios. The findings indicated that the addition of noisy data did not improve the generalization ability; however, the inclusion of a penalty term did.

In the research [22], the authors focused on studying Gaussian Noise Injections (GNIs), a widely used regularization technique in neural network training. They examined the "implicit effect" of GNIs on the dynamics of Stochastic Gradient Descent (SGD) and showed that it introduces asymmetric heavy-tailed noise in gradient updates. They formally demonstrated that GNIs lead to an "implicit bias" that varies based on the weight of the tails and the degree of asymmetry by developing a stochastic differential equation similar to Langevin dynamics to explain these modified dynamics. The authors also provided empirical evidence supporting the proposed dynamics model and demonstrated how the implicit effect of GNIs negatively impacts network performance.

The approach presented in [23] aimed to enhance the resilience of a multilayer perceptron (MLP) against link failures, multiplicative noise, and additive noise. The study utilized an MLP, an artificial neural network, for supervised learning. When implementing an MLP in hardware, the weights may be affected by multiplicative and/or additive noises. Furthermore, if an MLP is constructed using analogue circuits, it is susceptible to link failures known as stuck-at 0 issues. The suggested approach employed four terms to train the system: mean square error (MSE), the l2 norm of the weight vector, the sum of squares of the first-order derivatives of MSE with respect to weights, and the sum of squares of the second-order derivatives of MSE with respect to weights. These terms were utilized to address the impact of link failures and noise on the MLP. To evaluate the effectiveness of the proposed methodology, ten regression tests and ten classification tasks were conducted under scenarios involving link failure, multiplicative noise, and additive noise. The experimental results demonstrated the efficacy of the suggested regularization approach in achieving reliable MLP training and improving the system's robustness against various types of disturbances.

The paper [24] proposed a novel technique called STE (stochastically trained ensemble) layers for regularizing neural networks. The method aimed to enhance the averaging properties of other stochastic regularization techniques, such as dropout, by explicitly averaging outputs while applying stochastic regularization to an ensemble of weight matrices. Notably, this approach provided stronger regularization during testing without incurring additional computational costs. To evaluate the effectiveness of the STE layers, the authors conducted experiments using common network topologies and a variety of picture categorization tasks. The results consistently demonstrated continuous improvement in performance, highlighting the efficacy of the proposed technique for regularizing neural networks.

The research paper [25] put forward and examined two algorithms designed for training fault-tolerant and sparse multilayer perceptrons (MLPs) using the group lasso penalty. These algorithms involved incorporating noise into the weights during the learning process, either by addition or multiplication. This approach aimed to enhance the network's ability to generalize and prune unnecessary connections. To address the non-differentiability of the group lasso penalty, the study employed a smooth approximation technique. Additionally, the researchers provided rigorous evidence of the asymptotic convergence of the algorithms. Numerical simulations were conducted using both synthetic and benchmark datasets to evaluate the effectiveness of the proposed strategies. The results of these simulations demonstrated the efficacy of the suggested approaches in improving the performance of fault-tolerant and sparse MLPs.

3 Method

In the methodology section of our research, we incorporated two widely used techniques, namely Gaussian noise and Chaotic noise to enhance the performance of ANNs. These methods effectively improved the learning dynamics and generalization capabilities of the models by introducing controlled perturbations at multiple layers of the network, including data and network design. However, prior to applying noise to the network model, we conducted an initial evaluation without any noise application.

3.1 Evaluation Without Noise

The model architecture consisted of an input layer with 5 neurons, two hidden layers with 16 neurons each, and an output layer with a single neuron. The ReLU activation function was employed in the hidden layers to introduce non-linearity and aid the network in learning complex patterns. For binary classification tasks, the output layer utilized the sigmoid activation function, mapping the neuron's output to a probability value between 0 and 1, representing the likelihood of the input belonging to the positive class. During the model compilation stage, we employed the adam optimizer, binary crossentropy loss function, and accuracy metrics provided by Keras.

3.2 Applying Gaussian Noise

The utilization of Gaussian noise serves to replicate random fluctuations in data and is commonly employed in various applications. It follows a probability density function known as the Gaussian distribution or normal distribution. During the training of artificial neural networks (ANNs), Gaussian noise can be incorporated into different components such as input data, hidden layers, or network parameters. By introducing this noise, ANNs become more adept at generalizing to unseen data, as they are exposed to a wider range of input variations. This approach aims to enhance the model's adaptability and generalization capabilities by mitigating the effects of overfitting. Overfitting occurs when a model becomes excessively specialized to the training dataset, resulting in poorer performance on unseen data. In this research, it was focused on introducing Gaussian noise specifically to the hidden layer of the network model. The model architecture follows a sequential design, where layers are added in a sequential manner using the tf.keras.models.Sequential class. Each layer's output serves as the input for the subsequent layer, establishing a flow of information throughout the network. The architecture consists of two hidden layers, each composed of 16 neurons, positioned between the input and output layers. After the first hidden layer, the addition of Gaussian noise contributes to the introduction of controlled perturbations within the network activations, ultimately enhancing generalization and robustness. The input layer of the model is defined using tf.keras.layers.Dense, encompassing 16 neurons. The Rectified Linear Unit (ReLU) activation function is applied to introduce non-linearity. Additionally, the input shape is specified as (5,), indicating the presence of five features in the input data. A Gaussian noise layer is incorporated following the first hidden layer, utilizing tf.keras.layers.GaussianNoise. This layer introduces Gaussian noise with a standard deviation of 0.01 to the activations of the preceding layer. Its purpose is to enhance the model's robustness and improve generalization by selectively injecting controlled noise into the activations of the hidden layer. The second hidden layer is defined using tf.keras.layers.Dense and consists of 16 neurons, employing the ReLU activation function. The output layer is established with tf.keras.layers.Dense, comprises a single neuron and utilizes the sigmoid function, suitable for addressing binary classification problems. The output of this layer represents the predicted probability of the input belonging to the positive class. During model compilation, the adam optimizer, binary crossentropy loss function and accuracy metrics provided by Keras are employed.

3.3 Applying Chaotic Noise

An alternative approach involves incorporating Chaotic noise ANNs as a means of enhancing their performance. Chaotic noise, unlike Gaussian noise, is generated by Chaotic dynamical systems rather than a random distribution. The inherent complexity and unpredictability of these systems have the potential to nonlinearly alter the activations within the network. By introducing Chaotic noise, ANNs can benefit from increased effectiveness and resilience. This can be achieved by incorporating Chaotic noise into the input or weights of the ANN, enabling improvements in generalization, prevention of overfitting and avoidance of local minima. Regarding the model architecture, we implemented a custom layer that integrates a Chaotic map function using TensorFlow operations to inject Chaotic noise. The model begins with an input layer designed to accept a feature vector of length 5, denoted by an input shape of (5,). The first dense layer consists of 6 neurons and employs the 'gelu' activation function. It takes the input from the previous layer and applies the activation function to generate an output. To introduce Chaotic noise, a custom ChaoticNoiseLayer is utilized. This layer injects Chaotic noise into the output of the first dense layer, leveraging the perturbations generated by the Chaotic dynamical system. The second dense layer, also composed of 6 neurons with the 'gelu' activation function, takes the output of the Chaotic Noise Layer 1 as its input and produces an output using the activation function. Similarly, Chaotic Noise Layer 2 is employed to inject Chaotic noise into the output of the second dense layer, providing additional randomization. The third dense layer, comprising 6 neurons with the 'gelu' activation function, takes the output of Chaotic Noise Layer 2 as input and generates an output using the activation function. The fourth dense layer consists of 3 neurons with the 'gelu' activation function. It takes the output of the third dense layer as its input and produces an output using the activation function. The final layer of the model, referred to as Dense Layer 5 or the output layer, comprises a single neuron. It utilizes the sigmoid activation function to generate a binary classification output of either 0 or 1, based on the input received from Dense Layer 4. During model compilation, the adam optimizer, binary crossentropy loss function and accuracy metrics provided by Keras are utilized.

4 Result and Discussion

The performance of the suggested bio-inspired noise profiling approach was evaluated using an artificial neural network (ANN) model trained on the dataset related to insects' visual perception and it was modelled as a classification problem. The dataset, we used was split into training and testing sets with a ratio of 80:20. As stated in the method section, we applied three methods to test our datasets. There were two datasets: Animal and Shaded.

Animal dataset: The Animal dataset consisted of 288 samples with 5 features and 1 label. The 5 features were the data collected by the 5 photoreceptors (qMR1, qMR7p, qMR7y, qMR8p, qMR8y). Our goal was to classify whether it was animal or non-animal.

Shaded dataset: Similarly, The Shaded dataset consisted of 288 samples with 5 features and 1 label. The 5 features were the data collected by the 5 photoreceptors (qMR1, qMR7p, qMR7y, qMR8p, qMR8y). Our goal was to classify whether it was shaded or non-shaded.

4.1 Hyperparameter Selection

The hyperparameters, including the learning rate, batch size, and other necessary parameters were fine-tuned using the training and validation data which are shown in the Table 1. It shows the hyperparameter values selected.

4.2 Results

The datasets were first trained and tested without noise, resulting in the best accuracy of 72.41% in the Animal dataset and 100% in the Shaded dataset.

Table 1. Hyperparameter selection

Parameter	Animal	Shaded
Batch size	32	32
Number of epochs	100	100
Optimizer	Adam	Adam
Metrics	Accuracy	Accuracy
Number of training samples	230	230
Number of test samples	58	58
Loss function	binary_crossentropy	binary_crossentropy
Activation function (without noise/Gaussian noise)	relu	relu
Activation function (Chaotic noise)	gelu	gelu

When Gaussian noise was introduced, better accuracy was achieved compared to the noiseless scenario, with the Animal dataset reaching 74.14% accuracy and the Shaded dataset remaining at 100%. Finally, by utilizing Chaotic noise, the best results were obtained, yielding an accuracy of 75.86% in the Animal dataset and 100% in the Shaded dataset, as depicted in the Table 2.

Table 2. Test sets results

Without noise		With Gaussian noise		With Chaotic noise	
Animal	Shaded	Animal	Shaded	Animal	Shaded
0.7241	1	0.7414	1	0.7586	1

Most significantly, Gaussian and Chaotic noise additions had a beneficial effect on the networks and as a result, the overfitting of the network during the training period was decreased. Accuracy and loss per epoch for both datasets are shown in the Figs. 1, 2, 3, 4, 5, 6, 7, 8, 9, 10, 11 and 12). It is seen in Figs. 1 and 2, how the overfitting occurred in the network model without using noise in the Animal dataset. However, it is seen the effect of Gaussian in Figs. 3 and 4 and Chaotic noises in Figs. 5 and 6 respectively on the same dataset. Similarly, Figs. 7 and 8 shows the model behaviour on the Shaded dataset without noise and the effects of Gaussian noise on this dataset can be seen in Fig. 9, in Fig. 10, and finally the Chaotic noise on the same dataset had the impact as in Figs. 11 and 12 respectively.

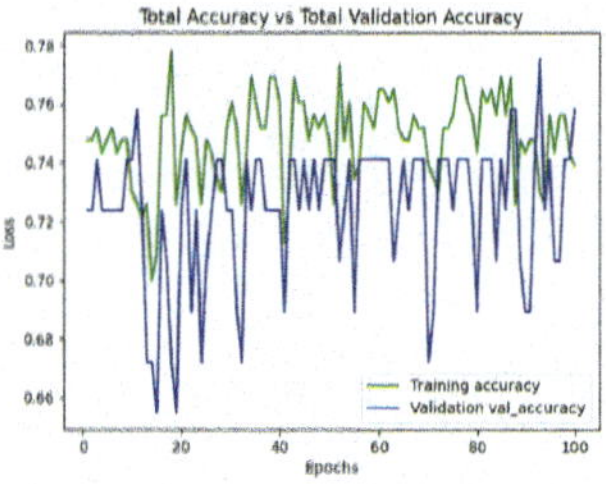

Fig. 1. Without noise on Animal

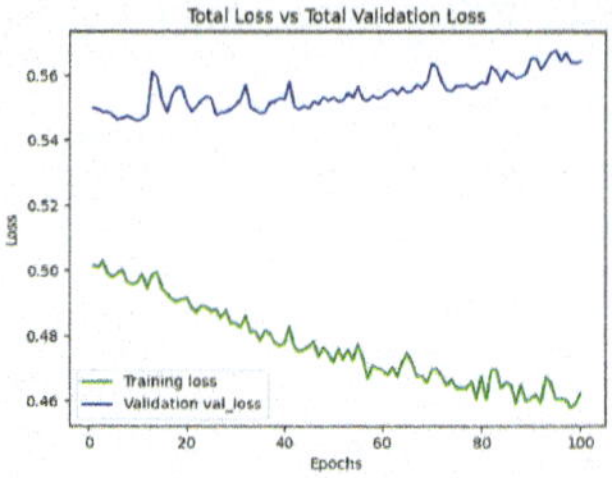

Fig. 2. Without noise on Animal

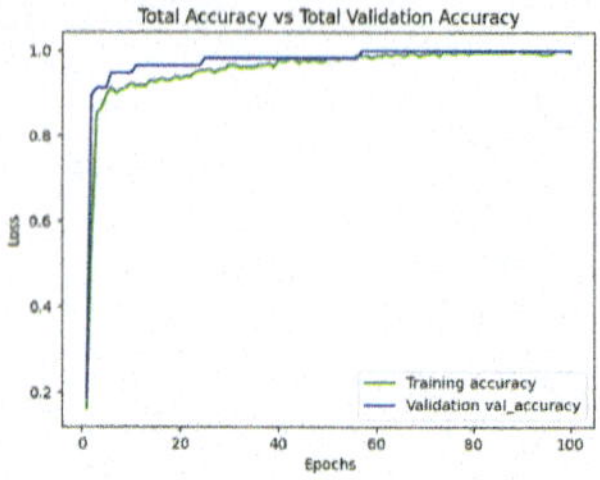

Fig. 3. Gaussian on Animal

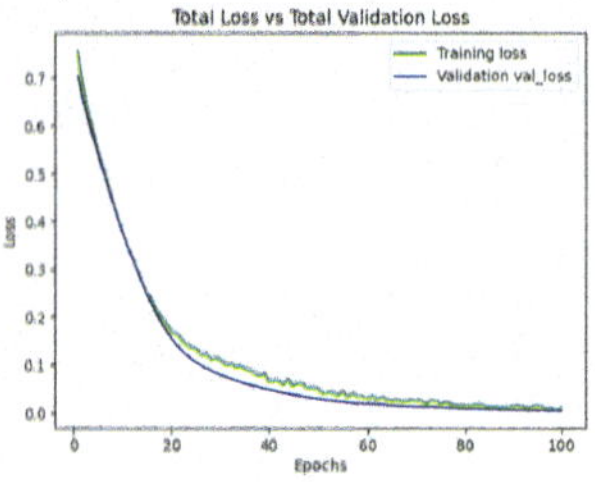

Fig. 4. Gaussian on Animal

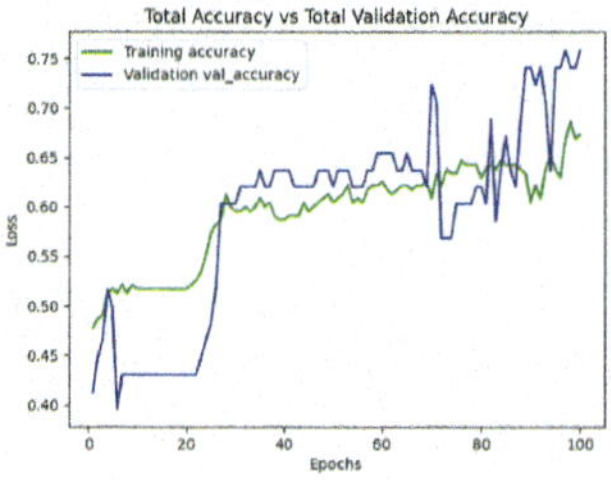

Fig. 5. Chaotic on Animal

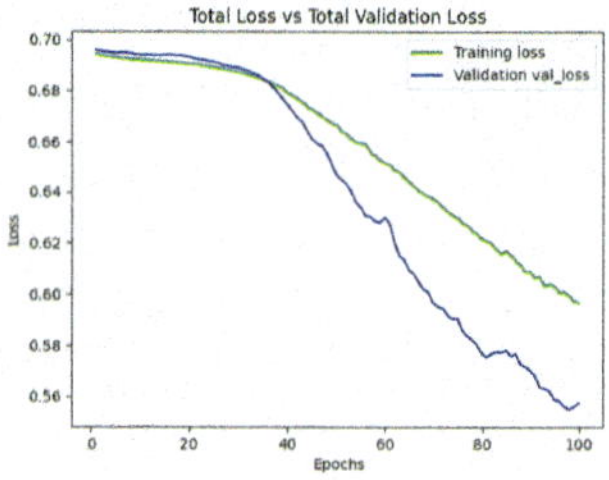

Fig. 6. Chaotic on Animal

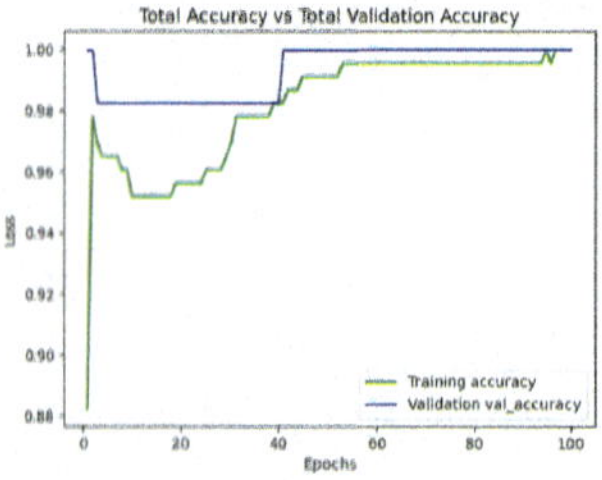

Fig. 7. Without noise on Shaded

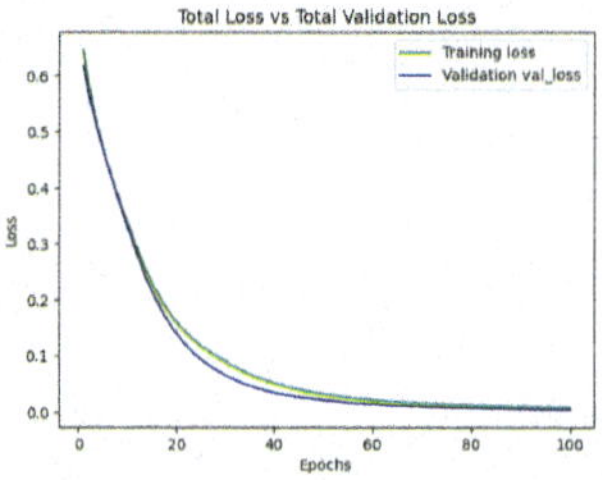

Fig. 8. Without noise Shaded

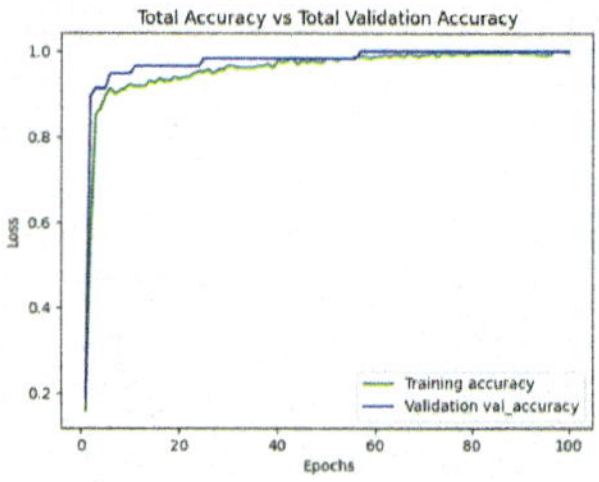

Fig. 9. Gaussian on Shaded

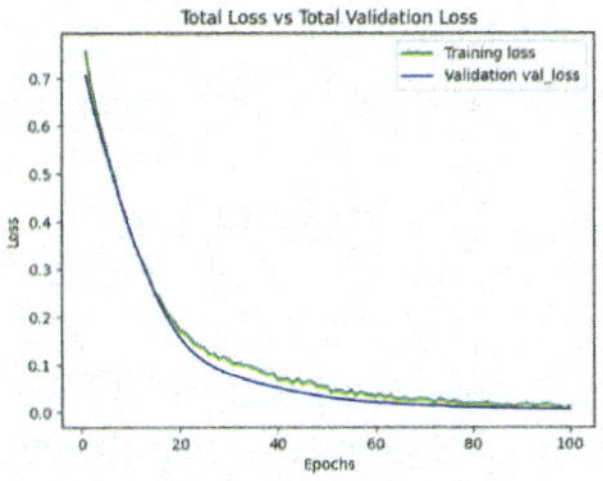

Fig. 10. Gaussian on Shaded

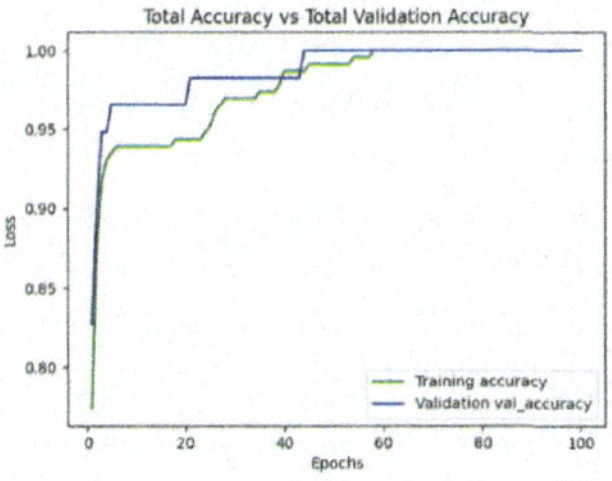

Fig. 11. Chaotic on Shaded

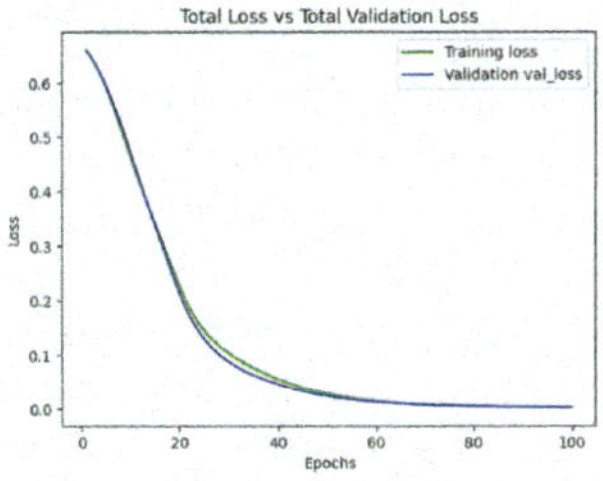

Fig. 12. Chaotic on Shaded

5 Conclusion

In this study, we investigated the use of bio-inspired noise profiling techniques to enhance Artificial Neural Networks' (ANNs') performance in the visual perception of insects. We concentrated on introducing Gaussian and Chaotic noises into the hidden layers of the ANN, taking into account the difficulties given by the native data-collection mechanism and the requirement for robust visual modelling. Our research has demonstrated that adding Gaussian and Chaotic noises to the hidden layers of ANNs improves their ability to generalise and adapt. By introducing controlled perturbations to the network, the ANN learns more generalized features, as a result, it, becomes resilient to noise and disturbances in visual stimuli, effectively mitigating overfitting and improving accuracy on unseen data. By bridging the gap between ANNs and the field of biological research, this method expands the use of Gaussian and Chaotic noises to insect visual perception. Our research highlights the significance of taking noise into account when constructing ANN structures, resulting in more precise models that mimic insect sensory systems. Further investigation into additional noise profiling methods and particular noise distributions in the biological science field may yield knowledge that will improve ANN training for tasks involving perception. In conclusion, our experiment showed how ANN performance is enhanced by bio-inspired noise profiling, particularly both Gaussian and Chaotic noises, making them more durable and adaptive in the setting of insect visual perception.

References

1. Abiodun, O.I., Jantan, A., Omolara, A.E., Dada, K.V., Mohamed, N.A., Arshad, H.: Heliyon **4**(11), e00938 (2018). https://doi.org/10.1016/j.heliyon.2018.e00938. https://www.sciencedirect.com/science/article/pii/S2405844018332067
2. Chen, M., Challita, U., Saad, W., Yin, C., Debbah, M.: IEEE Commun. Surv. Tutor. **21**(4), 3039 (2019). https://doi.org/10.1109/COMST.2019.2926625
3. Yang, G.R., Wang, X.J.: Neuron **107**(6), 1048 (2020)

4. Zhang, Q., Yu, H., Barbiero, M., Wang, B., Gu, M.: Light Sci. Appl. **8**(1), 42 (2019). https://doi.org/10.1038/s41377-019-0151-0. https://www.nature.com/articles/s41377-019-0151-0
5. Isaev, I., Dolenko, S.: Procedia Comput. Sci. **123**, 171 (2018). https://doi.org/10.1016/j.procs.2018.01.028. https://www.sciencedirect.com/science/article/pii/S1877050918300292
6. Wang, X., Wang, J., Zhang, K., Lin, F., Chang, Q.: Neurocomputing **452**, 796 (2021). https://doi.org/10.1016/j.neucom.2020.03.119. https://www.sciencedirect.com/science/article/pii/S0925231220310365
7. Mutasa, S., Sun, S., Ha, R.: Clin. Imaging **65**, 96 (2020). https://doi.org/10.1016/j.clinimag.2020.04.025. https://www.sciencedirect.com/science/article/pii/S0899707120301376
8. Ying, X.: J. Phys: Conf. Ser. **1168**(2), 022022 (2019). https://doi.org/10.1088/1742-6596/1168/2/022022
9. Hu, T., Wang, W., Lin, C., Cheng, G.: Proceedings of the 24th International Conference on Artificial Intelligence and Statistics, Proceedings of Machine Learning Research, vol. 130, ed. by A. Banerjee, K. Fukumizu (PMLR, 2021), Proceedings of Machine Learning Research, vol. 130, pp. 829–837. https://proceedings.mlr.press/v130/hu21a.html
10. Xiao, L., Zhang, Z., Peng, Y.: Noise Optimization for Artificial Neural Networks (2021). https://doi.org/10.48550/arXiv.2102.04450. ArXiv:2102.04450 [cs]
11. Van Der Kooi, C.J., Stavenga, D.G., Arikawa, K., Belušič, G., Kelber, A.: Ann. Rev. Entomol. **66**(1), 435 (2021). https://doi.org/10.1146/annurev-ento-061720-071644. https://www.annualreviews.org/doi/10.1146/annurev-ento-061720-071644
12. Sapkal, A., Kulkarni, U.V.: Procedia Comput. Sci. **143**, 309 (2018). https://doi.org/10.1016/j.procs.2018.10.401. https://www.sciencedirect.com/science/article/pii/S1877050918320982
13. Reid, S.: Adaptive chaotic injection to reduce overfitting in artificial neural networks. https://mspace.lib.umanitoba.ca/server/api/core/bitstreams/0c99d012-e5af-4077-9b25-3c147ccac83f/content. Accessed 20 June 2023
14. Reid, S., Ferens, K., Kinsner, W.: 2022 IEEE 21st International Conference on Cognitive Informatics and Cognitive Computing (ICCI*CC), pp. 22–31 (2022). https://doi.org/10.1109/ICCICC57084.2022.10101500
15. Anderson, J.C., Laughlin, S.B.: Vis. Res. **40**(1), 13 (2000). https://doi.org/10.1016/S0042-6989(99)00171-6. https://www.sciencedirect.com/science/article/pii/S0042698999001716
16. Zhang, Z., Jiang, J., Chen, M., Wang, Z., Peng, Y., Yu, Z.: A Novel Noise Injection-Based Training Scheme for Better Model Robustness (2023). https://doi.org/10.48550/arXiv.2302.10802. ArXiv:2302.10802 [cs]
17. Benedetti, M., Ventura, E.: Training neural networks with structured noise improves classification and generalization (2023). https://doi.org/10.48550/arXiv.2302.13417. ArXiv:2302.13417 [cond-mat]
18. Camuto, A., Willetts, M., Simsekli, U., Roberts, S.J., Holmes, C.C.: Advances in Neural Information Processing Systems, vol. 33, pp. 16603–16614. Curran Associates Inc. (2020). https://proceedings.neurips.cc/paper/2020/hash/c16a5320fa475530d9583c34fd356ef5-Abstract.html
19. Adilova, L., Paul, N., Schlicht, P.: ECML PKDD 2018 Workshops, ed. by A. Monreale, C. Alzate, M. Kamp, Y. Krishnamurthy, D. Paurat, M. Sayed-Mouchaweh, A. Bifet, J. Gama, R.P. Ribeiro (Springer International Publishing, Cham, 2019),

Communications in Computer and Information Science, pp. 37–48. https://doi.org/10.1007/978-3-030-14880-5_4
20. Sum, J., Leung, C.S., Ho, K.: IEEE Trans. Neural Netw. Learn. Syst. **31**(6), 2227 (2020). https://doi.org/10.1109/TNNLS.2019.2927689
21. Huiying, Y., Zhibin, H., Feng, Z.: 2017 of the 16th International Symposium on Distributed Computing and Applications to Business. Engineering and Science (DCABES), pp. 93–96 (2017). https://doi.org/10.1109/DCABES.2017.27. ISSN: 2473-3636
22. Camuto, A., Wang, X., Zhu, L., Holmes, C., Gurbuzbalaban, M., Simsekli, U.: Proceedings of the 38th International Conference on Machine Learning (PMLR, 2021), pp. 1249–1260. https://proceedings.mlr.press/v139/camuto21a.html
23. Dey, P., Nag, K., Pal, T., Pal, N.R.: IEEE Trans. Syst. Man, Cybern. Syst. **48**(8), 1255 (2018). https://doi.org/10.1109/TSMC.2017.2664143
24. Labach, A., Valaee, S.: 2020 IEEE 30th International Workshop on Machine Learning for Signal Processing (MLSP) (2020), pp. 1–6. https://doi.org/10.1109/MLSP49062.2020.9231761. ISSN: 1551-2541
25. Wang, J., Chang, Q., Chang, Q., Liu, Y., Pal, N.R.: IEEE Trans. Cybern. **49**(12), 4346 (2019). https://doi.org/10.1109/TCYB.2018.2864142

Machine Learning

AI Generated Art: Latent Diffusion-Based Style and Detection

Jordan J. Bird[1(✉)], Chloe M. Barnes[2], and Ahmad Lotfi[1]

[1] Department of Computer Science, Nottingham Trent University, Nottingham, UK
{jordan.bird,ahmad.lotfi}@ntu.ac.uk
[2] Department of Computer Science, Aston University, Birmingham, UK
c.barnes1@aston.ac.uk

Abstract. AI-generated artworks are rapidly improving in quality, and bring many ethical issues to the forefront of discussion. Data scarcity leaves many individuals under-represented due to aspects such as age and ethnicity, which can provide useful context when transferring artistic styles to an image. In this study, we consider current issues through the engineering of an AI art model trained on work inspired by Vincent van Gogh. The model is fine-tuned from a dataset of nearly 6 billion images and thus enables style transfer to individuals and entities not present in the art dataset given the knowledge of context. All models in this work are trained on consumer-level computing hardware with presented hyperparameters and configurations. Finally, we explore the application of computer vision models that can detect when an artwork has been created by human or machine with 98.14% accuracy. The dataset and models are open-sourced for future work.

Keywords: Computational creativity · Latent diffusion · Image classification

1 Introduction

Art has always been a reflection of human creativity, imagination, and emotion. However, with the advent of Artificial Intelligence (AI), the definition and creation of art is rapidly changing. Latent diffusion models, a type of generative model, have emerged as a promising tool for generating art with AI [1]. These models have the ability to generate unique and diverse outputs, making them suitable for exploring new forms of artistic expression.

Recent rapid developments in synthetic media, such as the use of latent diffusion models to learn artistic styles, have enabled Machine Learning (ML) models to go from creating interesting images to winning art competitions in just a few years [2]. This cutting-edge technology challenges the traditional concept of creativity, raising important ethical and philosophical questions. These new approaches have potential to democratise access to the art world and inspire new forms of creative expression, but also raise a number of professional, social,

N. Naik et al. (Eds.): UKCI 2023, AISC 1453, pp. 157–169, 2024.
https://doi.org/10.1007/978-3-031-47508-5_13

ethical, and legal issues that must be addressed. The rate at which these developments are taking place have superseded our current understanding of ethics and the law given the uncharted territory in which these new approaches exist.

Neural style transfer enables users to create new images inspired by an artist. Incorporating fine-tuned latent diffusion, the large data requirements of traditional methods are overcome, leading to the possibility of learning an artist's style from a small number of paintings without catastrophic forgetting of other concepts. For example, if a traditional Generative Adversarial Network (GAN) is trained on paintings of a non-diverse group of people, applying such a model to an individual who has different characteristics will often transfer unwanted knowledge and transform physical characteristics such as skin tone and age. This study shows that fine-tuning from large image datasets with a small number of paintings (which feature little diversity) enables the transfer style to anyone regardless of their physical characteristics.

Traditionally, training of generative approaches is largely limited to those with the means to access complex computational resources, with modern methods of latent diffusion estimated to cost seven figures to train. This study considers those limitations and we perform all experiments on consumer-level hardware in a step towards making AI art more accessible to wider society. Detection of AI-generated artworks is another growing concern, in part due to ML models winning competitions. This raises questions regarding the potential for bad actors to claim that their machine-generated work is, in fact, created by a human. Such concerns are furthered by financial motives given the well-known report of an AI-generated artwork selling for $432,000 (USD) [3]. Given this, this study explores and provides additional ML models that can detect when a painting has been created using the proposed approach. Uses of this could be useful, for example, during a competition submission process.

In the remainder of this paper, relevant background and related work are presented in Sect. 2 followed by the proposed method in Sect. 3, before results are presented in Sect. 4. Finally, pertinent conclusions from this research are drawn in Sect. 5 and direction of future research is suggested.

2 Background and Related Work

Latent diffusion models are a new approach in the field of generative models; thus, the literature is young and few applications have been explored.

In ML research, diffusion models are a rapidly growing area of study with notable recent developments in the model space with OpenAI's Dall-E [4], Google's Imagen [5], and StabilityAI's open-source Stable Diffusion [1]. Since these developments are pushing the boundaries of image quality from realistic to arguably artistic, the public imagination has been captured and there is ongoing debate as to the professional, social, ethical and legal implications of such technology. There are several reasons why it is important to be able to distinguish between art generated by AI and art created by humans; first, the ability to identify AI-generated art helps to ensure that the authenticity and

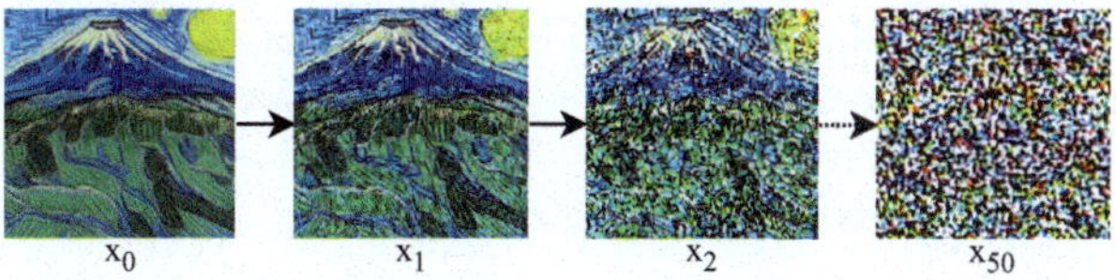

Fig. 1. An example of the forward diffusion process for 50 steps. Prompt: "Mount Fuji, in the style of van Gogh.". Note that the visual space is shown for readability purposes, but Stable Diffusion adds noise in the 768-dimensional latent space.

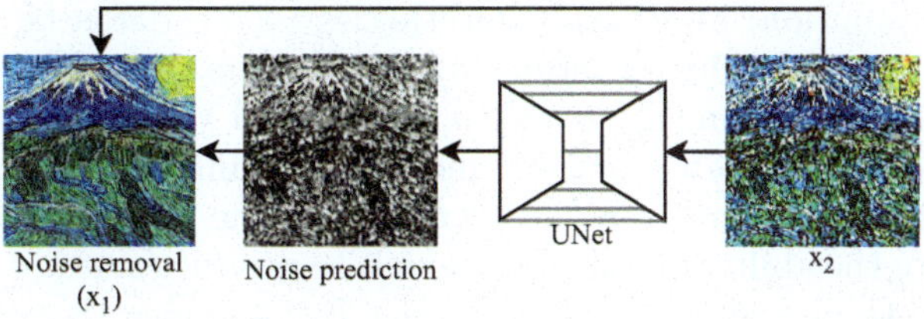

Fig. 2. An example of the backward diffusion process with text embedding omitted. Note that the visual space is shown for readability purposes, but Stable Diffusion removes noise in the 768-dimensional latent space.

originality of the art are accurately represented. That is, the knowledge of where such an image has come from, be that a machine or a human being. There are also legal implications, since ownership and copyright of AI-generated art are a complex and evolving area of law. It also helps to consider ethics, since art created by AI algorithms raises ethical questions about the role of technology in the creative process and the extent to which AI algorithms can truly be considered "creative". Knowing the origin of the art can help to contextualise these ethical questions. Furthermore, the ability to distinguish between AI-generated and human-created art is also important in terms of evaluating the artistic merit of the work to be attributed to the artist. Being able to identify AI-generated art is important for establishing transparency and accountability in the field of art and technology and for ensuring that the art is properly contextualised and valued.

Diffusion models aim to learn via modelling how image data diffuse through latent space. That is, given the stepwise addition of noise, which can be observed in Fig. 1, a model of how the function can be operated in reverse. If a model can add noise to an image until all pixels are completely random and then learn to reverse that process, the latter function could then effectively be used to generate images from random noise. This process is additionally directed by including an embedded representation of text, such as a descriptive caption.

An example of this function can be seen in Fig. 2 where the noise within the image x_i is predicted and removed to form the image x_{i-1}. To expand on the previous explanation, a noised image at timestep t, x_t, can be formed through the following by $x_t = \sqrt{\bar{\alpha}_t}x_0 + \sqrt{1 - \bar{\alpha}_t}\varepsilon$, with regards to original image x_0, noise

ε, and adjustments according to t, $\bar{\alpha}$. The reversal of this process is the task to be learnt by a neural network denoted ε_θ, with the loss function as:

$$Loss = E_{t,x_0,\varepsilon}\left[\left\|\varepsilon - \varepsilon_\theta\left(x_t, t\right)\right\|^2\right]. \tag{1}$$

For readability purposes, equations are given in the visual space, i.e. with regards to x. Stable diffusion deals with noise in the latent space, that is, the replacement of x_t with z_t in $\left\|\varepsilon - \varepsilon_\theta\left(x_t, t\right)\right\|^2$. Further technical details on the approach can be obtained from [1]. Stable diffusion applies the above theory to the LAION2B-en, LAION-high-resolution, and LAION-aesthetics v2.5+ datasets, subsets of the LAION-5B dataset [6], which contains 5.85 billion text-image pairs.

In terms of application, the research field is still young. [7] proposes a fine-tuning approach to generate synthetic medical imaging data via the model, noting an increase in quality from the standard Stable Diffusion weights. In [8], researchers note the ability of diffusion models to produce state-of-the-art quality paintings, with the results of a *visual Turing Test* showing that human subjects often predict a synthetic image is a real painting at a similar rate to that of real paintings. In [9], the results suggest that various latent diffusion models, similar to the approach used in this study, leave a type of digital fingerprint within images that can be detected and classified. Given that latent diffusion-based approaches for the synthesis of artworks is less explored, much of the relevant literature for the classification of such images is situated in deepfake detection. In 2019, researchers suggested an optical flow approach to the detection of synthetic human faces [10], achieving a classification rate of 81.61% accuracy on the FaceForensics dataset. The hybrid approach leveraging convolutional and recurrent methods as studied by [11] presented mixed results for three different publicly available deepfake detection datasets, the most promising performance was on the FaceForensics dataset at 91.21%. Related studies in [12] suggest that human beings have a limited ability to recognise manipulated images.

3 Method

This study has two main objectives. First, the fine-tuning of diffusion models on consumer-level hardware; the goal of this is to teach a new concept to a pre-trained model in the form of a new word that does not exist in the model vocabulary. Second, this work then explores a preliminary method of AI art detection given the growing issues and implications that technology has on society.

3.1 Data Collection and Fine-Tuning

Data is collected from the 2017 film, *Loving Vincent* [13]. Each frame of the film was handpainted by 125 classically trained artists in the style of Vincent van Gogh and runs at 24 frames per second. Frames from the artist-inspired film were chosen over original paintings due to additional visual information, such as the view of the same subjects from different angles and lighting conditions.

An example of this can be seen in Fig. 3. One frame is extracted for every two seconds of film (i.e. every 48th frame) and title/credit sequences are removed. Each extracted frame is then centre cropped and resized to 512px square using the Lanczos algorithm. This results in a final dataset of 1705 images.

Fig. 3. A comparison of van Gogh's original painting [14], to four similar frames from the film which show the subject at differing angles and facial expressions.

The fine-tuning method used is the DreamBooth approach from Google [15]. The original paper suggested providing training examples of subjects, for example fine-tuning the model on an individual to then enable the model to generate images of that person in different styles and settings. In this approach, the model is instead trained on each of the extracted frames from the film with the same unique identifier to generalise the stylistic qualities of the film. That is, training the model to learn the general concept of the similarities of the frames (the style) rather than individual entities such as cast members or physical objects.

The training of Stable Diffusion cost around $160,000 and took a total of 79,000 h on A100 GPUs ($32,000 USD) [16], and therefore it is important first to explore hyperparameters and options for realistic training on common hardware. The GPU used for this study is a consumer-level Nvidia RTX 3080Ti (MSRP $1,199), and training for one epoch is carried out at a batch size of 1.

Experiments in this work explore appropriate methods of training by combining hyperparameters with the goal of training on the selected hardware: (i) whether the model is frozen or not; (ii) the scheduler[1] (number of denoising steps, stochastic or deterministic, denoising algorithm) from a choice of Pseudo Numerical Methods for Diffusion Models on Manifolds (pndm) [17], Denoising Diffusion Implicit Models (ddim) [18], or Elucidating the Design Space of Diffusion-Based Generative Models (euler, euler-ancestral, dpm) [19]; (iii) use of the Low-Rank Adaptation (LoRA) approach [20]; (iv) The use of Exponentially-weighted Moving Averages (EMA); (v) Use of 8-bit precision in the ADAM optimiser [21]

[1] Further details on schedulers can be found at: https://huggingface.co/docs/diffusers/using-diffusers/schedulers.

and mixed precision from None, FP16, or BF16; and, finally, (vi) the implementation of memory-efficient attention approaches from *xformers* [22] or flash attention [23]. Every combination of the noted hyperparameters are attempted to discern the most appropriate methods for training on consumer-level hardware, aiming to overcome a main barrier that both researchers and enthusiasts encounter when implementing state-of-the-art approaches. The time taken to train the model on a clean system along with the VRAM usage is recorded.

3.2 Synthetic Art Detection

This section describes data generation and the learning process for the synthetic art detection model. The goal of this approach is to produce a computer vision model that can discern between real and generated artwork.

Initially, a dataset of synthetic art images is generated via the model produced by the experiments in the previous section. The generation of these works is unguided with the prompt ("painting in the style of UNIQUE-TOKEN".[2]); 1705 images are generated to match the size of the style dataset and balance both classes equally.

The method selected for the detection of synthetic artworks in this study is the Convolutional Neural Network (CNN). They are generally divided into two types of layer (with additional layers that transform data, such as flattening n-dimensional blocks); first, the convolutional layers, which are designed to learn to extract visual features from the input images using filters. Secondly, the classical dense layers of neurons, which aim to learn to extract further information from the learnt filters prior to prediction. The learning process of the proposed approach is based on binary cross-entropy $H_p(q) = -\frac{1}{N}\sum_{i=1}^{N} y_i \cdot \log(p(y_i)) + (1 - y_i) \cdot \log(1 - p(y_i))$. Where y is the class label (AI-Generated: 0, Real Painting: 1) and $p(y)$ is the probability predictions on all data. Initially, a network without dense layers is trained with $16, 32, 64, 128$ filters and $1, 2, 3$ layers. The best feature extractor is then used to optimise the dense layers, where $1, 2, 3$ layers of $32, 64, 128, \ldots, 4096$ rectified linear units are implemented.

4 Results and Observations

This section discusses the results of the two sets of experiments and their implications. First, the results for the fine-tuning of latent diffusion are presented; namely, hyperparameter exploration for execution on consumer-level hardware, metrics observed during the training process, and finally observations when prompting the trained model in several ways. Second, computer vision models for the detection of AI-generated artwork are detailed.

[2] A unique token is used to add a new term to the dictionary without interfering with the base knowledge.

Table 1. The combinations of hyperparameters that enabled DreamBooth fine-tuning on consumer-level computer hardware.

Hyperparameter	Attribute
Frozen model	Yes
Scheduler	All
LoRA	Yes
EMA	No
8-bit ADAM	Yes
Mixed precision	FP16
Memory attention	None

It was noted that few combinations of hyperparameters could execute on consumer-level hardware. These results are communicated to the research community in Table 1. As shown by the literature, training these algorithms is a computationally expensive process, but it was possible to tune the models. Specifically, LoRA with FP16 precision must be used.

Figure 4 shows the losses observed during fine-tuning for the viable models (first 100 steps omitted for readability purposes). The lowest average loss after one epoch was the dpm model at 0.167. Table 2 shows the amount of VRAM used and the time it took each of the schedulers to fine-tune the model. All approaches except Euler-ancestral used 6.5 GB of VRAM, well below the suggested requirements. All models trained successfully on the RTX 3080Ti GPU.

Figure 5 shows random examples of images generated by the model. The prompt given was the same as the training process, "painting in the style of UNIQUE-TOKEN". As can be observed, the style of the training data is successfully transferred, although lacking sharpness compared to the dataset. An interesting aspect of this approach can be observed in Fig. 7. The film had a small cast of 20 actors, mostly of white British descent. The under-representation of groups of people in ML data is a long-standing problem [24], and this figure

Table 2. GPU memory required and the time taken for different schedulers to fine-tune the model for one epoch.

Scheduler	VRAM usage (GB)	Tuning time (s)
ddim	6.5	693.6
dpm	6.5	676.8
euler	6.5	691.2
lms	6.5	670.2
pndm	6.5	656.4
euler-ancestral	7.7	644.4

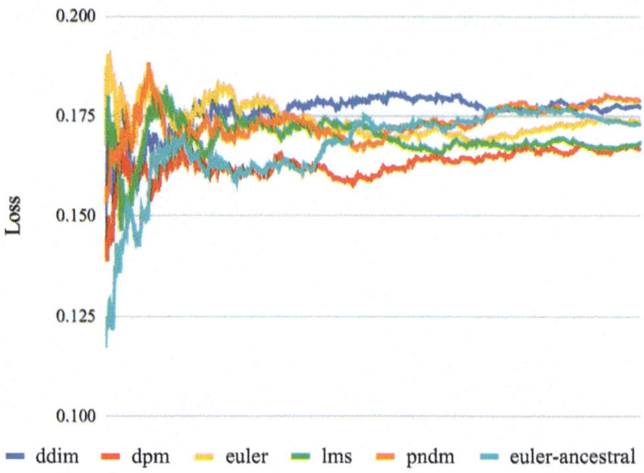

Fig. 4. Observed average losses for the fine-tuning process of one epoch.

Fig. 5. Examples of random images generated by our model.

shows that the proposed approach enables the possibility of generating stylised paintings of people who were not represented within the training data. More traditional style transfer approaches, such as the GAN, are known to transfer additional aspects that are unwanted, such as skin tone which was noted out by [25]. This is a problem of context, given that the models are provided a set of images that contain entities, but the ideal outcome is to only transfer the styles that encapsulate all images.

To provide further examples, Fig. 6a shows how knowledge of individuals (commonly found in the LAION dataset) can be generated in the new style. The nature of the proposed approach also extends to unseen entities and locations. An example of random entities that were not present in the training data can be observed in Fig. 6b, and similarly Fig. 6c shows locations in the film style.

(a)

(b)

(c)

Fig. 6. Examples of individuals, entities, and locations present within the LAION dataset with style transferred from our model. (a) Individuals from left-top; Barack Obama, Queen Elizabeth II, Elon Musk and left-bottom; Nicolas Cage, Morgan Freeman, Emilia Clarke. (b) Entities from left-top; Teapot, VW Golf, Chair and left-bottom; Black cat, Border Collie Dog, Chicken. (c) Locations from left-top; London (UK), Paris (France), Barcelona (Spain) and left-bottom; Tokyo (Japan), New York (USA), Mount Fuji (Japan).

The generated data and model are made open source and available to the public for online download.[3] The model is embedded as a standard Stable Diffusion checkpoint, and thus is compatible with the original code.

4.1 Detection of Synthetic Artwork

This section details the findings of the CNN approach to the detection of synthetic images. The results in this section are related to the binary classification of real versus AI-generated artworks. Accuracy metrics are presented since the dataset is perfectly balanced.

Table 3 presents results for the feature extractor model. It seems that the detection of AI artwork is relatively easy for a CNN model, which is interesting given that the literature review suggested that this is difficult for humans. The

[3] The dataset from this study can be downloaded from https://www.kaggle.com/datasets/birdy654/detecting-ai-generated-artwork.

Fig. 7. Generative samples showing that the nature of fine-tuning leads to the ability of generating stylised images of individuals who were not represented in the style training data.

Table 3. Validation accuracy for the tuning of the feature extraction model. The results show the ability of a computer vision model to discern human art from AI-generated art.

Filters	Layers		
	1	*2*	*3*
16	96.87	96.68	97.75
32	97.46	97.36	98.03
64	98.14	97.65	98.04
128	97.85	97.17	97.75

Table 4. Validation accuracy for the tuning of the CNN's dense layers. Results show the ability of a computer vision model to discern human art from AI-generated art.

Neurons	Layers		
	1	*2*	*3*
32	97.65	98.44	98.53
64	98.04	97.56	98.04
128	98.53	98.04	98.14
256	98.04	97.56	96.97
512	98.44	96.87	96.77
1024	98.14	97.56	97.56
2048	97.65	95.6	97.07
4096	97.36	98.04	96.09

best feature extractor consisted of one layer that produced 64 filters, leading to an overall classification accuracy of 98.14%. Table 4 then shows the final results of the deep neural network for AI art detection. The best overall model contained either one layer of 128 neurons or three layers of 32 neurons. With the data at hand, these results suggest that AI-generated artworks can be detected with 98.53% accuracy from a dataset of real paintings versus latent diffusion outputs of the same.

5 Conclusion and Future Work

The line between human and machine-generated art is becoming increasingly blurred. This study has led to several interesting findings. Firstly, results showed that it is possible to fine-tune state-of-the-art latent diffusion models for the generation of AI artwork on consumer-level hardware with hyperparameters presented; hyperparameter benchmarking found that the highest average performance was with the dpm scheduler. The results from this work demonstrated the remarkable capabilities of ML models to generate artwork in a given style, and then to transfer that style to images of people, objects, and locations not present in the original training data due to the nature of fine-tuning. However, as the reach and influence of AI-generated artwork grows, so too do the increasingly relevant issues that need to be addressed. This work highlights the need for ongoing discussion and development in this area, with the aim of creating a future where machine-generated art can co-exist with human-created art in a responsible and sustainable way; towards this, this study considered the implications that realistic AI art generation has by implementing computer vision to detect when an artwork is machine generated. In the future, models could be further trained on more data to attempt to overcome limitations, such as the lack of sharpness. This study along with the state-of-the-art shows that AI-generated art is becoming more realistic, and the ability for people to distinguish what is created by humans or machines is likewise becoming increasingly ineffective. Additionally, to enhance the detection of AI art, explainable AI approaches could be considered to present the visual attributes in the image that suggest that the work was not created by a human being to increase the trustworthiness of predictions as an expert system.

References

1. Rombach, R., Blattmann, A., Lorenz, D., Esser, P., Ommer, B.: High-resolution image synthesis with latent diffusion models. In: Proceedings of the IEEE/CVF Conference on Computer Vision and Pattern Recognition, pp. 10,684–10,695 (2022)
2. Roose, K.: An AI-generated picture won an art prize. artists aren't happy. The New York Times **2**, 2022 (2022)
3. Epstein, Z., Levine, S., Rand, D.G., Rahwan, I.: Who gets credit for AI-generated art? Iscience **23**(9), 101515 (2020)

4. Ramesh, A., Pavlov, M., Goh, G., Gray, S., Voss, C., Radford, A., Chen, M., Sutskever, I.: Zero-shot text-to-image generation. In: International Conference on Machine Learning, pp. 8821–8831. PMLR (2021)
5. Saharia, C., Chan, W., Saxena, S., Li, L., Whang, J., Denton, E., Ghasemipour, S.K.S., Ayan, B.K., Mahdavi, S.S., Lopes, R.G., et al.: Photorealistic text-to-image diffusion models with deep language understanding. arXiv:2205.11487 (2022)
6. Schuhmann, C., Beaumont, R., Vencu, R., Gordon, C., Wightman, R., Cherti, M., Coombes, T., Katta, A., Mullis, C., Wortsman, M., et al.: Laion-5b: an open large-scale dataset for training next generation image-text models. arXiv:2210.08402 (2022)
7. Chambon, P., Bluethgen, C., Langlotz, C.P., Chaudhari, A.: Adapting pre-trained vision-language foundational models to medical imaging domains. arXiv:2210.04133 (2022)
8. Yi, D., Guo, C., Bai, T.: Exploring painting synthesis with diffusion models. In: 2021 IEEE 1st International Conference on Digital Twins and Parallel Intelligence (DTPI), pp. 332–335. IEEE (2021)
9. Sha, Z., Li, Z., Yu, N., Zhang, Y.: De-fake: detection and attribution of fake images generated by text-to-image diffusion models. arXiv:2210.06998 (2022)
10. Amerini, I., Galteri, L., Caldelli, R., Del Bimbo, A.: Deepfake video detection through optical flow based CNN. In: Proceedings of the IEEE/CVF International Conference on Computer Vision Workshops (2019)
11. Saikia, P., Dholaria, D., Yadav, P., Patel, V., Roy, M.: A hybrid CNN-LSTM model for video deepfake detection by leveraging optical flow features. In: 2022 International Joint Conference on Neural Networks (IJCNN), pp. 1–7. IEEE (2022)
12. Nightingale, S.J., Wade, K.A., Watson, D.G.: Can people identify original and manipulated photos of real-world scenes? Cogn. Res. Princ. Implic. **2**(1), 1–21 (2017)
13. Kobiela, D., Welchman, H.: Loving Vincent. Universal Pictures. https://lovingvincent.com/ (2017)
14. van Gogh, V.: Self-portrait (1889)
15. Ruiz, N., Li, Y., Jampani, V., Pritch, Y., Rubinstein, M., Aberman, K.: Dreambooth: fine tuning text-to-image diffusion models for subject-driven generation. arXiv:2208.12242 (2022)
16. Stephenson, C., Seguin, L.: Training stable diffusion from scratch costs $160k. https://www.mosaicml.com/blog/ (2023). Accessed 03 February 2023
17. Liu, L., Ren, Y., Lin, Z., Zhao, Z.: Pseudo numerical methods for diffusion models on manifolds. arXiv:2202.09778 (2022)
18. Song, J., Meng, C., Ermon, S.: Denoising diffusion implicit models. arXiv:2010.02502 (2020)
19. Karras, T., Aittala, M., Aila, T., Laine, S.: Elucidating the design space of diffusion-based generative models. arXiv:2206.00364 (2022)
20. Hu, E.J., Shen, Y., Wallis, P., Allen-Zhu, Z., Li, Y., Wang, S., Wang, L., Chen, W.: Lora: low-rank adaptation of large language models. arXiv:2106.09685 (2021)
21. Dettmers, T., Lewis, M., Belkada, Y., Zettlemoyer, L.: Llm.int8(): 8-bit matrix multiplication for transformers at scale. arXiv:2208.07339 (2022)
22. Lefaudeux, B., Massa, F., Liskovich, D., Xiong, W., Caggiano, V., Naren, S., Xu, M., Hu, J., Tintore, M., Zhang, S., Labatut, P., Haziza, D.: xformers: a modular and hackable transformer modelling library. https://github.com/facebookresearch/xformers (2022)

23. Dao, T., Fu, D.Y., Ermon, S., Rudra, A., Ré, C.: Flash attention: fast and memory-efficient exact attention with IO-awareness. In: Advances in Neural Information Processing Systems (2022)
24. Mehrabi, N., Morstatter, F., Saxena, N., Lerman, K., Galstyan, A.: A survey on bias and fairness in machine learning. ACM Comput. Surv. (CSUR) **54**(6), 1–35 (2021)
25. Ba, Y., Wang, Z., Karinca, K.D., Bozkurt, O.D., Kadambi, A.: Style transfer with bio-realistic appearance manipulation for skin-tone inclusive RPPG. In: 2022 IEEE International Conference on Computational Photography (ICCP), pp. 1–12. IEEE (2022)

Object Detection in Heritage Archives Using a Human-in-Loop Concept

Surya Kasturi[1(✉)], Alex Shenfield[1], Chris Roast[1], Danny Le Page[2], and Alice Broome[2]

[1] Sheffield Hallam University, Sheffield, UK
s.kasturi@shu.ac.uk
[2] British Online Archives, Leeds, UK

Abstract. The use of object detection has become common within the area of computer vision and has been considered essential for a numerous applications. Currently, the field of object detection has undergone significant development and can be broadly classified into two categories: traditional machine learning methods that employ diverse computer vision techniques, and deep learning methods. This paper proposes a methodology that incorporates the human-in-loop feedback concept to enhance the deep learning object detection capabilities of pre-trained models. These Deep Learning models were developed using a custom humanities and social science dataset that was obtained from the British Online Archives collections database.

Keywords: Object detection · Human-in-loop · Deep learning

1 Introduction

Machine learning (ML) is a widely known concept that has gained significant interest in various domains, such as computer vision, pattern recognition, and data retrieval. ML allows computers to learn from data without explicit programming, improving themselves through experience. ML algorithms analyze historical data, identify patterns, and establish mathematical relationships between inputs and outputs. This technique relies on large training databases and computational power. While ML is fascinating, artificial intelligence (AI) is an even more advanced and intriguing technology. AI involves computer systems simulating human cognitive processes, including learning and problem-solving.

Human involvement plays a crucial role in every step of the machine learning (ML) pipeline, starting from data preparation to result inference. Before constructing a model, data scientists dedicate substantial time to data preprocessing [13]. This involves tasks such as data extraction, integration, and cleaning. The data is then categorized and divided into separate training and test sets. Throughout the entire development process of training and testing the ML model, human participation is evident. The following sections of this paper explore existing knowledge on human involvement across different phases of ML

N. Naik et al. (Eds.): UKCI 2023, AISC 1453, pp. 170–181, 2024.
https://doi.org/10.1007/978-3-031-47508-5_14

development. Additionally, we present a methodology and corresponding results that showcase improvements in object detection techniques applied to archival documents.

British Online Archives faces challenges due to the time-consuming publication process, limiting the volume and richness of curated collections and metadata. Humanities researchers rely on curated collections around a topic of interest to inspire and facilitate their work, but time constraints result in limited metadata provided. Researchers in humanities require consideration of both written and graphical content, but searching graphical content remains challenging compared to textual content. This complexity hinders systematic search and analysis of graphical material. In order to expedite the process of curating and publishing archives while also generating detailed and easily searchable metadata, we propose a machine learning pipeline as to produce comprehensive metadata about the elements within the collection. This extensive metadata, which describes various aspects of the curated collection, is automatically generated. This automation allows editors to concentrate on validating, organizing, and refining the contents of the collection. Once the collection is published, users can access the metadata, which provides detailed information. This enhanced accessibility enables users to systematically search for graphical content using both keywords and free-text queries, improving their overall experience. The object detection is crucial component of our research and a part of this research has been proposed in this paper. This research is a part of KTP (Knowledge Transfer Partnership) project which is funded by UKRI through Innovate UK.

2 Prior Work

Utilising pre-existing knowledge into the learning framework is a viable strategy for addressing data sparsity, as it obviates the need for the learner to derive the knowledge solely from the available data [3]. Humans possess extensive prior knowledge as specialised agents. The developer has the potential to facilitate machine learning through the incorporation of human wisdom and knowledge, which can aid in addressing the issue of sparse data, particularly in domains where there is insufficient training data [24]. To address these challenges a new concept named as Human-in-Loop (HIL) has been proposed. This approach primarily focuses on involving human expertise into the modelling procedure [7].

A conventional machine learning algorithm generally comprises of three components [21]: data pre-processing, data modelling, and process optimisation via developer modifications to enhance the performance of the model. In the typical process of model development, human intervention is required during the data pre-processing stage to transform unstructured data into structured label data. This practise has been identified by some researchers as an application of the Human-in-the-Loop (HIL) concept [1]. Usually, the efficiency of deep learning is dependent upon the quality of the data. To obtain effective performance in a novel task, a substantial quantity of accurately labelled data is required. The process of annotating extensive sets of data necessitates significant effort and time investment. This can pose a challenge for tasks that require multiple iterations and cannot accommodate the associated costs and delays. In contrast to data annotation, iterative data labelling places greater emphasis on user experience enabling users to engage in the data annotation process directly. So, here the objectives has been divided into two primary areas: first, improving the learning system through iterative labeling and second involves giving importance to engaging and communicating with users. This means actively involving users in the learning process, gathering their feedback, and incorporating their insights to enhance the system's performance.

Yu et al. [23] employed a labelling scheme that was partially automated, utilising deep learning techniques with human-in-the-loop to reduce the need for manual labour in the annotation process. This represents the fundamental model of uncomplicated iterative annotation. Several domains within the realm of Artificial Intelligence, including Natural Language Processing (NLP) and Computer Vision (CV), employ diverse methodologies that utilise human intelligence for the purposes of training and inferring experimental outcomes. Research related for both NLP and CV covers a range of techniques that combine human and machine intelligence. The utilisation of heuristic methods has considered the varied nature of human creativity in order to attain outcomes of superior quality.

The utilisation of Deep Learning techniques, specifically neural network-based methods, has become the leading approach for executing various computer vision tasks, as evidenced by recent studies [20]. In order to enhance the efficiency of stated techniques, human feedback has been incorporated into the deep learning framework to improve the system's overall intelligence in addressing difficult scenarios that are beyond the model's capacity to handle. Object detection, which is considered to be a fundamental and challenging problem in the field of computer vision, has drawn substantial interest in recent times [4]. Yao et al. [22] highlight that the repeated cycles of queries can incur significant costs and consume substantial time, rendering it impractical to engage in interactions with end-users. They proposed an interactive architecture for object detection that enables users to rectify a limited number of annotations suggested by a model for an unannotated image or test dataset with the highest predicted annotation cost. Madono et al. [12] proposed a proficient framework for object detection that involves human-in-the-loop. The framework is comprised of bi-directional deep SORT [19] and annotation-free segment identification (AFSID). The responsibility of humans within this architecture pertains to the verification of object candidates that cannot be automatically detected by bi-directional deep SORT. Subsequently, the model should be trained on the supplementary objects that have been annotated by individuals.

Numerous researchers have been dedicating their efforts towards enhancing the performance of object detection models. These models can be classified into two categories: one-stage object detectors and two-stage object detectors. One-stage object detection models execute classification and regression operations on closely spaced anchor boxes, without generating a sparsely populated Region of Interest (RoI) set. The YOLO algorithm [14], represents an initial foray into the direct detection of objects on a feature map with high density. The utilisation of multi-scale features has been proposed by SSD [11] as a means of detecting objects with varying scales. Later, RetinaNet [10] introduced the use of focal loss as a solution to tackle the issue of imbalanced classes in the context of dense object detection.

Currently, two-stage detectors exhibit superior performance in terms of detection accuracy. The detectors employ a two-stage approach wherein the initial stage generates sparse region proposals, followed by a subsequent stage that performs regression and classification on the proposed regions. The RCNN model [5] employed computer vision techniques such as Selective Search [18] and Edge Boxes [25] at a low level to produce proposals. Subsequently, a CNN was utilised to extract features for the purpose of training an SVM classifier and bounding box regressor. Fast R-CNN [4] then proposed a method of feature extraction for individual proposals on a feature map that is shared, through spatial pyramid pooling. Later, building on this, Faster R-CNN [15] incorporated the region proposal process within the deep ConvNet architecture, resulting in a detector that can be trained end-to-end.

The authors of R-FCN [2] introduced a region-based fully convolutional network as a means of producing features that are sensitive to regions for the purpose of detection which traditional methods lacked. By directly producing region-sensitive features using a fully convolutional network, R-FCN achieves faster inference times and better localization accuracy. FPN (Featured Pyramid Network) [9] an architectural approach that employs top-down processing and lateral connections to produce a feature pyramid suitable for detecting objects at multiple scales. FPN preserves both semantic information and spatial details, improving object detection across various scales. This approach has become widely adopted and has advanced the accuracy and robustness of object detection models. The EfficientDet model [16] utilises a compound scaling technique to simultaneously increase the dimensions of depth, width, and resolution for the backbone, BiFPN, and box/class prediction networks. The compound scaling technique used in EfficientDet enhances the model's capacity, improves feature representation, and allows for more precise object detection across different scales, contributing to its success in the field of object detection.

The current research emphasises on the development of a pipeline that is defined by ease of use and robustness. Even though involving humans in model inference incurs additional costs [22], we believe that human in loop techniques such as interactive machine learning will actually provide significant improvements in the process where there is a scarcity of data for training the model.

3 Object Detection

Our implementation of Human-in-Loop for object detection in archival documents involves six fundamental steps:

1. Dataset collection and annotation using the Label Studio [8] tool
2. Object detection model training using a transfer learning approach (which also entails selecting the appropriate model)

3. Inference on validation data
4. Modification or correction of the model's inference outcomes (using customised Label Studio user interface)
5. Retraining the model with new learning parameters after collecting a few newly annotated samples
6. Evaluation of the results in a held out test set.

3.1 Dataset for Base Model

The models have undergone training on a dataset comprising 146 images containing of 180 objects and 3 classes. The valid dataset during training has 35 images containing 53 objects. The test dataset, on the other hand, consists of 37 images containing 54 objects and the same classes as the other datasets. Some dataset samples are illustrated in Fig. 2.

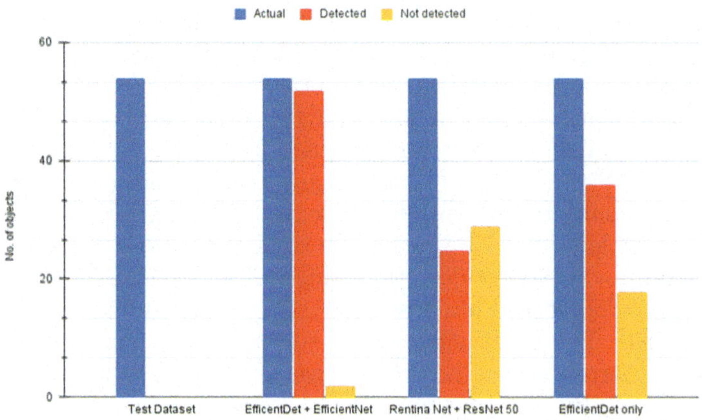

Fig. 1. Comparisons of various base models performance

3.2 Model Configuration

In this study our baseline object detection model is a two-stage fine-tuned EfficientDet architecture with a second stage EfficientNet classfier. Initial evaluation work indicates that this combination outperforms a single stage EfficientDet model and a two stage model based on RetinaNet and ResNet50. The comparison of different base models is presented in Fig. 1. It is evident that the two-stage model, namely EfficientDet + EfficientNet, outperforms the other two models. The two stage EfficientDet + EfficientNet model configuration is then further tuned using the Human-in-the-Loop (HIL) implementation discussed in the Sect. 4 to improve the overall performance of the system.

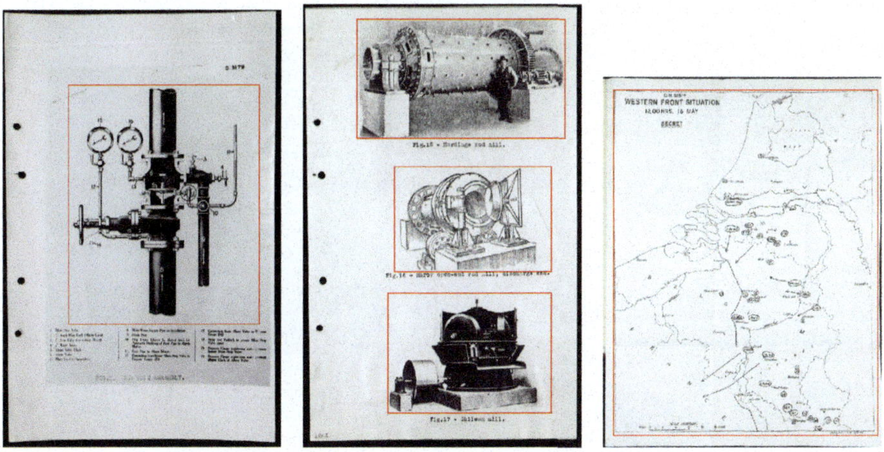

Fig. 2. Examples with bounding boxes

4 Implementation of Human-in-Loop

The essential elements of the Human-in-Loop framework entail the development of a user interface to facilitate user inputs and the establishment of a pipeline to enable automatic model retraining in response to human feedback. The subsequent sections will elaborate on the utilisation of Label Studio [8] as a user-facing interface for the purpose of rectifying or altering the outcomes generated by the model.

4.1 The Interface

The Human-in-Loop system requires an interface component that must exhibit simplicity in order to ensure ease of use for all users. The dataset employed in Sect. 3.1 was curated through the utilisation of Label Studio, a tool that enables the importing of extensive image datasets from cloud storage platforms like S3. All the data utilised in our study was obtained from the British Online Archives. The visual representation depicted in Fig. 3 provides an overview of the interface design intended for the user's perspective. This shows how users are able to access the predictions generated by the model and provide feedback to the pipeline.

4.2 The Pipeline

The effective implementation of machine learning pipeline integration constitutes another significant element of human-in-the-loop. The incorporation of this integration enhances the model's ability to acquire knowledge from user feedback. During the initial stage of the pipeline, the data undergoes pre-processing, which involves the creation of annotated data and the removal of abnormal data. The implementation of augmentation and normalisation techniques on the training

Fig. 3. Screenshot of the Label Studio Interface

dataset is utilised to enhance the quality of the training process. The transfer learning [17] methodology is employed in order to create a baseline model for the implementation of our Human-in-Loop (HIL) process. Non-Maximum Suppression (NMS) [6] is employed during post-processing to eliminate redundant bounding boxes and facilitate the selection of optimal bounding boxes. A distinct test dataset was generated, which was not exposed to the model during the HIL training phase, in order to assess its efficiency during evaluation. Finally the prediction of model on the test dataset available to the user on the Label Studio platform. Users then have the opportunity to review the predicted images and make necessary adjustments to the bounding boxes. The adjusted images data will then be collected and fed back into the model, thereby enhancing its performance.

The parameters utilised for our object detection model are as follows:

1. Image size: 256×256
2. Learning rate: 0.005 (for initial training) & 0.00005 (For re-training based on user feedback)
3. Batch size: 13
4. IOU threshold: 0.45
5. Prediction confidence = 0.50.

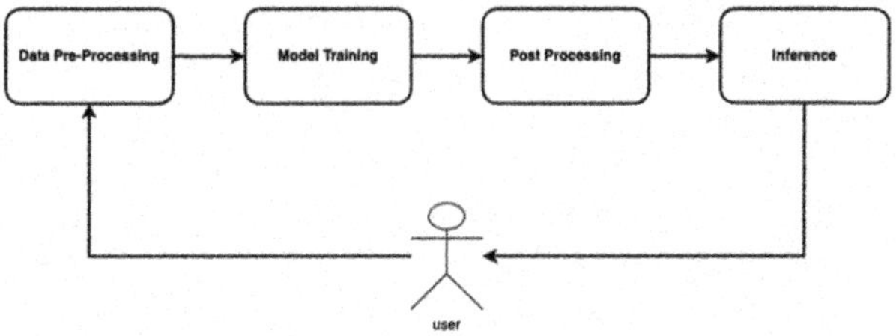

Fig. 4. Human-in-loop process

4.3 Evaluation Method

Following the integration of interface and pipeline, an essential aspect of the human-in-the-loop process is result evaluation. In this section, we present our evaluation method. Initially, we refer to the results obtained from the first inference as HIL-0%, indicating that no human feedback was involved in generating these results. Subsequently, we introduce HIL-10%, HIL-15%, and HIL-20%, which signify that users have corrected 10%, 15%, and 20% of poorly performing predicted bounding boxes in the test dataset, respectively.

The evaluation employs two distinct test datasets, namely test-1 and test-2. One of these datasets will be utilised for the purpose of rectifying the predictions, while the other dataset will be exclusively utilised for evaluating the model's performance across varying levels of HIL. Upon the user's modification of the bounding boxes on one of the test dataset, the corresponding corrected images will replace some random images in the initial training dataset which then becomes a new training dataset. A small learning rate of 0.00005 (in our case) is employed to retrain the model using the new training dataset. To retain previously learned information while incorporating the user-provided data, we utilize transfer-learning techniques that load the pre-trained weights of the model from HIL-0%. This approach allows us to make gradual adjustments to the model's weights, ensuring the assimilation of the new data without compromising the existing knowledge. Subsequently, we will conduct a comparison of the mean Intersection over Union (mIOU), mean Average Precision (mAP) scores, Precision and Recall specifically at IOU values of 0.5 and 0.75, for each of the distinct stages of models involved in this procedure. The comprehensive examination and outcomes can be found in the following Sect. 5

5 Results

The diverse outcomes of the model's performance at different percentages of HIL (Human-in-the-Loop) corrections are evident from the provided metrics in Table 1. An example object detection at different HIL levels are shown in Fig. 6

The Table 1 represents the model performance at various stages of the HIL process. The results show that the model's performance improves as the percentage of images corrected by the user increases. This is because a newer training

Table 1. Table of evaluation results

		mAP		Precision		Recall		FNs		FPs	
HIL	mIOU	@0.5	@0.75	@0.5	@0.75	@0.5	@0.75	@0.5	@0.75	@0.5	@0.75
0%	0.789	0.966	0.920	0.925	0.807	0.832	0.726	161	263	64	166
10%	0.805	0.966	0.933	0.927	0.817	0.856	0.754	138	236	64	162
15%	0.801	0.963	0.929	0.920	0.815	0.855	0.757	139	233	71	165
20%	**0.810**	**0.967**	**0.935**	**0.933**	**0.831**	**0.87**	**0.776**	**124**	**215**	**60**	**151**

dataset allows the model to learn more about the different types of objects that it is likely to encounter. Specifically, the mIOU and mAP scores for HIL-20 are higher than the scores for HIL-0, HIL-10, and HIL-15. This suggests that using 20% of the corrected images might provide a better results. However, it is worth noting that HIL-15 had a higher number of false positives and false negatives compared to the other models except HIL-0. This increase in false positives and false negatives could be attributed to various factors, including human errors during the correction process, imbalanced distribution of objects in the dataset, or the complexity and small size of the objects leading the model to predict bounding boxes for non-existent objects or missing some objects. Overall, the results show promise, indicating that using a higher percentage of corrected images (such as 20%) for training might yield better performance for this dataset. Nevertheless, it is essential to continue evaluating the model on different datasets to assess its adaptability and performance across various object types. Figure 7 presents an overview of precision-recall curves at various stages of HIL. The key observation is that as the HIL percentage increases, there is a less pronounced decrease in precision at the initial increase in recall which concludes that HIL can be used to improve the accuracy of object detection models, without sacrificing too much precision.

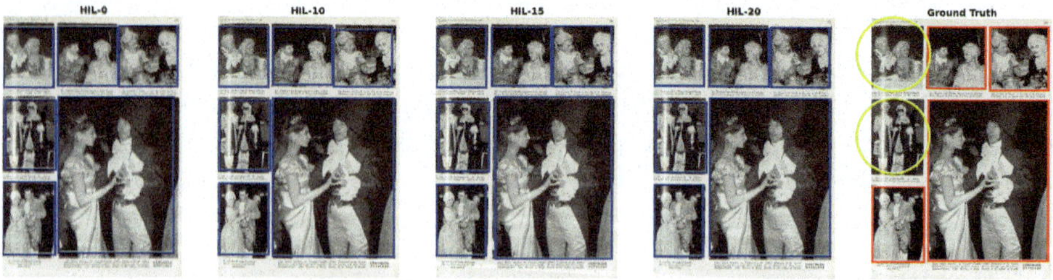

Fig. 5. Human error (missed annotation highlighted in yellow circle)

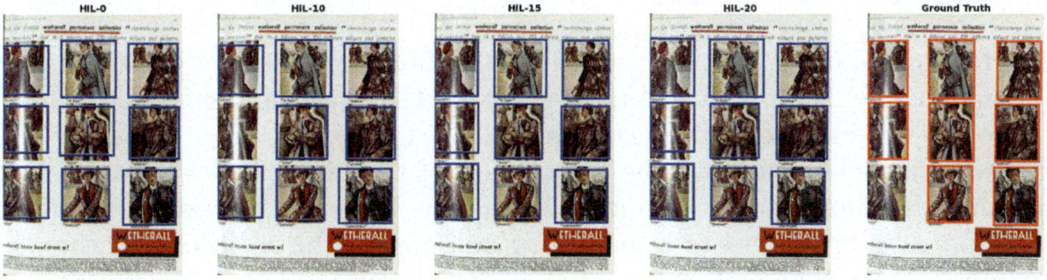

(a)

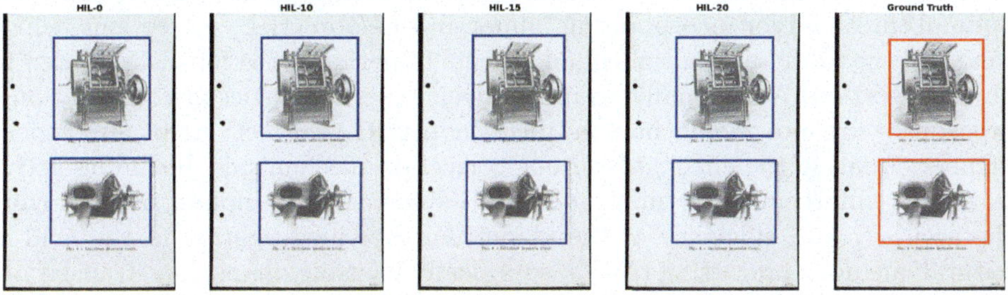

(b)

Fig. 6. Test dataset HIL progress

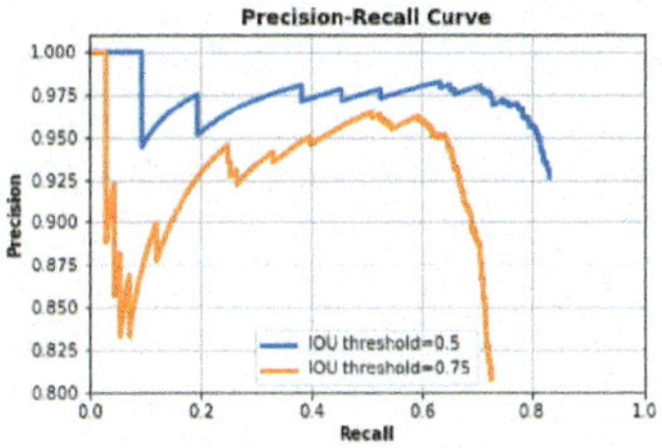

(a) HIL-0%

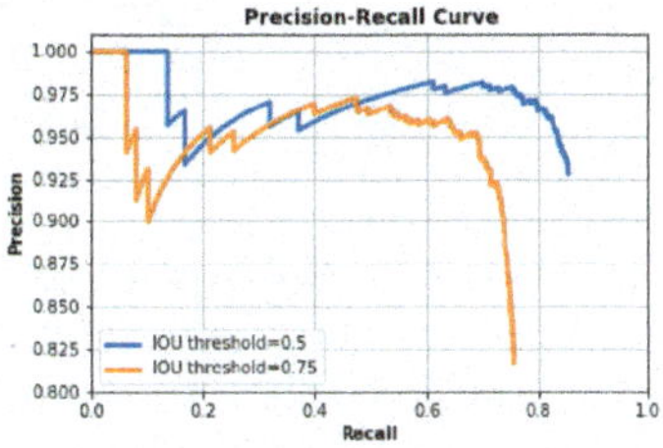

(b) HIL-10%

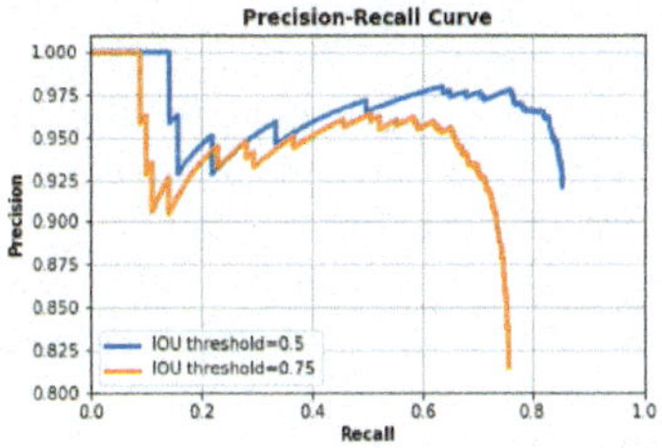

(c) HIL-15%

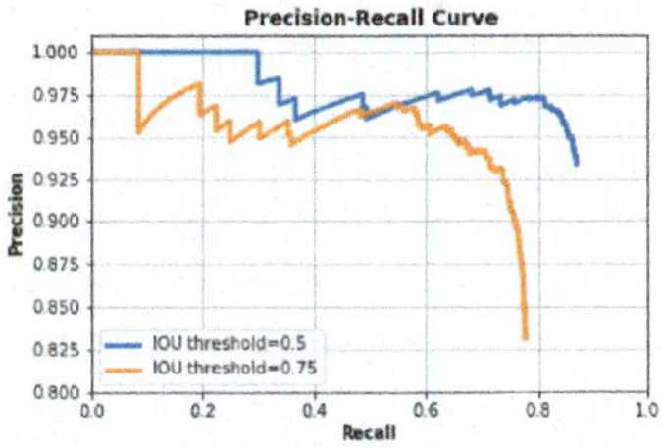

(d) HIL-20%

Fig. 7. Precision-recall curves at different % of HIL

6 Conclusion

The findings indicate that the inclusion of HIL corrections at a moderate level (approximately 10–20%) can improve the performance of the model in tasks related to object detection. Furthermore, based on the findings in Sect. 5 HIL helps the model to improve the localisation of the objects. Nevertheless, augmenting the dependence on human corrections beyond a particular threshold could potentially give rise to incongruities and impede the precision of the model. Striking a balance between automated predictions and human corrections is crucial for achieving optimal performance in these tasks.

One potential avenue for further investigation and analysis to determining the optimal threshold for incorporating human-in-the-loop (HIL) corrections, which can yield the most substantial enhancements in performance for tasks related to object detection. Additionally, examine diverse methodologies or computational procedures for integrating human-in-the-loop (HIL) corrections in an efficient manner. Analyse the effects of various correction mechanisms, including active learning, reinforcement learning, and selective correction sampling, on improving the accuracy and efficiency of the model. Further investigation in the field of natural language processing (NLP), specifically focusing on machine translation, presents promising opportunities for significant advancements. An area worth investigating is the possibility of utilising human-in-the-loop (HIL) corrections as a means of improving the calibre of machine translation results.

References

1. Chai, C., Li, G.: Human-in-the-loop techniques in machine learning. IEEE Data Eng. Bull. **43**(3), 37–52 (2020)
2. Dai, J., Li, Y., He, K., Sun, J.: R-FCN: object detection via region-based fully convolutional networks. Adv. Neural Inf. Process. Syst. **29** (2016)
3. Diligenti, M., Roychowdhury, S., Gori, M.: Integrating prior knowledge into deep learning. In: 2017 16th IEEE International Conference on Machine Learning and Applications (ICMLA), pp. 920–923 (2017)
4. Girshick, R.: Fast R-CNN. In: Proceedings of the IEEE International Conference on Computer Vision, pp. 1440–1448 (2015)
5. Girshick, R., Donahue, J., Darrell, T., Malik, J.: Rich feature hierarchies for accurate object detection and semantic segmentation. In: Proceedings of the IEEE Conference on Computer Vision and Pattern Recognition, pp. 580–587 (2014)
6. Hosang, J., Benenson, R., Schiele, B.: A convnet for non-maximum suppression. In: Pattern Recognition: 38th German Conference, GCPR 2016, Hannover, Germany, September 12–15, 2016, Proceedings, vol. 38, pp. 192–204. Springer (2016)
7. Kumar, V., Smith-Renner, A., Findlater, L., Seppi, K., Boyd-Graber, J.: Why didn't you listen to me? Comparing user control of human-in-the-loop topic models. arXiv:1905.09864 (2019)
8. Label Studio contributors: Label Studio. https://labelstud.io/ (2021). Accessed September 2021
9. Lin, T.Y., Dollár, P., Girshick, R., He, K., Hariharan, B., Belongie, S.: Feature pyramid networks for object detection. In: Proceedings of the IEEE Conference on Computer Vision and Pattern Recognition, pp. 2117–2125 (2017)

10. Lin, T.Y., Goyal, P., Girshick, R., He, K., Dollár, P.: Focal loss for dense object detection. In: Proceedings of the IEEE International Conference on Computer Vision, pp. 2980–2988 (2017)
11. Liu, W., Anguelov, D., Erhan, D., Szegedy, C., Reed, S., Fu, C., Berg, A.: SSD: single shot multibox detector. In: European Conference on Computer Vision (ECCV) (2016)
12. Madono, K., Nakano, T., Kobayashi, T., Ogawa, T.: Efficient human-in-the-loop object detection using bi-directional deep sort and annotation-free segment identification. In: 2020 Asia-Pacific Signal and Information Processing Association Annual Summit and Conference (APSIPA ASC), pp. 1226–1233. IEEE (2020)
13. Obaid, H.S., Dheyab, S.A., Sabry, S.S.: The impact of data pre-processing techniques and dimensionality reduction on the accuracy of machine learning. In: 2019 9th Annual Information Technology, Electromechanical Engineering and Microelectronics Conference (IEMECON), pp. 279–283. IEEE (2019)
14. Redmon, J., Divvala, S., Girshick, R., Farhadi, A.: You only look once: unified, real-time object detection. In: Proceedings of the IEEE Conference on Computer Vision and Pattern Recognition, pp. 779–788 (2016)
15. Ren, S., He, K., Girshick, R., Sun, J.: Faster R-CNN: towards real-time object detection with region proposal networks. Adv. Neural Inf. Process. Syst. **28** (2015)
16. Tan, M., Pang, R., Le, Q.V.: Efficientdet: scalable and efficient object detection. In: Proceedings of the IEEE/CVF Conference on Computer Vision and Pattern Recognition, pp. 10781–10790 (2020)
17. Torrey, L., Shavlik, J.: Transfer learning. In: Handbook of Research on Machine Learning Applications and Trends: Algorithms, Methods, and Techniques, pp. 242–264. IGI Global (2010)
18. Uijlings, J.R., Van De Sande, K.E., Gevers, T., Smeulders, A.W.: Selective search for object recognition. Int. J. Comput. Vis. **104**, 154–171 (2013)
19. Wojke, N., Bewley, A., Paulus, D.: Simple online and realtime tracking with a deep association metric. In: 2017 IEEE International Conference on Image Processing (ICIP), pp. 3645–3649. IEEE (2017)
20. Wu, X., Xu, B., Zheng, Y., Ye, H., Yang, J., He, L.: Fast video crowd counting with a temporal aware network. Neurocomputing **403**, 13–20 (2020)
21. Xin, D., Ma, L., Liu, J., Macke, S., Song, S., Parameswaran, A.: Accelerating human-in-the-loop machine learning: challenges and opportunities. In: Proceedings of the Second Workshop on Data Management for End-to-End Machine Learning, pp. 1–4 (2018)
22. Yao, A., Gall, J., Leistner, C., Van Gool, L.: Interactive object detection. In: 2012 IEEE Conference on Computer Vision and Pattern Recognition, pp. 3242–3249. IEEE (2012)
23. Yu, F., Seff, A., Zhang, Y., Song, S., Funkhouser, T., Xiao, J.: LSUN: construction of a large-scale image dataset using deep learning with humans in the loop. arXiv:1506.03365 (2015)
24. Zhang, R., Torabi, F., Guan, L., Ballard, D.H., Stone, P.: Leveraging human guidance for deep reinforcement learning tasks. arXiv:1909.09906 (2019)
25. Zitnick, C.L., Dollár, P.: Edge boxes: locating object proposals from edges. In: Computer Vision–ECCV 2014: 13th European Conference, Zurich, Switzerland, September 6–12, 2014, Proceedings, Part V, vol. 13, pp. 391–405. Springer (2014)

Semi-supervised Semantic Segmentation with Complementary Reconfirmation Mechanism

Yifan Xiao[1], Jing Dong[1(✉)], Qiang Zhang[1,2], Pengfei Yi[1], Rui Liu[1], and Xiaopeng Wei[2(✉)]

[1] The Key Laboratory of Advanced Design and Intelligent Computing, Ministry of Education, School of Software Engineering, DaLian University, Dalian, China
xiaoyifan@s.dlu.edu.cn, dongjing@dlu.edu.cn, zhangq@dlu.edu.cn, istrrr@163.com, liurui@dlu.edu.cn

[2] School of Computer Science and Technology, Dalian University of Technology, Dalian, China
xpwei@dlut.edu.cn

Abstract. Pseudo-labeling based methods are commonly employed for effectively utilizing unlabeled data in semi-supervised semantic image segmentation. It tends to select high-confidence pixels in images as pseudo labels and discard most of the pixels predicted with low confidence, but each pixel is valuable for accurate segmentation. Therefore, we propose a semi-supervised semantic image segmentation algorithm based on complementary reconfirmation mechanism (CR-Seg) to constrain the low-confidence pixels. Firstly, the predictions are divided into high-confidence and low-confidence pixels by a dynamic threshold. The High-confidence pixels supervise the predictions of the student model and guide the classifier to learn what they are. While the low-confidence pixels are equally important for model training, they are used to generate complementary reconfirmation loss for providing additional information as complementary labels. Our method achieves mIoU of 69.12, 73.84, 74.03, and 76.91% under 1/16, 1/8, 1/4, and 1/2 partitions of the classic PASCAL VOC 2012. The experimental results demonstrate that low-confidence pixels can provide more information to the model as complementary labels, thereby improving the model's segmentation performance.

Keywords: Semi-supervised semantic segmentation · Pseudo-labeling · Complementary learning

1 Introduction

Fully supervised semantic segmentation requires the classification of learned pixel semantic labels from many pixel-level annotated images, which is a fundamental task in computer vision. With the improvement of deep neural

N. Naik et al. (Eds.): UKCI 2023, AISC 1453, pp. 182–194, 2024.
https://doi.org/10.1007/978-3-031-47508-5_15

networks[1], we can use many labeled images to train and obtain accurate segmentation models. However, obtaining these labeled images is expensive, which limits the application of semantic segmentation in real-life applications. To alleviate this problem, semi-supervised semantic segmentation uses a small number of labeled images and many unlabeled images to learn segmentation models.

The core of semi-supervised semantic segmentation is how to efficiently utilize unlabeled images, and common approaches use consistency regularization [2–4] and pseudo-labeling [5,6] for unlabeled images. Consistency regularization-based methods compute the difference in predictions of the same unlabeled image after different perturbations as loss values, which are used to enhance the similarity of predictions from different perturbations. The common perturbations are data augmentation [2], and network perturbations [5]. Pseudo-labeling is a specific method of entropy minimization in semi-supervised learning, where the hard pseudo-label generated by the model is used to supervise the prediction of another model. When the model is updated during training, the quality of the pseudo-label is continuously improved, and the quality of the model is enhanced finally.

Recent work [7] combines consistency regularization with pseudo-labeling and achieves performance gains. However, these methods have a limitation of setting confidence thresholds in advance. In the early stage of training, a low threshold can generate many low-quality samples, while a high threshold leads to too few samples and make it difficult for the model to converge. More importantly, low-confidence samples usually contain some categories that are easily confused by the model, they cannot participate in training because of some ambiguous predictions. Therefore, how to improve model performance from low-confidence labels is an important issue for semi-supervised semantic segmentation.

This paper proposes a semi-supervised semantic segmentation algorithm based on complementary reconfirmation mechanism to make use of low-confidence unlabeled data. For the same unlabeled samples with different data augmentation, there are varying degrees of prediction overlap, although they are not identical. The consistent predictions can be given complementary labels to help the model learn the correct categories. Our main contributions can be summarized as follows:

(1) We propose a new training model based on Mean-Teacher, where unlabeled samples are fed into the student and teacher networks via Cutmix, and the predictions are calculated with unsupervised losses and complementary reconfirmation losses, respectively.
(2) We propose a complementary reconfirmation mechanism that imposes consistency regularization on the same low confidence samples through different data augmentation, which makes efficient use of unlabeled data.

The rest of this paper is organized as follows. The related works on semi-supervised semantic segmentation are briefly reviewed in Sect. 2. Section 3 presents the semi-supervised semantic segmentation with complementary reconfirmation mechanism. Section 4 shows the experimental results. Our work is summarized in Sect. 5.

2 Related Work

In this section, we review the work related to semi-supervised learning and semi-supervised semantic segmentation.

2.1 Semi-supervised Learning

The goal of semi-supervised learning is to use the information provided by a small amount of labeled data to represent the distribution of unlabeled data. Consistency regularization [9] and entropy minimization [8,10] are the two main methods of semi-supervised learning. Consistent regularization encourages the model to output the same probability distribution on the same sample for different data augmentation (e.g., color and shape). Entropy minimization-based methods assign pseudo-labels to unlabeled data and jointly label the data to train the model. Recent work [9] combines consistency regularization and entropy minimization to propose a framework to take full advantage of unlabeled data. In addition, we refer to Mean Teacher [8] and FlexMatch [10] in designing the framework.

2.2 Semi-supervised Semantic Segmentation

Early approaches of semi-supervised semantic segmentation used generative adversarial networks [11] to supervise unlabeled data and determine the differences between the generated predictions and the true labeled values. But GAN-based approaches [11] require a lot of time for training and may encounter situations of gradient vanishing, leading to model collapse. Inspired by recent work on semi-supervised learning, methods based on consistency regularization and entropy minimization have achieved good performance. CCT [3] introduces consistency between the outputs of different decoders. CutMix-Seg [2] uses the CutMix [12] data augmentation on unlabeled data to produce greater perturbation on the input and constrains the consistency of the output. CPS [5] uses two models with different initialization parameters to generate two pseudo-labels for cross-supervision and achieve consistency regularization. Reco [7] trains a teacher-student model to achieve entropy minimization using hard pseudo-labels for semantic comparison learning.

The above methods usually use all pseudo-labels or partial pseudo-labels obtained by threshold filter for supervised training. However, using all pseudo-labels for training increases the uncertainty of the model, likewise using only high-confidence labels to participate constraint causes the waste of low-confidence samples. Our proposed method tries to make use of unlabeled data to obtain advanced performance.

3 Approach

3.1 Overview

Semi-supervised semantic segmentation is one of the common vision tasks. The dataset x consisting of labeled images x^l with label y and unlabeled images x^u is used to train the segmentation network. This method uses data augmentation on the unlabeled images to obtain the augmented images x^{aug}. The labeled image x^l, the unlabeled image x^u and the augmented image x^{aug} are used as inputs to the segmentation network for training.

The semi-supervised semantic segmentation model based on complementary reconfirmation mechanism follows a self-training strategy, and the network architecture is shown in Fig. 1. The teacher and student models have the same network architecture with different weights. The weights θ_1 of the student model are updated in backpropagation, and the weights θ_2 of the teacher model are updated by an exponential moving average EMA of the student model weights:

$$\theta_2^t = \alpha\theta_2^{t-1} + (1-\alpha)\theta_1^t, \tag{1}$$

t is the current training epoch, α is the smoothing factor.

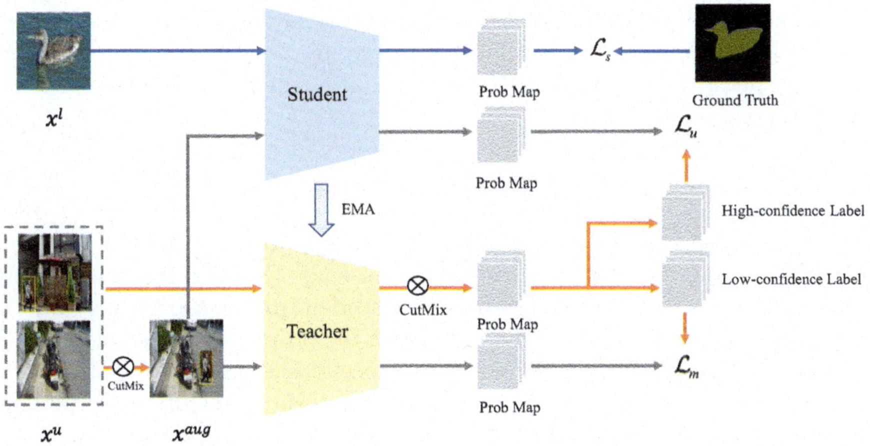

Fig. 1. Semi-supervised semantic segmentation based on complementary reconfirmation mechanism

The teacher model and the student model consist of a CNN-based encoder and a decoder with a segmentation header. Following the training strategies of Fixmatch [6] and Flexmatch [10], N labeled images and unlabeled images are selected for training respectively. For the labeled images, the supervised loss $\mathcal{L}_s$ is used to optimize student model. For unlabeled images, x^u is fed into the teacher model after weak data augmentation to obtain N probability feature maps, which are divided into high-confidence labels and low-confidence labels after Cutmix. From another branch of x^u, x^{aug} is obtained after CutMix and input into the student model and the teacher model for prediction respectively. The unsupervised loss $\mathcal{L}_u$ is calculated with the prediction results of the student model and high confidence pseudo-labels, which will be introduced in Sect. 3.2. The complementary reconfirmation loss $\mathcal{L}_m$ is calculated with the prediction results of the teacher model and low confidence pseudo-label, which will be presented in Sect. 3.3. In summary, the overall loss is:

$$\mathcal{L} = \mathcal{L}_s + \lambda \mathcal{L}_u + \gamma \mathcal{L}_m, \tag{2}$$

where λ is 0.2 and γ is 0.1, they are weighted hyper-parameters used to balance the importance of the corresponding losses. The supervisory loss $\mathcal{L}_s$ is the cross-entropy between the labels y and the predictions of the student model. It is calculated as follows:

$$\mathcal{L}_s = \frac{1}{N_l} \sum_{i=1}^{N_l} \frac{1}{W \times H} \sum_{j=1}^{W \times H} l_{ce} \left(f_{stu} \left(x_{ij}^l ; \theta_1 \right), y_{ij} \right), \tag{3}$$

N_l denotes the number of labeled images in training, W and H are the width and height of the input image. The encoder and decoder form the segmentation network f, f_{tea} and f_{stu} denote the teacher model segmentation network and the student model segmentation network, respectively. x_{ij}^l denotes the predicted value of the j-th pixel in the i-th labeled image, y_{ij} denotes the annotated label of the j-th pixel in the i-th labeled image, and l_{ce} denotes the standard cross-entropy loss.

3.2 Dynamic Classification of Pseudo Labels

In the early stages of training, the teacher model produces a large number of inaccurate pseudo-labels with noise, which misleads the model and interferes in its optimization direction. Furthermore, the student model learns noisy information passed from the teacher model, leading to the propagation and accumulation of errors. Therefore, filtering low confidence pseudo-labels is essential. However, FixMatch [6] and other work [7] use a fixed threshold to filter pseudo-labels, and only high-quality unlabeled data are usable for loss. It ignores some of the unlabeled data.

Drawing on previous work [4], the Shannon entropy is used as a basis for discerning pseudo-label uncertainty. The entropy $U(p_{ij})$ is spread as a one-dimensional array and its quantile of $(1 - a_t)$ is taken as the threshold of the t-th epoch γ_t. A linear function is used to generate the scale value a_t with the following equation:

$$a_t = a_0 \cdot \left(1 - \frac{t}{total\ epoch} \right). \tag{4}$$

The pseudo-labels are divided into high confidence pseudo-labels y_{ij}^h and low confidence pseudo-labels y_{ij}^l by means of a threshold γ_t. The high confidence region is considered as the true label to guide the model training. The high confidence pseudo-label y_{ij}^h is defined as:

$$y_{ij}^h = \begin{cases} argmax p_{ij}(c), & \text{if } U(p_{ij}) < \gamma_t, \\ \text{ignore}, & \text{otherwise}, \end{cases} \tag{5}$$

p_{ij} denotes the predicted vector feature, and when the Shannon entropy of the pixel in row i and column j of an image is lower than the threshold γ_t, we consider it to have less variation in pixel values. Confusing pixels usually refer to high frequency information such as edges and textures in an image. Therefore, if the pixel has a low Shannon entropy, it means that the pixel's grey value does not vary much, which means that the pixel is in a flat region or low frequency region and has a higher reliability and certainty compared to high frequency information. The high confidence pseudo-label y_{ij}^h is used for unsupervised constraints:

$$\mathcal{L}_u = \frac{1}{N_u} \sum_{i=1}^{N_u} \frac{1}{W \times H} \sum_{j=1}^{W \times H} l_{ce} \left(f_{stu} \left(x_{ij}^u; \theta_1 \right), y_{ij}^h \right), \tag{6}$$

N_u represents the number of unlabeled images in the training.

For low confidence regions, it also contains valuable information for segmentation, the complementary reconfirmation mechanism utilizes low-confidence pseudo-labels to prevent information loss. Low confidence pseudo-labels y_{ij}^l is defined as:

$$y_{ij}^l = \begin{cases} \text{ignore} , & \text{otherwise}, \\ argmaxp_{ij}(c), & \text{if } U\left(p_{ij}\right) > \gamma_t. \end{cases} \tag{7}$$

3.3 Complementary Reconfirmation Mechanism

Recent semi-supervised semantic segmentation algorithms based on consistent regularization only use samples with high confidence predictions to train the model, which can lead to inefficient use of unlabeled data, especially in the early stages of the training process. However, this wastage of low confidence predictions can affect the final performance of the model. Therefore, the central question is how to use low confidence samples for learning.

As shown in Fig. 2, on the PASCAL VOC 2012 dataset, the unlabeled image x^u is input to the segmentation model f_{tea} to obtain a confidence segmentation map for generating high and low confidence labels. After the CutMix operation, the blended image x^{aug} is also input to the segmentation model f_tea to obtain the predicted probability map. The label categories are ranked in order of prediction probability, and the top two categories (Dog, Sofa) are labeled as high confidence labels to calculate the unsupervised loss $\mathcal{L}_u$. The predictions of the top-k categories in the low confidence labels are used to complement the probability map generated by x^{aug} to reconfirm the loss constraint. The top-k categories such as Horse, Sheep and Table are easily confused categories, the probability maps generated by x^{aug} also contain predictions from Cow, Sheep and Table. Based on the above findings, some overlap in the order of similarity can be shared for strongly augmented versions of the images. We devised a paradigm where high confidence samples could guide the model to learn what it is, and low confidence samples could guide the model to distinguish its confusable categories for complementary labeling.

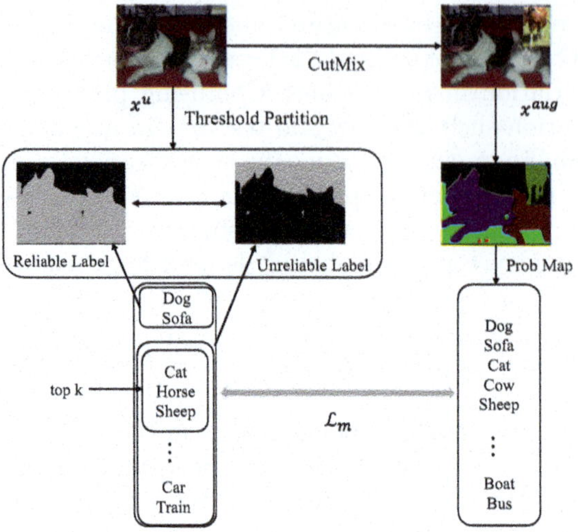

Fig. 2. Complementary reconfirmation mechanism

Similar to the consistency loss used for common high-confidence labels, we impose consistency predictions on unlabeled samples from two different data-augmented versions for complementary learning. These available low-confidence labels can facilitate segmentation model learning with less error information. The complementary reconfirmation loss $\mathcal{L}_m$ is defined as follows:

$$\mathcal{L}_m = \frac{1}{N_u} \sum_{i=1}^{N_u} \frac{1}{W \times H} \sum_{j=1}^{W \times H} l_{ce} \left(f_{tea} \left(x_{ij}^{aug}; \theta_2 \right), G_k(y_{ij}^l) \right). \tag{8}$$

The strongly data-augmented x^{aug} is fed into the teacher model with weight θ_2. The low confidence label y_{ij}^l is obtained from thresholded by Eq. (7). $G_k(m)$ denotes the largest m values in each row of the return x, i.e. *torch.topk*(x). In this paper, only the top-3 predictions of the low confidence pseudo-label are selected for training, which focuses on those more easily confused regions and are more beneficial for complementary learning.

4 Experiment

4.1 Experiment Settings

Datasets. PASCAL VOC 2012 [13] contains 20 foreground classes and 1 background. The training and validation sets consist of 1464 and 1449 images respectively. As in previous work [4], the SBD [15] was used as a supplementary dataset with 9118 additional training images. Cityscapes [14] is commonly used for real-world scene understanding, there are 19 categories. It consists of 2975 training images, 500 validation images and 1525 test images.

Experiment Details. We used DeepLabv3+ [16] and ResNet-101 [17] as the segmentation model. When training on the PASCAL VOC 2012, the initial learning rate of 0.0005, weight decay of 0.0001, batch size of 2, cropped image size of 321 × 321 and training period of 100. When training on the Cityscapes, the initial learning rate is 0.005, the weight decay is 0.0005, the batch size is 2, the cropped image size is 769 × 769, and the training period is 250.

Evaluation Indicators. Mean Intersection over Union (mIoU) is one of the most used metrics. Each pixel in the image is determined the semantic class to which it belongs, and then the intersection between the prediction and the label is calculated.

4.2 Results of Semi-supervised Semantic Segmentation

Our method was compared with recent semi-supervised semantic segmentation methods: models such as Mean Teacher [8], CutMix-Seg [2], Reco [7], CPS [5] and U2PL [4]. The classic and blender PASCAL VOC differ in the training set, and they both have the same validation set of 1449 images.

Results on the Classic PASCAL VOC 2012. Figure 3 shows the results of different algorithms on the classic PASCAL VOC 2012 dataset. The accuracy curve transformation of CR-Seg is smoother, and when training with 1/16 and 1/8 labeled samples, and CR-Seg has a high accuracy gain compared to other works. When there are only a small number of labeled samples, the knowledge used to guide the learning of the segmentation model is limited. Using low confidence pseudo-labels for complementary reconfirmation can effectively provide additional information for the model to learn, thus improving the accuracy of the model. The experimental results demonstrate that CR-Seg is more suitable for situations where the number of labels is sparse.

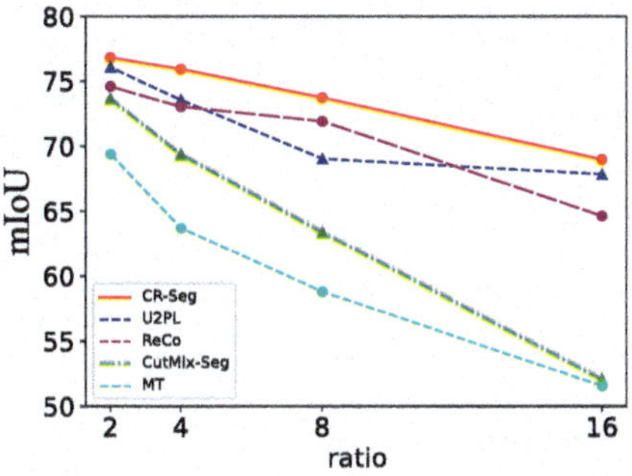

Fig. 3. Results of different methods on the classic PASCAL VOC 2012

Table 1 compares the results of CR-Seg with other methods on the classic PASCAL VOC 2012. It improved the most under 1/16 and 1/8 partitioning. Under 1/4 and 1/2 partitioning, the accuracy gains of CR-Seg and the other methods become smaller. As the models tend to train more towards fully supervised as the input labeled samples increase, the effectiveness of the semi-supervised algorithm becomes limited.

Table 1. Results on the model (ResNet-101) on the classic PASCAL VOC 2012

Method	1/16(92)	1/8(183)	1/4(366)	1/2(732)
SupOnly	46.21	54.37	66.46	71.89
MT [8]	51.72	58.93	63.86	69.51
CutMix-Seg [2]	52.16	63.47	69.46	73.73
ReCo [7]	64.78	72.02	73.14	74.69
U2PL [4]	67.98	69.15	73.66	76.16
CR-Seg	69.12	73.84	74.03	76.91

Results on the Blender PASCAL VOC 2012. Table 2 shows the results of the comparison on the blender PASCAL VOC 2012 dataset. Under the 1/16 and 1/8 partitioning protocols, CR-Seg improves by 1.47 and 1.84% compared to CPS.

Table 2. Results of the model (ResNet-101) on the blender PASCAL VOC 2012

Method	1/16(662)	1/8(1323)	1/4(2646)	1/2(5291)
SupOnly	67.97	71.66	76.23	77.22
MT [8]	70.51	71.53	73.02	76.58
CutMix-Seg [2]	71.66	75.51	77.33	78.21
CCT [3]	71.86	73.68	76.51	77.40
CPS [5]	74.48	76.44	77.68	78.64
CR-Seg	75.95	78.28	78.72	79.16

Results on the Cityscapes. Table 3 shows the results of the experiments on the Cityscapes dataset. It can be seen that the accuracy improvement of CR-Seg and other methods over SupOnly is limited because the Cityscapes dataset has a long-tailed distribution, causing the model's predictions to be biased towards the category with a larger sample size.

Table 3. Results of the model (ResNet-101) on Cityscapes

Method	1/16(186)	1/8(372)	1/4(744)	1/2(1488)
SupOnly	64.28	71.56	73.27	77.16
MT [8]	69.03	72.06	74.20	78.15
CutMix-Seg [2]	67.06	71.83	76.36	78.25
CCT [3]	69.32	74.12	75.99	78.10
CPS [5]	69.78	74.31	74.58	76.81
CR-Seg	69.78	74.25	75.38	78.94

4.3 Ablation Experiment

To demonstrate that pseudo-labeling of low-confidence samples is beneficial for training semi-supervised semantic segmentation models, we set up experiments with different confidence levels of labeling. We selected labeled samples at 1/16 and 1/8 scale for the ablation experiments. Table 4 shows the results on the classical PASCAL VOC 2012 validation set, where unsupervised loss improves the baseline model accuracy by 19.68% at 1/16 and 13.09% at 1/8 partitioning. The model improves by 3.23 and 6.38% under 1/16 and 1/8 partitioning using complementary reconfirmation loss compared to supervised and unsupervised loss, demonstrating the effectiveness of complementary reconfirmation with low confidence samples.

Table 4. Ablation experiment using samples of different loss functions

$\mathcal{L}_m$	$\mathcal{L}_u$	$\mathcal{L}_m$	mIoU(1/16)	mIoU(1/8)
√			46.21	54.37
√	√		65.89	67.46
√	√	√	69.12	73.84

4.4 Visualization Results

Visualization results are shown to further analyze our approach. The model was trained on 183 labeled samples and approximately 10,400 unlabeled samples. As shown in Fig. 4, training the model in a supervised only (SupOnly) appears to be fragile with limited labeled data. The CR-Seg showed much better performance, whether it was a car on the street or a pedestrian on the pavement. Our method is almost equivalent to the segmentation results of the labels in terms of determining a clear outline of the object and identifying the corresponding class.

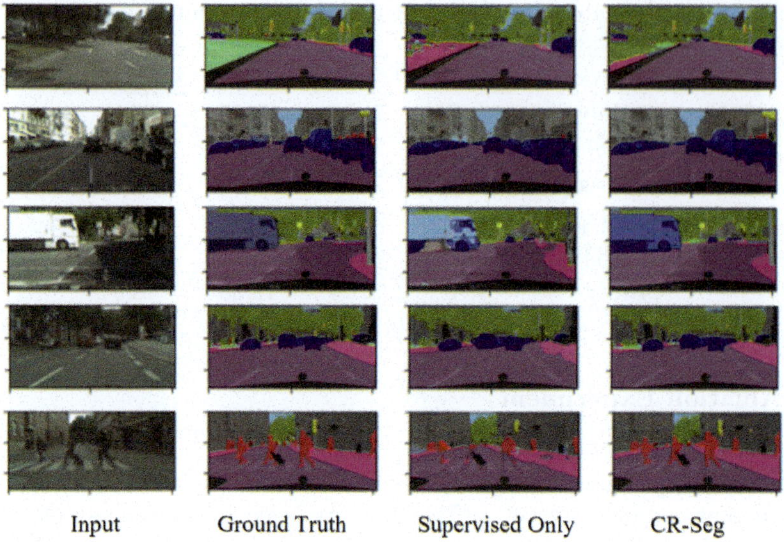

Fig. 4. Visualization results in Cityscapes

5 Conclusion

This paper proposes a semi-supervised semantic segmentation method based on complementary reconfirmation, which effectively reduces the use of labeled data and makes efficient use of pseudo-labels. High confidence pixels are used to supervise the predictions of the student model and low confidence pixels is used as complementary labels to allow the model to learn other categories for generating complementary reconfirmation loss, thus improving the segmentation performance of the model. Experiments on PASCAL VOC 2012 and Cityscapes were conducted to demonstrate the effectiveness of the CR-Seg model.

Acknowledgment. This work is supported in part by the Key Program of NSFC (No. U1908214), 111 Project (No. D23006), the Scientific Research Fundation of Education Department of Liaoning Province (No. LJKMZ20221839), the Science and Technology Innovation Fund of Dalian (No. 2020JJ25CY001), Program for Innovative Research Team in University of Liaoning Province (LT2020015), Support Plan for Key Field Innovation Team of Dalian (2021RT06).

References

1. Wang, J., Sun, K., Cheng, T., Jiang, B., Deng, C., Zhao, Y., Liu, D., Mu, Y., Tan, M., Wang, X.: Deep high-resolution representation learning for visual recognition. IEEE Trans. Pattern Anal. Mach. Intell. **43**(10), 3349–3364 (2020)
2. French, G., Laine, S., Aila, T., Mackiewicz, M., Finlayson, G.: Semi-supervised semantic segmentation needs strong, varied perturbations. In: British Machine Vision Conference, pp. 1–21 (2019)

3. Ouali, Y., Hudelot, C., Tami, M.: Semi-supervised semantic segmentation with cross-consistency training. In: The IEEE/CVF Conference on Computer Vision and Pattern Recognition, pp. 12,674–12,684 (2020)
4. Wang, Y., Wang, H., Shen, Y., Fei, J., Li, W., Jin, G., Wu, L., Zhao, R., Le, X.: Semi-supervised semantic segmentation using unreliable pseudo-labels. In: The IEEE/CVF Conference on Computer Vision and Pattern Recognition, pp. 4248–4257 (2022)
5. Chen, X., Yuan, Y., Zeng, G., Wang, J.: Semi-supervised semantic segmentation with cross pseudo supervision. In: The IEEE/CVF Conference on Computer Vision and Pattern Recognition, pp. 2613–2622 (2021)
6. Sohn, K., Berthelot, D., Carlini, N., Zhang, Z., Zhang, H., Raffel, C.A., Cubuk, E.D., Kurakin, A., Li, C.L.: Fixmatch: simplifying semi-supervised learning with consistency and confidence. Adv. Neural. Inf. Process. Syst. **33**, 596–608 (2020)
7. Liu, S., Zhi, S., Johns, E., Davison, A.J.: Bootstrapping semantic segmentation with regional contrast. In: International Conference on Learning Representations, pp. 774–791 (2021)
8. Tarvainen, A., Valpola, H.: Mean teachers are better role models: Weight-averaged consistency targets improve semi-supervised deep learning results. Adv. Neural. Inf. Process. Syst. **30**, 774–790 (2017)
9. Duan, Y., Zhao, Z., Qi, L., Wang, L., Zhou, L., Shi, Y., Gao, Y.: Mutexmatch: semi-supervised learning with mutex-based consistency regularization. IEEE Trans. Neural Netw. Learn. Syst. (2022)
10. Zhang, B., Wang, Y., Hou, W., Wu, H., Wang, J., Okumura, M., Shinozaki, T.: Flexmatch: boosting semi-supervised learning with curriculum pseudo labeling. Adv. Neural. Inf. Process. Syst. **34**, 18408–18419 (2021)
11. Souly, N., Spampinato, C., Shah, M.: Semi supervised semantic segmentation using generative adversarial network. In: Proceedings of the IEEE International Conference on Computer Vision, pp. 5688–5696 (2017)
12. Yun, S., Han, D., Oh, S.J., Chun, S., Choe, J., Yoo, Y.: Cutmix: Regularization strategy to train strong classifiers with localizable features. In: The IEEE/CVF International Conference on Computer Vision, pp. 6023–6032 (2019)
13. Everingham, M., Eslami, S.A., Van Gool, L., Williams, C.K., Winn, J., Zisserman, A.: The pascal visual object classes challenge: a retrospective. Int. J. Comput. Vision **111**, 98–136 (2015)
14. Cordts, M., Omran, M., Ramos, S. et al.: The cityscapes dataset for semantic urban scene understanding. In: The IEEE Conference on Computer Vision and Pattern Recognition, pp. 3213–3223 (2016)
15. Hariharan, B., Arbeláez, P., Bourdev, L., Maji, S., Malik, J.: Semantic contours from inverse detectors. In: 2011 International Conference on Computer Vision, pp. 991–998. IEEE (2011)

16. Chen, L.C., Zhu, Y., Papandreou, G., Schroff, F., Adam, H.: Encoder-decoder with atrous separable convolution for semantic image segmentation. In: The European Conference on Computer Vision (ECCV), pp. 801–818 (2018)
17. He, K., Zhang, X., Ren, S., Sun, J.: Deep residual learning for image recognition. In: The IEEE Conference on Computer Vision and Pattern Recognition, pp. 770–778 (2016)

Towards the Use of Machine Learning Classifiers for Human Activity Recognition Using Accelerometer and Heart Rate Data from ActiGraph

Matthew Oyeleye[1], Tianhua Chen[1](✉), Pan Su[2], and Grigoris Antoniou[1]

[1] Department of Computer Science, School of Computing and Engineering, University of Huddersfield, Huddersfield, UK
t.chen@hud.ac.uk

[2] School of Control and Computer Engineering, North China Electric Power University, Baoding, China

Abstract. Human Activity Recognition (HAR) aims at detecting human physical activities such as eating, running, laying down, sitting, etc., through sensor-generated data. With the ubiquitous nature and utilization of sensor-enabled devices such as smartphones, smartwatches, and wristbands in daily life, numerous modern applications have been developed and implemented in HAR around the world. In this study, rather than using only accelerometry data generated from smartphones which are more commonly adopted in recent literature, we aim to predict human activities using an accelerometer and heart rate (HR) data generated by Actigraph, as they can accurately measure moderate-to-vigorous intensity physical which is mostly affected by body composition and also better suited for self-monitoring. For this purpose, we explored the effectiveness of these features through the application of machine learning classifiers. A very recently publicly available Actigraph-generated data (MMASH) that contains accelerometer and HR recordings were used in the experiments. To evaluate the effectiveness of different indicators for recognising human activities, we performed a series of four experiments. In working towards recognising four activities, the best-performing machine learning models achieved an averaged accuracy value of 67±11% through using HR as a significant feature. The result shows that HR provides more information that can be used to predict better human activity recognition.

Keywords: Human activity recognition · Machine learning · Heart rate · Accelerometer · Actigraph

N. Naik et al. (Eds.): UKCI 2023, AISC 1453, pp. 195–208, 2024.
https://doi.org/10.1007/978-3-031-47508-5_16

1 Introduction

The use of sensors has greatly helped in Human Activity Recognition (HAR), which has played an important role in establishing the user's relationship with its immediate environment. They have therefore been adopted as aiding technology for chronic illnesses and medical diagnosis, healthcare, active and assisted living, home automation, and building an adaptive human-computer interaction system [1–3] which in turn has made HAR a more dynamic and challenging research area.

Human daily activities are mostly impulsive. Due to the fact that individual predilections and life practices, people tend to display dissimilar behavioral traits even in similar activities. HAR can be a challenging and arduous task in extracting a subject's behavioral characteristics when the acquisition sensor is unable to distinctively differentiate between the subject's major features.

There are two broad classifications of HAR: vision-based and sensor-based activity recognition [4]. In the Video-based HAR, human activities and behavior are recognized using the images and videos captured by cameras. In Computer Vision (CV) based HAR, human-body monitoring technology must be adopted in such a way that the subject must remain in the field of view. This has subjected the CV-based HAR to certain limitations such as—suitable environmental conditions (such as illumination, light rays, etc.), complex background obstruction, multiple individual tracking in a single image, and concurrent recognition as well as obstructed targets. The operating process of CV-based also requires voluminous data processing that requires enormous computational power and storage area, hence increasing its gross overhead [5]. The sensor-based HAR on the other hand is found more feasible in capturing the human activity data than the vision-based HAR. Wearable sensors and smartphone sensors are mostly used in the field of HAR due to their low cost and invasive feature and easy installation. Also, wearable sensor-based HAR methods are not restricted to the effect of environmental factors, huge device deployment costs, and computational power.

Nowadays, sensors such as accelerometers, gyroscopes, and magnetometers, which can be used to track a user's 3-axial linear acceleration, and also 3-axial angular velocity are broadly integrated with mobile devices such as smartphones, smartwatches, and arm and wristbands that can be easily carried around by people. Although wearable sensors can record individual information more accurately, they also possess disadvantages. For instance, an individual can feel uncomfortable having a wearable device on for a long period, and it can be unfitting for identifying complex behaviors. However, in this study, we focused on a wearable sensor-based device called ActiGraph wGT3X-BT.

An actiGraph is a wearable device that is integrated with sensors that measure external forces along a reference axis for accurate monitoring of human activities. Rather than just capturing only the user's linear acceleration data, the device can also capture the user's heartbeat/heart rate, total number of movements, energy expenditure, and even sleep data. Actigraph generates data in a time-series streaming format which can be processed on a row-by-row basis by time progression [6].

This study is motivated by the indispensable process of real-time assessing individuals' health risks such as cardiovascular diseases, musculoskeletal disorders, and stress by monitoring and recognizing their daily physical activity [7] taking the advantage of rise in advanced technology such as wearable monitoring most especially in the use of heart rate for monitoring patient well being [6,8,9]. For example, real-time assessment and monitoring can significantly enhance establishing the doctor-patient connection while also saving healthcare costs. Also, it provides great support and functionalities in smart home healthcare applications that involve frequent need and constant monitoring of the environment and resident behavior through the use of an intelligent system. Furthermore, it can also provide aid to a patient's recovery by giving him feedback.

Although many studies have been conducted in the area of human activity recognition using several approaches, to the best of our knowledge, none of the existing studies used data on the tri-axial and heart rate generated by an actigraph as a feature for predicting or classifying human activities. Rather, they commonly adopt the use of only tri-axial (x, y, and z) sensor data collected from a smartphone [4,10]. For this reason, in this study, we explored the processing power of ML classifiers to investigate the use of actigraph-generated time-series tri-axial (x, y, and z), heart rate (HR), and step count data in the prediction and classifying of 4 daily human activities.

The rest of the paper is organised as follows. Section 2 introduces the related work and motivations for this research. Section 3 presents the experimental outcomes with discussion. Section 4 concludes the paper and points out future research direction.

2 Related Work

Several studies have been directed at HAR in numerous facets in the recent past using data recorded by wearable sensors data [11]. Various ML-based models have been applied in human activities recognition and these models performed excellently well.

Generally, there are four major steps involved in the HAR conceptual framework: data collection and sensors are used to collect the data in time-series format. The second phase is data preprocessing and segmentation, this involves using an established fixed-length sliding window approach to split the time series sensor data into segments of the same length. The third and most pivotal phase in HAR is the feature extraction phase. This is aimed at extracting features from the data segments derived from phase two. And finally, the last phase is the activity classification process. This involves the use of Machine Learning (ML) based methods like Gradient Boosting (GB), Decision Tree (DT), Support Vector Machine (SVM), Multi-layer Perceptron (MLP), etc. to infer human activities. Nevertheless, the performance of these classifiers can be stirred by large intra-class and little inter-class similarities [12].

A study by [13] applied multi-class SVM on triaxial linear acceleration and angular velocity signals recorded by smartphone to classify six activities (e.g., standing, sitting, laying down, walking, walking downstairs and upstairs) performed by 30 people in a controlled environment. They released the dataset from their experiment for public use. In another study by [14], the authors developed an ensemble of classifiers combining J48 DT, MLP, and LR techniques. The ensemble of classifiers was validated on public WISDM data consisting of six daily activities mentioned earlier. [15] integrated three classifiers (i.e., MLP, SVM, LogitBoost) for HAR using accelerometer data recorded by smartphones. The experiment was carried out on four people who engaged in six activities (slow walking, fast walking, running, stairs-up, stairs-down, and dancing) and they outlined that the combination performed excellently well when the phone is in-pocket position.

Also, in a study by [16], statistical features (such as mean and standard deviation) are extracted from the raw sensor (accelerometer) data generated from a smartphone. Several multiple classifiers (i.e., logistic regression (LR), DT, and MLP) are analyzed, and MLP performed best. A novel HAR system with multiple combined features to monitor human physical movements from continuous sequences via tri-axial inertial sensors was proposed by [17] using a DT classifier for recognition. The proposed system was experimented with and tested on three benchmark datasets, the result showed an extraordinary level of performance when compared to conventional solutions.

The study by [18] aimed at improving the quality of life most especially in elderly people with disabilities by accurately monitoring their daily activities. The authors explored 12 classification models. The classifiers are validated on 3 datasets containing activities in indoor environments. The results from their experiments show that two classifiers i.e., LR and OneR have quality metrics higher than 90%. [17] proposed a granular computing-based approach for classifying human activities. Their approach was evaluated by applying 5 ML classifiers i.e., K-nearest neighbors (KNN), MLP, SVM, Naive Bayes, and random forest (RF) on a public smartphone dataset containing 6 activities. They reported that their approach performed better than the traditional classifiers.

The accuracy of steps measured by ActiGraph-GT3X was compared with the steps measured by WeRun (a smartphones-based) by [19], the results from their experiment show that ActiGraph-GT3X is not constrained to the body composition, hence measures user's steps more accurately. Also, [20] pragmatically compared the effectiveness of using ActiGraph accelerometry and Consumer-Based activity trackers in estimating the sedentary behavior of people in a free-living environment. Their results indicate that both are excellently suited for self-monitoring.

Despite the numerous studies and research directed at predicting and classifying human activities, none of these studies had attempted the use of actigraph-generated accelerometry tri-axial and heart rate data to predict or classify daily human activities. Rather they mostly rely on only accelerometry tri-axial data collected using smartphones [4,10].

A Multilevel Monitoring of Activity and Sleep in Healthy People (MMASH) dataset was presented by [21] with the aim of providing physiological data which in turn can be used to acquire a comprehensible view of a person's medical condition and behavior through the use of wearable devices The MMASH dataset provides 24 h of continuous psycho-physiological data such as heart rate, wrist accelerometry, and physical activity data recorded by Actigraph; in addition with sleep quality index and psychological characteristics (e.g., anxiety status, stressful events, and emotion declaration) data. This further evidenced the advanced use of wearable devices in the analysis of human activities and health monitoring. However, limited analysis had been attempted with this dataset. For instance, a study by [22] explored the data to review stress detection and sleep quality prediction in an academic environment. Also, it was adopted by [23] to survey how to manage perceived loneliness and social isolation levels for older adults. In another study by [24], the dataset was used for personalized recognition of wake/state using shapelets and the K-means algorithm.

To the best of our knowledge, none of the existing studies have explored the MMASH dataset in predicting daily activities. For this reason, we are motivated to explore this state-of-the-art dataset more widely in the prediction of human daily activities in this study.

3 Materials and Methods

As mentioned earlier in Sect. 1, HAR is much of a pattern recognition task, that consists of data collection, data preprocessing and segmentation, feature extraction, and lastly activity classification. As seen in Fig. 1, the ActiGraph wGT3X-BT generated time series tri-axial, heart rate, and step data was extracted from the MMASH dataset. The extracted data are preprocessed and split into fixed-length sliding window segments in the second stage. Thirdly we used traditional feature extraction methods to transform the segmented dataset. This process manually pulls out statistically important feature vectors from the segmented data. And finally, we explored 10-fold StratifiedKFold cross-validation to split the transformed data into testing and training subsets. These data subsets are then used by the ML classifiers to predict and classify the daily human activities and the accuracy of each model is measured.

3.1 Data Collection

In this study, we used a publicly available Multilevel Monitoring of Activity and Sleep in Healthy People (MMASH) dataset [21] which provided 24-hour continuous actigraph data collected from 22 healthy young males. The participants wore the Actigraph (Actigraph wGT3X-BT-Actigraph LLC, Pensacola, FL, USA) continuously for 24 h: (between 9:00 a.m. and 9:00 p.m. on the next day) during both the day (during physical activities too) and at night. The dataset readings were recorded on a second-second basis for each participant. The Actigraph device captured each participant's accelerometer three x, y, and

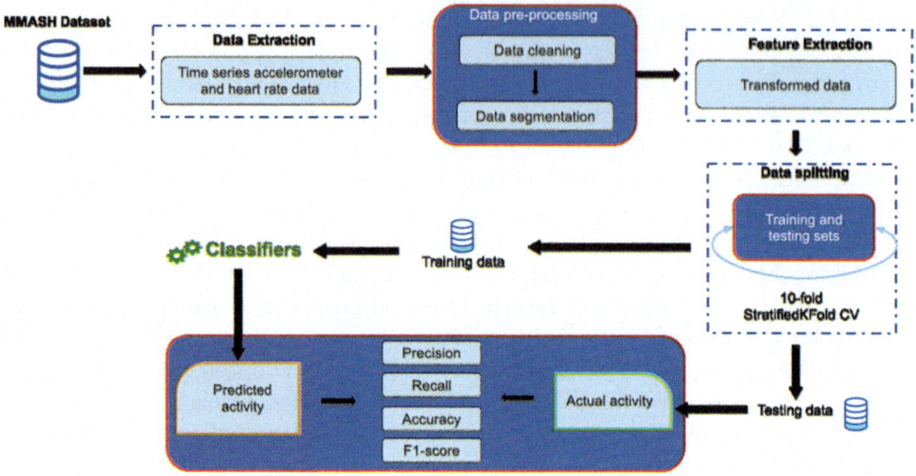

Fig. 1. Methodology design

z-directions, steps, and HR. Activities engaged by each participant were also recorded. However, the frequency at which the data was collected was not mentioned by the authors of the dataset. It was only reported on a second-second time series basis.

There are 12 overall activities recorded by the 22 participants which are: sleeping, laying down, sitting (e.g. studying, eating, and driving), light movement (e.g. slow/medium walk, chores and work), medium (e.g. fast walk and bike), heavy (e.g. gym, running), eating, small screen usage (e.g. smartphone and computer), large screen usage (e.g. TV and cinema), caffeinated drink consumption (e.g. coffee or coke), smoking, and alcohol consumption.

There is a total of 7 feature variables, 'Participant', 'Time', 'Axis1', 'Axis2', 'Axis3', 'Steps', and 'HR' (where 'Axis1' is the x-axis, and 'Axis2' is the y-axis, and 'Axis3' is the y-axis). Our target variable is 'activity' which we intend to predict.

3.2 Data Preprocessing and Segmentation

3.2.1 Data Preprocessing The first phase of our data preprocessing is data cleaning as described in Fig. 1, this is done by dropping null values. We performed an Exploratory Data Analysis (EDA) to analyze and understand the dataset. As seen in Fig. 2, we can observe that there is a significant class imbalance in the activity distribution with the majority of the samples having the class labels 'sitting', 'small screen usage', 'light movement' and 'sleeping', where 'smoking', 'medium movement', 'heavy movement', 'laying down', and 'alcohol consumption' are least represented in the dataset.

To further understand the dataset, we observed each user's contribution to each activity. As seen in Fig. 3, we observed that the participants didn't engage

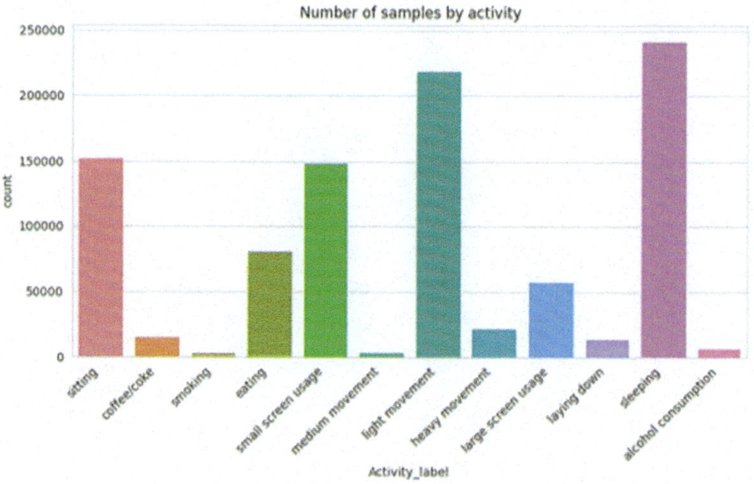

Fig. 2. Activity class distribution

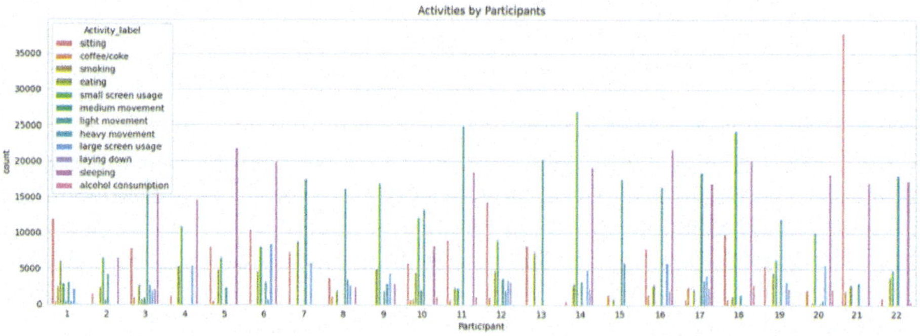

Fig. 3. Activity class distribution by participants

in all the activities. Participants 1 and 2 engaged in 10 of the activities. The highly significant class imbalance in the dataset will make the models biased towards the majority class only. For these reasons, we selected Participant 3 data for the experimental purpose having a high number of activities and a considerably minimal variance in activities distribution compared to Participant 1 data. Firstly, we dropped all activities with very high counts, we calculated the average count of the remaining activities and then selected activities whose counts are equal to or greater than the average activity. This process yielded 4 activities i.e., eating, heavy movement, laying down, and, large screen usage. We further investigated the separability of the 4 activity classes using the t-distributed Stochastic Neighbor Embedding (t-SNE) algorithm. As seen clearly in the TSNE cluster in Fig. 4, the four activities are not distinctively separated.

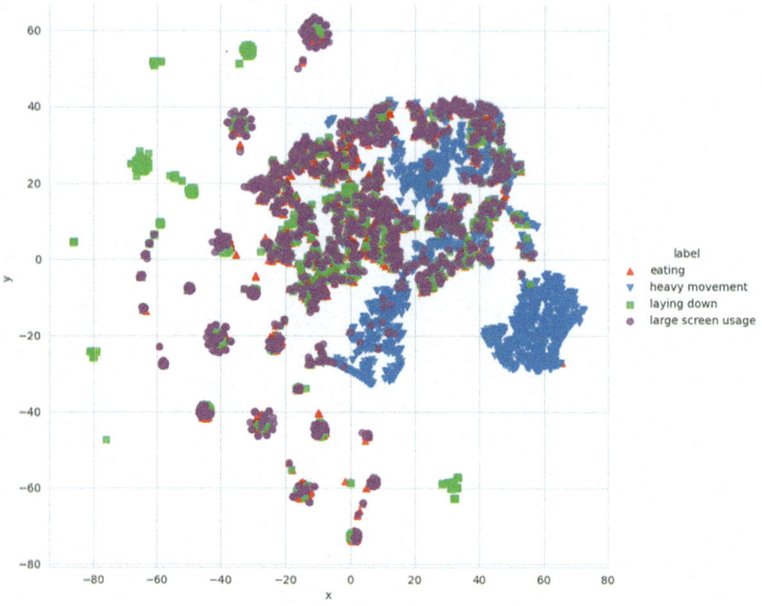

Fig. 4. TSNE activity separability cluster distribution

3.2.2 Time-Series Data Segmentation As mentioned in Sect. 1, the raw data generated by an actigraph is time-series data. Standard classification models are not theoretically appropriate to be directly applied to these raw time-series data [16]. For this reason, the time-series data need to be transformed into a format suitable for the ML classifiers. The raw time-series data is transformed using the 'windowing' technique. This technique allows us to divide the data into 5 s windows and produce new features from the 5-second segmentation. We assigned the class label for the transformed features by taking the most occurring activity in that window. A 5-second window size can be considered optimal for capturing the repeated movements involved in most of the activities as the movement may not be adequately captured with a limited window size while the transformed data may result in less data-point for the model training [16]. Rather than taking distinct windows, we also took overlapping windows with 3 s overlap which ensures that every succeeding row is provided with information from data in the preceding window.

3.3 Feature Extraction

And finally, for our feature engineering process, we explored the following 22 statistical features i.e. mean, weighted average, harmonic mean, geometric mean, variance, standard deviation, average absolute deviation, minimum value, maximum value, the difference of maximum and minimum values, median, median absolute deviation, interquartile range, negative values count, positive values count, number of values above mean, number of peaks, skewness, kurtosis, energy, average resultant acceleration, signal magnitude area, and applied it on the 'Axis1', 'Axis2', 'Axis3', 'Steps', and 'HR' features of our transformed dataset. This process aims to prepare the raw time-series data to best fit the standard classification models so that the models can uncover the hidden patterns in the time-series data and also improve the models' predictive power.

3.4 Machine Learning Classifiers

The machine learning classifiers learn the relationship function between sets of input and output sequences known as predictors and target variables, denoted by X and y respectively, in making predictions [6]. Artificial Intelligence and machine learning have been increasingly applied in the health and well-being domain [25–27], which has in turn demonstrated their broad applicability in this area and is therefore also considered in this task of human activity recognition. In particular, we applied support vector machine (SVM) (with both RBF and linear kernels) and k-Nearest Neighbour (kNN) which have been considered some of the top 10 popular algorithms in data mining [28]; in addition, we applied Logistic Regression (LR), Deccision Treee (DT), Random Forest (RF), Gradient Boost Machine (GB) and Multiple Layer Perceptron (MLP) which are also common choices in machine learning [29].

3.5 Model Evaluation

We evaluated the performance of the ML classifiers using precision, recall, accuracy, and F1-Score.

The precision which is the ratio between the true positives and all the positives is calculated as:

$$Precision = \frac{TP}{TP + FP} \tag{1}$$

The recall also refer to as sensitivity is the measure of a model correctly identifying true positives and is calculated as:

$$Recall = \frac{TP}{TP + FN} \tag{2}$$

The accuracy is the ratio of the total number of correct predictions and the total number of predictions and it is calculated as:

$$Accuracy = \frac{TP + TN}{TP + FP + TN + FN} \tag{3}$$

F1-score is the harmonic mean of the precision and recall. Instead of balancing precision and recall, a good F1-score indicates good Precision and a good Recall value and it is calculated as:

$$F1 - Score = 2 \times \frac{Precision \times Recall}{Precision + Recall} \tag{4}$$

where:
$TruePositive\,(TP)$: a positive value predicted as positive value
$FalsePositive\,(FP)$: a negative value predicted as positive value
$TrueNegative\,(TN)$: a negative value predicted as negative value
$FalseNegative\,(FN)$: a positive value predicted as negative value.

3.6 Result and Discussion

In this study, we performed five (5) categorical experiments with different feature variables from the dataset. We applied a 10-fold StratifiedKFold cross-validation on each data resulting from each data feature used to split the transformed data into test and training splits. The classifiers are trained with the training sets and are evaluated with the testing sets.

In the first experiment, we predicted HAR using 'Axis1', 'Axis2', 'Axis3', 'Steps', and 'HR' variable features which yielded 102 features after performing the feature extraction. The second experiment was performed using the 'Steps', and 'HR' variable features, and the feature extraction process yielded a dataset with 40 features. The third experiment was carried out using only the 'HR' variable feature and 20 features are generated from the feature extraction process. Fourthly, we only experimented using the 'Steps' variable feature and produced 20 features from the extraction process. And finally, we experimented using the triaxial accelerometer data 'Axis1', 'Axis2', and 'Axis3' variable features and 62 features as produced from the extraction process.

The results from the experiments as described below in Table 1, we observed that the models performed average well in Experiments 1, 2, and 3 having average precision, recall, accuracy, and f1-score values above 60 ± 12 %, but very poorly in experiment 4 and 5 with average precision, recall, accuracy, and f1-score values below 50 ± 05 %. It was also observed in Experiment 1 that model GB performed the best. Again, it is observed in all the experiments that both GB and MLP models had the best f1-score with average values of 61 ± 14 % in Experiments 1–3 and, 40 ± 7 % in Experiments 4 and 5. Since the models performed averagely well with experiments having HR as a feature, we deduced that HR provides the models with more information to predict better, hence, with HR, it is possible to predict human activities.

We believe that the average performance of the models for activity recognition on the MMASH dataset is a result of the reasons below:

- The activities collected in the MMASH dataset are recorded in real-life conditions rather than in a controlled environment, thereby, making it challenging to identify.
- There are huge inter-class activity similarities. For instance, eating while using a large screen as seen in Fig. 4.

Table 1. Models experimental results

Experiment	Metrics	LR (%)	DT (%)	RF (%)	GB (%)	KNN (%)	SVM-RBF (%)	SVM-Lib (%)	MLP (%)
1	Precision	60 ± 14	59 ± 09	62 ± 13	66 ± 11	59 ± 10	66 ± 17	62 ± 13	64 ± 11
	Recall	61 ± 14	59 ± 10	62 ± 14	64 ± 13	56 ± 10	64 ± 13	62 ± 13	63 ± 12
	Accuracy	64 ± 13	61 ± 10	64 ± 13	67 ± 12	59 ± 09	67 ± 13	65 ± 12	66 ± 11
	F1-Score	59 ± 14	59 ± 10	60 ± 14	62 ± 14	55 ± 10	61 ± 14	60 ± 13	62 ± 12
2	Precision	61 ± 15	60 ± 10.	61 ± 11	66 ± 12	63 ± 11	65 ± 18	64 ± 16	64 ± 17
	Recall	62 ± 14	61 ± 11	62 ± 12	63 ± 14	62 ± 11	63 ± 13	63 ± 13	64 ± 14
	Accuracy	66 ± 13	63 ± 10	64 ± 11	66 ± 13	65 ± 10	67 ± 12	67 ± 13	67 ± 14
	F1-Score	59 ± 14	59 ± 11	60 ± 12	62 ± 15	60 ± 12	60 ± 15	60 ± 14	62 ± 16
3	Precision	62 ± 14	59 ± 11	60 ± 10	63 ± 13	65 ± 08	64 ± 16	63 ± 15	62 ± 16
	Recall	62 ± 12	59 ± 11	60 ± 11	62 ± 13	61 ± 11	63 ± 12	63 ± 12	63 ± 13
	Accuracy	66 ± 12	62 ± 10	63 ± 10	65 ± 12	65 ± 10	68 ± 11	67 ± 12	67 ± 13
	F1-Score	58 ± 13	58 ± 12	59 ± 11	60 ± 14	58 ± 12	59 ± 14	58 ± 13	60 ± 14
4	Precision	36 ± 07	36 ± 06	36 ± 06	37 ± 07	27 ± 06	36 ± 07	36 ± 07	36 ± 06
	Recall	44 ± 07	43 ± 07	43 ± 07	43 ± 07	35 ± 10	44 ± 06	44 ± 06	43 ± 06
	Accuracy	45 ± 08	45 ± 07	45 ± 07	45 ± 07	39 ± 11	45 ± 07	46 ± 07	45 ± 07
	F1-Score	37 ± 07	37 ± 06	37 ± 06	37 ± 06	23 ± 11	37 ± 06	37 ± 06	37 ± 05
5	Precision	44 ± 07	43 ± 07	45 ± 08	44 ± 08	35 ± 08	43 ± 11	44 ± 08	43 ± 07
	Recall	47 ± 07	45 ± 07	47 ± 08	47 ± 07	35 ± 08	47 ± 08	47 ± 07	47 ± 07
	Accuracy	49 ± 07	46 ± 08	49 ± 09	49 ± 08	38 ± 09	50 ± 08	50 ± 07	48 ± 07
	F1-Score	43 ± 07	42 ± 07	43 ± 08	43 ± 07	30 ± 07	42 ± 08	42 ± 07	43 ± 07

- The time to switch between activities is also very small such as eating and large screen usage.
- Each activity has a different period, making the data to be relatively imbalanced and may result in low activity recognition.

4 Conclusion and Future Work

This study aimed to analyze the use of popular ML classifiers to classify human activities on data accelerometer and HR data generated by Actigraph rather than mostly used accelerometry data generated from smartphones. For our experimental purpose, we explored the Actigraph generated data provided by MMASH dataset [21] which contains accelerometer and HR recording and 8 ML classifiers. We preprocessed the dataset and explored 22 statistical features for the feature engineering process. Finally, we performed 4 experiments using different variable features. The experimental results show that HR can provide more information to accurately predict HAR. In the future, we plan to use deep learning models such as convolutional and recurrent neural networks which can automatically capture long-term dependencies in the data and perform feature extraction to improve the accuracy in predicting HAR. We are considering predicting all the activities by exploring a sampling technique that helps in overcoming the activity class imbalanced problem.

References

1. Li, H., Yang, G.: Dietary nutritional information autonomous perception method based on machine vision in smart homes. Entropy **24**(7) (2022)
2. Tsai, T.-H., Huang, C.-C., Zhang, K.-L.: Design of hand gesture recognition system for human-computer interaction. Multimedia Tools Appl. **79**, 5989–6007 (2020). Mar.
3. Uddin, M.Z., Hassan, M.M.: Activity recognition for cognitive assistance using body sensors data and deep convolutional neural network. IEEE Sens. J. **19**(19), 8413–8419 (2019)
4. Dua, N., Singh, S.N., Semwal, V.B., Challa, S.K.: Inception inspired CNN-GRU hybrid network for human activity recognition. Multimedia Tools Appl. **82**, 5369–5403 (2023). Feb
5. Lin, J., Li, Y., Yang, G.: Fpgan: face de-identification method with generative adversarial networks for social robots. Neural Netw. **133**, 132–147 (2021)
6. Oyeleye, M., Chen, T., Titarenko, S., Antoniou, G.: A predictive analysis of heart rates using machine learning techniques. Int. J. Environ. Res. Public Health **19**(4), 2417 (2022)
7. Dinarević, E.C., Husić, J.B., Baraković, S.: Issues of human activity recognition in healthcare. In: 2019 18th International Symposium INFOTEH-JAHORINA (INFOTEH), pp. 1–6 (2019)
8. Mohan, S., Thirumalai, C., Srivastava, G.: Effective heart disease prediction using hybrid machine learning techniques. IEEE Access **7**, 81542–81554 (2019)

9. Casalino, G., Castellano, G., Zaza, G.: On the use of FIS inside a telehealth system for cardiovascular risk monitoring. In: 2021 29th Mediterranean Conference on Control and Automation (MED), pp. 173–178. IEEE (2021)
10. Ghate, V., Sweetlin Hemalatha, C.: Hybrid deep learning approaches for smartphone sensor-based human activity recognition. Multimedia Tools Appl. **80**, 35585–35604 (2021)
11. Nweke, H.F., Teh, Y.W., Mujtaba, G., Al-garadi, M.A.: Data fusion and multiple classifier systems for human activity detection and health monitoring: review and open research directions. Inform. Fusion **46**, 147–170 (2019)
12. Li, Y., Yang, G., Su, Z., Li, S., Wang, Y.: Human activity recognition based on multi environment sensor data. Inform. Fusion **91**, 47–63 (2023)
13. Anguita, D., Ghio, A., Oneto, L., Parra, X., Reyes-Ortiz, J.L., et al.: A public domain dataset for human activity recognition using smartphones. ESANN **3**, 3 (2013)
14. Catal, C., Tufekci, S., Pirmit, E., Kocabag, G.: On the use of ensemble of classifiers for accelerometer-based activity recognition. Appl. Soft Comput. **37**, 1018–1022 (2015)
15. Bayat, A., Pomplun, M., Tran, D.A.: A study on human activity recognition using accelerometer data from smartphones. Procedia Comput. Sci. **34**, 450–457 (2014). The 9th International Conference on Future Networks and Communications (FNC'14)/The 11th International Conference on Mobile Systems and Pervasive Computing (MobiSPC'14)/Affiliated Workshops
16. Kwapisz, J.R., Weiss, G.M., Moore, S.A.: Activity recognition using cell phone accelerometers. SIGKDD Explor. Newsl. **12**, 74–82 (2011). Mar
17. Jalal, A., Batool, M., Kim, K.: Stochastic recognition of physical activity and healthcare using tri-axial inertial wearable sensors. Appl. Sci. 10(20) (2020)
18. Patricia, A.-C.P., Enrico, V., Shariq, B.A., Emiro, D.-l.-F., Alberto, P.-M.M., Isabel, O.-C.A., Tariq, M.I., Restrepo, J.K.G., Fulvio, P.: Machine learning applied to datasets of human activity recognition:data analysis in health care. Curr. Med. Imaging **19**(1), 46–64 (2023)
19. Wu, X., Kumar, V., Ross Quinlan, J., Ghosh, J., Yang, Q., Motoda, H., McLachlan, G.J., Ng, A., Liu, B., Yu, P.S., Zhou, Z.-H., Steinbach, M., Hand, D.J., Steinberg, D.: Top 10 algorithms in data mining. Knowl. Inform. Syst. **14**(1), 1–37 (2008)
20. Gomersall, S.R., Ng, N., Burton, N.W., Pavey, T.G., Gilson, N.D., Brown, W.J.: Estimating physical activity and sedentary behavior in a free-living context: a pragmatic comparison of consumer-based activity trackers and actigraph accelerometry. J. Med. Internet Res. **18**(9), e239 (2016)
21. Rossi, A., Da Pozzo, E., Menicagli, D., Tremolanti, C., Priami, C., Sirbu, A., Clifton, D., Martini, C., Morelli, D.: Multilevel monitoring of activity and sleep in healthy people. PhysioNet (2020)
22. Shanbhog, S.M., Medikonda, J.: A clinical and technical methodological review on stress detection and sleep quality prediction in an academic environment. Comput. Methods Programs Biomed. **235**, 107521 (2023)
23. Site, A., Lohan, E.S., Jolanki, O., Valkama, O., Hernandez, R.R., Latikka, R., Alekseeva, D., Vasudevan, S., Afolaranmi, S., Ometov, A., Oksanen, A., Martinez Lastra, J., Nurmi, J., Fernandez, F.N.: Managing perceived loneliness and social-isolation levels for older adults: a survey with focus on wearables-based solutions. Sensors **22**(3) (2022)
24. Geng, D., Qin, Z., Wang, J., Gao, Z., Zhao, N.: Personalized recognition of wake/sleep state based on the combined shapelets and k-means algorithm. Biomed. Signal Process. Control **71**, 103132 (2022)

25. Chen, T., Su, P., Shen, Y., Chen, L., Mahmud, M., Zhao, Y., Antoniou, G.: A dominant set-informed interpretable fuzzy system for automated diagnosis of dementia. Front. Neurosci. **16** (2022)
26. Su, P., Chen, T., Xie, J., Ma, B., Qi, H., Liu, J., Zhao, Y.: A density and reliability guided aggregation for the assessment of vessels and nerve fibres tortuosity. IEEE Access **8**, 139199–139211 (2020)
27. Chen, T., Carter, J., Mahmud, M., Khuman, A. (eds.): Artificial Intelligence in Healthcare: Recent Applications and Developments. Brain Informatics and Health. Springer, Singapore (2022). Oct.
28. Wu, X., Kumar, V., Ross Quinlan, J., Ghosh, J., Yang, Q., Motoda, H., McLachlan, G.J., Ng, A., Liu, B., Yu, P.S., Zhou, Z.-H., Steinbach, M., Hand, D.J., Steinberg, D.: Top 10 algorithms in data mining. Knowl. Inform. Syst. **14**, 1–37 (2008)
29. Géron, A.: Hands-on machine learning with scikit-learn and tensorflow: concepts. Tools Tech. Build Intell. Syst. (2017)

Effect of Financial News Headlines on Crypto Prices Using Sentiment Analysis

Ankit Limone(✉), Mahak Gupta, Nitin Nagar, and Shaligram Prajapat

International Institute of Professional Studies, Devi Ahilya Vishwavidyalaya, Indore, India
ankitlimone16@gmail.com

Abstract. Crypto prices are not solely based on historical data, dependent on the current state of the market, which can also fluctuate it significantly. A crypto's price can increase or decrease depending on financial news headlines. In order to understand the effect of a headline of news, sentiment analysis is one of the most popular techniques, whether it is positive or negative. The crypto market is not stable and fluctuates constantly, therefore, it is very challenging to predict crypto prices precisely. We used Vader (Valence Aware Dictionary for Sentiment Reasoning) model and TextBlob model of sentiment analysis combining with Support Vector Regressor (SVR) and Prophet (formally Fbprophet) to analyze that how much does news headline effect on the prediction of crypto price of Bitcoin.

Keywords: Sentiment Analysis · Crypto price prediction · Vader · SVM · Prophet · TextBlob · Natural Language Tool Kit (NLTK)

1 Introduction

Stock price prediction is very popular and interesting topic nowadays, but stock price prediction is done by not only using the past data, but also the daily news that affects the performance of the stock [1]. Sentiment analysis is the technique which can analyze news headlines and tells that if it is positive for the company's stock or negative [2]. There were many machine learning models and approaches to get the sentiment analysis of any given text such as TextBlob, Vader, RoBERTa, Transformer, Bag of Words, LSTM, Naïve Bayes, Natural Language Processing (NLP) [3–8].

Investors make their decision of selling or buying stock based on the market conditions [9]. Financial text-based news information majorly help investors to take decisions of buying, selling, and holding crypto and for how long. In machine learning, NLTK is one of the popular method by which we can analyze any text which is positive or negative for any cryptocurrency [10].

In this paper, we analyze that how much does sentiment analysis effects on the crypto price, and we used Bitcoin as crypto currency in this experiment, we used Vader and TextBlob for Sentiment analysis, for prediction of the Bitcoin price we used SVR and Prophet [11–13]. We took Bitcoin's ticker name BTC-USD for the analysis purpose from Yahoo Finance and for sentiment analysis we took data from Kaggle [14, 15]. Yahoo

N. Naik et al. (Eds.): UKCI 2023, AISC 1453, pp. 209–219, 2024.
https://doi.org/10.1007/978-3-031-47508-5_17

Finance is a stock screener website which provides the historical data and many other information like financial news.

2 Methodology

The methodology diagram in Fig. 1. Illustrates a systematic approach proposed for Bitcoin price prediction with the influence of financial news headlines. The diagram comprises nine interconnected components. The initial step, labeled "News Headlines," involves the collection of financial news headlines as the dataset. The second step, "Preprocessing using TextBlob," focuses on data filtering to enhance data quality. The third step, "Merge Headlines," combines multiple headlines into a unified one. The fourth step, "Calculated Subjectivity & Polarity," measures the positive and negative sentiment of the headlines. The fifth step, "Sentiment Analyzer," determines the emotional tone of the text, classifying it as positive, negative, or neutral. Similarly, for price prediction, the first step, "Crypto Price," represents the dataset containing cryptocurrency prices. The second step, "Remove empty columns and spaces," involves data preprocessing to enhance data quality. The third step, "Support Vector Regressor & Prophet Prediction," focuses on predicting future Bitcoin prices. The "Merge Dataset" step combines the sentiment analysis results and the predicted Bitcoin prices, enabling comprehensive analysis and insights. Based on this data, investors can make informed decisions.

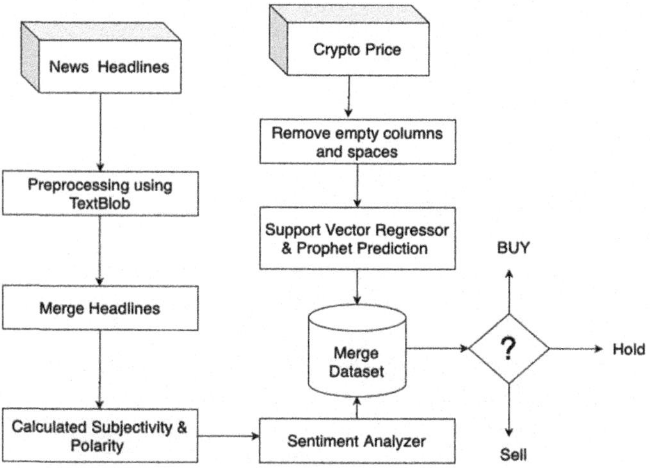

Fig. 1. Proposed Approach for Bitcoin price prediction

2.1 Data Collection and Preprocessing

Firstly, for sentiment analysis we took the news headlines dataset from Kaggle, dataset contains 4 columns namely URL, Name, Desc and Date and 690 rows. In pre-processing phase, We removed unnecessary data and eliminated timestamps from the date column in the dataset using TextBlob. We removed empty rows and incomplete rows from dataset using Pandas.

Then, for Bitcoin price data, we took from Yahoo Finance and pre-process the dataset and removed empty and incomplete data rows from dataset (see Table 1).

Table 1. Dataset of Bitcoin price

	Date	Open	High	Low	Close	Adj Close	Volume
0	08-07-2020	9253.020508	9450.335938	9249.500000	9428.333008	9428.333008	19702359883
1	07-07-2020	9349.161133	9360.617188	9201.815430	9252.277344	9252.277344	13839652595
2	06-07-2020	9072.849609	9375.474609	9058.664063	9375.474609	9375.474609	17889263252
3	05-07-2020	9126.090820	9162.183594	8977.015625	9073.942383	9073.942383	12903406143
4	04-07-2020	9084.233398	9183.295898	9053.629883	9132.488281	9132.488281	12290528515

2.2 Models and Supporting Tools

Following Table 2. Presents some candidate models or tool for the goal of proposed problem. Later on we have merged the results of sentiment and prediction dataset at the end and using these investors can take their financial decisions accordingly.

3 Experimental Analysis

The Experimental Analysis section comprises the methodology implemented in this paper, as well as the results obtained from the three distinct phases of the experiment that were conducted. We used multiple python libraries for sentiment analysis and crypto price prediction like as Pandas [17], NumPy [18], TextBlob, Regular Expression, Vader, Scikit-learn, Matplotlib, Seaborn, Plotly, and Prophet [19–23]. In the first phase for sentiment analysis, we took Vader's Sentiment Intensity Analyzer (SIA) and using SIA we calculated the negative, neutral, positive, and compound scores to get the sentiment of the news headlines. By Textblob we evaluated subjectivity and polarity and combined with the SIA values in the dataset (see Table 4).

In the second phase of our experiment, we employed SVR for Bitcoin(BTC-USD) price prediction, To ensure a reliable evaluation of the model's performance, we divided the dataset into two subsets: a training set and a testing set. The training set was utilized to train the SVR model, where we specifically utilized the 'Close' column data. Once the model was trained, we proceeded to forecast the future 5-day prices using the 'Close' column data from the testing set. We calculated the RMSE value of 379.062, MAE value of 309.201 and MPE value of 0.0209 and calculated correlation coefficient value is 0.1850.

In the above Fig. 2, the red curve represents the prediction of the Bitcoin price, while the blue curve represents the actual price of Bitcoin. Next, we took Prophet library to

Table 2. Dataset of Bitcoin price

Model/Tool	Feature for sentiment discovery and analysis	Dataset	Applicability for analysis of Financial News Headlines on Crypto Prices
Vader	• Vader (**V**alence **A**ware **D**ictionary for Sentiment **R**easoning) is tool for sentiment analysis • Vader is sensitive to intensity and polarity of the emotion of a text • Text data that is unlabelled can be processed directly by this technique [16]	*News headlines*	• ***Input***: News dataset from Kaggle • ***Preprocessing phase***: In News dataset from Kaggle, we Combined Name and Desc column • ***Output***: Classified words according to its positivity and negativity (Polarity of emotion) • ***Processing Module***: NLTK package • **Result**: Computed polarity as negative(−0.05), positive(0.05) and neutral(−0.05 to 0.05) scores of the combined news headline (Fig. 4)
TextBlob	• TextBlob is a python library to process textual data • TextBlob uses lexicon-based methods on input text files to label positive and negative	*News headlines*	• ***Input***: News data set from Kaggle • ***Preprocessing phase***: In News data set from Kaggle, we Combined Name and Desc column • ***Output***: Classified words according to its positivity and negativity (Polarity of emotion) • ***Processing Module***: Lexicon based Method • **Result**: Computed subjectivity and polarity as negative(0), positive(1) scores of the combined news headline (Fig. 5)

(*continued*)

Table 2. (*continued*)

Model/Tool	Feature for sentiment discovery and analysis	Dataset	Applicability for analysis of Financial News Headlines on Crypto Prices
Prophet	• Originally it was Fbprophet was an automated python library (Facebook) • It is format specific and needs certain columns to be customised like ds (Daily Seasonality), yhat (yhat is predicted value of target variable y) and cutoff (cutoff refers to date point)	*Bitcoin Price*	• ***Input***: Bitcoin Price dataset from Yahoo Finance • ***Preprocessing phase***: We trained and tested our Bitcoin dataset using this library and calculated the prediction and error values • ***Output*** : Prophet provides hourly, daily, weekly, and yearly seasonality values and graphs • ***Processing Module***: Prophet Prediction Library • **Result**: Computed Multiple Seasonality (Table 3)
SVR	• Support Vector Regressor is regression and classification supervised learning algorithm to predict discrete values and type of support vector machine	*Bitcoin Price*	• ***Input***: Bitcoin Price dataset from Yahoo Finance • ***Preprocessing phase***: We trained and tested our Bitcoin dataset using this library and calculated the prediction and error values • ***Output***: Calculated and predict Bitcoin price and root mean squared (RMSE), mean absolute error (MAE), and mean percentage error (MPE) • ***Processing Module***: It works on best fit line in hyperplane concept • **Result**: Actual Price vs Predicted Price before combining (Fig. 2)

Table 3. Prophet Dataset

	ds	yhat	yhat_lower	yhat_upper	y	cutoff
0	2020-04-07	23547.067008	23547.066975	23547.067026	9132.488281	2020-03-21
1	2020-05-07	26553.847489	26553.847390	26553.847556	9073.942383	2020-03-21
2	2020-06-07	-183412.028194	-183412.028197	-183412.028191	9375.474609	2020-05-12
3	2020-06-13	8983.412230	8983.412227	8983.412233	9475.277344	2020-05-12
4	2020-06-14	-183419.400242	-183419.400245	-183419.400240	9386.788086	2020-05-12

Table 4. Sentiment Analysis result Dataset

Date	combined_news	Subjectivity	Polarity	compound	negative	neutral	positive
08-07-2020	Animoca Brands achieves record growth in reve...	0.481818	0.375758	0.8225	0.000	0.815	0.185
08-07-2020	Energy company EDP Renewables confirms April ...	0.850000	0.000000	-0.6808	0.179	0.731	0.090
08-07-2020	A blockchain project builds a layer of securit...	0.290000	0.110000	0.7096	0.000	0.879	0.121
08-07-2020	Bitcoin Will Never Be Truly Private Says Andre...	0.458333	0.000000	0.2016	0.036	0.890	0.074
08-07-2020	Pennsylvania man admits to defrauding West Vir...	0.458333	-0.208333	0.2023	0.058	0.849	0.092

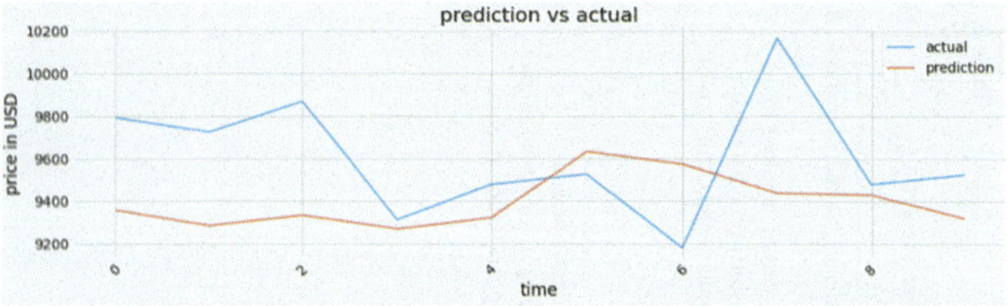

Fig. 2. Actual Price vs Predicted Price using SVR before merged dataset

predict Bitcoin price. We renamed date column to ds and close column to y and remove other columns after that we trained the model, and we evaluate performance matrix (see Table 5).

Figure 3 is for multiple seasonality of the prediction of Bitcoin daily, weekly and trend graphs separately. The seasonality component captures the repetitive patterns or cycles that occur at fixed intervals, such as daily, weekly, or hourly. In daily seasonality, the y-axis represents the average deviation from the overall trend for each day. Similarly, weekly seasonality, the y-axis represents the average deviation from the trend for each day of the week.

In the Final phase, we merged the sentiment analysis results with crypto price prediction using SVR and Prophet results to examine the effect of news headlines on the price

Table 5. Prophet performance metrics for Bitcoin price prediction

	horizon	mse	rmse	mae	mape	mdape	smape	coverage
0	9 days	27643.888746	166.264514	134.286729	0.014041	0.009385	0.014150	1.00
1	10 days	5285.455965	72.701141	68.800987	0.007394	0.007155	0.007384	1.00
2	11 days	31567.337529	177.671994	133.741457	0.014455	0.008708	0.014603	0.75
3	12 days	203415.723397	451.016323	316.604086	0.035485	0.021277	0.036834	0.50
4	13 days	408469.243541	639.115986	594.357908	0.065746	0.065924	0.066282	0.00

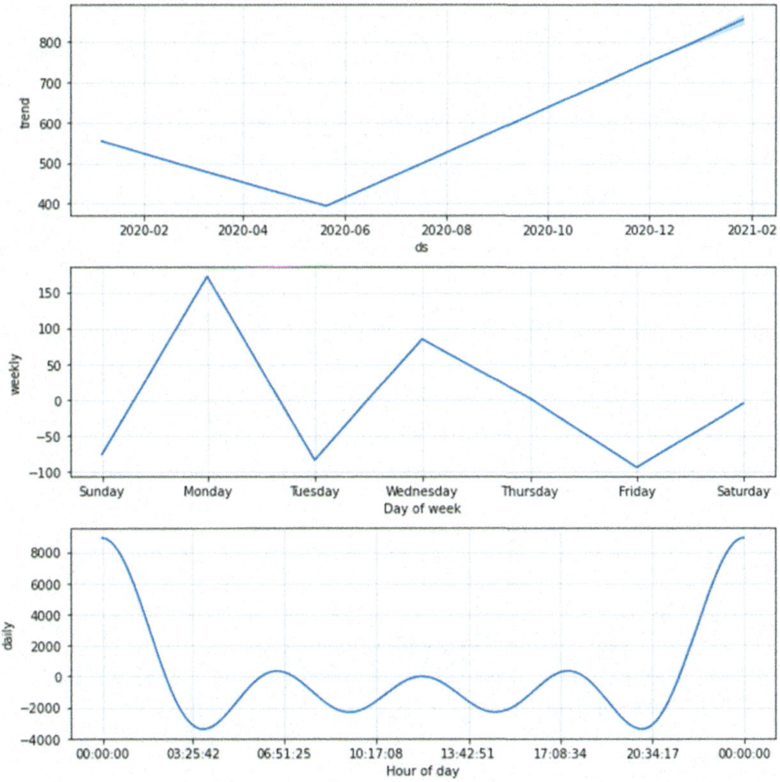

Fig. 3. Multiple seasonality graphs for predicted price of Bitcoin

of the crypto Bitcoin price in the market and how it can help in real world investors to take their financial decisions. For further evaluations we kept only some columns from merged dataset namely Open, High, Low, Close, Volume, Subjectivity, Polarity, compound, negative, neutral, positive and Label. Then we divide the dataset into testing and training part and trained the model using Linear Discriminant Analyzer for prediction (see Table 6).

Table 6. Merged Dataset of Predicted Price and Sentiment Analysis

	Open	High	Low	Close	Volume	Subjectivity	Polarity	compound	negative	neutral	positive	Label
0	9253.020508	9450.335938	9249.500000	9428.333008	19702359883	0.481818	0.375758	0.8225	0.000	0.815	0.185	1
1	9253.020508	9450.335938	9249.500000	9428.333008	19702359883	0.850000	0.000000	-0.6808	0.179	0.731	0.090	0
2	9253.020508	9450.335938	9249.500000	9428.333008	19702359883	0.290000	0.110000	0.7096	0.000	0.879	0.121	1
3	9253.020508	9450.335938	9249.500000	9428.333008	19702359883	0.458333	0.000000	0.2016	0.036	0.890	0.074	1
4	9253.020508	9450.335938	9249.500000	9428.333008	19702359883	0.458333	-0.208333	0.2023	0.058	0.849	0.092	1

Table 7. Classification Report of Predictions Results

	precision	recall	f1-score	support
0	0.91	1.00	0.95	69
1	1.00	0.90	0.95	69
accuracy			0.95	138
macro avg	0.95	0.95	0.95	138
weighted avg	0.95	0.95	0.95	138

In Table 7 we calculated the classification report of the merged data and predictions.

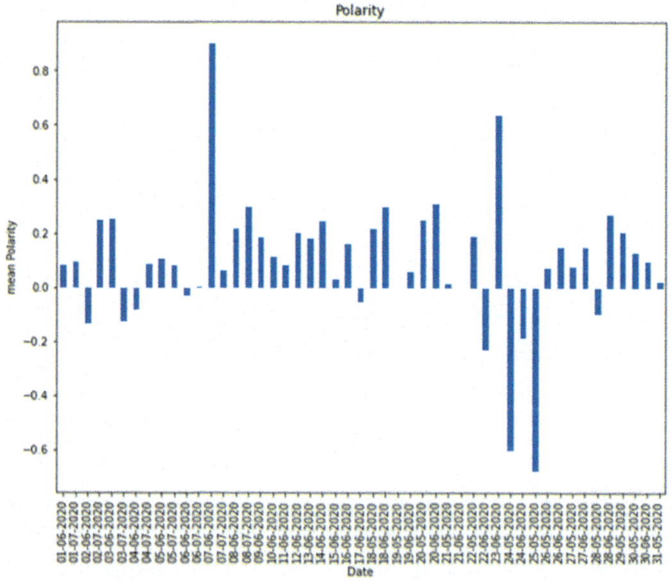

Fig. 4. Polarity Graph for Positivity and Negativity of News Headlines

Figure 4 is of polarity graph of the merged dataset of prediction of Bitcoin price date wise, it varies between 1 to −1.

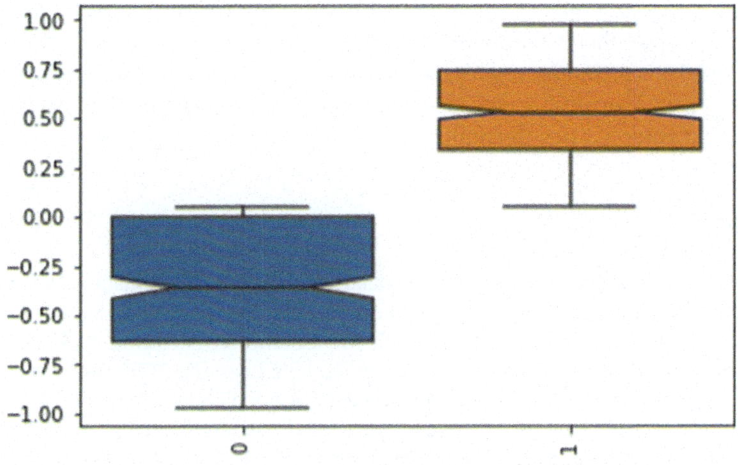

Fig. 5. Boxplot of Confusion Matrix for Sentiment Analysis

In Fig. 5. of confusion matrix, blue shows the neutral and negative sentiments and orange shows the positive sentiments. Negative and neutral are grouped and labelled as 0 similarly positive group is labelled as 1.

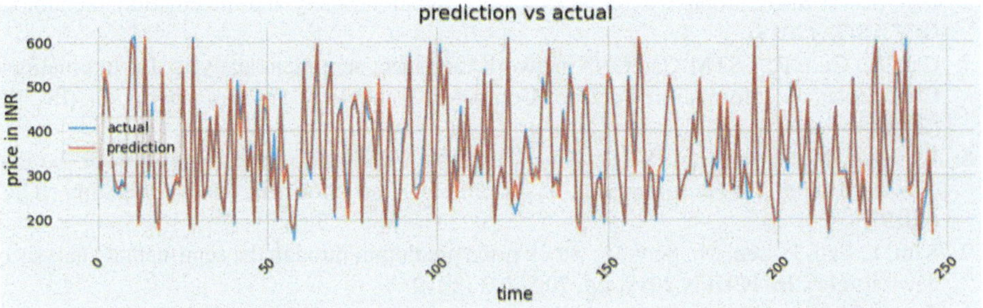

Fig. 6. Actual Price vs Predicted Price using SVR after Merged Dataset

In the above Fig. 6 depicts the prediction of the Bitcoin price represented by the red curve, while the actual price of Bitcoin is shown by the blue curve. This prediction and actual price comparison was conducted using a merged dataset with 94.9% accuracy.

4 Conclusion and Future Work

After conducting an experiment that involved combining the results of the sentiment analyzer and Bitcoin price prediction, we were able to achieve an impressive accuracy rate of 94.9%. Initially, we performed sentiment analysis and Bitcoin price prediction individually on a dataset consisting of 690 rows. However, by merging the results of both analyses, we observed a significant improvement in the effectiveness of our predictions. Notably, our analysis revealed that the sentiment conveyed in news headlines has a considerable impact on the price of Bitcoin. Moving forward, there is ample potential to expand upon this experiment by exploring alternative prediction algorithms and sentiment analysis techniques, with the aim of achieving even higher levels of accuracy. By doing so, we can further enhance our understanding of the relationship between sentiment and Bitcoin price dynamics.

References

1. Agarwal, A.: Sentiment analysis of financial news. In: 12th International Conference on Computational Intelligence and Communication Networks, pp. 312–315 (2020)
2. Mankar, T., Hotchandani, T., Madhawani, M., Chidrawar, A., Lifna, C.S.: Stock Market Prediciton based on Social Sentiments using Machine Learning. IEEE Xplore
3. Khan, R., Rustam, F., Kanwal, K., Mehmood, A., Choi, G.S.: US based COVID-19 tweets sentiment analysis using TextBlob and supervised machine learning algorithm. In: 2021 International Conference on Artificial Intelligence (ICAI), Islamabad, Pakistan (2021)
4. Amin, A., Hossain, I., Akther, A., Alam, K.M.: Bengali VADER: a sentiment analysis approach using modified VADER. In: 2019 International Conference on Electrical, Computer and Communication Engineering (ECCE) (2019)
5. Cui, Y., Zhu, C.: Fine-grained Chinese named entity recognition with RoBERTa and convolutional attention networks. In: 2020 IEEE 6th International Conference on Computer and Communications (2020)
6. Hao, Z., Kaidong, L., Feng, Q.: Improvement of word bag model based on image classification. In: 1st International Conference on Civil Aviation Safety and Information Technology (ICCASIT) (2019)
7. Cao, S., Gao, P.: LSTM-GateCNN network for aspect sentiment analysis. In: International Conference on Information Science, Computer Technology and Transportation (ISCTT) (2020)
8. Wongkar, M., Angdresey, A.: Sentiment analysis using naive bayes algorithm of the data crawler: Twitter. In: Fourth International Conference on Informatics and Computing (ICIC) (2019)
9. Kim, J., Seo, J., Lee, M., Seok, J.: Stock price prediction through the sentimental analysis of news articles. In: ICUFN 2019, pp. 700–702 (2019)
10. Gupta, R., Chen, M.: Sentiment analysis for stock price prediction. In: IEEE Conference on Multimedia Information Processing and Retrieval, pp. 213–218 (2020)
11. "VADER," Open Source. https://www.nltk.org/_modules/nltk/sentiment/vader.html. Accessed Feb 2023
12. "TextBlob: Simplified Text Processing," Open Source. https://textblob.readthedocs.io/en/dev/. Accessed Feb 2023
13. "Prophet," Meta. https://facebook.github.io/prophet/. Accessed Feb 2023
14. "Yahoo Finance". https://finance.yahoo.com/. Accessed Feb 2023
15. "Kaggle Bitcoin News," Google. https://www.kaggle.com/datasets/ffejgnaw/bitcoin-news-june-2020-onwards. Accessed Feb 2023
16. Mohan, S., Mullapudi, S., Sammeta, S., Vijayvergia, P., Anastasiu, D.C.: Stock price prediciton using news sentiment analysis. In: 5th International Conference on Big Data Computing Service and Applications, pp. 205–208 (2019)

17. "Pandas". https://pandas.pydata.org/
18. "Numpy". https://numpy.org/
19. "Regular Expression (RE)," Python Open Source. https://docs.python.org/3/library/re.html. Accessed Feb 2023
20. "Scikit-Learn". https://scikit-learn.org/
21. "Matplotlib". https://matplotlib.org/
22. "Seaborn," PyData. https://seaborn.pydata.org/. Accessed Feb 2023
23. "Plotly," Plotly. https://plotly.com/. Accessed Feb 2023
24. "Kaggle," Google. https://www.kaggle.com/. Accessed Feb 2023
25. Kale, A., Khanbilkar, O., Jivani, H., Kumkar, P., Madan, I., Sarode, D.T.: Forecasting Indian stock market using artificial neural network. In: International Conference on Computing Communication Control and Automation (ICCUBEA), vol. 4 (2018)
26. Shah, D., Isah, H., Zulkernine, F.: Prediciting the effects of news sentiment on the stock market. In: IEEE International Conference on Big Data, pp. 4705–4708 (2018)
27. Inamdar, A., Bhagtani, A., Bhatt, S., Shetty, P.M.: Predicting cryptocurrency value using sentiment analysis. In: International Conference on Intelligent Computing and Control System, pp. 932–933 (2019)
28. Sun, T., Wang, J., Zang, P., Cao, Y., Liu, B., Wang, D.: Predicting stock price returns using microblog sentiment for Chinese stock market. In: International Conference on Big Data Computing and Communications, pp. 87–96 (2017)
29. Wei, D.: Prediction of stock price based on LSTM neural network. In: International Conference on Artificial Intelligence and Advanced Manufacturing, p. 4 (2019)

Detection of Cyberbullying on Social Media Platforms Using Machine Learning

Mohammad Usmaan Ali(✉) and Raluca Lefticaru

Department of Computer Science, University of Bradford, Bradford, UK
{m.u.ali4,r.lefticaru}@bradford.ac.uk

Abstract. Cyberbullying is a prominent issue that affects many people within their lifetime. In this paper we explore the use of Machine Learning (ML) and Natural Language Processing (NLP) techniques to support automatic detection of cyberbullying. The paper discusses first the significance of cyberbullying and its relationship to cybersecurity. Then, in order to illustrate the automatic detection approach and its integration into a web application, we considered a benchmark dataset, the Offensive Language Identification Dataset (OLID). This is an annotated large-scale dataset with approximate 14,000 English tweets, that was used in various works for detecting offensive posts in social media, predicting their type and target. To solve the classification problems associated with OLID dataset, nine supervised models were developed for each task of the dataset. We used the TFIDF (Term Frequency Inverse Document Frequency) feature selection and GridSearchCV to find the optimum parameters for each of the ML algorithms. Evaluation metrics, such as build time, accuracy, precision, recall, and F1-score were used to compare the ML techniques. The Random Forest models achieved 82, 90, and 61% in accuracy for the tasks associated with the dataset, which were the best performing algorithms supported by the other metrics. The Random Forest models that achieved the best performance, were integrated within a Flask web application, publicly available. This serves as proof of concept, allowing users to test the classifier. The paper explains how a developed ML model can be integrated into a web application. Furthermore, one can develop an API for handling larger and more frequent data requests or integrate in a social media platform the classifier.

Keywords: Cybersecurity · Cyberbullying · Machine learning · Natural language processing · Twitter · Web development

1 Introduction

With the rise of the digital era, social media has become an integral part of everyone's life redefining how people communicate. Following the increase in users of various platforms such as Facebook, Instagram, Reddit, Twitter, and many more, users can voice their opinions. These platforms also allow malicious

N. Naik et al. (Eds.): UKCI 2023, AISC 1453, pp. 220–233, 2024.
https://doi.org/10.1007/978-3-031-47508-5_18

users to exploit their features to carry out a form of bullying called cyberbullying [21]. According to Waller et al. [24], the number of victims of cyberbullying has doubled from 18 to 36% in a 10 years period. Other statistics range from 3 to 72%, depending on the interpretation of cyberbullying and the measures used by researchers when gathering and analysing data [23]. This significant increase can be explained by technology being more accessible and integrated into people lives.

There are numerous forms of cyberbullying that bullies can exploit, exclusion, harassment, doxing, trickery, cyberstalking, fraping, dissing, trolling. The consequences of cyberbullying can be extreme; it can lead to depression, hopelessness, loneliness and even suicide [11]. Cyberbullying is a severe problem that must be addressed, and it was analysed by numerous psychology studies [11,22,24]. Currently, technical and non-technical preventive measures are in place to mitigate cyberbullying [22]. However, they each have their prescribed issues and drawbacks. The cyberbullying topic has received a lot of attention from Machine Learning (ML) and Natural Language Processing (NLP) communities which work on automating its detection.

One benefit of using *Artificial Intelligence* (AI) methods, in particular integrating ML models in social media to help with cyberbullying detection, is that they can run constantly. Moreover, these models do not have the issues that arise in human moderators of not reaching all the posts, nor suffer from mental distress. AI could be implemented to remove offensive posts before a user gets the chance to see them, ensuring the posts that users see on their feeds are free from disturbing content. However, as a recent study [18] highlights, the quality and accuracy of the classifiers should be improved. Moreover, many of the research works develop classifiers, evaluate them, but they do not present the classifier integration into a system, nor provide an interface or web application to interact with it [12,17]. Other works focus on developing suitable datasets for cyberbullying detection in various languages, which is also a challenging task [14,25].

The work presented in this paper uses the Offensive Language Identification Dataset (OLID) [25], that was made available for the research community in the OffensEval competition [26]. This dataset has been then used in teaching curricula in various universities in the UK and USA. A detailed description of this project is presented also in a Master's dissertation [4].

One of the papers contributions is to consider the relationship between cyberbullying and cybersecurity. Cyberbullying is a rapidly growing and significant issue [24]. However, the current literature doesn't discuss the severity of the issue and its relationship to cybersecurity, so this will be explored in the paper. Following this, using the OLID collection, the paper illustrates how NLP and ML techniques can be applied to detect cyberbullying. This is done transparently and all the source code is available online on GitLab. The paper provides practical guidance on how developers can determine the optimal accuracy of their model using GridSearchCV. Finally, once the ML models have been built, they can be applied in a real example. The models are incorporated within a web application developed in Flask, allowing users interaction. Compared to related works, which do not consider the application of an ML model once it has been created, this paper is integrating it into a web application.

The paper is structured as follows: Sect. 2 explores the cybersecurity and cyberbullying relation, it also covers the required knowledge of ML and NLP. Section 3 discusses related works, Sect. 4, explains the methodology and the implementation of ML and NLP. Section 5 evaluates the results and Sect. 6 presents the integration of the ML models into a web application. Finally, Sect. 7 concludes the paper, its limitations, and future work.

2 Background

2.1 Cyberbullying and Cybersecurity

Cybersecurity has multiple definitions. One definition described by the Dutch Ministry of Security and Justice is "cybersecurity is the state of being free from danger or harm caused by the malfunction or failure of ICT or its misuse." [9].

Firstly the term "ICT" encompasses information, technology, and communication. Hence, social media is encompassed within ICT as it contains all three. The term "free from misuse of ICT" is imperative as misusing can include many acts. Humans can use it to misuse social media platforms to act offensively and cause harm to individuals. Since cybersecurity is the state of "being free from danger or harm", as harm can include attacks on an individual's mental or physical well-being, we can conclude that cyberbullying fits within the cybersecurity definition or umbrella.

Cybersecurity traditionally had been associated with the CIA triad: Confidentiality, Integrity, and Availability [10]. Cyberbullying can potentially violate the CIA triad, for instance, if a cyberbully manages to get ahold of a victim's account. In this case, it violates confidentiality as they are not authorized to view the data on their account. It can also violate integrity as they could change their profile data and should not be able to. Finally, it violates the availability of the account, if the cyberbully changes the password, locking the victim out.

Generally, we take cybersecurity seriously as it is a misuse of ICT. However, we should consider cyberbullying under the same umbrella as cybersecurity due to the act of cyberbullying being as equivalent to violating cybersecurity. Singhal et al. [20] emphasise the dangers and significance of cyberbullying, although the relationship to cybersecurity is not mentioned.

2.2 Machine Learning

Machine learning is a subset of AI, that is heavily used to learn from data, identify patterns, make informed decisions or predictions, based on passive observations [7,19]. In this paper supervised learning is used. For more details, a survey on text classification algorithms can be consulted [16].

2.3 Natural Language Processing

NLP employs multiple disciplines such as computational linguistics, statistics, ML, and deep learning models to allow computers to process human language.

Text classification can be broken into rule-based and ML-based. Rule-based can be particularly time-consuming, as it is difficult to gather and manage the rules. It could also require a professional on the domain [5]. Hence, the ML approach is considered ideal, as not having the mentioned limitations. Text classification usually follows a certain workflow [16]:

1. **Data gathering**: can be done manually, or automatically, via an API. In some cases, online datasets are already available.
2. **Pre-processing**: requires cleaning up the dataset. This could include noise removal, stop-words removal, spelling corrections, slang word removal, non-ASCII characters removal, stemming, and lemmatization.
3. **Feature Extraction**: separating and assembling the most prominent features to use in the model. Methods such as TFIDF (Term Frequency Inverse Document Frequency), BoW (Bag of Words), and N-gram can be used [16].
4. **Model selection**: appropriate models are trained on the dataset once the text has been pre-processed and the most significant features have been extracted. Parameters can be tuned for obtaining the best results.
5. **Evaluation**: once trained and tested, specific metrics are used to evaluate the results of the models, such as confusion matrix, accuracy, precision, recall, F1-score and others. A previous stage can be revisited to improve the scores.

3 Related Work

Recent systematic reviews of automatic cyberbullying detection in social media are given in [3,18]. Rosa et al. summarise in [18] important studies and datasets up to 2019, among them the Formspring dataset. In [13] the authors survey a related topic, the automatic detection of hate speech.

In this paper, we consider the OLID Twitter dataset from Zampieri et al. [25]. The authors provide this annotated, large-scale dataset and perform first experiments, setting the baseline for future research. The tasks proposed are (a) identify offensive tweets; (b) categorize the offence type; (c) offence target identification. As ML techniques, Zampieri et al. use SVM (Support Vector Machine), BiLSTM (Bidirectional Long Short-Term-Memory) and CNN (Convolutional Neural Network). The OLID collection was used in the OffensEval challenge and a summary of the submissions is presented in [26] – more than 100 teams participated in the challenge and they used a plethora of ML and NLP techniques.

In [27] a challenge dataset that contains tweets in English, Arabic, Danish, and Greek is proposed, with three tasks: (a) identify the language; (b) classify the offence type; (c) identify the target. The results of the various teams performing the tasks and given in [27].

Ahuja et al. use in [2] a Twitter dataset for sentiment analysis, having 4242 tweets (1037 negative, 1952 neutral, and 1252 positive). Six different algorithms (Decision Tree, SVM, KNN, Random Forest, Logistic Regression, and Naive Bayes) are used and the F1-score, accuracy, precision, and recall are measured at the end. For the feature selection, TFIDF and N-grams were selected and compared.

A detailed study, aiming to reproduce the literature findings on various datasets from Wikipedia, Formspring and Twitter, is presented in [8]. The authors develop four different deep learning models: CNN, Long Short-Term Memory (LSTM), Bidirectional LSTM (BLSTM), and BLSTM with attention. Then they transfer learning on another dataset that has been collected from YouTube and conclude that deep learning models outperform the machine learning models for the YouTube dataset.

Although the authors of these studies have applied the various ML models to different datasets, they do not consider the practical application or integration of the developed models. There is a gap in applying the ML models once they have been developed and this paper aims to illustrate how to address this in Sect. 6. Also, there is a gap in the relationship between cyberbullying and cybersecurity that has not been explored before and this is discussed in Sect. 2.1.

4 Methodology

4.1 Preliminaries

It is worth noting that the project's source code is publicly available on GitLab at https://gitlab.com/muali4/project. This allows the project to be transparent and allows users to browse the code freely. The source code contains an exercises folder, aimed to familiarise the users with the technologies. An Exploratory Data Analysis (EDA) was performed on the training dataset, allowing to identify the total entries, class split, missing values, and the average word distribution per task. This is presented in the webpage of the web application https://uali.pythonanywhere.com/ under "Project Explained". Another preliminary analysis using WEKA[1] was performed and its results can be seen in the "results" folder in GitLab. The parameters used for WEKA are given in the README.md under "WEKA".

4.2 Dataset Description and Tasks

The OLID collection was acquired from https://github.com/joeykay9/offenseval/. It was a part of the OffensEval challenge proposed at the International Workshop on Semantic Evaluation in 2019 [25].

The dataset included multiple files when downloaded. In "olid-annotation.txt" the three tasks were described. **Task A** aimed to determine if the tweet was offensive (OFF) or not (NOT). For **Task B**, if the tweet was

[1] https://git.cms.waikato.ac.nz/weka/weka.

found offensive at task A, the aim was to determine if the tweet was targeted (TIN) or untargeted (UNT). For **Task C**, if the tweet was offensive and targeted, the purpose was to determine who it was targeted towards. The tweet could target an individual (IND), group (GRP), or other (OTH).

The dataset had a training file "olid-training-v1.0.tsv" which contained 13,240 tweets and the label for each of the tasks A, B, and C. There were three test files "testset-levelX", where "X" is "a", "b", or "c". These test files contained only the testing tweets for each of the tasks. The labels for the testing tweets were separated into three files "labels-levelX.csv", where"X" is "a", "b", or "c".

It is worth acknowledging that while progressing on the tasks, there is less data available. If a tweet is not offensive, it cannot be targeted or untargeted, nor can it be targetted at an individual, group, or other. This explains the missing values.

4.3 Pre-processing

Two of the pre-processing steps were already done. All the URLs in the tweets were converted into "URL" and anytime a user was tagged into a tweet their handle was converted to "@USER" for anonymity but still maintaining the integrity of the tweet [27]. It is worth mentioning that we used regular expression operations[2] and the NLTK[3] (Natural Language Processing Toolkit) library for some of the other pre-processing. The workflow followed was:

Merging Datasets and Renaming Columns The first step was to merge the datasets for ease of management. We merged the "testset-levelX.csv" with the corresponding "labels-levelX.csv". We also renamed the headings and the file names for readability. This step was optional, but done for clarity.

Removing Numbers and Punctuation The next step was to remove any noise in the tweets. In the training dataset, we used regular expressions to remove any numbers and punctuation as this will not add any benefit in any of the three tasks.

Removing Emojis and Non-ASCII Characters We removed any non-ASCII characters and emojis as these are considered noise and will not be used to determine the outcome. Emojis and non-ASCII characters are used for human understanding rather than computer understanding. Using regex we removed these from the tweets.

[2] https://docs.python.org/3/library/re.html.
[3] https://www.nltk.org/.

Removing Whitespaces and Tabs We removed any whitespaces and tabs as these are again considered noise and will not help determine the class of any of the three tasks. Whitespaces and tabs are used for human understanding rather than computer understanding.

Lowercasing Text Lowercasing all the text was done for the reduction in feature selection. In the tweets, users might use uppercase letters emphasizing certain words. For instance, "YES" and "yes" can mean the same thing, the model will understand this as the same word rather than a different word.

Removing Stopwords Stopwords are considered the most common words that add no meaning to the text. For instance, the word "the" is used for human understanding of the text but is not required for computer understanding. Removing these also reduces the number of irrelevant features we consider. This was done using the *NLTK* library in Python.

PoS Tagging and Lemmatizing Finally, we decided to use lemmatization rather than stemming due to it yielding more relevant results [6,15]. We used the PoS (Part of Speech) tagging to tag the words in the tweet which is then used to lemmatize the word down to its root form. This was done to reduce the number of features and was achieved by using the *NLTK* library.

4.4 Feature Selection

The most frequently used methods are TFIDF, BoW, and N-gram. We opted for the TFIDF due to it being more meaningful than BoW. In [2], TFIDF yields about 3–4% better results than N-gram and other authors report better results for TFIDF compared to BoW [12]. The TFIDF algorithm works as:

$$w_{i,j} = tf_{i,j} * log\left(\frac{N}{df_i}\right)$$

where N is the number of documents, i is a term, j is a document, $tf_{i,j}$ is the number of occurrences of i in j, and df_i is the number of documents containing the term i.

4.5 Model Selection

We decided to use multiple different models with different underlying algorithms using the *sklearn*[4] library in Python:

- Support Vector Machine (SVM)
- Logistic Regression

[4] https://scikit-learn.org/stable/.

- Naive Bayes
- Decision Tree
- Voted Perceptron
- Convolution Neural Network (CNN)
- Random Forest
- Gradient Boosting

A dummy classifier was used to act as baseline or success criterion. It classifies all the data points to the one with the most frequent label in the dataset. If the above models surpass the score of the Dummy classifier, it can be considered as a success. In this step, we also tuned the parameters of the models. For this we used the GridSearchCV, which is a part of the *sklearn* library, to iterate through the listed parameters for each model and determine the ones with the highest score. Other authors, when applying machine learning in various domains, such as medical diagnosis [1] evaluated the performance of ML techniques with and without GridSearchCV. It is worth mentioning the positive improvements obtained when tuning the hyper-parameter GridSearchCV.

4.6 Evaluation

Once the ML models have been developed, they were compared by measuring the time taken to build them, the accuracy, precision, recall, and F1-score using the *classification report* in the *sklearn* library.

The evaluation results are shown in Table 1. The build time was measured using the *time*[5] library in Python, and this gives a better idea of scalability when applying the ML models in a web application. All the metrics were compared for each task and for each algorithm, to determine the optimal model to use in a web application.

5 Results Discussion

Table 1 presents the evaluation results and shows in bold the selected (best) algorithm and in italic the baseline Dummy classifier. Obviously, the Dummy classifier is significantly quicker than the other algorithms, due to it having no complex underlying algorithm and instead grouping the datapoints with the most frequent label.

For task A, the Random Forest was selected. It performed the best in accuracy, recall, and F1-score. The build time was significantly larger than the others. However, the accuracy/time trade-off was accepted for a higher score.

For task B, the Random Forest was selected. Although the Voted Perceptron performed better in the F1-score, the accuracy score did not exceed the success criteria. The next best F1-score was the CNN and it also did not exceed the success criteria. The Random Forest was the only one that exceeded the baseline for accuracy. Hence, the Random Forest was selected.

[5] https://docs.python.org/3/library/time.html.

Table 1. Performance of ML algorithms

Task	ML model	Build time (s)	Accuracy	Precision	Recall	F1-score
Task A	Dummy classifier	0.1441	0.72	0.36	0.50	0.42
	Naive bayes	0.2156	0.76	0.82	0.58	0.58
	SVM	14.3179	0.81	0.80	0.70	0.72
	Logistic regression	4.9334	0.80	0.81	0.66	0.69
	Random forest	18.1346	0.82	0.81	0.71	0.73
	Decision tree	5.5087	0.75	0.70	0.71	0.71
	Gradient boosting	4.0435	0.80	0.82	0.66	0.68
	Voted perceptron	0.5229	0.75	0.68	0.68	0.68
	CNN	1.5948	0.72	0.36	0.50	0.42
Task B	Dummy classifier	0.091	0.89	0.44	0.50	0.47
	Naive bayes	0.1888	0.89	0.44	0.50	0.47
	SVM	1.1761	0.89	0.44	0.50	0.47
	Logistic regression	0.2006	0.89	0.95	0.52	0.51
	Random forest	3.0416	0.90	0.85	0.57	0.60
	Decision tree	1.1679	0.84	0.60	0.60	0.60
	Gradient boosting	1.7762	0.89	0.70	0.52	0.50
	Voted perceptron	0.1053	0.87	0.67	0.67	0.67
	CNN	30.8263	0.83	0.61	0.63	0.62
Task C	Dummy classifier	0.0828	0.47	0.16	0.33	0.21
	Naive bayes	0.3685	0.49	0.34	0.35	0.27
	SVM	2.0119	0.60	0.40	0.46	0.42
	Logistic regression	8.1116	0.58	0.39	0.44	0.40
	Random forest	1.896	0.61	0.41	0.47	0.44
	Decision tree	0.9044	0.62	0.50	0.50	0.49
	Gradient boosting	4.185	0.62	0.42	0.48	0.44
	Voted perceptron	0.1192	0.57	0.52	0.50	0.50
	CNN	0.8426	0.47	0.16	0.33	0.21

For task C, it seemed like the Decision Tree performed better than the Random Forest in accuracy and F1-score. However, when we run the function to train the model repeatedly, the score for Decision Tree oscillated up or down. The Random Forest is more consistent, hence it was selected. The Voted Perceptron had the best F1-score, but performed inadequately compared to the Random Forest in accuracy.

The F1-scores presented in Table 1 for Random Forest are comparable to similar results from [26], further solidifying the obtained results. The CNN algorithm performed poorly across all the tasks, further investigation suggesting to attempt a different CNN architecture, such as the one used in [25]. Although, it could also be attributed to the hyper-parameters selected for the CNN and altering these could lead to better results.

Comparing our results to the current ML models used within social media is not feasible due to not having access to the architecture or results of these models used by the organisations. They are concealed and considered as black-boxes, hence they cannot be compared. This reinforces the need for transparency of the ML models, hence why we developed our project.

Although the results of the project are comparable to similar results from [26], the focus of the paper lies in providing transparency of the ML models developed and integrating them into a web application.

5.1 Threats to Validity

We identified the following threats to validity.

Data sampling bias: the training data set may not be representative for the entire population of tweets or for other social media platforms, hence the results of the research may not be generalizable. However, the OLID dataset [25] contains approximately 14,000 tweets and it is to date the most comprehensive Twitter dataset for this problem, to the best of our knowledge, with sufficient variety. For future work, we will consider including in training datasets from other sources, especially from other social media platforms.

Model selection bias: to mitigate this threat, we used a variety of machine learning algorithms. Moreover, the OLID dataset had *training* and *testing* data and the ML algorithm that performed best on the *testing* dataset was selected.

Model tuning and hyper-parameters: improper or biased hyper-parameter tuning may impact the integrity of the study. We have documented this process and used GridSearchCV for tuning the hyper-parameters.

Overfitting: to mitigate this, we used a large dataset with variety of labels and we used ensemble models for the results. We did not have a separate validation dataset to use, however this is something we could consider in the future work.

In addition to these specific mitigation strategies, we have ensured *transparency*, regarding dataset, ML algorithms, evaluation metrics, and *replication* of the research: all data and code used are available online, such that other researchers can replicate the results.

6 Model Integration into a Web Application

Once the ML models have been trained and compared, the chosen best model was integrated into a web application. The selected algorithm was the Random Forest for all the tasks.

Since the ML models were developed in Python using the *sklearn* library, Django or the Flask web framework could be used. Due to the lightweight and microframework features of Flask,[6] it was the ideal choice.

[6] https://flask.palletsprojects.com/en/2.3.x/.

Since the Random Forest took a relatively long time to train, we opted for the idea of saving the Random Forest for each task and testing it on the inputted tweet from the user. It is faster to load in an ML model rather than train it again and timely responsiveness is imperative for user engagement.

To make the integration easier, all the preprocessing and the task models were refactored into standalone functions in a separate file. The functions could then be used within the web Flask *views.py* file.

The next step was to save the trained ML models into a file allowing for it to be loaded into the *views.py* file of the Flask web application. The *pickle*[7] library in Python was used for this. The TFIDF feature selection for each of the selected ML models was also required and saved.

Once the models and their corresponding TFIDF feature selections are integrated, the user can input the text to be analysed via a form, having a restriction of 280 characters to emulate Twitter. The user input is preprocessed via the functions created, then the ML models are loaded, and the inputted tweet is classified. The results and any metrics collected are presented. It is worth mentioning that, in order to use the inputted tweet from one function to another, a *session* is needed. Additional features, such as the exploratory data analysis performed on the training data, can be viewed on the web application; this is done in runtime due to its speed.

The web application was deployed on a free hosting service *Pythonanywhere*. The selected architecture is server-client, with all the processing occurring on the server. It can be accessed via https://uali.pythonanywhere.com/.

This web integration is a proof of concept, demonstrating the feasibility of this approach. As the ML model can be successful integrated in web sites or social media platforms, one can go further and develop an API for handling of larger and more frequent data.

7 Conclusion and Future Work

This paper explored current solutions for automatic detection of cyberbullying, discussed the significance of cyberbullying and its relationship to cybersecurity. It considered also text classification and its workflow to develop ML-based solutions. Multiple ML models were analysed to determine the best-scoring one. Metrics such as build time, accuracy, precision, recall, and F1-score were collected to determine the best-performing algorithm for the given tasks.

Among the algorithms considered, Random Forest was a clear winner across the three tasks, and this model was implemented within a web application for users to interact with. The research demonstrates the further application of an ML model beyond training and testing it; it considers the transparency of the algorithm by making it available on GitLab. It also considers using GridSearchCV to fine-tune the parameters to achieve a better score across the metrics.

[7] https://docs.python.org/3/library/pickle.html.

Future work includes researching on improving the quality and accuracy of the CNN model. The hyper-parameters could be explored and developed to fit the tasks, we could also utilise a CNN model with a different architecture to determine its suitability. We could use other deep learning algorithms, such as LSTM, RNN (Recurrent Neural Network), BLSTM etc. These could be compared to the traditional ML algorithms.

Multi-threading or parallel computation could be considered and incorporated for faster training times on Random Forest and other models. We could use GPU's when developing the deep learning models.

Another promising tool to use for cyberbullying detection is using sentiment analysis. This could potentially aid the detection, given the negative emotion that is expected in case of a cyberbullying.

Different research directions consist in expanding the models by incorporating other languages within the dataset or considering voice recordings and videos, processing the sound waves and classifying media for detection of offensive content.

For further development on the application of the models, an API can be developed which can be invoked to determine if the tweet is offensive or not. This would allow for developers to implement this solution into their code-base at real-time. Finally, for the users of the web application we could implement a feedback system. Such system could incrementally update the dataset, based on the users input and retrain the ML model, thus improving it.

References

1. Ahmad, G.N., Fatima, H., Ullah, S., Saidi, A.S., et al.: Efficient medical diagnosis of human heart diseases using machine learning techniques with and without GridSearchCV. IEEE Access **10**, 80151–80173 (2022)
2. Ahuja, R., Chug, A., Kohli, S., Gupta, S., Ahuja, P.: The impact of features extraction on the sentiment analysis. Procedia Comput. Sci. **152**, 341–348 (2019)
3. Al-Garadi, M.A., Hussain, M.R., Khan, N., Murtaza, G., Nweke, H.F., Ali, I., Mujtaba, G., Chiroma, H., Khattak, H.A., Gani, A.: Predicting cyberbullying on social media in the big data era using machine learning algorithms: review of literature and open challenges. IEEE Access **7**, 70701–70718 (2019)
4. Ali, M.U.: Detection of cyberbullying on social media platforms using machine learning. Master's thesis, University of Bradford (2023)
5. Atanassova, I., Bertin, M., Mayr, P.: Mining scientific papers: NLP-enhanced bibliometrics. Front. Res. Metrics Anal. **4**(2), 1–3 (2019)
6. Balakrishnan, V., Lloyd-Yemoh, E.: Stemming and lemmatization: a comparison of retrieval performances. Lecture Notes Softw. Eng. **2**(3), 262–267 (2014)
7. Berry, M.W., Mohamed, A., Yap, B.W.: Supervised and Unsupervised Learning for Data Science. Springer, New York, NY (2019)
8. Dadvar, M., Eckert, K.: Cyberbullying Detection in Social Networks Using Deep Learning Based Models; A Reproducibility Study. arXiv preprint arXiv:1812.08046 (2018)

9. Fichtner, L.: What kind of cyber security? Theorising cyber security and mapping approaches. Internet Policy Rev. **7**(2) (2018)
10. Ham, J.V.D.: Toward a better understanding of "cybersecurity". Digital Threats: Res. Practice **2**(3), 1–3 (2021)
11. Hinduja, S., Patchin, J.W.: Bullying, cyberbullying, and suicide. Arch. Suicide Res. **14**(3), 206–221 (2010)
12. Islam, M.M., Uddin, M.A., Islam, L., Akter, A., Sharmin, S., Acharjee, U.K.: Cyberbullying detection on social networks using machine learning approaches. In: 2020 IEEE Asia-Pacific Conference on Computer Science and Data Engineering (CSDE), pp. 1–6. IEEE (2020)
13. Jahan, M.S., Oussalah, M.: A systematic review of hate speech automatic detection using natural language processing. Neurocomputing 126232 (2023)
14. Khan, S., Qureshi, A.: Cyberbullying detection in Urdu language using machine learning. In: 2022 International Conference on Emerging Trends in Electrical, Control, and Telecommunication Engineering (ETECTE), pp. 1–6. IEEE (2022)
15. Khyani, D., Siddhartha, B., Niveditha, N., Divya, B.: An interpretation of lemmatization and stemming in natural language processing. J. Univ. Shanghai Sci. Technol. **22**(10), 350–357 (2021)
16. Kowsari, K., Jafari Meimandi, K., Heidarysafa, M., Mendu, S., Barnes, L., Brown, D.: Text classification algorithms: a survey. Information **10**(4), 150 (2019)
17. Perera, A., Fernando, P.: Accurate cyberbullying detection and prevention on social media. Procedia Comput. Sci. **181**, 605–611 (2021)
18. Rosa, H., Pereira, N., Ribeiro, R., Ferreira, P.C., Carvalho, J.P., Oliveira, S., Coheur, L., Paulino, P., Simão, A.V., Trancoso, I.: Automatic cyberbullying detection: a systematic review. Comput. Hum. Behav. **93**, 333–345 (2019)
19. Sindhu, V., Nivedha, S., Prakash, M.: An empirical science research on bioinformatics in machine learning. J. Mech. Continua Math. Sci. **7**, 86–94 (2020)
20. Singhal, M., Ling, C., Kumarswamy, N., Stringhini, G., Nilizadeh, S.: SoK: Content Moderation in Social Media, from Guidelines to Enforcement, and Research to Practice. arXiv preprint arXiv:2206.14855 (2022)
21. Slonje, R., Smith, P.K., Frisén, A.: The nature of cyberbullying, and strategies for prevention. Comput. Hum. Behav. **29**(1), 26–32 (2013)
22. Snakenborg, J., Van Acker, R., Gable, R.A.: Cyberbullying: prevention and intervention to protect our children and youth. Prevent. Sch. Fail. Altern. Educ. Child. Youth **55**(2), 88–95 (2011)
23. Truell, A.D., Zhao, J.J., Lazaros, E.J., Davison, C., Nicley, D.L.: Cyberbullying: important considerations. Issues Inform. Syst. **20**(2), 83–88 (2019)
24. Waller, A.P., Lokhande, A.P., Ekambaram, V., Deshpande, S.N., Ostermeyer, B.: Cyberbullying: an unceasing threat in today's digitalized world. Psychiatr. Ann. **48**(9), 408–415 (2018)
25. Zampieri, M., Malmasi, S., Nakov, P., Rosenthal, S., Farra, N., Kumar, R.: Predicting the type and target of offensive posts in social media. In: Proceedings of the 2019 Conference of the North American Chapter of the Association for Computational Linguistics: Human Language Technologies, NAACL-HLT 2019, pp. 1415–1420. Association for Computational Linguistics (2019)

26. Zampieri, M., Malmasi, S., Nakov, P., Rosenthal, S., Farra, N., Kumar, R.: SemEval-2019 Task 6: identifying and categorizing offensive language in social media (OffensEval). In: Proceedings of the 13th International Workshop on Semantic Evaluation, SemEval@NAACL-HLT 2019, pp. 75–86. Association for Computational Linguistics (2019)
27. Zampieri, M., Nakov, P., Rosenthal, S., Atanasova, P., Karadzhov, G., Mubarak, H., Derczynski, L., Pitenis, Z., Çöltekin, Ç.: SemEval-2020 Task 12: multilingual offensive language identification in social media (OffensEval 2020). In: Proceedings of the Fourteenth Workshop on Semantic Evaluation, SemEval@COLING 2020, pp. 1425–1447. International Committee for Computational Linguistics (2020)

Analyzing Supervised Learning Models for Predicting Student Dropout and Success in Higher Education

Shraddha Bhurre(✉) and Shaligram Prajapat

International Institute of Professional Studies, DAVV, Indore, India
shraddhabhurre@gmail.com, shaligram.prajapat@iips.edu.in

Abstract. The global education industry has experienced significant transformations, yet it continues to grapple with challenges, particularly in higher education, such as declining student success rates and course abandonment. Addressing these issues necessitates proactive identification of students at risk of failure and timely intervention, through appropriate models. This study focuses on a comparative analysis of various supervised learning models that effectively predict student success and dropout. Specifically, the performance of five models, namely MLP (Multilayer Perceptron), SL (Simple Logistic), DT (Decision Tree), RF (Random Forest), and REPTree (Reduced Error Pruning Tree), is evaluated using a Kaggle dataset comprising 35 attributes and 4424 instances. The experiment encompasses all attributes and evaluates model accuracy based on Precision, Recall, and F-measure for all 5 models. Additionally, the study also compares Correctly and Incorrectly Classified Instances of these Machine Learning models. The findings reveal that Random Forest achieves the highest percentage of correctly classified instances and surpasses other supervised learning methods in terms of accuracy.

Keywords: Supervised Learning Models · MLP(Multilayer Perceptron) · SL(Simple Logistic) · DT(Decision tree) · RF(Random Forest) · And REPTree (Reduced Error Pruning Tree)

1 Introduction

Universities are producing a vast amount of electronic student data as a result of the digitization of academic operations. It is essential for them to successfully convert this enormous data into information that can aid educators, administrators, and policymakers in analyzing it to improve decision-making. Furthermore, by giving pertinent information to many stakeholders, it may help improve the quality of educational procedures. The use of data mining, data science, and machine learning to save, retrieve, and use educational data has become necessary for this sector to determine the effectiveness of expected outcomes. Students' success depends upon various factors and there are various supervised learning methods that develop such models which can analyze students' success and dropout rate. This paper focuses on analysis of 5 Supervised Machine Learning Models with several factors/attributes that could affect student success rates. The list of

N. Naik et al. (Eds.): UKCI 2023, AISC 1453, pp. 234–248, 2024.
https://doi.org/10.1007/978-3-031-47508-5_19

attributes/factors is given in Table 1. The Models used in this study are MLP(Multilayer Perceptron), SL(Simple Logistic), DT(Decision tree), RF(Random Forest), and REP-Tree (Reduced Error Pruning Tree). A real dataset from Kaggle is used to test these algorithms and comparative analyze these Machine Learning Models for successfully predicting student retention rate.

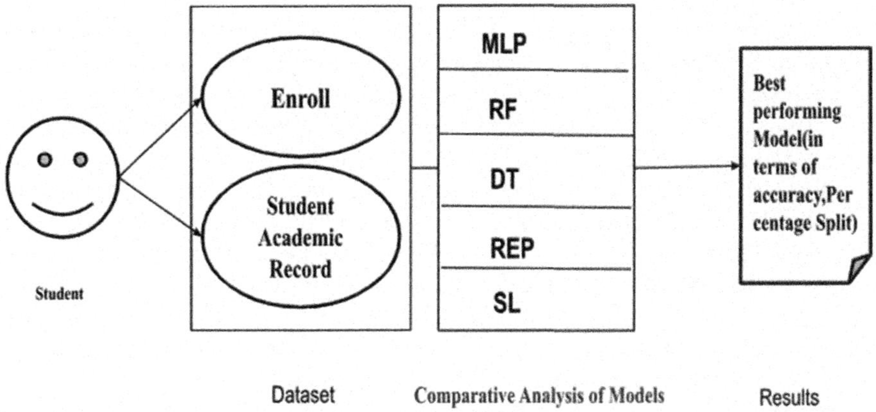

Fig. 1. Proposed Work

In Fig. 1 a simplified use case diagram for the proposed work. Here we have two primary use cases: "Enroll" and "Record Academic Performance. "Enroll: Represents the use case where a student enrolls in an educational institution or a course. It involves capturing information such as Marital status, Application mode, Application order, Nationality, Educational special needs, Debtor, Gender, Scholarship holder, Age at enrollment, and International. Record Academic Performance: This represents the use case where academic performance data is recorded. It involves capturing data related to Academic Records, including Curricular units 1st and 2nd semesters, evaluations, credits, grades, etc. A short description of the algorithms being used for this study has been presented in Sect. 3. The methodology and data set has been discussed in Sect. 4 with corresponding metrics, results, and analysis. Section 5 presents the conclusions of this study.

2 Related Work

Several Studies have been conducted on educational datasets to analyze and predict student performance.

The proper classification based on highly impacting variable need to be observed in order to accurately judge student success and dropout rate. Educational data is pre-processed and analyzed using multiple attributes and performance prediction methodologies.

It yielded an accuracy of 61.3 for the Decision tree, 73.8 for ANN, and 72.5 for NB. It also reveals that the Decision tree (C4.5/J48) has got only a 6% to 7% performance improvement using this method [1].

Regularly watching and monitoring the students can help them perform better. One study experimented on 120 students who took the same test as the subjects of the data collection. Course instructed by two unique lectures. The professors' internal quizzes, assignments, and mock exams served as the basis for their observations of the students. In this study, the students are divided into various groups based on their performance using Multilayer Perceptron (MLP), and many influences on the student's performance are also noted [2].

Using the Multilayer Perceptron (MLP) classifier, which is available in WEKA software implementations, one study intends to analyze the student's graduation predictions. This can be done in the fourth semester. Next, test MLP against Naive Bayes classification, IBk, and Tree J48 to compare its performance. In this study, the testing method is a cross-validation and percentage split. Root Mean Square Error (RMSE) and occurrences that were successfully classified throughout testing were the parameters. With an accuracy of J48 81.82% and the value of the least RSME, i.e. 0.273, MLP outperforms all competing methods in the Cross Validation mode. The accuracy of MLP mode on a percentage split is 92.31%, the same as that of Naive Bayes [3].

Based on both short- and long-term data, Tao Zhang et al.'s study [4] predicts changes in the student's performance over time. The long-term information, like a student's name or student ID, remains constant. This algorithm evaluated the results against the actual academic outcomes to predict the exam ranking for the 1995 students. Based on prediction accuracy, a comparison of step regression, decision tree, logistic regression, and SVM regression was made. The outcomes demonstrate that the SVM regression outperforms other methods (prediction 77%).

One study collected and looked at data from a public university in the Republic of Kosovo. The dataset includes information on student dropouts from the previous six academic years, along with demographic, academic, and enrollment information. The model and predictions were created using Logistic Regression, one of the most used Machine Learning and Artificial Intelligence techniques. With a percentage of 90% and an F1 score of 0.85, the results demonstrate that a high level of prediction accuracy was attained, demonstrating the model's excellent performance and the dependability of the results [5].

To see if improved classification performance could be attained, three decision strategies were employed to integrate the findings of the machine learning algorithms in different ways. Testing and training are the two steps of the experiment. These phases are carried out in three stages, each of which corresponds to a different semester stage. The number of attributes in the dataset was raised at each step until all attributes were added at the final stage. The dataset's most notable feature was that it exclusively included time-varying attributes rather than time-invariant ones like gender or age. This type of dataset has helped researchers figure out how much time-invariant data affect prediction accuracy. The accuracy and sensitivity of the experiment outcomes were assessed. The use of instance-based learning Classifier, Decision Tree, and Naïve Bayes in predicting students at risk are discussed in [6].

Dech Thammasiri et al. [8] proposed a model to provide an early classification of the poor academic performance of freshmen. Four classification methods with three balancing methods were applied to resolve the class imbalance problem. In a result, the combination of the support vector machine and SMOTE achieved the 90.24% highest overall accuracy.

A methodology for early classification of low freshman academic performance was put forth by Dech Thammasiri et al. [12]. To fix the class imbalance issue, four classification methods and three balancing methods were used. Results showed that the SMOTE and support vector machine combination had the highest overall accuracy of 90.24%.

Educational mining is concerned with establishing ways of obtaining knowledge from educational data, and it uses data mining techniques and tools to uncover hidden patterns and discover new knowledge from huge educational databases. Educational mining knowledge can be used for decision-making in higher education institutions.

According to [14], in Classification technique C5.0, Naïve Bayes Classification is the best algorithm in performance, and in Clustering Technique K-Mean clustering algorithm is the best algorithm or in the Association Rule Technique Apriori algorithm is the best and most accurate as compared to other algorithms.

To see if improved classification performance could be attained, three decision strategies were employed to integrate the findings of the machine learning algorithms in different ways. Testing and training are the two steps of the experiment. These phases are carried out in three stages, each of which corresponds to a different semester stage. The number of attributes in the dataset was raised at each step until all attributes were added at the final stage. The dataset's most notable feature was that it exclusively included time-varying attributes rather than time-invariant ones like gender or age. This type of dataset has helped researchers figure out how much time-invariant data affect prediction accuracy. The accuracy and sensitivity of the experiment outcomes were assessed. The use of instance-based learning Classifier, Decision Tree, and Naïve Bayes in predicting students at risk is discussed in [15].

3 Supervised Learning Models for Predicting Student Dropout and Success in Higher Education

3.1 Multilayer Perceptron

MLPs are neural network models that work as universal approximators, i.e., they can approximate any continuous function. MLPs are composed of neurons called perceptions. It consists of three types of layers—the input layer, output layer, and hidden layer. The required task such as prediction and classification is performed by the output layer. An arbitrary number of hidden layers that are placed in between the input and output layers are the true computational engine of the MLP [7].

3.2 Simple Logistic

Based on some dependent variables, logistic regression is used to forecast the probability of particular classes. It estimates the logistic of the outcome after computing the sum of the input features (there is typically a biased term).

For a binary classification problem, logistic regression's result is always between 0 and 1, which is appropriate. The likelihood that the current sample will be assigned to class = 1 increases with increasing value, and vice versa [9].

Table 1. Noteworthy observation from existing studies

Ref. No	Functionality	Algorithm Used	Scope of Prediction
[1]	Outlier detection in evaluating student performance	Not mentioned	Student performance
[2]	Analysis of student academic performance using the MLP model	Multilayer Perceptron (MLP)	Student academic performance
[3]	Prediction analysis of student graduation using MLP	Multilayer Perceptron (MLP)	Student graduation
[4]	Predicting the poor performance of college students based on behavior patterns	Not mentioned	College student performance
[5]	Application of logistic regression for predicting student Dropout	Logistic Regression	Student dropout
[6]	Identifying at-risk students using machine learning techniques	Instance-based learning Classifier, Decision Tree, Naïve Bayes	At-risk students
[7]	Performance comparison of MLP and Radial Basis Function Artificial Neural Networks in analyzing educational data	Multilayer Perceptron (MLP), Radial Basis Function Artificial Neural Networks	Educational data analysis
[8]	Introduction and explanation of the Random Forest algorithm	Random Forest	Not mentioned
[9]	Overview and Explanation of logistic regression in data analysis	Logistic Regression	Not mentioned
[10]	Prediction of student academic performance using decision tree algorithm	Decision Tree	Student academic performance
[11]	Comparative study of reduced error pruning methods in decision tree algorithms	Reduced Error Pruning Tree (REPTree)	Decision tree algorithms

(*continued*)

Table 1. (*continued*)

Ref. No	Functionality	Algorithm Used	Scope of Prediction
[12]	Prediction of freshmen student attrition and addressing the class imbalance problem	Support Vector Machine (SVM) with SMOTE	Freshmen student attrition
[13]	Discussion of data imbalance and solutions for achieving data democracy	Not mentioned	Not mentioned
[14]	Review of data mining techniques for educational data analysis	C5.0, Naïve Bayes Classification, K-Mean clustering algorithm, Apriori algorithm	Educational data analysis
[15]	Utilization of a real dataset from Kaggle to test algorithms for predicting student success and dropout rates	MLP, Simple Logistics, Decision Tree, Random Forest, REPTree	Student success and dropout rates

3.3 Decision Tree

The decision tree makes explicit all possible alternatives and traces each alternative to its conclusion in a single view, to make easy comparisons among the various alternatives.

Another main advantage is the ability to select the most biased feature and comprehensible nature. It is also easy to classify and Interpretable easily. Also used for both continuous and discrete data sets. Variable screening and feature section are good enough in the decision tree. By talking about its performance, non-linear does not affect any of the parameters of the decision tree [10].

3.4 Random Forest

Random Forest is a generic principle of classifier combination proposed in [8], which employs L tree-structured base classifiers h(X,n), N = 1,2,3,…L, where X is the input data and n is a family of identical and dependent distributed random vectors.

Every Decision Tree is created by picking data at random from the given data. For each Decision Tree, for example, a Random Forest (as in Random Subspaces) can be constructed by randomly picking a feature subset and/or randomly sampling a training data subset (the idea of Bagging). In a Random Forest, the features are randomly selected in each decision split. The correlation between trees is reduced by randomly selecting the features which improve the prediction power and result in higher efficiency.

3.5 REPTree

Reduced Error Pruning Tree ("REPT"), in its simplest form, is a quick decision tree learning method that constructs decision trees based on information gain or variance reduction. This algorithm's main pruning method involves using REP with back overfitting. It politely arranges the numerical attribute values once and handles the missing values in fractional situations using an embedded technique by C4.5. We can see that this algorithm used the C4.5 approach and included a basic REP count in its processing [11].

In the context of "Analyzing Supervised Learning Models for Predicting Student Dropout and Success in Higher Education," the following information about logistic regression, decision trees, random forest, and REPTree can be incorporated:

Logistic regression is employed to forecast the probability of specific classes based on dependent variables. It calculates the logistic of the outcome by summing the input features, typically including a biased term. For binary classification, logistic regression results always fall between 0 and 1, which is suitable. Higher values indicate an increasing likelihood of assigning the current sample to class = 1, and vice versa [9].

Decision trees present all possible alternatives and trace each alternative to its conclusion, facilitating easy comparisons. They offer advantages such as selecting the most influential feature and being easily interpretable and applicable to both continuous and discrete datasets. Decision trees also exhibit good performance in variable screening and feature selection. Furthermore, they remain unaffected by non-linear parameters [10].

Random Forest is a classifier combination technique that constructs L tree-structured base classifiers, where each Decision Tree is created by randomly selecting data from the given dataset. Random Forest employs features such as random feature subsets and training data sampling to create diverse decision trees (utilizing the concept of Bagging). In each decision split, Random Forest randomly selects features, reducing correlations between trees and improving prediction power and efficiency [8].

REPTree, short for Reduced Error Pruning Tree, is a rapid decision tree learning method that constructs decision trees based on information gain or variance reduction. This algorithm employs REP with back overfitting as its main pruning method. It also incorporates the C4.5 approach, using an embedded technique to handle missing values in fractional situations. REPTree enhances its processing by efficiently organizing numerical attribute values and employing a basic REP count [C4.5-inspired] [Reference for C4.5].

The characteristics of all the supervised learning algorithms are shown in Table 2.

4 Methodology for Analysis

4.1 Data Description

Dataset: [15].

This dataset contains data from a higher education institution on various variables related to undergraduate students, including demographics, social-economic factors, and academic performance. The dataset consists of 4424 students' data and 35 features. All features are listed in Table 3:

4.2 Experimental Setup

The Datasets with default features have been considered and were prepared using a percentage split, 90% of the data will be considered as a training set and 10% of the data will

Table 2. Characteristics of supervised learning algorithms

Model	Complexity	Accuracy	Interpretability
Multilayer Perceptron (MLP)	High	Medium	Complex, but can achieve high accuracy
Simple Logistic (SL)	Simple	Low	Simplest among five model
Decision Tree(DT)	Medium	Medium	More complex than Simple Logistic but still relatively easy to understand
Random Forest(RF)	Complex	High	The most complex
REPTree (REFP)	Medium	High	Good compromise between Simple Logistic and Random Forest

be considered as a testing set. The classification algorithms MLP(Multilayer Perceptron), SL(Simple Logistic), DT(Decision tree), RF(Random Forest), and REPTree (Reduced Error Pruning Tree), will then analyze the given dataset with 35 features described in Table 3. This will give an efficient analysis of given Supervised Learning Models on student success and dropout. The results of the Model were taken and compared.

4.3 Measures of Evaluation

A confusion matrix is a table that is used to define the performance of a classification algorithm. A confusion matrix visualizes and summarizes the performance of a classification algorithm (see Table 4).

Confusion matrices represent counts from predicted and actual values. The output "TN" stands for True Negative which shows the number of negative examples classified accurately. Similarly, "TP" stands for True Positive which indicates the number of positive examples classified accurately. The term "FP" shows a False Positive value, i.e., the number of actual negative examples classified as positive; and "FN" means a False Negative value which is the number of actual positive examples classified as negative. One of the most commonly used metrics while performing classification is accuracy. The accuracy of a model (through a confusion matrix) is calculated using the given formula below [13].

Accuracy

It represents the number of correctly classified data instances over the total number of data instances.

$$Accuracy = \frac{TN + TP}{TN + TP + FN + FP} \quad (1)$$

Table 3. Features Description

Attributes	Description	Value
Marital Status	Marital status of students	Categorical
Application Mode	The method of application used by the student	Categorical
Application Order	The order in which the student applied	Numerical
Course	The course is taken by the student	Categorical
Daytime/Evening Attendance	Whether the student attends classes during the day or in the evening	Categorical
Previous Qualification	The qualification obtained by the student before enrolling in higher education	Categorical
Nationality	The nationality of the student	Categorical
Mother's Qualification	The qualification of the student's mother	Categorical
Father's Qualification	The qualification of the student's father	Categorical
Mother's Occupation	The occupation of the student's mother	Categorical
Father's Occupation	The occupation of the student's father	Categorical
Displaced	Whether the student is a displaced person	Categorical
Educational Special Needs	Whether the student has any special educational needs	Categorical
Debtor	Whether the student is a debtor	Categorical
Tuition Fees are up to date	Whether the student's tuition fees are up to date	Categorical
Gender	The gender of the student	Categorical
Scholarship Holder	Whether the student is a scholarship holder	Categorical
Age at enrolment	The age of the student at the time of enrollment	Numerical
International	Whether the student is an international student	Categorical
Curricular units 1st sem (credited)	The number of curricular units credited by the student in the first semester	Numerical
Curricular units 1st sem (enrolled)	The number of curricular units enrolled by the student in the first semester	Numerical
Curricular units 1st sem (evaluations)	The number of curricular units evaluated by the student in the first semester	Numerical
Curricular units 1st sem (approved)	The number of curricular units approved by the student in the first semester	Numerical

Precision

Table 4. Confusion Matrix

Total Count		Predicted Value	
		Negative	Positive
Actual Value	Negative	TN(True Negative)	FP(False Positive)
	Positive	FN(False Negative)	TP(True Positive)

It is defined as the ratio of correctly classified positive instances (True Positive) to the total number of classified positive instances (either correctly or incorrectly).

$$Precision = \frac{TP}{TP + FP} \tag{2}$$

Recall

It is calculated as the ratio between the number of Positive samples correctly classified as Positive to the total number of Positive samples. The recall measures the model's ability to detect positive samples. The higher the recall, the more positive samples detected.

$$Recall = \frac{TP}{TP + FN} \tag{3}$$

F1 Score

It is defined as the harmonic mean of precision and recall.

$$F1Score = 2 \times \frac{(Precision * Recall)}{(Precision + recall)} \tag{4}$$

CCI(Correctly Classified Instances)

The sum of TP(True Positive) and TN(True Negative) is CCI.

$$\mathrm{CCI} = \mathrm{TP} + \mathrm{TN} \tag{5}$$

ICI(Incorrectly Classified Instances).

The sum of FP(False Positive) and FN(False Negative) is ICI.

$$\mathrm{ICI} = \mathrm{FP} + \mathrm{FN} \tag{6}$$

4.4 Computation

First, all 5 algorithms process the dataset for percentage split of 90:10 ratio, that is 90% data for training and 10% for testing. Here Data is of 4424 students, out of which 90%(3982) taken as training data and 10%(442) as testing data. Through Multilayer

Table 5. Confusion Matrix for MLP

a	b	c	
111	12	14	a = dropout
10	200	8	b = graduate
17	30	40	c = enrolled

Perceptron following Confusion Matrix is generated which further will be used to give TPR and FPR values.

For a,(dropout)

From Table 5

True Positive(TP) = 111
False Positive(FP) = 10 + 17 = 27
True Negative(TN) = 200 + 40 + 8 + 30 = 278
False Negative(FN) = 12 + 14 = 26

$$Precision = \frac{TP}{TP+FP} = \frac{111}{111+27} = \mathbf{0.804}$$

$$Recall = \frac{TP}{TP+FN} = \frac{111}{111+26} = \mathbf{0.810}$$

$$F1Score = 2 \times \frac{(Precision*Recall)}{(Precision+recall)} = 2 \times \frac{(0.804*0.810)}{(0.804+0.810)} = \mathbf{0.806}$$

For b,(graduate)

True Positive(TP) = 200
False Positive(FP) = 12 + 30 = 42
True Negative(TN) = 111 + 14 + 17 + 40 = 182
False Negative(FN) = 10 + 8 = 18

$$Precision = \frac{TP}{TP+FP} = \frac{200}{200+42} = \mathbf{0.826}$$

$$Recall = \frac{TP}{TP+FN} = \frac{200}{200+18} = \mathbf{0.917}$$

$$F1Score = 2 \times \frac{(Precision*Recall)}{(Precision+recall)} = 2 \times \frac{(0.826*0.917)}{(0.826+0.917)} = \mathbf{0.868}$$

For c,(enrolled)

True Positive(TP) = 40
False Positive(FP) = 14 + 8 = 22
True Negative(TN) = 111 + 12 + 10 + 200 = 333
False Negative(FN) = 17 + 30 = 47

$$Precision = \frac{TP}{TP+FP} = \frac{40}{40+22} = \mathbf{0.645}$$

$$Recall = \frac{TP}{TP+FN} = \frac{40}{40+47} = \mathbf{0.459}$$

$$F1Score = 2 \times \frac{(Precision*Recall)}{(Precision+recall)} = 2 \times \frac{(0.645*0.459)}{(0.645+0.459)} = \mathbf{0.536}$$

Overall accuracy, CCI and ICI for MLP model:

$$Accuracy = \frac{TN+TP}{TN+TP+FN+FP} = \frac{351}{278+111+26+27} = 0.794 = 79.41\%$$

Weighted Average,

Precision = 0.784
Recall = 0.794
F-measure = 0.785

CCI(Correctly Classified Instances) = 351.

ICI(Incorrectly Classified Instances) = 91.

In this way, calculations for the rest of the supervised models SL(Simple Logistic), DT(Decision tree), RF(Random Forest), and REPTree (Reduced Error Pruning Tree) have been done through their respective Confusion matrix. The detailed result is discussed in the following section.

4.5 Result Analysis

In order to compare the performance of all 5 Models (MLP (Multilayer Perceptron), SL (Simple Logistic), DT (Decision Tree), RF (Random Forest), and REPTree (Reduced Error Pruning Tree)), this study will compute the their Accuracy, Precision, Recall, F1 Score, CCI(Correctly Classified Instances) and ICI(Incorrectly Classified Instances) through Confusion Matrix for each Model as we given in Table 5.

The Precision will show the ratio between the number of students Dropout, Graduate, and Enrolled that are correctly predicted and the total number of students are Dropout, Enrolled and Graduate (both correctly and incorrectly predicted). In case of MLP it is 0.784.

Accuracy will be the ratio between number of correct predictions for dropout, graduate and enrolled student and total number of predictions. In case of MLP it is 79.41%.

Recall is a measure of completeness, which represent the ratio between number of dropout, graduate and success students that are correctly predicted and the sum of True Positive and False Negative Instances. In case of MLP it is 0.794.

F1 Score is harmonic mean of precision and recall. Which is 0.785 in case of MLP.

CCI will be computed as how many total True Positive and True Negative instances are there. In case of MLP it is 351.ICI will be Computed through how many False Positive and False Negative are there. In case of MLP it is 91.

Table 6. Performance Comparison of all Supervised Learning Models (90% split)

Model	Accuracy (%)	Precision	Recall	F1 Score	CCI	ICI
MLP	79.41%	0.784	0.794	0.785	351	91
SL	78.73%	0.774	0.787	0.77	348	94
DT	76.01%	0.747	0.76	0.75	336	106
RF	80.99%	0.8	0.81	0.797	358	84
REP Tree	76.47%	0.745	0.765	0.738	338	104

Table 6 show the experiment results of (MLP (Multilayer Perceptron), SL (Simple Logistic), DT (Decision Tree), RF (Random Forest), and REPTree (Reduced Error Pruning Tree)). In which Random Forest have highest accuracy among all. Figure 2 and 3 shows the graphical representation for Comparison of performances of all 5 Models.

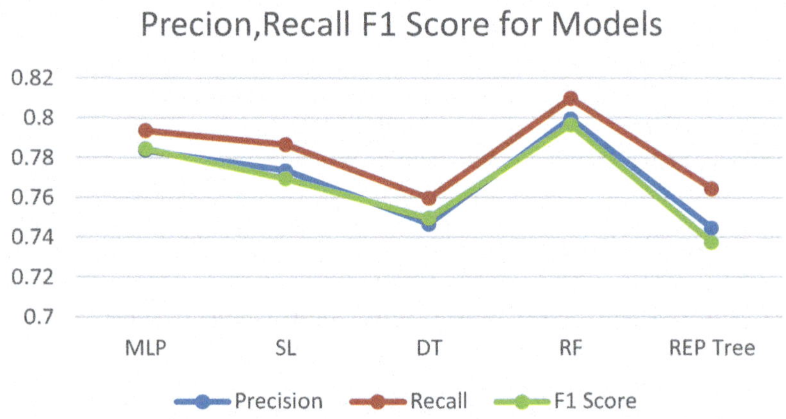

Fig. 2. Precision, Recall and F1 Score Comparison graph

Figure 4 represented the Correctly and Incorrectly Classified Instances. That is Correctly and incorrectly classified a(dropout), b(graduate) and c(Enrolled) students. Among all the Models here, Random Forest predict maximum correct instances that is 358.

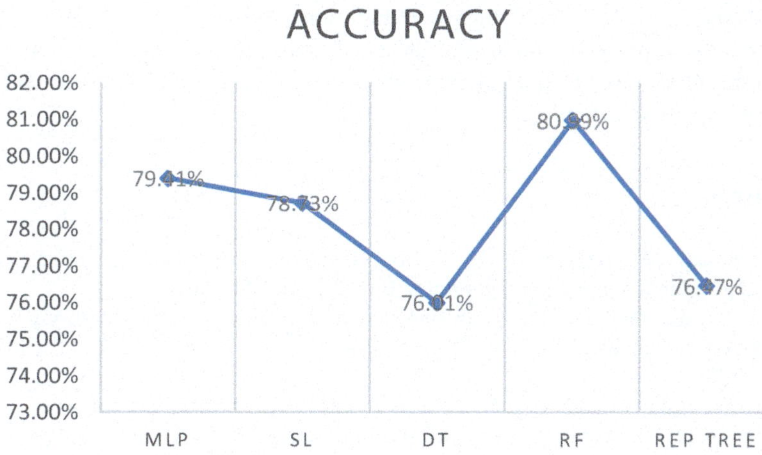

Fig. 3. Accuracy Graph for all Models

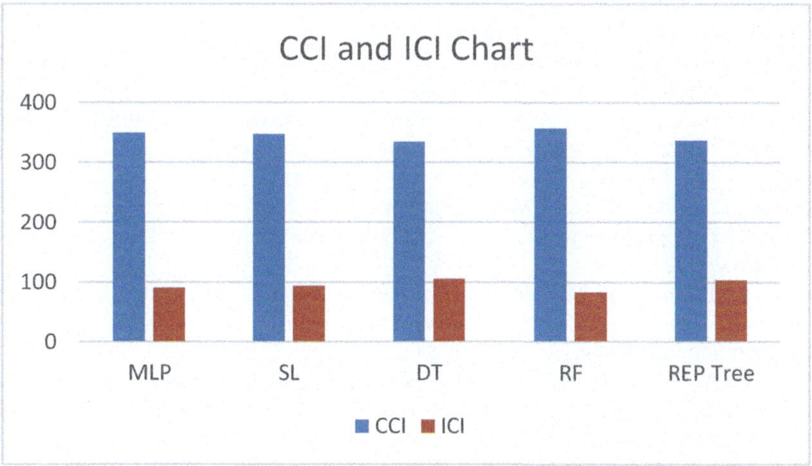

Fig. 4. Correctly and Incorrectly Classified Instances by all Models

5 Conclusion

This study explores the performance of various supervised learning algorithms, namely MLP (Multilayer Perceptron), SL (Simple Logistic), DT (Decision Tree), RF (Random Forest), and REPTree (Reduced Error Pruning Tree), using a dataset comprising 4424 students. The dataset includes 35 features, as described in Table 3.

The results obtained from the analyzed algorithms are presented in Table 6. The experiment was conducted a percentage split approach. The ratio taken was 90:10 that is 90% training data and 10% testing data. Among the evaluated algorithms, Random Forest demonstrated the highest accuracy 80.99%.

Based on these findings, it is suggested that future research should focus on applying this comparative analysis on Indian educational datasets. Such comparison will be useful in developing predictive and analytical mechanisms to assess student performance in the online learning environment.

References

1. Ajith, P., Sai, M., Tejaswi, B.: Evaluation of student performance: an outlier detection perspective. Int. J. Innov. Technol. Explor. Eng. **2**(2), 40–44 (2013)
2. Sinthia, G., Balamurugan, M.: Analyzing student's academic performance using multilayer perceptron model. Int. J. Recent. Technol. Eng. (IJRTE) **7**(5S3), 156–160 (2019). ISSN: 2277-3878
3. Windarti, M., Prasetyaninrum, P.T.: Prediction analysis student graduate using multilayer perceptron. In: International Conference on Online and Blended Learning, pp. 53–57 (2019)
4. Zhang, X., Sun, G., Pan, Y., Sun, H., Tan, J.: Poor performance discovery of college students based on behavior pattern. In: 2017 IEEE, pp. 1–8 (Aug. 2017)
5. Ujkani, B., Minkovska, D., Stoyanova, L.: Application of logistic regression technique for predicting student dropout. In: 2022 XXXI International Scientific Conference Electronics (ET), Sozopol, Bulgaria, pp. 1–4 (2022)
6. Er, E.: Identifying at-risk students using machine learning techniques: a case study with IS 100. Int. J. Mach. Learn. Comput. **2**(4) (2012)
7. Kayri, M.: An intelligent approach to educational data: performance comparison of the multilayer perceptron and the radial basis function artificial neural networks. Educ. Sci.: Theory Pract., 1247–1255 (2015)
8. Breiman, L.: Random forests. Mach. Learn. **45**, 5–32 (2001)
9. Maalouf, M.: Logistic regression in data analysis: an overview. Int. J. Data Anal. Tech. Strat. **3**(3), 281–299 (2011)
10. Hasan, R., Palaniappan, S., Raziff, A.R.A., Mahmood, S., Sarker, K.U.: Student academic performance prediction by using decision tree algorithm. In: 4th International Conference on Computer and Information Sciences (ICCOINS), pp. 1–5 (2018)
11. Mohamed, W.N.H.W., Salleh, M.N.M., Omar, A.H.: A comparative study of reduced error pruning method in decision tree algorithms. In: 2012 IEEE International Conference on Control System, Computing and Engineering, pp. 392–397 (2012)
12. Thammasiri, D., Delen, D., Meesad, P., Kasap, N.: A critical assessment of imbalanced class distribution problem: the case of predicting freshmen student attrition. Expert Syst. Appl. **41**(2), 321–330 (2014)
13. Kulkarni, A., Chong, D., Batarseh, F.A.: Foundations of data imbalance and solutions for a data democracy. In: Data Democracy, pp. 83–106. Academic Press (2020)
14. Sharma, P., Sharma, S.: Data mining techniques for educational data: a review. Int. J. Eng. Technol. Manag. Res., pp. 166–177 (2018). https://doi.org/10.29121/ijetmr.v5.i2.2018.641
15. Realinho, V., Machado, J., Baptis, L., Martins, M.V.: Predict students' dropout and academic success (1.0). Zenodo. https://doi.org/10.5281/zenodo.5777340

An Exploratory Ukraine Rising Commodities Price Analysis: Towards a Resilient Food System

Hiral Arora[1](✉), Ambikesh Jayal[2], and Edmond Prakash[3]

[1] Somerville School, Noida, India
hiralarora14@gmail.com
[2] Information Technology and Systems, University of Canberra, Bruce, ACT, Australia
ambi.jayal@canberra.edu.au
[3] Centre for Creative Technologies Research, University for the Creative Arts, Farnham, UK
edmond.prakash@uca.ac.uk

Abstract. Indeed, Ukraine has faced adverse situations that have significant implications for commodity price fluctuations which lead to regional and global poverty and food security crises. This study is concerned with the Ukraine commodity situation of the past 5 years due to the pandemic and the Russia-Ukraine war. Statistical analyses are performed to get insights into commodities and wavering prices. This study elicits knowledge from World Food Program (WFP) Ukraine dataset and demonstrates the stable price commodities and outliers price commodities corresponding to a particular city and time. Hence, exploratory commodity price is performed to present stability analysis, outlier analysis, adverse time commodity price analysis, and temporal vegetable price analysis. The linear regression (LR) model has also been used to predict the forecast of commodity prices based on historical data. The LR model fits well and has achieved less MSE, RMSE, and MAE loss concerning all training, validation, and test dataset.

Keywords: Ukraine Commodity · Food Price · Analysis · Visualization · Food Price Pre- diction · Linear Regression

1 Introduction and Related Literature

From earlier times, Ukraine's land is known for its fertile nature and fairly high volume of agricultural production. The country plays a vital role in global food stability, also acknowledged as the "Breadbasket of Europe". Ukraine is one of the world's top agricultural producers of various commodities such as wheat, corn, and barley. It has vast agricultural potential due to its fertile soil, pleasing climate, and extensive land bank. In recent years, Ukraine's commodities experienced several fluctuations due to reasons such as natural disasters, political instability, medical emergencies (COVID), Russia Ukraine invasion, etc. On 24 February 2023, an article published by World Food Program (WFP) [1] highlights that Ukraine and its neighborhood have been upended in the sense of food after a year of war began. This has enormously raised food price challenges for WFP humanitarian operations. The fallout of war has taken countries

N. Naik et al. (Eds.): UKCI 2023, AISC 1453, pp. 249–258, 2024.
https://doi.org/10.1007/978-3-031-47508-5_20

such as Lebanon, Sudan, Yemen, and many more a step closer to the hunger catastrophe because of Ukraine's agricultural ruination. The effects of war due to various military equipment's devastated the cultivation which resulted in significant setbacks for commodity prices in Ukraine. WFP humanitarian operations work towards the removal of war debris and improvise land for cultivation by restoring agricultural activities. A number of studies and explorations have been performed by researchers on diverse crises and risks of Ukraine's commodity shifts.

In 2023, Arndt et al. published work on the crisis of agriculture food systems, poverty, and food insecurity in 19 developing countries as an impact of the Russia-Ukraine war [3]. The study outcome confirms the adverse effect of crisis specifically poverty and hunger. Even, countries will be suffered from higher food prices. The study analysis depicts world commodity price changes from mid-2021 till mid-2022, estimated impacts on national and agrifood system GDP, and estimated impact on poverty and food security worldwide. Another relevant research in the same direction is presented by Mottaleb and fellow researchers [9], they examined the impact of Russia's invasion of Ukraine on wheat price, its consumption, and calorie intake from wheat. The study is performed on 160 countries' online data-bases taken from the Food and Agriculture Organization (FAO) of the United nation from the years 2016 to 2019. The study claims that due to a reduction in wheat export, developing countries need to explore new land areas for wheat production. Chepeliev et al. in their 2023 publication assess the Ukraine war implications for global and European agriculture. The insightful and useful aspect covered in this research paper is showcasing change in agriculture and food export across countries where the plot is shown to depict war-related agriculture shock, fertilizer-related shock, weather shocks, etc. [5]. A number of research works have been done on Russia-Ukraine war to find war implications for global and regional food security and corresponding policy responses [1]. Therefore, the data analysis and studies in this area are primarily done in the context of the global food crisis, food security, or food price escalation [2, 4, 6, 7].

The research work discussed in this manuscript is focused specifically on Ukraine's food price study and exploration in order to portray the Ukraine country food situation as the WFP article quotes that roughly one in three families in Ukraine i.e. approximately 11 million people are food insecure. This analysis aims to study commodities price fluctuation in Ukraine since January 2018. As per our knowledge and literature study, exploration of food price impact specifically on Ukraine country has not been explored before. This work studies a detailed analysis of commodities and presents data insights that might help Ukraine's humanitarian operations and support government organizations. This work can support drafting policies including near-term improvement of domestic commodity production to meet substantial and country- sufficient food production goals majorly to manage the forecasted food crisis in Ukraine. The contributions of research work are as follows:

- Ukraine Commodity price stability is performed to get to know the commodities with less or more fluctuating prices;
- Outliers corresponding to location and date are analyzed which show the highest or out-of-range commodity price;

- Adverse time commodity price analyses have been done and a comparison Pre-COVID, during COVID, and war period commodity price is depicted;
- Temporal vegetable price analysis is explored and results show price variation based on commodity requirements corresponding to Ukraine's situation;
- Finally, linear regression as a regression prediction model is applied to forecast the commodities' prices. Results depict the good fit of the model for commodity price prediction.

The organization of the rest of the study is as follows. Section 2 details about dataset used for the Ukraine region. Section 3 depicts exploratory Ukraine commodity analysis and data insights corresponding to each analysis. Section 4 details Ukraine's commodity price prediction model and model results are displayed in Sect. 5. Finally, Sect. 6 concludes the study and displays a few future research directions.

2 Dataset Information

The Ukraine food price [1] dataset used for this research is taken from Humanitarian Data Exchange (HDX) data repository. This repository sourced data from World Food Program (WFB) where 98 countries' food price datasets are accessible. Our targeted Ukraine food price dataset is available from 2004 onwards and consists of more than 100000 records. Although for this study, the dataset has been reduced and the 2018 onwards dataset is taken to widen the study for recent adverse situations such as COVID and War period. The dataset detailed on which the present study is shown is depicted in Fig. 1. Six commodity categories are available along with varying time price data records. The category and frequency of data in that specific category are depicted in the bar graph. The overall food data study is performed on the timely fluctuating price of six commodity categories. Maximum timely food price variations data are available in the cereals and tubers category, on the other end minimum data is available in the miscellaneous category. The dataset of 24 cities of Ukraine country is considered and those cities are highlighted by blue dot in Fig. 2.

	category	Frequency
0	cereals and tubers	15041
1	meat, fish and eggs	8962
2	milk and dairy	3842
3	miscellaneous food	1280
4	non-food	13962
5	oil and fats	3806
6	vegetables and fruits	6405

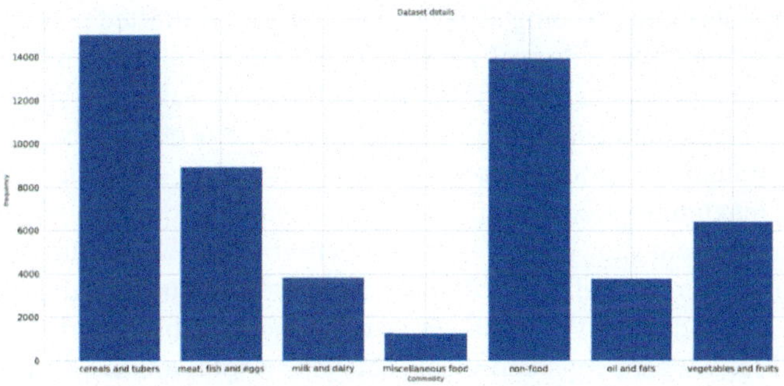

Fig. 1. Ukraine Food price Dataset taken for study

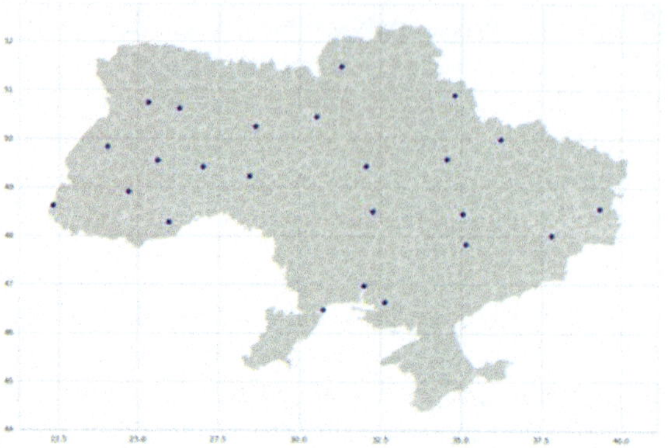

Fig. 2. Ukraine 24 countries taken for rising commodities Price Analysis highlighted by blue dots

3 Exploratory Ukraine Commodity Analysis

This research paper is written to portray insights into Ukraine's commodity fluctuating using visual relationships. The factors considered for analysis are time and location for all the commodities that exist in the dataset. This research paper discerns the commodity price variations in diverse cities of Ukraine according to adverse scenarios that Ukraine faced during the past 6 years period due to pandemics, war, and weather situations. Also, this is useful for government policy-making with respect to improvising agriculture decline due to land distortions and later adverse weather that lowered yield and increased commodity prices All the performed Commodity price analyses are discussed in further subsections.

3.1 Food Price Stability and Outlier Analysis

Commodities' price fluctuations, stability, and outlier detection are discussed to get insights into Ukraine's rising commodities analysis. This analysis is helpful for government, humanitarian service providers, non-profile government organizations, and even individual human being also [8]. The box plot is drafted in order to measure the stability of commodities. Figure 3 shows the plotted box plot for milk and dairy products where it is clearly depicted that milk price is the most stable out of the other two products. The sour cream and curd price interquartile range is almost the same but the minimum and maximum curd price is more in comparison. Also, curd price consists of outliers towards the maximum price i.e. curd price raised much more than the expected maximum price. The adjacent shown table is of all 6 outlier points when curd price touched a peak in two cities Kherson and Kyiv continuously for 5 months just after the war began.

Second, Non-Vegetarian Food commodity stability and analysis outcome is reflected in Fig. 4 where eggs and meat (chicken, whole) are the most stable items in the entire 5 years period. Although all the commodities under this category contain outliers, observation claims the peak commodity price period of outliers for food items under this commodity category is the same as was in the milk and dairy product commodity category. i.e. Non-Veg Commodity's price was maximum for the 6 months after the war began.

date	market	commodity	usdprice
15-03-2022	Kherson	Curd	6.3585
15-04-2022	Kherson	Curd	6.0124
15-04-2022	Kyiv city	Curd	5.8755
15-05-2022	Kherson	Curd	6.1333
15-05-2022	Kyiv city	Curd	5.8158
15-06-2022	Kherson	Curd	6.1926
15-07-2022	Kherson	Curd	6.2478
15-07-2022	Kyiv city	Curd	5.8715

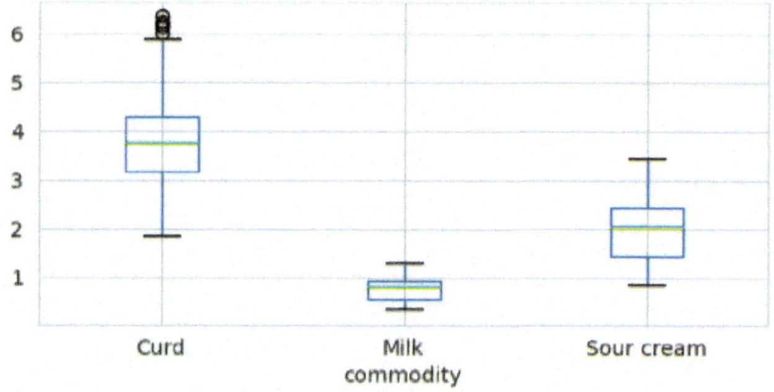

Fig. 3. Milk and Dairy Food Price Stability Analysis and Outliers

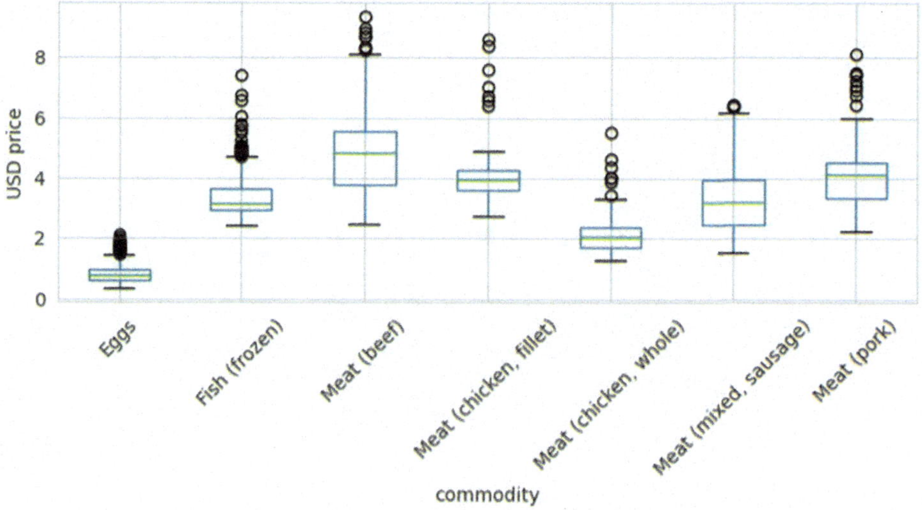

Fig. 4. Meat, Fish, and Eggs Food Price Stability Analysis and Outliers

The third commodity in the list consists of butter, fat, and oil prices. Figure 5 shows the variation in prices over the 5 years using a box plot diagram. The box plot results state that the median line of butter and oil prices is towards the minimum boundary i.e. maximum time price was less. Whereas, the maximum butter price is reaching to £7.2 without any outliers. Another observation about the outlier is the same as in the previous two categories i.e. price increase from Mar 2022. Overall, this commodity category data distribution is shown using a violin plot in Fig. 5 itself. This violin plot depicts the higher probability of these commodity observations in the range from £1 to £2 and the thin range shows the lower probability of observations in the dataset. The violin plot is narrowing down mainly after £2 and gets thinner up to £7.256.

One other interesting pattern observation is, while visualizing Non-Food commodities price analysis shown in Fig. 6. Almost all commodities' price is stable with less variation except imported antibiotics. According to observations, the minimum price of

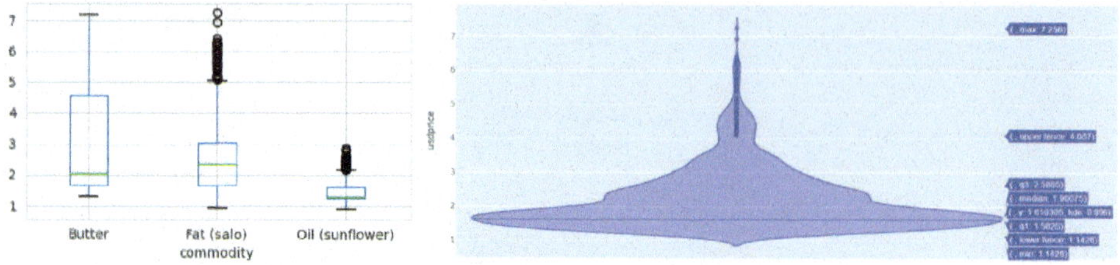

Fig. 5. Oil & Fat Price Stability and price variation Analysis

imported antibiotics is also much higher than any other commodity in this category. But, price escalations in terms of outliers are after the war period and in war areas. Actual data on the WFP website is available till April 2023 and this has been observed that usually, all the commodities prices are rapidly escalating.

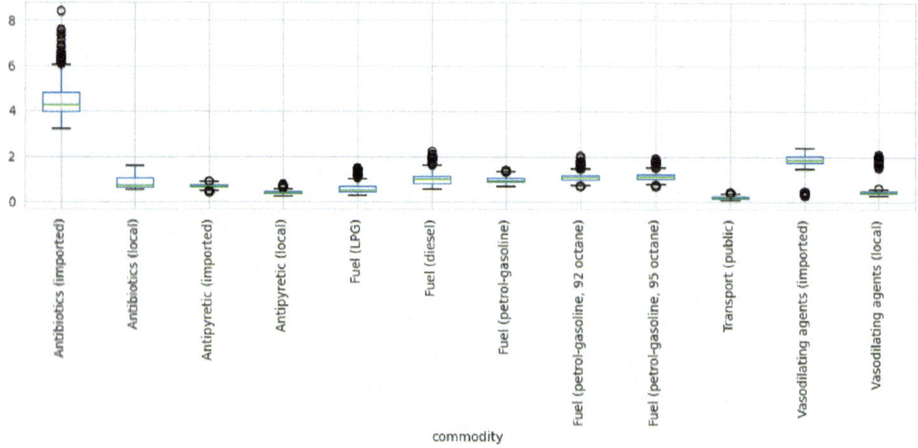

Fig. 6. Non-Food Price Stability Analysis

3.2 Adverse Time Commodity Analysis

Generally, the price of almost all commodities is pushed up in adverse conditions such as Russia-Ukraine War, the Pandemic, and extreme weather situations. To understand this, the price variation plot is drafted for three time periods pre-pandemic, during the pandemic, and war period those are shown in red, green, and blue color respectively (see Fig. 7).

Almost all the commodities are having fewer prices before the COVID period as compared to other considered two time spans (see red line in plot). Further, the green line i.e. Pandemic period price line is in the middle maximum amount of time. Finally, the commodities price is at its peak after the war period and records depict an increase in commodity prices constantly which needs to be tackled. The improvisation of food

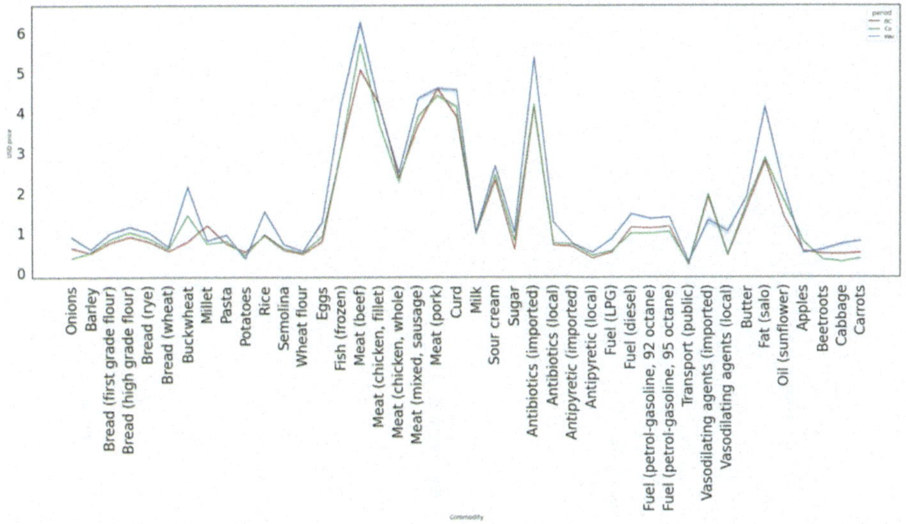

Fig. 7. Adverse Time Commodity Price Variation

price stability locally in Ukraine is a reflection of global (especially Europe) food crisis management, estimation, and successful recovery plan.

3.3 Temporal Vegetable Price Analysis

This analysis is performed to analyze the price variation on daily needs such as vegetables. The Analysis is quite interesting and shows the high price of "Apple" fruit during the pandemic period. The price of all vegetables was rapidly increasing during wartime and came to an almost stable situation after 8 months. This is seen in the plot shown in Fig. 8(a) that price is again at a peak for vegetable commodities such as Onions, Carrots, and a similar pattern is reflected for other commodities as well those are available in the overall commodity price dataset taken into consideration. Therefore, according to this study, Humanitarian operations need to consider these as a priority and find out an effective solution to handle this situation. Some Government agriculture policies can help in stabilizing commodities prices and save Ukraine from hunger situation. Figure 8(b) shows all outliers towards the extremely high end for all the vegetable commodities.

4 Ukraine Commodity Price Prediction Model

Statistical modeling technique linear regression has been used to analyze the relationship between two independent variables (location and date) and the dependent variable (USD price). Linear regression assumes a linear relationship between the independent and dependent variables i.e. changes in independent variables are associated with proportional changes in dependent variable. Regression model is a suitable model for predicting the commodity price. Linear regression model provides good fit if dependent and independent factors are having linear relationship.

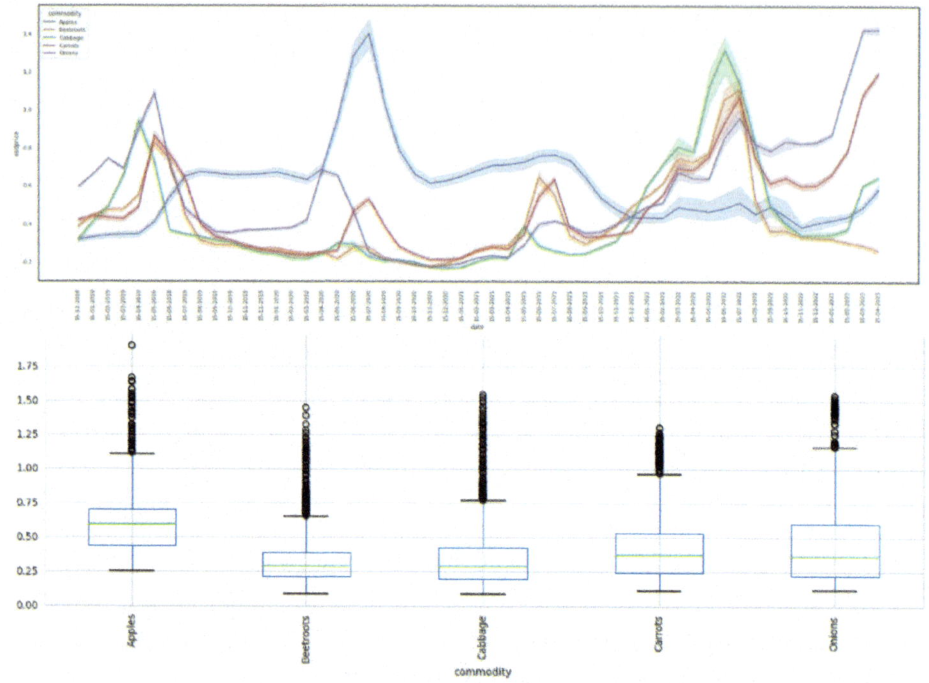

Fig. 8. (a) Temporal Vegetable price variation analysis (b) Box plot for vegetable commodity

The equation used for multiple linear regression is written as

$$USDPrice = X0 * Longitude + X1 * Latitude + X2 * date + Int + \varepsilon \quad (1)$$

where USDPrice is the dependent variable, X_0, X_1, X_2 are slope coefficient for the dependent variable is Longitude, Latitude, and date of historical commodity price respectively. *Int* is intercept coefficient which needs to be calculated based on training model and ε is error.

5 Ukraine Commodity Price Set up and Prediction Result

The performance of Ukraine commodity price prediction is validated using Mean Square Error (MSE), Root Mean Square Error (RMSE), and Mean Absolute Error (MAE). These performance measures are commonly used to measure the average difference between predicted and actual values in a regression problem. The equations used for these measures are as follows:

$$MSE = \frac{1}{n}\sum_{i=1}^{n}(y_i - \hat{y}_i)^2 \quad (2)$$

$$RMSE = \left(\frac{1}{n}\sum_{i=1}^{n}(y_i - \hat{y}_i)^2\right) \quad (3)$$

$$MAE = \frac{\sum_{i=1}^{n} |y_i - \hat{y}_i|}{n} \tag{4}$$

where, n is total number of records, Σ is sum of squared difference between actual value y and predicted value $\hat{y}$.

Results are validated on various partitioning of the dataset but here we have presented the result of partitioning of data using pareto principal where 70% of data in the training set, 20% in the validation set, and the remaining 10% in the test data set. For this partition achieved MSE, RMSE, and MAE are noted in Table 1. These interpretable performance measures show good prediction accuracy of linear regression model for food commodity price forecast.

Table 1. Price Prediction Performance Outcome

Performance Measure → Dataset	MSE	RMSE	MAE
Training Set	1.88	1.37	1.08
Validation Set	1.88	1.63	1.19
Test Set	2.38	1.54	1.16

6 Conclusion and Future Scope

This paper demonstrated the statistical analysis performed inside Ukraine commodity price analysis and prediction. Stability analysis showcased the stable and unstable commodities, Outliers i.e. price higher than maximum was extracted which shows the cities and date on which commodity price was radical. Adverse time i.e. pandemic time and war time commodity price analysis can reveal facts such as apple price was majorly high during the pandemic whereas onion, carrot, and wheat i.e. daily commodities price is high during the war period. Another interestingly high commodity price under the Non-Food category was imported antibiotics that too during Wartime. Several other analyses are performed to give future policy directions or base assumptions to government and humanitarian agencies. To predict the future price, linear regression was applied which gave interestingly good results for price forecasting using location and date as features. Forecast performance is validated using MSE, MAE, and RMSE performance measures. Future work is also needed to confront the global food Price exploration in accordance to Ukraine agricultural situation. Artifcial intelligence needs to be applied to suggest policies to handle and stabilize food prices locally and globally.

References

1. Abay, K.A., Breisinger, C., Glauber, J., Kurdi, S., Laborde, D., Siddig, K.: The Russia-Ukraine war: implications for global and regional food security and potential policy responses. Glob. Food Sec. **36**, 100675 (2023)

2. Abu Hatab, A.: Russia's invasion of Ukraine jeopardizes food security in Africa: shocks to global food supply chains could lead to social and political unrest. Nordiska Afrikainstitutet (2022)
3. Arndt, C., Diao, X., Dorosh, P., Pauw, K., Thurlow, J.: The Ukraine war and rising commodity prices: implications for developing countries. Glob. Food Sec. **36**, 100680 (2023)
4. Behnassi, M., El Haiba, M.: Implications of the Russia-Ukraine war for global food security. Nat. Hum. Behav.Behav. **6**(6), 754–755 (2022)
5. Chepeliev, M., Maliszewska, M., Pereira, M.F.S.E.: The war in Ukraine, food security and the role for Europe. EuroChoices **22**(1), 4–13 (2023)
6. Harrison, F.: Out of the ashes: Ukraine and the new social paradigm. Int. J. Environ. Stud. **80**(2), 517–524 (2023)
7. Lang, T., McKee, M.: The reinvasion of Ukraine threatens global food supplies. bmj, 376 (2022)
8. Mittal, V., Kaul, A., Gupta, S.S., Arora, A.: Multivariate features based instagram post analysis to enrich user experience. Procedia Comput. Sci. **122**, 138–145 (2017)
9. Mottaleb, K.A., Kruseman, G., Snapp, S.: Potential impacts of Ukraine-Russia armed conflict on global wheat food security: a quantitative exploration. Glob. Food Sec. **35**, 100659 (2022)
10. Taneja, A., Arora, A.: Identification of relevant contextual dimensions using regression analysis. In: 2018 Eleventh International Conference on Contemporary Computing (IC3), pp. 1–6. IEEE (Aug 2018)

Available Website Names Classification Using Naïve Baye

Kanokphon Kane[1], Khwunta Kirimasthong[1(✉)], and Tossapon Boongoen[2]

[1] Center of Excellence in AI and Emerging Technologies, School of Information Technology, Mae Fah Luang University, Chiang Rai, Thailand
6451501003@lamduan.mfu.ac.th, Khwunta.kir@mfu.ac.th

[2] Department of Computer Science, Aberystwyth University, Ceredigion SY23 3DB, UK
tob45@aber.ac.uk

Abstract. This paper presents a method for classifying website names using machine learning techniques based on the analysis of URLs from different websites on the internet. The primary objective is to categorize websites as either positive or negative, aiding in access permissions. The proposed method offers advantages such as improved content filtering, increased risk awareness, enhanced access control, and a comparative analysis with Decision Tree and Logistic Regression models. The experimental dataset includes training and testing data of website URLs, along with external datasets for sentiment analysis. The results demonstrate an impressive accuracy rate of 94%, validating the suitability of the method for website name classification. Future work can explore the application of the classification method in network security to detect and block negative websites by classifying them as malicious URLs. This extension would further enhance protection against harmful content and contribute to a more secure online environment.

Keywords: Multinomial naïve bayes · Website name classification · Machine learning · Malicious URL · Network security

1 Introduction

The internet has become an integral part of our lives, impacting various aspects such as work, education, social interaction, and communication. Its widespread usage has reached all age groups, from children to the elderly. The growth of social media platforms has been particularly significant, with billions of users worldwide [1]. However, along with the positive aspects, there is also a darker side to the internet. Negative and illegal websites, including gambling sites, pornography, and platforms facilitating inappropriate conversations, pose risks to internet users. These risks are particularly concerning for individuals with limited knowledge and awareness of online threats. Websites can contain hidden viruses and malware, which can harm users and their devices. It is crucial for users to exercise caution and be aware of the potential dangers that exist on the internet.

This study proposes a methodology for classifying websites into positive or negative text content categories by analyzing their URLs and utilizing machine learning techniques, specifically the Multinomial Naïve Bayes algorithm. The primary objective is to

N. Naik et al. (Eds.): UKCI 2023, AISC 1453, pp. 259–269, 2024.
https://doi.org/10.1007/978-3-031-47508-5_21

enhance content filtering capabilities and promote a safer browsing experience for internet users. By analyzing website URLs, the classification system raises risk awareness among users, empowering them to make informed decisions and mitigate potential dangers associated with accessing negative websites. Moreover, the study aims to provide a comparative analysis of the proposed classification method with other models, such as Decision Tree and Logistic Regression, to evaluate their strengths and weaknesses in the context of website name classification. The findings from this analysis contribute to the advancement of content filtering techniques and aid in improving internet security in the future.

2 Related Works

2.1 Uniform Resource Locator (URLs)

A URL is the address of a specific webpage on the internet. URLs play a crucial role in website navigation as they help users access specific pages on a website. URLs can be used for both good and bad purposes in online conversations. The URLs can be used to share valuable information, resources, or entertainment. On the other hand, they can also be used to spread malicious or dirty content such as malware, harsh online conversations, scams, dirty websites, and other attacks [2, 3]. To protect users from such malicious URLs, various web classification methods have been researched, which extract features from the content of web pages to determine if a URL is trustworthy or not [4, 5]. Therefore, it is important to be cautious while clicking on URLs and to verify the authenticity of the website before providing any personal or sensitive information.

Nowadays, the detection of malicious URLs is widely interested. Numerous studies have demonstrated the efficacy of machine learning and deep learning algorithms in detecting malicious URLs across diverse approaches [4]. Moreover, identifying malicious URLs could involve a content-based approach that examines the content of a matching web page associated with a URL to determine its malicious nature [6].

2.2 Text Classification

Text classification is a common tool that analyses an incoming message and tells whether the underlying sentiment is positive, negative or neutral. You can input a sentence of your choice and gauge the underlying sentiment.

In text sentiment analysis, the sentiment analysis task is usually modeled as a classification problem [7]. Classification methods that are based on vector space models, such as classification algorithm based on distance measurement, support vector machine algorithm [5] neural network method, maximum entropy method, and the Naive Bayesian method. Text classification and sentiment analysis is a part of Machine Learning [8].

2.3 Machine Learning (ML)

Machine learning, a branch of artificial intelligence (AI) and computer science, generates empirical-data-based models that can make decisions and judgments in new situations

[2, 7]. The method is often used to process and analyze large amounts of data and is widely used in finance, healthcare, education, and other fields.

Multinomial Naïve Bayes Classification: The URL Filter is a method used to analyze URLs for potentially harmful elements, and its accuracy can be improved with a large training sample set. The Multinomial Naïve Bayes algorithm has been identified as one of the most effective approaches in text categorization and identifying keywords in text messages [9, 10]. Compared to traditional web page classification methods, this approach is faster as it does not require fetching and evaluating the pages themselves [9, 11]. Achieved an overall accuracy of 95.12% in their studies. In comparison to other models proposed by different researchers, our model covers more security considerations [10].

Decision Tree: Decision Tree-based classification algorithms that have an advantages in terms of simplicity, interpretability, and the ability to handle feature interactions [10, 12]. The main idea of a Decision Tree is to create a tree based on attributes for categorized data points. Each internal node in the tree represents a test for an attribute (for example, if the value of a variable is more than 5), test results are represented by branches, class labels are represented by leaf nodes, and classification rules are represented by paths from root to leaf nodes. In developing an effective ML model to predict spam discovery in e-mail, researchers compared five major ML classification algorithms: LR, DT, NB, KNN, and SVM. Finally, they concluded the Random Tree algorithm as the best choice for performance [10] compared the performance between four classifiers: Decision Tree (DT), K-Nearest Neighbor (KNN), Naives Bayesian (NB) and Naives Bayes (NBM) in different situations using feature selection with stemming and without stemming [5, 13].

Logistic regression (LR): Logistic regression is a popular and fundamental machine learning algorithm that is extensively utilized. It stems from the concept that linear regression can also be applied to solve classification problems [13]. In logistic regression, the model aims to estimate the probability of the output based on the input. It can be utilized to construct a classifier, such as assigning inputs to different classes based on a selected cutoff probability. This approach offers a general approach for creating binary classifiers.

3 Methodology

The proposed website name classification method utilizes the Naive Bayes algorithm for classification, as depicted in Fig. 1. The process starts with data preparation and then proceeds to apply the Naive Bayes algorithm to categorize the website names.

3.1 Dataset

The experimental dataset of this work is divided into 3 datasets as follows: training, testing, and the external dataset, respectively. These datasets were obtained from publicly available data sources such as Kaggle and included positive and negative words by Spam Text Message Classification [14], Toxic Comment Classification [15], dirty words [16], Cleaned toxic comments [17], for train and test, Datasets specifically designed for natural

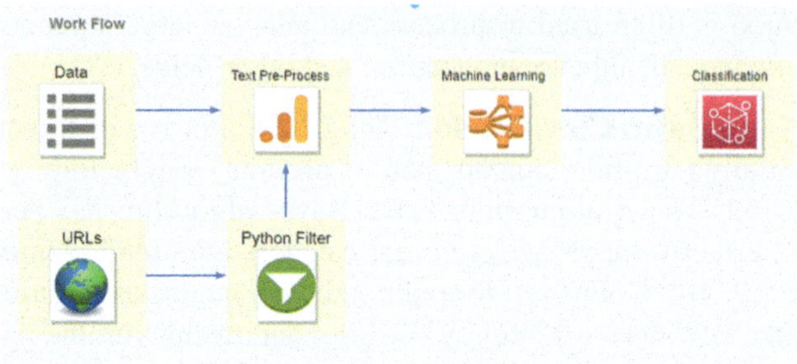

Fig. 1. The overview of website name classification based naive bayes algorithm.

language processing specially for testing unseen data that not related to training data [18] and the Twitter Sentiment dataset for external dataset that unseen data [5]. The analysis focuses on extracting information regarding positive and negative content from the text.

The experimental data is classified into two categories: negative and positive. The negative class encompasses instances of hate speech, inappropriate language, racism, and sexual harassment, while the positive class does not include any of these elements.

Training dataset: This dataset consists of 2 classes which are 0 is a negative sentence that is vulgar, or abusive, hate speech, dirty language, foul language, toxic comments, Spam message and 1 is a positive sentence in general without the use of vulgar language [14–16]. There are 10,181 records in each class for model training.

Testing dataset (Unseen): The testing dataset comprises 267,401 records and is categorized into two classes based on sentiment analysis. Class 0: the 151,461 records of negative sentiment and includes sentences that contain profanity, violence, toxic, or dirty words. Class 1: the 114,940 records of positive sentiment and including non-negative sentiment category [18]. The testing dataset is referred to as unseen data because it is not used during the training phase of the machine learning models. Instead, the models are trained on a separate training dataset to learn patterns and relationships between features and labels. The testing dataset is then used to evaluate the performance of the trained models and assess how well they can generalize to new, unseen data.

External dataset (Unseen): This work uses the Twitter messages about sexual harassment and racism [5] to test the use of different types of negative and common words to evaluate the performance of proposed model comparing with other classification that can predict external datasets after model training and model testing unknown data.

3.2 Text Pre-Processing

Text pre-processing is a critical first step in natural language processing to get raw text data ready for analysis. In this study, our text data is unorganized and contains unnecessary information. The pre-processing steps are crucial in transforming the text into a clean and organized format, making analysis easier and more accurate. The pre-processing steps that we use as follow,

- *Remove HTML Tags and URIs*: HTML tags and URIs are structural elements found in the raw text, like website links and formatting codes. However, for our website name classification, we only care about the actual text content. By getting rid of HTML tags and URIs, we can focus solely on the meaningful words, ensuring our machine learning algorithms work with relevant information.

- *Eliminate Punctuation*: Punctuation marks like commas and periods don't add to the meaning of website names. Keeping them could confuse our machine learning algorithms. By removing punctuation, we simplify the text and get rid of unnecessary noise, helping our algorithms focus better on the essential content.
- *Remove Stop Words*: Stop words are common words like "the" and "and" that don't carry much meaning. They appear frequently but don't help classify website names as positive or negative. Removing stop words reduces the text size and gets rid of irrelevant information, making our analysis more efficient and accurate.

By applying these pre-processing steps, we convert the messy raw text into a clean and organized version, ready for accurate analysis and classification of website names based on their positive or negative content. The structural features removed during pre-processing are detailed in Table 1.

Table 1. Structural features

Heading level	Example
Punctuation	%, ^,), &, #, $, @, *, (, &, ^, %, $, #,:; /,. &, |,}, {,], [, ?
HTML tag	<html>, <head>, <body>, <! --…-->
Image tags	<image>, <picture>, <map>
Link or navigation	<!DOCTYPE>,

3.3 Feature Extraction

Feature Extraction is a method for decreasing the quantity of elements in a dataset by producing new ones from existing ones (and afterward disposing of the first elements). The first arrangement of elements should then have the option to rundown most of the data in the new diminished arrangement of elements. Feature Extraction is done with the Bag of words technique [8, 13], Count Vectorizer [7, 8, 11, 19], and Tf–Idf technique [8, 10].

Bag of words technique: Input the data from CSV file and then the process of cleaning and transforming raw data prior to processing with clean numbers or special characters for data integrity before analysis. After that, the processing data will be made to Data Frame. Training data will only extract field text and use Count Vectorizer to tokenize and start wrapping and then keep [8, 13]. The number of positive and negative equal to balance the data for model training, this data has all the dataset 20362 records in train dataset, 267,401 records in test dataset and records in external data.

Count Vectorizer: The Count Vectorizer is a simple technique that converts text data into a numerical format. Each word in the dataset becomes a column, and the count of each word's occurrence in a document is placed in the corresponding position in the matrix. This representation allows machine learning algorithms to work with text data effectively [7, 8, 11, 19].

Term Frequency Inverse Document Frequency (TF-IDF): TF-IDF is a popular method for converting text into a meaningful mathematical representation used for machine learning tasks. It calculates the importance of each word in a document relative to the entire dataset. TF represents the frequency of a word in a document, while IDF represents the inverse document frequency, considering the word's importance across all documents. The product of TF and IDF gives a weight to each word, making it a valuable feature for classification [8, 10].

Term Frequency (TF): The purpose of TF is to find out how often each word appears in each document [7]. Finding the TF value is found by the formula of Term Frequency. The TF value is computed by Eq. (1) below,

$$TF(t, d) = \frac{\text{number of time t appears in d}}{\text{total number of terms in d}} \quad (1)$$

In this paper, The TF value is computed in every document, here are 41,333 features. Then, compute the IDF value. IDF stands for Inverse Document Frequency or its inversion in document frequency Which is derived from documents that have a lot of TF values, but still does not answer that it is an important word because TF will focus on each document, but IDF is the value calculated from all documents that can be General to some extent, the IDF value can be calculated by Eq. (2) below,

$$IDF(t) = \text{Log}\frac{1 + \text{n}}{1 + \text{df(t)}} + 1 \quad (2)$$

After that, the project multiplies the TF and IDF values so that the weight of the TF and the weight of the IDF, which are opposites, are weights that can separate the keyword.

Train and Test: After making the statement, this paper will model training by first preparing the class in an array-like arr_text, then train from the transformer. When the train is complete, the project will create a model using and create a test. Here The project will use array text [0] for testing. This paper divides the data into 75% of training and 25% of testing. Then convert it to a vector and store the value in one interpreter to predict that the prediction will come out as a show result Negative that matches Expected. Bringing all the data to the train to test it should be matched.

3.4 Python Filter

Python filters are commonly employed to extract and filter text from URLs when requested. By using the "get" command, one can retrieve the HTML code of a web page by providing its URL. The HTML code represents the structure and visual presentation of the entire website. You can view the HTML code of a web page by right-clicking on an empty area of the page and selecting "view source". In the context of researching malicious URLs, statistical methods can be applied to analyze character distribution features and structural characteristics. By extracting the text obtained from the web page, these statistical techniques can be employed to identify specific patterns or anomalies that may indicate malicious behavior.

3.5 Classifier Model

In this paper, section describes the Third popular machine learning algorithms which we have considered for the purpose of evaluation. Though there are several classification algorithms available to train and test the machine learning model, these algorithms yield better performance in terms of evaluation parameters like accuracy, mean squared error, Confusion Matrix and so on.

Naive Bayes Classification: Naive Bayes is an algorithm that is commonly used in natural language processing (NLP) tasks such as Spam filtering, Sentiment analysis, Classification, and recommendation. It is based on Bayes' Theorem as shown below [20] The formula takes the following form as shown equation below (3):

$$P(m|n) = \frac{P(n|m)P(m)}{P(n)} \tag{3}$$

From Eq. (3) above, P (m I n) is the probability of class x. Where x is the target and predictor are the attributes. P (m) is the prior probability of class. P (n I m) is the probability predictor of the given class. P (n) is the prior probability of predictor [20].

A Naive Bayes classifier gives us the conditional probabilities of events occurring related to each other by using Bayes' Theorem. Naive Bayes, it works with frequencies of features. It is used text classification or Basics of Natural Language Processing.

Decision Tree: Decision tree is the one of the useful and popular tools for classification and prediction. A Decision tree is a flowchart like tree structure, where the top node is called the root and the output node is leaf and each branch shows an outcome of the test. To build the decision tree for our classification problem, we must calculate information gain and entropy for the features. Entropy is the average rate at which information is given by the source data. The Entropy Eq. (4) is shown below,

$$Entropy = \sum_{i=1}^{C} -P_i \times log_2(P_j) \tag{4}$$

Logistic Regression: Logistic Regression is a classification algorithm which is based on the probability concept and its cost function lies between 0 and 1 [1, 12]. In this method sigmoid function as in Eq. (5) that is used to model the data shown below,

$$g(z) = \frac{1}{1 + e^{-z}} \tag{5}$$

Overall, the text pre-processing and feature extraction methods help in transforming the raw text data into a clean and organized format suitable for machine learning analysis and classification. The classifier models, Naive Bayes, Decision Tree, and Logistic Regression are then used to predict website URLs based on their content and classify them as positive or negative.

4 Experiment and Results

4.1 Functionality

The main objective of the functionality test is to confirm that the server's action, response, and URI functionality all function as expected. This study evaluates the effectiveness of the categorization based on the model's response when a client requests a package, and the model sends the output to the client along with URLs. The reply message includes:

- Positive: The analysis of the machine learning prediction URL on the website informs the user that the website is safe to browse and that it is positive.
- Negative: The study of machine learning prediction URLs on the website informs the user that the website may not be safe to visit and that they should exercise caution when using it.

4.2 Results

The functional test results shown in Table 2 represents the conclusion of the response from collecting experimental results followed by its status code and shows the resulting test.

Table 2. The functionality test table

Domain name	Method	Result
https://www.bbc.com	REQUEST/GET	Positive
https://www.merriam-webster.com/dictionary/sex	REQUEST/GET	Positive
https://Pornhub.com	REQUEST/GET	Negative
https://www.youporn.com	REQUEST/GET	Negative
https://miku-doujin.com	REQUEST/GET	Negative

Table 3. Accuracy of naive bayes classification

Machine learning: multinomial naïve baye				
	Accuracy (%)	Precision (%)	Recall (%)	F1 score (%)
Train dataset	94	94	94	94
Test dataset	94	94	94	94
External dataset	55.8	56	48	48

The confusion matrices that represents the accuracy, precision, recall, and F1 score to evaluate the optimal models are shown in Table. 6 (Tables 4 and 5).

Table 4. Accuracy of decision tree classification

Machine learning: decision tree				
	Accuracy (%)	Precision (%)	Recall (%)	F1 score (%)
Train dataset	60	78	61	54
Test dataset	88	89	89	89
External dataset	55	52	56	46

Table 5. Accuracy of logistic regression classification

Machine learning: logistic regression				
	Accuracy (%)	Precision (%)	Recall (%)	F1 score (%)
Train dataset	94	94	94	94
Test dataset	92	93	93	93
External dataset	56	54	57	45

Table 6. Confusion matrix

	Machine learning					
	Multinomial naïve baye		Decision tree		Logistic regression	
	Predicted positive	Predicted negative	Predicted positive	Predicted negative	Predicted positive	Predicted negative
Actual Positive	0.96	0.042	0.88	0.12	0.9	0.1
Actual Negative	0.08	0.92	0.096	0.9	0.036	0.96

5 Conclusion

The results of the classification models, namely Naive Bayes, Decision Tree, and Logistic Regression, are presented and discussed. Table 3 shows the accuracy, precision, recall, and F1 score for each model on the training, test, and external datasets.

From the results, it is evident that Multinomial Naïve Bayes achieved the highest accuracy of 94% on both the training and test datasets, indicating its efficiency in predicting website URLs based on their content. The precision, recall, and F1 score for Multinomial Naïve Bayes were also consistently high at 94% on the training and test datasets, highlighting its ability to accurately classify positive and negative website names.

On the other hand, Decision Tree demonstrated lower performance with an accuracy of 60% on the training dataset and 88% on the test dataset. While its precision and recall on the test dataset were 89%, the F1 score was lower at 54%, indicating that Decision Tree may struggle with balancing precision and recall.

Logistic Regression performed well with an accuracy of 94% on the training dataset and 92% on the test dataset. Its precision, recall, and F1 score on the test dataset were all 93%, showcasing its effectiveness in classifying website names.

However, when evaluating the models on the external dataset focused on sexual harassment from Twitter, the accuracy of all models dropped significantly. Multinomial Naïve Bayes achieved an accuracy of 55.8%, while Logistic Regression and Decision Tree achieved accuracies of 56% and 55%, respectively. This demonstrates the challenge of predicting content from social media platforms due to the diverse usage patterns of individual users.

The confusion matrices (Table 6) The confusion matrices provided valuable insights into each model's performance, highlighting their strengths and weaknesses in classifying positive and negative website names. Multinomial Naïve Bayes demonstrated a high true positive rate but had a relatively higher false positive rate, indicating that it occasionally misclassified negative website names as positive.

In conclusion, Multinomial Naïve Bayes emerged as the most effective classifier for predicting website URLs based on their content, showing a balance between accuracy and speed. While Decision Tree and Logistic Regression also performed reasonably well, they struggled to match the accuracy of Multinomial Naïve Bayes. The study's findings contribute to the development of safer online environments by identifying and blocking negative websites effectively.

However, the research has some limitations, including the exclusion of other potentially more efficient classification models and the inability to predict text within images. Future studies should explore a broader range of models and address the challenges posed by text in images for a more comprehensive analysis. Moreover, we can expand the testing dataset to classify the neutral websites that are the legal website that would be helpful for the classification evaluation.

Overall, the proposed classification method based on Multinomial Naïve Bayes shows promising results and has the potential to significantly enhance website URL classification, ensuring user safety and contributing to network security efforts.

Acknowledgement. This research work is supported by Mae Fah Luang University. This is a part of the project funded by the Ministry of Higher Education, Science, Research, and Innovation (Big security data fusion and analysis).

References

1. Bergeron, J., Debbabi, M., Desharnais, J., Erhioui, M.M., Lavoie, Y., Tawbi N.: Static detection of malicious code in executable programs. In: Proceedings of the Symposium on Requirements Engineering for Information Security, Indianapolis (2001)
2. Jordan, M.I., Mitchell, T.M.: Machine learning: trends, perspectives, and prospects. Science **349**(6245), 255–260 (2015)
3. Hotho, A., Maedche, A., Staab, S.: Ontology-based text document clustering. KI. **16**(4), 48–54 (2002)

4. Mishra, S., Soni, D.: SMISHING detector: a security model to detect smishing through SMS content analysis and URL behavior analysis. Futur. Gener. Comput. Syst. **108**, 803–881 (2020)
5. Sarlan, A., Nadam, C., Basri, S.: Twitter sentiment analysis. In: Proceedings of the 6th International Conference on Information Technology and Multimedia (2014)
6. Nair, S.M.: Detecting malicious URL using machine learning: a survey. Int. J. Res. Appl. Sci. Eng. Technol. **8**(5), 2670–2677 (2020)
7. Liu, H., Chen, X., Liu, X.: A study of the application of weight distributing method combining sentiment dictionary and TF-IDF for text sentiment analysis. IEEE Access **10**, 32280–32289 (2022)
8. Huang, D., Xu, K., Pei, J.: Malicious URL detection by dynamically mining patterns without pre-defined elements. World Wide Web **17**(6), 1375–1394 (2013)
9. Moshchuk, A., Bragin, T., Gribble, S.D., Levy, H.M.: A crawler-based study of spyware in the web. In: Proceedings of the Network and Distributed System Security Symposium (NDSS'06), San Diego, California, USA, The Internet Society (2006)
10. Nandhini, S., Marseline K.S.J.: Performance evaluation of machine learning algorithms for email spam detection. In: 2020 International Conference on Emerging Trends in Information Technology and Engineering (Ic-ETITE) (2020)
11. Wang, Z., Li, S., Wang, B., Ren, X., Yang, T.A.: Malicious URL detection model based on convolutional neural network. Commun. Comput. Inf. Sci. 34–40 (2020)
12. Jati, W.K., Kemas Muslim, L.: Optimization of decision tree algorithm in text classification of job applicants using particle swarm optimization. In: 2020 3rd International Conference on Information and Communications Technology (ICOIACT) (2020)
13. Rohmawati, U.A., Sihwi, S.W., Cahyani, D.E. Semar: an interface for Indonesian hate speech detection using machine learning. In: 2018 International Seminar on Research of Information Technology and Intelligent Systems (ISRITI) (2018)
14. AI, T.: Spam text message classification. Kaggle. https://www.kaggle.com/datasets/team-ai/spam-text-message-classification. Accessed 27 August 2017
15. G, A.: Toxic comment classification. Kaggle. https://www.kaggle.com/datasets/akashsuper2000/toxic-comment-classification?slect=train.csv&fbclid=IwAR1GBSocbVKDibH9sd-oD7-DSe2LuireEinS87CCyGJkq6Cyl9-kZrXIqfc. Accessed 12 October 2022
16. Hu, C.: Dirty_words. Kaggle. https://www.kaggle.com/datasets/chenghonghu/dirty-words?select=bad_words2.csv. Accessed 15 January 2022
17. Zafar.: Cleaned toxic comments. Kaggle. Accessed 12 March 2018
18. Tanyel, T.: Datasets for natural language processing. Kaggle. https://www.kaggle.com/datasets/toygarr/datasets-for-natural-language-processing. Accessed 04 January 2022
19. Xuan, C., Dinh, H., Victor, T.: Malicious URL detection based on machine learning. Int. J. Adv. Comput. Sci. Appl. **11**(1) (2020)
20. Huang, Y., Li, L.: Naive bayes classification algorithm based on small sample set. In: 2011 IEEE International Conference on Cloud Computing and Intelligence Systems (2011)

Evolutionary Computation

U2FSM: Unsupervised Square Finite State Machine for Gait Events Estimation from Instrumented Insoles

Luigi D'Arco, Haiying Wang, and Huiru Zheng(✉)

Ulster University, Belfast BT15 1ED, UK
{darco-l,hy.wang,h.zheng}@ulster.ac.uk

Abstract. Gait analysis is a research field that aims to assess and analyse a person's locomotion patterns. Traditional methods rely on visual evaluations by medical experts, but recent advances in biomechanics have introduced objective solutions such as motion capture systems, force plates, and pressure mats. However, these solutions are expensive, cumbersome, and limited to controlled environments. This paper proposes a novel hybrid model for gait event detection using instrumented insoles with pressure sensors. The model combines a finite state machine and a fuzzy c-mean algorithm to accurately identify gait events, including heel strike, foot flat, heel off, and toe-off. Random sampling was employed to evaluate the model's performance, ensuring representative results across the population. Nine parameters, including the duration of events, stride, stance and swing duration, and percentage of stance and swing phases were the main focus of the evaluation. The proposed system demonstrated accurate recognition of step counts and duration, with minimal variations compared to manually annotated data. Although a 0.1 s overall error in the duration of the gait events was identified, favouring longer heel strikes over shorter foot flat events, this was attributed to the amplitude-based annotation process's constraints. The proposed solution aligned with the optimal percentages of the stance and swing phases according to the gait cycle model. The results indicate the reliability and potential applicability of the proposed system in real-world scenarios. Future research will focus on refining the model, addressing observed errors, and exploring additional gait parameters to provide a comprehensive analysis of human locomotion patterns.

Keywords: Gait events · Instrumented insoles · Pressure sensors · Finite state machine · Fuzzy C-Means

1 Introduction

Gait analysis is a field of research that systematically analyses and assesses a person's locomotion. It entails assessing a person's movement in terms of different factors, such as their lower limb biomechanics, joint angles, muscle activity, and general coordination. Gait analysis' main objective is to find any irregularities,

N. Naik et al. (Eds.): UKCI 2023, AISC 1453, pp. 273–285, 2024.
https://doi.org/10.1007/978-3-031-47508-5_22

deviations, or inefficiencies in a person's gait pattern that might be responsible for discomfort, injury, or functional limitations. Throughout a person's life, a wide range of circumstances might have an impact on their ability to walk. Ageing [10], muscular atrophy, neurological or musculoskeletal diseases [3], and accidents [1] are some of them. As the disorders become more extensive, many people choose to lead sedentary lifestyles, which worsens their physical health and raises their risk for secondary health issues [11] like diabetes, obesity, and to some extent decline in organ functioning [14]. To date, the main method for determining the degree of impairment a person has during a walk is the visual evaluation by a medical expert, who evaluates the condition by subjecting the patients to different movement tests. With advances in biomechanics, several solutions have arisen to mitigate the bias derived from the subjective evaluation of such experts, including motion capture systems, force plates, and pressure mats, which are now considered gold-standard devices. These solutions provide objective data that helps analyse gait in its entirety, such as stride length, stride length, cadence, foot strike pattern, joint range of motion, and timing of muscle activation [13]. While these solutions produce reliable results, they are very expensive and cumbersome and limit the analysis to a controlled environment and specific exercises. With the miniaturisation of control units and sensors, new devices are emerging such as instrumented insoles and smart insoles, revolutionising the methodology to conduct gait analysis. These types of devices allow the combination of multiple sensors, maintaining a reduced small form factor and costs. The integrated sensors vary from solution to solution according to the needs but generally, pressure and inertia sensors are employed. Comprehensive gait monitoring can be made possible by such types of devices, whose applications can be extended either to multiple activities [5], due to their small size and few restrictions on the user's motion, and seamless daily life monitoring.

Multiple solutions have been presented in the literature which allow to determine the fundamental parameters for gait evaluation, however, they lack of a detailed analysis of the different events which form a normal gait cycle model. Considering the 4-event gait model, consisting of heel strike (HS), foot flat (FF), heel off (HO), and toe off (TO), determining the onset of these events allows for a throughout analysis of the individual's gait behaviour that can be comparable with the gold standard devices. This research aims to develop a new gait event detection approach that uses data extracted from instrumented insoles, consisting of 8 pressure sensors. A hybrid solution has been proposed, which combines a finite state machine, in which the states are the gait events, and a fuzzy c-mean algorithm, which is used to determine the transitions between states. A set of rules based on the general knowledge of the topic has been included to reduce the number of transitions to those allowed within a gait cycle while avoiding anomalous transitions. Various data preprocessing techniques have been used to improve system performance, including sensors aggregation, noise reduction and normalisation.

The rest of the paper is structured as follows: in Sect. 2 an analysis of the current state-of-the-art solution has been presented. The methodology is presented in Sect. 3 followed by the discussion of the results in Sect. 4. The research objectives are summarised in Sect. 5.

2 Related Work

In recent years, the field of biomechanics has witnessed significant advancements in the development of wearable sensor technologies for gait analysis. Focusing on instrumented or smart insoles, multiple studies can be found in the literature that aimed at the detection of gait events starting from raw data generated by pressure and inertial sensors. A cluster-based strategy for gait events detection from instrumented insoles, consisting of 16 force-sensing resistors, was proposed by Salis et al. [15]. Data were collected at a sampling rate of 1000 Hz, however, the recordings were later down-sampled to 100 Hz. Salis et al. normalised the raw data from the insoles, which were expressed in terms of voltage, and then smoothed the data using a nonlinear median filter with 5 contiguous points. The rising and falling edges of each of the pressure sensor data were identified using a method based on the first derivative, followed by a threshold-based peak recognition algorithm. The rising and falling peaks were organised into clusters corresponding to the same foot contact. The foot ground-contact interval was only calculated if the same cluster contained at least three activated pressure sensors. Therefore, the first rising minima of the activation cluster were used to identify the initial contact, while the last falling minima of the deactivation cluster identified the final contact. Nine participants were included in the study to validate the proposed solution, who were asked to walk on force pads for six minutes while wearing the instrumented insoles. The ground reaction force of the force platforms was used as a reference measure to calculate the root mean squared error, bias and standard deviation to evaluate the instrumented insoles, resulting in a mean error of less than 10 ms for foot contacts. Kim et al. [9] presented a heel-strike and toe-off detection algorithm using smart insoles with inertial measurement units. Seven healthy subjects were included in the study. The employed smart insoles had an inertial measurement unit composed of a tri-axes accelerometer, gyroscope and compass, and five pressure sensors placed at the outer toes, inner toe, inner heel, outer heel, and midfoot. The pressure sensors were used as a reference for the developed solution that relies entirely on inertial sensors. The sampling frequency during the experiments was set to 100 Hz. An adaptive zero-velocity update (ZUPT) algorithm consisting of three stages has been presented. The high peak was first identified for each step using an adaptive minimum-maximum threshold, followed by the determination of the ZUPT threshold based on the maximum speed of one step, and finally, if multiple ZUPT intervals were found within one step, the longest interval was chosen as the ZUPT interval. The heel-strike and toe-off events were then identified using a combination of the ZUPT technique and a continuous wavelet transform approach. In both walking and running, the suggested solution had an average error of 0.02 s. Furthermore, Kim et al. carried out a number of treadmill studies at speeds ranging from 3 to 5 km/h in order to validate the suggested approach.

In accordance with the findings of the literature, the results showed that as speed increased, cadence increased and single-support and double-support times decreased. Hoseini et al. [8] proposed a gradient fuzzy-based method for gait event detection at different walking speeds. A pair of instrumented shoes was involved, consisting of 8 pressure sensors per foot. In addition, to obtain a ground truth of initial contact and toe-off events, two switches were positioned at the beginning and at the end of the under part of the shoe, respectively. The sampling frequency involved in the data collection was set to 200 Hz. For each of the force-sensing resistors, the gradient was computed and the fuzzy set was defined as large or small according to an exponential function, allowing for the creation of the membership functions. The study aimed at recognising five gait events, including initial contact (IC), loading response (LR), mid-stance (MS), pre-swing (PS), and swing (SW), and for each of them, a fuzzy rule was defined and optimised through a genetic algorithm. Five healthy participants were involved in the experiments and were asked to walk on a treadmill at three different speeds (0.4, 0.85, and 1.3 m/s) for one minute. Overall a mean error of -14.3 ± 16.9 and 1.24 ± 17.0 ms were identified for the recognition of stance and swing phases, respectively.

While previous solutions have focused primarily on initial contact and final contact detection, a comprehensive approach including extended gait events can provide insights into human locomotion. By exploring the extraction of additional events such as flat foot and heel-off, it is possible to capture the intricate details of gait dynamics and improve the understanding of an individual's correctness of walking. Furthermore, existing solutions rely on threshold-based algorithms and expert judgement, presenting limitations in terms of accuracy and reliability if unanalysed cases arise. Machine learning bases its knowledge on previously collected data and can provide solutions that can handle a multiplicity of cases, however, the supervised approach requires annotated data which are not easily reconstructable on time series. For this reason, this study focuses on an unsupervised machine-learning solution that leverages large data sets gathered from a diverse cohort of participants. This approach can exploit the power of data-based analysis, discover hidden patterns and provide a more solid and objective detection of gait events, which combined with a fuzzy logic algorithm better manage the uncertainties of possible incoming data. In addition, this solution allows the application in real time since it does not need to know the entire data sequence but focuses on a small window of data.

3 Methodology

This research study proposes a novel gait event detection technique based on instrumented insoles. The overall architecture is depicted in Fig. 1 and consists of two primary modules: data preprocessing and gait event identification. The former enables the analysis of data taken from instrumented insoles, whilst the latter first uses the data to train a fuzzy c-means algorithm before incorporating it into a finite state machine to identify gait events. With the exception of the data utilised as input, the overall architecture is repeated for each foot.

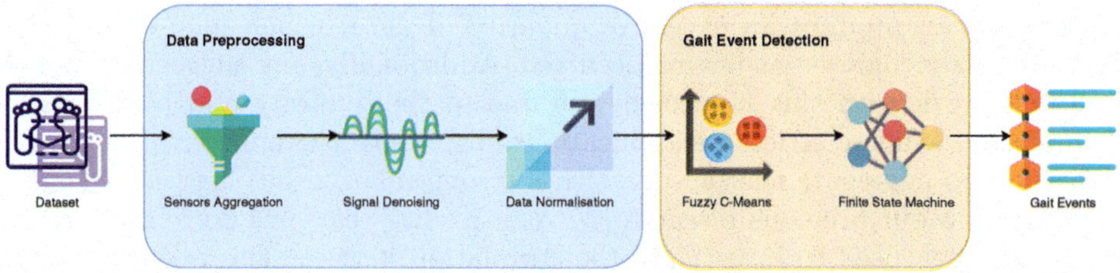

Fig. 1. Proposed system's architecture overview. To allow independent gait event extraction the architecture is repeated for each foot.

3.1 Dataset Description

In this study, a publicly available dataset has been employed, the "Gait in Parkinson's Disease" from Physiobank [6]. This database consists of gait measurements from 93 people diagnosed with idiopathic Parkinson's disease (PD) with an average age of 66.3 years, of whom 63% were men. Additionally, the database comprises 73 healthy controls with a mean age of 66.3 years, of which 55% were men. The recorded data within the database relates to the vertical ground reaction forces (vGRFs) observed while the subjects engaged in walking at a self-selected pace for approximately 2 min on a level surface. Each foot was equipped with 8 sensors (Ultraflex Computer Dyno Graphy, Infotronic Inc.) that precisely measure force in Newtons with respect to time. A sampling frequency of 100 Hz was chosen for recording and digitising the data. For each participant demographic information has been reported, as well as the measures of disease severity, using the Hoehn & Yahr staging and the Unified Parkinson's Disease Rating Scale (UPDRS). The dataset is comprised of data derived from three distinct studies, allowing for the virtual partitioning of the dataset into three subsets based on the respective source study. In the context of this research, the data from the Yogev et al. study [18] was utilised for training the predictive model, while the data from the Hausdorff et al. study [7] were reserved exclusively for testing purposes.

3.2 Data Preprocessing

The quality and reliability of the data used for analysis directly impact the accuracy and validity of research findings. To enhance the performance of the proposed solution and to address the data imperfections and redundancy, different data preprocessing techniques have been applied to the raw vGRFs, including sensors aggregation, noise reduction and normalisation.

Sensors Analysis and Aggregation The dataset under investigation offers data collected by eight pressure sensors placed throughout the foot's sole. Although, with these sensors, it is feasible to get a thorough study of the pressure applied to the ground by the user, it must be considered that the number of sensors can produce an excessive quantity of data, which may alter how a gait event is recognised as having occurred. Additionally, any suggested solution will only work with this kind of device due to the quantity and placement of the sensors, which restricts its applications. To reduce the number of sensors, it is feasible to aggregate the sensors with a strong relationship between them, as conducted in our previous research [4]. As a result, Pearson's correlation coefficient has been used to investigate the correlation between the various sensors. Pearson's correlation coefficient is a measure of the strength and direction of the linear relationship between two variables. It is denoted with the symbol r and can take values between -1 and $+1$, where $+1$ indicates a strong positive relationship (as one variable increases, the other tends to increase proportionally), -1 indicates a strong negative linear relationship (as one variable increases, the other tends to decrease proportionally), and 0 it indicates a weak or no linear relationship between the variables. It is calculated as:

$$r = \frac{\sum((x_i - \bar{x})(y_i - \bar{y}))}{\sqrt{\sum(x_i - \bar{x})^2 \sum(y_i - \bar{y})^2}} \tag{1}$$

where x_i and y_i are the i-th values of the two variables, $\bar{x}$ and $\bar{y}$ are the means of the two variables.

The study of the correlations revealed that the sensors in the same zone (front, middle, and back) presented high correlation values ($r > 0.5$). As a result, the sensors from the same zone have been combined using the Euclidean formula to calculate the magnitude signal, as follows:

$$G_i = \sqrt{\sum_{j=0}^{|G_i|} x_j^2}, \quad i = \{\text{front}, \text{middle}, \text{back}\} \tag{2}$$

where G_i is the ith group (front, middle and back), $|G_i|$ is the number of sensors belonging to the ith group, and x_j is the value of the jth sensor in the group.

Noise Reduction Technique Noise interference is a common challenge in data collection, especially in sensor-based systems. For this solution, a second-order Butterworth low-pass filter has been employed with a cut-off frequency of 10 Hz, since the human motion is typically less than 5 Hz [12].

Data Normalisation Data normalisation is a crucial step in preparing the dataset for analysis. By normalising the data, we aimed to bring all features within a consistent range, eliminating any biases introduced by varying scales or units. For each pressure sensor group (G_i), the following equation has been applied to convert the original range of values to a new range of values between 0 and 1.

$$G_{i_{SC}} = \frac{g - \min(G_i)}{\max(G_i) - \min(G_i)}, \quad g \in G_i, i = \{\text{front}, \text{middle}, \text{back}\} \tag{3}$$

where $G_{i_{SC}}$ is the sensor group with the scaled values, and g is an element in the group G_i.

3.3 Gait Event Detection Algorithm

In the literature many solutions can be identified for the recognition of gait events, however, they are mainly based on thresholds or on rules established on the basis of the researcher's experience. These types of solutions, although they allow a more detailed understanding of the reason for the occurrence of a specific event, are limited by a narrow coverage which can determine their functioning when borderline or anomalous cases are encountered. As a result, in this study, we propose a hybrid approach based on machine learning and on rules dictated by the understanding of the subject.

A gait cycle refers to a sequence of physical movements performed during walking that involve the motion of the lower limbs. It is formally defined as the time interval between two consecutive HS of the same foot, which is also known as a stride. The gait cycle consists of two primary phases: the stance phase and the swing phase. During the stance phase, the foot remains in contact with the ground, providing stability and support. Conversely, the swing phase occurs when the foot is lifted off the ground and moves through the air. According to the gait cycle model under consideration, different events can occur during the stance and swing phases. TO is an event of the swing phase, whereas HS, FF, and HO have been considered stance events for the purposes of this study.

Fuzzy C-Means Algorithm Fuzzy C-Means (FCM) is a clustering algorithm that enables soft or fuzzy clustering. Unlike traditional hard clustering algorithms, FCM assigns membership degrees to data points, representing the extent to which they belong to each cluster [16]. This allows for a more flexible representation of data patterns and accommodates situations where data points may exhibit overlapping characteristics. The FCM algorithm iteratively updates the cluster centroids and the fuzzy membership degrees to achieve convergence. The primary objective of the algorithm is to minimise the fuzzy partition coefficient, an objective function that captures the fuzziness or uncertainty of the clustering [2]. The fuzzy partition coefficient incorporates both the distances between data points and cluster centroids, as well as the membership degrees. The key parameter in FCM is the fuzzifier parameter, denoted as m. The value of m ($m > 1$) controls the degree of fuzziness in the clustering process. A larger value of m leads to a fuzzier partition, where data points can have significant membership degrees in multiple clusters. On the other hand, a smaller value of m makes the partition more crisp, resembling traditional hard clustering.

In this study, the FCM was used to determine the belonging of each observation to a certain event. The number of clusters has been set to 4 which represents the number of events that can occur during a gait cycle. The fuzzifier parameter m has been set to 2. Two FCMs were developed to independently process data from the right and left foot, using the same parameters with the only difference in the input data. In addition, to provide adherence between consecutive predictions a sliding window approach has been included to smooth the prediction. The prediction for an observation at time t, $\hat{y}_t$, is given by the argument of the maxima of the prediction of the elements in the window ($W = [x_{t-10}, x_{t+10}]$).

Finite State Machine A Finite State Machine (FSM) is a mathematical model used to describe and analyse systems that exhibit discrete behaviour. It consists of a set of states (Q), a set of input symbols (Σ), and a transition function (δ) that maps the current state and an input symbol to the next state. The behaviour of the FSM is determined by the initial state, the sequence of inputs, and the transition function [17].

In this study, an FSM has been implemented for estimating the gait event from the pressure sensors data. The set of states Q has been defined as the sets of possible gait events (HS, FF, HO, TO) and an init state that is used when no walking is detected. The input symbols Σ include all the possible values obtained by the observations from the pressure sensor groups (front, middle, back). The transition function δ that maps $Q \times \Sigma \longrightarrow P(Q)$ is defined as the prediction of the inner Fuzzy C-Means algorithm prediction for the observation x_i ($\hat{y}_i(x_i)$). However, a set of predefined knowledge-based rules have been identified to restrict anomalies transitions between the states, as shown in Fig. 2. The combination of the FSM and the FCM has made it possible to obtain a prediction of the gait events based on an unsupervised machine learning model, which is based on the characteristics of the signals while maintaining the general knowledge of the subject, thus avoiding anomalous transitions which are not part of a correct gait cycle flow. For each foot, an FSM was developed with the related FCM algorithm. The two FSMs were then joined by an agent which allows the extraction of the gait parameters.

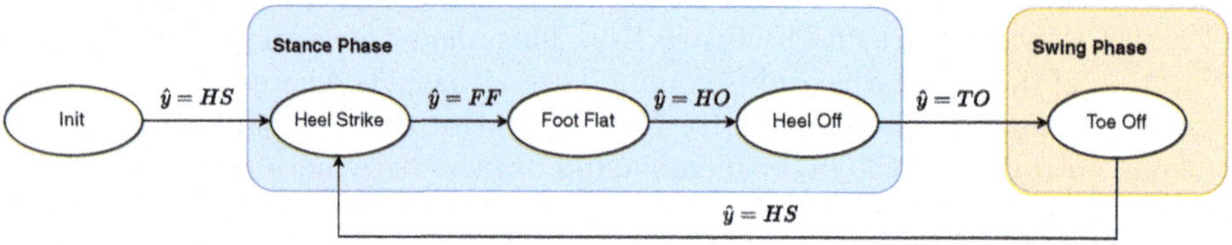

Fig. 2. Proposed finite state machine in which states reflect gait events and the transitions between them are based on the predictions of the fuzzy c-means algorithm, $\hat{y}$, with constraints based on the gait cycle model's knowledge.

4 Results and Discussion

To assess the proposed gait event detection model, a random sampling approach from the testing population has been employed. This approach ensures that the evaluation process is representative of the true diversity and characteristics of the population under consideration reducing any potential bias. A total of six sessions were chosen for evaluation, with each session involving a different participant. The researchers manually labelled the data by carefully inspecting the raw signals obtained from the gait recordings of the selected testing participants, identifying the gait events starting points.

Table 1. Summary of the comparison analysis between the extracted gait event using the proposed solution and the annotated gait events.

Session	Metric	Annotations				Proposed method				MAE	
		Left		Right		Left		Right		Left	Right
		Mean	Std	Mean	Std	Mean	Std	Mean	Std		
JuCo01_01 (HY: 0)	Step Count	21		20		21		20			
	HS_d	0.107	0.016	0.093	0.012	0.326	0.046	0.279	0.047	0.219	0.187
	FF_d	0.445	0.039	0.424	0.032	0.193	0.026	0.197	0.036	0.251	0.226
	HO_d	0.211	0.019	0.216	0.019	0.143	0.016	0.180	0.018	0.068	0.036
	TO_d	0.385	0.017	0.412	0.014	0.485	0.019	0.488	0.015	0.100	0.075
	STR_d	1.178	0.031	1.175	0.022	1.178	0.032	1.175	0.023	0.004	0.005
	STA_d	0.783	0.031	0.753	0.019	0.683	0.027	0.677	0.018	0.100	0.076
	SWI_d	0.385	0.017	0.412	0.014	0.485	0.019	0.488	0.015	0.100	0.075
	STA_p	66.481	1.441	64.056	0.996	57.980	1.323	57.617	1.029	8.500	6.439
	SWI_p	32.670	1.435	35.093	0.995	41.170	1.320	41.532	1.029	8.499	6.439
JuCo09_01 (HY: 0)	Step Count	19		18		19		18			
	HS_d	0.084	0.010	0.080	0.008	0.181	0.015	0.187	0.010	0.097	0.106
	FF_d	0.314	0.020	0.309	0.013	0.131	0.014	0.146	0.018	0.183	0.163
	HO_d	0.189	0.011	0.177	0.009	0.166	0.011	0.157	0.014	0.023	0.020
	TO_d	0.273	0.018	0.292	0.017	0.382	0.015	0.368	0.016	0.108	0.076
	STR_d	0.889	0.026	0.888	0.022	0.889	0.026	0.887	0.020	0.004	0.003
	STA_d	0.606	0.012	0.586	0.011	0.497	0.015	0.509	0.011	0.108	0.077
	SWI_d	0.273	0.018	0.292	0.017	0.382	0.015	0.368	0.016	0.108	0.076
	STA_p	68.190	1.178	66.038	1.186	55.955	0.874	57.404	1.114	12.235	8.633
	SWI_p	30.685	1.202	32.835	1.205	42.919	0.880	41.467	1.126	12.234	8.632
JuCo15_01 (HY: 0)	Step Count	18		17		18		17			
	HS_d	0.098	0.010	0.092	0.008	0.247	0.019	0.246	0.014	0.148	0.154
	FF_d	0.377	0.015	0.358	0.014	0.177	0.018	0.165	0.012	0.200	0.192
	HO_d	0.224	0.012	0.224	0.016	0.176	0.015	0.192	0.017	0.047	0.032
	TO_d	0.351	0.014	0.372	0.008	0.446	0.011	0.444	0.009	0.095	0.071
	STR_d	1.079	0.021	1.076	0.016	1.076	0.018	1.076	0.016	0.007	0.002
	STA_d	0.718	0.009	0.694	0.016	0.619	0.011	0.623	0.013	0.098	0.071
	SWI_d	0.351	0.014	0.372	0.008	0.446	0.011	0.444	0.009	0.095	0.071
	STA_p	66.557	0.737	64.476	0.750	57.595	0.672	57.866	0.683	8.961	6.609
	SWI_p	32.517	0.751	34.595	0.744	41.475	0.676	41.204	0.680	8.958	6.609
JuPt03_07 (HY: 2.5)	Step Count	21		20		21		20			
	HS_d	0.086	0.007	0.121	0.010	0.206	0.036	0.219	0.031	0.120	0.098
	FF_d	0.324	0.028	0.294	0.018	0.229	0.031	0.156	0.035	0.100	0.138
	HO_d	0.272	0.033	0.247	0.020	0.158	0.018	0.205	0.021	0.113	0.041
	TO_d	0.341	0.011	0.360	0.016	0.431	0.014	0.440	0.019	0.089	0.081
	STR_d	1.054	0.029	1.050	0.026	1.054	0.032	1.050	0.025	0.002	0.005
	STA_d	0.702	0.022	0.681	0.014	0.613	0.022	0.600	0.012	0.089	0.081
	SWI_d	0.341	0.011	0.360	0.016	0.431	0.014	0.440	0.019	0.089	0.081
	STA_p	66.648	0.737	64.836	0.850	58.170	0.884	57.128	1.017	8.478	7.707
	SWI_p	32.402	0.733	34.212	0.861	40.881	0.880	41.919	1.029	8.478	7.707
JuPt15_01 (HY: 2.5)	Step Count	22		21		22		21			
	HS_d	0.069	0.009	0.087	0.008	0.173	0.013	0.215	0.022	0.103	0.128
	FF_d	0.334	0.019	0.369	0.020	0.216	0.016	0.187	0.024	0.117	0.182
	HO_d	0.229	0.020	0.183	0.013	0.152	0.007	0.151	0.008	0.076	0.031
	TO_d	0.326	0.008	0.316	0.009	0.415	0.007	0.402	0.009	0.089	0.086
	STR_d	0.988	0.021	0.985	0.017	0.986	0.020	0.985	0.018	0.003	0.002
	STA_d	0.651	0.016	0.659	0.017	0.561	0.016	0.573	0.016	0.090	0.086
	SWI_d	0.326	0.008	0.316	0.009	0.415	0.007	0.402	0.009	0.089	0.086
	STA_p	65.944	0.547	66.887	0.902	56.860	0.664	58.138	0.859	9.084	8.748
	SWI_p	33.043	0.544	32.098	0.895	42.126	0.653	40.847	0.852	9.083	8.748
JuPt25_01 (HY: 2.5)	Step Count	19		18		19		18			
	HS_d	0.190	0.030	0.127	0.009	0.220	0.025	0.193	0.021	0.030	0.065
	FF_d	0.205	0.016	0.263	0.024	0.218	0.015	0.238	0.025	0.014	0.029
	HO_d	0.309	0.025	0.342	0.034	0.167	0.019	0.214	0.025	0.141	0.128
	TO_d	0.394	0.017	0.362	0.024	0.488	0.025	0.451	0.016	0.094	0.088
	STR_d	1.127	0.022	1.125	0.025	1.124	0.025	1.126	0.025	0.006	0.002
	STA_d	0.724	0.015	0.753	0.029	0.626	0.016	0.665	0.021	0.097	0.087
	SWI_d	0.394	0.017	0.362	0.024	0.488	0.025	0.451	0.016	0.094	0.088
	STA_p	64.200	1.114	66.913	2.017	55.681	1.565	59.050	1.212	8.519	7.863
	SWI_p	34.912	1.120	32.198	2.017	43.429	1.574	40.062	1.211	8.516	7.864

FF_d: foot flat duration; HO_d: heel off duration; HS_d: heel strike duration; HY: hoehn and yahr scale; MAE: mean absolute error; STA_d stance duration; STA_p stance percentage; STR_d: stride duration; SWI_d swing duration; SWI_p swing percentage; TO_d: toe of duration

To evaluate the model's effectiveness, a comparison analysis with the annotated samples has been carried out focusing on a wide range of parameters, including the duration of HS, FF, HO, TO, stride, stance and swing, as well as the stance and swing percentage. The duration of the event (HS, FF, HO, TO) has been calculated as follows:

$$E_d = n_E/fs \tag{4}$$

where E_d is the event's duration in seconds, n_E is the number of consecutive samples that make up the event, and fs is the sampling frequency. The interval between two consecutive HS events divided by the sampling frequency has been used to calculate the stride duration, as follows:

$$STR_d = (HS_i + 1 - HS_i)/fs \tag{5}$$

where STR_d is the stride duration in seconds, fs is the sampling frequency, and HS_i+1 and HS_i are the beginning timestamps of the second and first HS events, respectively. The stance duration has been calculated as the interval between the HS events and the start of the next TO event divided by the sampling frequency, as follows:

$$STA_d = (TO - HS)/fs \tag{6}$$

where STA_d is the stance duration in seconds, fs is the sampling frequency, TO is the beginning timestamp of the TO event, and HS is the timestamp of the HS event. The time between the start of the TO event and the start of the subsequent HS event has been divided by the sampling frequency to calculate the swing duration, as follows:

$$SWI_d = (HS - TO)/fs \tag{7}$$

where SWI_d is the swing duration in seconds, fs is the sampling frequency, and HS and TO are the start timestamps of the HS and TO events, respectively. The percentage of the stance and swing phases has been calculated as the ratio between the respective duration and the stride duration multiplied by 100.

The findings of the comparison are summarised in Table 1, reporting for each subject the mean and standard deviation for each parameter and the mean absolute error (MAE) across the steps. The proposed method provided accurate recognition of the number of steps performed by each individual across all test samples. Additionally, the suggested approach showed minimal variation in the data obtained from manual annotations when it came to measuring the duration of each step. The average variation was specifically 0.004 s for the left foot and 0.003 s for the right foot. With an error of 0.06 s for the right foot and 0.08 s for the left foot, the stance duration and swing duration error can be considered minimal. Similarly, for the duration of the events the MAE values differ from 0.067 to 0.124 s for the left foot, and from 0.041 to 0.133 s for the right foot. Analysing the individual sessions, however, it can be seen that the behaviour of the proposed solution favours longer HS to the detriment of the duration of the FF, which is the cause of the main variation analysed in the comparison analysis. Although there is a difference in behaviour, this is not to be considered incorrect because the researchers manually annotated the data using only the signal amplitude as a reference. Without actual visual proof of the intervals when the events occurred, there is not enough information to choose the most accurate solution, even though there is minimal difference between them. In addition, the proposed solution is closely aligned with the optimal gait cycle model, where the stance and swing percentages are 60% and 40%, respectively. This once again confirms the consistency of the proposed solution with the gait cycle model. Furthermore, the comparison analysis provides evidence that the proposed solution is not affected by the differences in gait impairments of the participants (differences on the Hoehn and Yahr scale), which lays the foundations for using such a solution in cases of gait disorders. However, this requires further investigation.

Overall, the findings demonstrate the effectiveness of the proposed system in recognising gait events, which results in accurate recognition of step counts and duration, with only negligible deviations from manually annotated data. The combination of an unsupervised machine learning algorithm, fuzzy c-means, and a set of knowledge-based rules has produced a solution with low average variability, indicating the system's consistency. Additionally, focusing the prediction of the gait event of a specific sample to a small window of data allowed the development of a solution that can be integrated into daily living applications, given that it can produce nearly real-time predictions without the need for predetermined training on the subject profile. However, further investigation is needed to analyse the solution's performance on complex activities besides walking.

5 Conclusion

In this research paper, a novel hybrid model for gait event detection that combines an unsupervised learning algorithm and the general knowledge rules of the gait cycle model has been proposed. The proposed model was evaluated using a random sampling approach to ensure representative results. The analysis focused on nine parameters related to gait events and duration. The findings of the comparison between the proposed model and manually annotated data demonstrated accurate recognition of step counts and duration. The system showed minimal variation in measuring the duration of each step, with errors considered negligible. The overall MAE values in stance and swing duration were marginal with an error of 0.06 s for the right foot and 0.08 s for the left foot. The proposed solution closely aligned with the optimal percentages of the stance and swing phases according to the gait cycle model. While the comparative analysis revealed minimal variation favouring longer heel strikes over foot flat duration, this behaviour does not indicate a flaw in the proposed solution, as the annotation process was performed based on the amplitude of the signals with a lack of visual evidence, making it difficult to determine the closest solution to reality.

Although only a few parameters were extracted for validation purposes in this study, it should be noted that multiple parameters can be extracted for each event, such as balance and stability among others. In this regard, the solution is necessary for an in-depth analysis of the gait patterns of a subject. Further research will focus on refining the model to address the observed errors and investigate additional gait parameters for a comprehensive analysis of human locomotion patterns.

Acknowledgements. This research is supported by the European Union's Horizon 2020 research and innovation programme under the Marie Skłodowska-Curie grant agreement No. 823978, and Invest Northern Ireland Proof of Concept project (PoC809). Luigi D'Arco is funded by Ulster University Beitto Research Collaboration Programme.

References

1. Antwi-Afari, M.F., Li, H., Anwer, S., Yevu, S.K., Wu, Z., Antwi-Afari, P., Kim, I.: Quantifying workers' gait patterns to identify safety hazards in construction using a wearable insole pressure system. Saf. Sci. **129**, 104855 (2020). https://doi.org/10.1016/j.ssci.2020.104855
2. Bezdek, J.C.: Pattern Recognition with Fuzzy Objective Function Algorithms. Springer Science & Business Media (2013)
3. Das, R., Paul, S., Mourya, G.K., Kumar, N., Hussain, M.: Recent trends and practices toward assessment and rehabilitation of neurodegenerative disorders: Insights from human gait. Front. Neurosci. **16** (2022). https://doi.org/10.3389/FNINS.2022.859298
4. D'Arco, L., Wang, H., Zheng, H.: A rapid detection of Parkinson's disease using smart insoles: A statistical and machine learning approach. In: IEEE International Conference on Bioinformatics and Biomedicine (BIBM), pp. 2985–2992 (2022). https://doi.org/10.1109/BIBM55620.2022.9995237
5. D'Arco, L., Wang, H., Zheng, H.: Deephar: a deep feed-forward neural network algorithm for smart insole-based human activity recognition. Neural Comput. Appl. **35**, 13547–13563 (2023). https://doi.org/10.1007/S00521-023-08363-W
6. Goldberger, A.L., Amaral, L.A.N., Glass, L., Hausdorff, J.M., Ivanov, P.C., Mark, R.G., Mietus, J.E., Moody, G.B., Peng, C.K., Stanley, H.E.: PhysioBank, PhysioToolkit, and PhysioNet: components of a new research resource for complex physiologic signals. Circulation **101**(23), e215–e220 (2000). https://doi.org/10.1161/01.CIR.101.23.e215
7. Hausdorff, J.M., Lowenthal, J., Herman, T., Gruendlinger, L., Peretz, C., Giladi, N.: Rhythmic auditory stimulation modulates gait variability in Parkinson's disease. Eur. J. Neurosci. **26**(8), 2369–2375 (2007). https://doi.org/10.1111/J.1460-9568.2007.05810.X
8. Hoseini, A., Hosseini-Zahraei, S., Akbarzadeh, A.: Fuzzy-based gait events detection system during level-ground walking using wearable insole. In: 29th National and 7th International Iranian Conference on Biomedical Engineering (ICBME), pp. 333–339 (2022). https://doi.org/10.1109/ICBME57741.2022.10052821
9. Kim, J.K., Bae, M.N., Lee, K.B., Hong, S.G.: Gait event detection algorithm based on smart insoles. ETRI J. **42**, 46–53 (2020). https://doi.org/10.4218/ETRIJ.2018-0639
10. Krishnan, C., Washabaugh, E.P., Reid, C.E., Althoen, M.M., Ranganathan, R.: Learning new gait patterns: age-related differences in skill acquisition and interlimb transfer. Exp. Gerontol. **111**, 45–52 (2018). https://doi.org/10.1016/j.exger.2018.07.001

11. Myers, J., Lee, M., Kiratli, J.: Cardiovascular disease in spinal cord injury: an overview of prevalence, risk, evaluation, and management. Am. J. Phys. Med. Rehabil. **86**, 142–152 (2007). https://doi.org/10.1097/PHM.0B013E31802F0247
12. Pandit, S., Godiyal, A.K., Vimal, A.K., Singh, U., Joshi, D., Kalyanasundaram, D.: An affordable insole-sensor-based trans-femoral prosthesis for normal gait. Sensors **18**(3) (2018). https://doi.org/10.3390/s18030706
13. Rani, V., Kumar, M.: Human gait recognition: a systematic review. Multimedia Tools Appl. **2023**, 1–35 (2023). https://doi.org/10.1007/S11042-023-15079-5
14. Rosso, A.L., Sanders, J.L., Arnold, A.M., Boudreau, R.M., Hirsch, C.H., Carlson, M.C., Rosano, C., Kritchevsky, S.B., Newman, A.B.: Multisystem physiologic impairments and changes in gait speed of older adults. J. Gerontol. Ser. A, Biol. Sci. Med. Sci. **70**, 319–324 (2015). https://doi.org/10.1093/GERONA/GLU176
15. Salis, F., Bertuletti, S., Bonci, T., Croce, U.D., Mazzà, C., Cereatti, A.: A method for gait events detection based on low spatial resolution pressure insoles data. J. Biomech. **127**, 110687 (2021). https://doi.org/10.1016/J.JBIOMECH.2021.110687
16. Suganya, R., Shanthi, R.: Fuzzy c-means algorithm-a review. Int. J. Sci. Res. Publ. **2**(11), 1 (2012)
17. Ying, M.: A formal model of computing with words. IEEE Trans. Fuzzy Syst. **10**(5), 640–652 (2002). https://doi.org/10.1109/TFUZZ.2002.803497
18. Yogev, G., Giladi, N., Peretz, C., Springer, S., Simon, E.S., Hausdorff, J.M.: Dual tasking, gait rhythmicity, and Parkinson's disease: which aspects of gait are attention demanding? Eur. J. Neurosci. **22**(5), 1248–1256 (2005). https://doi.org/10.1111/J.1460-9568.2005.04298.X

Graph Attention Based Spatial Temporal Network for EEG Signal Representation

James Ronald Msonda[1,2](✉), Zhimin He[3], and Chuan Lu[2]

[1] Department of Computer Science, Aberystwyth University, Wales, UK
jam139@aber.ac.uk
[2] Malawi University of Business and Applied Sciences, Blantyre, Malawi
[3] Department of Psychology, Aberystwyth University, Wales, UK

Abstract. Graph attention networks (GATs) based architectures have proved to be powerful at implicitly learning relationships between adjacent nodes in a graph. For electroencephalogram (EEG) signals, however, it is also essential to highlight electrode locations or underlying brain regions which are active when a particular event related potential (ERP) is evoked. Moreover, it is often important to identify corresponding EEG signal time segments within which the ERP is activated. We introduce a GAT Inspired Spatial Temporal (GIST) network that uses multilayer GAT as its base for three attention blocks: edge attentions, followed by node attention and temporal attention layers, which focus on relevant brain regions and time windows for better EEG signal classification performance, and interpretability. We assess the capability of the architecture by using publicly available Transcranial Electrical Stimulation (TES), neonatal pain (NP) and DREAMER EEG datasets. With these datasets, the model achieves competitive performance. Most importantly, the paper presents attention visualisation and suggests ways of interpreting them for EEG signal understanding.

Keywords: EEG models · Electroencephalography · Graph neural networks · Attention mechanism · Interpretable machine learning

1 Introduction

The success of deep neural networks (DNN) at learning from data in areas such as image classification, natural language processing, audio classification and speech generation etc., has been attributed to the applicability of convolution operations to a common system of coordinates in the Euclidean space [1]. However, representing electroencephalogram (EEG) signals within the same n-dimensional linear space fails to capture vital information including strengths and directionality of relationships between electrode locations within and between underlying brain regions.

Geometric deep learning (GDL) was conceptualized to replicate the achievement of erstwhile vanilla deep learning in the non-Euclidean space, often dealing with graph structured data [2]. A graphical representation of EEG data encapsulates interlinks and structural organization between EEG electrodes (channels).

N. Naik et al. (Eds.): UKCI 2023, AISC 1453, pp. 286–298, 2024.
https://doi.org/10.1007/978-3-031-47508-5_23

This paper introduces a GAT Inspired Spatial Temporal (GIST) network, which draws inspiration from the success of Graph Attention (GAT) network at expressing the strength of the connections between nodes through self-attention. The architecture also has a node attention layer to learn the importance of individual nodes (electrodes), followed by a temporal attention layer which focusses on informative time windows. This completes the EEG feature representation learning part of the architecture. For EEG signal classification, a multilayer perceptron is added to the top of the model. In this paper, we make the following contributions: 1) propose a novel nearly transparent graphical model for high level EEG signal feature learning; 2) assess the feature learning ability of the model by using 3 different EEG datasets; 3) suggest ways through which the learned attention weights can be leveraged for model diagnosis and interpretation; and 4) demonstrate the practical use of the model on real world problems (a) identifying regions associated with physical pain; and (b) channel selection for emotion classification.

2 Related Work

2.1 Graph Neural Networks

In a graph structure, the nodes, or vertices, are linked together by edges. Nodes and edges often have multi-dimensional features. While static forms of graphs provide informative spatial representations, better insight emanates from the time-varying structural changes of such graphs e.g., adding/removing a node and creating/modifying an edge or their features or weights.

Graph neural networks (GNN) were devised to bring convolutional neural network (CNN) like operations to the domain of geometric deep learning and thrive on the principle of message passing. In the message passing operation, node features of all of the target node's neighbours are aggregated to create new features for the said node. For the current node to learn from features of a node two hops away (neighbour of a neighbour), two iterations would be needed. Stacking these message passing layers enables the current node to learn even from the entire graph. It follows therefore, that the resulting embedding in a GNN encodes both the node features and existing node to node relations. Common GNN architectures include Graph Convolution Networks (GCN) [3] and Graph ATtention networks (GAT) [4].

2.2 Graph Attention Networks

Unlike GCN, where all neighbours of a node are given equal importance, in GAT, the features of the target node's neighbours are given learned weights before aggregation. Weighting is done through an attention mechanism.

To illustrate this point, assume that the input to a GAT layer is a collection of node features $\mathrm{h} = \overrightarrow{\mathrm{h}_1}, \overrightarrow{\mathrm{h}_2}, \overrightarrow{\mathrm{h}_3}, \ldots, \overrightarrow{\mathrm{h}_\mathrm{N}}$, $\overrightarrow{\mathrm{h}_\mathrm{i}} \in \mathrm{R}^\mathrm{F}$ where N denotes the total number of nodes in a graph, F the number of features available in each node. Equations (1–4) summarise how node embeddings $h^{(l+1)}$ are obtained given features h^l at a lower layer l.

$$z_i^{(l)} = W^{(l)} h_i^{(l)} \tag{1}$$

$$e_{ij}^{(l)} = LeakyReLU\left(\vec{a}^{(l)^T}\left(z_i^{(l)}|z_j^{(l)}\right)\right) \tag{2}$$

$$\alpha_{ij}^{(l)} = \frac{\exp\left(e_{ij}^{(l)}\right)}{\sum_{k\in\mathcal{N}(i)}\exp\left(e_{ik}^{(l)}\right)} \tag{3}$$

$$h_i^{(l+1)} = \sigma\left(\sum_{j\in\mathcal{N}(i)}\alpha_{ij}^{(l)}z_j^{(l)}\right) \tag{4}$$

As shown in (1), a learnable weight, $W^{(l)}$ is used to convert input features from a low level to a higher level to improve their expressivity. The outputs of (1) from two adjacent nodes, i and j are concatenated $\left(z_i^{(l)}\right)\Big\|z_j^{(l)}$, and an additive attention score $e_{ij}^{(l)}$ is obtained by taking the dot product of the result and a learnable weight, $\vec{a}^{(l)}$ before applying LeakyRelu activation function. In (3), SoftMax is used to normalize the scores attained in (2) across all single hop neighbours of i, $\mathcal{N}(i)$. Finally, normalized attention scores $\alpha_{ij}^{(l)}$ are used to provide weights to adjacent node embeddings ($z_j^{(l)}$) which are further aggregated as shown in (4), to get new embeddings for the target node where σ is an activation function such as Rectified Linear Unit (ReLU).

The attention score as calculated in GAT indicates the importance of a node to its neighbour, thus it can also be regarded as learned edge weight between concerned nodes. We refer to this as a form of edge attention. In case of multiple attention heads, node embeddings can be obtained by combining the outputs of the heads by concatenation or averaging, as shown in (5), with H as the number of heads.

$$h_i^{(l+1)} = \sigma\left(\frac{1}{H}\sum_{k=1}^{H}\sum_{j\in\mathcal{N}(i)}\alpha_{ijk}^{(l)}z_j^{(l)}\right) \tag{5}$$

2.3 EEG Graph Models

Zhang et al. [5] proposed a Graph based Hierarchical Attention Model (G-HAM) which encodes channel connectivity as either Euclidean distance between electrodes' spatial positions or structural neighbourhood. The node features are the raw signals. These node signals are sliced before conventional CNN is applied to them for feature extraction. This layer is followed by attention mechanisms which isolate important time slices and nodes. However, with the edge weights remaining constant across temporal slices and trials during graph formation, the model does not capture the dynamic nature of relations between brain regions, and there is no guarantee for a link between spatial and functional relationships.

Dynamical GCNs were proposed in [6] for EEG emotion classification, which adaptively adjust edge weights during model training. A similar approach is taken in [7] where a spatial temporal GCN is used to learn important edges and eventually estimate a latent graph structure. In [8], layers of GCN are used to extract features from temporal portions of EEG signals. Thereafter, long short-term memory (LSTM) is used to learn temporal changes across time slices.

As opposed to GAT process described in Eqs. (1)–(4), GCN's convolution operation results in (6) below, where $c_{ij} = \sqrt{|\mathcal{N}(i)||\mathcal{N}(j)|}$. Clearly, c_{ij} is a function of the structural configuration of the graph. Thus, owing to their dependence on graph structure, GCN based methods suffer from limited generalisability.

$$h_i^{(l+1)} = \sigma\left(\sum_{j\in\mathcal{N}(i)} \frac{1}{c_{ij}} z_j^{(l)}\right) \tag{6}$$

On the other hand, GAT replaces c_{ij} with attention mechanism. This ensures that different edge weights are implicitly learned, which if applied to an EEG graph model would offer a good approximation of the functional brain connectome. Besides, the learned edge weights can be visualized for model interpretation. While interactivity between nodes is necessary, it is not sufficient for EEG signal understanding. In EEG signal localisation and lateralisation, for example, it is essential to isolate relevant nodes associated with a particular brain activation. Identifying a time window within which an event related potential (ERP) occurs is also relevant e.g., when measuring signal propagation speed or the time it takes for a brain to react to a painful stimulus.

We propose a GAT Inspired Spatial Temporal model which learns relationship between brain areas (edge attention), detects significant channels (node attention) and identifies relevant temporal segments (temporal attention) responsible for the ERP under study. This architecture allows for interpretability and explainability of brain activities following an ERP. Moreover, node attention can also be used for channel selection for production affordable portable devices for specific applications, e.g., seizure detection, neural marketing, emotion recognition etc.

3 GIST Network Architecture

The GIST network architecture, shown in Fig. 1, takes its input as windowed signals and outputs predicted labels. Before the classifier, there are three attention blocks namely: edge attention, node attention and temporal attention. This section discusses these building blocks.

3.1 Input Segmentation

The input EEG signals are segmented into a fixed number of time windows, Q. Slicing signals enables learning from the temporal dynamics of the recorded signal and apportioning importance values to each slice which facilitates identification of a window within which relevant ERPs occurred. One graph is created from each time segment. Thus, there are Q graphs per recording.

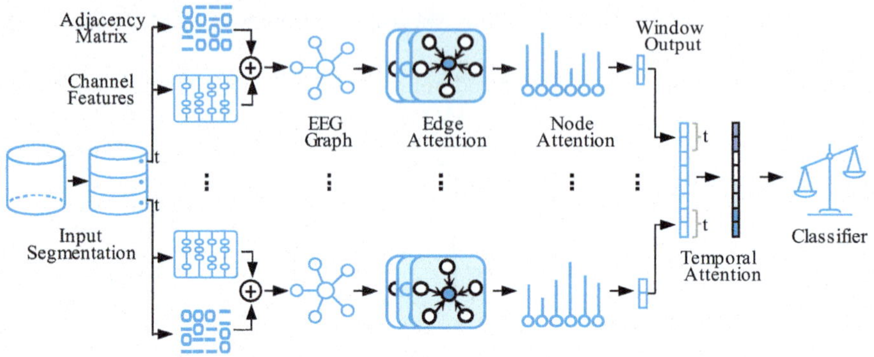

Fig. 1: GIST network architecture

3.2 Graph Representation

To formulate a graph, each electrode position (channel) in a temporal slice becomes a node. A set of F features is generated from the raw signal at each of the N nodes. Connectivity between nodes is encoded using a so-called adjacency matrix, A, such that a 1 indicates that an edge exists between nodes i and j. Otherwise, a 0 is inserted.

Adjacency Matrix. For the human brain, it is believed that measures of temporal and/or spectral oscillatory synchrony of recorded EEG signals define the functional connectivity between its regions. Common measures include coherence, transfer entropy, phase locking value, phase-slope index, and Granger causality [9]. Of these methods, coherence is the most popular because it is easy to interpret [10]. It reveals the magnitude of oscillatory frequency coupling between signals. To produce an adjacency matrix from coherence scores between signals in a time window, a threshold (k) value was used to determine if an edge existed between nodes of concerned signals.

Features. Node features extracted from raw EEG signals ranged from conventional statistical features (mean, kurtosis etc.) to nonlinear and nonstationary measures of entropy and fractal dimension. Details of these are available in [11].

3.3 Edge Attention

For edge attention, multi-layer GAT network was used to learn the interdependence between nodes. As it has been highlighted in Sect. 2.3, in a GAT network, attention is defined as the importance of adjacent nodes to a central node - in a way, quantifying the strength of the edges between nodes.

It must be noted that at any given point, there are multiple electrical activities taking place in the human brain. Thus, the adjacency matrix cannot distinguish between brain connectivity related to an ERP and that due to other background activities. The role of this layer is, therefore, to tune edge weights in response to a target ERP. As the model learns to classify whether a desired event took place or not, edge attention weights are automatically adjusted accordingly. Given that in the architecture, we have chopped our signal into Q time windows, the multi-head node update function in (5) can be modified to factor in this temporal slice (t) element as shown in (7),

$$h_i^{\{t,(l+1)\}} = \sigma\left(\frac{1}{H}\sum_{k=1}^{H}\sum_{j\in\mathcal{N}(i)} \alpha_{ijk}^{(t,l)} z_j^{(t,l)}\right) \tag{7}$$

where $h_i^{\{t,(l+1)\}}$ is the output of ith node for tth time window, whose value depends on $z_j^{(t,l)}$, the output of its connected node (with index j) at layer l and tth time window, as well as the corresponding edge attentions $\alpha_{ijk}^{(t,l)}$ for kth attention head. σ is an activation function usually a LeakyReLU.

3.4 Node Attention

The output of edge attention layer at temporal window, t is an updated set of node embeddings $h^{\{t,(l+1)\}} \in R^{N\times F'}$, where the number of features, F' do not necessarily have to be the same as the original feature size, F. This becomes an input to the node attention layer within the same time slice, t. The purpose of node attention is to identify nodes whose electrical activities can be associated with the presence of a particular external stimulation. The node embedding vector is transposed to get an $F' \times N$ matrix, which is multiplied with a learned attention weight, $a_{node_i}^{(t)}$ as depicted in (8). After applying an activation function, the importance value, $d_i^{(t)}$ for node i in window t, is determined by taking the mean across the F' dimension. Again, to make these attention values comparable, we apply a softmax function (9) to normalize the values. The normalised attention is then used as weights for the node embeddings to produce the node attention block output $h_i{\prime}^{(t)}$ (10). For N nodes, the output is then $h^{'(t)} \in R^{N\times F'}$.

$$d_i^{(t)} = \frac{1}{F'}\left\{LeakyReLU\left(a_{node_i}^{(t)} h_i^{\{t,\ (l+1)\}T}\right)\right\} \tag{8}$$

$$\alpha_{node_i}^{(t)} = \frac{\exp\left(d_i^{(t)}\right)}{\sum_{i\in N^{(t)}} \exp\left(d_i^{(t)}\right)} \tag{9}$$

$$h_i{\prime}^{(t)} = \alpha_{node_i}^{(t)} h_i^{\{t,(l+1)\}} \tag{10}$$

3.5 Temporal Attention

The temporal attention block attempts to focus on certain time windows depending on their relevance to the end classification task. It also follows the pattern of additive attention as described above. Equations (11)–(15) specify how the final features of the GIST network are obtained. In this case, $a_{temp}^{(t)}$ is a learnable weight while $b^{(t)}$ and $\alpha_{temp}^{(t)}$ represent temporal slice importance and normalised attention score respectively.

$$\mathrm{b}^{(t)} = LeakyReLU\left(a_{temp}^{(t)}\ h^{'(t)}\right) \tag{11}$$

$$\alpha_{temp}^{(t)} = \frac{\exp(b^{(t)})}{\sum_{t \in Q} \exp(b^{(t)})} \tag{12}$$

$$o^{(t)} = \alpha_{temp}^{(t)} h'^{(t)} \tag{13}$$

$$out = \Big\Vert_{t=1}^{Q} o^{(t)} \tag{14}$$

$$\tilde{y} = MLP(\boldsymbol{out}) \tag{15}$$

For Q time windows, the operator $\Big\Vert_{t=1}^{Q} O^{(t)}$ is used to chain together Q slice outputs to feed into the classification block. Here, $\Big\Vert_{t=1}^{Q} o^{(t)} = \left(o^{(1)}||o^{(2)}||...||o^{(Q)}\right)$

3.6 Classifier

The classification block comprises of a multilayer perceptron (MLP), or a fully connected feed forward neural network. This takes a flattened output of (14) to produce class predictions. In our experiments we empirically opted for an MLP comprising of 3 dense layers interleaved with dropout layers for regularisation and ReLU for activation.

4 Experiments

The main objective behind the GIST network is to model EEG signals in a way that facilitates understanding and interpretability. To this end three datasets were used: Transcranial Electrical Stimulation (TES) [12], Neonatal Pain (NP) and a Database for Emotion Recognition through EEG and ECG Signals (DREAMER) [13].

The TES dataset has known node positions which were stimulated and hence used to demonstrate the roles of edge and node attention blocks. NP signals were time locked to an ERP and hence, this is used to explore the capability of temporal attention in the network. We further investigate the usefulness of the model by applying it to two real world problems: brain regions associated with physical pain (NP), and emotion classification (DREAMER).

Thus, TES data was used to predict which part of the brain (frontal or motor region) was stimulated, while in NP the task was to classify if a given EEG record was done during painful heel lance or not. Finally, in DREAMER, the goal was to distinguish between emotions experienced by subjects while watching video clips on a binary scale of positive or negative valence. In TES and NP, model performance was evaluated by group stratified K-Fold validation where subjects formed the groups. On the other hand, leave one subject out (LOSO) cross validation was used in DREAMER. A summary of these datasets and time windows is shown in Table 1.

Table 1. Summary of datasets used to test GIST network

Database	No. subjects	Windows	No. channels	Sampling rate (Hz)	Stimuli
TES	20 (7 F)	5	30	2000	30Hz 1mA current
NP	112 (52 F)	4	20	2000	Heel lance
DREAMER	23 (9 F)	12	14	128	Video clips

4.1 Experimental Settings

DGLGraph [14] package with PyTorch backend was used to create the models. The hyperparameter settings for our experiments were: GAT output feature size, 8; GAT hidden layer feature size, 8; number of GAT layers, 3; number of attention heads per layer, 3; attention drop rate, 0.1; LeakyReLU negative slope, 0.1; and 0.4 as the drop rate for the first two MLP layers. The learning rate was 0.001 using Adam optimiser with cross entropy as the loss function. A threshold, $k = 0.6$, was used to produce an adjacency matrix from coherence scores. These were chosen following a series of prior Bayesian optimisation experiments on subsets of the datasets. Bayesian optimisation is a relatively quick probabilistic method of progressively narrowing down hyperparameter choices based on previous evaluations.

It was observed that the most influential parameters were GAT output feature size (GOFS), GAT hidden layer feature size (GHLFS) and threshold (k). For GHLFS with possible values as 2, 4 and 8, an increase in the value produced a corresponding improvement in the accuracy. On the other hand, with GOFS, lower values were better. An exploration of the threshold value revealed that accuracy was low when k was either too low (highly dense matrix) or too high (highly sparse matrix).

5 Results and Discussion

5.1 Feature Learning

A dimensionality reduction strategy called Uniform Manifold Approximation and Projection (UMAP) [15] was used to visualise how well feature learning took place across the GIST model.

Figure 2 shows a projection of the low-level input features and outputs of the three attention blocks onto two dimensional spaces. The plots demonstrate that there is increasing separability between classes moving across the edge attention, node attention and temporal attention blocks. Thus, a trained GIST model is good at feature transformation for classification purposes. It is also worth noting that for a simpler problem such as TES, it is possible to separate the classes at the node attention level which enables node attention to easily identify the active electrode positions.

5.2 The Role of Attention Blocks

From the temporal point of view, the datasets used are of three different characteristics: EEG recordings in which the time at which an ERP is induced is unknown e.g.,

DREAMER where relevant parts of the clips at which emotions were evoked are not identified; synchronized data where the stimulation time point is fixed such as NP in which heel lance was done 2 s after the start of the recording; and a dataset like TES where a stimulus was applied in the entire duration of the recording.

From Fig. 3(a) we observe that highest temporal attention scores are in time windows 3. This is in line with what was expected since window 3 comes immediately after a heel lance and hence it being the most informative in as far as classification between lance and no lance is concerned. By extension, this also demonstrates that the brain's response to a noxious stimulus is within 1 s. Figure 3(b) displays strip plots of TES temporal attention by label. It can be observed that both windows 1 and 3 can distinguish between frontal and motor stimulation. This is also supported by a temporal output visualisation of the 5 windows in Fig. 4 where windows 1 and 3 show data points in nearly perfect clusters.

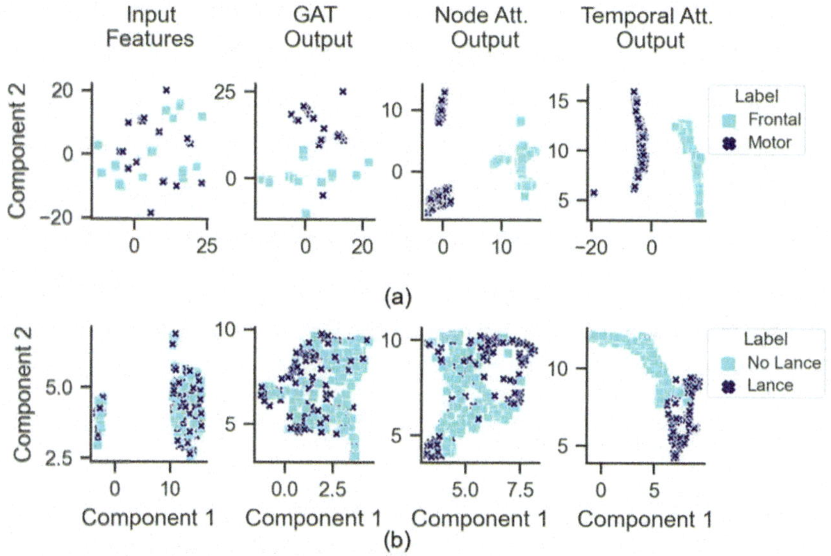

Fig. 2: UMAP visualisation of feature learning across the GIST network input, edge attention, node attention and temporal attention layers for **a** TES and **b** NP

Moreover, if we consider the edge attention at these two windows as depicted in Fig. 5(a) and (b), it is evident that the two windows attend to different brain regions. Window 1 focusses on the sensorimotor region while window 2 concentrates on the frontal area. Thus, for this data type, where the entire duration of EEG recording comprised of ERP, the time segments act in a manner similar to multi-head attention layers.

The node attention distribution information conveyed through topographic plots in 5(c) and (d) can be interpreted as either highlighting relevant electrode positions for classification or associating concerned brain areas with certain functions e.g. pain, 5(c) and emotion processing, 5(d). From 5(c) the highest intensity is around the premotor and sensorimotor region, followed by the parietal region. This observation is supported

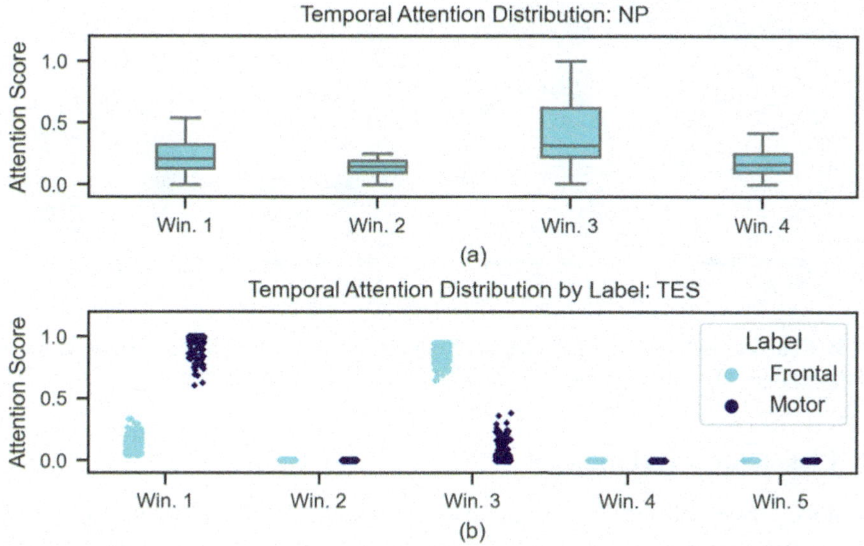

Fig. 3: **a** Temporal attention distribution box plots in TES and **b** Strip plots showing temporal attention by label in TES

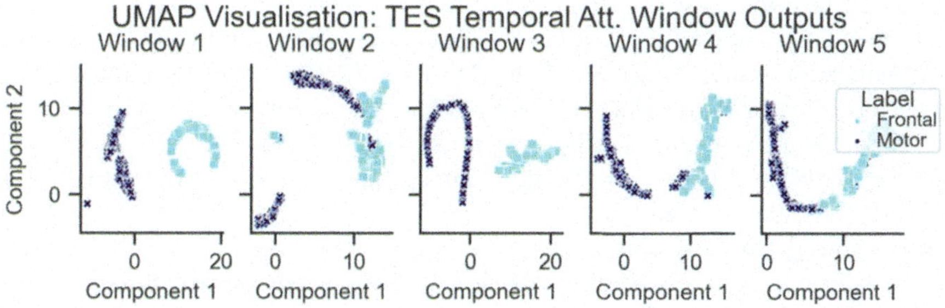

Fig. 4: UMAP visualisation of TES of the temporal attention block output of each individual window

by Tayeb et al. who established that 'noxious stimulation activates the pre-motor (Cz electrode) and moderately intense stimulation was found in the parietal lobe (P2, P4, and P6 electrodes)' [16]. 5(d) suggests that the frontal, temporal, and parietal regions are active during emotion processing. This is a view also shared by [17–19] among others. It follows therefore, that node attention in Figs. 5(d) can potentially be used to select channels for emotion recognition. However, applying this model for channel selection purposes must be done with extreme caution. This is because information contained in a node at the node attention layer is an aggregation of feature data from neighbouring nodes. The number of message passing rounds (how far wide the node goes to fetch for information) is controlled by the number of GAT layers.

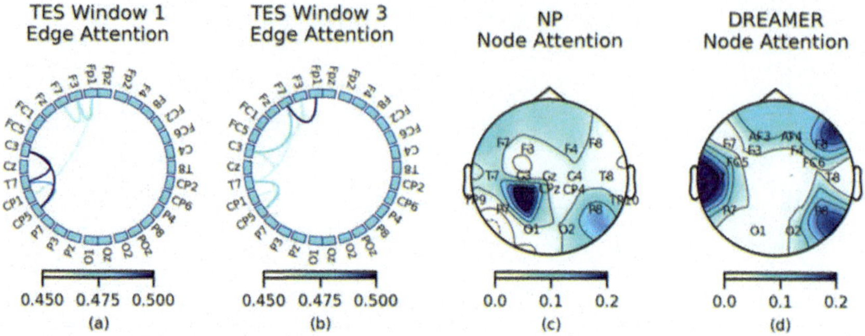

Fig. 5: Connectivity plots for **a** TES window 1 edge attention and **b** TES window 3 edge attention, and topographic plots for **c** NP node attention **d** DREAMER node attention

5.3 Classification Performance

The model's performance was assessed using accuracy, sensitivity, and specificity. Accuracy refers to the ratio of the number of correct predictions to the total number of predictions. Sensitivity, also called recall, hit rate, or true positive rate is calculated as the number of instances which were predicted to be positive which are truly positive divided by the number of all positive instances available in the dataset. The equivalent of sensitivity for the negative class is specificity (selectivity or true negative rate). Thus, it is the ratio of the number of true negatives to the sum of the true negatives and false positives.

Classification scores for TES, NP and DREAMER are provided in Table 2 below. The performance of the model on TES was very good, with scores of 99.4%, 100% and 98.8% for accuracy, specificity, and sensitivity respectively. On the other hand, specificity for NP was 67% even though accuracy and sensitivity were relatively higher at 76 and 96% in that order.

It is only DREAMER which has been found to have been used in other studies, albeit with different cross validation strategies. Table 3 compares performance of GIST network with other models. The within subject (WS) protocol utilised subject dependent leave one session out cross validation strategy in which a subject's recordings are split such that one session is used for testing while the rest are used for training. Thus, an accuracy score of 79.18% obtained using GIST network compares fairly with other models even though LOSO, a more challenging cross validation strategy was used.

Table 2. Model classification performance

Database	Accuracy (%)	Specificity (%)	Sensitivity (%)
NP	76	67	96
TES	99.40	100	98.80
DREAMER	79.18	75.40	80.20

Table 3. Performance on DREAMER

Model	Validation	Accuracy (%)
daSPDnet [20]	LOSO	67.99
CNN [21]	LOSO	75.93
DGCNN [6]	WS	86.23
GCB-net [22]	WS	86.99
GIST network	**LOSO**	**79.18**

6 Conclusion and Future Work

In this paper, a GIST network has been presented. The graph model thrives on the principle of attention to create a rich feature representation for classification purposes. The additive multi-level attention mechanisms used in the network facilitates understanding of EEG signals. From the attention scores, it is possible to visualise interdependence between brain regions, importance of individual electrode positions and the significance of temporal slices.

Future work could include adding an automatic feature extraction block to the GIST model. Modification of the architecture to accept heterogeneous graphs could also make the model accommodate multimodal data. To improve the model's channels selection capability, a dedicated channel attention layer could be added to the architecture just before the edge attention layer.

References

1. Bronstein, M.M., Bruna, J., LeCun, Y., Szlam, A., Vandergheynst, P.: Geometric deep learning: going beyond euclidean data. IEEE Signal Process. Mag. **34**, 18–42 (2017). https://doi.org/10.1109/MSP.2017.2693418
2. Zhou, J., Cui, G., Hu, S., Zhang, Z., Yang, C., Liu, Z., Wang, L., Li, C., Sun, M.: Graph neural networks: a review of methods and applications. AI Open. **1**, 57–81 (2020). https://doi.org/10.1016/j.aiopen.2021.01.001
3. Kipf, T.N., Welling, M.: Semi-supervised classification with graph convolutional networks. arXiv1609.02907 (2016)
4. Veličković, P., Cucurull, G., Casanova, A., Romero, A., Lio, P., Bengio, Y.: Graph attention networks. arXiv1710.10903 (2017)
5. Zhang, D., Yao, L., Chen, K., Wang, S., Haghighi, P.D., Sullivan, C.: A graph-based hierarchical attention model for movement intention detection from EEG signals. IEEE Trans. Neural Syst. Rehabil. Eng. **27**, 2247–2253 (2019). https://doi.org/10.1109/TNSRE.2019.2943362
6. Song, T., Zheng, W., Song, P., Cui, Z.: EEG emotion recognition using dynamical graph convolutional neural networks. IEEE Trans. Affect. Comput. **11**, 532–541 (2020). https://doi.org/10.1109/TAFFC.2018.2817622
7. Li, X., Qian, B., Wei, J., Li, A., Liu, X., Zheng, Q.: Classify EEG and reveal latent graph structure with spatio-temporal graph convolutional neural network. In: 2019 IEEE International Conference on Data Mining (ICDM). pp. 389–398 (2019). https://doi.org/10.1109/ICDM.2019.00049

8. Yin, Y., Zheng, X., Hu, B., Zhang, Y., Cui, X.: EEG emotion recognition using fusion model of graph convolutional neural networks and LSTM. Appl. Soft Comput. **100**, 106954 (2021). https://doi.org/10.1016/j.asoc.2020.106954
9. Cao, J., et al.: Brain functional and effective connectivity based on electroencephalography recordings: a review. Hum. Brain Mapp. **43**, 860–879 (2022). https://doi.org/10.1002/hbm.25683
10. Rocca, D.L., et al.: Human brain distinctiveness based on EEG spectral coherence connectivity. IEEE Trans. Biomed. Eng. **61**, 2406–2412 (2014). https://doi.org/10.1109/TBME.2014.2317881
11. Msonda, J.R., He, Z., Lu, C.: Feature reconstruction based channel selection for emotion recognition using EEG. In: 2021 IEEE Signal Processing in Medicine and Biology Symposium (SPMB). pp. 1–7 (2021). https://doi.org/10.1109/SPMB52430.2021.9672258
12. Gebodh, N., Esmaeilpour, Z., Datta, A., Bikson, M.: Dataset of concurrent EEG, ECG, and behavior with multiple doses of transcranial electrical stimulation. Sci. Data. **8**, 274 (2021). https://doi.org/10.1038/s41597-021-01046-y
13. Katsigiannis, S., Ramzan, N.: DREAMER: a database for emotion recognition through EEG and ECG signals from wireless low-cost off-the-shelf devices. IEEE J. Biomed. Heal. Informatics. **22**, 98–107 (2018). https://doi.org/10.1109/JBHI.2017.2688239
14. Wang, M., Zheng, D., Ye, Z., Gan, Q., Li, M., Song, X., Zhou, J., Ma, C., Yu, L., Gai, Y., Xiao, T., He, T., Karypis, G., Li, J., Zhang, Z.: Deep graph library: a graph-centric, highly-performant package for graph neural networks. https://arxiv.org/abs/1909.01315 (2019). https://doi.org/10.48550/ARXIV.1909.01315
15. McInnes, L., Healy, J., Melville, J.: UMAP: Uniform manifold approximation and projection for dimension reduction. https://arxiv.org/abs/1802.03426 (2018). https://doi.org/10.48550/ARXIV.1802.03426
16. Tayeb, Z., Bose, R., Dragomir, A., Osborn, L.E., Thakor, N.V., Cheng, G.: Decoding of pain perception using EEG signals for a real-time reflex system in prostheses: a case study. Sci. Rep. **10**, 4–8 (2020). https://doi.org/10.1038/s41598-020-62525-7
17. Phillips, M.L., Drevets, W.C., Rauch, S.L., Lane, R.: Neurobiology of emotion perception I: the neural basis of normal emotion perception. Biol. Psychiatry **54**, 504–514 (2003). https://doi.org/10.1016/S0006-3223(03)00168-9
18. Zheng, W.L., Lu, B.L.: Investigating critical frequency bands and channels for EEG-based emotion recognition with deep neural networks. IEEE Trans. Auton. Ment. Dev. **7**, 162–175 (2015). https://doi.org/10.1109/TAMD.2015.2431497
19. Zheng, W.: Multichannel EEG-based emotion recognition via group sparse canonical correlation analysis. IEEE Trans. Cogn. Dev. Syst. **9**, 281–290 (2017). https://doi.org/10.1109/TCDS.2016.2587290
20. Wang, Y., Qiu, S., Ma, X., He, H.: A prototype-based SPD matrix network for domain adaptation EEG emotion recognition. Pattern Recognit. **110**, 107626 (2021). https://doi.org/10.1016/j.patcog.2020.107626
21. Pandey, P., Seeja, K.R.: A one-dimensional CNN model for subject independent emotion recognition using EEG signals. In: Khanna, A., Gupta, D., Bhattacharyya, S., Hassanien, A.E., Anand, S., and Jaiswal, A. (eds.) International Conference on Innovative Computing and Communications. pp. 509–515. Springer Singapore, Singapore (2022)
22. Zhang, T., Wang, X., Xu, X., Chen, C.L.P.: GCB-Net: graph convolutional broad network and its application in emotion recognition. IEEE Trans. Affect. Comput. **13**, 379–388 (2022). https://doi.org/10.1109/TAFFC.2019.2937768

Hybridizing Lévy Flights and Cartesian Genetic Programming for Learning Swarm-Based Optimization

Jörg Bremer(✉) and Sebastian Lehnhoff

University of Oldenburg, 26129 Oldenburg, Germany
joerg.bremer@uni-Oldenburg.de
https://uol.de/ei

Abstract. Cartesian Genetic Programming is a well-established version of Genetic Programming and has meanwhile been applied to many use cases. The case of learning swarm behavior for optimization recently showed some fitness landscape characteristics that make program evolution harder due to the intrinsic barrier structure that is hard to pass by using standard mutation. In this paper, we explore possible improvements by replacing the standard uniform mutation by Lévy flights when training with a (μ+λ)-evolution strategy. We demonstrate the superiority of the new variation operation for training instances of the optimization learning problem and compare success rates and minimal computational effort.

Keywords: Cartesian genetic programming · Lévy flights · Mutation · Swarm-based optimization

1 Introduction

Cartesian Genetic Programming (CGP) is an efficient version of Genetic Programming (GP), introduced in [27]. It has become very popular and has meanwhile been broadly adopted [28] to many different use cases [5,22,25]. CGP already demonstrated its capabilities in synthesizing complex functions in several different use cases for example for image processing [11], or neural network training [18]. Some additions have been developed to CGP, e. g. recurrent CGP [36] or self-modifying CGP [10]. In this paper, we will focus on standard CGP as evolving the programs is often identical.

CGP uses an integer-based representation of a directed graph to encode the program. Integers encode addresses in data or functions by addresses in a look-up table. But also versions with float-based representations exist [4].

Cartesian genetic programs are often evolved using a $(1 + \lambda)$-evolution strategy (ES) with mutation only [36]. Mutation works on all or just on active genes [7]. But, also genetic algorithms with crossover schemes have been explored, and are—in the first place—useful when multiple chromosomes have independent fitness assessments [27,40]; what is not the case in our use case.

N. Naik et al. (Eds.): UKCI 2023, AISC 1453, pp. 299–310, 2024.
https://doi.org/10.1007/978-3-031-47508-5_24

A Lévy flight mimics a movement that resembles marine predators searching for prey with many small movements for searching the local neighborhood with due diligence but also allowing for quickly changing to another region if nothing is found. We harness this movement characteristics for jumping into another basin of attraction without running (too often) into non-evaluable solution on a barrier without trading the exploration quality of the current region.

Here, we focus on mutation and possible improvements. For some problems, CGP has difficulties to properly converge due to numerous, broad barriers in the fitness landscape. An example can be found in [2], where the authors try to evolve a swarm-based optimization algorithm with CGP. In [1], it had been observed that obviously a series of infinitely high barriers exists on the fitness landscape that separate basins of attraction of different local optima. These barriers are explained by intermediate solutions that encode nonsense control programs leading to no evaluable swarm optimization behavior at all. These barriers are observed to be rather frequent and obviously surround rugged funnels to a wide variety of local optima. This has been found by investigating random paths on the fitness landscape. The barriers often also exhibit a considerable width what makes them hard to pass without running into non-evaluable solution [1]. We propose to use Lévy flights for mutation to overcome this problem.

The rest of the paper is organized as follows. We start by an overview on Lévy flights in optimization and recap the basics of CGP and Lévy flights. We propose a hybridization of both methods and demonstrate the applicability and usefulness with some first results.

2 Related Work

The term Lévy flight first appeared in [23]. In general, a Lévy flight belongs to the class of random walks with steps made in isotropic direction and a step size following some heavy-tailed probability distribution [23]. For the case of a Lévy distribution, these walks are called Lévy flight.

Lévy flights can be used for modeling foraging of marine predators like sharks [30,37]. Lévy flights are characterized by frequent tiny and rare long step lengths. Moreover, same sites are revisited much less frequently than in other diffusion processes like Brownian motion. Lévy flights have been experimentally observed in nature in biological foraging [39], though the reasons behind are still not fully understood. One observation is that a Lévy distribution for step sizes in movement is advantageous because it reduces the probability of returning to previously visited site significantly—compared to a Gaussian distribution [20, 32]. This effect is independent of the chosen value of μ [33]. It is argued that n walkers on a random path visit a much larger number of new sites when using a Lévy distribution in contrast to a Brownian movement [38]. The n walker diffuse much faster and explore a larger portion of the landscape with a reduction of the relative competition. Therefore, it seems to be a superior strategy.

For optimization, the concept of Lévy flights has already been hybridized with several algorithms [14,17]. The most straightforward use is for generating search trajectories directly [41]. A meta-heuristic inspired by using particles simply moving on Lévy flight trajectories is given in [13]. A modification of particle swarm optimization is given in [9]. By using a Lévy distribution for relocation of particles that are detected to show a stall behaviour, escape from local minima is enhanced. In [15] an improvement is presented that integrates the Lévy distribution directly into the movement of particles. A similar approach is taken in some other population based algorithms [12,16,21].

A general problem arises for all GP problems as soon as the optimization algorithm reaches a region with solutions that encode programs that are not executable and thus a region with invaluable solution candidates.

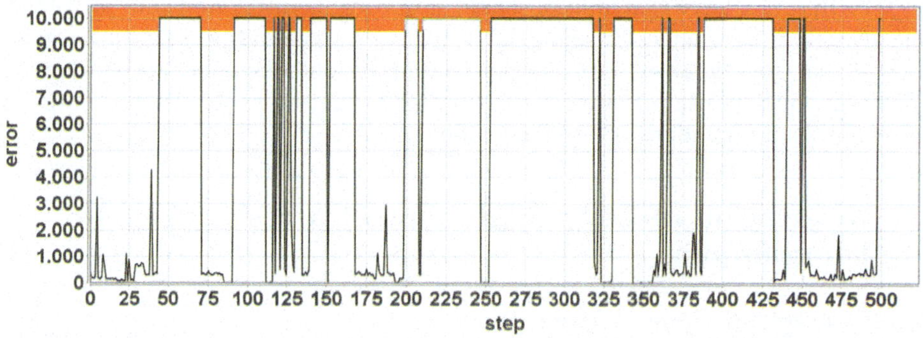

Fig. 1. Example series of residual error along a random path on the fitness landscape of the optimization learning problem.

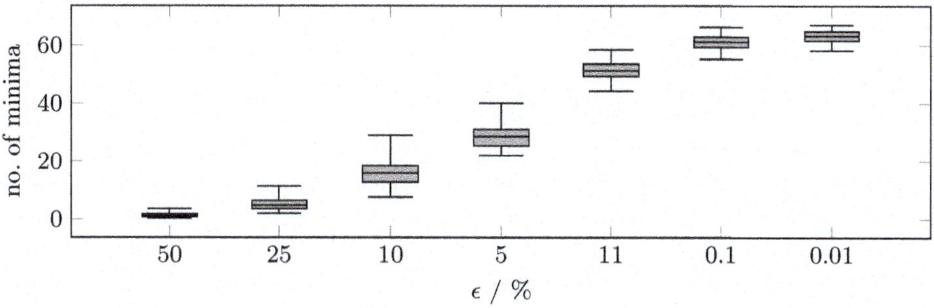

Fig. 2. Number of different local optima along a random path of length 500 on the fitness landscape of the optimization learning problem. The results are averaged over 100 paths respectively and show the presence of a significant number of local optima with a hight of more than 50% of the range of the objective function and the logistic growth of this number when smaller magnitudes are incorporated; modified after [1].

Figure 1 reproduces an example from [1] and displays the residual error (minimization problem) of succeeding solutions encountered on a random path of length 500 on the fitness landscape. The displayed error has been cut at a value of 10k in order to not run into numerical problems. It had been empirically found that as soon as the algorithmic search enters such regions (orange bar), the search gates stuck on this barrier if it is too broad and the mutation rate is too low. On the other hand, if the mutation rate is set too high, the search does not properly explore and degenerates to a mostly random search. Apart from these dead end barriers, several local minima of different magnitude are also present. This can be seen in Fig. 2.

We integrated Lévy flights into (μ+λ)-evolution strategies as mutation operator. Surprisingly, few research can be found on using Lévy distributions for mutation of offspring solutions and/or sampling new offspring solutions for the case of GP. In [8], a Lévy distribution-based mutation is integrated into an islanding model. An example for using Lévy flights for generating samples is given in [34]. For genetic or evolutionary programming one example can be found in [31]. The authors use a Lévy distribution for tuning of crossover and mutation probabilities in standard, tree-based code encoding.

3 Integration of Lévy Flights into Cartesian Genetic Programming

3.1 Cartesian Genetic Programming

CGP encodes computer programs as graph representation [35] and is an enhancement of a method originally developed for evolving digital circuits [24,26]. During evolution, connections in between the nodes are rewired, but using an intermediate output of an inner node is not mandatory. Thus, computation nodes may be left unused.

A chromosome in CGP encodes function as well as output genes and the connections in between the nodes. Together they encode the computational graph that represents an executable program. Figure 3 shows an example graph with six computational nodes, two inputs and two outputs. The encoding works as follows. The allele of a function gene represents the index in an associated function lookup-table ($\underline{0}$ to $\underline{3}$ in the example). Each computation node is encoded by a gene sequence consisting of the function look-up index and the connected input (or output of another computation node). The length of each gene sequence depends on the arity of the encoded function and thus varies. The graph in traditional CGP is acyclic. Parameters that are fed into a computation node may only be collected from previous nodes or from the inputs into the system. In this way, the execution order is predefined. Outputs can be connected to any computation node, or directly to any input. Nodes with output that is not connected to any other function or system output are considered inactive and are not used in the resulting program [24].

Finding a good program is achieved by evolving the chromosome e. g. with an ES. Originally, often a $(1 + \lambda)$-ES is used with mutation only. Mutation is

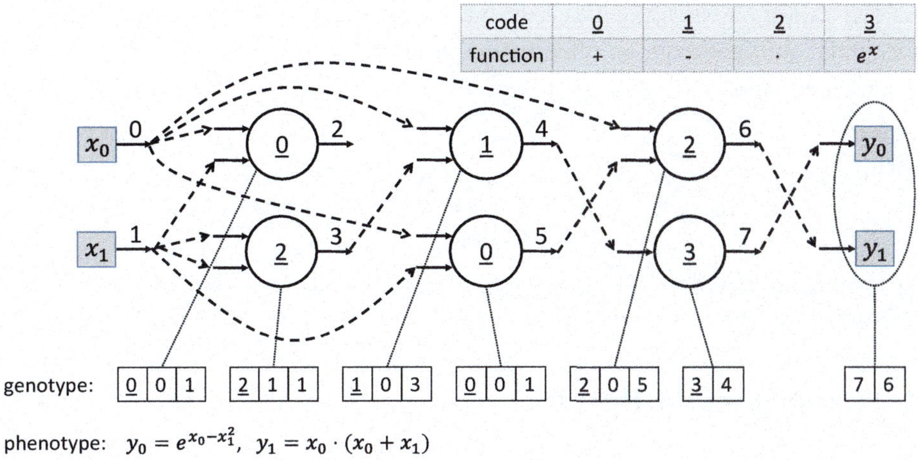

Fig. 3. Computational graph structure and its genotype and phenotype representation in Cartesian Genetic Programming; modified after [2].

done by [29]:

$$x_i' = \begin{cases} x_r \sim U(0,1) & \text{, if } r_i \sim U(0,1) < m \\ x_i & \text{, otherwise,} \end{cases} \tag{1}$$

i. e. by setting each gene of a parent solution to some random value in the search space (scaled to $[0,1]^d$) with likelihood m. The mutation rate m is fixed for all genes.

3.2 Sampling with Lévy Fights

The Lévy distribution is a continuous probability distribution for a non-negative random variable. The Lévy distribution is a heavy-tailed distribution with the right tail not exponentially bounded, meaning that in a random series of mostly small values occasionally large values occur by random. The probability density function is given by:

$$f_L(x;\mu,\gamma) = \frac{\sqrt{\frac{\gamma}{2\pi}} e^{-\frac{\gamma}{2(x-\mu)}}}{(x-\mu)^{3/2}}, \tag{2}$$

with location parameter μ describing location and scale parameter γ controlling dispersion and thus the range of occasional large random values. We tuned the Lévy distribution and used $\mu = 0$ and $\gamma = 0.01$ which reduces the occurrence of large values and obviously fosters the local exploration of fitness regions as soon as the optimization process reaches a new basin. The location parameter is set to zero because we want to use the old solution position in search space as starting point for sampling new solutions in the vicinity. The occasional occurrence of large values is still high enough to initiate occasional transitions to neighboring fitness basins.

For sampling offspring solutions, we take a selected elitist parent solution $\boldsymbol{x} \in [0,1]^n$ and generate an offspring solution $\boldsymbol{x}'$ by

$$x_i' = x_i + d_i \cdot l_i, \tag{3}$$

with

$$d_i = \begin{cases} -1 & \text{, if } r_i \sim U(0,1) < 0.5 \\ 1 & \text{, otherwise} \end{cases}, \quad 1 \leq i \leq n \tag{4}$$

with uniform distributed r_i to ensure isotropy of the direction and a Lévy distributed step size $l_i \sim f_L$.

4 Experimental Results

The general Lévy -based mutation concept should be applicable to a broad range of CGP use cases. On the other hand, it was our intention to cope with the specific barrier problem in the swarm problem class described above. Therefore, for the moment we restricted evaluation to the performance on learning intrinsic behavior for solving optimization problems with population-based solvers.

In [2], a first approach for automated design of emergent behavior using Cartesian Genetic Programming has been proposed. The goal of the problem is to learn optimization with a population-based algorithm, i. e. to evolve a swarm behavior that is capable of finding good solutions to arbitrary problem instances of global optimization.

By using CGP, a control program is evolved that guides individual particles through the search space of previously unseen optimization problems. In this way, a set of different particles jointly (interacting just by observation) performs a swarm-based optimization. Interaction takes place merely by mutual observation, as it is for example the case in particle swarm optimization.

As function set, the four basic arithmetic operations, a generator for normal distributed random numbers, the classical if-then-else-statement, and the set of standard order relations have been proposed in [2] and are also used here. As input, a particle gets the current position (in search space), the current objective value, and the position of the best particle. The output of the program was set to be the new particle position.

In [2], a float representation of the chromosome has proven useful. As advantage of the float representation, a larger range of operators is applicable for mutation. Prior to solution evaluation, the chromosome is mapped back to the integer domain. For evaluation of solutions, we used the same objective function as [2]:

$$\begin{aligned} &f(\{\mathcal{P}_i\}_1^n, n_{dec}, n_{inc}, \{\delta_i\}_1^n, t) \\ &= \frac{1}{n \cdot t} \sum_{i=1}^{n} f_{\mathcal{P}_i}(x) - \frac{1}{\ln \rho(1)} + \sum_{i=1}^{t \cdot n} \delta_i + \left(1 + \frac{n_{\text{dec}} - n_{\text{inc}}}{n_{\text{dec}} + n_{\text{inc}}}\right)^{-2}. \end{aligned} \tag{5}$$

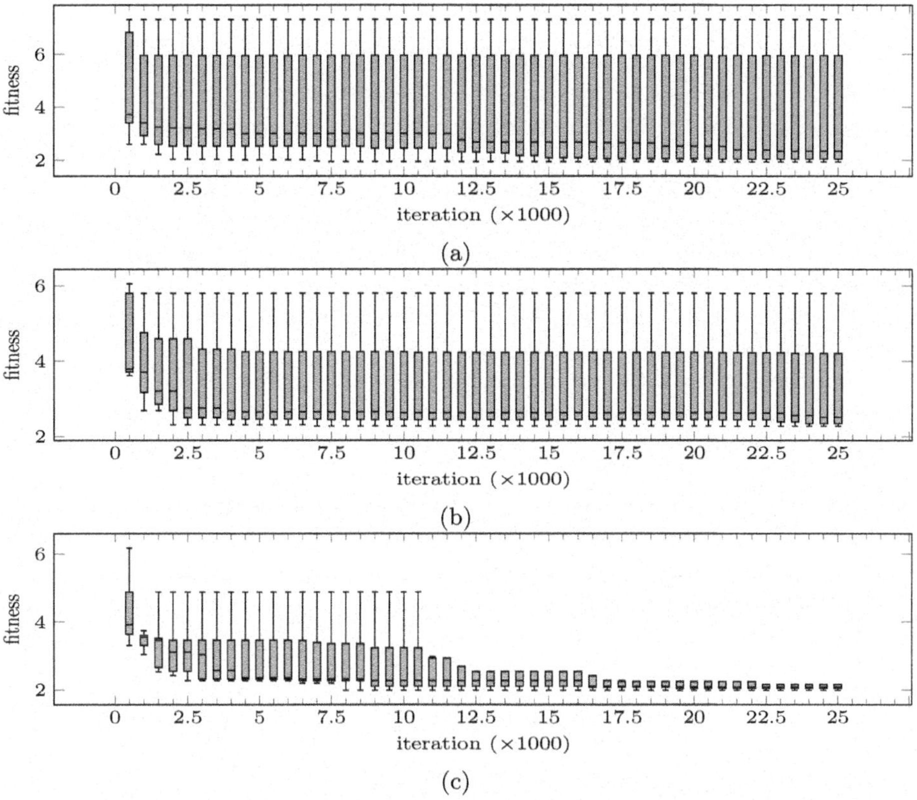

Fig. 4. Achieved mean fitness during the first 25000 iterations for different types of mutation (from top to bottom: original, Gaussian, Lévy flight). The resulting statistics were calculated from 25 runs for each mutation type.

Equation (5) aggregates in a scalarization approach four criteria that are all to be minimized: swarm efficiency, targeted behaviour, swarm diameter, and improvement. The swarm that is equipped with a control program encoded by a solution candidate has to solve a set $\{\mathcal{P}_i\}$ of n optimization problems. Each problem is solved t times (because of the stochastic character) and the mean achieved fitness value is incorporated in Eq. (5). To assess targeted behavior, correlation distance $-\frac{1}{\ln \rho(1)}$ and final swarm diameters δ_i are evaluated. Finally, the last term assesses the improvement of the swarm by the relation between total successful optimization steps n_{inc} and deteriorative n_{dec} steps. We used five standard benchmark problems [2] with $t = 10$ repetitions each. Please note, that Eq. (5) is not scaled and thus the domain is $[0, \infty]$. With this problem setting, we compared the original mutation operator with the Lévy flight-based one.

For comparison, we used statistical measures as introduced by Koza [19]. The cumulative probability of success for a budget of i objective evaluations is given by

$$P(M, i) = \frac{n_{\text{success}}(i)}{n_{\text{total}}}, \tag{6}$$

with n_{success} denoting the number of successful runs at i objective function calls and n_{total} denoting the total number of runs. M denotes the number of individuals. In our use case, M—although interpretable as number of agents—is of no use as the agent system works asynchronously and not in terms of generations with a constant number of evaluations per iteration. Instead, we simply use the budget i of the maximum number of objective functions calls allowed by all agents together and set $M := 1$. This approach is consistent with the generalization in [3].

From the success rate one can derive the mean number of independent runs that are required to achieve a minimum rate of success when the budget is fixed to a maximum of i evaluations per rum. Let z denote the wanted success rate, then

$$R(z) = \left\lceil \frac{\log(1-z)}{\log(1-P(M,i))} \right\rceil \tag{7}$$

gives the number of necessary runs. The computational effort $I(M, i, z) = M \cdot i \cdot R(z)$ gives the number of individual function evaluations that must be performed to solve a problem to a proportion of z [3]. As i is a matter of parametrization, Koza defines the minimum computational effort as

$$I_{\min}(M, z) = \min_i \; M \cdot R(z). \tag{8}$$

Table 1. Comparison of original mutation and Lévy flight sampling using Koza's statistics for a $(1+4)$-ES.

a	Original			Lévy		
M	$P(M,i)$	$R(z)$	$I(M,i,z)$	$P(M,i)$	$R(z)$	$I(M,i,z)$
25000	0.24	17	1.70×10^6	0.48	8	8.00×10^5
20000	0.20	21	1.68×10^6	0.48	8	6.40×10^5
15000	0.16	27	1.62×10^6	0.44	8	4.80×10^5
10000	0.12	37	1.48×10^6	0.32	12	4.88×10^5
5000	0.08	56	1.12×10^6	0.28	15	3.00×10^5
1000	0.04	113	4.52×10^5	0.04	113	4.52×10^5

Figure 4 shows the results of a first experiment. We compare the performance of the original mutation (Fig. 4a) [29] with Gaussian mutation (Fig. 4b) and the proposed Lévy flight sampling (Fig. 4c).

In addition to the original mutation implementation, we used Gaussian mutation which has already long time been used for evolutionary optimization [6]. Gaussian mutation samples locally around previously found solutions by adding

Table 2. Comparison of original mutation and Lévy flight sampling Koza's statistics for a $(1 + 12)$-ES.

a	Original			Lévy		
M	$P(M,i)$	$R(z)$	$I(M,i,z)$	$P(M,i)$	$R(z)$	$I(M,i,z)$
25000	0.52	7	7.00×10^5	0.56	6	6.00×10^5
20000	0.4	10	8.00×10^5	0.56	6	4.80×10^5
15000	0.32	12	7.20×10^5	0.48	8	4.80×10^5
10000	0.28	15	6.00×10^5	0.48	8	3.20×10^5
5000	0.12	37	7.40×10^5	0.32	12	2.40×10^5
1000	0.04	113	4.52×10^5	0.04	113	4.52×10^5

Table 3. Comparison of original mutation and Lévy flight sampling using Koza's statistics for a $(3 + 12)$-ES.

a	Original			Lévy		
M	$P(M,i)$	$R(z)$	$I(M,i,z)$	$P(M,i)$	$R(z)$	$I(M,i,z)$
25000	0.68	5	5.00×10^5	0.92	2	2.00×10^5
20000	0.64	5	4.00×10^5	0.88	3	2.40×10^5
15000	0.56	6	3.60×10^5	0.88	3	1.80×10^5
10000	0.32	12	4.80×10^5	0.72	4	1.60×10^5
5000	0.16	27	5.40×10^5	0.60	6	1.20×10^5
1000	0.0	N/A	N/A	0.24	17	6.80×10^4

a random value from a Normal distribution $N(0, \sigma)$ to an old solution position (element wise). All three mutation variations have merely one parameter each that is to be tuned. Parameter tuning in our study requires finding a good parametrization of the respective distribution for a fair comparison. The rest of the algorithm is kept constant for all mutation variants and is parametrized as proposed in [2]. These are the mutation rate m for the uniform mutation, σ for the Gaussian mutation, and $gamma$ for the case of Lévy flights. All these parameters have been tuned using a grid search with step size 0.01.

Each whisker in Fig. 4 shows the aggregated statistics over 500 iteration. While Gaussian mutation already achieve a slight improvement, the convergence is significantly faster with the proposed Lévy flight; which also reduces variance in fitness.

Tables 1, 2, 3 compare the original mutation with Lévy sampling using Koza's Miz statistics for different settings for μ and λ (see Eq. 2) of the evolution strategy. As a successful run, we counted a run that achieved an objective value of less than 2. This threshold has empirically shown to be an indicator for breeding swarms with good optimization behavior [2]. The Lévy method wins already for small ES population sizes with a higher success rate and less computational effort already with medium sized numbers of generations. The advantage grows with larger populations. Thus, Lévy can also take advantage from mating more parent solutions. The winner with the total minimum computational effort (120,000 evaluations) is also the Lévy method.

Although, the results are so far preliminary, they are promising and encourage further research.

5 Conclusion

Some problem classes like learning swarm-based behavior are hard to solve for CGP, because they result in exploration problems due to many and broad barriers on the fitness landscape that separate different basins of interest. This holds especially for algorithms that are mainly driven by mutation in Cartesian Genetic Programming evolved by $(\mu + \lambda)$-evolution strategies. Apart from huge barriers that demand occasional larger jumps for transition into neighboring basins of attraction, also intermediate regions in the fitness landscape with unevaluable solutions hinder smooth trajectories through search space.

We explored using Lévy flights as mutation operator. So far, the results are already promising as the convergences is significantly improved in our test cases. Further research will now consider different variants of integration, further test cases, and also different (heavy-tailed) distributions for further improvements.

References

1. Bremer, J.: Learning to Optimize, pp. 1–19. Springer International Publishing, Cham (2022). https://doi.org/10.1007/978-3-031-06839-3_1
2. Bremer, J., Lehnhoff, S.: Towards Evolutionary Emergence. Ann. Comput. Sci. Inform. Syst. **26**, 55–60 (2021)
3. Christensen, S., Oppacher, F.: An analysis of Koza's computational effort statistic for genetic programming. In: Genetic Programming: 5th European Conference, EuroGP 2002 Kinsale, Ireland, April 3–5, 2002 Proceedings 5. pp. 182–191 (2002)
4. Clegg, J., Walker, J.A., Miller, J.F.: A new crossover technique for cartesian genetic programming. In: Proceedings of the 9th Annual Conference on Genetic and Evolutionary Computation, pp. 1580–1587 (2007)
5. Diveev, A.: Cartesian genetic programming for synthesis of optimal control system. In: Proceedings of the Future Technologies Conference, pp. 205–222. Springer (2020)
6. Fogel, D.B., Atmar, J.W.: Comparing genetic operators with gaussian mutations in simulated evolutionary processes using linear systems. Biol. Cybern. **63**(2), 111–114 (1990)
7. Goldman, B.W., Punch, W.F.: Reducing wasted evaluations in cartesian genetic programming. In: European Conference on Genetic Programming, pp. 61–72. Springer (2013)
8. Gupta, R., Pal, R.: Biogeography-based optimization with Lévy-flight exploration for combinatorial optimization. In: 2018 8th International Conference on Cloud Computing, Data Science Engineering (Confluence), pp. 664–669 (2018)

9. Haklı, H., Uğuz, H.: A novel particle swarm optimization algorithm with Lévy flight. Appl. Soft Comput. **23**, 333–345 (2014)
10. Harding, S., Banzhaf, W., Miller, J.F.: A survey of self modifying cartesian genetic programming. In: Genetic Programming Theory and Practice VIII, pp. 91–107. Springer (2011)
11. Harding, S., Leitner, J., Schmidhuber, J.: Cartesian genetic programming for image processing. In: Genetic Programming Theory and Practice X, pp. 31–44. Springer (2013)
12. Heidari, A.A., Pahlavani, P.: An efficient modified grey wolf optimizer with lévy flight for optimization tasks. Appl. Soft Comput. **60**, 115–134 (2017)
13. Houssein, E.H., Saad, M.R., Hashim, F.A., Shaban, H., Hassaballah, M.: Lévy flight distribution: a new metaheuristic algorithm for solving engineering optimization problems. Eng. Appl. Artif. Intell. **94**, 103731 (2020)
14. Jamil, M., Zepernick, H.J.: Lévy flights and global optimization. In: Yang, X.S., Cui, Z., Xiao, R., Gandomi, A.H., Karamanoglu, M. (eds.) Swarm Intelligence and Bio-Inspired Computation, pp. 49–72. Elsevier, Oxford (2013). https://www.sciencedirect.com/science/article/pii/B978012405163800003X
15. Jensi, R., Jiji, G.W.: An enhanced particle swarm optimization with Lévy flight for global optimization. Appl. Soft Comput. **43**, 248–261 (2016)
16. Kaidi, W., Khishe, M., Mohammadi, M.: Dynamic Lévy flight chimp optimization. Knowl.-Based Syst. **235**, 107625 (2022)
17. Kamaruzaman, A.F., Zain, A.M., Yusuf, S.M., Udin, A.: Lévy flight algorithm for optimization problems—a literature review. In: Applied Mechanics and Materials, vol. 421, pp. 496–501. Trans Tech Publ (2013)
18. Khan, M.M., Ahmad, A.M., Khan, G.M., Miller, J.F.: Fast learning neural networks using cartesian genetic programming. Neurocomputing **121**, 274–289 (2013)
19. Koza, J.R., Koza, J.R.: Genetic Programming: On the Programming of Computers by Means of Natural Selection, vol. 1. MIT press (1992)
20. Levandowsky, M., Klafter, J., White, B.: Swimming behavior and chemosensory responses in the protistan microzooplankton as a function of the hydrodynamic regime. Bull. Mar. Sci. **43**(3), 758–763 (1988)
21. Liu, Y., Cao, B.: A novel ant colony optimization algorithm with Lévy flight. IEEE Access **8**, 67205–67213 (2020)
22. Manazir, A., Raza, K.: Recent developments in cartesian genetic programming and its variants. ACM Comput. Surv. (CSUR) **51**(6), 1–29 (2019)
23. Mandelbrot, B.B., Mandelbrot, B.B.: The Fractal Geometry of Nature, vol. 1. WH Freeman New York (1982)
24. Miller, J.: Cartesian Genetic Programming, vol. 43 (2003)
25. Miller, J.F., Mohid, M.: Function optimization using cartesian genetic programming. In: Proceedings of the 15th Annual Conference Companion on Genetic and Evolutionary Computation. pp. 147–148. GECCO '13 Companion, Association for Computing Machinery, New York, NY, USA (2013). https://doi.org/10.1145/2464576.2464646
26. Miller, J.F., Thomson, P., Fogarty, T.: Designing electronic circuits using evolutionary algorithms. arithmetic circuits: a case study. Genetic Algorithms and Evolution Strategies in Engineering and Computer Science, pp. 105–131 (1997)
27. Miller, J.F., et al.: An empirical study of the efficiency of learning boolean functions using a cartesian genetic programming approach. In: Proceedings of the Genetic and Evolutionary Computation Conference, vol. 2, pp. 1135–1142 (1999)
28. Miller, J.F.: Cartesian genetic programming: its status and future. Genet. Program Evolvable Mach. **21**(1), 129–168 (2020)

29. Oranchak, D.: Cartesian Genetic Programming for the Java Evolutionary Computing Toolkit (CGP for ECJ) (2010). http://www.oranchak.com/cgp/doc/
30. Reynolds, A.: Lévy flight movement patterns in marine predators may derive from turbulence cues. Proc. Roy. Soc. A: Math. Phys. Eng. Sci. **470**(2171), 20140408 (2014)
31. dos Santos Coelho, L., Bora, T.C., Klein, C.E.: A genetic programming approach based on lévy flight applied to nonlinear identification of a poppet valve. Appl. Math. Model. **38**(5–6), 1729–1736 (2014)
32. Schuster, F., Levandowsky, M.: Chemosensory responses of acanthamoeba castellanii: visual analysis of random movement and responses to chemical signals. J. Eukaryot. Microbiol. **43**(2), 150–158 (1996)
33. Shlesinger, M.F., Klafter, J.: Lévy walks versus lévy flights. On Growth and Form: Fractal and Non-fractal Patterns in Physics, pp. 279–283 (1986)
34. Shukla, S., Kumar, L., Bera, T., Dasgupta, R.: A Lévy Flight based Narrow Passage Sampling Method for Probabilistic Roadmap Planners. arXiv preprint arXiv:2107.00817 (2021)
35. Sotto, L.F.D.P., Kaufmann, P., Atkinson, T., Kalkreuth, R., Basgalupp, M.P.: A study on graph representations for genetic programming. In: Proceedings of the 2020 Genetic and Evolutionary Computation Conference. pp. 931–939. GECCO '20, Association for Computing Machinery, New York, NY, USA (2020), https://doi.org/10.1145/3377930.3390234
36. Turner, A.J., Miller, J.F.: Recurrent cartesian genetic programming. In: Bartz-Beielstein, T., Branke, J., Filipič, B., Smith, J. (eds.) Parallel Problem Solving from Nature—PPSN XIII, pp. 476–486. Springer International Publishing, Cham (2014)
37. Viswanathan, G.M.: Fish in lévy-flight foraging. Nature **465**(7301), 1018–1019 (2010)
38. Viswanathan, G.M., Afanasyev, V., Buldyrev, S.V., Murphy, E.J., Prince, P.A., Stanley, H.E.: Lévy flight search patterns of wandering albatrosses. Nature **381**(6581), 413–415 (1996)
39. Viswanathan, G., Afanasyev, V., Buldyrev, S.V., Havlin, S., Da Luz, M., Raposo, E., Stanley, H.E.: Lévy flights in random searches. Phys. A **282**(1–2), 1–12 (2000)
40. Walker, J.A., Völk, K., Smith, S.L., Miller, J.F.: Parallel evolution using multichromosome cartesian genetic programming. Genet. Program Evolvable Mach. **10**(4), 417 (2009)
41. Zhou, Y., Ling, Y., Luo, Q.: Lévy flight trajectory-based whale optimization algorithm for engineering optimization. Eng. Comput. (2018)

Strategies to Apply Genetic Programming Directly to the Traveling Salesman Problem

Darren M. Chitty(✉)

Faculty of Environment, Science and Economy, University of Exeter, Exeter EX4 4QF, UK
darrenchitty@googlemail.com

Abstract. Genetic Programming (GP) is an evolutionary methodology for generating programs typically applied to classification and symbolic regression problems. GP is not ordinarily applied directly to solve combinatorial optimisation problems. However, GP can be considered similar to hyper-heuristic methods which apply simple heuristics sequentially to a given solution to a problem, a set of operations or a program. Consequently, this paper will present a novel implementation of GP which can directly solve optimisation problems. Similar to hyper-heuristics, a hill-climbing method to GP is presented whereby programs are constructed in small parts or phases, Phased-GP. Furthermore, acceptance strategies for the use of Phased-GP are explored to improve its performance. When Phased-GP is applied directly to Traveling Salesman Problems of up to 1000 cites solutions within 6% of optimal can be derived using only simple operators, a significant improvement over standard GP.

Keywords: Genetic programming · Combinatorial optimisation · Hyper-heuristics

1 Introduction

Solving complex combinatorial problems such as the Traveling Salesman Problem (TSP) is difficult due the size of the problem landscape. Indeed, the number of feasible solutions make these types of problems $\mathcal{NP}$-Hard in nature. Typically, the TSP is solved using exact or heuristic methods such as the state of the art Chained Lin-Kernighan [1]. Meta-heuristic methods are also commonly used such as Genetic Algorithms (GAs) [9] and Ant Colony Optimisation (ACO) [7]. These methods search within the *solution* problem space, combining or modifying solutions to improve them. An alternative method is to search within the space of *operations* that can manipulate a given solution. An example is 2-opt local search [5] which iteratively searches for inversion operations that improve a solution to local optima. Hyper-heuristics go further searching a range of available operations or heuristics that improve a solution to the optimal state [4].

N. Naik et al. (Eds.): UKCI 2023, AISC 1453, pp. 311–324, 2024.
https://doi.org/10.1007/978-3-031-47508-5_25

A set of operations applied sequentially to a solution could also be considered a *program* whereby each program line is an operation to be applied. A well known evolutionary methodology used to automatically generate programs is Genetic Programming (GP) [12]. GP uses the principles of Darwinian evolution namely natural selection, crossover and mutation, to successively improve a population of candidate programs. Typically, GP is applied to classification problems to correctly label data instances or symbolic regression to uncover variable relationships. GP is not generally *directly* applied to optimisation problems such as the TSP. However, given the hyper-heuristics methodology of finding heuristic operations being similar to a program, GP is likely equally applicable. Indeed, given a simple swap operation it should be possible to evolve a set of swaps that would result in the optimal TSP solution.

Consequently, this paper investigates the use of GP to solve TSP instances by evolving programs consisting of basic operations. This enables the power of GP itself to solve TSP instances to be fully evaluated rather than rely on sophisticated heuristics such as 2-opt. However, evolving long programs of operations and associated parameters is possibly too difficult for GP, more complex than searching the solution space with a GA. Therefore, a *hill-climbing* method will be explored whereby best programs are periodically saved and subsequent evolved programs operate on the output of the saved program, a *phased* approach to GP. The paper is laid out as follows: Sect. 2 provides background to the TSP and methods to solve it including hyper-heuristic methods. Section 3 describes the direct application of GP to the TSP and a Phased-GP methodology. In Sect. 4 results from applying GP and Phased-GP to a range of TSP instances will be presented and Sect. 5 considers differing acceptance strategies for Phased-GP. Finally Sect. 6 summarises the work and future research directions.

2 Background and Related Work

The goal of the Travelling Salesman Problem (TSP) is to visit all cities once minimising traversal cost. The symmetric TSP is represented as a complete weighted graph $G = (V, E, d)$ where $V = \{1, 2, \ldots, n\}$ are vertices defining cities and $E = \{(i, j)|(i, j) \in V \times V\}$ edges with distance d between pairs of cities such that $d_{ij} = d_{ji}$. The aim is to find a Hamiltonian cycle in G of minimal length. The TSP is $\mathcal{NP}$-Hard in nature, as vertices increase cycles increase exponentially.

Exact methods can solve small TSP instances but do not scale to larger problems. Consequently, meta-heuristic methods which can only guarantee near optimal solutions are popular. Genetic Algorithms (GAs) [9] use Darwinian evolution to successively improve a population of candidate solutions. Each chromosome consists of cities in the order to be visited and these undergo selection, crossover and mutation to generate new candidate orderings. Key to GAs applied to the TSP is crossover without repetition of cities thus a range of sophisticated crossovers have been proposed such as Partition crossover (PX) [21]. Alternatively, Ant Colony Optimisation (ACO) [7] applied to the TSP uses simulated ants traversing graph G visiting each vertex depositing pheromone defined by solution quality on edges taken. Ants probabilistically decide vertices to visit using pheromone levels on the edges of graph G and heuristic information. Whilst ACO can derive solutions within a few percent of optimal for small TSP instances ACO does not scale well. These meta-heuristics can be improved by local search methods such as 2-opt [5] which iteratively disconnect all possible edge pairs and reconnect them in opposite, a sub-tour reversal accepting if an improvement. A GA with Generalised Partition Crossover [20] combined with Lin-Kernigan local search has produced highly competitive results on TSP instances.

Exact and meta-heuristic methods operate directly on candidate solutions to the TSP, searching within the space of solutions. An alternative search is find a set of operations to a given solution that will derive the optimal solution. Hyper-heuristics [4] are a method used to search this problem space [3]. Hyper-heuristics do not operate with problem domain knowledge. Instead, low level heuristics are utilised ranging from simple methods such as swapping two vertices in a solution to deploying a local search method such as 2-opt. Cowling et al. [4] considered a range of simple hyper-heuristic methods to select heuristics. For instance, Simple Random (SR) which uses uniform probability of selection or Random Descent (RD) which continues to use the same heuristic until no longer successful.

These low-level heuristics applied can be considered as a set of steps or a sequence. Indeed, advanced hyper-heuristic methods search for sequences of heuristics to be applied in a single iteration such as the Selection-Based Sequence Hyper-Heuristic (SSHH) [11]. Sequences of heuristics to be applied can also be considered as a *program* of operations. An evolutionary method used to automatically design programs is Genetic Programming (GP) [12]. GP uses the principles of Darwinian evolution, natural selection, crossover and mutation, to evolve a population of programs to solve a given task. These tasks typically comprise classification, regression or forecasting problems. GP is not commonly used to directly solve combinatorial problems. However, there are some exceptions for instance Dimopoulus and Zalzaha [6] used GP to evolve dispatching rules for the one-machine tardiness optimisation problem. Tay and Ho [19] also use GP to evolve dispatching rules for the flexible job-shop problem with the evolved rules representing a heuristic or rule of thumb. Within civil engineering Soh and Yang [17] use GP to evolve optimised structures in terms of size and geometry. However, although GP is not used for combinatorial optimisation it is used within a hyper-heuristic framework but not directly. Instead, GP operates in a *generative* form of hyper-heuristics whereby the goal is to synthesize a novel heuristic method that can be used to solve an underlying problem such as the TSP directly within an evolutionary algorithm or hyper-heuristic framework. For instance, Ryser-Welch et al. [16] use a cartisian form of GP to evolve TSP solvers that use advanced operators such as 3-opt local search. Tavares and Pereira [18] use strongly typed GP to evolve communication strategies for ants within ACO for application to the TSP. Keller and Poli [10] use Linear GP (LGP) to generate parsimonious meta-heuristics to solve the TSP. Oltean [15] uses GP to create a bespoke GA for solving the TSP whereby GP decides which specific population members are crossed over and mutated. Duflo et al. [8] use GP to generate heuristics to solve TSP instances with improved results over nearest neighbour. In an alternative context, Nguyen et al. [14] use GP to decide heuristic selection within a hyper-heuristic framework to solve combinatorial optimisation problems. Given that GP does not seem to be used to solve combinatorial problems directly, this paper will explore the prospect of using GP in this context.

3 Applying Genetic Programming Directly to the TSP

To apply GP to a combinatorial optimisation problem such as the TSP requires evolving a set of operations to apply to a given initial solution that achieves optimality. Given a TSP instance and a random solution it can be considered that there exists a program that can reorder this to the optimal solution. Moreover, given a simple swap operation and a ten city TSP it is conceivable there exists a program of swaps of ten or less steps. With an optimal program the first swap could place the correct city at position one, the second swap the correct city at position two and so forth. An advantage of GP over a hyper-heuristic is that GP can learn both the best sequence of operations and their associated parameters.

GP is commonly used in a tree-based form whereby lower branches provide the input values to higher level operations. However, for combinatorial optimisation a sequential sequence of operations is required, essentially a form of traditional program whereby each line constitutes an operation to be performed. A variant of GP known as Linear GP (LGP) [2] generates programs of this type. Using LGP a program or sequence of operations can be evolved whereby the first element of each operation refers to which operation to perform and the following elements consist of the input values or parameters to this operation. For applying LGP to the TSP using a simple swap operation these parameters would be constant values and refer to positions within the solution, a value between 1 and the number of cities. Algorithm 1 provides an example of a program that could be evolved by LGP for application to the TSP using simple operators.

Algorithm 1 Exemplar GP Program to Solve the TSP

S' = initial global starting solution S	
S' = SWAP(3,8)	{swap city at position 3 with city at position 8}
S' = INSERT(7,2)	{insert city at position 7 into position 2}
S' = INSERT(9,1)	{insert city at position 9 into position 1}
S' = INVERT(3,6)	{invert cities between positions 3 and 6}
S' = SWAP(4,5)	{swap city at position 4 with city at position 5}
S' = INVERT(1,3)	{invert cities between positions 1 and 3}
S' = INVERT(4,7)	{invert cities between positions 4 and 7}

3.1 Phased-GP

However, it could be considered that finding an optimal program to solve the TSP is in fact more difficult than solving the TSP using a meta-heuristic such as a GA. This is due to evolving both operations and their associated parameters. If the number of operations required is equal to the number of cities and each operation has two parameters the solution space to be searched is three times larger. Consequently, a form of hill-climbing or *phased* approach is proposed for use within the LGP evolutionary process. Instead of evolving a fully complete large program that generates the optimal result from an initial randomly generated solution, a program can be evolved in stages or *phases*. Consider that at the early stages of the evolutionary process small programs have evolved that improve upon the initial solution but not to optimality. Rather than attempting to grow or evolve these small programs into increasingly larger programs, the best improvement could be *locked in*. With an optimisation problem such as the TSP, this simple program can be applied to the given solution to improve it and this becomes the new solution for subsequent evolved programs to be applied to. In effect, this intermediate program is *saved* and subsequent programs add to this. This is similar in respect to Automatically Defined Functions (ADFs) [13] in GP whereby both functions and programs are evolved and functions reused within the program. With a phased approach to GP it could be considered that a single use ADF is generated and applied at the beginning of subsequent programs. This methodology repeats until the conclusion of the evolutionary process. This is in effect hill-climbing to consistently improved solutions. This process can be termed as a *phased* approach to evolving a good program. Repeatedly doing this allows a program to be evolved *piecemeal*, each saved program can be output to provide the entire program that can optimise from the initial solution.

Consider, the exemplar program in Algorithm 1 if evolved in phases. The code within lines 2–4 is the best program evolved in phase 1. At this point the program is *saved* and all subsequent programs operate on the output of this program. This is achieved by updating the initial solution S with current solution S' from line 4. Program lines 5–6 constitute the best program evolved by building upon the output program from phase 1. This too is *saved* and solution S updated with S' and phase 3 evolves the final 2 lines. The best programs from each phase constitute the complete solution.

The phased variant of GP (Phased-GP) is shown in Algorithm 2. Note, an extra inner loop in contrast to standard GP on line 7. This *phase* loop evolves a program on the current solution after which the best found program, if an improvement, updates the current solution *saving* the program. A new population of random programs is generated at the beginning of each phase and the evolutionary process begins again. If no improving program has been evolved within a phase in effect a restart occurs with a new population using the same solution. The number of phases equates to the total generations over phase generations.

Algorithm 2 Phased-GP Applied to the TSP

```
S = current solution to TSP instance generated randomly
S' = new solution                          {current solution modified by a program}
S_best = best solution                     {best new solution found in phase}
P_best = best program                      {best program within phase}
while generation less than max generations do
    P = generate population of random programs          {beginning of GP phase}
    for number of phase generations do
        increment generation
        for each program p in P do
            S' = executed program p on S
            if S' better than S_best then
                S_best=S'
                P_best=p
            end if
        end for
        P = generate new program population using selection, crossover, mutation
    end for
    if S_best better than S then
        S = S_best          {if best program improves current solution then update}
    end if
    reset S_best and P_best                             {completion of GP phase}
end while
```

4 Results

To measure the effectiveness of standard LGP and Phased-GP for combinatorial optimisation they will be tested using six medium TSP instances from the TSPLIB library described in Table 1 using the parameters in Table 2. A larger degree of evolution is used as in this instance the objective is $\mathcal{NP}$-Hard optimisation rather than generalisation. A high degree of mutation is used since when using GP to optimise operations and associated parameters, the introduction of new values is necessary. Four TSP operators are used, swap, insertion, inversion and a 3-opt move whereby three edges are disconnected and reconnected. The first three require two parameters, the fourth operator four. Thus, chromosomes use an arity of five, each operator using only the parameters it requires. Single point crossover is used such that child chromosome arity remains constant. Three mutation operators are utilised with uniform probability, crossover, addition and modification. Crossover mutation performs crossover with a random chromosome. Addition adds up to five random operators and modification makes up to five random changes to operators or parameters. A synchronous parallel implementation of GP is used with an eight core AMD Ryzen 2700 processor. Experiments are conducted over 25 random seeded execution runs.

Table 1. TSP instances utilised

TSP instance	Number of cities	Optimal solution	TSP instance	Number of cities	Optimal solution
d198	198	15780	a280	280	2579
lin318	318	42029	pcb442	442	50778
rat783	783	8806	pr1002	1002	259045

To provide a baseline for the use of GP applied directly to combinatorial optimisation problems standard LGP will be applied without any phases. Two

Table 2. GP parameters used in experiments

Population size -	512	Max. iterations -	100 k
Crossover prob. -	90%	Mutation prob. -	33%
Elitism rate -	10%	Random rate -	10%
Tournament size -	4	Terminal set -	Integers pertaining
Operators -	Swap, Insert, Invert, 3-opt move		to cities

Table 3. Average relative errors, runtimes and program lengths when applying LGP to range of TSP instances using both random and 2-opt improved initial solutions.

TSP	LGP from random solution			LGP From 2-opt improved solution		
	Error (%)	Runtime (s)	Prog. length	Error (%)	Runtime (s)	Prog. length
d198	200.24 ± 27.22	319.23 ± 13.34	113.09 ± 1.17	3.41 ± 0.95	61.26 ± 11.82	10.52 ± 6.07
a280	366.27 ± 47.23	485.99 ± 29.09	160.43 ± 1.81	11.04 ± 2.51	63.24 ± 5.88	9.55 ± 2.28
lin318	451.23 ± 77.32	553.51 ± 47.04	182.91 ± 1.23	9.10 ± 1.47	63.65 ± 8.55	8.26 ± 4.96
pcb442	508.03 ± 54.68	870.87 ± 64.04	253.70 ± 1.53	10.26 ± 1.39	71.00 ± 9.40	8.65 ± 2.52
rat783	755.08 ± 51.48	2166.09 ± 118.38	445.82 ± 4.10	11.83 ± 0.88	84.54 ± 7.81	8.17 ± 2.83
pr1002	1026.72 ± 126.01	3268.73 ± 257.90	570.49 ± 4.98	11.95 ± 1.25	93.01 ± 11.37	7.39 ± 3.06

approaches will be considered. Firstly, the ability of GP to optimise from a randomly generated solution is tested for each of the TSP instances and secondly, when 2-opt is used to improve these initial random solutions. These results are shown in Table 3. Firstly, it can be clearly observed that LGP does not perform well when optimising from a random solution. The relative error after 100 k generations is still several orders larger than the known optimal solutions. As hypothesised, using GP in its standard form is unlikely to find near optimal solutions. A key reason is the long length of the programs evolved hence improving programs of this size is difficult. Naturally, from a random solution a large number of operations are necessary. However, when using 2-opt to improve the initial solution a significant gain is observed. This is to be expected as 2-opt can improve random solutions greatly. Indeed, it was observed that when applying GP to these 2-opt improved initial solutions GP was not able to improve upon them much further. Note the difference in average program length when using 2-opt to improve the initial solution, even with lengthy evolution the program lengths have not grown. This is reflected in the runtimes with faster times due to the shorter programs evolved. Given the simple operators used always change a solution, if this is close to optimal longer programs are more likely to be increasingly detrimental to the solution quality than improvements.

The next stage is to consider using Phased-GP from random initial solutions. A fixed number of generations of program evolution will occur before the best program in this phase is used to update the current solution if an improvement. Results from using a range of evolutionary generations within phases are shown in Table 4. The first observation is when contrasting to standard LGP the results from Phased-GP are significantly improved even when using 2-opt to improve the initial solutions for LGP. This demonstrates the benefit of evolving small programs that improve upon the current solution and locking this in, hill-climbing to better solutions. The issue of finding improving modifications to increasingly larger programs is avoided. These results reinforce the hypothesis that GP needs to operate in *phases* in a hill-climbing manner to perform combinatorial optimisation.

A second observation is that whilst using phases within GP is beneficial, the degree of evolution within phases is important. A very small degree of evolution of five to ten generations seems best. Lower than this and the GP programs are barely evolved. Higher than this level and there are less hill-climbing opportunities as the total number of phases is lower. In terms of average program lengths Phased-GP has smaller programs due to bloat being unable to occur within a single phase. A low degree of evolution within a phase has larger programs due to the greater degree of new random populations. Runtimes are also lower for Phased-GP due to the shorter programs when compared to standard LGP.

5 Acceptance Strategies

Phased-GP has successfully demonstrated the capacity for GP to be used for combinatorial optimisation. This approach though uses a singular acceptance strategy. At the end of a given phase the best program found is applied to the current solution to generate a new solution only if it improves upon solution quality, a *greedy* method. If there is no improving program the next phase evolves a new population on the existing solution, a restart. It was observed in experiments that as optimality was approached many phases could occur without improvement. If the solution is trapped in a local optima it could be difficult for Phased-GP to both exit the local optima and find an improvement. Consequently, it is hypothesised that a differing acceptance strategy is required.

A simple alternative option is to consider always implementing the best found GP program within a phase to update the current solution. This could seem to be a poor acceptance strategy to consider but note that by using evolution within a phase and selecting the best found program this program will be unlikely to cause a significant decrease in current solution quality. However, repeatedly implementing programs from phases that reduce current solution quality may be problematic. A further acceptance strategy to consider is Simulated Annealing (SA). SA is based on an analogy from thermodynamics, when the temperature is high at the beginning of the process, the acceptance of solutions is more likely. Therefore, a program that reduces solution quality is much more likely to be accepted and implemented. As the temperature reduces this likelihood reduces. The probability of acceptance is dependant on both the temperature and the degree to which solution quality reduces. The SA probability p of accepting a program that reduces solution quality is defined as:

$$p_S = \frac{1}{1 + e^{\frac{eval(S') - eval(S_b)}{T}}} \tag{1}$$

where S' is the new solution, S_b is the best found solution and T is the current temperature defined as the remaining generations across phases.

Table 4. Relative errors, runtimes and program lengths when applying Phased-GP to TSP instances using a range of generations within phases and random initial solutions.

TSP	Phase gens.	Relative error (%)			Runtime	Program
		Average	Best	Worst	(secs)	Length
d198	3	2.12 ± 0.85†	0.71	4.03	57.44 ± 0.39	2.76 ± 0.01
	5	**1.93 ± 0.94**†	0.51	3.57	52.23 ± 0.44	2.26 ± 0.01
	10	1.96 ± 0.74†	1.05	3.64	49.84 ± 0.49	1.99 ± 0.01
	20	2.21 ± 0.92†	0.77	4.70	49.45 ± 0.37	2.01 ± 0.01
	30	2.53 ± 1.00†	0.95	4.92	49.63 ± 0.38	2.04 ± 0.05
a280	3	**7.56 ± 1.88**†	3.51	11.56	61.41 ± 0.30	2.69 ± 0.02
	5	7.70 ± 1.68†	3.93	10.36	55.75 ± 0.39	2.21 ± 0.02
	10	7.67 ± 1.60†	4.21	11.41	53.07 ± 0.37	1.97 ± 0.03
	20	8.01 ± 1.87†	3.93	10.77	52.55 ± 0.36	2.00 ± 0.02
	30	8.28 ± 1.59†	4.08	10.38	52.70 ± 0.43	2.04 ± 0.02
lin318	3	5.89 ± 1.31†	2.66	8.06	64.63 ± 0.48	2.68 ± 0.00
	5	5.50 ± 1.28†	3.40	8.38	58.66 ± 0.50	2.21 ± 0.03
	10	**5.04 ± 1.18**†	2.39	7.70	55.79 ± 0.54	1.97 ± 0.03
	20	5.56 ± 1.76†	2.92	9.48	55.32 ± 0.45	2.00 ± 0.01
	30	6.49 ± 1.62†	3.35	9.78	55.06 ± 0.46	2.06 ± 0.04
pcb442	3	7.31 ± 1.31†	4.94	10.24	72.69 ± 0.32	2.68 ± 0.01
	5	**6.30 ± 1.21**†	4.09	8.91	65.71 ± 0.50	2.20 ± 0.02
	10	6.48 ± 1.19†	4.41	9.11	62.37 ± 0.78	1.96 ± 0.03
	20	7.48 ± 1.23†	4.45	10.07	61.57 ± 0.79	2.01 ± 0.02
	30	7.29 ± 1.31†	4.86	9.99	61.33 ± 0.44	2.08 ± 0.05
rat783	3	8.37 ± 0.94†	7.14	10.15	94.58 ± 1.44	2.69 ± 0.00
	5	**7.87 ± 0.65**†	6.48	9.29	85.18 ± 1.40	2.21 ± 0.01
	10	8.08 ± 0.79†	6.41	10.18	79.46 ± 0.75	1.97 ± 0.03
	20	8.47 ± 0.87†	6.73	10.43	78.16 ± 1.02	2.06 ± 0.03
	30	8.76 ± 0.79†	7.28	10.54	78.31 ± 1.24	2.18 ± 0.05
pr1002	3	8.86 ± 0.99†	7.03	10.90	109.17 ± 2.57	2.68 ± 0.00
	5	**8.18 ± 0.90**†	6.60	9.85	97.13 ± 2.02	2.20 ± 0.02
	10	9.14 ± 0.95†	7.74	11.65	90.64 ± 1.44	1.98 ± 0.03
	20	9.52 ± 1.12†	6.72	11.57	89.03 ± 1.02	2.09 ± 0.03
	30	10.21 ± 0.96†	8.17	12.09	88.68 ± 0.45	2.25 ± 0.04

†Statistically significant improvement of Phased-GP over both LGP results with a $p < 0.01$ t-test, two-sided significance level and 24 degrees of freedom

Table 5. Average relative errors when applying Phased-GP to TSP instances for range of generations within phases and using always and SA acceptance strategies.

TSP	Phase gens.	Always acceptance			Simulated annealing acceptance		
		Av. error (%)	Best (%)	Worst (%)	Av. error (%)	Best (%)	Worst (%)
d198	3	0.97 ± 0.34†	0.55	1.72	0.93 ± 0.27†	0.49	1.66
	5	**0.69 ± 0.44**†	0.31	2.43	0.70 ± 0.45†	0.27	2.34
	10	0.90 ± 0.56†	0.19	2.34	0.86 ± 0.53†	0.36	2.41
	20	1.19 ± 0.57†	0.40	2.72	1.16 ± 0.51†	0.45	2.19
	30	1.43 ± 0.88†	0.53	4.57	1.47 ± 0.87†	0.41	5.02
a280	3	5.55 ± 1.52†	3.29	8.22	5.52 ± 1.42†	3.22	8.75
	5	3.35 ± 1.51†	1.12	5.67	3.74 ± 1.68†	0.80	6.61
	10	**3.23 ± 1.27**†	0.79	5.73	3.45 ± 1.18†	1.28	6.27
	20	4.64 ± 1.67†	1.21	7.55	4.37 ± 1.92†	0.72	7.77
	30	4.13 ± 1.52†	0.75	7.52	4.39 ± 1.70†	1.21	7.59
lin318	3	4.27 ± 1.16†	2.57	6.95	3.81 ± 0.91†	2.48	6.23
	5	3.81 ± 0.80†	2.15	5.56	**3.43 ± 0.98**†	1.76	5.86
	10	4.23 ± 1.45†	2.01	7.25	4.13 ± 1.21†	2.26	6.99
	20	4.61 ± 1.50†	1.76	7.80	4.85 ± 1.43	2.55	7.47
	30	5.30 ± 1.29†	3.33	7.86	5.25 ± 1.50†	3.33	10.04
pcb442	3	6.95 ± 1.20	4.76	10.82	5.80 ± 0.88†	3.93	7.61
	5	3.48 ± 0.82†	1.70	5.05	**3.21 ± 0.81**†	2.06	5.41
	10	3.72 ± 0.92†	2.25	5.69	3.66 ± 0.91†	1.60	5.17
	20	4.62 ± 1.08†	2.90	7.48	4.75 ± 1.23†	3.12	7.38
	30	4.87 ± 1.55†	2.02	7.88	4.38 ± 1.51†	2.30	9.09
rat783	3	14.25 ± 1.48	10.13	16.45	13.96 ± 1.23	11.08	16.13
	5	**6.36 ± 0.94**†	4.45	7.74	6.43 ± 0.96†	4.58	8.80
	10	6.63 ± 1.07†	4.66	8.75	6.61 ± 0.90†	5.15	8.38
	20	7.78 ± 0.77†	6.19	9.17	7.54 ± 0.53†	6.76	8.78
	30	8.30 ± 0.91	6.51	10.51	8.15 ± 0.97	6.74	10.61
pr1002	3	12.93 ± 1.40	10.18	15.93	10.29 ± 0.97	8.34	12.27
	5	7.12 ± 1.06†	5.40	9.21	**6.67 ± 0.87**†	5.15	8.48
	10	7.69 ± 1.28†	5.56	11.23	7.83 ± 1.20†	5.19	10.49
	20	8.99 ± 0.72	7.90	10.20	8.46 ± 0.57†	7.38	9.79
	30	9.98 ± 1.08	8.06	11.86	9.38 ± 0.91†	7.72	10.88

†Statistically significant improvement over equivalent greedy acceptance strategy with a $p < 0.01$ t-test, two-sided significance level and 24 degrees of freedom

The results from these acceptance strategies are shown in Table 5 where in contrast to the greedy acceptance strategy results in Table 4 an improvement in the average relative error has been achieved by both acceptance strategies. However, for the smallest degree of phase evolution lower quality results are achieved for larger TSPs using either strategy. With such little evolution the programs evolved are relatively poor hence the strategy of often accepting the best non-improving program from a phase is not beneficial in this case. With greater evolution of five or greater generations though the best found program is not likely to cause a significant step back if implemented. Furthermore, once a step back in solution quality is taken greater evolution increases the likelihood a subsequent solution improving program is found in the next phase. The results are mixed as to which is the best strategy to use between always accept and an SA strategy. Given that higher evolution within phases will generate an improving program or one that only slightly reduces quality it is likely SA mostly accepts the best program in any case.

It is possible to consider the theory that the always accept and SA acceptance strategies are broadly similar in that they mostly accept programs by using a fixed probability of acceptance. Instead of always accepting a non-improving program, a probability of 1.0, a set of fixed lower probabilities can be considered irrespective of the program effect on current solution quality. Furthermore, reducing the probability of acceptance will most likely be beneficial for low evolution within GP phases when less fit programs are considered for acceptance.

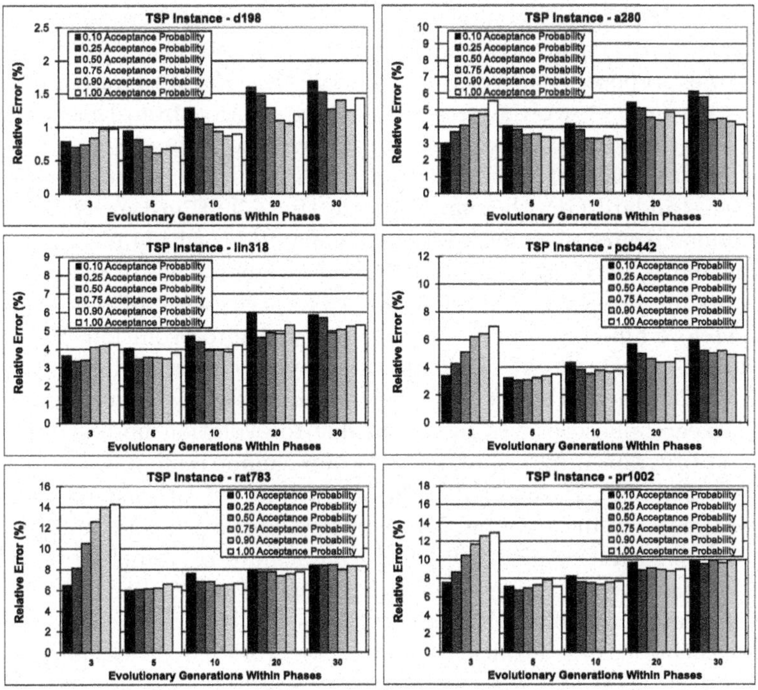

Fig. 1. Average TSP relative errors for Phased-GP when using a range of fixed acceptance probabilities and evolutionary generations within phases.

The results from using a set of fixed acceptance probabilities irrespective of program quality are shown in Fig. 1. Observe that with a low degree of evolution within phases, a high acceptance rate results in lower solution quality. Clearly, the best found lowly evolved phase program can reduce solution quality greatly. When acceptance probability is reduced to as low as 0.1 less backward steps are taken which is beneficial for low evolution phases. The opposite is true for high degrees of evolution within phases, a high acceptance rate gives the best results as the best found program within a phase is not too detrimental to the current solution. However, the overall best results are achieved with a configuration of a low degree of evolution within phases and a low to medium acceptance rate if no improving program is found. This configuration of Phased-GP achieves results superior to that of using an SA acceptance strategy.

6 Conclusions

This paper considered a novel use of Genetic Programming (GP) to directly solve combinatorial optimisation problems such as the Traveling Salesman Problem (TSP). Due to hyper-heuristics creating sequences of operations being akin to a program, GP is equally applicable. Moreover, it was hypothesised that standard GP evolving a single complete TSP program in one step is too difficult, a hill-climbing approach would be much more beneficial. Therefore, a *Phased-GP* approach was proposed which periodically saves the best program and *locks* this in. A new program is evolved on the resulting solution from this program and so forth. Programs could be evolved in parts within phases.

Experiments demonstrated that Phased-GP achieves considerably improved results over standard GP when applied directly to TSP instances. Significantly, only a minimal degree of evolution of 3–10 generations within phases is required. Key is the ability for Phased-GP to hill-climb in small steps or programs. Lower evolution within phases enables more phases to occur with a fixed set of generations. Therefore hill-climbing is of equal importance to evolution. Phased-GP is also capable of breaking out of local optima by accepting and *saving* programs that have a detrimental effect on solutions. In fact, *always* accepting the best program in each phase proves beneficial due to evolution finding a minimially detrimental program if an improving program cannot be found.

Overall, Phased-GP can find solutions within 6% of optimal for TSPs of up to 1,000 cities using only simple operators. However, compared to state of the art solvers such as GPX2 [20] which can achieve results close to optimality for the TSP instances considered in this paper, Phased-GP requires significant improvements to be viable. GPX2 though uses highly sophisticated local search methods to achieve these results. Hence, further work will investigate using more advanced operators with Phased-GP and improved hill-climbing strategies.

References

1. Applegate, D., Cook, W., Rohe, A.: Chained Lin-Kernighan for large traveling salesman problems. Informs J. Comput. **15**(1), 82–92 (2003)
2. Brameier, M., Banzhaf, W.: A comparison of linear genetic programming and neural networks in medical data mining. IEEE Trans. Evol. Comput. **5**(1), 17–26 (2001)
3. Burke, E.K., Gendreau, M., Hyde, M., Kendall, G., Ochoa, G., Özcan, E., Qu, R.: Hyper-heuristics: a survey of the state of the art. J. Oper. Res. Soc. **64**(12), 1695–1724 (2013)
4. Cowling, P., Kendall, G., Soubeiga, E.: A hyperheuristic approach to scheduling a sales summit. In: International Conference on the Practice and Theory of Automated Timetabling, pp. 176–190. Springer (2000)
5. Croes, G.A.: A method for solving traveling-salesman problems. Oper. Res. **6**(6), 791–812 (1958)
6. Dimopoulos, C., Zalzala, A.M.: Investigating the use of genetic programming for a classic one-machine scheduling problem. Adv. Eng. Softw. **32**(6), 489–498 (2001)
7. Dorigo, M., Gambardella, L.M.: Ant colony system: a cooperative learning approach to the traveling salesman problem. IEEE Trans. Evol. Comput. **1**(1), 53–66 (1997)
8. Duflo, G., Kieffer, E., Brust, M.R., Danoy, G., Bouvry, P.: A GP hyper-heuristic approach for generating TSP heuristics. In: 2019 IEEE International Parallel and Distributed Processing Symposium Workshops (IPDPSW), pp. 521–529. IEEE (2019)
9. Holland, J.H.: Adaptation in Natural and Artificial Systems: An Introductory Analysis with Applications to Biology, Control, and Artificial Intelligence. U Michigan Press (1975)
10. Keller, R.E., Poli, R.: Linear genetic programming of parsimonious metaheuristics. In: 2007 IEEE Congress on Evolutionary Computation, pp. 4508–4515. IEEE (2007)
11. Kheiri, A., Keedwell, E.: A hidden markov model approach to the problem of heuristic selection in hyper-heuristics with a case study in high school timetabling problems. Evol. Comput. **25**(3), 473–501 (2017)
12. Koza, J.R.: Genetic Programming (1992)
13. Koza, J.R.: Genetic programming II: Automatic Discovery of Reusable Programs. MIT press (1994)
14. Nguyen, S., Zhang, M., Johnston, M.: A genetic programming based hyper-heuristic approach for combinatorial optimisation. In: Proceedings of the 13th Annual Conference on Genetic and Evolutionary Computation, pp. 1299–1306 (2011)
15. Oltean, M.: Evolving evolutionary algorithms using linear genetic programming. Evol. Comput. **13**(3), 387–410 (2005)
16. Ryser-Welch, P., Miller, J.F., Swan, J., Trefzer, M.A.: Iterative cartesian genetic programming: creating general algorithms for solving travelling salesman problems. In: Genetic Programming: 19th European Conference, EuroGP 2016, Porto, Portugal, March 30-April 1, 2016, Proceedings 19, pp. 294–310. Springer (2016)
17. Soh, C.K., Yang, Y.: Genetic programming-based approach for structural optimization. J. Comput. Civ. Eng. **14**(1), 31–37 (2000)
18. Tavares, J., Pereira, F.B.: Designing pheromone update strategies with strongly typed genetic programming. In: Genetic Programming: 14th European Conference,

EuroGP 2011, Torino, Italy, April 27–29, 2011. Proceedings 14, pp. 85–96. Springer (2011)
19. Tay, J.C., Ho, N.B.: Evolving dispatching rules using genetic programming for solving multi-objective flexible job-shop problems. Comput. Indus. Eng. **54**(3), 453–473 (2008)
20. Tinós, R., Whitley, D., Ochoa, G.: A new generalized partition crossover for the traveling salesman problem: tunneling between local optima. Evol. Comput. **28**(2), 255–288 (2020)
21. Whitley, D., Hains, D., Howe, A.: Tunneling between optima: partition crossover for the traveling salesman problem. In: Proceedings of the 11th Annual Conference on Genetic and Evolutionary Computation, pp. 915–922 (2009)

A Method for Load Balancing and Energy Optimization in Cloud Computing Virtual Machine Scheduling

Kamlesh Chandravanshi[1], Gaurav Soni[1](✉), and Durgesh Kumar Mishra[2]

[1] School of Computing Science and Engineering, VIT Bhopal University, Bhopal-Indore Highway, Kothrikalan, Sehore Madhya Pradesh 466114, India
{kamlesh.chandravanshi,gaurav.soni}@vitbhopal.ac.in

[2] SCSIT, Symbiosis University of Applied Sciences, Indore, India

Abstract. In today's world, cloud computing delivers the payable on demand resources like platform, application, software, and infrastructure as a service. Because it has a variety of advantages, such as adaptability, scalability, dependability, capability, safety, swiftness, and all-time supportability, it is quite popular in all over the world. In this research, proposes a Load-balanced and Energy Optimization method (user to server) in Cloud Computing (LEOCC). The user sends the request to the core switch, where first check if the service is available in the core switch (check the load of each core switch) and if not then forward the request to the next level of switch, the aggregator switch. Here also measure the load of each aggregator switch and select a lightly loaded switch for the service request, which forwards the request to the next level for access switch while the service is unavailable. The same load and service are checked in the access layer switch also, if the requested service is unavailable, it is forwarded to a lightly loaded server for response generation and sent to the user via the reverse path. Otherwise, the response is sent by the access switch. The performance measured by average energy consumption, load, average delay, data received and task handling by virtual machine. The results of LEOCC are compared with various scheduling approaches, i.e., round robin, random, and dens. Finally, LEOCC provides better load balancing and energy optimization in cloud computing as compared to other approaches.

Keywords: Cloud · Energy · LEOCC · Load balancing · Scheduling

1 Introduction

The term "cloud computing" refers to a technique that provide resources, such as processing power, data storage, and application-oriented services, available as on demand. One of the most popular way by that new technologies are generating money in today's industry of cloud computing [1]. Cloud computing popularity is growing steadily as a result of the cost-effectiveness and simplicity of the service. It is also covering business sector of all sizes. Thus, more data centers and servers are setup to enhance the cloud

N. Naik et al. (Eds.): UKCI 2023, AISC 1453, pp. 325–335, 2024.
https://doi.org/10.1007/978-3-031-47508-5_26

infrastructure. To keep their services current, cloud providers need to incorporate these new technologies, but this will raise heterogeneity. Both providers and customers need to take advantage of the growing heterogeneity, if they are to achieve their aims of efficient resource utilization and cost-effectiveness. In addition, the cloud environment is highly unstable because many users utilize various computer resources without having much insight into the other users or the underlying infrastructure. There is a significant possibility of users interfering with one another due to resource sharing and conflict, despite the efforts of many cloud providers like Amazon to isolate individual users' operations to assure a certain degree of speed. The cloud concept can use in different networks for efficient communication [2, 3].

Sometimes, cloud services will purposefully migrate to less predictable resource containers in order to provide lower pricing. For instance, the "spot" instances provided by Amazon's EC2 service are bid on by customers at a fraction of the cost of the more commonplace "regular" instances. Eviction may occur at any time if EC2's load increases and the price of spot instances exceed the bid. For example, micro instances provide increased CPU capacity for short periods of time when additional cycles are available, but this is not guaranteed. Amazon EC2 offers a wide variety of platforms and performance levels, even within a single instance type [4]. As a result of this confluence of circumstances, estimates of resource availability and the efficiency of applications are highly subject to variation.

This article is divided into seven sections. In Sect. 1 gives a brief overview of cloud computing and in Sect. 2 discusses Literature survey on load balancing and energy efficiency. In Sect. 3 mentioned the proposed LEOCC algorithm and in Sect. 4 mentioned simulation's parameters and in Sect. 5 mentioned result comparison with different schemes and in Sect. 6 finally conclude the research work with future scope.

2 Literature Survey

In this section, discussing about recent cloud scheduling techniques. Cloud computing is described here as work in the field of improving cloud services, which is useful for efficient storage and other service provision to customers.

Dzmitry Kliazovich et al. [5] "DENS: data center energy-efficient network-aware scheduling" is a phrase that stands for "data center energy-efficient network-aware scheduling." The significance of communication fabric in data centre energy usage is highlighted in this book, which also introduces DENS, method of scheduling that takes into account both energy savings and network conditions. Rahmeh et al. [6] proposed a cloud computing, active clustering mechanism for balancing the load. The active clustering approach's principal goal is to group related nodes together. The cluster is formed by the clustered node. After that, the algorithm focus on to work on these groupings. The matchmaker node may be detached. Yi Lua et al. [7] proposed a Join-Idle-Queue method. This method is employed to evenly disperse data loads across a wide network. The idle processors are load balanced using this strategy. It primarily affects the dispatchers' availability of idle processors. Galloway et al. [8] Power Aware Load Balancing (PALB) Algorithm was proposed by The PALB method was developed to provide the cluster controller more say over the computing process. The three main parts of PALB are the balancing component, the upscale part, and the downscale part. In the balancing phase, the health of the virtual machine is examined. If the current usage of the nodes is higher than 75%, the upscale portion is responsible for powering up the extra nodes. In order to solve the issue of random load in cloud computing, the ESAE approach was developed [9]. This technique uses a queue to store the jobs in order of submission. Tin-Yu Wu et al. [10] proposed an Index Name Server (INS) approach by that seeks to reduce instances of duplicate information in a database. An ideal selection point is generated using this algorithm. The choice is based on a number of factors, including the server's location, maximum bandwidth, data hash code, weight factor, and so on. The other important option is the busy level, which determines whether or not the connection can tolerate extra nodes. Rich Lee et al. [11] proposed a Weighted Least Connection (WLC proposed a technique to identify the node with the fewest connections. Once the node with the fewest connections has been identified, the job is allocated to it. However, the fundamental drawback of this technique is that it ignores key parameters such as processing speed, bandwidth, and so on. This limitation can be greatly enhanced by using a method known as Exponential Smooth Forecast based on Weighted Least Connection (ESWLC). The time series and node capabilities are taken into account by ESWLC [12]. This is accomplished by making decisions depending on CPU power, the number of connections, memory, and other factors. The node is chosen using an exponential smoothing algorithm. Al-Jaroodi et al. [13] proposed a Dual Direction FTP method. A file of size "n" is partitioned into "n/2" subsets. The task is delegated to the nodes, and they process it in batches. Each node has an independent entity and functioning. As an illustration, one node may start processing from the first block and advance gradually, whereas another node may start processing at the last node and advance negatively. Sharma et al. [14] proposed scheme to approach to allocate workload on virtual machines in a way that maximizes resource utilization, devised the Throttled method. In 2013, however, a new Throttled algorithm [15] was introduced, which reduced response time and resource utilization. The updated technique showed how the index is maintained for the virtual machines. The user gives the load balancer its first direction, which is to find the right virtual machine. This approach, disregards the significance of the work at hand. Methods were developed for ranking the jobs in order of importance [16]. It considers work that in queue for processing. The job's queue time, which should be as short as possible, has been considered in research.

3 Load Balancing and Energy Optimization Algorithm for Virtual Machine Scheduling in Cloud Computing (LEOCC)

In the recent network architecture cloud environment play important role for low cost-based service provide to users, due to higher demand of cloud service load of the cloud devices i.e., switch and sever are proportionally increases. In this research focus on the development of a heterogeneous cloud load balancing and low energy consumption approach that reduces the energy consumption of edge servers and all switches in a cloud system while also balancing their respective workloads. Here, we'll go over the formal description and talk about the technique in detail. The suggested LEOCC technique outperforms the current scheduling system in terms of load balancing and energy usage.

Algorithm: Load balancing Energy Optimization algorithm for Virtual machine scheduling in Cloud Computing (LEOCC).

Input: CN: Cloud Network
U_k: Number of users
S_n: n servers in ∈ CN
Q_l: Queue in server
T_p: Task Priority
$Load_S$: Device Load
T_{h_l}: Load Threshold
S_{wi_Core}: Core Switches in CN
S_{wi_Agr}: Aggregator Switches ∈ CN
S_{wi_Access}: Access Switches ∈ CN
n_{ports}: Number of Port in switch
P_r: Power consumption by active port
$L_{chassis}$: Power consumption by switch hardware
$L_{linecards}$: Power consumption by an active line card

Output: Power Consume (core, agg, access) switch, server energy consumes, average load of server, delay, data receives, virtual machine task schedule.

Procedure:

CN deploy for service provide to U_k
While
U_k request to S_n for service
Request come in S_{wi_Core}

$$T_{h\_l\_Core} = \frac{\sum_{i=1}^{n} S_{wi\_Core(i)_{load}}}{\text{Number } S_{wi\_Core}}$$

If S_{wi_Core} is available & $S_{wi\_Core_{load}} < T_{h\_l}$ **Then**
If Service available in S_{wi_Cor} **Then**
S_{wi_Cor} response to U_k
Else
S_{wi_Core} forward Request to S_{wi_Agr}

$$T_{h\_l\_Agr} = \frac{\sum_{i=1}^{n} S_{wi\_Agr(i)_{load}}}{\text{Number } S_{wi\_Agr}}$$

If S_{wi_Agr} is available & $S_{wi\_Agr_{Load}} < T_{h\_l\_Agr}$ **Then**
If Service available in S_{wi_Agr} **Then**
S_{wi_Agr} response to U_k
Else
S_{wi_Agr} forward Request to S_{wi_Access}

$$T_{h\_l\_Access} = \frac{\sum_{i=1}^{n} S_{wi\_Access(i)load}}{\text{Number } S_{wi\_Access}}$$

If S_{wi_Access} is available & $S_{wi\_Access_{Load}} < T_{h\_l\_Acces}$ **Then**
If Service available in S_{wi_Access} **Then**
S_{wi_Access} response to U_k
Else
S_{wi_Access} forward Request to S_n
S_n take U_k task and response sent by reverse route
S_{wi_Core}, S_{wi_Agr}, S_{wi_Access} use n_{ports} to broadcast request all connected S_i
$L_{switch(Core,\ Agr,\ Access)} = L_{chassis} + L_{linecards} + \sum_{i=0}^{R} n_{ports} \cdot P_r$
S_n Calculate $T_{h\_l(Core,\ Agr,\ Access)} = \frac{\sum_{i=1}^{n} s_n}{i}$
If S_n load $< T_{h\_l}(S_n)$ **than**
Select S_n the Load of device is lower S_{wi_access} forward the U_k request to service S_n
S_n task in Q_l for service
S_n priority assign P_t to U_k task
S_n Execute U_k request and send reply to U_k
Calculate $Load_S$ = Total Service / second
Else
Wait for response by server
End if
End Procedure

4 Simulation Parameters

The efficiency of the protocols in the cloud based on their performance across these four parameters. Now that the entries in the table accurately reflect the consumption of energy by nodes, the switch energy, server energy, and data center load that are specified in Table 1 are as follows: In this case, each of the four protocols that share the same parameters is identical, but the approach that is taken to access the clod will be different.

5 Simulation Result

In this section, we describe the outcome of various scheduling strategies and compare the performance with respect to average energy consumption, load on the server, average delay, percentage of data receives and task scheduling. The outcome of proposed LEOCC provides a better result as compared to DENS, random and round robin.

5.1 Average Energy Consumption and Load Analysis on Server

The conventional schemes like DENS, Random and Round Robin handled the load of servers properly. The requests of users on the cloud are efficiently handled on all three

Table 1: Cloud simulation parameter

Scheduling technique	Round robin	Random	DENS	LEOCC
Simulation duration (sec.)	101.5	101.5	101.5	101.5
Datacenter architecture	Three tier heterogeneous debug			
Switches (core)	2	2	2	2
Switches (aggregator)	4	4	4	4
Switches (access)	4	4	4	4
Number of servers	48	48	48	48
Users	3	3	3	3
Datacenter load	47.1%	46.9%	46.9%	47.1%
Total tasks	30629	30629	30629	30629
Average tasks/server	638.1	638.1	638.1	638.1
Total energy	718.9 W/h	711.8 W/h	712.8 W/h	650.1 W/h
Switch energy (Core)	162.3 W/h	162.3 W/h	162.3 W/h	162.3 W/h
Switch energy (aggregator)	324.6 W/h	324.6 W/h	324.6 W/h	324.6 W/h
Switch energy (access)	33.6 W/h	33.6 W/h	33.6 W/h	33.6 W/h
Server energy	198.4 W/h	191.3 W/h	192.3 W/h	190.1 W/h

levels. If more energy is consumed by a server, it suggests the server is working harder than necessary since there is no good method for balancing the load. As a result of its effective load management, the suggested LEOCC strategy enhances cloud responsiveness. In the given figure, it is clearly visible that the energy consumption of LEOCC is only 125w*h, but the rest of the schemes show energy consumption of more than 190w*h. This is a huge energy consumption difference (Figs. 1 and 2).

There are many people that utilize the cloud and rely on data centers for their storage, processing, and other resource needs. The performance of the earlier DENS, Random, and Round Robin schemes is better if the present server has a suitable amount of demand and has minimal energy for communication, however the LEOCC shows less load % as compared to the others. The correct switching between services improves performance and lessens the network's long-term resource consumption problem.

5.2 Percentage of Data Received and Average Delay Analysis

Packet receiving ratio percentage is one of the important parameters to measure the performance of LEOCC, DENS, Random and Round Robin. Here we clearly show that the performance of the proposed LEOCC is showing an improvement in data packets received because of proper switching of load on servers. The proper switching is handled by the first two layers, and the last layer request is forwarded to the server (Fig. 3).

The delay performance clearly shows that the performance of LEOCC is better due to efficiently handling load on servers and switch load on lightly loaded servers. Here

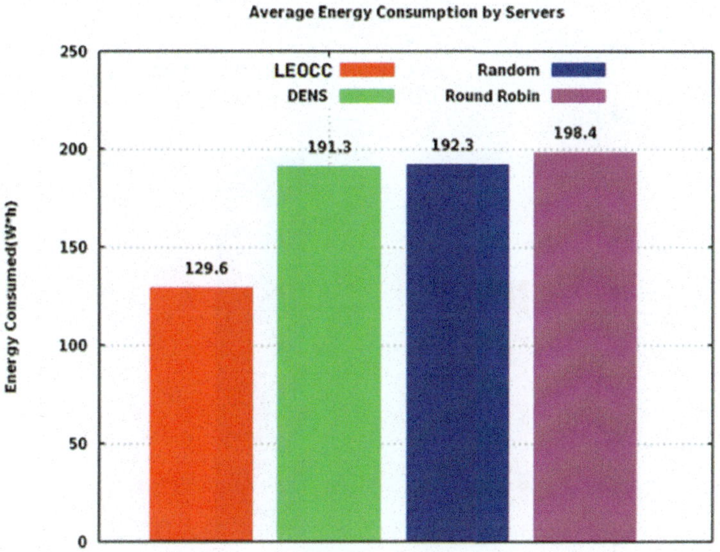

Fig. 1: Server average energy consumption

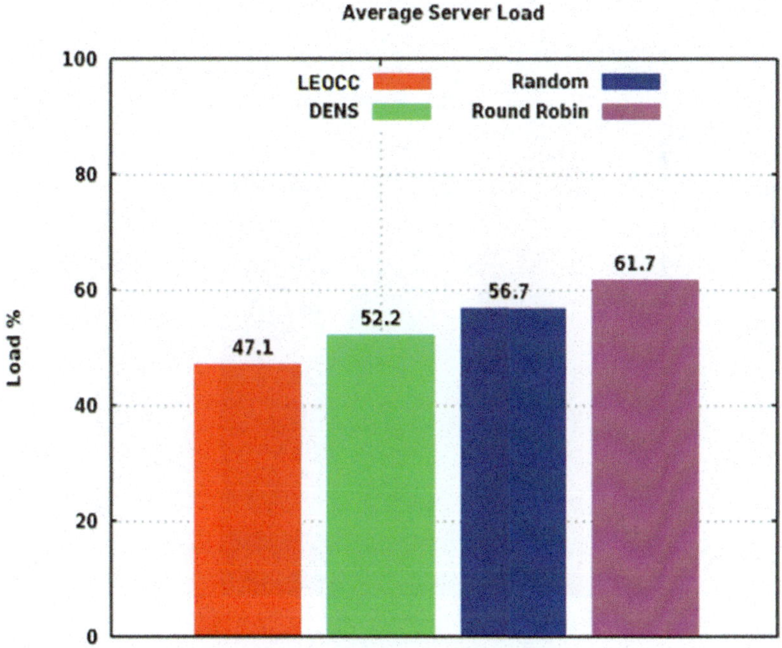

Fig. 2: Analysis of average server load

we clearly show that the Round Robin delay performance is very poor. Random scheme performance is better, but DENS performance is better than Random (Fig. 4).

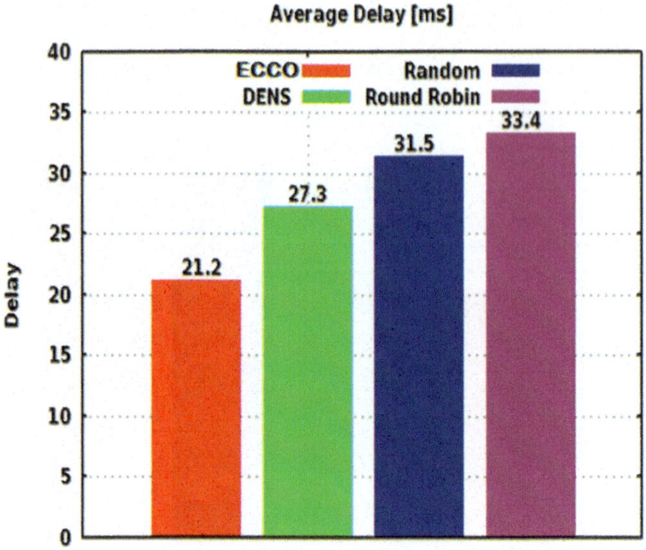

Fig. 3: Analysis of average delay [ms]

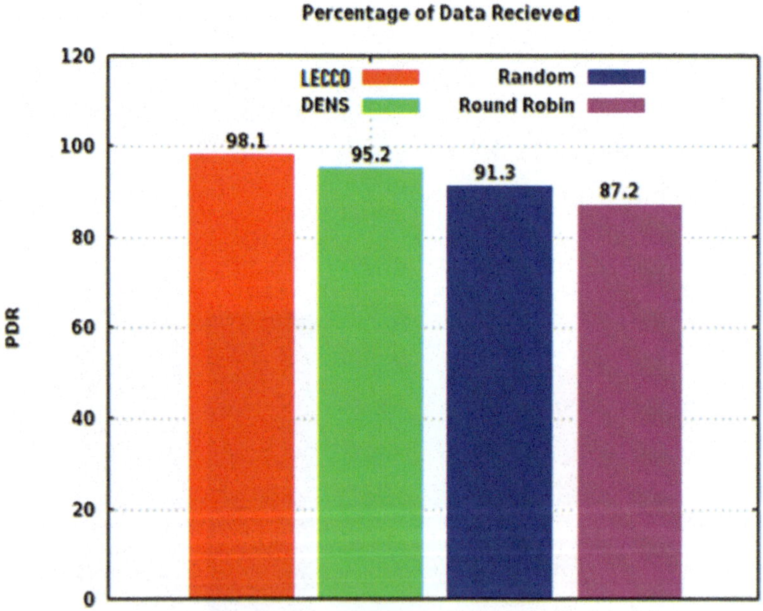

Fig. 4: Packets receiving analysis

5.3 Analysis of Virtual Machine Task Schedule

Switches play an important role in load handling on servers. The main aim of an efficient method is to switch the load on the other server. If one of the servers is busy in scheduling, then the other server is a better choice to handle the request. In this scenario, only the LEOCC technique achieves a good load balance on the virtual machine; the other three approaches all suffer from performance deterioration. Keeping the available servers evenly distributed is another useful function served by the switching mechanism. The execution time or response time is decreased because to the new technique, which correctly schedules user requests. The server is responding appropriately to all network clients. The increased network traffic caused by a larger user population can be mitigated by careful planning of request and response times (Fig. 5).

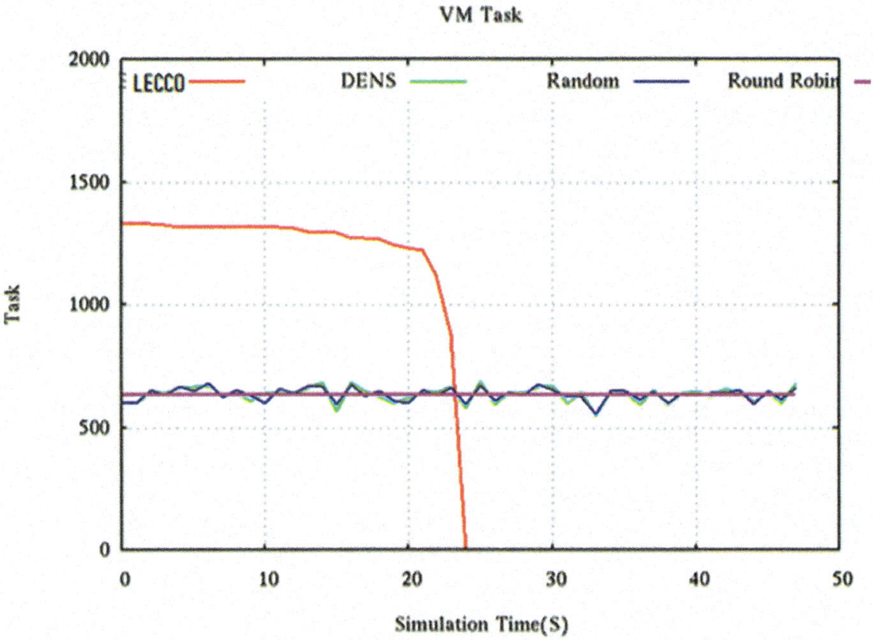

Fig. 5. Analysis of virtual machine task handle

6 Conclusion

Cloud computing is one solution for providing network services to users while also meeting their needs for data storage, platforms, software, and so on. Cloud computing is now very useful for larger organizations, increasing the load on the cloud system. To overcome the load on servers and energy consumption of servers and intermediate switches, we proposed a Load balancing and Energy Optimization in Cloud Computing (LEOCC) that enhances performance with number of user increases and decreases. In today's cloud environment, various virtual scheduling techniques, such as round robin, HEROS, DENS and random are used for efficient task scheduling at server end, but no one can completely estimate the energy utilization and load balancing from user end to server end (not including intermediate switch's load). In this research focus on the need of improve and minimize network load and energy consumption. The proposed LEOCC virtual scheduling would identify the load on each intermediate switch (core, aggregator, access), as well as the load on each server, and assign work based on current load and service availability that optimizing cloud system performance. The results shows that LEOCC scheduling reduced 4ms average delay(minimum), energy consumption, and average server load while increasing percentage of data delivery (98.1%).

Cloud computing is a very wide field of open research, so in the future we will apply some AI based techniques for future prediction-based task scheduling approach and vulnerability monitoring for security system development.

References

1. Subba Rao, B.V., Sharma, V., Rathore, N., Prasad, D., Anandaram, H., Soni, G.: A secure framework to prevent three-tier cloud architecture from malicious malware injection attacks. Int. J. Cloud Appl. Comput. (IJCAC). **13**(1) (2023)
2. Chandravanshi, K., Soni, G., Mishra, D.K.: An efficient energy aware for reliable route discovery using energy with movement detection technique in MANET. In: Lecture Notes in Electrical Engineering, vol. 929. Springer (2022)
3. Soni, G., Chandravanshi, K., Kaurav, A.S., Dutta, S.R.: A bandwidth-efficient and quick response traffic congestion control qos approach for VANET in 6G. In: Innovations in Computer Science and Engineering. Lecture Notes in Networks and Systems, vol. 385. Springer (2022)
4. Michael, A., Lee. G.: Evaluating amazon's ec2 as a research platform. http://radlab.cs.berkeley.edu/w/upload/0/0c/EC2_Performance (2023)
5. Kliazovich, D., Bouvry, P.: DENS: data centre energy-efficient network-aware scheduling. Springer Science Received (2011)
6. Rahmeh, O.A., Johnson, P., Bendiab, A.T.: A dynamic biased random sampling scheme or scalable and reliable grid networks. INFOCOMP-J. Comput. Sci. **7**(4), 1−10 (2008)
7. Lua, Y., Xiea, Q., Kliotb, G.: Join-idle-queue: a novel load balancing algorithm for dynamically scalable web services. In: 29th International Symposium on Computer Performance, Modeling, Measurements and Evaluation, pp.1056–1071 (2011)
8. Galloway, J.M., Smith, K.L., Vrbsky, S.S.: Power aware load balancing for cloud computing. In: IEEE Proceedings of the World Congress on Engineering and Computer Science (WCECS), pp. 19–21 (2011)
9. Kaur, J.: Comparison of load balancing algorithms in a Cloud. Int. J. Eng. Res. Appl. (IJERA) **2**(3), 169–173 (2012)
10. Whu, T.Y., Lee, W.T., Lin, Y.S.: Dynamic load balancing mechanism based on cloud storage. In: proceedings of IEEE Computing, Communications and Applications Conference (ComComAp), pp. 102–106 (2012)
11. Lee. R., Jeng, B.: Load-balancing tactics in cloud. In: IEEE Proceeding of International Conference on Cyber-Enabled Distributed Computing and Knowledge Discovery (CyberC), pp. 447–454 (2011)
12. Ren, X., Lin, R., Zou, H.: Dynamic load balancing strategy for cloud computing platform based on exponential smoothing forecast. In: IEEE Proceedings of International Conference on Cloud Computing and Intelligent Systems (CCIS), pp. 220–224 (2011)
13. Jaroodi, A., Mohamed, J., Mohamed, N.: DDFTP: dual-direction FTP. In: the Proceeding of 11th IEEE/ACM International Symposium on Cluster, Cloud and Grid Computing (CCGrid), pp. 504–503 (2011)

14. Sharma, M., Sharma, P.: Efficient load balancing algorithm in vm cloud environment, M.Tech. Dissertation, Information Technology Department, Dharmsinh Desai University (2012)
15. Domanal, S.G., Reddy, G.R.: Load balancing in cloud computing using modified throttled algorithm. In: Proceeding of IEEE Cloud Computing in Emerging Markets (CCEM), pp. 1–5 (2013)
16. Babu, L.D., Krishna, P.V.: Honey bee behaviour inspired load balancing of tasks in cloud computing environments. Proceeding Appl. Soft Comput. **13**(5), pp. 2292–2303 (2013)

A Dynamic Hyper Heuristic Approach for Solving the Static Frequency Assignment Problem

Khaled Alrajhi(✉)

King Khalid Military Academy, Riyadh, Saudi Arabia
kanr2008@hotmail.com

Abstract. This study proposes a novel approach to solve the minimum-order frequency assignment problem. This problem involves assigning a frequency to each request while satisfying a set of constraints and minimizing the number of used frequencies. This approach solves the static problem by modeling it as a dynamic problem through dividing this static problem into smaller sub-problems, which are then solved in turn in a dynamic process. The proposed approach can be thought of as an algorithm that combines multiple heuristics to solve hard combinatorial optimization problems. Such heuristics are called low level heuristics and are managed by hyper heuristic algorithm. Different mechanisms for selecting the low level heuristics are investigated. Several techniques are also used to make this approach in this study more efficient. One of these is using a lower bound on the number of frequencies that are required for a feasible solution to exist in each sub-problem, based on the underlying graph coloring model. These lower bounds ensure that the search focuses on parts of the solution space that are likely to contain feasible solutions. Overall, this approach showed competitive performance compared with other algorithms in the literature.

1 Introduction

The frequency assignment problem (FAP) is related to wireless communication networks, which are used in many applications such as mobile phones, TV broadcasting and Wi-Fi. The aim of the FAP is to assign frequencies to wireless communication connections (also known as requests) while satisfying a set of constraints, which are usually related to prevention of a loss of signal quality. Note that the FAP is not a single problem. Rather, there are variants of the FAP that are encountered in practice. The static minimum order FAP (MO-FAP) is the first variant of the FAP that was discussed in the literature. In the MO-FAP, the aim is to assign frequencies to requests in such a way that no interference occurs, and the number of used frequencies is minimized.

In this paper, a dynamic hyper heuristic (DHH) approach is applied to MO-FAP. Several novel and existing techniques are used. One of these is using a lower bound on the number of frequencies that are required from each domain for a feasible solution to exist in each sub-problem, based on the underlying graph coloring model. (see [1] for example). This ensures that we never waste time trying to find a solution with a set of

N. Naik et al. (Eds.): UKCI 2023, AISC 1453, pp. 336–347, 2024.
https://doi.org/10.1007/978-3-031-47508-5_27

frequencies that do not satisfy the lower bounds in each sub-problem. Another technique is to apply simple and advanced LLHs associated with an independent tabu list for each LLH for each sub-problem. This is different from other approaches for the static FAP in the literature (e.g. [11, 12]). Different mechanisms for selecting LLHs are investigated.

This paper is organized as follows: the next section gives an overview of the static MO-FAP. Section 3 explains how the underlying graph coloring model for the static MO-FAP can be used to provide a lower bound on the number of frequencies and how this information can then be used to assist the search. In Sects. 4 and 5, the description of the DHH approach for the static MO-FAP is given. In Sect. 6, the results of this approach are given and compared with those of existing approaches in the literature before this paper finishes with conclusions.

2 Overview of the Static MO-FAP

The main concept of the static MO-FAP is assigning a frequency to each request while satisfying a set of constraints and minimizing the number of used frequencies. The static MO-FAP can be defined formally as follows: given.

- A set of requests $\mathrm{R} = \{r_1, r_2, \ldots, r_{\mathrm{NR}}\}$, where NR is the number of requests,
- A set of frequencies $\mathrm{F} = \{f_1, f_2, \ldots, f_{\mathrm{NF}}\} \subset \mathbb{Z}^+$, where NF is the number of frequencies,
- A set of constraints related to the requests and frequencies (described below).

The goal is to assign one frequency to each request so that the given set of constraints are satisfied and the number of used frequencies is minimized. The frequency that is assigned to requests r_i is denoted as f_{r_i} throughout of this study. The static MO-FAP has four variants of constraints as follows:

1. Bidirectional Constraints: this type of constraint forms a link between each pair of requests $\{r_{2i-1}, r_{2i}\}$, where $i = 1, \ldots, NR/2$. In these constraints, the frequencies $f_{r_{2i-1}}$ and $f_{r_{2i}}$ should be distance $d_{r_{2i-1}r_{2i}}$ apart. These constraints can be written as follows:

$$\left|f_{r_{2i-1}} - f_{r_{2i}}\right| = d_{r_{2i-1}r_{2i}} \quad for \ \ i = 1, \ldots, \mathrm{NR}/2 \tag{1}$$

2. Interference Constraints: this type of constraint forms a link between a pair of requests $\{r_i, r_j\}$, where the pair of frequencies f_{r_i} and f_{r_j} should be more than distance $d_{r_ir_j}$ apart. These constraints can be written as follows:

$$\left|f_{r_i} - f_{r_j}\right| > d_{r_ir_j} \quad for \ \ 1 \leq i < j \leq \mathrm{NR} \tag{2}$$

3. Domain Constraints: the available frequencies for each request r_i are denoted by the domain $D_{r_i} \subset F$, where $\cup_{r_i \in R} D_{r_i} = F$. Hence, the frequency which is assigned to r_i must belong to D_{r_i}.
4. Pre-assignment Constraints: for certain requests, the frequencies have already been pre-assigned to given values i.e. $f_{r_i} = p_{r_i}$, where p_{r_i} is a given value.

3 Modeling the Static MO-FAP as a Dynamic Problem

In the DTS approach, the static MO-FAP is broken down into smaller sub-problems, each of which is considered at a specific time period. To achieve this, each request is given an integer number between 0 and n (where n is a positive integer) indicating the time period in which it becomes known. In effect, the problem is divided into $n + 1$ smaller sub-problems$P_0, P_1, \ldots, P_n$, where n is the number of sub-problems after the initial sub-problemP_0. Each sub-problem P_i contains a subset of requests which become know at time period$i$. The initial sub-problem P_0 is solved first at time period0. After that, the next sub-problem P_1 is considered at time period 1 and the process continues until all the sub-problems are considered. In this study, we found that the number of sub-problems does not impact on the performance of the DTS approach for solving the static MO-FAP, so the number of sub-problems is fixed at 21 (i.e. $n = 20$).

Based on the number of the requests known at time period 0 (belonging to the initial sub-problem P_0), 10 different versions of a dynamic problem are generated. These versions are named using percentages which indicate the number of requests known at time period 0. These 10 versions are named 0%, 10%, 20%, 30%, 40%, 50%, 60%, 70%, 80%, 90% (note that 100% means all the requests are known at time period 0 and so corresponds to the static MO-FAP).

An example of how a static MO-FAP is modeled as a dynamic problem is illustrated in Fig. 1, where each node represents a request, each edge a bidirectional or interference constraint and each color a time period in which a request becomes known for the first time.

After breaking the static MO-FAP into smaller sub-problems, these sub-problems will be solved in turn.

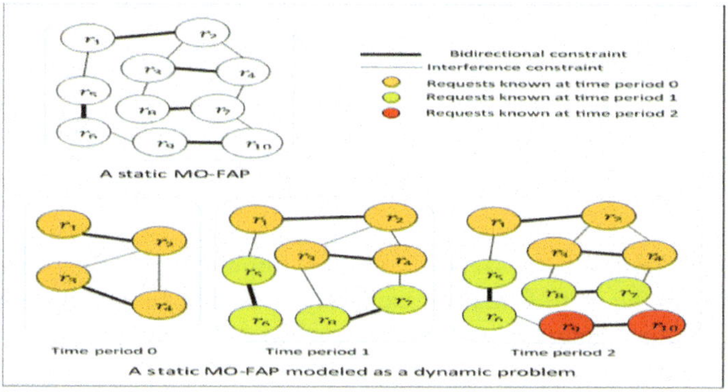

Fig. 1. An example of modeling a static MO-FAP as a dynamic problem over 3 time periods.

4 Graph Coloring Model for the Static MO-FAP

The graph coloring problem (GCP) can be viewed as an underlying model of the static MO-FAP [9]. The GCP involves allocating a color to each vertex such that no adjacent vertices are in the same color class and the number of colors is minimized. The static

MO-FAP can be represented as a GCP by representing each request as a vertex and a bidirectional or an interference constraint as an edge joining the corresponding vertices.

One useful concept of graph theory is the idea of cliques. A clique in a graph can be defined as a set of vertices in which each vertex is linked to all other vertices. A maximum clique is the largest among all cliques in a graph. Vertices in a clique have to be allocated to a different color in a feasible coloring. Therefore, the size of the maximum clique acts as a lower bound on the minimum number of colors.

As the requests belong to different domains, the graph coloring model for each domain can be considered separately and then a lower bound on the number of frequencies that is required from each domain can be calculated. An overall lower bound on the total number of frequencies for a whole instance can also be calculated in a similar way. A branch and bound algorithm is used to obtain the set of all maximum cliques for each domain within each sub-problem.

5 The Dynamic Hyper Heuristic Approach

5.1 Solution Space and Cost Function

It was found that the interference constraints are the most difficult constraints to be satisfied ([8]). So, the solution space is defined here as the set of all possible solutions that satisfy bidirectional, domain and pre-assignment constraints, and the cost function is defined as the number of broken interference constraints, also known as the number of violations. This is different from other configurations in DHH for the static FAP in the literature where no constraints are relaxed (e.g. [11, 12]). Using the solution space which relaxes the interference constraints gives the following sub-problem: minimize the number of violations with a fixed number of used frequencies. Requests and frequencies are considered as pairs based on the bidirectional constraints.

5.2 Structure of the Dynamic Hyper Heuristic Approach

Our Approach starts by solving the sub-problem P_0 in the initial solution phase in the same way as [1]. If this gives a feasible initial solution, then the creating violations phase (see [1]) is used to reduce the number of used frequencies (otherwise, the creating violations phase is skipped and all available frequencies are allowed). After that, DHH is applied using LLHs (see Sect. 5.3) to find a feasible solution with a fixed number of used frequencies. One of the LLHs is selected in each iteration based on the selection mechanism (see Sect. 5.4) to find a new solution. This solution is accepted or rejected based on the move acceptance criteria (see Sect. 5.5), which accepts worse solutions a limited number of times to diversify the search. The process continues until one of the stopping criteria is met (see Sect. 5.6). Then, the sub-problem P_1 is solved in a similar way. DHH continues in the same way until all the sub-problems are considered.

5.3 The Low Level Heuristics

Here, 13 LLHs are applied. Some of them are simple (and previously used, see e.g. [12]) and the others more advanced. As this algorithm accepts neighbour solutions with the same cost as the current solution, cycling is a problem which may be faced in each LLH. To avoid this, each LLH has an independent local tabu list. Any change made to the solution by each LLH is added to the local tabu list. All the local tabu lists are cleared when a feasible solution is achieved, i.e. the sub-problem is solved.

Each LLH starts by choosing either a frequency f_k to be removed or a request r_i to be re-assigned. Such a frequency or request should satisfy two conditions: (i) it is not in the local tabu list, and (ii) it is involved in most violations. If more than one frequency or request satisfies these conditions, then one of them is chosen randomly.

A frequency f_j to be assigned in place of the removed frequency f_k or to be assigned to the chosen request r_i should satisfy the following conditions: (i) f_j is not in the local tabu list, and (ii) f_j results in the minimum number of violations. If more than one frequency satisfies these conditions, then one of them is chosen randomly.

The LLHs which start by choosing a frequency f_k are given as follows:

- LLH_1: the set of requests that are currently assigned to the chosen frequency f_k is swapped with its partner (based on the bidirectional constants).
- LLH_2: all requests assigned to f_k are re-assigned to either the chosen unused frequency f_j or one of the used frequencies. If the assignment to f_j results in zero violations, then this is always made, otherwise each request is assigned to a used frequency that results in the smallest number of violations. In the event of a tie, the requests are assigned randomly.
- LLH_3:a request is randomly selected from the set of requests that are currently assigned to the chosen frequency f_k to be re-assigned to the used frequency f_j.

The LLHs which start by choosing a request r_i is given as follows:

- LLH_4: the selected request r_i is assigned to the chosen used frequency f_j.
- LLH_5: the selected request r_i is swapped with its partner (based on the bidirectional constraints).
- LLH_6: it is similar to LLH_2 but here the frequency of the selected request r_i is chosen to be removed.

The following LLHs are not required to satisfy the condition (ii) for selecting f_k, or r_i, or f_j.

- LLH_7: it is similar to LLH_2 but here the frequency f_k which is assigned to the fewest requests is removed.
- LLH_8: a request r_i is chosen randomly to be re-assigned to a used frequency f_j which is also chosen randomly.
- LLH_9:a request r_i is chosen randomly to be swapped with its partner (based on the bidirectional constraints).
- LLH_{10}: a used frequency f_k is chosen randomly, and then one of the requests that is assigned to f_k, say r_i, is randomly selected. After that, a used frequency f_j which results in the minimum number of violations is assigned to r_i. In case of a tie, one of them is chosen randomly.
- LLH_{11}: each request is swapped with its partner (based on the bidirectional constraints) as long as this does not increase the number of violations.
- LLH_{12}: for each request r_i, the used frequency f_j that results in the minimum number of violations is chosen. In case of a tie, one of them is chosen randomly.

- LLH_{13}: assign an unused frequency f_j in place of a used frequency f_k. All the possible choices of f_j and f_k are considered. Then, the choice that results in the lowest number of violations is chosen.

LLHs may sometimes give worse solutions (i.e. a greater number of violations), which the DHH may choose to accept in order to diversify the search. Thus, LLHs can be divided into two groups: intensification and diversification LLHs. Here, the former group contains LLH_1, LLH_3, LLH_4, LLH_5, LLH_8, LLH_9, LLH_{10}, LLH_{11} and LLH_{12}. The latter group contains LLH_2, LLH_6, LLH_7 and LLH_{13}.

5.4 LLH Selection Mechanisms

The LLH selection mechanisms can be executed in two ways: a non-adaptive selection method such as random selection, and an adaptive selection method such as probabilistic selection [4].

Random Selection. This is the oldest, the simplest and the most commonly used selection mechanism of the LLHs in DHH [5]. This type of selection mechanism was previously used in DHH for the static FAP in [12].

Probabilistic Selection. Probabilistic selection was used previously in [14]. Here, the LLHs are selected probabilistically based on their performance. At the beginning, each intensification LLH has an equal chance of being selected, whereas the diversification LLHs are not selected. Then, the probabilities are updated according to their effect on the number of violations. The following three approaches of updating the probabilities are considered:

- **Approach 1**: If the selected LLH_j decreases the number of violations, its probability is increased using Formula (3) (N is a parameter, which is 50 in this study).

$$P(LLH_j) \leftarrow P(LLH_j) + \frac{1}{N+1} \tag{3}$$

In contrast, if LLH_j increases the number of violations, the probability of selecting LLH_j is decreased using Formula (4).

$$P(LLH_j) \leftarrow P(LLH_j) - \frac{1}{N+1} \tag{4}$$

- **Approach 2**: If the selected LLH_j decreases the number of violations by M, the probability of LLH_j is increased using Formula (5). So, its probability is increased by a greater amount if it causes a larger improvement in the number of violations.

$$P(LLH_j) \leftarrow \frac{P(LLH_j) + M}{N+1} \tag{5}$$

In contrast, if the number of violations increases, the probability is unchanged.

- **Approach 3**: This is a mixture of the first two approaches. If the selected LLH_j decreases the number of violations, its probability is increased by Formula (5). In contrast, if it increases the number of violations, its probability is decreased according to Formula (4).

After increasing or decreasing the probability of the selected LLH using one of the above approaches, the probabilities are then normalised by Formula (6).

$$P(LLH_j) \leftarrow \frac{P(LLH_j)}{\sum_{i=1,i\neq 2,6,7}^{12} P(LLH_i)} \tag{6}$$

Limitation on probabilities. If the probabilities can increase or decrease without limits, some LLHs may be ignored as their probabilities approach zero. To ensure a balance between selecting all the LLHs, we use a minimum limit of the probability of each LLH, i.e. it is updated using Formula (7) (the limit is 0.08 in this study).

$$P(LLH_j) \leftarrow Max(0.08, P(LLH_j)) \tag{7}$$

Using a limit for the probability of selecting each LLH was previously used in [29], but this technique is implemented differently in our DHH approach by considering two stages of removing the extra probability, which is given by Formula (8).

$$Extra\ probability = 0.08 - P(LLH_j) \tag{8}$$

a) *Removing the extra probability by equivalent division.* The extra probability is equally removed from those LLHs which do not reach the limit of the probability value. Assume that there are N such LLHs. Then, the reduction probability, given by Formula (9), is subtracted from each probability of the n LLHs.

$$Reduction\ probability = \frac{Extra\ probability}{n+1} \tag{9}$$

b) *Removing the extra probability by proportional division.* The probability of some LLH may become less than the limit after the previous stage. This happens when its probability is narrowly above 0.08. In this stage, the extra probability is removed proportionally from the probabilities of LLHs which do not reach the limit. Say there are msuch LLHs. Their probabilities are reduced based on Formula (10).

$$Reduction\ probability(LLH_j) = \frac{P(LLH_j)}{\sum_{i=1}^{m} P(LLH_i)} \times Extra\ probability \tag{10}$$

5.5 Acceptance Criteria

A combination of two types of acceptance criteria is applied. This concept was previously used in e.g. [14]. The first type is applied when an intensification LLH is selected: only neighbour solutions that are not worse than the current solution are accepted. This is commonly used and one of the successful acceptance criteria in the literature [13]. The second type is for a diversification LLH: all neighbour solutions are accepted even if the number of violations increases. Note that a diversification LLH is selected when no better neighbour solution has been found by intensification LLHs for a certain number of iterations. Each diversification LLH is allowed to give a worse solution for no more than a given number of times.

5.6 Stopping Criteria

The DHH approach has three stopping criteria: (i) a feasible solution whose number of frequencies is equal to the lower bound is found (as this is the optimal solution), (ii) the number of iterations equals a given number without successfully solving the sub-problem (see Sect. 5.1), i.e. a feasible solution could not be achieved (note that the number of iterations is reset to zero each time the sub-problem is solved), and (iii) the number of violations remains unchanged for more than a given number of iterations and the number of executions of each diversification LLH reaches a given number.

6 Experiments and Results

This section presents results of DHH for the static MO-FAP using CELAR and GRAPH datasets. These datasets (and known optimal solutions) are available on the FAP website http://fap.zib.de/problems/CALMA/ (last accessed 25 February 2015). The parameters of the DHH approach are set based on experimentations as follows: The maximum number of iterations is 2,500. The tabu tenure of a local tabu list for each LLH is 5. The diversification LLHs are used when no better neighbour solution can be achieved using the intensification LLHs for 50 consecutively iterations. Each diversification LLH is allowed to give a worse solution for no more than 6 times. The algorithm was coded using FORTRAN 95. All experiments were conducted on a 3.0 GHz Intel Core I3–2120 Processor (2nd Generation) with 8GB RAM and a 1TB hard drive.

6.1 Results Comparison of the Dynamic Hyper Heuristic Approach

This section compares variants of the DHH approach for the static FAP. For each instance, DHH is run 5 times. The selected number of runs is chosen based on preliminary experiments.

Probabilistic Selection of the LLHs. Here three approaches of updating the probabilities of LLHs without and with limits are compared. The total average numbers of used frequencies over all instances for each approach are shown in Table 1.

Table 1. Total average number of used frequencies for each probabilistic selection approach.

	Approach 1	Approach 2	Approach 3
Without a limit	62.0	57.6	56.0
With a limit	56.8	57.6	54.4

It is found by the Wilcoxon signed-rank test at the 0.05 significance level that the performance of DHH with a limit on the probabilities of LLHs is significantly better than DHH without a limit. Hence, the three approaches are compared on selected instances as shown in Table 2. The selected instances represent different instances with different numbers of requests and constraints.

Table 2. Average number of used frequencies for each probabilistic selection approach.

Instance	Approach 1	Approach 2	Approach 3
CELAR 01	20.0	20.0	18.8
CELAR 03	17.2	16.4	16.4
GRAPH 08	19.6	21.2	19.2

It is found by the Wilcoxon signed-rank test at the 0.05 significance level that there is no significant difference between the average results over the selected three instances of these approaches. Hence, these approaches are compared based on the average run time as shown in Fig. 10 (Table 3).

Table 3. Average run time (min) in each instance for each probabilistic selection approach.

Instance	Approach 1	Approach 2	Approach 3
CELAR 01	37.6	33.6	26.0
CELAR 03	9.2	9.4	6.6
GRAPH 08	26.4	25.2	20.8

It is found by the Wilcoxon signed-rank test at the 0.05 significance level that there is a significant difference between the average run times of these approaches on these instances. Thus, the best probabilistic selection mechanism is approach 3.

Comparison of the LLH Selection Mechanisms. The results of the DHH algorithm with the random and the probabilistic selection mechanisms are shown in Table 4.

It is found by the Wilcoxon signed-rank test at the 0.05 significance level that there is no significant difference between the average results of DHH based on the LLHs selection mechanisms. Therefore, these approaches are compared based on the run time as also shown in Table 4. It is found by the Wilcoxon signed-rank test at the 0.05 significance level that the average run time using the random selection is significantly better than using the probabilistic selection of the LLHs. Hence, the random selection of the LLHs is selected for the DHH approach to be compared with other algorithms in the literature.

6.2 Results Comparison with Other Algorithms

This section compares the performance of the DHH approach with other algorithms in the literature. The comparison is shown in Table 4, where a bold number means that the optimal solution was achieved, and a dash "-" means the result is not available (Table 5).

Table 4. Average number of used frequencies and average run time of the DHH algorithm with each LLH selection mechanism.

Instance	Random selection		Probabilistic selection	
	No. Freq	Time (min)	No. Freq	Time (min)
CELAR 01	19.2	18.0	18.8	36.0
CELAR 02	14.0	0.0	14.0	0.0
CELAR 03	16.8	3.2	16.4	7.6
CELAR 04	46.0	0.9	46.0	0.9
CELAR 11	40.0	10.4	43.2	19.2
GRAPH 01	18.4	0.8	18.8	1.1
GRAPH 02	14.8	3.0	14.8	4.6
GRAPH 08	20.0	13.2	19.2	28.8
GRAPH 09	22.0	19.2	22.4	38.4
GRAPH 14	10.8	15.6	10.8	38.4

Table 4 shows that DHH approach achieved the optimal solution to the majority of the instances. In fact, it achieved the optimal solution for all the instances except for CELAR 11, GRAPH 08 and GRAPH 14. Overall, this approach shows competitive results compared with other approaches in the literature.

7 Conclusions

In this paper, we have presented a novel approach for solving the static MO-FAP. This approach solves this problem by modeling it as a dynamic problem through dividing this problem into smaller sub-problems, which are then solved in turn in a dynamic process using DHH approach. This approach includes 13 simple and advanced LLHs, some of which are used for diversification. Furthermore, each LLH has an independent tabu list to avoid cycling. Two different methods of the LLHs selection were compared: random and probabilistic selection. The probabilistic selection gives a higher probability to the LLHs which reduce the number of violations. Moreover, two types of the LLH probabilistic selection were tested: without and with a limit. It was found that the random selection performed better than the probabilistic selection. Overall, DHH showed competitive performance compared with other algorithms in the literature.

Table 5. Results of DHH and other algorithms in the literature.

Instance	GENET [3]	Genetic algorithm [10]	Potential reduction [18]	A nonlinear approach [17]	Evolutionary search [7]	Simulating annealing [16]	Variable depth search [16]	Tabu search [2]	Tabu search [16]	Tabu search [1]	The DHH approach	Optimal solution
CELAR 01	16	20	16	16	–	16	16	18	16	16	16	16
CELAR 02	14	14	14	–	14	14	14	14	14	14	14	14
CELAR 03	14	16	16	16	14	14	14	14	14	14	14	14
CELAR 04	46	46	46	–	–	46	46	46	46	46	46	46
CELAR 11	24	32	–	–	–	24	24	24	22	38	36	22
GRAPH 01	18	20	18	18	18	–	–	18	18	18	18	18
GRAPH 02	14	16	14	14	14	–	–	16	14	14	14	14
GRAPH 08	22	–	18	18	–	–	–	24	20	18	20	18
GRAPH 09	22	28	18	18	–	–	–	22	22	18	18	18
GRAPH 14	–	14	10	10	–	–	–	12	10	8	10	8

References

1. Alrajhi, K., Thompson, J., Padungwech, W.: Tabu search hybridized with multiple neighborhood structures for the frequency assignment problem. In: International Workshop on Hybrid Metaheuristics, pp. 157–170. Springer, Cham (2016)
2. Bouju, A., Boyce, J.F., Dimitropoulos, C.H.D., Vom Scheidt, G., Taylor, J.G.: Tabu search for the radio links frequency assignment problem. Applied Decision Technologies (ADT'95), London (1995a)
3. Bouju, A., Boyce, J.F., Dimitropoulos, C.H.D., Vom Scheidt, G., Taylor, J.G., Likas, A., Papageorgiou, G., Stafylopatis, A.: Intelligent search for the radio link frequency assignment problem. In: Proceedings of the International Conference on Digital Signal Processing, Cyprus (1995b)
4. Burke, E.K., Hyde, M., Kendall, G., Ochoa, G., Özcan, E., Woodward, J.R.: A Classification of Hyper-Heuristic Approaches. In: Handbook of metaheuristics, pp. 449–468. Springer, US (2010)
5. Chakhlevitch, K., Cowling, P.: Hyperheuristics: rrrecent developments. In: Adaptive and Multilevel Metaheuristics, pp. 3–29. Springer, Berlin Heidelberg (2008)
6. Chaves-González, J.M., Vega-Rodríguez, M.A., Gómez-Pulido, J.A., Sánchez-Pérez, J.M.: Optimizing a realistic large-scale frequency assignment problem using a new parallel evolutionary approach. Eng. Optim.Optim. **43**(8), 813–842 (2011)
7. Crisan, C., Mühlenbein, H.: The frequency assignment problem: A look at the performance of evolutionary search. In: Artificial Evolution, pp. 263–273. Springer, Berlin Heidelberg (1997)
8. Dorne, R., Hao, J.K.: Constraint handling in evolutionary search: a case study of the frequency assignment. In: Parallel Problem Solving from Nature—PPSN IV, pp. 801–810. Springer, Berlin Heidelberg (1996)
9. Hale, W.K.: Frequency assignment: theory and applications. Proc. IEEE **68**(12), 1497–1514 (1980)
10. Kapsalis, A., Chardaire, P., Rayward-Smith, V.J., Smith, G.D.: The radio link frequency assignment problem: A case study using genetic algorithms. In: Evolutionary Computing, pp. 117–131. Springer, Berlin Heidelberg (1995)
11. Kendall, G., Mohamad, M.: Channel assignment in cellular communication using a great deluge hyper-heuristic. In: Networks, 2004 (ICON 2004). Proceedings. 12th IEEE International Conference on, vol. 2, pp. 769–773. IEEE (2004a)
12. Kendall, G., Mohamad, M.: Channel assignment optimisation using a hyper-heuristic. In: Cybernetics and Intelligent Systems, 2004 IEEE Conference on, vol. 2, pp. 791–796. IEEE (2004b)
13. Özcan, E., Bilgin, B., Korkmaz, E.E.: A comprehensive analysis of hyper-heuristics. Intell. Data Anal. **12**(1), 3–23 (2008)
14. Rattadilok, P., Gaw, A., Kwan, R.S.: Distributed choice function hyper-heuristics for timetabling and scheduling. In: Practice and Theory of Auto-mated Timetabling V, pp. 51–67. Springer, Berlin Heidelberg (2004)
15. Segura, C., Miranda, G., León, C.: Parallel hyperheuristics for the frequency assignment problem. Memetic Comput. **3**(1), 33–49 (2011)
16. Tiourine, S.R., Hurkens, C.A.J., Lenstra, J.K.: Local search algorithms for the radio link frequency assignment problem. Telecommun. Syst.. Syst. **13**(2–4), 293–314 (2000)
17. Warners, J.P.: A nonlinear approach to a class of combinatorial optimization problems. Stat. Neerl.t. Neerl. **52**(2), 162–184 (1998)
18. Warners, J.P., Terlaky, T., Roos, C., Jansen, B.: A potential reduction approach to the frequency assignment problem. Discret. Appl. Math.. Appl. Math. **78**(1), 251–282 (1997)

Cybersecurity

Cyberattack Analysis Utilising Attack Tree with Weighted Mean Probability and Risk of Attack

Nitin Naik[1(✉)], Paul Jenkins[2], Paul Grace[1], Shaligram Prajapat[3], Dishita Naik[4], Jingping Song[5], Jian Xu[5], and Ricardo M. Czekster[1]

[1] School of Computer Science and Digital Technologies, Aston University, Birmingham, UK
{n.naik1,p.grace,r.meloczekster}@aston.ac.uk
[2] Cardiff School of Technologies, Cardiff Metropolitan University, Cardiff, UK
pjenkins2@cardiffmet.ac.uk
[3] International Institute of Professional Studies, Devi Ahilya University, Indore, India
shaligram.prajapat@iips.edu.in
[4] Birmingham City University, Birmingham, UK
dishita.naik@mail.bcu.ac.uk
[5] Software College, Northeastern University, Shenyang, China
songjp@swc.neu.edu.cn, xuj@mail.neu.edu.cn

Abstract. As technology advances and AI becomes embedded and accepted into everyday life, the risk of cyberattacks by adversaries increases. These cyberattacks are ubiquitous affecting both businesses and individuals alike, and causing financial and reputational loss as a result. Numerous cyberattack analysis methods are available to analyse the risk of cyberattacks and offer the appropriate mitigation strategy. Nonetheless, several cyberattack analysis methods may not be effective and applicable in all cyberattack conditions due to several reasons such as their cost, complexity, resources and expertise. Therefore, this paper builds on an economical, simple and adaptable method for cyberattack analysis using an attack tree with weighted mean probability and risk of attack. It begins with an examination of a weighted mean approach followed by an investigation of the different types of weighted mean functions. Utilizing a series of orderly steps to perform a cyberattack analysis and assess its potential risk in an easy and effective manner. This method provides the means to calculate the potential risk of attack and therefore any mitigation that can be employed to minimise its effect.

Keywords: Cyberattack analysis · Attack tree · Weighted mean probability of attack · Weighted mean risk of attack · Information theft attack

1 Introduction

As technology develops and evolves the challenge for all is to protect where possible and mitigate cyberattacks. There are a number of reasons for the increase

N. Naik et al. (Eds.): UKCI 2023, AISC 1453, pp. 351–363, 2024.
https://doi.org/10.1007/978-3-031-47508-5_28

in these challenges such as advancement in Artificial Intelligence (AI), weaker cybersecurity and unprotected data [13]. Therefore, it is often difficult to assess the risk of attack, especially with attackers utilising sophisticated attack methods, requiring equally complex means to prevent or mitigate the attack [11]. A number of methods exist to alleviate these attacks, however, they are often limited due to their cost, complexity, resources and expertise required to use them [10,12]. Furthermore, these methods are often specific in nature, targeting precise attacks. Therefore, there is a continuing necessity to explore new methods for cyberattack analysis satisfying the aforementioned criteria.

The method proposed by the paper is a cost-effective, uncomplicated and adaptable method to assess cyberattack risks based upon the design of an attack tree and a range of weighted mean probability and risk calculations [8]. The design of an attack tree is a diagrammatic approach used to describe an attack, its features and their analysis [6]. These possible attacks can be represented diagrammatically as a tree structure, where the attack goal is denoted by the root node of the tree with leaf nodes representing actions required to achieve the attack goal [15]. One of the key advantages of the proposed method is that it is not resource hungry and can be utilised by most users without specialised cybersecurity knowledge [7].

Furthermore, utilising a weighted mean function minimises the effects of outliers in the data by including the weighted value, whilst maintaining simplicity of the method. In this paper three different types of weighted means, namely weighted arithmetic, geometric and harmonic mean, will be assessed to determine the most accurate depending on the data. The proposed method combines the attack tree with weighted mean probability and risk to present a series of orderly steps to perform a cyberattack analysis for any given cyberattack, assessing its potential risk in an uncomplicated and adaptable manner. It proposes the necessary parameters and formulas to be used in calculating the weighted mean probability and risk of attack utilising an attack tree, which can be further utilised to mitigate the risk [9].

This paper is organised into the following sections: Sect. 2 describes the attack tree model; Sect. 3 presents the proposed method for cyberattack analysis utilising an attack tree with weighted mean probability and risk of an attack, and its application for an information theft attack on an organisation; Sect. 4 presents the conclusion and future work.

2 Attack Tree Model

An attack tree is a graphical means of representing and depicting a possible attack on a system. This graphical representation allows users to analyse the attack in a simple and effective way. The aim of the attack is set as the root node, with the leaf nodes used to represent the different methods or actions required to realise the attack [14–16]. The attack tree method represents an efficient and cost effective means of examining a possible attack on any IT system, which is due to the low level of resource required [4]. The attack tree is formulated in a similar way to that of a structure chart, where, as each level of leaf nodes is added, more detail is appended in terms of actions required to support the goal of an attack node, designated as the root node. These leaf nodes and subnodes each represent an attack-vector until finally the last level depicts an atomic action exploiting a specific vulnerability as outlined in Fig. 1. Each level of the tree is connected using one of two logical operators, namely OR and AND (disjunction $\vee$ and conjunction $\wedge$). The AND ($\wedge$) operation denotes that all child actions necessarily must be complete to achieve the desired action of the parent node. Moreover, the OR ($\vee$) operator indicates that anyone of the child nodes can be completed to achieve the objective of the parent node, this is the default mode. Depending on the scenario, the attack vector and vulnerability can be expressed on multiple levels of the diagram. The AND ($\wedge$) operator is usually noted on the diagram, whilst the OR ($\vee$) operator is not indicated as it is the default relationship.

One of the key advantages of attack trees is that it enables security experts to assess different scenarios of an attack, allowing different stakeholders with different backgrounds and skills to provide their feedback, and hence improve and tailor mitigations for the attack. The attack tree method can be used to perform various types of attack analysis depending on the types of attack trees and their connecting operators. For example, an attack tree utilising sequential AND operator (denoted as SAND) can be used to analyse time-dependent attacks by describing sequential nodes as conjunctive nodes with a notion of progress of time [2]. Similarly, an attack tree utilising sequential AND operator can also be used to perform risk analysis with conditional probabilities [4,5]. Another attack tree utilising ordered AND operator (denoted as OAND) can be used to represent temporal dependencies between various attack components [3]. This attack tree method offers several benefits over other attack analysis methods as it is illustrative, understandable, economical, efficient, customisable, scalable, reusable; and it helps develop mitigation strategies at granular levels [1,15].

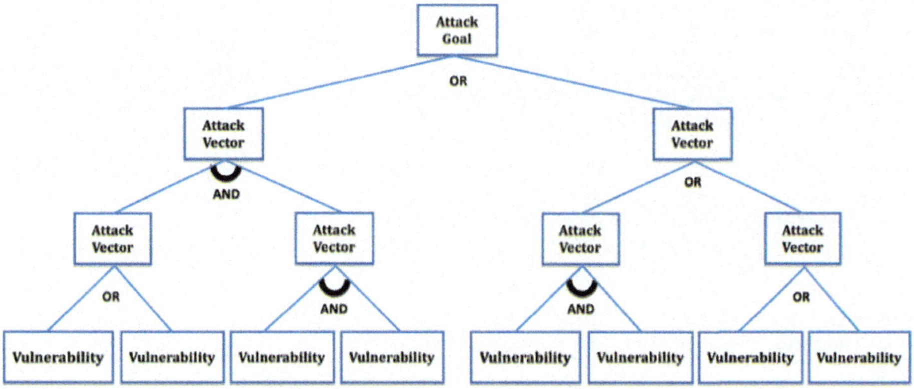

Fig. 1. Template of an attack tree for cyberattack analysis

3 Proposed Method for Cyberattack Analysis Using Attack Tree with Weighted Mean Probability and Risk of Attack

This section proposes a method for cyberattack analysis based on an attack tree and a range of weighted mean functions to calculate probability and risk of an attack, which is the further enhancement of the earlier proposed method for cyberattack analysis as illustrated in Fig. 2, that provided an excellent cost-effective means of assessing the risk of cyberattack and proposing attack mitigation [8]. In order to perform a cyberattack analysis and assess its potential risk in an easy and effective manner, the necessary parameters and formulas for calculating the weighted mean probability and risk of an attack based on its attack tree, and on that basis, its potential mitigations can be determined. Naik et al. (2022) considered a given scenario of an information theft attack on an organisation, which is a very common attack and applicable to any IT system. This cyberattack analysis of an information theft attack on an organisation follows the proposed method stages illustrated in Fig. 2, demonstrating how the method can be easily applied to any attack scenario.

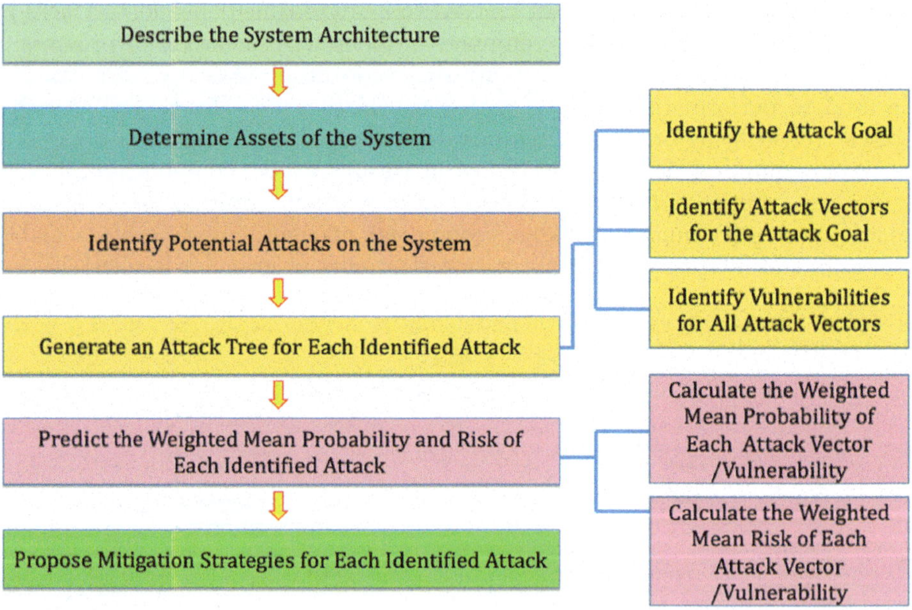

Fig. 2. Steps of the proposed method for cyberattack analysis based on attack tree with weighted mean probability and risk of attack

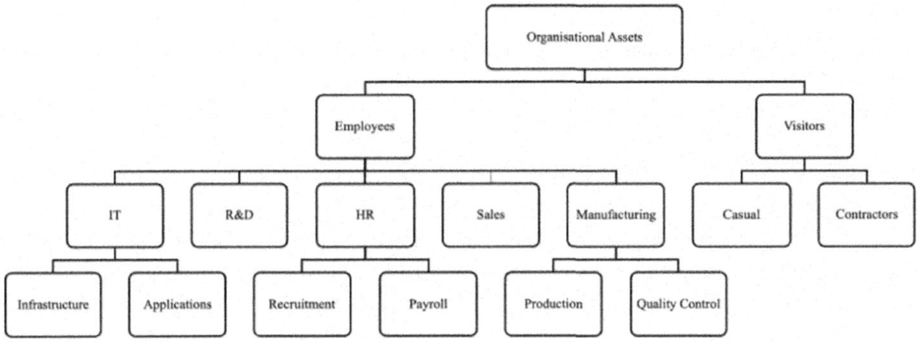

Fig. 3. Identified assets of an organisation for an application of the proposed method

3.1 Describe the System Architecture

In considering a cyberattack analysis, initially a description of the system architecture is prepared and constructed by identifying the relevant assets. Here, a general architecture of an organisation is considered, which includes various departments IT, Research & Development, HR, Sales and Manufacturing as shown in Fig. 3. Therefore, it can be easily mapped to most of similar organisational structures to perform an attack analysis of an information theft attack.

3.2 Determine the Assets of the System

For any system, assets are the most crucial component from the cyberattack viewpoint, an attacker either directly performs an attack on these assets or they are affected by an attack. The identified assets of an organisation are shown in Fig. 3, covering those entities which are relevant to an information theft attack. Again, these assets are very generic and can be easily mapped to most of the similar assets in any organisation; however, the selected assets can be customised depending on the specific organisation.

3.3 Identify Potential Attacks on the System

In this paper an information theft attack on an organisation is considered to demonstrate the method. This attack was selected as it is very common attack type and can be applied to a number of scenarios. However, other potential attacks on the system can also be identified and analysed in similar way.

3.4 Generate an Attack Tree for Each Identified Attack

In generating an attack tree, it is dependent upon on the number of attacks identified in the previous step, if there are a number possible attacks, an attack tree is generated for each attack. Figure 4 illustrates the created attack tree for the information theft attack, where the goal of a malicious user is to steal information using a number of different identified attack vectors that exploit specific vulnerabilities of assets within an organisation. Here, each illustrated path (i.e., from each leaf node to the root node) to steal information has to be evaluated for its potential success and risk by using a suitable method. The attack tree facilitates the graphical and hierarchical illustration of an attack for its analysis; however, it does not provide a specific method to determine parameter values for each node in an attack tree such as its probability, severity and risk. Therefore, a suitable method is proposed in the next step to calculate the weighted mean probability and risk of attack using different weighted mean functions.

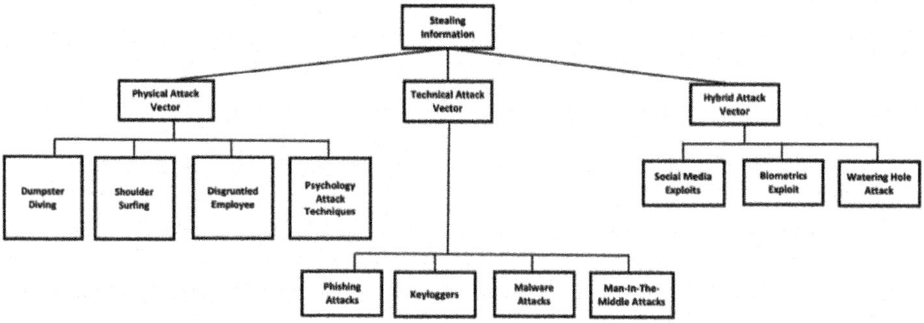

Fig. 4. An attack tree to analyse information theft attack and its associated risks in an organisation

3.5 Predict the Weighted Mean Probability and Risk of Attack Using the Proposed Parameters and Formulas

Utilising the developed attack tree of the information theft attack as shown in Fig. 4, a weighted mean probability and risk is calculated for each attack vector of the information theft attack tree using the proposed formulas. These values will assist security analysts to assess the overall risk of each attack vector and develop their possible mitigation strategies. In this attack tree of an information theft attack as shown in Fig. 4, there are three main attack vectors identified: a physical attack vector, a technical attack vector and a hybrid attack vector. However, this analysis will demonstrate the calculation of the weighted mean probability and risk of only physical attack vector, and a similar calculation can be performed for the other two attack vectors: the technical attack vector and the hybrid attack vector. The proposed formulas for the weighted mean probability and risk of attack in the following two subsections can be used at the desired level of an attack tree. In this method, four weighted functions are utilised to calculate probability and risk of attack: weighted sum, weighted arithmetic mean, weighted geometric mean, and weighted harmonic mean, which are explained as follows:

Weighted Sum The generic formula of weighted sum is:

$$W_S = \sum_{i=1}^{i=n} (w_i * x_i) = w_1 * x_1 + w_2 * x_2 + \cdots + w_n * x_n \tag{1}$$

Where
W_S is the weighted sum;
n is the number of terms;
x_i is the ith data value to be averaged;
w_i is the weight to be applied on the data x_i.

Weighted Arithmetic Mean The generic formula of weighted arithmetic mean is:

$$W_A = \frac{\sum_{i=1}^{i=n} (w_i * x_i)}{\sum_{i=1}^{i=n} w_i} = \frac{w_1 * x_1 + w_2 * x_2 + \cdots + w_n * x_n}{w_1 + w_2 + \cdots + w_n} \tag{2}$$

Where
W_A is the weighted arithmetic mean;
n is the number of terms;
x_i is the ith data value to be averaged;
w_i is the weight to be applied on the data x_i.

Weighted Geometric Mean The generic formula of weighted geometric mean is:

$$W_G = \sqrt[n]{\prod_{i=1}^{i=n} x_i^{w_i}} = \sqrt[n]{x_1^{w_1} * x_2^{w_2} * x_3^{w_3} \cdots * x_n^{w_n}} \tag{3}$$

where
W_G is the weighted geometric mean;
n is the number of terms;
x_i is the ith data value to be averaged;
w_i is the weight to be applied on the data x_i.

Weighted Harmonic Mean The generic formula of weighted harmonic mean is:

$$W_H = \frac{\sum_{i=1}^{i=n} w_i}{\sum_{i=1}^{i=n} \frac{w_i}{x_i}} = \frac{w_1 + w_2 + \cdots + w_n}{\frac{w_1}{x_1} + \frac{w_2}{x_2} + \cdots + \frac{w_n}{x_n}} \tag{4}$$

where
W_H is the weighted harmonic mean;
n is the number of terms;
x_i is the ith data value to be averaged;
w_i is the weight to be applied on the data x_i.

Calculate the Weighted Mean Probability of Attack Vector/Vulnerability Using the Proposed Formula This subsection proposes several parameters for calculating different mean probabilities of an attack [6, 7]. An expression for the probability P of an attack is derived based upon the four selected parameters: cost of attack, technical difficulty in performing an attack, number of times an attack occurred/unit time and total number of attacks occurred/unit time as shown in Eq. 5.

$$P = \frac{\frac{(C*D)}{SF1} + \frac{N}{T}}{SF2} \tag{5}$$

Where
P is the probability of an attack;
C is the cost of an attack;
D is the technical difficulty;
N is the number of times an attack occurred per unit of time;
T is the total number of attack occurred per unit of time;
SF is the scaling factor;
$SF1 = Max(C) * Max(D)$;
$SF2$ is the number of terms in the formula.

All these values are normally derived from the data or obtained from security experts; for example, the selected range for the first two parameters for this research work, cost of attack and technical difficulty in performing an attack are shown in Table 1, where the highest value is 1 and the lowest value is 5, which can be adapted depending on the specific analysis requirements. The specific value in the range can be selected for an attack by security experts based on the available data or their own assessment. The remaining two parameters, number of times an attack occurred/unit time and total number of attacks occurred/unit time can be obtained from the available data for a particular time period (e.g., day, week, month or year). At the initial stage, if these two values are not available then the probability of attack can still be calculated, which will require readjustment once the data is available to obtain these two values.

Table 2 presents some example scenarios related to each individual vulnerability linked to the physical attack vector to demonstrate the calculation of the probability of attack based on different values of the four selected parameters in the proposed formula. Here, the total number of attacks occurred/unit time is 10 for all the example scenarios, and all other values are randomly selected within the given range to demonstrate a wide range of conditions. Once the probability of all individual attack vectors (or vulnerabilities) linked to a particular attack is calculated then their weighted probability is calculated by multiplying the corresponding weights as shown in Table 3.

Table 3 demonstrates some example values of the probability of each individual vulnerability linked to the physical attack vector for the calculation of their weighted probability of attack. These weights can be derived from the data or provided by security experts. Finally, based on the weighted probability of all vulnerabilities linked to the physical attack vector, the weighted mean probability of the physical attack vector is calculated as shown in Table 3.

Table 1. Proposed Cost, technical difficulty and severity levels of attack (or attack vector/vulnerability)

Cost of attack (C)	Technical difficulty in derforming an dttack (D)	Severity of attack (S)
1 = Highest	1 = Highest	1 = Lowest
2	2	2
3	3	3
4	4	4
5 = Lowest	5 = Lowest	5 = Highest

Calculate the Weighted Mean Risk of Attack Vector/Vulnerability Using the Proposed Formula After calculating the weighted probability of all individual attack vectors (or vulnerabilities), linked to a particular attack and deriving the severity of attack from the data or obtained from security experts, their weighted risk can be calculated using the proposed formula for the weighted risk of attack as shown in Eq. 6. Here, the suggested range for the severity of attack for this research work is shown in Table 1, where the lowest value is 1 and the highest value is 5, which can be adapted depending the specific analysis requirements.

Table 4 demonstrates the calculation of the weighted mean risk of a physical attack based on different values of the two selected parameters in the proposed formula. Here, the value of the weighted probability of each individual attack is obtained from Table 3, and their corresponding severity is selected within the

Table 2. Example scenarios for calculating the probability of an attack (or vulnerability) based on the proposed formula

Attack	Cost of an attack (C)	Technical difficulty in performing an attack (D)	Number of times an attack occurred/unit time (N)	Total number of attacks occurred/unit time (T)	Probability of an attack P = [(C*D)/SF1 + N/T]/SF2 (Here SF1 = 5 * 5 = 25, and SF2 = 2)
Dumpster diving	1	5	4	10	P = [(1 * 5) / 25 + (4/10)]/2 = [0.2 + 0.4]/2 = 0.6/2 = 0.3
Shoulder surfing	5	1	2	10	P = [(5 * 1) / 25 + (2/10)]/2 = [0.2 + 0.2]/2 = 0.4/2 = 0.2
Disgruntled employee	3	5	2	10	P = [(3 * 5) / 25 + (2/10)]/2 = [0.6 + 0.2]/2 = 0.8/2 = 0.4
Psychology attack techniques	4	5	4	10	P = [(4 * 5) / 25 + (2/10)]/2 = [0.8 + 0.2]/2 = 1/2 = 0.5

Table 3. Example scenarios for calculating the different weighted mean probabilities of a physical attack

Attack	Probability of an attack P_i	Weighting factor W_i	Weighted sum probability W_S	Weighted arithmetic mean probability W_A	Weighted geometric mean probability W_G	Weighted harmonic mean probabilityW_H
Dumpster diving	0.3	0.1	P1*W1 + P2*W2 + P3*W3 + P4*W4 = 0.3*0.1 + 0.2*0.1 + 0.4*0.4 + 0.5*0.4 = 0.03 + 0.02 + 0.16 + 0.20 = 0.41	(P1*W1 + P2*W2 + P3*W3 + P4*W4)/(W1 + W2 + W3 + W4) = (0.3*0.1 + 0.2*0.1 + 0.4*0.4 + 0.5*0.4)/(0.1 + 0.1 + 0.4 + 0.4) = (0.03 + 0.02 + 0.16 + 0.20)/(0.1 + 0.1 + 0.4 + 0.4) = 0.41	$(P1^{W1} * P2^{W2} * P3^{W3} * P4^{W4})^{(1/(W1 + W2 + W3 + W4))} = [0.3^{0.1} * 0.2^{0.1} * 0.4^{0.4} * 0.5^{0.4})^{(1/(0.1 + 0.1 + 0.4 + 0.4))} = [(0.3 * 0.2)^{0.1} * (0.4*0.5)^{0.4}] = 0.40$	1/((W1/P1 + W2/P2 + W3/P3 + W4/P4)/(W1 + W2 + W3 + W4)) = 1/((0.1/0.3 + 0.1/0.2 + 0.4/0.4 + 0.4/0.5)/(0.1 + 0.1 + 0.4 + 0.5)) = 30/79 = 0.38
Shoulder surfing	0.2	0.1				
Disgruntled employee	0.4	0.4				
Psychology attack techniques	0.5	0.4				

given range (see Table 1) to demonstrate a wide range of conditions. Finally, the weighted mean risk of a physical attack vector is calculated based on different weighted mean functions as shown in Table 4. The comparative analysis of four weighted mean functions show a minor difference in their results. However, this could be more effectively evaluated if the ground truth is known for the given data and selected a suitable weighted mean function for the given condition.

$$R_w = \frac{S * P_w}{SF3} \tag{6}$$

where
R_w is the weighted mean risk of physical attack vector;
S is the severity;
P_w is the weighted probability;
$SF3$ is the scaling factor = Max(S).

Table 4. Example scenarios for calculating the different weighted mean risks of a physical attack

Severity of an attack (S)	Weighted sum risk of an attack R_w = (S * P_w)/SF3 (Here SF3 = 5)	Weighted arithmatic mean risk of an attack R_w = (S * P_w)/SF3 (Here SF3 = 5)	Weighted geometric mean risk of an attack R_w = (S * P_w)/SF3 (Here SF3 = 5)	Weighted harmonic mean risk of an attack R_w = (S * P_w)/SF3 (Here SF3 = 5)
1	R_w = (1 * 0.41)/5 = 0.082	R_w = (1 * 0.41)/5 = 0.082	R_w = (1 * 0.40)/5 = 0.08	R_w = (1 * 0.38)/5 = 0.076
2	R_w = (2 * 0.41)/5 = 0.16	R_w = (2 * 0.41)/5 = 0.16	R_w = (2 * 0.40)/5 = 0.16	R_w = (2 * 0.38)/5 = 0.15
3	R_w = (3 * 0.41)/5 = 0.25	R_w = (3 * 0.41)/5 = 0.25	R_w = (3 * 0.40)/5 = 0.24	R_w = (3 * 0.38)/5 = 0.23
4	R_w = (4 * 0.41)/5 = 0.33	R_w = (4 * 0.41)/5 = 0.33	R_w = (4 * 0.40)/5 = 0.32	R_w = (4 * 0.38)/5 = 0.30
5	R_w = (5 * 0.41)/5 = 0.41	R_w = (5* 0.41)/5 = 0.41	R_w = (5 * 0.40)/5 = 0.40	R_w = (5 * 0.38)/5 = 0.38

Similarly, the weighted mean probability and risk of the other two attack vectors technical attack vector and hybrid attack vector in the attack tree can be calculated using the above formulas and calculations. After calculating the weighted mean risk of all attack vectors in the attack tree, the total weighted mean probability and risk of attack can also be calculated in a similar manner if required.

3.6 Propose Mitigation Strategies for Each Identified Attack

Given that the attack tree has identified the different attack vectors, and weighted mean risks have been assigned to them, it should be easy to implement well known mitigation strategies or develop customised mitigation strategies for these three types of attack vectors, and thus for the information theft attack. Here, these mitigation strategies should be selected by developers or security experts according to their acceptance level of the previously measured weighted mean risk of all attack vectors. For example, the physical attack vector is a low-risk attack and hence security experts may wish to mitigate or avoid such risk depending on their organisational security policy.

4 Conclusion

This paper presented an economical, simple and adaptable method for cyberattack analysis utilising an attack tree with weighted mean probability and risk of attack. The proposed method comprises orderly steps to perform a cyberattack analysis for a cyberattack and assess its potential risk in an easy and effective manner. It proposed the required parameters and formulas for calculating the weighted mean probability and risk of a cyberattack based on its attack tree, deriving its potential mitigation strategy. The proposed method was explained based on an assumed scenario of an information theft attack on an organisation, which is a very common attack and applicable to any IT system. This cyberattack analysis of an information theft attack on an organisation followed the steps illustrated in the proposed method to clearly demonstrate how the proposed method can be easily applied to any cyberattack scenario. This proposed cyberattack analysis method is a systematic and generalised method for analysing cyberattacks and their security risks, and can be applied to the majority of cyberattacks. However, this is a preliminary realisation of the proposed method in an assumed scenario of an information theft attack on an organisation, and requires more comprehensive analysis and testing for its further improvement.

References

1. Amenaza.com: The SecurITree advantage (2021). https://www.amenaza.com/SS-advantage.php
2. Arnold, F., Hermanns, H., Pulungan, R., Stoelinga, M.: Time-dependent analysis of attacks. In: International Conference on Principles of Security and Trust, pp. 285–305. Springer (2014)
3. Camtepe, S.A., Yener, B.: Modeling and detection of complex attacks. In: 2007 Third International Conference on Security and Privacy in Communications Networks and the Workshops-SecureComm 2007, pp. 234–243. IEEE (2007)
4. Jhawar, R., Kordy, B., Mauw, S., Radomirović, S., Trujillo-Rasua, R.: Attack trees with sequential conjunction. In: IFIP International Information Security and Privacy Conference, pp. 339–353. Springer (2015)
5. Jiang, R., Luo, J., Wang, X.: An attack tree based risk assessment for location privacy in wireless sensor networks. In: 2012 8th International Conference on Wireless Communications, Networking and Mobile Computing, pp. 1–4. IEEE (2012)
6. Naik, N., Grace, P., Jenkins, P.: An attack tree based risk analysis method for investigating attacks and facilitating their mitigations in self-sovereign identity. In: IEEE Symposium Series on Computational Intelligence (SSCI). IEEE (2021)
7. Naik, N., Grace, P., Jenkins, P., Naik, K., Song, J.: An evaluation of potential attack surfaces based on attack tree modelling and risk matrix applied to self-sovereign identity. Comput. Secur. **120**, 102808 (2022)
8. Naik, N., Jenkins, P., Grace, P.: Cyberattack analysis based on attack tree with weighted average probability and risk of attack. In: UK Workshop on Computational Intelligence (UKCI). Springer (2022)
9. Naik, N., Jenkins, P., Grace, P., Naik, D., Prajapat, S., Song, J., Xu, J., Czekster, R.M.: Analysing cyberattacks using attack tree and fuzzy rules. In: UK Workshop on Computational Intelligence (UKCI). Springer (2023)

10. Naik, N., Jenkins, P., Grace, P., Song, J.: Comparing attack models for IT systems: lockheed Martin's cyber kill chain, MITRE ATT&CK framework and diamond model. In: 2022 IEEE International Symposium on Systems Engineering (ISSE). IEEE (2022)
11. Naik, N., Jenkins, P., Savage, N., Yang, L., Boongoen, T., Iam-On, N.: Fuzzy-import hashing: a static analysis technique for malware detection. Forensic Sci. Int. Digital Invest. **37**, 301139 (2021)
12. Naik, N., Jenkins, P., Savage, N., Yang, L., Boongoen, T., Iam-On, N., Naik, K., Song, J.: Embedded YARA rules: strengthening YARA rules utilising fuzzy hashing and fuzzy rules for malware analysis. Complex Intell. Syst. **7**(2), 687–702 (2021)
13. Naik, N., Shang, C., Jenkins, P., Shen, Q.: D-FRI-Honeypot: A secure sting operation for hacking the hackers using dynamic fuzzy rule interpolation. IEEE Trans. Emerg. Topics Comput. Intell. **5**(6), 893–907 (2020)
14. Salter, C., Saydjari, O.S., Schneier, B., Wallner, J.: Toward a secure system engineering methodolgy. In: Proceedings of the 1998 Workshop on New Security Paradigms, pp. 2–10 (1998)
15. Schneier, B.: Attack trees. Dr. Dobb's J. **24**(12), 21–29 (1999)
16. Weiss, J.D.: A system security engineering process. In: Proceedings of the 14th National Computer Security Conference, vol. 249, pp. 572–581 (1991)

Analysing Cyberattacks Using Attack Tree and Fuzzy Rules

Nitin Naik[1(✉)], Paul Jenkins[2], Paul Grace[1], Dishita Naik[3], Shaligram Prajapat[4], Jingping Song[5], Jian Xu[5], and Ricardo M. Czekster[1]

[1] School of Computer Science and Digital Technologies, Aston University, Birmingham, UK
{n.naik1,p.grace,r.meloczekster}@aston.ac.uk

[2] Cardiff School of Technologies, Cardiff Metropolitan University, Cardiff, UK
pjenkins2@cardiffmet.ac.uk

[3] Birmingham City University, Birmingham, UK
dishita.naik@mail.bcu.ac.uk

[4] International Institute of Professional Studies, Devi Ahilya University, Indore, India
shaligram.prajapat@iips.edu.in

[5] Software College, Northeastern University, Shenyang, China
songjp@swc.neu.edu.cn, xuj@mail.neu.edu.cn

Abstract. Understanding the development and execution of a cyberattack is intrinsic in its prevention and mitigation. A suitable cyberattack analysis method can be utilised in analysing cyberattacks. However, not every analysis method can be utilised for analysing every type of cyberattack due to the specific aim, strategy, requirements and skills of an analysis method. Therefore, deciding on a simple and suitable analysis method is always a challenging task, which requires a continuous exploration of new analysis methods. This paper presents a simple and generic method for cyberattack analysis using an attack tree and fuzzy rules. The attack tree provides a graphical and granular relationship between a cyberattacker and a victim to understand the taxonomy of an attack. Subsequently, the probability and risk of each leaf node in the attack tree is calculated using the proposed formulas. Finally, fuzzy rules formalise human reasoning to manage the approximation and uncertainty of the data to determine the overall risk of attack. This method proposes a process consisting of a sequence of steps to perform a step-by-step analysis of a cyberattack and evaluate its potential risk in a simple and efficient manner, hence its prevention and mitigation can be determined beforehand. Furthermore, the paper presents a case study of an information theft attack on an organisation and its analysis using the proposed analysis method, which can be beneficial in the analysis of other similar attacks.

Keywords: Cyberattack · Attack tree · Fuzzy rules · Fuzzy logic · Probability of attack · Severity of attack · Risk of attack · Information theft attack

N. Naik et al. (Eds.): UKCI 2023, AISC 1453, pp. 364–378, 2024.
https://doi.org/10.1007/978-3-031-47508-5_29

1 Introduction

Cybersecurity involves protecting IT infrastructure by detecting, responding, mitigating and preventing cyberattacks [13]. Understanding the development and execution of a cyberattack is intrinsic in detecting, responding, mitigating and preventing that cyberattack [7]. A suitable cyberattack analysis method can be utilised in analysing cyberattacks for understanding the development and execution of it. However, not every cyberattack analysis method can be utilised for analysing every type of cyberattack due to the specific aim, strategy, requirements and skills of an analysis method [10,12]. Therefore, deciding on a simple and suitable cyberattack analysis method is always a challenging task, which requires a continuous exploration of new simple and generic analysis methods [11].

This paper presents a simple and generic method for cyberattack analysis using an attack tree and fuzzy rules. This method proposes a process consisting of a sequence of steps to perform a step-by-step analysis of a cyberattack and evaluate its potential risk in a simple and efficient manner [8,9]. The proposed sequence of steps is to: describe the system architecture, determine assets of the system, identify potential attacks on the system, generate an attack tree for each identified attack, predict the risk of each identified attack using fuzzy rules, and propose mitigation strategies for each identified attack. In this proposed method, the attack tree provides a graphical and granular relationship between a cyberattacker and a victim to understand the taxonomy of an attack [6,15]. Subsequently, the probability and risk of each leaf node in the attack tree is calculated using the proposed formulas. Finally, fuzzy rules formalise human reasoning to manage the approximation and uncertainty of the data to determine the overall risk of attack. This paper also presents a case study of an information theft attack on an organisation and its analysis using the proposed analysis method, which can be beneficial in the analysis of other similar attacks.

This paper consists of the following sections: Sect. 2 explains the attack tree model; Sect. 3 presents the proposed method and its stages for analysing cyberattack using attack tree and fuzzy rules. Section 4 presents an application of the proposed method in analysing information theft attack on an organisation including some numerical examples. Section 5 renders the summary of the current and future work.

2 Attack Tree Model

An attack tree is a systematic and illustrative method of describing an attack on a system and analysing its features, where potential attacks against a system are represented in a tree structure, with the attack goal being represented as the root node and different methods or actions of achieving the attack goal as leaf nodes [14–16]. An attack tree method is an efficient and economical method to perform an analysis for potential attacks on any IT system, as it does not require significant resources and a fully implemented IT system [4].

In this research work, the attack tree structure is designed in such a way that each attack tree comprises a root node representing the attack goal, with several levels of subnodes representing attack vectors to perform that attack, and finally, leaf node representing an atomic action exploiting a vulnerability to achieve the attack goal as shown in Fig. 1. The different levels of the tree are structured and connected using two main operators: conjunction (denoted as AND) and disjunction (denoted as OR). The AND relationship represents that all child nodes' actions should to be performed in order to achieve the action of the parent node; and the OR relationship represents that any one child node's action should be performed in order to achieve the action of the parent node. The attack vector and vulnerability can have multiple levels depending on the specific attack scenario. In attack tree diagrams, the AND relationship should be indicated, whereas OR relationship is normally a default relationship and does not require explicit indication.

The attack tree enables security analysts to implement a process where different stakeholders with different backgrounds and skills provide their feedback to help analyse potential attacks and facilitate their mitigations. The attack tree method can be used to perform various types of attack analysis depending on the types of attack trees and their connecting operators. For example, an attack tree utilising sequential AND operator (denoted as SAND) can be used to analyse time-dependent attacks by describing sequential nodes as conjunctive nodes with a notion of progress of time [2]. Similarly, an attack tree utilising sequential AND operator can also be used to perform risk analysis with conditional probabilities [4,5]. Another attack tree utilising ordered AND operator (denoted as OAND) can be used to represent temporal dependencies between various attack components [3]. This attack tree method offers several benefits over other attack analysis methods as it is illustrative, understandable, economical, efficient, customizable, scalable, reusable; and it helps develop mitigation strategies at granular levels [1,15].

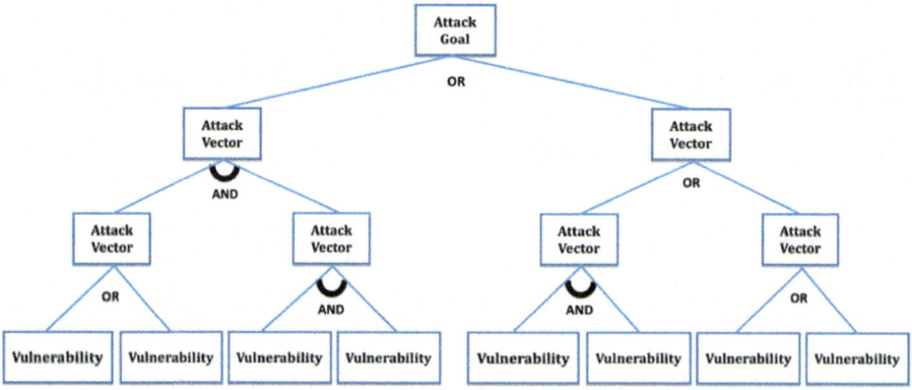

Fig. 1. Template of an attack tree for cyberattack analysis

3 Proposed Method for Analysing Cyberattack Using Attack Tree and Fuzzy Rules

This section proposes a simple and generic method for cyberattack analysis using an attack tree and fuzzy rules. This proposed method consists of a sequence of steps (see Fig. 2) to perform a step-by-step analysis of a cyberattack and evaluate its potential risk in a simple and efficient manner. The proposed sequence of steps is to: describe the system architecture, determine assets of the system, identify potential attacks on the system, generate an attack tree for each identified attack, predict the risk of each identified attack using fuzzy rules, and propose mitigation strategies for each identified attack. Additionally, it proposes the necessary parameters and formulas for calculating the probability and risk of attack based on its created attack tree. Subsequently, fuzzy rules are created utilising these values to predict the overall risk of attack (or attack vector), and on that basis, its potential mitigations can be determined.

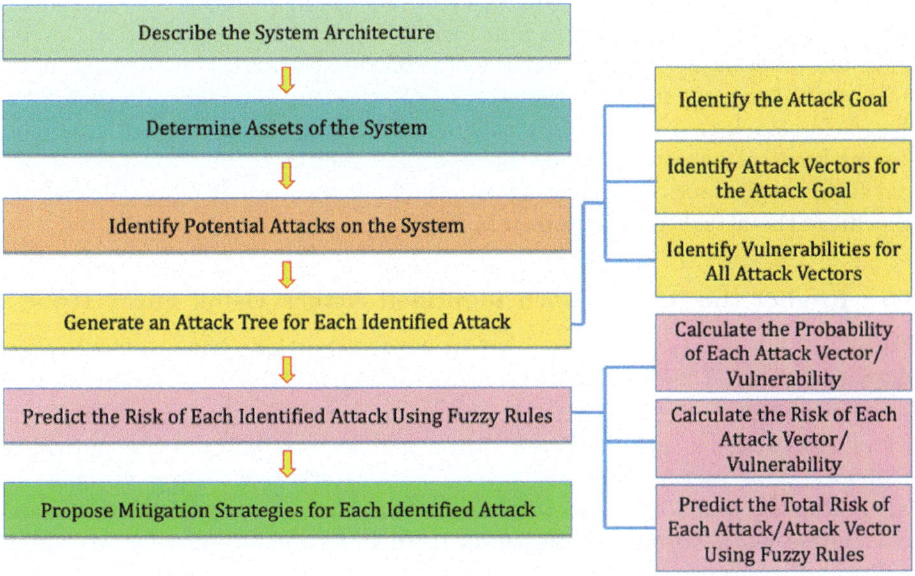

Fig. 2. Steps of the proposed method for cyberattack analysis based on attack tree and fuzzy rules

3.1 Describe the System Architecture

For any cyberattack analysis, the first step of the proposed method is to describe the system architecture and functionality of the system for which an attack analysis is to be performed.

3.2 Determine the Assets of the System

Assets are the most important component of the system for any cyberattack analysis, as an attacker performs an attack on these assets. Therefore, determining those assets which are the potential target of an attack is the most crucial step for the further analysis of an attack.

3.3 Identify Potential Attacks on the System

After determining the assets of the system, potential attacks on the system should be identified for analysing their risk. The identification of potential attacks can be performed on the basis of the existing knowledge of attacks and historical data available for the system, and/or empirical research/data available for the similar types of systems.

3.4 Generate an Attack Tree for Each Identified Attack

Depending on the number of attacks identified in the previous step, an attack tree is generated for each attack. Here, the attack goal is represented as the root node and different methods or actions of achieving the attack goal as leaf nodes [15]. In this proposed method, the attack tree structure is designed in such a way that each attack tree comprises a root node representing the attack goal, with several levels of sub-nodes representing attack vectors to perform that attack, and finally, leaf nodes representing an atomic action exploiting a vulnerability to achieve the attack goal as shown in Fig. 1.

3.5 Predict the Risk of Each Identified Attack Using Fuzzy Rules

Utilising the developed attack tree of all identified attacks, the probability and risk are calculated for each attack vector (or vulnerability) in the attack tree. Later, these values will be utilised in generating fuzzy rules in order to assess the overall risk of each attack (or attack vector) and develop their possible mitigation strategies. This section proposes several parameters and general formulas for calculating the probability and risk of attack based on a comprehensive research and analysis [6,7]. Notably, these generic parameters and formulas for probability and risk of attack can be adapted and applied at any level of the attack tree (i.e., attack vector/vulnerability), given that the proposed parameters can be defined and obtained at that level of the attack tree.

Calculate the Probability of Each Attack Vector/Vulnerability Using the Proposed Formula A formula for the probability of attack is derived based on the four selected parameters: cost of attack, technical difficulty in performing an attack, number of times an attack occurred/unit time and total number of attacks occurred/unit time as shown in Table 1.

The values of all these proposed parameters are generally derived from the data or obtained from security experts. In this paper, the suggested values or ranges of these parameters are provided for simulation purposes; for example, the selected range for the first two parameters, cost of an attack and technical difficulty in performing an attack are shown in Table 2, where the highest value is 1 and the lowest value is 5, which can be adapted depending on the requirements of a specific analysis. The specific value in the range can be selected for an attack by security experts depending on the available data or their own assessment.

The rest of the two parameters, number of times an attack occurred/unit time and total number of attacks occurred/unit time can be obtained from the available data for a particular time period (e.g., day, week, month or year). At the initial stage, if these two values are not available then the probability of attack can still be calculated; however, it will require readjustment once these two values are known or the data is available to obtain these two values.

Table 1. Proposed parameters and formula for calculating the probability of attack (or attack vector/vulnerability)

Cost of attack	Technical difficulty in performing an attack	Number of times an attack occurred/unit time	Total number of attacks Occurred/unit time	Probability of Attack
C	D	N	T	P = [(C*D)/SF1 + N/T]/SF2 (SF = Scaling Factor, Here SF1 = Max(C) * Max (D), and SF2 = Number of terms in the formula)

Table 2. Suggested cost, technical difficulty and severity levels of attack (or attack vector/vulnerability)

Cost of attack (C)	Technical difficulty in performing an attack (D)	Severity of attack (S)
1 = Highest	1 = Highest	1 = Lowest
2	2	2
3	3	3
4	4	4
5 = Lowest	5 = Lowest	5 = Highest

Calculate the Risk of Each Attack Vector/Vulnerability Using the Proposed Formula After calculating the probability of all individual attack vectors (or vulnerabilities), linked to a particular attack and deriving the severity of an attack from the data or obtained from security experts, their risk can be calculated using the proposed formula for the risk of attack as shown in Table 3. Here, the suggested range for the severity of attack (or attack vector/vulnerability) for this research work is shown in Table 2, where the lowest value is 1 and the highest value is 5, which can be adapted depending on the requirements of a specific analysis.

Table 3. Proposed parameters and formula for calculating the risk of attack (or attack vector/vulnerability)

Severity of attack	Probability of attack	Risk of attack
S	P	R = (S * P)/SF3 (SF3 = Scaling Factor = Max(S))

Predict the Total Risk of Attack/Attack Vector Using Fuzzy Rules Fuzzy rules can be created for utilising the above approximate values effectively and predicting the overall risk of attack (or attack vector) in a user-friendly manner. These fuzzy rules are created based on the previously calculated probability and risk of all individual attack vectors (vulnerabilities). Once the probability and risk of all individual attack vectors (or vulnerabilities), linked to a particular attack (or attack vector) is calculated, the overall risk of attack (or attack vector) can be determined using the created fuzzy rules. The main benefit of using fuzzy rules is that the value of all the parameters in the proposed formulas are approximate assessment from security experts or based on the data and may not be precise, therefore, it requires an approach to manage approximation and uncertainty which is the strength of fuzzy logic.

3.6 Propose Mitigation Strategies for Each Identified Attack

Once the overall risk of attacks or attack vectors is calculated, it should be straightforward to implement well known mitigation strategies or develop customised mitigation strategies for these attacks or attack vectors according to their acceptance level of risk for a particular organisation.

4 Application of the Proposed Method for Analysing Information Theft Attack

The proposed cyberattack analysis method is applied to a given scenario of an information theft attack on an organisation, which is a very common attack and applicable to any IT system. This cyberattack analysis of an information theft attack on an organisation follows the proposed method stages as illustrated in Fig. 2 to clearly demonstrate how the proposed method can be easily applied to any attack scenario. Therefore, the proposed method is a systematic and generalised method for analysing cyberattacks and their security risks, and can be applied to the majority of cyberattacks.

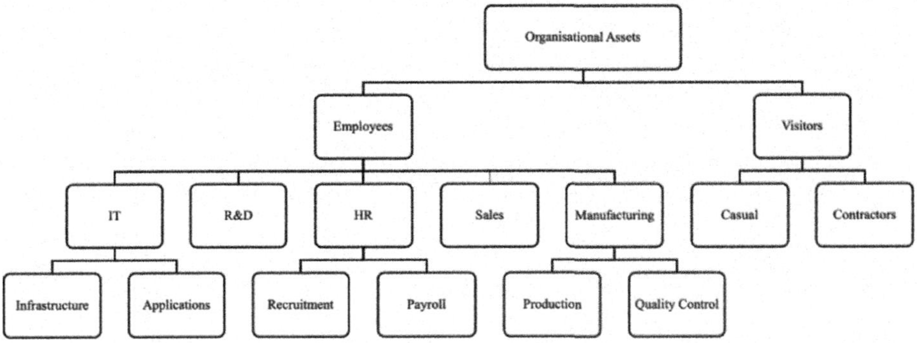

Fig. 3. Identified assets of an organisation for an application of the proposed method

4.1 Describe the System Architecture

In this analysis of an information theft attack on an organisation, a general architecture of an organisation is considered, which includes various departments such as IT, Research & Development, HR, Sales and Manufacturing as shown in Fig. 3. Therefore, it can be easily mapped to most of the similar organisational structures to perform an attack analysis of an information theft attack. Additionally, departments can be added or removed according to the nature and requirements of an organisation.

4.2 Determine the Assets of the System

For the analysis of an information theft attack on an organisation, the identified assets of an organisation are shown in Fig. 3, covering those entities which are relevant to an information theft attack. Again, these assets are very generic and can be easily mapped to most of the similar assets in any organisation to perform an attack analysis of an information theft attack. Additionally, assets can be added or removed according to the nature and requirements of an organisation.

4.3 Identify Potential Attacks on the System

As mentioned earlier, this cyberattack analysis will examine only one attack scenario that of an information theft attack on an organisation to demonstrate its successful implementation. This attack has been selected due to the fact that it is a very common attack type and applicable to most IT systems. However, the other potential attacks on IT systems can also be identified and analysed in similar way.

4.4 Generate an Attack Tree for Each Identified Attack

For this analysis of an information theft attack on an organisation, only one attack tree is generated for this attack. Figure 4 illustrates the generated attack tree for the information theft attack, where the goal of an attack is to steal information using a number of different identified attack vectors that exploit specific vulnerabilities of assets within an organisation. Here, each illustrated path (i.e., from each leaf node to the root node) to steal information has to be evaluated for its potential success and risk by using the proposed method in the next subsection.

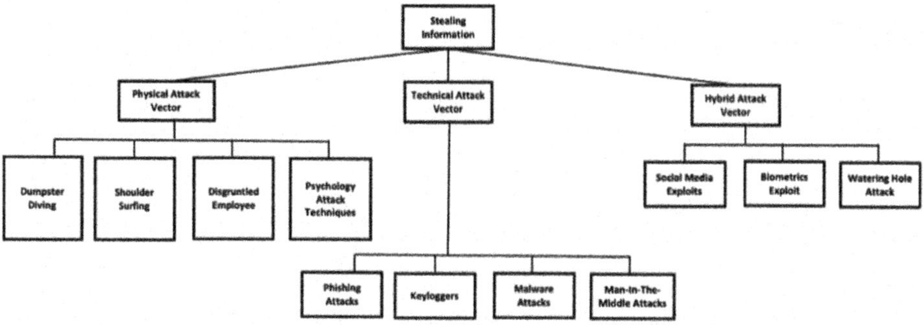

Fig. 4. An attack tree to analyse information theft attack and its associated risks in an organisation

4.5 Predict the Risk of Each Identified Attack Using Fuzzy Rules

For this analysis of an information theft attack on an organisation, the probability and risk is calculated for each attack vector in the attack tree. In this attack tree of an information theft attack as shown in Fig. 4, there are three main attack vectors identified: physical attack vector, technical attack vector and hybrid attack vector. However, this analysis will demonstrate the calculation of the probability and risk of only physical attack vector, and a similar calculation can be performed for the other two attack vectors: technical attack vector and hybrid attack vector.

Calculate the Probability of Attack Vector/Vulnerability Using the Proposed Formula The probability of all individual attack vectors (or vulnerabilities) linked to a particular attack (or attack vector) is calculated based on the proposed formula of the probability of attack in the previous section. Table 4 presents several example scenarios to demonstrate the calculation of the probability of attack based on different values of the four selected parameters in the proposed formula. Here, the total number of attacks occurred/unit time is assumed 10 for all the example scenarios, and all other values are randomly selected within the given range to demonstrate a wide range of conditions. For example, cost of an attack and technical difficulty in performing an attack are selected randomly within the given range from Table 2.

Similarly, for this analysis of an information theft attack on an organisation, the probability of all attack vectors (or vulnerabilities) linked to the physical attack vector can be calculated based on the above formula, and some example values are shown in Table 6.

Calculate the Risk of Attack Vector/Vulnerability Using the Proposed Formula Once the probability of all individual attack vectors (or vulnerabilities) linked to a particular attack (or attack vector) is calculated then their risk can be calculated utilising the corresponding value of their probability and severity. The severity can be determined by the security expert from the given range. Table 5 presents several example scenarios to demonstrate the calculation of the risk of attack based on different values of the probability and severity in the proposed formula. Here, the value of the probability of each individual attack vector (or vulnerabilities) can be calculated based on Table 4, and their corresponding severity is selected randomly within the given range from Table 2 to demonstrate a wide range of conditions.

Similarly, for this analysis of an information theft attack on an organisation, Table 6 demonstrates an example scenario for the calculation of the risk of all individual attack vectors (or vulnerabilities) linked to the physical attack based on different values of the probability and severity in the proposed formula.

Furthermore, the probability and risk of the other two attack vectors, technical attack vector and hybrid attack vector in the attack tree can be calculated using the above formulas and calculations.

Predict the Total Risk of Attack/Attack Vector Using Fuzzy Rules For this analysis of an information theft attack on an organisation, a sample of fuzzy rules (see Fig. 5) are created for only a physical attack vector to determine its overall risk. However, similar rules can be created for the other two attack vectors, technical attack vector and hybrid attack vector to determine their overall risk. These fuzzy rules are based on four input variables derived from four lower-level attack vectors (or vulnerabilities) of a physical attack vector: Dumpster Diving, Shoulder Surfing, Disgruntled Employee, and Psychology Attack Techniques. These four input variables are called DDE (Dumpster Diving Risk), SSR (Shoulder Surfing Risk), DER (Disgruntled Employee Risk) and

Table 4. Example scenarios for calculating the probability of attack (or vulnerability) based on the proposed formula

Cost of attack (C)	Technical difficulty in performing an attack (D)	Number of times an attack occurred/unit time (N)	Total number of attacks occurred/unit time (T)	Probability of attack P = [(C*D)/SF1 + N/T]/SF2 (Here SF1 = 5 * 5 = 25, and SF2 = 2)
2	3	0	10	P = [(2 * 3)/25 + 5/10]/2 = [6/25 + 0/10]/2 = [0.24 + 0]/2 = 0.24/2 = 0.12
2	3	5	10	P = [(2 * 3)/25 + 5/10]/2 = [6/25 + 5/10]/2 = [0.24 + 0.5]/2 = 0.74/2 = 0.37
2	3	10	10	P = [(2 * 3)/25 + 10/10]/2 = [6/25 + 10/10]/2 = [0.24 + 1]/2 = 1.24/2 = 0.62
1	1	0	10	P = [(1 * 1)/25 + 0/10]/2 = [1/25 + 0/10]/2 = [0.04 + 0]/2 = 0.04/2 = 0.02
1	1	5	10	P = [(1 * 1)/25 + 5/10]/2 = [1/25 + 5/10]/2 = [0.04 + 0.5]/2 = 0.54/2 = 0.27
1	1	10	10	P = [(1 * 1)/25 + 5/10]/2 = [1/25 + 10/10]/2 = [0.04 + 1.04]/2 = 1.04/2 = 0.52
5	5	0	10	P = [(5 * 5)/25 + 0/10]/2 = [25/25 + 0/10]/2 = [1 + 0]/2 = 1/2 = 0.5
5	5	5	10	P = [(5 * 5)/25 + 10/10]/2 = [25/25 + 5/10]/2 = [1 + 0.5]/2 = 1.5/2 = 0.75
5	5	10	10	P = [(5 * 5)/25 + 10/10]/2 = [25/25 + 10/10]/2 = [1 + 1]/2 = 2/2 = 1

PATR (Psychology Attack Techniques Risk) respectively. All the input variables are divided into three fuzzy sets: Low= 0.1 to 0.4, Medium = 0.3 to 0.7 and High = 0.6 to 1.0. These four lower-level attack vectors (or vulnerabilities) are the mechanism to perform a physical attack vector, therefore, based on their four input variables, the output variable PAVR (Physical Attack Vector Risk) is derived which is the overall risk of a physical attack vector. The output variable is also divided into three fuzzy sets: Low= 0.1 to 0.4, Medium = 0.3 to 0.7 and High = 0.6 to 1.0.

It should be noted that these fuzzy input variables assume equal weight, and on that basis fuzzy rules are generated. However, a weight can be assigned to each input variable based on the data or assessment from security experts, which

Table 5. Example scenarios for calculating the risk of attack

Severity of attack vector (S)	Probability of attack vector (P)	Risk of attack R = (S * P)/SF3 (Here SF3 = 5)
2	0.1	R = (2 * 0.1)/5 = 0.2/5 = 0.04
2	0.3	R = (2 * 0.3)/5 = 0.6/5 = 0.12
2	0.5	R = (2 * 0.5)/5 = 1.0/5 = 0.20
2	0.7	R = (2 * 0.7)/5 = 1.4/5 = 0.28
2	0.9	R = (2 * 0.9)/5 = 1.8/5 = 0.36
1	0.2	R = (1 * 0.2)/5 = 0.2/5 = 0.04
1	0.4	R = (1 * 0.4)/5 = 0.4/5 = 0.08
1	0.6	R = (1 * 0.6)/5 = 0.6/5 = 0.12
1	0.8	R = (1 * 0.8)/5 = 0.8/5 = 0.16
1	1	R = (1 * 1)/5 = 1/5 = 0.20
5	0.1	R = (5 * 0.1)/5 = 0.5/5 = 0.10
5	0.3	R = (5 * 0.3)/5 = 1.5/5 = 0.30
5	0.5	R = (5 * 0.5)/5 = 2.5/5 = 0.50
5	0.7	R = (5 * 0.7)/5 = 3.5/5 = 0.70
5	0.9	R = (5 * 0.9)/5 = 4.5/5 = 0.90

Table 6. Example scenarios for calculating the risk of physical attack vector

Attack vector/vulnerability	Severity of attack vector/vulnerability (S)	Probability of attack vector/vulnerability (P)	Risk of attack R = (S * P)/SF3 (Here SF3 = 5)
Dumpster diving	3	0.03	R = (3 * 0.03)/5 = 0.09/5 = 0.018
Shoulder surfing	4	0.02	R = (4 * 0.02)/5 = 0.08/5 = 0.016
Disgruntled employee	1	0.16	R = (1 * 0.16)/5 = 0.16/5 = 0.032
Psychology attack techniques	5	0.20	R = (5 * 0.20)/5 = 1/5 = 0.20

will create customised fuzzy rules depending on the requirement of a specific cyberattack analysis.

1. If DDR is Low	AND	SSR is Low	AND	DER is Low	AND	PATR is Low	Then PAVR is Low
2. If DDR is Low	AND	SSR is Low	AND	DER is Low	AND	PATR is Medium	Then PAVR is Medium
3. If DDR is Low	AND	SSR is Low	AND	DER is Low	AND	PATR is High	Then PAVR is High
4. If DDR is Low	AND	SSR is Low	AND	DER is Medium	AND	PATR is Low	Then PAVR is Medium
5. If DDR is Low	AND	SSR is Low	AND	DER is High	AND	PATR is Low	Then PAVR is High
6. If DDR is Low	AND	SSR is Medium	AND	DER Low	AND	PATR is Low	Then PAVR is Medium
7. If DDR is Low	AND	SSR is High	AND	DER is Low	AND	PATR is Low	Then PAVR is High
8. If DDR is Medium	AND	SSR is Low	AND	DER Low	AND	PATR is Low	Then PAVR is Medium
9. If DDR is High	AND	SSR is Low	AND	DER is Low	AND	PATR is Low	Then PAVR is High
10. If DDR is Low	AND	SSR is Low	AND	DER is Medium	AND	PATR is Medium	Then PAVR is Medium
11. If DDR is Low	AND	SSR is Low	AND	DER is High	AND	PATR is High	Then PAVR is High
12. If DDR is Low	AND	SSR is Medium	AND	DER is Medium	AND	PATR is Medium	Then PAVR is Medium
13. If DDR is Low	AND	SSR is High	AND	DER is High	AND	PATR is High	Then PAVR is High
14. If DDR is Medium	AND	SSR is Medium	AND	DER is Medium	AND	PATR is Medium	Then PAVR is Medium
15. If DDR is Medium	AND	SSR is High	AND	DER is High	AND	PATR is High	Then PAVR is High
16. If DDR is High	AND	SSR is High	AND	DER is High	AND	PATR is High	Then PAVR is High

Fig. 5. Sample fuzzy rules to determine the overall risk of physical attack vector

4.6 Propose Mitigation Strategies for Each Identified Attack

For this analysis of an information theft attack on an organisation, the mitigation strategies should be selected or matured by developers or security experts according to their acceptance level of the previously measured risk of all attack vectors. For example, the physical attack vector is a low-risk attack and hence security experts may wish to mitigate or avoid such risk depending on their organisational security policies.

5 Conclusion

This paper presented a simple and generic method for cyberattack analysis using an attack tree and fuzzy rules. This proposed method consists of a sequence of steps to perform a step-by-step analysis of a cyberattack and evaluate its potential risk in a simple and efficient manner. The attack tree provides a graphical and granular relationship between a cyberattacker and a victim to understand the taxonomy of an attack. However, it does not provide a specific method to determine parameter values for each node of each path in an attack tree such as the probability, severity and risk of attack (or attack vector/vulnerability). Therefore, this paper proposed necessary parameters and formulas to calculate the probability and risk of attack (or attack vector/vulnerability). Subsequently, fuzzy rules were created utilising these values to predict the overall risk of attack (or attack vector), and on that basis, its potential mitigations can be determined. Furthermore, the paper presented a case study of an information theft attack on an organisation and its analysis using the proposed cyberattack analysis method, which can be beneficial in the analysis of other similar attacks. Therefore, the proposed method is a simple and generic method for analysing cyberattacks and their security risks, and can be applied to the majority of cyberattacks. However, it has only presented the initial concept of applying fuzzy rules to predict the overall risk of attack, and requires further design and implementation of these fuzzy rules to test their successful working. Additionally, this paper presented an initial application of the proposed method in a given scenario of an information theft attack on an organisation, and requires further analysis and testing for its refinement.

References

1. Amenaza.com: The SecurITree advantage (2021). https://www.amenaza.com/SS-advantage.php
2. Arnold, F., Hermanns, H., Pulungan, R., Stoelinga, M.: Time-dependent analysis of attacks. In: International Conference on Principles of Security and Trust, pp. 285–305. Springer (2014)
3. Camtepe, S.A., Yener, B.: Modeling and detection of complex attacks. In: 2007 Third International Conference on Security and Privacy in Communications Networks and the Workshops-SecureComm 2007, pp. 234–243. IEEE (2007)
4. Jhawar, R., Kordy, B., Mauw, S., Radomirović, S., Trujillo-Rasua, R.: Attack trees with sequential conjunction. In: IFIP International Information Security and Privacy Conference, pp. 339–353. Springer (2015)
5. Jiang, R., Luo, J., Wang, X.: An attack tree based risk assessment for location privacy in wireless sensor networks. In: 2012 8th International Conference on Wireless Communications, Networking and Mobile Computing, pp. 1–4. IEEE (2012)
6. Naik, N., Grace, P., Jenkins, P.: An attack tree based risk analysis method for investigating attacks and facilitating their mitigations in self-sovereign identity. In: IEEE Symposium Series on Computational Intelligence (SSCI). IEEE (2021)
7. Naik, N., Grace, P., Jenkins, P., Naik, K., Song, J.: An evaluation of potential attack surfaces based on attack tree modelling and risk matrix applied to self-sovereign identity. Comput. Secur. **120**, 102808 (2022)
8. Naik, N., Jenkins, P., Grace, P.: Cyberattack analysis based on attack tree with weighted average probability and risk of attack. In: UK Workshop on Computational Intelligence (UKCI). Springer (2022)
9. Naik, N., Jenkins, P., Grace, P., Prajapat, S., Naik, D., Song, J., Xu, J., Czekster, R.M.: Cyberattack analysis utilising attack tree with weighted mean probability and risk of attack. In: UK Workshop on Computational Intelligence (UKCI). Springer (2023)
10. Naik, N., Jenkins, P., Grace, P., Song, J.: Comparing attack models for IT systems: lockheed Martin's cyber kill chain, MITRE ATT&CK framework and diamond model. In: 2022 IEEE International Symposium on Systems Engineering (ISSE). IEEE (2022)
11. Naik, N., Jenkins, P., Savage, N., Yang, L., Boongoen, T., Iam-On, N.: Fuzzy-import hashing: a static analysis technique for malware detection. Forensic Sci. Int. Digit. Invest. **37**, 301139 (2021)
12. Naik, N., Jenkins, P., Savage, N., Yang, L., Boongoen, T., Iam-On, N., Naik, K., Song, J.: Embedded YARA rules: strengthening YARA rules utilising fuzzy hashing and fuzzy rules for malware analysis. Complex Intell. Syst. **7**(2), 687–702 (2021)

13. Naik, N., Shang, C., Jenkins, P., Shen, Q.: D-FRI-Honeypot: a secure sting operation for hacking the hackers using dynamic fuzzy rule interpolation. IEEE Trans. Emerg. Top. Comput. Intell. **5**(6), 893–907 (2020)
14. Salter, C., Saydjari, O.S., Schneier, B., Wallner, J.: Toward a secure system engineering methodology. In: Proceedings of the 1998 Workshop on New Security Paradigms, pp. 2–10 (1998)
15. Schneier, B.: Attack trees. Dr. Dobb's J. **24**(12), 21–29 (1999)
16. Weiss, J.D.: A system security engineering process. In: Proceedings of the 14th National Computer Security Conference, vol. 249, pp. 572–581 (1991)

Malware Prediction Using Tabular Deep Learning Models

Ahmad Alzu'bi[1](✉), Abdelrahman Abuarqoub[2], Mohammad Abdullah[1], Rami Abu Agolah[1], and Moayyad Al Ajlouni[1]

[1] Department of Computer Science, Jordan University of Science and Technology, Irbid, Jordan
agalzubi@just.edu.jo

[2] Cardiff School of Technologies, Cardiff Metropolitan University, Cardiff, UK

Abstract. As technology progresses, malware evolves, becoming increasingly perilous and posing a significant challenge in combating cybercriminals. With the abundance of massive data on vulnerabilities, Deep learning techniques present a chance to further boost data and system security. This paper introduces a deep neural network model that automatically generates embedding layers for each categorical feature. Its foundation lies primarily in training neural oblivious decision ensembles and TabNet model on malware data, benefiting from both end-to-end gradient-based optimisation and the power of multi-layer hierarchical representation learning. These deep architectures possess the capacity to learn numerous parameters and identify patterns within large-scale datasets. The proposed models were evaluated using the Microsoft malware prediction dataset, which includes nine million labelled subjects and 83 features. This work marks one of the early attempts to utilise deep tabular architectures for malware prediction. The experimental results demonstrate the model's effectiveness, achieving an accuracy of 66.1% and AUC of 72.8%.

Keywords: Malware prediction · Tabular neural network · Deep learning

1 Introduction

Malware is disruptive malicious software designed to inflict undesired or harmful consequences on a computer system. Detecting malware has become an essential requirement in computer security due to the considerable costs and damage inflicted by these malicious programs [1]. Black-hat hackers have several motivations for targeting victims, with financial gain being among the most common incentives. An approach used for this purpose involves utilizing malware like Adware, which autonomously presents advertisements on the compromised system [2]. Additional common ways or types of malware include Crypto-Mining malware [3], Ransomware [4], Corporate espionage and personal information acquisition [5], and Spyware that monitors and collects users' information, including email addresses and credit card numbers. Several evasion and mitigating techniques have been developed to identify this infection. These techniques have evolved

N. Naik et al. (Eds.): UKCI 2023, AISC 1453, pp. 379–389, 2024.
https://doi.org/10.1007/978-3-031-47508-5_30

from early-stage signature-based detection to the utilisation of machine learning methods [6]. By employing machine learning approaches with carefully optimized hyper-parameters, we can develop extended models capable of predicting previously unseen malware samples [7]. This adaptability in malware detection empowers the system to keep pace with the ever-increasing volume and diversity of threats.

Conventional machine learning approaches typically rely on structured data, while deep learning, employing neural networks, is effectively able to handle large amounts of unstructured data. Machine learning algorithms are relatively straightforward to configure and utilise, but their results may be constrained in certain cases or applications [8, 9]. Due to the categorical nature of the extensive data involved in the task of malware detection, most machine learning algorithms are unable to process it directly without converting it into numerical values. Moreover, the effectiveness of traditional ML algorithms largely relies on the encoding technique utilised for categorical variables. Most importantly, earlier research studies [10, 11] introduced the No Free Lunch (NFL) theorem, which proposes that all learning algorithms achieve equivalent performance when averaged across all potential datasets. This seemingly counterintuitive idea implies the impracticality of discovering a universally effective predictive algorithm. An accurate prediction of malware attacks that can alter their signatures over time, known as polymorphic malware [12], is also challenging. In such scenarios, heuristic-based malware detectors prove insufficient for accurately predicting malware. Hence, addressing this issue effectively necessitates the development of well-tailored deep learning-based prediction algorithms capable of detecting or predicting polymorphic malware attacks based on machine specifications.

As a result, the challenges posed by the characteristics of tabular or categorical malware data have motivated us to explore the effectiveness of deep learning models in malware detection. This study places significant emphasis on various aspects, including data processing, content analysis, data cleansing, and feature selection, specifically tailored for categorical malware data. In addition, we show how deep learning algorithms can be used to produce top-of-the-line results on tabular data, as they have been shown to do so when applied to both huge categorical data [13]. To the best of our knowledge, this is the first work to employ tabular deep learning architectures for malware prediction, in which we conducted experiments using three distinct scenarios.

First, we trained a fine-tuned TabNet [14] model on malware data, which is a contemporary architecture that leverages ensembles of oblivious decision trees [15]. The architecture design of oblivious decision trees is a deep neural network tailored for tabular problems, which applies gradient boosting and ensures highly efficient inference while demonstrating significant resistance to overfitting. Second, a conventional two-dimensional CNN architecture was implemented to reshape the tabular malware data in accordance with the attribute parameters. Third, we built a deep neural network capable of handling the embeddings of categorical features, which were subsequently concatenated with numerical features to facilitate discriminating malware data through mathematical modelling.

The remaining part of this paper is structured as follows. Section 2 reviews the related works; the methodology is presented in Sect. 3; the experimental results and evaluations are discussed in Sect. 4; and Sect. 5 concludes this paper.

2 Related Work

Machine learning algorithms have been widely employed in malware classification. Malware classification can be associated with the categorisation of binaries as harmful or benign, as well as the categorisation of malware samples into known various malware families. However, this paper focuses on determining whether the device is going to be infected or not.

For conventional ML techniques, a variety of high-accuracy classifiers have recently been proposed. Zhang et al. [16] introduced a novel approach, called soft relevance value (s-value), to evaluate the feature soft relevance of malware prediction. This method leverages the mixed distance criterion, commonly used in pattern recognition, to differentiate testing samples as a new family that was not labelled in the training set. When compared to the results of the Microsoft Malware Prediction Competition [17], the training and prediction time costs account for only 16.7% and 3.8% of the winner's respective time costs. Narayanan et al. [18] suggested a method for enhancing malware categorisation performance, by which they extracted the required characteristics for classification using Principal Component Analysis (PCA). They evaluated several classifiers, including kNN and SVM.

To categorise malware samples into known various malware families, Zhang et al. [19] proposed a malware classification method that assigns malware to its relevant family. The classifier mainly relies on two efficient ensemble learning models, XGBoost and ExtraTreeClassifer, in addition to a stacking approach. Based on malware visualisation methodology, Nataraj et al. [20] suggested a malware categorisation method with Gray-scale pictures created from the malware samples. The KNN technique was used to classify 25 families within a malware corpus. R. 20. Bahtiyar et al. [21] also extracted information for predicting and detecting advanced malware based on features of conventional and advanced malware instances seen in the wild. To predict the specific type of malware, they utilised correlations between features associated with conventional malware and advanced malware.

Pan et al. [22] selected the best 42 features through Chi-square testing to overcome the problem of high dimensionality. After employing various machine learning methods, they determined that LightGBM emerged as the most effective solution, achieving the highest accuracy while requiring less time. Through analysing the feature importance using the LightGBM algorithm, they further concluded that anti-virus software with vulnerabilities or pitfalls is more susceptible to increased attacks.

Among existing deep learning models, CNN-based approaches are highly effective for most tasks involving textual data and computer vision applications [23]. Rhode et al. [24] investigated the feasibility of identifying the potential threat level of an executable based on a brief snapshot of behavioural data. They found that an ensemble of RNNs can accurately predict whether a program is malicious or benign within the first few seconds of its execution.

Kolosnjaji et al. [25] built a neural network with convolutional and recurrent layers to collect the best characteristics for classification expressed as a series of API function calls, and this combined neural network architecture gives them an 85.6 percent precision and 89.4 percent recall. Kalash et.al [26] also used a CNN-based architecture to classify malware by converting Malimg and Microsoft malware data to grayscale images, and they reported remarkable accuracies.

Many recent studies [27] have increasingly emphasised leveraging deep learning techniques for detecting or predicting malware across diverse domains, such as mobile malware detection [28], malware analysis in cloud IAAS [29], malware detection in IoT environments [30], and quantum computing environments [31]. However, our work is distinguished from the existing work by utilising the power of tabular deep architectures to predict malware after a careful procedure of feature preprocessing and selection, allowing the use of the most discriminating data. In addition, we evaluated a deep learning architecture that handles the embeddings of malware features combined with its numerical data.

3 Methodology

The pipeline of the proposed deep-learning architectures for tabular malware prediction is depicted in Fig. 1. The dataset of malware row data is first preprocessed, and then the most discriminating features are empirically chosen to be kept for training and testing the model. Then, as illustrated in the following subsections, three distinct deep architectures are built and trained on the processed dataset to predict malware at the beginning of a device operation.

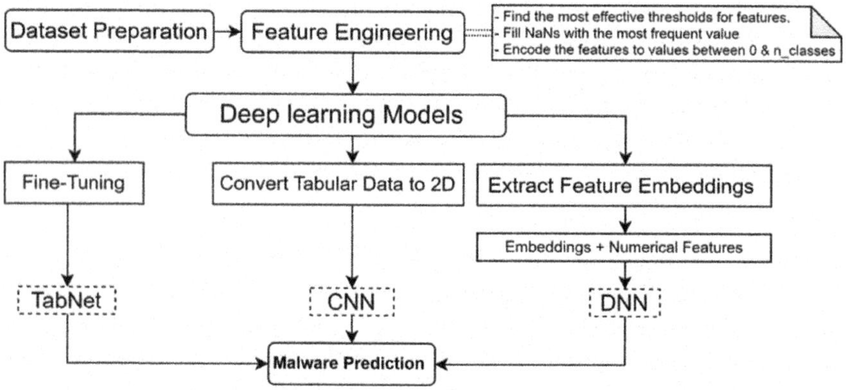

Fig. 1. The pipeline of the proposed tabular deep architecture for malware prediction.

3.1 Microsoft Malware Prediction Dataset

The dataset used for evaluation in this paper is the Microsoft Malware Prediction dataset (Kaggle), and it contains roughly 9 million rows and 83 features with the aim of estimating the likelihood of a system getting infected by various malware based on various machine's features. In this dataset, each machine has its own row, which is uniquely recognised by a "Machine Identifier". The property "HasDetections" indicates whether malware was discovered on the computer or not, where a label value 0 means the device is not infected, and a label value 1 marks the device as infected.

The malware features in this data corpus are generally divided into six categories, each of which provides detailed information about every aspect of the device, including Geographic Location, e.g., country identifier, Hardware Information, e.g., whether it's a touch device or pen capable, Firmware Information, e.g., manufacturer identifier, Operating System Information, e.g., version and build number, Device Status and Activities, e.g., whether the device is AlwaysOnAlwaysConnected capable, and Windows Defender Information, e.g., product name and application version. When categorising the features, 65% are of categorical features, while 10% and 25% pertain to numeric and binary features, respectively.

3.2 Feature Engineering

Accurately representing data structures using existing deep learning models is considerably more challenging due to the complexity and diversity of real-world malware datasets. Hence, in order to construct a reliable predictive malware model, it becomes imperative to employ a robust feature engineering process. This procedure involves carefully selecting and transforming the most crucial features from raw data, which has been demonstrated to enhance the performance of machine learning algorithms in terms of accuracy and training time [32].

In this work, extensive experiments were applied to the malware dataset with thorough analysis and feature selection. Firstly, we need to find the best thresholds for the amount of NaN entries to drop columns at, the number of unique values, and the percentage of the most prominent value for each categorical feature. The malware features were initially categorised into binary, numeric and categorical. Then we calculated the percentage of missing values for each feature and sorted them decreasingly. Three percentage values, which are 20, 30, and 40%, were empirically evaluated to find the best amount of NaN values to drop out. Moreover, having a feature with high unique values provides no useful information for the prediction model. Therefore, we empirically dropped the categorical features that have unique values greater than three thresholds 800, 400, and 100, evaluated separately.

Finally, we calculated the most prominent value for all categorical features to identify the features that have imbalanced values. We tested the values 70, 80, and 90%. As a result, any features with missing values were filled with the most frequent value for categorical and binary features and the median value for numeric features. After that, we used a label encoder for the remaining categorical features, which transforms the labels into values between 0 and the number of classes. The results of these feature processing procedures will be discussed consecutively.

We also calculated the importance of each malware feature included in the final features corpus using LightGBM [9], which includes two techniques for dealing with huge numbers of data instances and features: gradient-based one-side sampling and exclusive feature bundling. Each feature is used to split the training data across all trees in the LightGBM model. Let w_i is the weight of feature x_i, the feature importance score $\mathrm{FIS_i}$ is calculated for each split, as follows:

$$FIS_i = \{s | s = w_i x_i\} \tag{1}$$

Figure 2 shows the importance of the scores that demonstrate the effect of each feature on the trained model, which are obtained by training the LightGBM model on the remaining 42 features as a result of the data preprocessing.

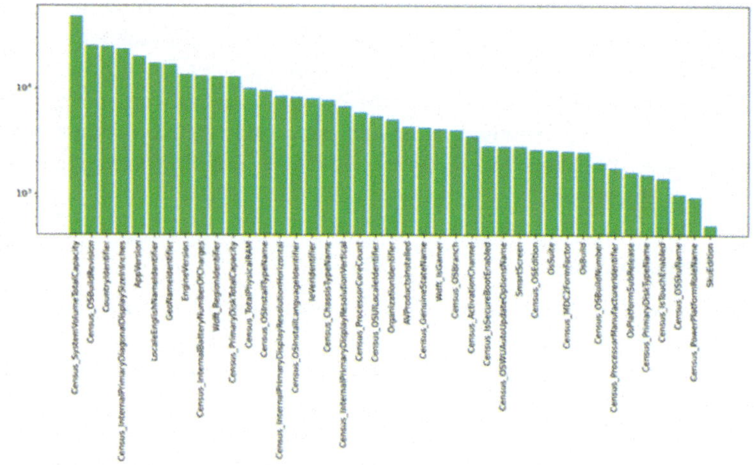

Fig. 2. The importance scores of malware categorical features.

3.3 Deep Learning Models

The following deep learning models are used to classify the tabular malware data.

TabNet. The deep architecture of TabNet [14] was fine-tuned on the dataset prepared for the evaluation of malware prediction. The TabNet baseline incorporates sequential attention, a learning technique that facilitates the selection of relevant model properties at each level. This approach enables the model to provide explanations for its predictions and contributes to the development of more accurate models, which is the core value of TabNet. Additionally, TabNet utilises sparse example feature extraction, employing a serial multi-step architecture where each phase influences the decision based on the selected features. It enhances learning capacity through non-linear processing of the extracted features and emulates ensembling through higher dimensions and multiple steps. This enabled us to efficiently train the prediction model on the tabular data.

CNN. By leveraging their hierarchical structure, local receptive fields, and weight-sharing mechanisms, CNN models can effectively recognise distinctive characteristics and spatial dependencies within malware samples. This enables them to learn discriminative representations of malware features, enhancing the accuracy and efficacy of malware detection systems. However, it is necessary to change the shape of malware data from one dimension to 2D to make it compatible with the CNN model. This enables the CCN model to receive the input features as images, in which the pixels were fed to the input layer in the form of arrays. The complete structure of the CNN layers we constructed in this work is shown in Fig. 3.

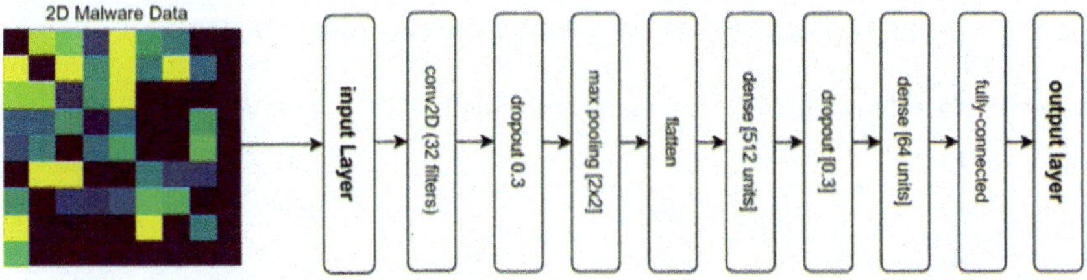

Fig. 3. The generic CNN architecture designed for malware categorical features.

DNN Embeddings. We developed a deep neural architecture to extract the embedding of the categorical features which were then concatenated with the existing numerical features. This model facilitates advanced mathematical modelling. Three dense layers with 64, 32, and 16 units were used. Given the number of hidden units, the main steps adopted for extracting feature embeddings start by creating an input layer for each malware feature. If the feature type is categorical then an embedding layer is created for it; otherwise, the feature data is concatenated with the numerical features. Then, a concatenation layer is added to the DNN architecture to aggregate all the resulting outputs, followed by a flatten layer. A dense layer in accordance with the number of hidden units is also added before the top layers that end with the output layer, providing the decision of malware infection.

4 Experiments and Results

4.1 Experiments Setup

All experiments were performed on the same cloud processing configurations. The evaluation metrics employed in our study are accuracy and area under the curve (AUC). Accuracy measures the model's capability to identify correlations and patterns among variables, and a higher accuracy indicates a stronger ability to generalise to unseen data, leading to better predictions and insights. It can be calculated by dividing the number of correct predictions by the total number of predictions. On the other hand, AUC is a metric that evaluates a classifier's ability to differentiate between classes. It quantifies how well the model distinguishes between positive and negative classes, with a higher AUC indicating superior performance.

4.2 The Results of Feature Engineering

The final set of malware features was generated by applying the feature preprocessing and selection procedure, as illustrated in Subsect. 3.2. Table 1 lists the accuracy results obtained by the LightGBM with default parameters under different feature configurations. The best accuracy is achieved when dropping the categorical features of unique values greater than a threshold of 400 and a threshold of 40% for NaN values. To obtain the most prominent values for categorical features, the amount of 90% scored the best accuracy result. As a result, the generated collection of malware features is evaluated by the three deep learning models under this configuration with data splits of 75% for training and 25% for testing.

Table 1. The experimental results of malware feature engineering.

Scenario	NaN threshold (%)	Prominent threshold (%)	Unique values	Accuracy
S1	40	90	800	0.6588
S2	40	80	800	0.6589
S3	40	70	800	0.6583
S4	40	90	400	0.6607
S5	40	80	400	0.6600
S6	40	70	400	0.6599
S7	40	90	100	0.6491
S8	40	80	100	0.6495
S9	40	70	100	0.6491
S10	30	90	800	0.6388
S11	30	80	800	0.6385
S12	30	70	800	0.6383
S13	30	90	400	0.6407
S14	30	80	400	0.6396
S15	30	70	400	0.6398
S16	30	90	100	0.6280
S17	30	80	100	0.6282
S18	30	70	100	0.6509
S19	20	90	800	0.6385
S20	20	80	800	0.6380
S21	20	70	800	0.6382
S22	20	90	400	0.6408
S23	20	80	400	0.6399
S24	20	70	400	0.6396
S25	20	90	100	0.6294
S26	20	80	100	0.6278
S27	20	70	100	0.6280

4.3 The Results of Tabular Deep Models

The hyperparameter configured for the fine-tuned TabNet model is summarised in Table 2. The three deep learning models, i.e., fine-tuned TabNet, CNN, and DNN-Embeddings, are evaluated using the resulting set of malware data with 42 features for each record.

Table 2. The hyperparameters of fine-tuned TabNet.

Hyperparameter	Value
Optimizer_fn	Torch.optim.Adam
Learning rate	2e2
Scheduler	StepLR
Scheduler params	Step size = 10, gamma = 0.9
Batch size	256
Virtual batch size	128
Mask type	Entmax
Max epochs	10
Weights	1
Drop last	False

As shown in Table 3, among all the models evaluated, the TabNet model achieved the highest accuracy of 0.661 and AUC of 0.701, which demonstrates its superior performance. This can be attributed to the utilisation of a sequential attention technique, which enhances classification performance by effectively capturing interdependencies among features, particularly in tabular data where feature interactions are prominent. Deep learning offers the advantage of extracting embeddings from categorical features, which contributes to improved results compared to traditional machine learning approaches, especially when dealing with large-scale datasets like the one employed in this research. Most importantly, this research work represents an initial demonstration of the effectiveness of tabular deep learning models in malware prediction. Due to resource constraints, implementing the DNN-embedding model with a high number of epochs was challenging, mainly because the model includes a large number of layers.

Table 3. The experimental results of tabular malware prediction.

Model	Accuracy	AUC
Fine-tuned TabNet	0.642	0.701
DNN-embeddings	0.614	0.529
Baseline CNN	0.620	0.672

5 Conclusion

In this research paper, we have developed and explored a series of deep learning models for the purpose of malware detection. The experimental findings present highly promising and compelling evidence regarding the effectiveness of tabular deep learning architectures, particularly when fine-tuned on specific-domain data, such as malware prediction. Our approach also involves the construction of the DNN-embeddings model that automatically determines the appropriate number of layers and extracts embeddings from categorical data, making it well-suited for training on tabular data with a substantial number of features. Leveraging deep learning for tabular data, particularly in tasks like malware prediction, offers the advantage of extracting valuable information from categorical features and maximising the model's potential. Notably, tabular deep learning models achieved comparable performance results to those reported in the Microsoft malware prediction competition. As part of future endeavors, the scope of this study can be broadened by examining the effectiveness of integrating the DNN-embeddings with the TabNet-based architecture. A potential avenue for further analytical investigation involves exploring more pre-trained deep learning models to enhance the procedure for effectively handling more complex malware.

References

1. Yuxin, D., Siyi, Z.: Malware detection based on deep learning algorithm. Neural Comput. & Applic. **31**, 461–472 (2019)
2. Ye, Y., Li, T., Adjeroh, D., Iyengar, S.S.: A survey on malware detection using data mining techniques. ACM Comput. Surv. (CSUR) **50**(3), 1–40 (2017)
3. Pastrana, S., Suarez-Tangil, G.: A first look at the crypto-mining malware ecosystem: a decade of unrestricted wealth. In: Proceedings of the Internet Measurement Conference, pp. 73–86 (2019)
4. McIntosh, T.R., Jang-Jaccard, J., Watters, P.A.: Large scale behavioral analysis of ransomware attacks. In: Neural Information Processing: 25th International Conference, ICONIP 2018, Siem Reap, Proceedings, Part VI 25, pp. 217–229. Springer (2018)
5. Button, M.: Economic and industrial espionage. Secur. J. **33**, 1–5 (2020)
6. Sharma, A., Sahay, S.K.: Evolution and detection of polymorphic and metamorphic malwares: a survey. arXiv:1406.7061 (2014)
7. Anderson, H.S., Kharkar, A., Filar, B., Roth, P.: Evading machine learning malware detection. Black Hat, 1–6 (2017)
8. Jiao, Z., Hu, P., Xu, H., Wang, Q.: Machine learning and deep learning in chemical health and safety: a systematic review of techniques and applications. ACS Chem. Health & Saf. **27**(6), 316–334 (2020)
9. Ke, G., Meng, Q., Finley, T., Wang, T., Chen, W., Ma, W., Ye, Q., Liu, T.Y.: Lightgbm: a highly efficient gradient boosting decision tree. Adv. Neural Inf. Process. Syst. **30** (2017)
10. Wolpert, D.H.: The existence of a prior distinctions between learning algorithms. Neural Comput. **8**, 1391–1420 (1996)
11. Gomez, D., Rojas, A.: An empirical overview of the no free lunch theorem and its effect on real-world machine learning classification. Neural Comput. **28**, 216–228 (2016)
12. Akhtar, M.S., Feng, T.: Malware analysis and detection using machine learning algorithms. Symmetry **14**(11), 2304 (2022)
13. Hayashi, Y.: Does deep learning work well for categorical datasets with mainly nominal attributes? Electronics **9**(11), 1966 (2020)
14. Arik, S.Ö., Pfister, T.: Tabnet: Attentive interpretable tabular learning. Proc. AAAI Conf. Artif. Intell. **35**(8), 6679–6687 (2021)
15. Popov, S., Morozov, S., Babenko, A.: Neural oblivious decision ensembles for deep learning on tabular data. arXiv:1909.06312 (2019)

16. Zhang, Y., Liu, Z., Jiang, Y.: The classification and detection of malware using soft relevance evaluation. IEEE Trans. Reliab. **71**(1), 309–320 (2020)
17. Ronen, R., Radu, M., Feuerstein, C., Yom-Tov, E., Ahmadi, M.: Microsoft malware classification challenge. arXiv:1802.10135 (2018)
18. Narayanan, B.N., Djaneye-Boundjou, O., Kebede, T.M.: Performance analysis of machine learning and pattern recognition algorithms for malware classification. In: 2016 IEEE National Aerospace and Electronics Conference (NAECON) and Ohio Innovation Summit (OIS), pp. 338–342. IEEE (2016)
19. Zhang, Y., Huang, Q., Ma, X., Yang, Z., Jiang, J.: Using multi-features and ensemble learning method for imbalanced malware classification. In: 2016 IEEE Trustcom/BigDataSE/ISPA, pp. 965–973. IEEE (2016)
20. Nataraj, L., Karthikeyan, S., Jacob, G., Manjunath, B.S.: Malware images: visualization and automatic classification. In: Proceedings of the 8th International Symposium on Visualization for Cyber Security, pp. 1–7 (2011)
21. Bahtiyar, Ş, Yaman, M.B., Altınîğne, C.Y.: A multi-dimensional machine learning approach to predict advanced malware. Comput. Netw. **160**, 118–129 (2019)
22. Pan, Q., Tang, W., Yao, S.: The application of LightGBM in microsoft malware detection. J. Phys. Conf. Ser. **1684**(1), 012041 (2020)
23. Younis, L.B., Sweda, S., Alzu'bi, A.: Forensics analysis of private web browsing using android memory acquisition. In: 2021 12th International Conference on Information and Communication Systems (ICICS), pp. 273–278. IEEE (2021)
24. Rhode, M., Burnap, P., Jones, K.: Early-stage malware prediction using recurrent neural networks. Comput. & Secur. **77**, 578–594 (2018)
25. Kolosnjaji, B., Zarras, A., Webster, G., Eckert, C.: Deep learning for classification of malware system call sequences. In: AI 2016: Advances in Artificial Intelligence: 29th Australasian Joint Conference, Hobart, TAS, Australia, December 5–8, 2016, Proceedings 29, pp 137–149. Springer International Publishing (2016)
26. Kalash, M., Rochan, M., Mohammed, N., Bruce, N.D., Wang, Y., Iqbal, F.: Malware classification with deep convolutional neural networks. In: 2018 9th IFIP International Conference on New Technologies, Mobility and Security (NTMS), pp. 1–5. IEEE (2018)
27. Gopinath, M., Sethuraman, S.C.: A comprehensive survey on deep learning based malware detection techniques. Comput. Sci. Rev. **47**, 100529 (2023)
28. Wang, Z., Liu, Q., Chi, Y.: Review of android malware detection based on deep learning. IEEE Access **8**, 181102–181126 (2020)
29. McDole, A., Gupta, M., Abdelsalam, M., Mittal, S., Alazab, M.: Deep learning techniques for behavioral malware analysis in cloud iaas. In: Malware Analysis Using Artificial Intelligence and Deep Learning, pp. 269–285 (2021)
30. Khan, A.R., Yasin, A., Usman, S.M., Hussain, S., Khalid, S., Ullah, S.S.: Exploring lightweight deep learning solution for malware detection IoT constraint environment. Electronics **11**(24), 4147 (2022)
31. Abuarqoub, A., Abuarqoub, S., Alzu'bi, A., Muthanna, A.: The impact of quantum computing on security in emerging technologies. In: The 5th International Conference on Future Networks & Distributed Systems, pp. 171–176. ACM (2021)
32. Kasongo, S.M., Sun, Y.: A deep learning method with filter based feature engineering for wireless intrusion detection system. IEEE access. **7**, 38597–38607 (2019)

An Intrusion Detection System Using the XGBoost Algorithm for SDVN

Adi El-Dalahmeh[1(✉)], Jie Li[1], Ghaith El-Dalahmeh[2], Mohammad Abdur Razzaque[1], Yao Tan[3], and Victor Chang[4]

[1] Teesside University, TS1 3BX Middlesbrough, UK
{A.El-Dalahmeh,jie.li}@tees.ac.uk
[2] MARS Robotics, Irbid, Jordan
[3] College of Computer Science and Engineering, Chongqing University of Technology, 400054 Chongqing, China
[4] Aston University, B4 7ET Birmingham, UK

Abstract. Vehicular ad hoc networks (VANETs) rely on Software-Defined Networking (SDN) to enable continuous exchange of information and messages about vehicle and road conditions. This facilitates convenience for users and improves decision-making and safety. However, the communication of Electronic Control Units (ECUs) through the Control Area Network (CAN) poses security risks. The CAN is vulnerable to a range of security attacks, including Denial of Service (DoS), fuzzy attacks, and spoofing RPM, which can cause traffic congestion, fatal accidents, or disrupt the network services provided to users. To address these security challenges, we propose an Intrusion Detection System (IDS) that uses the XGBoost machine learning algorithm. Our IDS leverages a car-hacking dataset to detect traffic patterns and classify them as normal or attack patterns. Specifically, our research examines three types of attacks: DoS, fuzzy, and spoofing RPM, which are present in the car-hacking dataset. We show that our proposed IDS outperforms external infiltration systems that use KNN and LSTM-AE algorithms. By enhancing the security of SDN-based VANETs, our proposed framework contributes to safer and more reliable vehicular communication.
.

Keywords: IDS · Security · XGBoost

1 Introduction

A Vehicular ad hoc networks (VANET), a subset of the mobile ad hoc network (MANET), consists of mobile nodes that act as both hosts and routers. However, unlike MANET, VANET's constant vehicle movement causes the network topology to continuously change [7]. VANET employs different communication types, including vehicle-to-vehicle (V2V), vehicle-to-infrastructure unit (V2I), and vehicle-to-pedestrian (V2P) communication. Although these connections improve information about roads and network status, they are short and

N. Naik et al. (Eds.): UKCI 2023, AISC 1453, pp. 390–402, 2024.
https://doi.org/10.1007/978-3-031-47508-5_31

intermittent, making them vulnerable to attacks like Denial of Service (DoS) that aim to reduce the quality of service or stop the network entirely [10].

In the case of VANETs, although they have unique characteristics due to the vehicular environment and specific communication needs, they still adhere to the underlying principles of traditional networks. This includes concepts such as addressing, routing, packet delivery, network protocols, and data transmission. However, it requires innovation and flexibility in its infrastructure to improve its information collection, security and privacy, complete network control, and connection time [11]. In recent years, Software-Defined Networking (SDN) has been combined with VANET for its advantages in network management and programmable infrastructure. SDN addresses most VANET issues, such as effective resource utilization for all network users and compatibility with the network's constantly changing nature. SDN also meets specific VANET requirements like scalability, short latency, network heterogeneity, and providing a network overview [4]. Figure 1 illustrates the SDN-based VANET architecture. In an SDN-based VANET architecture, the control plane is managed by a centralized controller, while the data plane is handled by the network nodes. The controller has a global view of the network and can make real-time decisions about routing, traffic management, and security based on the network conditions.

Despite these benefits, the SDN-based VANET network faces several problems, as shown in the following points [6]:

- Security and privacy: the control unit in this network assumes responsibility for all instructions, making it a target for various attacks such as denial-of-service (DoS), Sybil, or man-in-the-middle attacks. Therefore, safeguarding central control is essential.
- Reliability, a vehicle's dependability reduces its risk rate; however, it is crucial to continuously monitor vehicle behavior to identify any malicious activity that can lead to fuzzy attacks. Such attacks involve sending incorrect messages to create traffic jams or fatal accidents. Considering the significant impact of security attacks on human life, an IDS must be put in place to protect users effectively.

Several Machine Learning-based IDSs have been proposed in previous studies, including TSK+ fuzzy inference-based IDS [9,15], Neural Network supported IDS [8], and, in [14] proposes a hybrid IDS approach to reduce Sybil attacks in vehicles. The proposed approach uses a combination of a clustering algorithm and a DNN. However, these approaches are mostly designed for TCP/IP networks and only a few have been developed for SDN-based VANETs. The current methods mainly focus on detecting specific types of attacks and may not be suitable for detecting attacks in VANETs due to their unique characteristics. Additionally, the use of SDN in VANETs poses specific security challenges, such as the need to protect the central control from various attacks, and the continuous monitoring of vehicle behaviour to detect malicious activities. In this study, we propose an XGBoost-based IDS to detect DoS, fuzzy, and spoofing RPM attacks in VANET networks. Our approach aims to address the limitations of existing

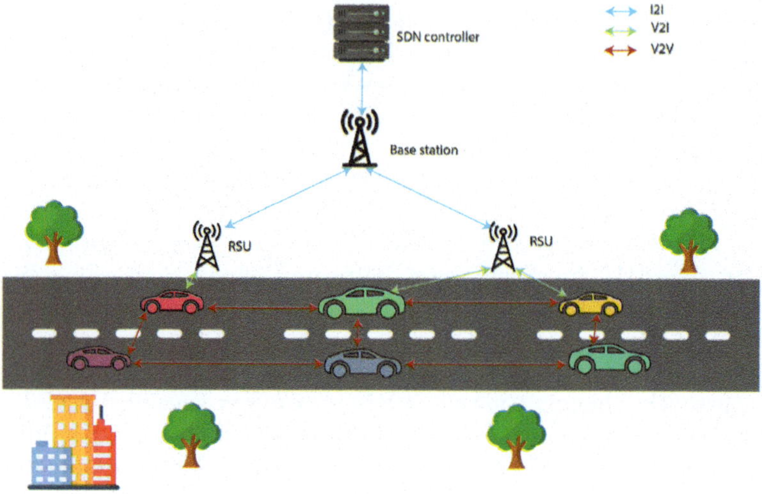

Fig. 1. SDN-based VANET architecture

methods and reduce the risk associated with the use of SDN in VANETs by utilizing a car-hacking dataset.

1.1 Research Contribution

- We propose an IDS based on the XGBoost learning algorithm to protect communications in an SDN based on VANET to increase the security and privacy of communications.
- The proposed system detects and classifies three types of attacks: DoS, Fuzzy and spoofing RPM attacks. A car-hacking data set was adopted, produced from an experienced Hyundai vehicle.

The rest of the paper is structured as follows: Sect. 2 presents the security attacks and learning algorithms. Section 3 details the proposed IDS. Section 4 presents the results and discussion, while Sect. 5 delivers the research conclusions.

2 Background

In this section, we will discuss the security attacks that target SDN-based VANET. Security attacks aim to disrupt the network's services, cause traffic congestion, accidents, or steal user information. The three types of attacks we will discuss are DoS, fuzzy, and spoofing RPM.

- A DoS attack occurs when a malicious node sends numerous packets to the target node, making it difficult to distinguish legitimate messages, resulting in network slowdown or complete network failure. The malicious equipment can also send wireless signals to interfere with communications between vehicles and infrastructure, reducing or preventing communication quality [1].

- A fuzzy attack is an injection attack in which a malicious vehicle injects the network with a large number of fake spam messages. Vehicles continuously communicate in the network to improve traffic safety on the roads. If an attacker injects false information, vehicles may make poor decisions leading to traffic congestion or accidents. Fake messages may include false emergency messages, incorrect road conditions that cause traffic jams, and data that cause confusion [12].
- A spoofing RPM attack tampers with the gear and RPM values, transmitting incorrect messages to nearby vehicles, leading to wrong decisions such as sudden braking or changing lanes to avoid collisions.[13].

The IDS monitors and analyzes networks to detect potential attacks. Machine or deep learning algorithms enhance the detection and classification accuracy of security attacks. These systems train machine learning algorithms using a large set of normal and abnormal data to identify natural patterns in the network and other patterns that could indicate a potential security threat. There are various types of learning algorithms used, such as KNN, LSTM-AE, and XGboost.

- KNN: The KNN algorithm is a supervised machine learning algorithm used for classification tasks. In this algorithm, objects are classified based on their proximity to data points in the training dataset. The study cited in [2] used KNN to classify traffic patterns into normal and unusual by calculating the distance between the traffic pattern received and the known traffic patterns in the training set. The classification was then based on which category was more common in its neighbors. Although KNN is effective with small and medium datasets, it is not the best choice for extensive data due to the high computational costs.
- LSTM-AE: It is a type of recurrent neural network (RNN) that processes sequential data. In the study mentioned in [3], researchers proposed an intrusion detection system (IDS) based on LSTM-AE to detect traffic patterns in a network. The LSTM-AE processes traffic and converts it into a series of feature vectors. These vectors are then classified as 'normal' or 'extraordinary'. The proposed framework achieved good results with the databases (UNSW-NB15). However, when applied to the car-hacking database, its results were more limited due to the difference in data between the databases.
- XGboost: This is a machine learning method that creates a series of decision trees to learn each tree on a data category based on its advantage. The creation of trees is continuous, and the mistakes made by the trees need to be corrected [5]. In IDS, XGBoost has many suitable characteristics for intrusion detection activities. It can deal with the numerical and categorical features essential for dealing with traffic data, and it is scalable and can handle big data effectively [5].

3 Proposed Work

The objective of security attacks like DoS, fuzzy, and spoofing RPM is to disable or partially disrupt a system, cause traffic congestion, or endanger users by producing a large number of messages or creating fraudulent ones. To counteract these threats, an IDS was developed to identify any malicious activity in the network. In communication protocols, messages generated by users or controllers are marked with a timestamp, which changes when the vehicle sends a message. In this study, we put forward an IDS based on the XGBoost algorithm and a car-hacking dataset to detect DoS, fuzzy, and spoofing RPM attacks. The structure of the proposed framework is illustrated in Fig. 2, and the following steps outline the procedure of the system:

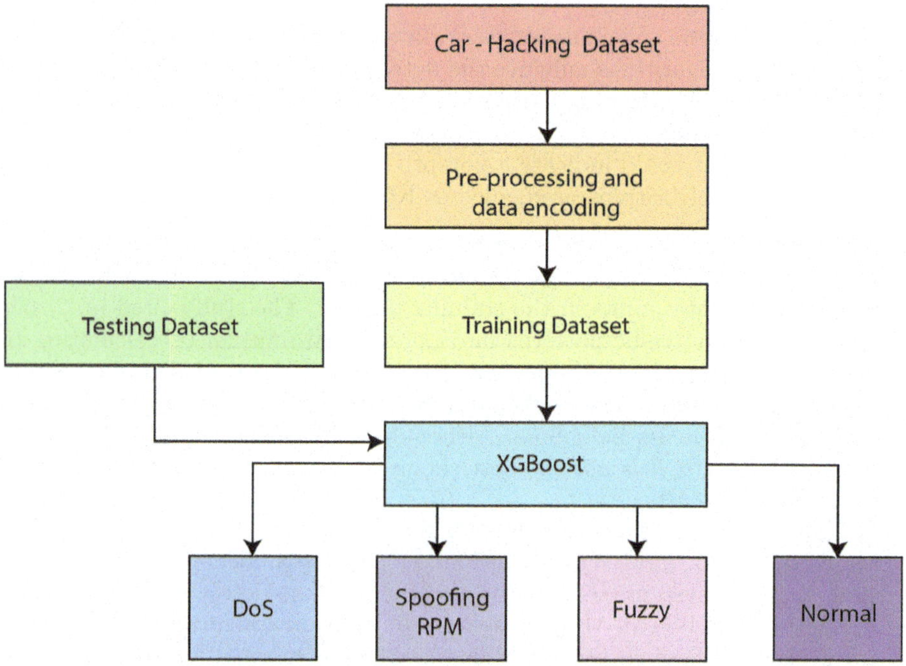

Fig. 2. Intrusion detection system diagram

3.1 Car-Hacking Dataset

The system utilized a car-hacking dataset [12,13] to detect DoS, fuzzy, and spoofing RPM attacks. To gather the necessary data, a Hyundai vehicle was used as a test vehicle, and its traffic and CAN messages were observed in real-time using OBD2 and a Raspberry Pi connected to a computer. The regular data was then subjected to attacks by injecting a DoS attack every 0.3 milliseconds, a fuzzy attack every 0.5 milliseconds, and a spoofing RPM attack every 1 millisecond through the CAN ID. The dataset comprised 300 interpolations of message injection, and each attack lasted for 3–5 seconds, resulting in a total of 30–40 minutes of traffic per set. The dataset contained 12 columns with unique timestamps, representing the time interval for each message, the message identifier, and the number of bits in each message (DLC). The data length varied between 0–7 bytes, and the last column indicated whether the message was normal code (R) or attack code (T). The three datasets were merged into a single file for ease of use. Table 1 illustrates the number of messages in each dataset [12,13]:

Table 1. Dataset overview.

Attack	Normal messages	Injected messages
DoS	3,078,250	587,521
Fuzzy	3,347,013	491,847
Spoofing RPM	2,290,185	654,897

3.2 Pre-processing and Data Encoding

In order to deal with the data better and accurately by the XGBoost learning method, we pre-processed the data as follows:

- Remove empty rows from the data set.
- Remove the extra column of headings that do not specify any data analysis information.
- Merge all the columns into one and remove spaces from it, as it contains eight columns, each representing 1 byte of data.
- Convert hexadecimal values to decimal numbers only.
- Set addresses to separate DOS, fuzzy, spoofing RPM, and normal from each other. The data set was pre-processed for each of the three attacks before the XGBoost training process using the processed data, as described next.

3.3 Training Phase

After pre-processing the data, the data set was separated into two groups. The first group for training contained 70% of the data set and the second contained 30% to test the effectiveness of the proposed IDS. XGBoost helps to improve memory and hardware resources through additional training, as each training stage depends on the results of the previous one. The proposed framework was designed in Python to visualize traffic and send and receive CAN. Table 2 shows the numbers of normal and attack messages used in the training phase.

3.4 Testing Phase

In the testing stage, we sent normal messages of different ID and size. After this, we sent messages for all three attack types, with random information and ID. Table 3 shows the numbers of normal and attack messages used in the testing phase.

Table 2. Training dataset.

Attack	Total messages	Normal messages	Injected messages
DoS	560,000	460,000	100,000
Fuzzy	518,842	420,183	98,659
Spoofing RPM	702,765	550,974	151,791

Table 3. Training dataset.

Attack	Total messages	Normal messages	Injected messages
DoS	240,000	198,461	41,539
Fuzzy	100,00	80,731	19,269
Spoofing RPM	264,716	238,952	25,764

4 Evaluation

In this section, we present of the proposed system compared to the external works [2] on KNN and [3] on LSTM-AE against DoS, Fuzzy, and Spoofing RPM attacks.

4.1 Evaluation Criteria

Several criteria were used to evaluate the proposed framework: accuracy, detection, recall rate, and F1-SCORE:

- Accuracy: the rate of effectiveness of the proposed framework in detecting cases classified as normal or attack. The accuracy equation is represented as follows:

$$Accuracy = \frac{TP + TN}{TP + TN + FP + FN} \tag{1}$$

- Precision: the rate of correctly identifying attacks among those detected. It is represented as follows:

$$Precision = \frac{TP}{TP + FP} \tag{2}$$

- Recall: the ratio of correctly detected attacks to the total attacks in the test dataset. The third equation represents recovery.

$$Recall = \frac{TP}{TP + FN} \tag{3}$$

- F1-SCORE: The weighted average of accuracy and recovery is calculated from the fourth equation:

$$F1 - SCORE = 2\frac{(Precision * Recall)}{(Precision + Recall)} \tag{4}$$

True positive (TP) represents real attacks correctly classified as attacks, while a true negative (TN) is the total number of normal data correctly classified as normal. False positive (FP) is the number of normal data incorrectly classified as an attack. A false negative (FN) represents the number of attack data incorrectly classified as normal.

4.2 Results

Figure 3 presents the accuracy results of the proposed framework compared with the external frameworks KNN [2] and LSTM-AE [3] against a DoS attack. Figure 3 shows the accuracy rates for each of the three systems; the proposed system achieved an accuracy rate of 99.89% compared to 97.4% for KNN and 99% for LSTM-AE against DoS attack. Also, presents the results of the three IDS against fuzzy attacks, the results were: 99.93% for XGBoost, 98.5% for KNN and 99.4% for LSTM-AE. In regard to the spoofing RPM attack, our accuracy rate was 99.93%, while KNN achieved 98.5% and LSTM-AE achieved 99.4%. In addition, we achieved 99.89%, KNN 99.7%, LSTM-AE 99.6% in a normal case.

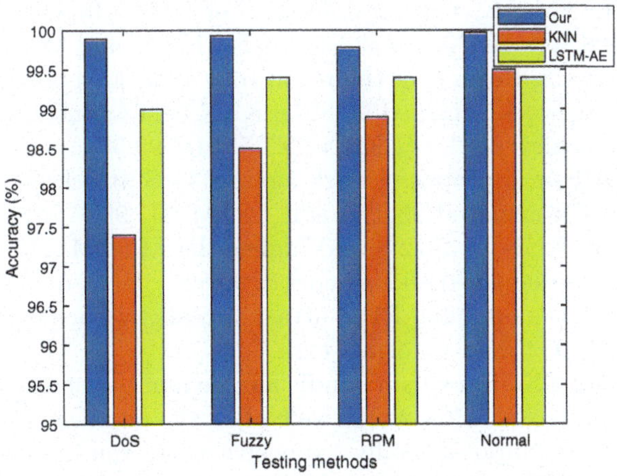

Fig. 3. Accuracy rate against attacks

From Fig. 4 the precision rate against DoS attack was 99.61% for our system, 96.3% for KKN, and 99% for LSTM-AE. The precision rate of the proposed framework was 99.55% compared to 97.7% for KNN and 99.2% for LTSM-AE against Fuzzy attack. For precision against Spoofing RPM, XGBoost achieved 99.55%, KNN 97.7% and LSTM-AE 99.2%. In a normal case, we achieved 99.9%, KNN 99.5%, and LSTM-AE 99.3%.

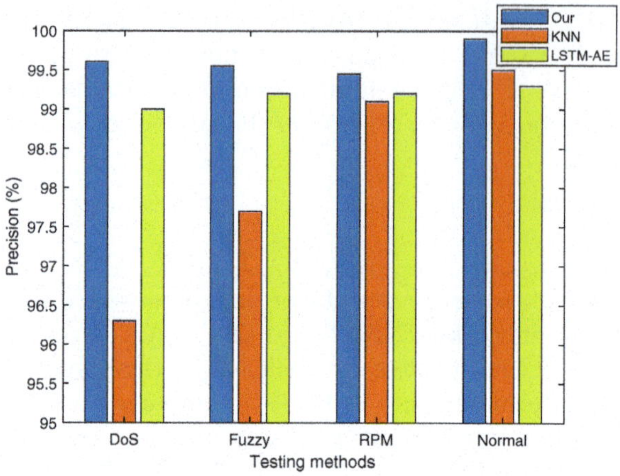

Fig. 4. Precision rate against attacks

The recall rate is represented in Fig. 5; our rate was 0.73% higher than KNN and less than LSTM-AE, with a difference of 0.05% In DoS case, while the recall rate was 98.86% for XGBoost, 98.4% for KNN, and 98.5% for LSTM-AE against Fuzzy attack. Regarding the recall rate against Spoofing RPM attacks, the proposed framework achieved 98.86%, KNN 98.4%, and LSTM 98.5%. In a normal case, we achieved 99.9%, KNN 99.7%, and 99.5%.

For F1-SCORE, the proposed work against DoS attack scored the highest, 99.71%, while KNN achieved 93.4 % and LSTM-AE 99%. We achieved an F1-Score higher than KNN by 2.72% and higher than LSTM-AE by 1.12% in the Fuzzy case. Also, we achieved an F1-Score of 99.82%, KNN of 97.1%, and LSTM-AE of 98.7% against a Spoofing RPM attack. Finally, we achieved 99.89%, KNN 99.7%, LSTM-AE 99.6% in a normal case.

From the results, XGBoost can handle a large number of features and missing values better than KNN. KNN works well when the number of features is small, but it can become computationally expensive when the number of features is large. Moreover, KNN requires the imputation of missing values, which can lead to sub-optimal results. On the other hand, XGBoost can handle missing values by learning the best split at each node of the decision tree. XGBoost can handle complex non-linear relationships between features better than LSTM-AE. LSTM-AE is a type of neural network that is often used for sequence data, such as time series. While LSTM-AE can capture non-linear relationships between features, it may not be as effective in handling complex non-linear relationships between features as XGBoost. XGBoost uses decision trees to model non-linear relationships between features, which can be more effective in certain scenarios. Also, XGBoost is faster to train than LSTM-AE. LSTM-AE is a deep neural network that can be computationally expensive to train, especially on large datasets. XGBoost, on the other hand, is a gradient-boosting algorithm that can

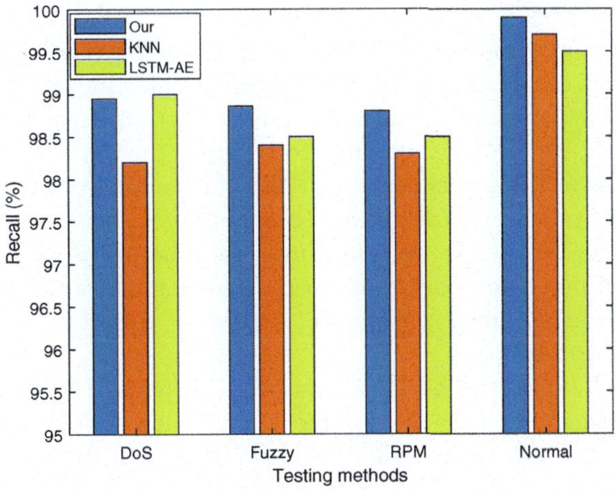

Fig. 5. Recall rate

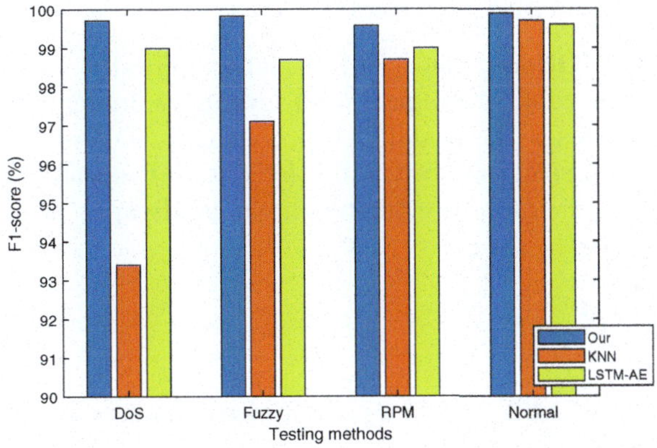

Fig. 6. F1-score rate

be trained relatively quickly on large datasets. Moreover, XGBoost can be easily parallelized, which makes it scalable to larger datasets.

In summary, XGBoost may be preferred over KNN and LSTM-AE in scenarios where the number of features is large, and complex non-linear relationships between features need to be modeled quickly and effectively. However, the choice of algorithm ultimately depends on the dataset, and it's important to evaluate multiple algorithms to find the best one for a given task.

In our proposed work, we address the scalability problem in SDVNs by utilizing the XGBoost algorithm and a carefully curated car-hacking dataset. Here's how we tackle the scalability challenge:

- Efficient Learning Algorithm: We employ the XGBoost algorithm, which is known for its scalability and efficiency in handling large-scale datasets.

XGBoost leverages parallel processing and optimized tree construction algorithms to handle a significant volume of data effectively. By using XGBoost, we ensure that our IDS can efficiently process and analyze the extensive car-hacking dataset, enabling scalability in detecting security attacks.
- Pre-processing and Data Encoding: To enhance the scalability of our IDS, we employ pre-processing techniques on the car-hacking dataset. This involves removing empty rows, unnecessary columns, and spaces, as well as converting hexadecimal values to decimal numbers. The pre-processed data is then encoded to differentiate between normal and attack messages. These steps optimize the data representation and prepare it for effective learning by the XGBoost algorithm.
- Incremental Learning: The XGBoost algorithm used in our IDS supports incremental learning. This means that the training process can be performed in stages, where each stage builds upon the results of the previous one. Incremental learning enables the IDS to handle large datasets without requiring them to be processed all at once. It also allows for efficient use of memory and hardware resources, enhancing the scalability of the IDS in processing and learning from extensive data.

By employing the XGBoost algorithm, utilizing a curated car-hacking dataset, employing pre-processing techniques, and leveraging incremental learning, our proposed IDS addresses the scalability problem in SDVNs. These approaches ensure efficient processing, analysis, and detection of security attacks, enabling the IDS to scale and adapt to evolving threat landscapes in real-world SDVN environments.

5 Conclusion

In the SDN-based VANET, vehicles communicate with each other and roadside units. In the process of communication inside the car, the CAN channel is an essential element in the communication process. Therefore, it must be protected because it is vulnerable to attacks that aim to disrupt the communications process or create traffic jams or fatal accidents. In this paper, we proposed an IDS to detect DoS, fuzzy and spoofing RPM attacks using the XGBoost learning method. In the system, we used a car-hacking dataset, where the data were divided after the pre-treatment process into 70% for training and 30% for testing. By analyzing the results, the proposed framework achieved better results than the external works that used KNN and LSTM-AE to build IDS.

In future work, it is crucial for IDS to broaden their scope in SDVNs by incorporating a wider range of security threats. This expansion will enhance the effectiveness and applicability of IDS in addressing evolving cybersecurity risks. Areas of focus include:

- Additional Attack Types: IDS should adapt to new attack vectors, encompassing threats specific to SDVNs such as control plane attacks, virtualization layer attacks, and attacks on SDN controllers.

- Insider Threats: Mitigating insider threats becomes imperative in SDVNs, where multiple parties with different levels of trust have access to the network infrastructure. IDS should incorporate techniques to detect and mitigate unauthorized access, data exfiltration, and malicious activities initiated by trusted users.
- Zero-Day Exploits: IDS should explore advanced techniques like anomaly detection, behavior analysis, and machine learning to detect zero-day exploits and unknown threats that lack existing security patches or signature-based detection methods.
- There is a need for a more diverse and comprehensive dataset that captures various attack scenarios commonly encountered in SDVNs. By incorporating such a dataset, the IDS can improve its ability to detect and mitigate a broader range of attacks, enhancing its effectiveness and applicability in real-world SDVN environments.
- In the future, there is a growing need to deploy and utilize IDS in real-world environments rather than simulation environments.

References

1. Ahmed, W., Elhadef, M.: Dos attacks and countermeasures in vanets. In: Advanced Multimedia and Ubiquitous Engineering: MUE/FutureTech 2018, vol. 12, pp. 333–341. Springer (2019)
2. Alshammari, A., Zohdy, M.A., Debnath, D., Corser, G.: Classification approach for intrusion detection in vehicle systems. Wirel. Eng. Technol. **9**(4), 79–94 (2018)
3. Ashraf, J., Bakhshi, A.D., Moustafa, N., Khurshid, H., Javed, A., Beheshti, A.: Novel deep learning-enabled lSTM autoencoder architecture for discovering anomalous events from intelligent transportation systems. IEEE Trans. Intell. Transp. Syst. **22**(7), 4507–4518 (2020)
4. Di Maio, A., Palattella, M.R., Soua, R., Lamorte, L., Vilajosana, X., Alonso-Zarate, J., Engel, T.: Enabling SDN in Vanets: What is the impact on security? Sensors **16**(12), 2077 (2016)
5. Gad, A.R., Nashat, A.A., Barkat, T.M.: Intrusion detection system using machine learning for vehicular ad hoc networks based on ton-iot dataset. IEEE Access **9**, 142206–142217 (2021)
6. Ghonge, M.M.: Software-defined network-based vehicular ad hoc networks: a comprehensive review. Software Defined Networking for Ad Hoc Networks, pp. 33–53 (2022)
7. Goumiri, S., Riahla, M.A., Hamadouche, M.: Security issues in self-organized adhoc networks (manet, vanet, and fanet): a survey. In: Artificial Intelligence and Its Applications: Proceeding of the 2nd International Conference on Artificial Intelligence and Its Applications (2021), pp. 312–324. Springer (2022)
8. Li, J., Qu, Y., Chao, F., Shum, H.P., Ho, E.S., Yang, L.: Machine learning algorithms for network intrusion detection. AI in Cybersecurity, pp. 151–179 (2019)
9. Li, J., Yang, L., Qu, Y., Sexton, G.: An extended Takagi-Sugeno-Kang inference system (tsk+) with fuzzy interpolation and its rule base generation. Soft. Comput. **22**, 3155–3170 (2018)
10. Malhi, A.K., Batra, S., Pannu, H.S.: Security of vehicular ad-hoc networks: a comprehensive survey. Comput. Secur. **89**, 101664 (2020)

11. Mchergui, A., Moulahi, T., Zeadally, S.: Survey on artificial intelligence (AI) techniques for vehicular ad-hoc networks (Vanets). Veh. Commun. **34**, 100403 (2022)
12. Seo, E., Song, H.M., Kim, H.K.: Gids: Gan based intrusion detection system for in-vehicle network. In: 2018 16th Annual Conference on Privacy, Security and Trust (PST), pp. 1–6. IEEE (2018)
13. Song, H.M., Woo, J., Kim, H.K.: In-vehicle network intrusion detection using deep convolutional neural network. Veh. Commun. **21**, 100198 (2020)
14. Velayudhan, N.C., Anitha, A., Madanan, M.: Sybil attack detection and secure data transmission in vanet using cmeha-dnn and md5-ecc. J. Ambient Intell. Humanized Comput. 1–13 (2021)
15. Yang, L., Li, J., Fehringer, G., Barraclough, P., Sexton, G., Cao, Y.: Intrusion detection system by fuzzy interpolation. In: 2017 IEEE International Conference on Fuzzy Systems (FUZZ-IEEE), pp. 1–6. IEEE (2017)

Privacy and Security Landscape of Metaverse

Vibhushinie Bentotahewa, Shadan Khattak, Chaminda Hewage, Sandeep Singh Sengar, and Paul Jenkins(✉)

Cybersecurity and Information Networks Centre, Cardiff School of Technologies, Cardiff Metropolitan University, Cardiff, UK
{vibentotahewa,skhattak,chewage,sssengar,pjenkins2}@cardiffmet.ac.uk

Abstract. The Metaverse will create an immersive and interoperable virtual universes for user interaction. The advancement of AR/VR technology, brain–computer interfaces (BCIs) via sensor technologies and 5G/6G links, has important implications for the Metaverse as different application domains will be realized sooner than predicted. However, its immersiveness and interoperability creates significant privacy and security issues beyond Web 2.0 technology. This position paper advances the existing knowledge in the space of privacy and security implications of the proposed platforms with particular focus on the friction between the Metaverse and existing data protection laws such as EU GDPR. Furthermore, this elaborates on Artificial Intelligence (AI) usage in the Metaverse, potential technical solutions for identified privacy and security challenges and future research directions as recommendations.

Keywords: Metaverse · Privacy · Security · GDPR · Artificial Intelligence · Encryption · Digital Forensics · Blockchain

1 Introduction

The Emerging Metaverse based technologies will collect vast amounts of personal information to provide an improved Quality of Experience (QoE) in real-time for its users. However, the collection of large amounts of personal data poses a significant privacy and security risk. Moreover, the notion that the Metaverse should be a personalized experience additionally raises potential privacy and security risks to users. Therefore, to effectively manage future threats, this position paper aims to assess the potential privacy and security issues associated with the Metaverse and proposes both policy and technical recommendations to address these issues in the early stages.

Security and privacy aspects will have serious challenges and concerns in the Metaverse, specifically for social media platforms [15]. Recently, these concerns have been a major focal point of research in Metaverse related technologies [7,15, 39,50,52]. Some of this research focussed on security and privacy concerns in the Metaverse [7,15,50]. Malicious users can monitor and assemble Metaverse users'

N. Naik et al. (Eds.): UKCI 2023, AISC 1453, pp. 403–417, 2024.
https://doi.org/10.1007/978-3-031-47508-5_32

activities (e.g., purchase actions and interactions with other users), biometrics (e.g., vocal inflections and facial expressions), all in real-time, which might be employed to identify the user later (linked or linkable information to personal identity). Thus, to offer users with appropriate services efficiently and securely, it is important to consider the privacy and cybersecurity concerns from the design stage, since the Metaverse is made in the digital (or cyber) environment while collecting the data from the physical world. Furthermore, these actions performed on the Metaverse could have significant impact on the individuals and their rights, society and the economy.

Although the Metaverse offers an exciting and immersive digital society, it could lead to several side effects such as cyberbullying, harassment, and hate speech to be confronted, yet these are not new, society has been living with such malice for years. According to the reports published by Researchers from the Centre for Countering Hate, incidents of harassment and abuse on Facebook's VR chat occurred every 7 minutes [51] over a 12-hour period [51]. One consolation is that Meta has stated that it would continue to improve their user experience and ensure that safety tools are easy to find and use [51].

In the Metaverse worlds, hateful sentiments and slurs can be orally or textually expressed in one-to-one or one-to-many contexts [13] Concerns have been expressed that such hate speech assaults would affect the dignity of those who were targeted and undermines their emotional and psychological health, even promote violence towards marginalised groups [13]. The most concerning scenario is that the assaulted become marginalised and result in self-isolation. The impact for individuals may be significant due to the immersive experience compared to the 2D user experiences of other social media platforms (e.g. the use of hate speech in Facebook).

In the following sections, significant privacy and security issues in the Metaverse are presented and a discussion of the potential solutions. Section 2 discusses the privacy and security issues in the Metaverse while Sect. 3 presents the AI usage in the Metaverse. Section 4 presents key recommendations and future research directions. Finally, conclusions are presented in Sect. 5.

2 Privacy and Security Challenges

2.1 Impact of Metaverse Technologies on Privacy of the Users

Sensors in the Metaverse collect large amounts of data from their respective environments [61], which has come under scrutiny and has raised serious concerns regarding the privacy of the data subjects. The collection of specific types of data on brain wave patterns, facial expressions, eye movements, hand movements, speech and biometric features [50], and the surrounding environment related physiological, physical, biometric, and social interactions using sensors is in excess of generic data collection practices [6]. The amount and type of data collected by such a vast platform carries substantial privacy and security concerns, with incidental data breaches having a severe impact on the security and privacy of the users. Unlike passwords, biometric data breaches will have a greater impact on individuals since they cannot be replaced. The content (including users' personal information) stored in a Metaverse platform can be leaked and forged [13,51,57]. For example, an avatar's information (such as audio and video recording) might be hacked while the user is using the platform, or an attacker might forge the avatar and misuse it (in a similar way to taking control of a social media account on Facebook). However, it is hard to establish when to correctly check the security actions required to exercise control over personal information since complex services in the Metaverse share various types of private information in real-time instead of at a particular moment.

Human resemblance through an avatar can increase concerns about the complexity of identifying the perpetrators. The challenge is to allay any fears about the avatar world and to establish safeguards in the design stage. To ensure that proper safeguards are in place, the policy makers should ensure that there should be a proper mechanism in the form of a legal document that would function as a safety net for the individuals against the Metaverse associated privacy and security challenges.

Further research is required on the impact of a vulnerable population, such as underage users of Metaverse. The easy access for underage users to venture virtually into venues that would be prohibited in the physical world would be detrimental for certain age groups [36]. For instance, ChatGPT and TikTok came under severe criticism recently since these platforms were unable to verify the age of its users [19]. Due to the impact it can create on vulnerable populations, there should be secure frameworks to verify the users and activities.

2.2 The Tension Between the Metaverse Technology and Data Protection Laws and Regulations

The Metaverse developers would require a vast amount of personal information to create a truly immersive and personalized Metaverse with high accuracy and QoE (e.g., create an immersive and personalise space). However, the collection of such personal data conflicts with the data protection principles for example the data minimisation principle of the General Data Protection Regulation (GDPR) [29]. Therefore, to uphold the privacy factor, taking adequate measures to minimise the amount of data collected is as important as the purpose for which the data is collected. Furthermore, Data protection laws such as GDPR stipulates that data collected for a specific purpose cannot be used for any other purpose [29]. Therefore, it is incumbent on the creators of Metaverse technologies to take into consideration that the collection and use of data is undertaken responsibly and prudently to avoid contravention of data protection regulations. A number of researchers are investigating the use of self-sovereign identity as a development of federated identity management systems, where users have total control over their identity and where it is stored. In these systems, when a user wishes to use a service their credentials are verified by a verifier who has a trust relationship with the issuer of the credentials. Only the minimum credentials are shared. All transactions are conducted on Distributed Ledger Technolgy (DLT), to ensure immutability and transparency [26].

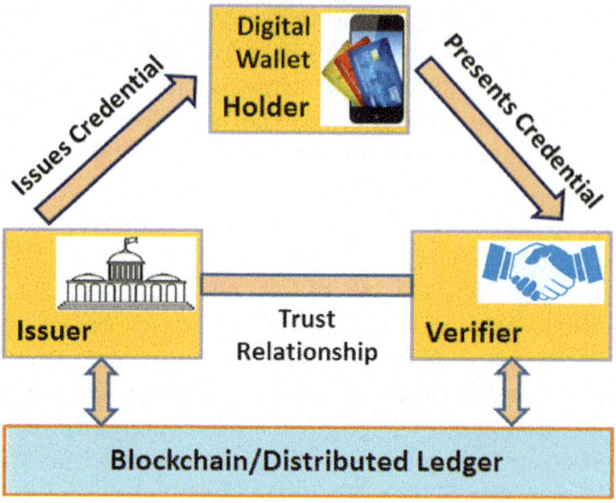

Fig. 1. Self-sovereign ecosystem [26]

In the Metaverse environment, the consumers are permitted to move digital assets and avatars between platforms and across the Metaverse [32]. On that basis, software developers are required to establish bilateral or multilateral data sharing agreements to improve the seamlessness of the consumer experience [32]. For example, when transferring data between territories, all the countries involved in the process should have adequate data protection mechanisms as stated in the GDPR regulatory framework. These conditions will become increasingly difficult to comply with in the Metaverse where data exchange is rapid and involves a multitude of participants. This will conflict with data localization guidelines of different data protection regulations across the world. For example, the Data Protection Bill in India encourages data localization whereas GDPR allows transfer of data provided that the data processes are in compliance with detailed GDPR principles.

The rights of the individuals under data protection regulations are diverse in different regions. For example, the GDPR has clearly stated the rights of the individual residing in the EU/EEA region [30]. However, the Metaverse environment is borderless and it is unclear how it will conform to rights of the individuals. Furthermore, GDPR allows the data subject to request the organisation to delete his/her data except for legal reasons [28], and in such circumstances, the organisations that process European citizen's data must abide by the stipulated requirements. In the borderless nature of the Metaverse environment, the data subject may not be able to enjoy the same rights they have in real life. Therefore, it is important to have alternative arrangements to revisit the GDPR and add measures to protect the rights of the data subjects in the Metaverse.

Due to GDPR regulations, in the real world, the data controller, data processor, and data protection officer are held accountable for the information the organisation collects [31]. However, the lack of control in the Metaverse platform raises questions about the security of the data collected from the users. GDPR clearly states and establishes the rights of the data subject [30], one of which is the right to have the data erased [28] when the collection or other processing of personal data is no longer necessary, or on withdrawal of consent given by the data subject for processing their data. However, in the operation of the Metaverse, adherence to the rights of the individuals had not been feasible in some instances. For example, the right to be forgotten is a fundamental principle of the GDPR [28], however, it may not be possible to honour that requirement in the Metaverse due to difficulties in identifying the parties responsible for data security. Naik and Jenkins [27] discuss the issue of privacy and sovereignty in an environment where the internet is boundless, suggesting a number of key features that all states would be required adhere to in co-operating with one another ensure the privacy and security of individuals.

There are obstacles to practically identifying the characteristic of the Metaverse. For example, if an avatar is verbally abusive in the Metaverse, it is not clear if it is the same as subjecting someone to verbal abuse in real life. Therefore, it is important that the Metaverse lawmakers and regulators adapt clear definitions for the crimes and introduce punishments that fit these crimes. In addition, there are legal issues that surround the application of the law in a way it considers a threat to an avatar in the Metaverse as a threat to an individual, it is not clear in which court the prosecution of the perpetrators would take place. The anonymity and creation of fake avatars would further add to the dilemma and the perpetrators would attempt to evade punishment. Therefore, there should be a mechanism to identify/trace the individuals behind the avatar in the conceptual and design stage of the avatar. In light of the above, it is important to revisit and review the current data protection mechanisms as they were not designed to address some of the challenges and complexities arising from the inception of the Metaverse. Otherwise cyber bullying and crimes will thrive on this platform. The emphasis must be on the requirement to identify the shortcomings and introduce new clauses to ensure the personal privacy of the data subject.

Some countries do not yet have meaningful data protection mechanisms, while others are making progress in developing appropriate mechanisms to strengthen their statute. However, the requirement is for all countries to have unified, universal data protection mechanisms to meet the challenges of upcoming technologies such as the Metaverse. The creation of an international organisation in the form of an association of states wishing to have access to the Metaverse should take collective decisions on the collection, processing, and storing of vast quantities of sensor generated personal data.

2.3 Security Issues of the Metaverse

The Metaverse will be a combination of number of complex technologies. This will add further security issues to the Metaverse platforms and increase the attack surface naturally. They key security issues can arise due to platform automation, authentication, and integrity.

2.3.1 Automation Security Risks As discussed in previous sections, Metaverse will deal with abundance of data from both users and the platform. The scalability and sustainability of these environments will be a significant challenge. Therefore, most of these information flows and processes need to be automated in order to operate the system in real-time. Some of these algorithms will be based on AI and Machine Learning algorithms to provide efficiency, performance and scalability. This could lead to bias decision making which is detrimental when it comes to social and personal data (unfair outcomes and lack of transparency) [1]. Furthermore, the complex AI algorithms such as Deep learning approaches can consume vast amount of energy as well which could create other environmental problems.

2.3.2 Authentication and Integrity Similar to overtaking social media accounts, future Metaverse accounts can be overtaken by either humans or bots. This will create significant personal, societal and economical impact to individuals involved. Advanced AI algorithms could be used to mimic the original avatar in place of the owner similar to its 2D counterpart [4]. Therefore, advanced authentication and integrity checking algorithms should be in place to tackle these issues. In the future users may need to prove that they are not a 3D bot by performing certain actions.

2.4 Implementation of Technical Solutions

The existing management policies and protection measures are inadequate to shield the virtual environment of the Metaverse from cyber-attacks [40] and cybercrime [18], thus it must be enhanced to outfit the features of the Metaverse. For illustration, pseudonymization [37] of private information or a more dynamic access control and fine-grained authentication policy for data is needed. Additionally, sensitive data must be stored securely using encryption to limit the impact of illegal access. Metaverse systems can accumulate more sensitive information than traditional systems, and this can knowingly interrupt user privacy. For illustration, HMDs (Head-mounted displays) with always-on cameras might record video in private spaces and Metaverse headsets with live microphones might record all discussions. Additionally, eye-tracking approach might record what the user looks at [10].

Privacy issues and countermeasures in the Metaverse have been proposed particularly for Avatars [38]. Here, authors highlighted the significance of protecting the privacy of users, suggesting solutions such as multiple cloned avatars, teleportable avatars, and physically invisible avatars, which could recognize user behaviour patterns and execute similar behaviours. The protection of this data should utilise a variety of techniques such as pseudonymization [37], fine-grained authentication [22], encryption [18], and dynamic access control [53] to avoid privacy issues.

During this communication process, only the legitimate receiver should know the content of communication, a third party should not capable of stealing the information. For this purpose, encryption techniques [12,42,62] can be applied where the sender transfers the information after encrypting with the use of a secret key and the authentic receiver decrypts it by using the same key following it's receipt. Thumbnail-preserving encryption can be implemented to reduce the contradiction between visual observability and privacy security [56], which preserves the coarse details of the visual content, but removes the fine one. A cryptographic asset non-fungible token (NFT) [48] cannot be replicated and that can offer a user with proof of ownership of a unique digital asset, e.g., an audio file, an image, or a video. Many domains are investigating the use of NFT by offering digital items that have actual value in real and/or virtual worlds, similar to cosmetic items for a digital image, avatar, video content, music file, and even property or land sold in virtual worlds.

Invisible watermarking is a useful technique in which a specific mark to identify the artefacts is embedded within them [11]. This embedding mark does not affect the visual results of its personal goods in the Metaverse due to it's invisibility and can be removed or detected whenever required. Thus, some approaches, such as authentication [3], content protection [59], and tamper-resistance [54] are realized, which further prevents malicious avatars from illegally copying and stealing goods. Additionally, in comparison to the real-world situation, watermarking is a better "fit" for the Metaverse. In the real world, while applying watermarking it is mandatory to modify the visual content and destroy some physical features. Therefore, the presence of watermarking can be illegitimately detected using some technical means. However, this is not the case in the generative digital scenario [58], where it is difficult to detect by existing means even if the watermarking is applied over it.

An AI technique, a Generative Adversarial Network (GAN) [2], can be used to generate context images for high-quality dynamic game scenarios in the Metaverse. However, it poses security threats such as poisoned and adversarial samples which are difficult to detect. In the existing literature, by employing adversarial samples as a component of training data, numerous efforts have been performed to resist adversarial samples via adversarial reinforcement learning [44], virtual adversarial learning [60], adversarial transfer learning [60] and adversarial representation learning [41], etc., this can be useful to resist adversarial threats in the building of the Metaverse.

In privacy preserving interactive Metaverse game design, Corcoran et al. [17] differentiated the group privacy and individual privacy. The former refers to the privacy related with a group of individuals (for example, an organization, a social group, and a nation) while the latter refers to the behavioural traits, purchasing patterns, image/video data, communication, and location/space associated to an individual. AI algorithms could be employed to identify those harassing others and spreading hate speech and legal mechanisms could possibly be levied to bring perpetrators to justice with suitable fines for ignoring/disobeying privacy regulations. If the Metaverse can develop advanced tools, such offences can be dealt with by restricting or banning people's access after all other measures had been taken.

Images and videos have been extensively examined in Digital forensics, enabling accountability in the Metaverse under consideration. Swaminathan et al. [45] proposed a general forensic technique for digital camera images, in this method post-camera and in-camera image processing produces a series of diverse fingerprint traces. The assessed post-camera fingerprints may be used to validate image authenticity (i.e., whether a particular digital image is from a computer graphics program, particular camera, or scanner). Though, the usage of anti-forensics makes trusted digital forensics complicated, therefore to remove this problem, Stamm et al. [24] presented an automatic video frame addition or deletion forensic technique.

One way to verify the authenticity of digital content in a Metaverse is through the use of a Distributed Ledger Technology (DLT) [43]. For example, Microsoft, the BBC, CBC, and the New York Times are using DLT for content authenticity in their joint project, Project Origin [34]. Similarly, Streambed [35] uses blockchain technology to map content to creators (e.g., people, brands etc.) which assists the content creators to hold greater control over how their content is used/reused and by whom. It is expected that in the coming years, blockchain will be one of the major tools for news and video content authentication.

3 AI in the Metaverse

Privacy and cybersecurity must deliver several measures, techniques and solutions to confirm that systems and users are protected from varied vulnerabilities (e.g., unfair outcomes of Artificial Intelligence (AI) based algorithms) and threats in the Metaverse. Since AI is one of the key technologies to construct the Metaverse [16], its intrinsic vulnerabilities may be inherited by the Metaverse and its constituent parts. For example, AI algorithms have been used to generate and spread fake news, which is particularly undemanding with the emergence of software such as ChatGPT and Goodle Bard. The use of best practice guidelines would similarly assist researchers and developers of Metaverse technologies to comply with standards and provide secure and privacy preserving services over the platform [25].

Distributed Ledger Technology (DLT), offers promising solutions to many security and privacy issues in the Metaverse where AI can be used to detect cyberattacks used in protecting digital assets in the Metaverse. For example, Tanwar et al. (2019) [46], utilised Machine Learning (ML) algorithms and DLT architectures to identify cyberattacks in blockchain networks. Similarly, Fan et al. (2021) [8], employed a Federated Learning (FL) based framework with deep model using average aggregation for privacy preservation of heterogeneous IoT devices while Liu et al. (2021)[23], exploited an FL based framework with Convolutional Neural Network (CNN) model averaging and training to detect malicious attacks in transportation systems.

Many systems are protected by cybersecurity techniques such as the use of an Intrusion Detection System (IDS) or Intrusion Prevention System (IPS), both monitor network traffic on the system searching for know malware patterns. The

Metaverse would be no exception to this, however, as malware products become more complex, they employ a range of stealth techniques to avoid detection. This is especially difficult with completely new virus signatures, which are unknow to many signature databases. As a result, researchers and malware detection systems have turned to the use if ML and AI. As an example, Thockchom et al. (2023) [47], proposed a lightweight ML solution for an IDS, where the ML algorithm examines the network traffic for particular malware profiles. Most of these system utilise a single classifier, using a single ML algorithm, whilst Thockchom et al. [47] propose a hybrid system using a various combinations of ML algorithms, for example using one for pre-processing and another for training and testing the data.

The proposed model uses a combination of well-known and standard ML techniques, namely, Gaussian naive Bayes, logistic regression and decision tree and stochastic gradient descent. Initially a Chi-Square test is used to select the features and ensure that features are independent of one another:

$$x^2 = \sum \frac{(O_i - E_i)^2}{E_i} \tag{1}$$

Following this the base classifiers are applied, for each ML technique, beginning with Gaussian Naïve Bayes, then Logistic Regression, Decision Tree and finally Stochastic Gradient Descent. Making use of the formulas:

$$P(y|x_1 \ldots x_i) = \frac{(Px_1|y)(Px_2|y)\ldots(Px_n|y)(P_y)}{P(x_1)P(x_2)\ldots P(x_n)} \tag{2}$$

In terms of the Decision Tree approach, the Gini index is utilised:

$$G_{ini} = 1 - \sum_{i=1}^{n}(P_i)^2 \tag{3}$$

and finally, Logistic Regression, which is an ML predictive analysis classification algorithm centered on probability. As with all probability function (sigmoid), maps real values onto new values in the range 0 to 1. It is a linear model that is capable of working with large datasets, but less so with non-linear problems.

$$0 <= h0(x) <= 1 \tag{4}$$

Once these stages are complete a meta classifier using a Stochastic Gradient Descent (SGD) function.

These types of ML and AI techniques are beginning to be used in various IDS systems to protect existing platforms, such as Facebook, Twitter and Instagram protecting users from malignant malware and other software aiming to use exfiltration to exploit user data.

One of the most challenging issues is to transform these systems using AI and ML to operate on smaller platforms and in real-time.

4 Recommendations

This section summarises the main recommendations for the Metaverse as discussed in this paper. The recommendations are categorised into four sections as illustrated in the Fig. 2.

4.1 Data

The vast amount of personal data collected through sensors in the Metaverse presents a privacy and security threat not only to the users but additionally to anyone else who might be present in the surroundings. While frameworks [5,14,21] have been proposed which allow users and developers to control the privacy of their sensor-collected data, there is still a requirement for further limiting the collection and distribution of personal data in the Metaverse. Furthermore, it is important that users should be aware of when data relating to them is being collected. One potential solution for this is the use of visual cues when data is collected or transmitted [9]. The data for the Metaverse should be obtained via consent and access to such sensitive data must be minimized. If possible the data used in the Metaverse for interactions should not be linked to living individuals directly or indirectly. This will guarantee that the personal data used within the Metaverse platforms will minimally affect individual rights.

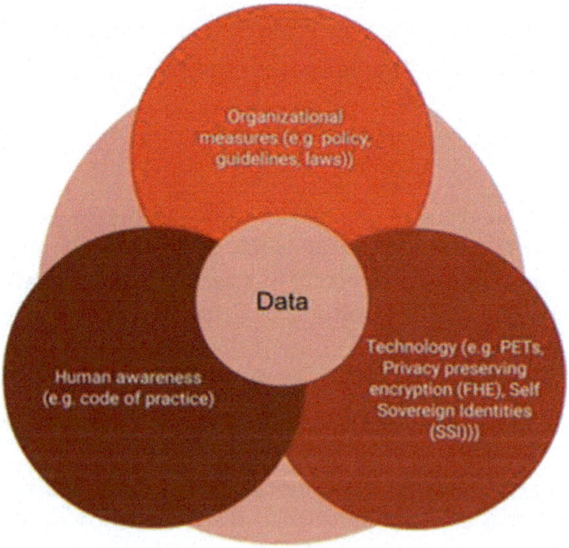

Fig. 2. Main recommendations for the Metaverse

The personal data oriented Metaverse applications should be maintained as private networks to avoid any misuse of personal data. Clear and concise guidelines (e.g. Privacy notice) should be provided on what data is collected, with whom the data is shared and who will process the data upon collection.

4.2 Organisations

In order to protect children and adults from illegal and harmful content online, the UK is set to introduce a new set of laws through the UK Online Safety Bill [33]. The bill will force social media platforms to ensure underage children are not able to access illegal or illicit social media platforms, for example, by using age assurance or verification technologies. Similarly, adults could be provided tool sets by social media platforms which will assist them to reduce the likelihood of seeing harmful or illegal content. Ofcom, which is the communications regulator in the UK, will have the power to act against any violating technology company, even if they are based outside the UK, if their services are accessible to UK users. This has implications for the Metaverse providing services in the UK. For example, in order to fully comply with the Online Safety Bill, the Metaverse creators must provide necessary privacy preserving tools to its users. Similarly, the Data Protection Act (DPA) 2018 controls how users' personal information is used by organisations, businesses or the government. It permits users to establish what information the government or organisation stores about them. In a Metaverse, it may be useful to extend the DPA to include the digital avatars of the users in addition to these bodies.

4.3 Technology

Technological measures are necessary to ensure privacy and security in the Metaverse. For example, in order to minimize the sharing of personal information collected through sensors, either Privacy Enhancing Technologies (PETs) should be used to obscure sensitive data before it is shared with cloud services [9] or in-sensor processing of data should be advocated. Similarly, DLT technologies such as blockchain should be used to provide a decentralized architecture for the Metaverse which can assist minimisation of the risks associated with centralized architectures (such as low transparency and control and by a few entities [51]). Metaverse developers should allow users to configure the privacy settings of their own personal spaces, or privacy bubbles [9].

Most of existing Web 2.0 based social media platforms such as Twitter and Facebook are susceptible to attacks since they hold an abundance of personal data about their users, leading to the creation of data honeypots that are lucrative for hackers. While these centralized infrastructures have been critical in today's web, a decentralized web with Self-Sovereign Identities (SSI), decentralized identifiers (DIDs) and verifiable claims, operating on a backbone infrastructure of distributed ledgers technology, may be the key to a secure online future [25,26]. SSI and DIDs will enable users to hold personal data while the requited information is only passed to the platforms such as the Metaverse. These technologies will return the control of their personal data back to the end users.

The novel encryption algorithms such as Fully Homomorphic Encryption (FHE) [49] will provide privacy preserving data sharing in emerging platforms such as the Metaverse. When the processing takes place in the encrypted domain, the processors or the end users will not be able to see personal data in plain text. In this regard, further advancement in many-to-many encryption approaches (e.g., [20,55]) require further investigation to support multi-user Metaverse environments.

4.4 People

Like any other technology, humans are likely to be the weakest link in the privacy/security chain. The end users should have good awareness of the risks in using the Metaverse technology and mitigation actions to reduce the impact from potential risks. In a Metaverse, people should be permitted to define codes of practice (and the ability to restrict access) within their own spaces. Further research is required in defining best practice guidelines for users. Finally, users should be made aware of their rights and responsibilities when using the Metaverse. Recommendations related to people are summarized below:

- Code of practice: This should be led by the users and supported by the technology providers.
- Best practice guidelines: The best practices should be shared among the users of the Metaverse to avoid any common exposures or vulnerabilities. This will create a community of safe users with the technology.
- User awareness: User awareness is the key. If the users are aware of the ways their privacy is going to be exposed through technology, they will be proactive and take precautionary actions.

5 Conclusions

Metaverse is a web of evolving relationships in a complex environment, and it is difficult to identify who is responsible for privacy in the Metaverse and, who has control over it whether in part or in its entirety. Then the question arises as to who would be responsible for protecting data and what would happen to the vast amount of data collected through the sensors. Furthermore, given the complex relationships that entangle the Metaverse, the dilemma is to ascertain how data protection will function. Therefore, it is important that the Metaverse environment creators establish procedural and technical criteria for securing data, before introducing the Metaverse to users.

References

1. Metaverse: Security and Privacy Issues (2021). https://doi.org/10.1109/TPSISA52974.2021.00032
2. Association of Computing Manufacturers: NFTGAN: Non-Fungible Token Art Generation Using Generative Adversarial Networks (2022). https://doi.org/10.1145/3529399.3529439
3. Bhattacharyya, D., Ranjan, R., Alisherov, F., Choi, M.: Bio-metric authentication: a review. Int. J. U-and E-Serv. Sci. Technol. **2**(3), 13–28 (2009)
4. Boneh, D., Grotto, A., McDaniel, P., Papernot, N.: How relevant is the turing test in the age of sophisbots? IEEE Secur. Priv. **17**(6), 64–71 (2019)
5. De Guzman, J.A., Thilakarathna, K., Seneviratne, A.: Security and privacy approaches in mixed reality: a literature survey. ACM Comput. Surv. (CSUR) **52**(6), 1–37 (2019)
6. Di Pietro, R., Cresci, S.: Metaverse: security and privacy issues. In: 2021 Third IEEE International Conference on Trust, Privacy and Security in Intelligent Systems and Applications (TPS-ISA), pp. 281–288. IEEE (2021)
7. Dwivedi, Y.E.A.: Metaverse beyond the hype: Multi-disciplinary perspectives on emerging challenges, opportunities and agenda for research practice and policy. Int. J. Inform. Manage. **66** (2022)
8. Fan, S., Zhang, H., Zeng, Y., Cai, W.: Hybrid blockchain-based resource trading system for federated learning in edge computing. IEEE Internet Things J. **8**, 2252–2264 (2021)
9. Fernandez, C., Hui, P.: Life, the metaverse and everything: an overview of privacy, ethics, and governance in metaverse. In: 2022 IEEE 42nd International Conference on Distributed Computing Systems Workshops (ICDCSW), pp. 272–277. IEEE (2022)
10. Fineman, B., Lewis, N.: Securing your reality: addressing security and privacy in virtual and augmented reality applications. https://er.educause.edu/articles/2018/5/securing-your-reality-addressing-security-and-privacy-in-virtual-andaugmented-reality-applications
11. Garg, P., Jain, A.: A robust technique for biometric image authentication using invisible watermarking. Multimedia Tools Appl. **82**(2), 2237–2253 (2023)
12. Hasan, M., et al.: Lightweight encryption technique to enhance medical image security on internet of medical things applications. IEEE Access **9**, 47731–47742 (2021)
13. Hinduja, S.: The Metaverse: Opportunities, Risks and Harms (2022). https://cyberbullying.org/metaverse
14. Hu, J., Iosifescu, A., LiKamWa, R.: Lenscap: split-process framework for fine-grained visual privacy control for augmented reality apps. In: Proceedings of the 19th Annual International Conference on Mobile Systems, Applications, and Services, pp. 14–27. ACM (2021)
15. Huang, Y., Li, Y.J., Cai, Z.: Security and privacy in the metaverse: a comprehensive survey. Big Data Mining Anal. **6**(2), 234–247 (2023)
16. Huynh-The, T., Pham, Q., Pham, X., Nguyen, T., Han, Z., Kim, D.: Artificial intelligence for the metaverse: a survey (2022)
17. IEEE: A Privacy Framework for Games and Interactive Media (2018)
18. Kaur, M., Kumar, V.: A comprehensive review on image encryption techniques. Arch. Comput. Methods Eng. **27**(1), 15–43 (2020)

19. Kreitmeir, D., Raschky, P.: The unintended consequences of censoring digital technology evidence from Ialy's chatgpt ban (2023)
20. Kumar, S., Hu, Y., Andersen, M., Popa, R., Culler, D.: Jedi: Many-to-many end-to-end encryption and key delegation for Iot. In: Proceedings of the 28th USENIX Security Symposium (USENIX Security), pp. 1519–1536. USENIX Association (2019). https://www.usenix.org/system/files/sec19-kumar-sam.pdf
21. Lebeck, K., Ruth, K., Kohno, T., Roesner, F.: Securing augmented reality output. In: 2017 IEEE Symposium on Security and Privacy (SP), pp. 320–337. IEEE (2017)
22. Lin, C., He, D., Huang, X., Choo, K., Vasilakos, A.: Bsein: A blockchain-based secure mutual authentication with fine-grained access control system for industry 4.0. J. Netw. Comput. Appl. **116**, 42–52 (2018)
23. Liu, H., Zhang, S., Zhang, P., Zhou, X., Shao, X., Pu, G., Zhang, Y.: Blockchain and federated learning for collaborative intrusion detection in vehicular edge computing. IEEE Trans. Veh. Technol. **70**, 6073–6084 (2021)
24. Stamm, M.C., Lin, W., Liu, K.: Temporal forensics and anti-forensics for motion compensated video. IEEE Trans. Inform. Forensics Secur. **7**(4), 1315–1329 (2012)
25. Naik, N., Jenkins, P.: Governing principles of self-sovereign identity applied to blockchain enabled privacy preserving identity management systems. In: IEEE International Symposium on Systems Engineering (ISSE), pp. 1–6 (2020)
26. Naik, N., Jenkins, P.: Your identity is yours: Take back control of your identity using gdpr compatible self-sovereign identity. In: 7th International Conference on Behavioral and Social Computing pp. 1–6 (2020)
27. Naik, N., Jenkins, P.: Is self-sovereign identity really sovereign? In: 2022 IEEE International Symposium on Systems Engineering (ISSE). IEEE (2022)
28. n.d.: Art. 17 gdpr- right to erasure ('right to be forgotten') (online). https://gdpr-info.eu/art-17-gdpr/
29. n.d.: Art. 5 gdpr-principles relating to the processing of personal data. https://gdpr-info.eu/art-5-gdpr/
30. n.d.: Chapter 3- rights of the data subject (online). https://gdpr-info.eu/chapter-3/
31. n.d.: Chapter 4 controller and processor (online). https://gdpr-info.eu/chapter-4/
32. n.d.: The metaverse: The evolution of a universal digital platform (2021). https://www.nortonrosefulbright.com/en-de/knowledge/publications/5cd471a1/the-metaverse-the-evolution-of-a-universal-digital-platform
33. n.d.: Online safety bill (2022). https://bills.parliament.uk/bills/3137
34. n.d.: Protecting trusted media. https://www.originproject.info/
35. n.d.: Protecting trusted media. https://www.streambedmedia.com/
36. n.d.: Metaverse app allows kids into virtual strip clubs - bbc news (2023). https://www.youtube.com/watch?v=3n9iXlA7HJk
37. Neubauer, T., Heurix, J.: A methodology for the pseudonymization of medical data. Int. J. Med. Informatics **80**(3), 190–204 (2011)
38. Neubauer, T., Heurix, J.: The social metaverse: battle for privacy. IEEE Technol. Soc. Mag. **37**(2), 52–61 (2018)
39. Park, S., Kim, Y.: A metaverse: taxonomy, components, applications and open challenges. IEEE Access **10**, 4209–4251 (2022)
40. Rid, T., Buchanan, B.: Attributing cyber attack. J. Strateg. Stud. **38**, 4–39 (2014)
41. Mai, S., Hu, H., Xing, S.: Modality to modality translation: An adversarial representation learning and graph fusion network for multimodal fusion. Proc. AAAI Conf. Artif. Intell. **34**(1), 164–172 (2020)

42. Seth, B., Dalal, S., Jaglan, V., Le, D., Mohan, S., Srivastava, G.: Integrating encryption techniques for secure data storage in the cloud. Trans. Emerg. Telecommun. Technol. **33**(4), e4108 (2022)
43. Springer: Distributed ledger technology. In Internet Computing (2020)
44. Sun, J., Zhang, T., Xie, X., Ma, X., Zheng, Y., Chen, L., Liu, Y.: Stealthy and efficient adversarial attacks against deep reinforcement learning. In: Proceedings of the AAAI Conference on Artificial Intelligence, vol. 34, pp. 5883–5891 (2022)
45. Swaminathan, A., Wu, M., Liu, K.: Digital image forensics via intrinsic fingerprints. IEEE Trans. Inf. Forensics Secur. **3**(1), 101–117 (2008)
46. Tanwar, S., Bhatia, Q., Patel, P., Kumari, A., Singh, P., Hong, W.: Machine learning adoption in blockchain-based smart applications: the challenges, and a way forward. IEEE Access **8**, 474–488 (2019)
47. Thockchom, N., Singh, M., Nand, U.: A novel ensemble learning-based model for network intrusion detection (2023). https://doi.org/10.1007/s40747-023-01013-7
48. Wang, Q., Li, R., Wang, Q., Chen, S.: Non-fungible token (nft): Overview, evaluation, opportunities and challenges
49. Wang, W., Hu, Y., Chen, L., Huang, X., Sunar, B.: Exploring the feasibility of fully homomorphic encryption. IEEE Trans. Comput. **64**(3), 698–706 (2015)
50. Wang, Y.: A survey on metaverse: fundamentals, security and privacy. IEEE Commun. Surv. Tutor. **25**(1), 319–352 (2023)
51. webfoundation.org: Online abuse in the metaverse untangled (online) (2022). https://webfoundation.org/2022/02/online-abuse-in-the-metaverse-untangled
52. Xu, M.: A full dive into realizing the edge-enabled metaverse: visions, enabling technologies and challenges. IEEE Commun. Surv. Tutor. **25**(1), 656–700 (2023)
53. Xu, S., Yang, G., Mu, Y., Deng, R.: Secure fine-grained access control and data sharing for dynamic groups in the cloud. IEEE Trans. Inf. Forensics Secur. **13**(8), 2101–2113 (2018)
54. Xu, S.E.A.: A secure EMR sharing system with tamper re-sistance and expressive access control. IEEE Trans. Depend. Secure Comput. **20**(1), 53–67 (2023)
55. Zhang, J., Zhong, H., Cui, J., Xu, Y., Liu, L.: Smaka: secure many-to-many authentication and key agreement scheme for vehicular networks. IEEE Trans. Inf. Forensics Secur. **16**, 1810–1824 (2021)
56. Zhang, Y., Zhao, R., Xiao, X., Lan, R., Liu, Z., Zhang, X.: Hf-tpe: High-fidelity thumbnail-preserving encryption. IEEE Trans. Circ. Syst. Video Technol. **32**(3), 947–961 (2022)
57. Zhang, Z., Ning, H., Shi, F.: Artificial intelligence in cybersecurity: research advances, challenges and opportunities. Artif. Intell. Rev. **55**, 1029–1053 (2022)
58. Zhao, R., Zhang, Y., Zhu, Y., Lan, R., Hua, Z.: Metaverse: Security and Privacy Concerns
59. Zhaofeng, M., Weihua, H., Hongmin, G.: A new blockchain-based trusted DRM scheme for built-in content protection. EURASIP J. Image Video Process. **1**, 1–12 (2018)
60. Zheng, H., Zhang, Z., Gu, J., Lee, H., Prakash, A.: In: Efficient adversarial training with transferable adversarial examples. In: 2020 IEEE/CVF Conference on Computer Vision and Pattern Recognition (CVPR), pp. 1178–1187 (2020)
61. Zhou, Y., Xiao, X., Chen, C., Zhao, X., Chen, J.: Self-powered sensing technologies for human metaverse interfacing. Joule **6**(7), 1381–1389 (2022)
62. Zia, U., M. McCartney, M., Scotney, B., Martinez, J., AbuTair, M., Memon, J., Sajjad, A.: Survey on image encryption techniques using chaotic maps in spatial, transform and spatiotemporal domains. Int. J. Inform. Secur. **21**(4), 917–935 (2022)

Machine Learning Based XSS Attacks Detection Method

Korrawit Santithanmanan[1], Khwunta Kirimasthong[1](✉), and Tossapon Boongoen[2]

[1] Center of Excellence in AI and Emerging Technologies, School of Information Technology, Mae Fah Luang University, Chiang Rai, Thailand
6451501004@lamduan.mfu.ac.th, Khwunta.kir@mfu.ac.th

[2] Department of Computer Science, Aberystwyth University, Ceredigion SY23 3DB, UK
tob45@aber.ac.uk

Abstract. Cross-Site Scripting (XSS) attacks pose a significant threat to web applications by allowing the attacker to use the XSS to inject malicious code, typically JavaScript, and send it to other users in the form of a URL to identify which URLs are malicious. The attackers inject and execute arbitrary code within a user's browser, potentially leading to unauthorized access and data theft. Therefore, the objective of this paper is to propose machine learning-based methods for detecting XSS attacks focusing on URLs that follow domain names were non-alphanumeric characters that can appear in Javascript by using k-NN, Decision Tree, SVM, and Gaussian Naive Bayes classification model. This information aids in selecting the most suitable model for real-world deployment, ensuring efficient and reliable detection of XSS attacks in web applications. By training the models on a diverse dataset containing both benign and malicious scripts, they learn to differentiate between safe and malicious code, enhancing the accuracy of detection to find the best model for detecting websites that will inject scripts or 33 non-alphanumeric characters and characters that can appear in Javascript that possibly steal sensitive information about victims. The evaluation results reveal the performance of each model in terms of its ability to identify and classify malicious URLs accurately.

Keywords: Cross-site scripting · XSS attack · Machine learning · k-NN · Decision tree · SVM · Gaussian naive bayes

1 Introduction

Web security is a critical concern in the digital age, with cyber-attacks posing a constant threat to the confidentiality and integrity of online systems. One prevalent vulnerability problem is Cross-Site Scripting (XSS), where attackers inject malicious code or scripts into websites to steal sensitive user information [1]. Hacker is popularly used by taking advantage of web vulnerabilities such as web boards, social media, and emails are particularly dangerous [2]. By tricking users into entering important information such as usernames, passwords and sending them back to hackers. When hackers gain information, they will spoof information by using the user's information to log in to various websites, causing a lot of damage to the user.

N. Naik et al. (Eds.): UKCI 2023, AISC 1453, pp. 418–429, 2024.
https://doi.org/10.1007/978-3-031-47508-5_33

Machine learning-based vulnerability prediction models have gained popularity in web security to combat this issue. XSS attacks exploit vulnerabilities in web applications or browsers, tricking unsuspecting users into executing malicious code by visiting compromised websites [3]. And can be triggered by unsuspecting users simply by visiting a compromised website [4], such as information theft, sensitive information (e.g., passwords, and credit card numbers) [5], website redirection, or even seizing control of the user's device [6]. The consequences of XSS attacks can be severe for individuals and organizations [4]. XSS attacks can be categorized into three main types: [7].

Reflect XSS. Cross-Site Scripting (XSS) is a common security vulnerability in web applications, particularly in search boxes. The principle behind XSS attacks in search boxes is that the attacker injects malicious JavaScript code into an input field or URL, which is then executed by the victim's browser [7, 8]. This can result in sensitive information being stolen or manipulated. It is important for developers to properly validate and sanitize user input to prevent XSS attacks.

Stored XSS. In a stored cross-site scripting (XSS) attack, the attacker injects malicious JavaScript code into a web application's database. When other users access the vulnerable web application, they inadvertently execute the malicious code, resulting in sensitive information being stolen or manipulated. Stored XSS is commonly found in forums or web boards, where an attacker may insert malicious JavaScript into a reply to a topic. When other users click to view the thread, their browser will execute the JavaScript code in the response, as specified by the attacker [7, 8].

DOM-Based XSS. This is an XSS vulnerability that is harder to detect than Reflected XSS and Stored XSS because DOM-based XSS does not need to rely on the back end to render JavaScript to users. When a web browser successfully loads the HTML of the webpage creates what is known as the Document Object Model (DOM), the document object on that page, JavaScript can modify the object in the HTML. An attacker exploits a JavaScript insertion vulnerability, and this type of attack is known as a DOM-based XSS [7, 9].

Machine learning has emerged as a powerful approach in the field of web security, offering the potential to detect XSS vulnerabilities with greater accuracy and efficiency compared to manual inspection methods [10, 11]. However, by leveraging machine learning models trained on extensive datasets, it becomes feasible to predict the susceptibility of a website to XSS attacks with a high level of precision. This application of machine learning holds significant value in protecting users from cyber-attacks and preserving the security of their sensitive information [11].

The objective of this paper is to develop and evaluate a machine learning-based method for detecting XSS attacks, primarily focusing on the URL that will be related to the type of Reflect XSS and DOM-Based XSS because the attacker will try to inject scripts into URLs. The method should possess the ability to accurately identify the danger of URLs and demonstrate robustness in handling various attack types and patterns. A comprehensive analysis, including a comparison with existing methods, should be conducted to evaluate the proposed method's performance, effectiveness, and efficiency. Ultimately, the thesis aims to provide a reliable and efficient solution for detecting XSS attacks by utilizing machine learning models.

2 Related Works

2.1 XSS Attack

Cross-site scripting (XSS) is a severe form of cyber-attack where malicious code is injected into a website, enabling its execution on other users' browsers [12]. Such attacks can lead to significant consequences, including the theft of sensitive information, website defacement, and malware propagation [12]. Understanding the characteristics and techniques of XSS attacks is crucial for effective prevention and defense strategies.

There exist different types of XSS attacks. Reflective XSS attacks involve injecting malicious code into a website, which is then reflected to the user, often in error messages or search results [13]. Persistent XSS attacks, on the other hand, involve injecting code that remains on the server, such as in forum posts or comments [13]. DOM-based XSS attacks manipulate the website's Document Object Model (DOM) to inject malicious code [14]. XSS attacks can be executed through various means, including phishing emails, malicious websites, exploiting vulnerabilities in web applications, or social engineering attacks [15]. They can also be utilized as components of larger attack campaigns, such as distributed denial of service (DDoS) attacks or data exfiltration [12].

To prevent XSS attacks, it is important for websites to properly sanitize user input and implement appropriate input validation controls [12]. It is also important to regularly update and patch the website to fix any identified vulnerabilities [16].

2.2 XSS of Detection

Cross-site scripting (XSS) is a critical web security vulnerability that enables attackers to inject malicious code into websites, posing risks such as information theft, website defacement, and malware propagation [12]. Effective detection and prevention of XSS vulnerabilities are therefore paramount.

Various methods exist for detecting XSS vulnerabilities. Static analysis tools examine website source code to identify potential vulnerabilities, while dynamic analysis tools monitor website behavior during execution to detect vulnerabilities [17, 18]. These automated approaches offer broad coverage but may generate false positives. Manual testing, involving human testers injecting code to identify vulnerabilities, offers higher accuracy but is time-consuming and may not cover all potential vulnerabilities [19].

This paper builds upon the idea of using machine learning to detect XSS attacks, as presented in "Detecting Cross-Site Scripting Attacks Using Machine Learning" [20]. The study utilizes Linear Kernel Support Vector Machines (SVM), Polynomial Kernel SVM, K-Nearest Neighbors (k-NN), and Random Forest classifiers to predict two classes: benign and malicious [20, 22]. The experiment's results demonstrate excellent performance, with k-NN, Polynomial Kernel SVM, Random Forest, and Linear Kernel SVM classifiers achieving accuracies of 99.75%, 99.60%, 99.50%, and 96.32% respectively [20, 21] (refer to Table 1).

Overall, a combination of static and dynamic analysis tools, along with manual testing, can provide the most comprehensive approach to detecting XSS vulnerabilities [17]. It is also important to regularly update and patch the website to fix any identified vulnerabilities [16].

Table 1. The results of this experiment by focusing on each performance were great the accuracy of k-NN, polynomial kernel SVM, random forest, and linear kernel SVM classifiers [20, 21].

Classifier	Linear (%)	Polynomial (%)	k-NN (%)	Random Forest (%)
Accuracy rate	96.32	99.60	99.75	99.50
Precision rate	98.33	99.69	99.88	99.84
Sensitivity (TPR)	94.53	99.22	99.61	99.15
Specificity (TNR)	98.26	99.69	99.88	99.84

3 Methodology

3.1 Overview of the Detection Method for XSS Attacks

Data collection is an important step for training and evaluating machine learning models. The data used for this purpose is typically a dataset of web pages and payloads, both legitimate and malicious, which are used to train the machine learning models to recognize patterns and features that indicate an XSS attack. The resulting dataset is then used to train and evaluate the machine learning models, with the goal of developing a detection method that can accurately identify XSS attacks in real-world web applications. Therefore, after training machine learning until finding the best machine learning model, the model was combined with the Python program. According to the step shown in Fig. 1.

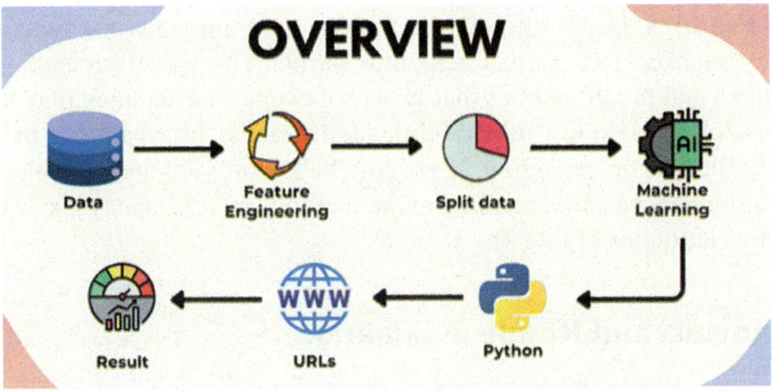

Fig. 1. Overview of the detection method for XSS attacks

3.2 Experimental Dataset

The dataset contains 65 features of non-alphanumeric (Boolean) of non-alphanumeric as shown in Table 2. The dataset in this paper consists of two gathered datasets that include both malicious and benign URLs. The total number of this dataset is 43,218 records, 28,068 records for benign scripts, and 15,150 records for malicious scripts. For the malicious (obfuscated) scripts, most of them came from XSSed.com, the world's largest online library of XSS-vulnerable websites. Dmoz and ClueWeb09 provided safe scripts [23]. The combination of those two sources of data was used in the training and test set.

3.3 Data Pre-processing

This step is important because it can help improve the performance of the machine learning model by transforming the data into a more suitable format. Using one-hot encoding to transform 33 non-alphanumeric characters to binary with values 0 or 1 only. Machine learning algorithms typically work with numerical data, so categorical features need to be transformed into a numerical format for these algorithms to process them effectively.

Structural Features. The entire set of non-alphanumeric characters that can appear in JavaScript code and are known for their specific structural properties. These characters can be present in any JavaScript script, but attackers may use specific strategies to circumvent the security measures of web applications, altering the character range used in their malicious scripts. For instance, malicious scripts may include blank lines or meaningless symbols, such as " < \ sc ri pt >," inserted between commands or tags to evade detection. In JavaScript, there are 33 non-alphanumeric characters, along with 5 possible combinations, resulting in a total of 38 structural features [20]. These features play a significant role in the context of Cross-Site Scripting (XSS) attacks, where attackers attempt to inject malicious scripts into web pages viewed by other users as shown in Table 4.

Behavioral Features. JavaScript refers to the various commands and functions that can be utilized by attackers to construct malicious scripts. These features encompass a wide range of actions and functionalities that JavaScript supports, and they play a crucial role in enabling attackers to exploit vulnerabilities and steal the information of the victim. The present study focuses on analyzing 21 specific behavioral features, detailed in Table 5, to gain a comprehensive understanding of the methods attackers may use in carrying out their malicious activities (Table 3).

4 Experiments and Results Evaluation

4.1 Experiments

The experiments were conducted using a Jupyter notebook, providing an effective working environment for this data science project. The dataset comprised 43,218 records, with 10,000 set aside as unseen data to rigorously test model accuracy. Approximately 10% of noisy data (3,322 records) was added to the training set using noise augmentation, a widely employed technique in machine learning and data analysis to enhance dataset diversity and model robustness. Careful consideration was given to achieve an optimal noise level, as excessive noise can negatively impact model performance. The remaining 36,540 records were used for training and testing, employing a 70:30 split. Four

Table 2. Dataset

Feature name	Value type
Contains &lt	Boolean
ScripTag	Boolean
Readable	Boolean
Contains " ><	Boolean
Contains ' ><	Boolean
Contains And	Boolean
Contains Percentage	Boolean
Contains BackSlash	Boolean
Contains Plus	Boolean
Contains Document	Boolean
Contains Window	Boolean
Contains Onload	Boolean
Contains Onerror	Boolean
Contains DIV	Boolean
Contains iframe	Boolean
Contains img	Boolean
Contains SRC	Boolean
Containss Var	Boolean
Contains Eval	Boolean
Contains href	Boolean
Contains Cookie	Boolean
Contains StringfromCharCode	Boolean
Contains Single Quote	Boolean
Contains Question Mark	Boolean
Contains Exclamation Mark	Boolean
Contains Semicolon	Boolean
Contains HTTP	Boolean
Contains JS	Boolean
Contains Equal	Boolean

classifiers (k-NN, Decision tree, Support Vector Machine, and Gaussian Naive Bayes) and 10-fold cross-validation was utilized to determine optimal model settings. Confusion matrices obtained from the test phase using the four techniques were analyzed to assess model performance and predictive capabilities. Table 5 shows confusion matrices obtained for the test phase, with four different machine learning techniques.

Table 3. Dataset (cont.)

Feature name	Value type
Contains Open Bracket	Boolean
Contains Close Bracket	Boolean
Contains Duble Bracket	Boolean
Contains Dollar	Boolean
Contains Open Parenthesis	Boolean
Contains Close Parenthesis	Boolean
Contains Asterisk	Boolean
Contains Comma	Boolean
Contains Hyphen	Boolean
Contains Less Than	Boolean
Contains Greater Than	Boolean
Contains At	Boolean
Contains Underscore	Boolean
Contains location	Boolean
Contains Search	Boolean
Contains &#	Boolean
Contains Colon	Boolean
Contains Dots	Boolean
Contains Open Brace	Boolean
Contains Close Brace	Boolean
Contains tilde	Boolean
Contains Spase	Boolean
Contains Qutions	Boolean
Contains Grave	Boolean
Contains Duble Equals	Boolean
Contains Duble Slash	Boolean
Contains Vertical Bar	Boolean
Contains Power	Boolean
Contains Broken Bar	Boolean
Contains Alert	Boolean
Contains Break Line	Boolean
Letters Ratio	2-degits decimal number

(continued)

Table 3. *(continued)*

Feature name	Value type
Numbers Ratio	2-degits decimal number
Symbols Ratio	2-degits decimal number
Class	Binary

Table 4. Structural Features

Feature	Terms
Punctuation	&, %, /, \, +,', ?, !,;, #, =, [,], $, (,), ^, *,,, -,, @, _,:, {,}, ~,., space, |, ¦,”
Punctuation combination	><, [], = =, &#

Table 5. Behavioral features

Feature	Description
Readability	Is the script readable - the number of alphabetical characters
Objects	Document, window, iframe, location, This
Events	createelement, String.fromCharCode, Search
Tags	DIV, IMG, < script
Attributes	SRC, Href, Cookie
SRC, Href, Cookie	Var
Functions	eval()
Protocols	HTTP
External File	.js file

Those confusion matrices presented in the previous table can indeed be used to calculate accuracy, precision, recall and F1 scores, which are provided in Table 6 (Fig. 2).

Table 6. Behavioral features

	Machine learning models							
	KNN		Decision tree		SVM		Gaussian NB	
	Benign	Malicious	Benign	Malicious	Benign	Malicious	Benign	Malicious
Benign	7356	135	7211	280	7321	170	6700	791
Malicious	173	3298	402	3069	318	3153	154	3317

Table 7. Data Evaluation

	Machine learning models			
	KNN (%)	Decision tree (%)	SVM (%)	Gaussian NB (%)
Accuracy	97.2	93.8	95.5	91.4
Precision	96.1	91.6	94.9	80.7
Recall	95.0	88.4	90.8	95.6
F1 score	95.5	90.0	92.8	87.5

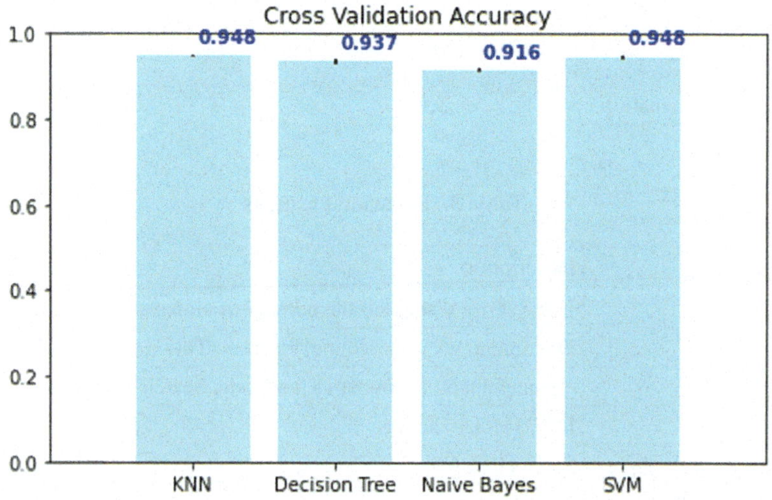

Fig. 2. Cross validation accuracy of 4 models that show accuracy and forecast time, these findings suggest that the k-NN classifier could be beneficial for detecting XSS attack in Table 7.

The experimental results presented in Fig. 3 showcase the evaluation of the best model, k-NN, using unseen data consisting of 10,000 records. The outcome demonstrates a favorable performance as both false positive and false negative instances are minimal, indicating the k-NN prediction model's proficiency in minimizing errors. Consequently, the k-NN model exhibits an impressive accuracy rate of 99.99%.

4.2 Results Evaluation

Upon completion of the model testing, it was determined that the k-NN model exhibited the highest accuracy among the four models. Consequently, the k-NN model was integrated into a Python program. The program operates by allowing users to input a URL in the designated "Put URL here" field. Upon submission, the program performs calculations to determine whether the website is classified as malicious or benign. Additionally, special characters present in the URL are displayed (Fig. 4).

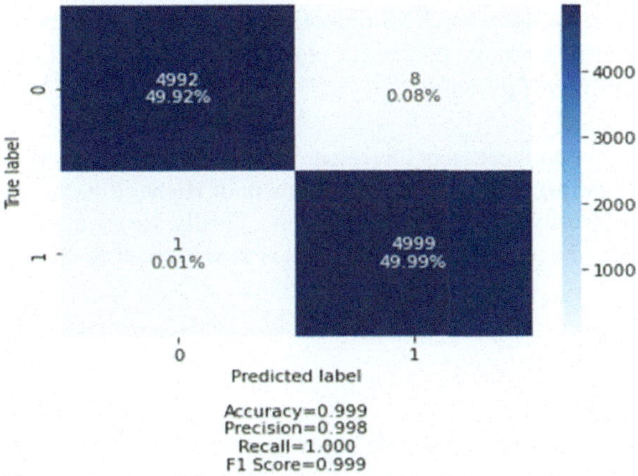

Fig. 3. This confusion matrix is k-NN because from the result k-NN has the highest accuracy of the 4 models.

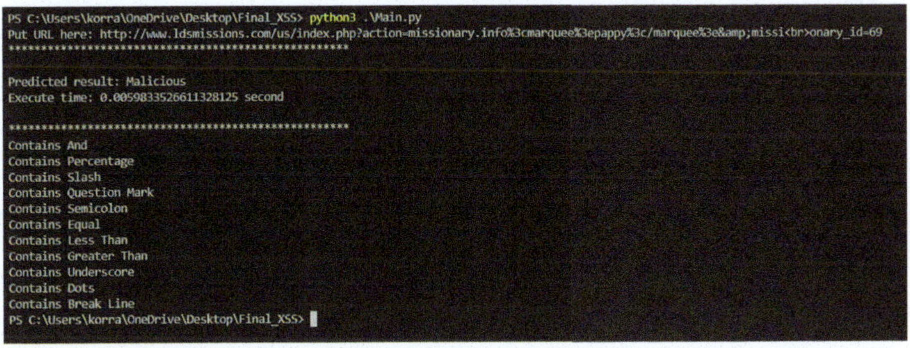

Fig. 4. The insecure URL will show as malicious as shown in the picture.

5 Conclusion

Cross-Site Scripting (XSS) poses a significant threat to the security of Internet web pages, exploiting vulnerabilities to inject malicious JavaScript code into URLs. This paper focuses on enhancing detection accuracy. Four machine learning models (k-NN, Decision tree, Gaussian NB, and SVM) were employed to analyze current insecurity URLs and steal sensitive user information. The study's findings demonstrate the effectiveness of these models, particularly the k-NN model, which achieved a remarkable accuracy rate of 99.9% and a rapid prediction time of 3.3689 s. However, it was found from the experimental results that all classification algorithms give a high accuracy and similar values which are caused by a small amount of experimental dataset. In future work, several avenues of research and improvement can be explored to further enhance the accuracy and efficiency of XSS detection systems such as developing real-time XSS detection systems that can analyze URLs on the fly as they are accessed by users or consider exploring deep learning techniques, such as recurrent neural networks (RNNs) or transformer-based models, for XSS detection. These approaches have shown significant success in natural language processing tasks and could potentially capture complex patterns in URLs more effectively.

Acknowledgment. The publication of this research work is supported by Mae Fah Luang University. This is a part of the project funded by the Ministry of Higher Education, Science, Research, and Innovation (Big security data fusion and analysis). Finally, we are most thankful for Mereani, F. A. and Howe, J. M. who provided the dataset that is used in creating a machine learning model.

References

1. Zhang, Y., Liu, Y.: A machine learning based approach for cross-site scripting vulnerability detection. Secur. Commun. Netw. **9**(14), 2678–2690 (2016)
2. Cai, M., Wang, L.: Cross-site scripting vulnerability prediction using machine learning techniques. Comput. Sci. Inf. Syst. **12**(2), 559–573 (2015)
3. Du, C.D., Zhang, Y., Wang, L., Ren, K.: Cross-site scripting (XSS) attack detection: a survey. IEEE Access **7**, 1–20 (2019)
4. Shears, T.R., Smith, C.L.: Web application security: the state of the art. ACM Comput. Surv. **52**(3), 1–30 (2019)
5. Li, X.J., Chen, L., Zhang, Y.: Cross-site scripting (XSS) attack detection: a review. ACM Comput. Surv. **53**(3), 1–34 (2020)
6. Magharei, S., Gorji, M., Rezaie, H.M.: A survey of cross-site scripting (XSS) attack detection techniques. J. Comput. Sci. **15**(7), 839–854 (2019)
7. Hiremath, O.: The different types of cross-site scripting (XSS). Sqreen Blog. https://blog.sqreen.com/types-of-cross-site-scripting-xss/. Accessed 18 December 2021
8. McKeever, O.: The different types of cross-site scripting (XSS). Sqreen Blog. https://blog.sqreen.com/types-of-cross-site-scripting-xss/. Accessed 18 December 2021
9. Nidecki, T.A.: Dom XSS: an explanation of dom-based cross-site scripting. Acunetix. https://www.acunetix.com/blog/articles/dom-xss-explained//. Accessed 18 December 2021
10. Gupta, R., Rana, R.: A machine learning approach for detecting cross-site scripting vulnerabilities. J. Inf. Secur. Appl. **37**, 69–78 (2018)
11. Wang, L., Cui, H., Liu, J.: A hybrid feature selection and classification model for cross-site scripting vulnerability prediction. Expert Syst. Appl. **87**, 167–176 (2017)
12. W3C.: Cross-Site Scripting (XSS), https://www.w3.org/OWASP/TopTen/2017-Top-10/A3-Cross-Site-Scripting. Accessed 17 April 2022
13. Ravishankar, C., Kumar, A., Lokesh, B.: A survey on cross-site scripting (XSS) attack and prevention techniques. J. King Saud Univ.-Comput. Inf. Sci. **30**(3), 277–287 (2018)
14. DOM-based cross-site scripting. https://www.owasp.org/index.php/DOM_Based_XSS. Accessed 11 July 2022
15. Nash, D.: Cross-site scripting (XSS) attacks. https://www.owasp.org/index.php/Cross-site_Scripting_(XSS). Accessed 18 December 2022
16. Houde, S., Tremblay, S., Ksentini, A.: A survey of cross-site scripting (XSS) prevention techniques. IEEE Access **6**, 52788–52804 (2018)
17. Sajjad, M., Aqeel, H., Muhammad, R.: Security testing of web applications: a survey. IEEE Access **5**, 23197–23212 (2017)
18. Güzel, B., Can, E., Akkuş, M.: A survey of dynamic analysis techniques for web applications. Comput. Secur. **61**, 21–37 (2016)

19. Krukowski, M., Zaremba, M., Dabrowski, D.: A review of XSS prevention techniques for web applications. Comput. Sci. Rev. **22**, 1–18 (2017)
20. Mereani, F.A., Howe, J.M.: Detecting cross-site scripting attacks using machine learning. In: The International Conference on Advanced Machine Learning Technologies and Applications (AMLTA2018), pp. 200–210 (2018a). https://doi.org/10.1007/978-3-319-74690-6_20
21. Mereani, F.A., Howe, J.M.: Preventing cross-site scripting attacks by combining classifiers. In: Proceedings of the 10th International Joint Conference on Computational Intelligence (2018). https://doi.org/10.5220/0006894901350143
22. Gogoi, B., Ahmed, T., Saikia, H.K.: Detection of XSS attacks in web applications: a machine learning approach. Int. J. Innov. Res. Comput. Sci. & Technol. **9**(1), 1–10 (2021). https://doi.org/10.21276/ijircst.2021.9.1.1
23. Falaka, B., Saputra, R.E., Setianingsih, C., Murti, M.A.: Sea wave detection system using web-based naive bayes algorithm. In: 2021 3rd International Conference on Electronics Representation and Algorithm (ICERA) (2021). https://doi.org/10.1109/icera53111.2021.9538697

Image Processing

Pre-image Calculation for Random Fourier Feature Kernel Machines

Bernard Tiddeman(✉) and Will Robinson

Department of Computer Science, Aberystwyth University, Aberystwyth, UK
bpt@aber.ac.uk, wgr@aber.ac.uk
https://www.aber.ac.uk/en/cs/staff-profiles/listing/profile/bpt/

Abstract. In this paper we address the pre-image problem for random Fourier feature based kernel machines. Full kernel methods require the calculation of the Gram matrix of scalar products between all training samples. This is not feasible for large sample sizes, and various approximation methods have been proposed previously. Calculation of the inverse transform, known as finding the pre-image, is frequently required, and has been tackled previously for the full kernel approach, but has been less well explored for approximation methods. In this paper we investigate two approaches suitable for finding the pre-image from random Fourier features. The first is a direct adaptation of a learning-based approach to the random features case. This requires the solution of a completely separate optimisation problem, with additional model parameters and a second pass through the data. The second approach only requires solving a single optimisation problem with an augmented vector and a single pass through the data samples. We compare the two methods on four datasets and demonstrate their effectiveness.

Keywords: Kernel pre-image · Radial basis functions · Random Fourier features · Kernel PCA

1 Introduction

1.1 Kernel Methods and Approximations

Kernel methods were first popularised by Boser et al. [2] with their use in Support Vector Machines (SVM) and were later extended to other problems, such as Kernel PCA [15]. Kernel methods work in a Reproducing Kernel Hilbert Space (RKHS) which maps points $\mathbf{x}$ into a higher dimensional space $\Phi(\mathbf{x})$, but avoids explicit calculation of the mapping by introducing a kernel function, $k(\mathbf{x}, \mathbf{y})$ such that:

$$k(\mathbf{x}, \mathbf{y}) = < \Phi(\mathbf{x}), \Phi(\mathbf{y}) > = \Phi(\mathbf{x})^T \Phi(\mathbf{y}) \quad (1)$$

The aim is to find a feature space $\Phi(\mathbf{x})$ such that the variables that are not linearly separable in the original space can be separated in the new feature space. In standard kernel methods, we avoid working in the feature space $\Phi(\mathbf{x})$ by only computing the inner products, which are given by $k(\mathbf{x}, \mathbf{y})$. This means that the feature space can in theory be very high (even infinite) dimensional.

N. Naik et al. (Eds.): UKCI 2023, AISC 1453, pp. 433–444, 2024.
https://doi.org/10.1007/978-3-031-47508-5_34

Standard kernel methods are not appropriate for very large sample sizes, because they involve computing and storing the Gramian (or Gram) matrix of inner products of all pairs of training samples. When we are dealing with potentially billions of samples, this problem becomes intractable. Various approximate solutions have been proposed for dealing with large datasets, including the Nyström sampling method [18], Random Fourier Features (RFF) [14] and many enhancements and extensions to these methods [10]. Here we build on the work of Rahimi et al. [14], who show that radial basis function (RBF) kernels can be approximated by mapping into an alternate feature space $z_\omega(\mathbf{x})$ constructed so that the expected value $\mathbb{E}[z_\omega(\mathbf{x})^T z_\omega(\mathbf{y})] = k(\mathbf{x}, \mathbf{y})$. They suggest two possible options for the function $z_\omega(\mathbf{x})$:

$$z_\omega(\mathbf{x}) = (\cos(\omega^T \mathbf{x}), \sin(\omega^T \mathbf{x})) \tag{2}$$

or

$$z_\omega(\mathbf{x}) = (\sqrt{2}\cos(\omega^T \mathbf{x} + b)) \tag{3}$$

where ω is drawn from the Fourier transform of the kernel function and b is drawn uniformly from $[0, 2\pi]$. Averaging over N such functions lowers the variance of the resulting approximation to the kernel function. This is equivalent to taking the scalar product of the two sample vectors constructed from the input vectors and a set of random samples from the Fourier transform of the kernel function.

The use of these random features solves the problem of applying kernel methods, such as kernel PCA, on very large samples, as we can convert each sample into the random feature space and perform standard PCA [11]. Such methods have also been integrated with state of the art techniques such as Transformers [4,13], neural tangent kernels [21], and graph neural networks [5,20]. It still leaves the problem that the above functions are not invertible, due to the aliasing introduced by the cos function. Invertibility is required in many algorithms, such as denoising, where data are projected into the kernel space and then back into the original space.

Reconstructing the input is known as finding the pre-image in kernel methods [1,6,8,12]. Earlier work used an iterative fixed point approach to minimize the error [12], but this is slow and can suffer from numerical instability. A multidimensional scaling approach was proposed by Kwok et al. [8], based on distances in the original and feature space. Honeine et al. [6] suggested a direct method based on dot products in original and feature space. All of the preceding methods require the full kernel matrix to be stored and processed, which is not feasible when the number of samples is very large.

Learning the parameters of the inverse mapping by ridge-regression in kernel PCA space was suggested by Bakir et al. [1]. In the work of Lopez-Paz et al. [11] they perform reconstruction from an RFF kernel PCA representation using a number of nonlinear regressors to map from the principal component weights back to the original feature space, but the precise nature of the regressors is not made clear.

In the following sections we develop two algorithms that allow recovery of the pre-image from samples projected into kernel PCA space and evaluate the approaches on 4 datasets of varying sizes.

2 Methods

In this work we develop two methods, one is based on [1] and requires a second pass through the dataset to build the reconstruction model. The second pass through the data, once the kernel PCA model has been built, is not only time consuming but also involves an additional set of parameters to tune. The second method we investigate is a single pass algorithm using augmented vectors. We append additional data to the mapped data vectors to allow reconstruction, and adapt the PCA to correctly handle the augmented vectors. The results indicate that both algorithms can allow accurate reconstruction using a number of datasets, while also maintaining the class separation capability of the kernel PCA approach. The preferred approach is likely to be problem dependent, and would need to take into account the overhead in each data pass (such as loading the data samples) as well as the overhead incurred from using longer sample vectors in the PCA algorithm.

2.1 Learnt Inversion

The method of [1] can be adapted to the approximate RBF kernel-PCA scenario. In their method, a linear model is learnt in the Reproducing Kernel Hilbert Space (RKHS) that maps the kernel PCA scores back to the original scores. When working in the full Kernel space a regularising term is required due to the high risk of multicolinearity in the RKHS. In contrast, for our problem, there is low risk of multicolinearity and we can instead use the Moore-Penrose pseudo inverse to learn the mapping parameters.

To train the model we first project each training sample into the kernel PCA component scores. We then construct RFF samples from a suitable RBF kernel space according to the method of Rahimi and Recht [14]. The kernel function and samples do not need to be the same as for the forward model. We then look for a set of parameters $\mathbf{A}$ such that:

$$\mathbf{AY} = \mathbf{X} \tag{4}$$

where $\mathbf{Y}$ is the matrix with the mapped kernel PCA scores for the training data as columns, with 1 appended for the constant term, and $\mathbf{X}$ is the corresponding matrix of original training data. This is solved in the least-squares sense using:

$$\mathbf{A} = \mathbf{XY}^T(\mathbf{YY}^T)^{-1} \tag{5}$$

An individual point in the kernal PCA space, $\mathbf{y}$, can then be mapped onto it's pre-image, by first mapping to it's RFF location and appending 1, to give vector, $\mathbf{y}'$ and then multiplying by $\mathbf{A}$.

It is also possible to include a regularising term in Eq. 5, leading to a ridge regression solution of the form:

$$\mathbf{A} = \mathbf{X}\mathbf{Y}^T(\mathbf{Y}\mathbf{Y}^T + \alpha\mathbf{I})^{-1} \tag{6}$$

where α is the ridge regression parameter.

2.2 Augmentation Method: Construction for Invertibility

To adapt the method of Rahimi and Recht [14] to be directly invertible, we propose augmenting the random vectors by appending a set of values to z_ω formed from an invertible mapping, f of the input points, i.e. the combined vector $\eta(x)$ takes the form:

$$\eta(x) = \begin{pmatrix} \cos(\omega_1^T x + b_1)/\sqrt{N} \\ \cos(\omega_2^T x + b_2)/\sqrt{N} \\ \vdots \\ f(\omega_1^T x + b_1) \\ f(\omega_2^T x + b_2) \\ \vdots \end{pmatrix} \tag{7}$$

Taking dot products between pairs of these vectors as the kernel approximation would require the expected value of the products $\mathbb{E}[f(\omega' x + b)f(\omega' y + b)]$ to be zero, or the scalar products between the appended parts of the vectors to be approximately zero for all mapped points. This is not possible in general for non-zero vectors e.g. the scalar product of a mapped vector with itself cannot be zero for a non-zero vector.

To get around this problem we replace the scalar product with a bilinear form $\eta(x)^T\mathbf{R}\eta(y)$, and use the matrix $\mathbf{R}$:

$$\mathbf{R} = \begin{pmatrix} \mathbf{I} & \mathbf{I} \\ \mathbf{I} & \mathbf{0} \end{pmatrix} \tag{8}$$

This can recover the same approximate kernel function provided that the expected value $\mathbb{E}[\cos(\omega^{\mathbf{T}}\mathbf{x} + b)f(\omega^{\mathbf{T}}\mathbf{y} + b)]$ is zero (and sufficient samples are used). This is true for any odd function f of a variable drawn from an even distribution, because cos is an even function. To this end, we also draw the values b from $[-\pi, \pi]$, rather than $[0, 2\pi]$. This random rotation ensures that the dot product provides a sample mean of the correct inverse Fourier function. The specific direction of the rotation is not critical to achieve the correct sample mean, but a symmetric interval around zero is required to ensure an expected value of zero for the $\cos(\omega^{\mathbf{T}}\mathbf{x} + b)f(\omega^{\mathbf{T}}\mathbf{y} + b)$ terms. We also need to ensure that rounding errors due to the sample size don't dominate, and that the inverse function f^{-1} exists and does not amplify errors in the reconstruction. Hence we recommend a scaled identity function $f(x) = \alpha x$ where α is used to control the variance of mapped values, to ensure that the deviation from the sample mean does not dominate the cosine terms.

Random Feature Kernel PCA with Augmented Vectors: The Eigenvalues λ_i and corresponding Eigenvectors $\mathbf{u}_i$ of the approximate kernel matrix satisfy:

$$\mathbf{K}\mathbf{u}_i \approx \mathbf{X}^T\mathbf{R}\mathbf{X}\mathbf{u}_i = \lambda_i\mathbf{u}_i \tag{9}$$

This holds even though $\mathbf{R}$ is not positive semi-definite (PSD), due to the nature of the sampled functions. Empirically we have found that, although this approximation to the kernel matrix does have negative eigenvalues, they are typically much smaller than the positive ones, and the difference increases with larger RFF sample sizes.

Ideally we would like to take the square root of the matrix $\mathbf{R}$ to provide a mapping into a space that is suitable both for forming the approximate covariance matrix and inverting the mapping. The presence of negative eigenvalues complicates matters slightly, but we show here that it is possible to calculate the principal component scores without recourse to complex variables. To derive the modified kernel algorithm, firstly, we can factor $\mathbf{R}$ into a real orthogonal matrix $\mathbf{P}$ and a diagonal matrix $\mathbf{D}$ with entries either purely real (for positive eigenvalues) or purely imaginary (for negative eigenvalues) such that $\mathbf{R} = \mathbf{P}^T\mathbf{D}\mathbf{D}\mathbf{P}$ so that Eq. (9) becomes:

$$\mathbf{X}^T\mathbf{P}^T\mathbf{D}\mathbf{D}\mathbf{P}\mathbf{X}\mathbf{u}_i = \lambda_i\mathbf{u}_i \tag{10}$$

If we were to use this complex mapping $\mathbf{Y} = \mathbf{D}\mathbf{P}\mathbf{X}$, multiplying both sides of Eq. (10) by $\mathbf{D}\mathbf{P}\mathbf{X}$ would give:

$$\mathbf{D}\mathbf{P}\mathbf{X}\mathbf{X}^T\mathbf{P}^T\mathbf{D}\mathbf{D}\mathbf{P}\mathbf{X}\mathbf{u}_i = \lambda_i\mathbf{D}\mathbf{P}\mathbf{X}\mathbf{u}_i \tag{11}$$

And substituting $\mathbf{v}_i = \mathbf{D}\mathbf{P}\mathbf{X}\mathbf{u}_i$ gives:

$$\mathbf{D}\mathbf{P}\mathbf{X}\mathbf{X}^T\mathbf{P}^T\mathbf{D}\mathbf{v}_i = \mathbf{Y}\mathbf{Y}^T\mathbf{v}_i = \lambda_i\mathbf{v}_i \tag{12}$$

where $\mathbf{Y}\mathbf{Y}^T/(N-1)$ is the covariance matrix of the mapped vectors $\mathbf{Y}$, $\mathbf{v}_i$ are the eigenvectors, and λ_i are the associated eigenvalues.

To analyse data samples, we would first convert the input sample into the form given by Eq. (7) to give $\mathbf{x}$, then multiply by $\mathbf{D}\mathbf{P}$ to give the mapped vector and then take the scalar product with each principal component used, $\mathbf{v}_i$, formed into the columns of matrix $\mathbf{V}$, to give the output scores $\mathbf{y}$ i.e.

$$\mathbf{y} = \mathbf{V}^T\mathbf{D}\mathbf{P}\mathbf{x} \tag{13}$$

Recalling that $\mathbf{V} = \mathbf{D}\mathbf{P}\mathbf{X}\mathbf{U}$ this gives:

$$\mathbf{y} = \mathbf{U}^T\mathbf{X}^T\mathbf{P}^T\mathbf{D}\mathbf{D}\mathbf{P}\mathbf{x} = \mathbf{U}^T\mathbf{X}^T\mathbf{R}\mathbf{x} \tag{14}$$

This shows the relationship between the eigenvectors of the full kernel matrix and the PCA component scores. Clearly, the scores are real, provided the eigenvectors of the full kernel matrix are real. Rather than working with the full kernel matrix approximation, we can see that multiplying both sides of Eq. (9) by $\mathbf{R}\mathbf{X}$ gives:

$$\mathbf{R}\mathbf{X}\mathbf{X}^T\mathbf{R}\mathbf{X}\mathbf{u}_i = \lambda_i\mathbf{R}\mathbf{X}\mathbf{u}_i \tag{15}$$

by replacing $\mathbf{w}_i = \mathbf{RXu}_i$ we can see that:

$$\mathbf{RXX}^T\mathbf{w}_i = \lambda_i\mathbf{w}_i \tag{16}$$

so by calculating the right eigenvalues and eigenvectors of $\mathbf{RXX}^T$ we can calculate the PCA scores using:

$$\mathbf{y} = \mathbf{U}^T\mathbf{X}^T\mathbf{Rx} = \mathbf{W}^T\mathbf{x} \tag{17}$$

where $\mathbf{W}$ is the matrix with columns $\mathbf{w}_i$, the principal components of $\mathbf{RXX}^T$. This resulting single pass algorithm is shown in Algorithms 1 and 2.

Direct Inversion from Augmented Vectors: If we consider our vectors ω_i as the columns of the matrix $\boldsymbol{\Omega}$, we can invert the above representation by extracting the half of the combined vector η comprising the $f(x)$ terms, and then use the appropriate inverse function i.e.:

$$\omega_i^T\mathbf{x} + b_i = f^{-1}(f(\omega_i^T\mathbf{x} + b_i)) \tag{18}$$

we can then recover $\mathbf{x}$ by subtracting the b_i values and using $\boldsymbol{\Omega}^{-1}$ or the pseudo inverse for a non-square $\boldsymbol{\Omega}$. The resulting algorithm is given in Algorithm 3.

Algorithm 1 Augmented PCA calculation

---

procedure AUGMENTEDPCA($\{\mathbf{x}_i\}, n, \gamma$)
  $\boldsymbol{\Omega}, \mathbf{b} \leftarrow$ INITIALIZERFF(n) ▷ According to [14]
  $\mathbf{C} \leftarrow \mathbf{0}$
  $\bar{\eta} \leftarrow 0$
  for all $\{\mathbf{x}_i\}$ **do**
    $\eta \leftarrow$ MAP($\mathbf{x}_i, \boldsymbol{\Omega}, \mathbf{b}$) ▷ According to equation 7
    $\bar{\eta} \leftarrow \bar{\eta} + \eta$
    $\mathbf{C} \leftarrow \mathbf{C} + \eta^T\eta$
  end for
  $\mathbf{R} \leftarrow \begin{pmatrix} \mathbf{I} & \mathbf{I} \\ \mathbf{I} & \mathbf{0} \end{pmatrix}$
  $\mathbf{C} \leftarrow \mathbf{C} - \bar{\eta}^T\bar{\eta}$
  $\mathbf{C} \leftarrow \mathbf{RC}$
  $\lambda, \mathbf{W} \leftarrow$ EIGENANALYSIS($\mathbf{C}$)
  $\lambda = \mathrm{Re}(\lambda)$
  $\lambda, \mathbf{W} = $ SORT($\lambda, \mathbf{W}$) ▷ Sort eigen values and vectors according to λ
  return $\boldsymbol{\Omega}, \bar{\eta}, \lambda, \mathbf{W}$
end procedure

Algorithm 2 Augmented PCA analysis

procedure APPLYAUGMENTEDPCA($\mathbf{x}, \mathbf{\Omega}, \mathbf{b}, \mathbf{W}, \bar{\eta}$) ▷ $\mathbf{W}$ with largest PCs
  $\eta \leftarrow \text{MAP}(\mathbf{x}_i, \mathbf{\Omega}, \mathbf{b})$ ▷ According to equation 7
  $\eta \leftarrow \eta - \bar{\eta}$
  $\mathbf{y} \leftarrow \mathbf{W}^T \eta$
  return y
end procedure

Algorithm 3 Augmented PCA reconstruction

procedure AUGMENTEDPCAPREIMAGE($\mathbf{y}, \mathbf{\Omega}, \mathbf{b}, \mathbf{W}, \bar{\eta}, \alpha$) ▷ $\mathbf{W}$ with largest PCs
  $\eta \leftarrow \mathbf{W}\mathbf{y} + \bar{\eta}$
  $\mathbf{z} \leftarrow \eta_{N/2..N}/\alpha$ ▷ extract the second half of the vector and rescale
  $\mathbf{z} \leftarrow \mathbf{\Omega}^{-1}(\mathbf{z} - \mathbf{b})$
  return z
end procedure

Computational Complexity. The complexity of both RFF training algorithms is given by $O(N(o + L^2) + L^3)$, where N is the number of samples, o is the overhead per sample (e.g. loading and pre-processing), and L is the length of the sample vectors. The L^2 terms come from forming the covariance matrix, and the L^3 term from the eigen analysis step. In the case $N >> L$ that we are addressing here, this reduces to $O(N(o + L^2))$. In cases where $o > L^2$ the complexity is then dominated by $O(No)$ and the augmented version is preferred (as o is reduced by half), otherwise the complexity is dominated by $O(NL^2)$ and the learnt algorithm is preferred (as L is reduced by half).

3 Experimental Evaluation

In this section we evaluate the two proposed reconstruction algorithms on a number of example datasets, including two small "toy" datasets, a medium sized dataset where it is still possible to calculate the full kernel PCA solutions, and a large dataset, where it is not practicable to directly calculate the full kernel PCA solution or inverse. The small datasets are useful to visualise the distribution of samples in the original and kernel PCA spaces, including the reconstructed test set. The medium sized dataset demonstrates a well-known denoising application, including the full kernel PCA approach. The larger dataset (MNIST) demonstrates that the methods are applicable where calculation of the full kernel PCA solution is not possible. Example results for each dataset are given in Table 1.

For the small datasets we use the circles and moons binary classification datasets from scikit-learn,[1] with the training and testing data shown in Fig. 1. Example results are shown in Fig. 2. All three methods (full kernel, random features with learnt inversion, and augmented random features with direct inversion) are capable of both linearly separating the classes, and also approximately

[1] https://scikit-learn.org/stable/datasets/sample_generators.html.

reconstructing the input data. In all three cases we use the same value for the Gaussian kernel parameter, γ, for both the mapping into kernel PCA space and for the learnt reconstruction model. In these examples the random feature based methods perform slightly worse than the full kernel method, which is as expected.

For the medium sized dataset, we extend the denoising example from [1] also used in the scikit-learn demo[2] using the USPS dataset [7]. Here, the training data is corrupted with normally distributed, zero mean random noise applied to each pixel, with a standard deviation of 0.25. We use the value for gamma suggested in the demo, and use all 6000 images in the training set for training and all 1000 images in the test set for testing. For the unregularised experiment we approximate an unregularised full kernel-PCA using a ridge regression parameter of 10^{-11}, which is just sufficient to avoid problems with colinearity. In the unregularised approach the augmented method narrowly outperforms the learnt method, but the difference between the two RFF methods is not significant ($p = 0.1265$) and none of the methods outperform linear PCA with an mse of 0.0124. Introducing regularisation, by increasing the full kernel ridge regression parameter to 10^{-3} and using Eq. 6 for the regularised learnt RFF pre-image with a regularisation parameter of 1.2 improves both algorithms to outperform linear PCA. In this case the full kernel PCA approach outperforms the regularised RFF approach. We know of no straightforward way to incorporate regularisation into the augmented approach. Example results are shown in Fig. 3.

For the larger dataset, we use the MNIST dataset [9]. We use the gamma value based on the approach suggested by [19] for the full MNIST dataset, which avoids the need for grid search, but assumes a Gaussian distribution for the clusters. Here we use the original full training set to build the model and the full test set for evaluation. Example results are shown in Fig. 4. In this (noise free) case, the augmented approach was better able to reconstruct the test images than either the learnt inversion with random Fourier features, or a linear model, but the difference between the two RFF methods was not significant ($p = 0.4005$). We also used the SSIM metric [17], and again the augmented method performed marginally better than the learnt method, but on this measure the linear approach outperformed the RFF kernel PCA methods.

4 Conclusion and Future Work

In this paper we have addressed the problem of finding the pre-image for Random Fourier Features (RFF) for approximate RBF kernel PCA. We have detailed two algorithms, and tested them on 4 datasets. The first is a simple adaptation of a learning based method [1] to the RFF case, that requires a second pass through the data and a second model requiring tuning of a set of additional parameters. The second method uses augmented vectors, and allows direct calculation of the inverse. The results demonstrate that both methods are capable of finding the

[2] https://scikit-learn.org/stable/auto_examples/applications/plot_digits_denoising.html.

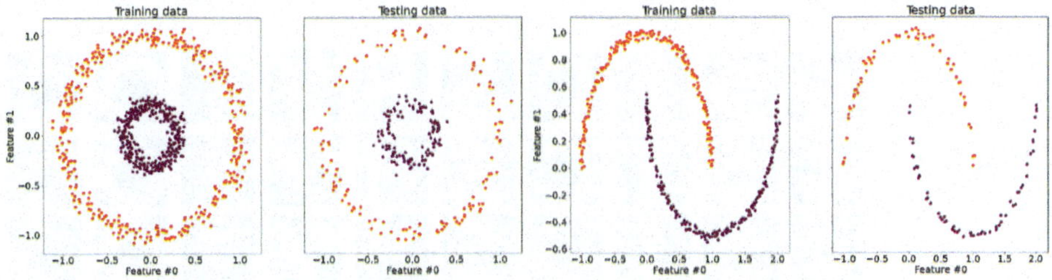

Fig. 1. Example toy datasets used in the experiments. From left to right: circles training data, circles test data, moons training data and moons test data.

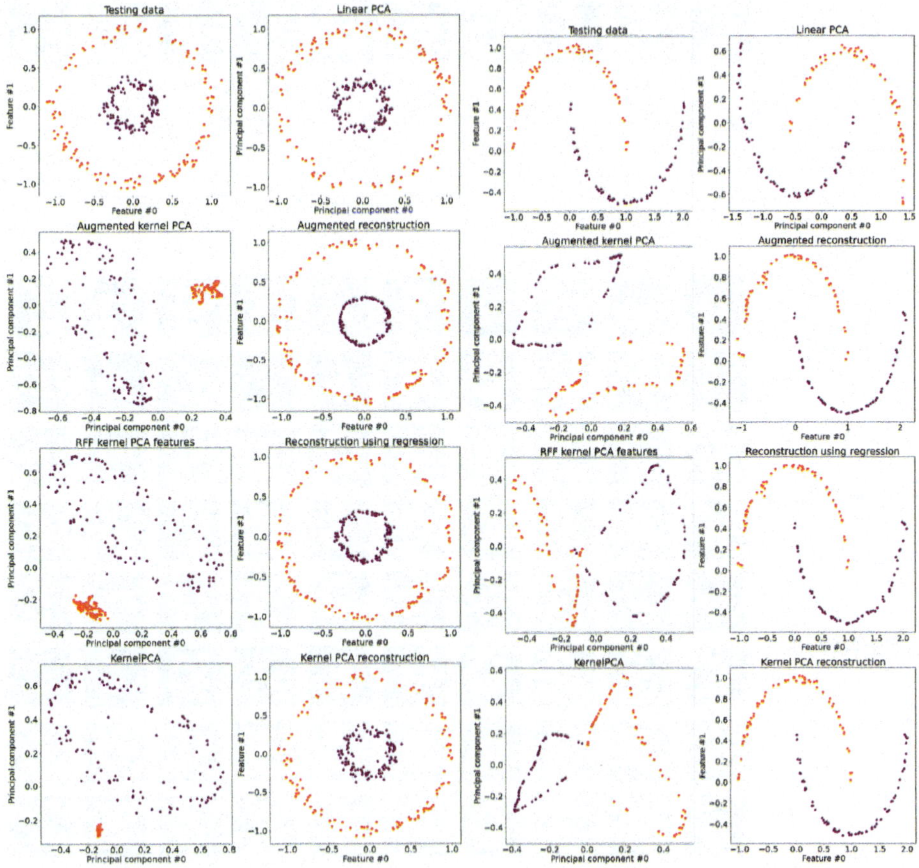

Fig. 2. Example results for the circles (left two columns) and moons data (right two columns). In reading order: test data; linear PCA; augmented RFF kernel PCA; augmented RFF kernel PCA reconstruction; RFF kernel PCA; RFF kernel PCA reconstruction; full kernel PCA; full kernel PCA reconstruction

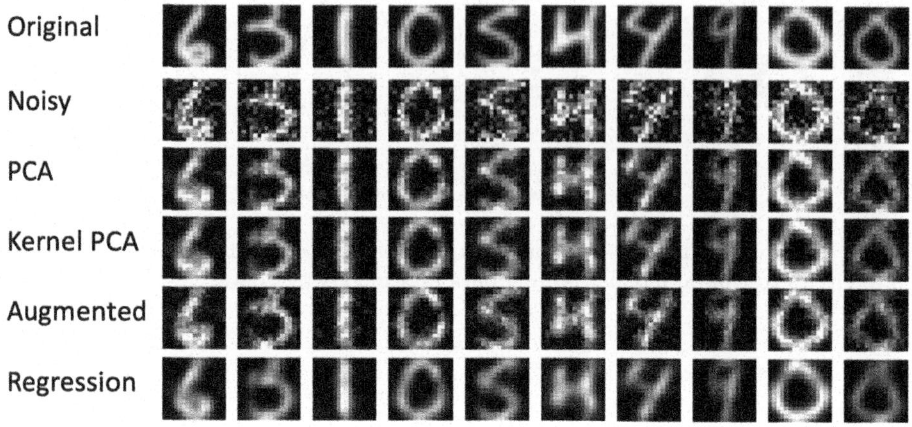

Fig. 3. Example denoising results for USPS digits data

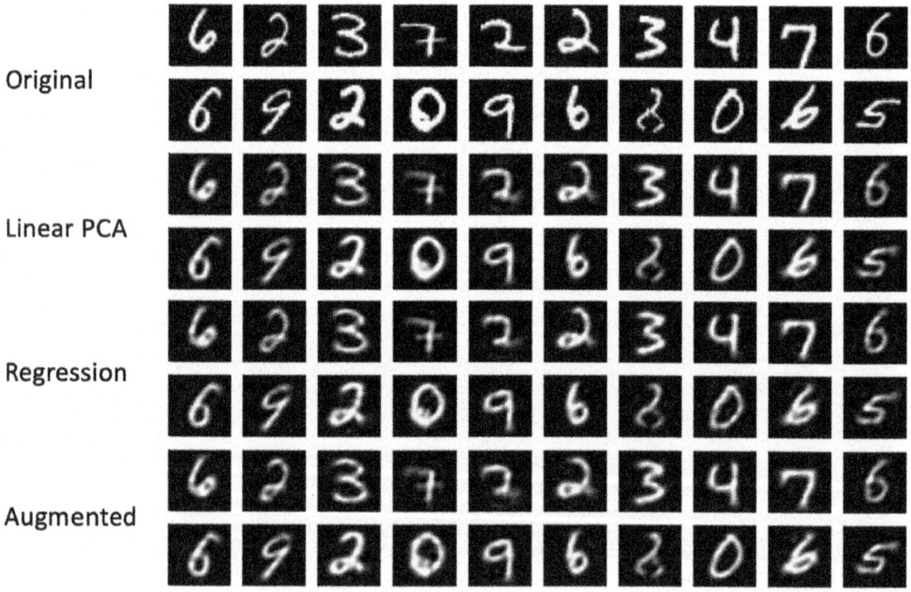

Fig. 4. Example results for MNIST digits data. Rows 1–2: original examples; rows 3–4: linear PCA reconstructions; rows 5–6: approximate RBF kernel PCA with learnt inversion; rows 7–8: approximate RBF kernel method with augmented samples.

Table 1. Table showing means squared error results (and SSIM for MNIST) on a variety of datasets and problems. We used 1000 random features for the two toy datasets and 2000 for the other datasets. For the augmented and regression algorithms we use averages over 100 repetitions for the circles, moons and USPS datasets and 10 repetitions for the MNIST dataset.

Dataset	γ	Components	Augmented	Regression	Full (* linear)
Circles	5	25	1.38×10^{-3}	1.36×10^{-3}	$\mathbf{2.67 \times 10^{-4}}$
Moons	5	25	2.55×10^{-4}	2.55×10^{-4}	$\mathbf{5.47 \times 10^{-5}}$
USPS digits	10^{-3}	32	$\mathbf{1.42 \times 10^{-2}}$	1.42×10^{-2}	1.58×10^{-2}
USPS regularised	10^{-3}	32	–	1.23×10^{-2}	$\mathbf{1.07 \times 10^{-2}}$
MNIST digits	7.5×10^{-8}	50	1.148×10^{6}	$\mathbf{1.147 \times 10^{6}}$	1.154×10^{6}*
MNIST SSIM	7.5×10^{-8}	50	0.6981	0.6978	**0.7377***

pre-image for the 4 datasets evaluated, while also maintaining the attractive features of the kernel methods.

In the future we intend to work towards extending Principal Component Wavelet Networks (PCWN) [16] to use kernel PCA, rather than linear PCA. PCWN combines an over complete wavelet basis with PCA to learn 1×1 convolution filters in a convolutional neural network (CNN) framework. The network is used to decompose and approximately reconstruct images, and has shown good results for a number of inverse linear problems when combined with the Alternating Direction Method of Multipliers (ADMM) algorithm [3] and can reduce overfitting problems seen in some adversarial CNN solutions. The method involves calculating PCA from samples that are proportional in number to the number of pixels in the set of training images, which could be of the order of billions with modern image datasets, and hence is not amenable to direct kernel PCA based approaches, and requires the use of approximations such as RFF. The method also requires the PCA step to be invertible, in order to reconstruct the image. Hence we will use the results from this paper to develop kernel-PCA version of the PCWN algorithm.

References

1. Bakır, G.H., Weston, J., Schölkopf, B.: Learning to find pre-images. Adv. Neural. Inf. Process. Syst. **16**, 449–456 (2004)
2. Boser, B.E., Guyon, I.M., Vapnik, V.N.: A training algorithm for optimal margin classifiers. In: Proceedings of the Fifth Annual Workshop on Computational Learning Theory, pp. 144–152 (1992)
3. Boyd, S., Parikh, N., Chu, E., Peleato, B., Eckstein, J., et al.: Distributed optimization and statistical learning via the alternating direction method of multipliers. Found. Trends Mach. Learn. **3**(1), 1–122 (2011)
4. Choromanski, K., Likhosherstov, V., Dohan, D., Song, X., Gane, A., Sarlos, T., Hawkins, P., Davis, J., Mohiuddin, A., Kaiser, L., et al.: Rethinking attention with performers. arXiv:2009.14794 (2020)

5. Du, S.S., Hou, K., Salakhutdinov, R.R., Poczos, B., Wang, R., Xu, K.: Graph neural tangent kernel: fusing graph neural networks with graph kernels. Adv. Neural Inf. Process. Syst. **32** (2019)
6. Honeine, P., Richard, C.: Solving the pre-image problem in kernel machines: a direct method. In: 2009 IEEE International Workshop on Machine Learning for Signal Processing, pp. 1–6. IEEE (2009)
7. Hull, J.M.: A database for handwritten text recognition research. IEEE Trans. Pattern Anal. Mach. Intell. **16**(5), 550–554 (1994)
8. Kwok, J.T.-Y., Tsang, I.W.-H.: The pre-image problem in kernel methods. IEEE Trans. Neural Netw. **15**(6), 1517–1525 (2004)
9. LeCun, Y., Cortes, C., Burges, C.J.: Mnist handwritten digit database. ATT Labs 2. http://www.yann.lecun.com/exdb/mnist (2010)
10. Liu, F., Huang, X., Chen, Y., Suykens, J.A.K.: Random features for kernel approximation: a survey on algorithms, theory, and beyond. IEEE Trans. Pattern Anal. Mach. Intell. **44**(10), 7128–7148 (2021)
11. Lopez-Paz, D., Sra, S., Smola, A., Ghahramani, Z., Schoelkopf, B.: Randomized nonlinear component analysis. In: Xing, E.P., Jebara, T. (Eds.) Proceedings of the 31st International Conference on Machine Learning, volume 32 of Proceedings of Machine Learning Research, pp. 1359–1367. Bejing, China, 22–24 Jun 2014 PMLR
12. Mika, S., Schölkopf, B., Smola, A., Müller, K.-R., Scholz, M., Rätsch, G.: Kernel PCA and de-noising in feature spaces. Adv. Neural Inf. Process. Syst. **11** (1998)
13. Peng, H., Pappas, N., Yogatama, D., Schwartz, R., Smith, N., Kong, L.: Random feature attention. In: International Conference on Learning Representations (ICLR 2021) (2021)
14. Rahimi, A., Recht, B.: Random features for large-scale kernel machines. Adv. Neural Inf. Process. Syst. **20** (2007)
15. Schölkopf, B., Smola, A., Müller, K.-R.: Kernel principal component analysis. In: 7th International Conference Lausanne Artificial Neural Networks-ICANN'97, Switzerland, October 8–10, 1997 Proceedings, pp. 583–588. Springer (2005)
16. Tiddeman, B., Ghahremani, M.: Principal component wavelet networks for solving linear inverse problems. Symmetry **13**(6) (2021)
17. Wang, Z., Bovik, A.C., Sheikh, H.R., Simoncelli, E.P.: Image quality assessment: from error visibility to structural similarity. IEEE Trans. Image Process. **13**(4), 600–612 (2004)
18. Williams, C., Seeger, M.: Using the Nyström method to speed up kernel machines. Adv. Neural Inf. Process. Syst. **13** (2000)
19. Yin, J.-B., Li, T., Shen, H.-B.: Gaussian kernel optimization: complex problem and a simple solution. Neurocomputing **74**(18), 3816–3822 (2011)
20. Zambon, D., Alippi, C., Livi, L.: Graph random neural features for distance-preserving graph representations. In: International Conference on Machine Learning, pp. 10968–10977. PMLR (2020)
21. Zandieh, A., Han, I., Avron, H., Shoham, N., Kim, C., Shin, J.: Scaling neural tangent kernels via sketching and random features. Adv. Neural. Inf. Process. Syst. **34**, 1062–1073 (2021)

Impact Characterization on Reinforced Aerospace Structures via Machine Learning

F. Dipietrangelo(✉), F. Nicassio, and G. Scarselli

Department of Engineering for Innovation, University of Salento, Via Monteroni, Lecce, Italy
flavio.dipietrangelo@unisalento.it

Abstract. In the present manuscript Machine Learning algorithms are compared in order to implement a model able to predict the coordinates of impacts on panels used in aerospace field. Experimental activities are conducted to build a proper impacts' dataset. Polynomial regression and artificial neural network are applied and optimised to panels without stringer to test their capability to identify the location of the impacts. Subsequently, the algorithms are applied to reinforced panels, with much more complexity of dynamic features of the system to test: the goal is not only on the impact position's detection but also on the event's severity. The work demonstrates the validity and the computing efficiency of the Machine Learning application to the impact localisation on realistic structures, regardless of specimen complexity.

Keywords: Structural health monitoring · Machine learning · Artificial neural network

1 Introduction

Structural health monitoring (SHM) in the aerospace field has reached a certain level of maturity for what concerns the methods and the sensors used for the damage detection in the different structural components of an aircraft. In [1] a review is provided of the main SHM techniques applied in the aviation based on the propagation of guided waves. Numerical, analytical and experimental characterization of damaged structures is discussed and the main conclusion is that, despite the different geometries involved with variable levels of irregularities, the tomographic method consisting of an unsupervised threshold definition and a multi-parameter processing is the most effective in detecting and locating the impact damage. Significant results are obtained even when thickness increase, stiffeners or both are located close to the monitored area. The Multi-Parameter Approach taken into consideration, indeed, may be generalised to different scenarios where the features mostly influenced by the defect may be specialized according to the different damage occurrence's generation (i.e., impact, fatigue or others). In [2] the state of the art on SHM of aerospace composite structures is represented with a special focus on the most effective sensors, commercially available and developed in laboratory. Working principles, properties, embedding capabilities, their interaction with host materials are discussed. General considerations about the influence of the aerospace environment on

N. Naik et al. (Eds.): UKCI 2023, AISC 1453, pp. 445–458, 2024.
https://doi.org/10.1007/978-3-031-47508-5_35

the development of SHM systems are also formulated: one of the conclusions of the manuscript is that piezo-sensors play still today a relevant role in the damage detection of lightweight structures. Other recent works, related to SHM applied to composites and aerospace field, focus on possible methods that can be adopted for the damage identification of typical structures [3–5], on the need to implement a wireless sensors' network on operating aircrafts in order to provide information about the health of primary structures [6], on the possible ways to feed the SHM sensors exploiting aircraft structural behaviour for energy harvesting [7], or address key aspects for the actual embedding of SHM devices into composite aerospace structures, like the need for a holistic damage assessment of the aircraft's structural health and an actual upscaling of these systems to realistic in-service structures [8]. Impacts can be detrimental to the aerospace structures and scientists pay significant efforts, on one side, to improve the materials' impact properties [9] and, on the other side, to detect impact related damage [10]. Machine learning algorithms are nowadays attractive and powerful tools in different fields of the engineering because they allow solutions of problems without a *a priori* knowledge of the whole physical frame behind the problems [11, 12]. In the present work, different machine learning algorithms are compared in order build a model able to locate on a thin aluminium panel the impact position with a remarkable level of accuracy. The present work extends a previous work carried out by authors [13] with the following new features: (i) the geometry of the tested samples was modified by reinforcements, responsible of waves' scattering through the panel; (ii) the built model is also capable to detect the level of the impact energy. In the present paper, after the description of the experimental setup and machine learning features, the chosen algorithms are explained in terms of math formulas. The results are reported with respect to the different specimens, while particular attention is paid to the explanation of physical phenomenon of speed increase of modes within the reinforced zone.

2 Experimental Setup

2.1 Hardware and Software

Impacts with low speed were performed on three squared specimens: 100 × 100 cm (Fig. 1 (a)); 100 × 100 cm with a "L-shape" stringer (Fig. 1(b)); 100 × 100 cm with three "L-shape" stringers (Fig. 1 (c)). Each panel was made of aluminium with the following features: density = 2700 kg/m^3, elastic modulus E = 72 GPa, Poisson's ratio $\nu = 0.33$, thickness = 1.2 mm. Four PZT $Pb\left[Z_{rx}Ti_{1-x}\right]O_3$ sensors [14] (diameter equal to 10 mm) were glued to the surface, at the vertices of a 50 × 50 cm area (Fig. 1 (a)). The impacts were performed inside this square zone on a grid of 0.25 mm of step, by dropping a ball of 1.2 g weight from the top of a 40 cm height tower [15].

The waves generated by 167 impacts were acquired by the PZT sensors, connected to the four channels of a Picoscope 6402D oscilloscope [16], in order to trace the voltage signals (Fig. 2).

The signals were buffered and sampled by Picoscope 6 software and exported as CSV file, in order to import them in MATLAB for data processing (Fig. 3).

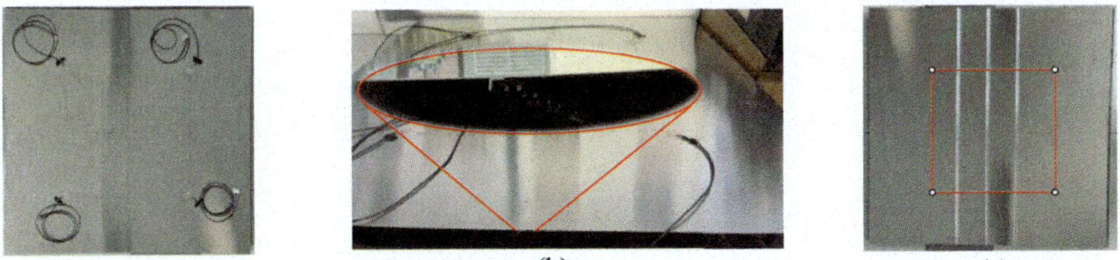

(a) (b) (c)

Fig. 1. Specimens: 100 × 100 cm (**a**); 100 × 100 cm with "L-shape" single stringer (**b**); 100 × 100 cm with three "L-shape" stringers (**c**).

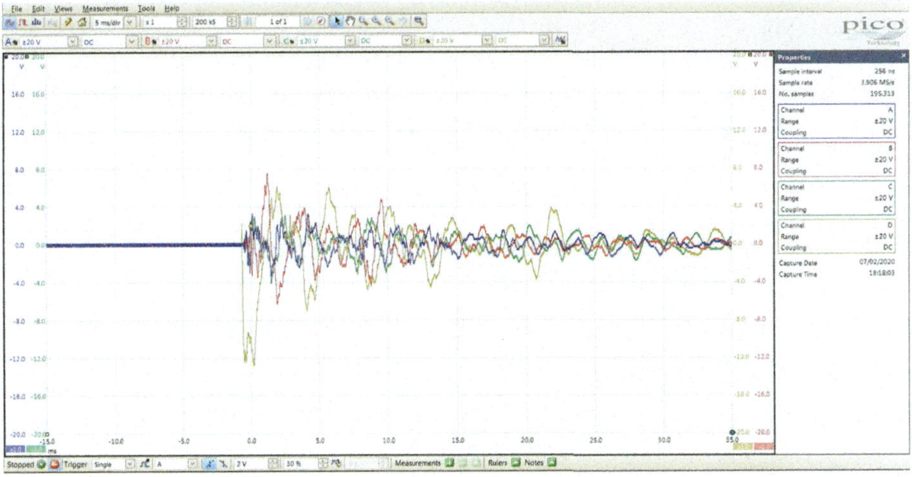

Fig. 2. Signal acquisition in Picoscope software.

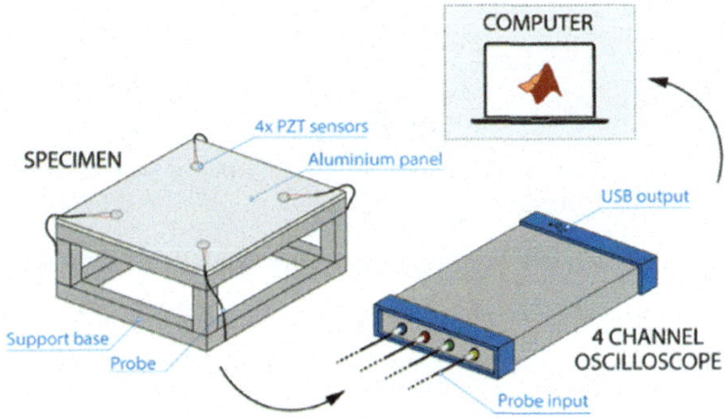

Fig. 3. Experimental setup.

2.2 Time of Flight Calculation

An impact produces Lamb waves, characterised by several modes and relative frequencies. In (0–700 kHz) range, only Symmetric and Antisymmetric zero-modes (S0, A0) are generated [17, 18].

The A0 and S0 tuning curves, excited on a similar aluminium plate [17], show that it is reasonable to identify in 40 kHz the maximum frequency at which the S0 mode is neglectable. In (0–40 kHz) range it is therefore equally reasonable to calculate the Time of Flight (ToF), that is the arrival time of the wave to a sensor, as the A0 ToF. For this reason, after the acquisition by the oscilloscope, the ToFs were evaluated according to the procedure explained in [19] for each sensor, by calculating the Short Time Fourier Transform in MATLAB, considering (0–40 kHz) range, a Hamming window and 40 Hz resolution (Fig. 4).

A sensor was chosen as a reference and the differences t_1, t_2, t_3 between the ToF at this sensor and the ToFs at the other ones were calculated. The actual coordinates (x,y), t_1, t_2 and t_3 became the columns of a dataset made of 167 rows (samples), corresponding to the 167 impacts points.

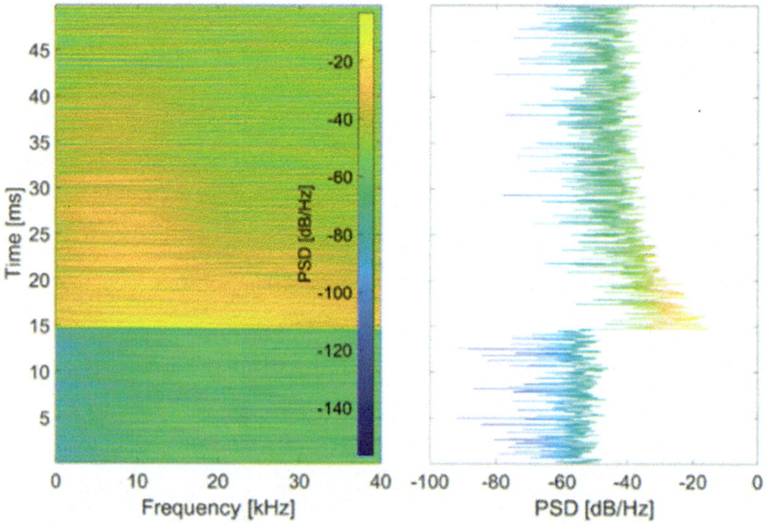

Fig. 4. Short Time Fourier Transform calculation in MATLAB.

3 Machine Learning Algorithms

3.1 Polynomial Regression

In this work the impact position (*x*,*y*) is function of t_1, t_2, t_3. Considering polynomial regression, (*x*,*y*) can be evaluated as polynomials of degree *d* and coefficients θ, as expressed in the following formulas:

$$x = \theta_{x0} + \sum_{i=1}^{3} \theta_{xi} t_i + \sum_{j=1}^{3}\sum_{k=1}^{3} \theta_{xjk} t_j t_k + \ldots \tag{1}$$

$$y = \theta_{y0} + \sum_{i=1}^{3} \theta_{y_i} t_i + \sum_{j=1}^{3}\sum_{k=1}^{3} \theta_{y_{jk}} t_j t_k + \ldots \tag{2}$$

The construction of the model consists of calculating *d* and θ in order to get closer to known data. Fixing *d* and considering *L* impacts, the following matrix form can be used:

$$\boldsymbol{U} = \boldsymbol{T}\boldsymbol{B} \tag{3}$$

$\boldsymbol{U}$ is $L \times 2$ matrix, in which the columns contain the *x* and *y* coordinates for the *L* impacts respectively, $\boldsymbol{T}$ is $L \times p$ matrix, in which the columns contain the polynomial features (t_i, power of the t_i up to grade *d* and their cross multiplication), $\boldsymbol{B}$ is $p \times 2$ Design Matrix of weight coefficients θ. $\boldsymbol{B}$ was calculated by using a subset of *M* samples as *training* data [12] and MATLAB *fitlm* [20] (a QR decomposition-based function):

$$\boldsymbol{B} = \left(\boldsymbol{T}^T\boldsymbol{T}\right)^{-1}\boldsymbol{T}^T\boldsymbol{U} \tag{4}$$

$\boldsymbol{T}$ is an ill-conditioned matrix, due to the correlation between its columns, and the inverse matrix calculated in (4) returned a rank-deficient error. Therefore, the *Z-Score scaling* was applied, getting new values of the features, with mean value 0 and standard deviation 1. It was possible to validate the model by calculating the Mean Radial Error (*MRE*) over *M* training samples and over *N* test samples, where the Radial Error about an impact is the Euclidean distance between the predicted and actual coordinates.

3.2 Artificial Neural Network

In a second phase, the Artificial Neural Network (ANN) was considered. An ANN is composed of layers of connected *artificial neurons* [12], that process input signals and send the output to the following layer. Each neuron performs two computations (5):

1. an intermediate variable *z* using the vectors *t* (input), *w* (weights) and *b* (biases);
2. an output *a* through the activation function *g*.

$$z = w^T t + b;\ a = g(z) \tag{5}$$

The first layer of an ANN is the input layer and contains the input variables. The last layer is the output layer and contains the output of the iterative calculation performed by the neurons of the internal *hidden* layers.

A Shallow Neural Network (SNN) is a feed-forward ANN consisting of one or two hidden layers and is able to solve both data fitting (regression) and pattern recognition (classification) problems. Several methods can initialise weights and biases, such as that proposed by Xavier [21], as well as different algorithms can be chosen in order to train the network through iterative procedures [22]-[23].

4 Results and Discussions

4.1 Panel 100 x 100 cm

In a first analysis the dataset was obtained by the impacts on the 100 × 100 cm specimen. The ***B*** calculation and the *MRE* estimating were performed considering different combinations of training and test sets with an 80/20 ratio (*K-Fold* cross validation with $K = 5$).

Different polynomial degrees were considered: Fig. 5 (a) shows *MRE* trends for (i) entire dataset (yellow curve), (ii) training set (blue curve) and (iii) test set (red curve) of it.

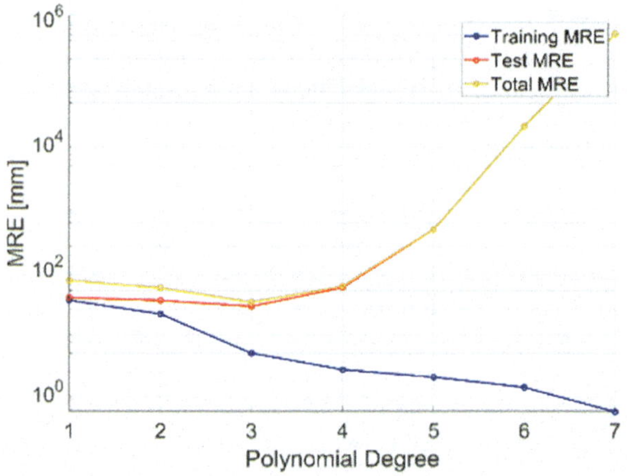

Fig. 5. Radial Error vs Polynomial Degree for 100 × 100 cm specimen.

The polynomial with $d = 3$ corresponded to the best model for generalisation because it minimised *MRE* and the gap between *MRE*s of training and test (Fig. 5). Therefore, the subsequent analysis, related to the best training size, was performed using $d = 3$ (Fig. 6). By keeping the same test set size of the previous analysis (20% of the database), it is clear that it is necessary to train the model with all the remaining data (80%).

A bias-variance diagnosis was performed for confirming $d = 3$ [12].

After the bias-variance diagnosis and then of the best choice of *d* and training size, a 50 test cases campaign was performed, with the calculation of mean *MRE*, $d = 3$ and

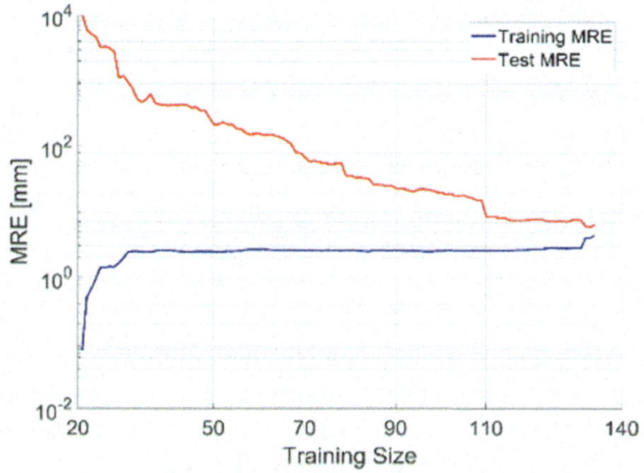

Fig. 6. Radial Error vs Training size for 100 × 100 cm specimen.

a casual splitting of data with 80/20 training/test ratio. Different test cases returned very different *MREs*, from 5.48 mm (the best one) to 27.45 mm (the worst one). The reason lies in the fact that the ToFs (used as the basis for the machine learning features) are very dissimilar and the model is not able to build a function that can fits well with such different values.

In a second phase, the SNN was considered, setting it with the following configuration: no features scaling; a sigmoid function in the hidden layer; a linear function in the output layer; three different learning algorithms; 50 test cases campaign with 70/30 training/test ratio. Each configuration was trained increasing the size of the hidden layer from 10 to 50 neurons. The best *MRE,* 3.08 mm, corresponded to a Bayesian Regularization (BR) trained model, containing 30 neurons, as shown in Table 1.

Table 1. Shallow neural network 100 × 100 cm panel best results

Number of neurons	Training method	*MRE* [mm]
30	Levenberg-Marquardt	6.74
30	Bayesian Regularization	**3.08**
20	Scaled conjugate Gradient	26.06

4.2 Panel with "L-shape" Single Stringer

Subsequently, a "*L-shape*" aluminium stiffener 1 mm thick was riveted in the middle of the 100 × 100 specimen and 167 impacts were performed on it, building a new dataset. Applying the same SNN and the same comparisons to these data, the best *MRE*, 4.24 mm, corresponded to a Bayesian Regularization (BR) trained model, containing 30 neurons, as shown in Table 2.

Table 2. Shallow neural network 100 × 100 cm specimen with "L-shape" single stiffener best results

Number of neurons	Training method	*MRE* [mm]
30	Levenberg-Marquardt	9.86
30	Bayesian Regularization	**4.24**
20	Scaled Conjugate Gradient	28.10

Therefore, the Bayesian Regularization training confirms its effectiveness and precision [22], at the cost of more complex algorithm and longer execution time. In Table 3 a deep comparison between the two scenarios (100 × 100 cm plate with and without stringer), with the same configuration and analysis, is shown.

Table 3. Error comparison between panels

Without Stringer			With Stringer		
Mean Error	Min Error	Max Error	Mean Error	Min Error	Max Error
3.09 mm	1.93 mm	6.18 mm	4.24 mm	3.16 mm	6.44 mm

The stiffener worsens the error, even if the mean one (4.24 mm) falls between the minimum (3.09 mm) and the maximum error (6.18 mm) of the case without reinforcement.

The reason for the deterioration in performance lies in the fact that the ML features are not coherent in the two setups: with respect to the first scenario, in the second one, because of the stiffener, two out of four ToFs are calculated from the detection of «transmitted» waves, that have passed through a discontinuity.

In order to quantify the phenomenon, an impact-sensor pair was chosen, by identifying the significant feature (ToF difference between the sensor and a reference one). The M2 impact location and the 4th sensor were chosen (impact number 105/167), with a distance (i.e., the wave path) of 49 cm (Fig. 7). The following calculation were obtained:

- ToF without stringer = 14.650 ms;
- ToF with stringer = 14.648 ms;
- ToF difference = 2.07 microseconds.

Therefore, ToF is shorter and the wave (orange plot) is anticipated in time with respect to the propagation without reinforcement (blue plot), as shown in MATLAB representation (Fig. 8). The behaviour described below was confirmed by finite elements simulations with dynamic characterization run in ANSYS, as shown in Fig. 9.

By concentrating the analysis on A0 and S0, the impact on the reinforcement generates reflected and transmitted modes, that travel at the same speed as the direct modes in the propagation inside the plate [24], but faster within the stiffened zone, because of the doubled thickness. Further confirmation was provided by the MATLAB calculation

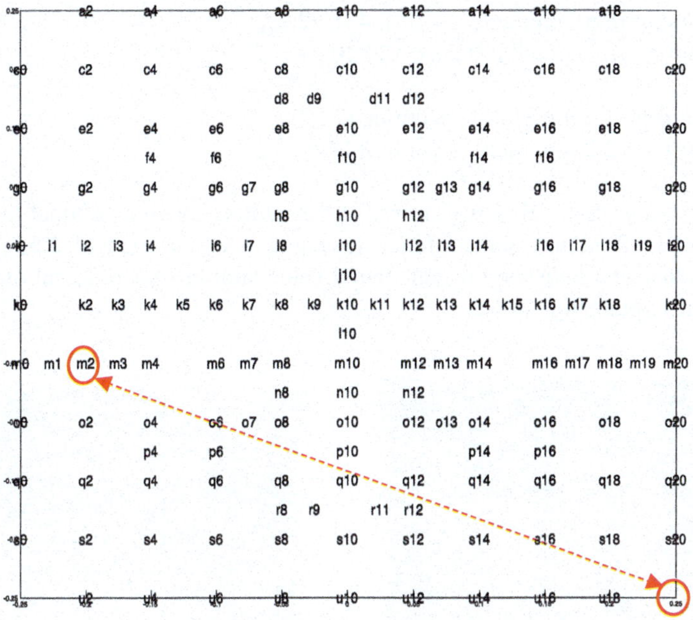

Fig. 7. Impact-sensor pair.

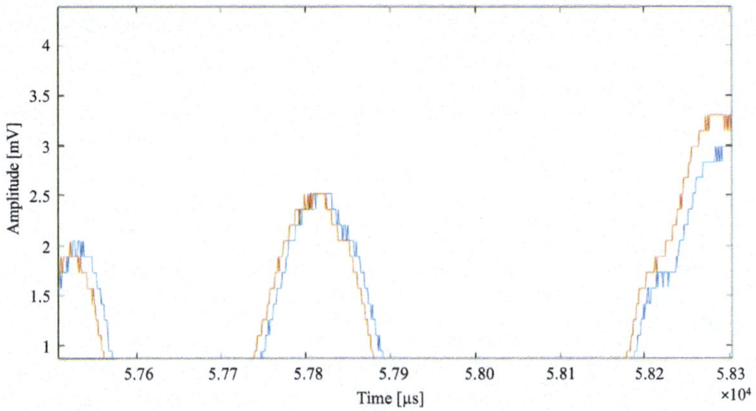

Fig. 8. Zoom of differences between signals traveling on simple plate (blue) and on reinforced plate (orange).

of dispersion curves at 40 kHz, at which the ToF was evaluated. The following A0 mode speeds were calculated (at 40 kHz):

- in the zone 1.2 mm thick = 1135 m/s;
- in the zone 2.2 mm thick = 1590 m/s.

Considering the M2-4th sensor path, the reinforced area is only 1 cm length. According to the speeds indicated above, the following propagation times were obtained.

- Plate without reinforcement (49 cm): 431.72 μs.
- Plate with reinforcement (49 cm): 429.20 μs.
  - In the zone 1.2 mm thick: 48 cm in 422.91 μs.
  - In the zone 2.2 mm thick: 1 cm in 6.29 μs.

Summarizing, the total arrival time without reinforcement was equal to 431.72 μs, while the total arrival time with stiffener was equal to 429.20 μs, with **2.52 μs** of difference. The closeness between this calculated value and the experimental one (**2.07 μs**) confirms the explanation above.

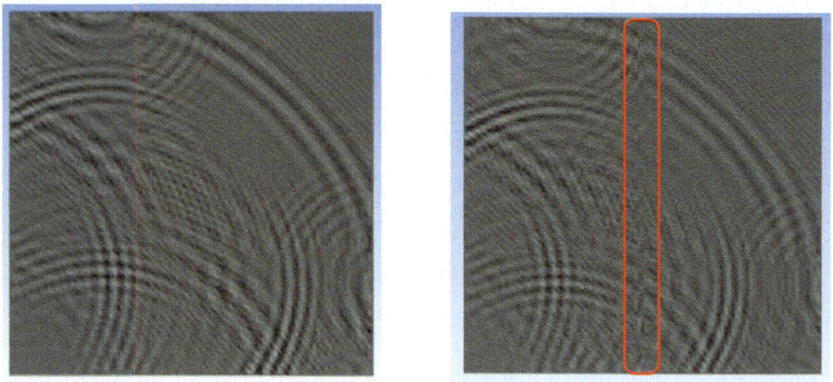

(a) (b)

Fig. 9. Finite elements simulations to highlight the different wave propagations without (a) or with (b) the stringer.

4.3 Plate with Three "L-shape" Stringers

4.3.1 Localisation

Subsequently, other two "L-shape" stiffeners 1 mm thick were fixed to the panel at an equally distance from the first stringer and 167 impacts were performed on it, building a new dataset. Applying the SNN to these data, using Scaled Conjugate Gradient and Levenberg-Marquardt training methods, the obtained *MRE* was greater than 15 mm. The best *MRE* corresponded to a Bayesian Regularization trained model, but with 40 neurons, as detailed in Table 4.

Therefore, in presence of three stiffeners, the phenomenon explained before is "multiplied", causing about twice the error. In order to reduce the *MRE*, a more complex SNN model was built with the following configuration: Z-Score scaling; 2 hidden layers with a sigmoid function; 40 neurons in the first layer; an output layer with a linear function; Bayesian Regularization method; maximum 1000 epochs; 50 test cases campaign with 70/30 training/test ratio.

Table 4. Shallow neural network error for a 100 × 100 plate with three "L-shape" stringers with Bayesian Regularization training

Number of neurons	Mean Error	Min Error	Max Error
30	8.88 mm	6.29 mm	14.81 mm
40	**8.80 mm**	5.75 mm	13.94 mm

Each configuration was trained (MATLAB *train* function [20]) increasing the size of the second hidden layer from 10 to 40 neurons. The best *MRE*, 6.88 mm, corresponded to a model containing 20 neurons in the second hidden layer, as shown in Table 5. Therefore, the increasing in network complexity improved the *MRE* of about 2 mm.

Table 5. Custom neural network performances

2ND layer neurons	Mean error	Min error	Max error
10	7.43 mm	5.06 mm	11.54 mm
20	**6.88 mm**	5.18 mm	8.74 mm
30	7.08 mm	5.23 mm	9.21 mm
40	7.34 mm	4.73 mm	11.33 mm

4.3.2 Energy Classification

The last analysis was to apply machine learning algorithm in order to predict the impact energy. New 167 impacts were performed on the 100 × 100 plate with three "L-shape" stringers, with a dropping ball from three different increasing heights, 10, 20, 40 cm. This choice allowed to consider three values of energy with each equal to double the previous one.

Processing the acquired waveforms for each of the 501 impacts, the power integration of the waves was calculated in order to build a new database with the energy as an additional feature. In a first step all previous regression algorithms were applied to this database, but the implemented models were not able to predict the correct energy levels. It is noteworthy that the presented scenario (with only three energy levels) can not be solved by a regression approach, because it represents a pattern recognition problem. Therefore, a pattern recognition Bayesian Regularization trained SNN, was considered, with the following configuration: Z-Score scaling; a hidden layer with a sigmoid function; an output layer with *softmax* function; maximum 1000 epochs; 50 test cases campaign with 70/30 training/test ratio. Each configuration was trained (MATLAB *train* function [20]) increasing the size of the hidden layer from 10 to 50 neurons. The best *Classification Error* (*CE*), 1.2%, corresponded to a Bayesian Regularization trained model, containing 30 neurons. In Fig. 10 the confusion matrix is related to the best test case.

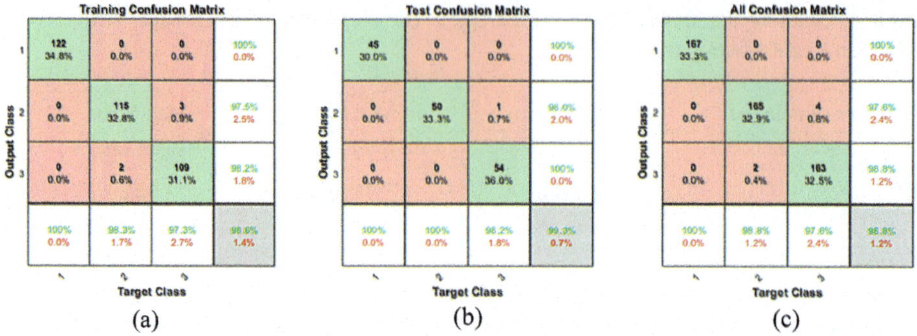

Fig. 10. Confusion matrix for a pattern recognition SNN with 30 neurons.

5 Conclusions

This work focused on the building of ML models that are able to predict the location of impacts with low speed, regardless of the material. The present study just needs four sensors and considers three 100 × 100 cm aluminium specimens: a simple panel, a panel with a "L-shape" stiffener and one with three "L-shape" stringers. After carrying out experimental activities, a 167 impacts database was built and polynomial regression and SNN were applied to it, limited to the panel without stringer. In order to get a good trade-off between the *MRE* (mean radial error between the predicted and actual coordinates) and the generalization, the polynomial degree $d = 3$ was identified as the best choice. At the same time, the Bayesian Regularization algorithm was the best choice for the SNN training. The *MRE* was calculated as mean over all the samples and over 50 test cases. For the 100 × 100 panel, the minimum polynomial regression error was 5.48 mm, while, applying the SNN, the best *MRE,* 3.08 mm, corresponded to a BR trained model, containing 30 neurons in the hidden layer. Subsequently, another dataset was built from the generation of 167 impacts on the panel with a single "L-shape" stiffener. In this scenario, applying the SNN, the best *MRE,* 4.24 mm, corresponded to a BR trained model, containing 30 neurons. The stringer makes the error worse, because of the different propagation speed within the reinforced zone and the resulting inconsistency of the ToFs, as confirmed by ANSYS simulations. Subsequently, another dataset was built from the generation of 167 impacts on the plate with other two "L-shape" stringers. In this scenario, applying the SNN, the error was double the previous one, because of the "multiplied" ToF inconsistency. A more complex SNN model was built, adding another hidden layer: the best *MRE* was 6.88 mm. The last analysis was to apply a pattern recognition SNN on a new dataset, in order to predict the impact energy on the plate with 3 "L-shape" stringers, dropping a ball from three different increasing heights. The best classification error, 1.2%, corresponded to a Bayesian Regarization trained model, containing 30 neurons.

In conclusion, this work can contribute in terms of novelty to the development of SHM systems, as it demonstrates with experimental data the machine learning approach validity, even with reinforcements and "discontinuities". The excellent agreement between the actual and predicted values makes it suitable for SHM systems, regardless of panel complexity and encourage the authors to verify the method on more complex structures (thickness variable thin structures, different materials and so on).

References

1. Boffa, N.D., Arena, M., Monaco, E., Viscardi, M., Ricci, F., Kundu, T.: About the combination of high and low frequency methods for impact detection on aerospace components. Prog. Aerosp. Sci. **129**, 100789 (2022). https://doi.org/10.1016/J.PAEROSCI.2021.100789
2. Rocha, H., Lafont, U., Nunes, J.P.: Optimisation of through-thickness embedding location of fibre Bragg grating sensor in CFRP for impact damage detection. Polymers **13**, 3078. 13 (2021) 3078. https://doi.org/10.3390/POLYM13183078
3. Nicassio, F., Carrino, S., Scarselli, G.: Elastic waves interference for the analysis of disbonds in single lap joints. Mech. Syst. Signal Process. **128**, 340–351 (2019). https://doi.org/10.1016/J.YMSSP.2019.04.011
4. Hassani, S., Mousavi, M., Gandomi, A.H.: Structural health monitoring in composite structures: a comprehensive review. Sensors 2022 **22**, 153 (2021). https://doi.org/10.3390/S22010153
5. Seifoori, S., Mahdian Parrany, A., Mirzarahmani, S.: Impact damage detection in CFRP and GFRP curved composite laminates subjected to low-velocity impacts. Compos. Struct. **261**, 113278 (2021). https://doi.org/10.1016/J.COMPSTRUCT.2020.113278
6. Sofi, A., Jane Regita, J., Rane, B., Lau, H.H.: Structural health monitoring using wireless smart sensor network – an overview. Mech. Syst. Signal Process. **163**, 108113 (2022). https://doi.org/10.1016/J.YMSSP.2021.108113
7. Elahi, H.: The investigation on structural health monitoring of aerospace structures via piezoelectric aeroelastic energy harvesting. Microsyst. Technol. **27**(7), 2605–2613. https://doi.org/10.1007/S00542-020-05017-Y
8. Broer, A.A.R., Benedictus, R., Zarouchas, D.: The need for multi-sensor data fusion in structural health monitoring of composite aircraft structures. Aerospace **9**, 183. https://doi.org/10.3390/AEROSPACE9040183
9. B. Kurian, R. Liyanapathirana, Machine Learning Techniques for Structural Health Monitoring, in: Lecture Notes in Mechanical Engineering, Pleiades Publishing, 2020: pp. 3–24. https://doi.org/10.1007/978-981-13-8331-1_1
10. Ai, L., Soltangharaei, V., Bayat, M., van Tooren, M., Ziehl, P.: Detection of impact on aircraft composite structure using machine learning techniques. Meas. Sci. Technol. **32**, 084013 (2021). https://doi.org/10.1088/1361-6501/ABE790
11. Shi, S., Du, D., Mercan, O., Kalkan, E., Wang, S.: A novel unsupervised real-time damage detection method for structural health monitoring using machine learning. Struct. Control. Health Monit. **29**, e3042 (2022). https://doi.org/10.1002/STC.3042
12. C.M. Bishop, Pattern Recognition and Machine Learning - Springer 2006
13. Dipietrangelo, F., Nicassio, F., Scarselli, G.: Structural Health Monitoring for impact localisation via machine learning. Mech. Syst. Signal Process. **183**, 109621 (2023). https://doi.org/10.1016/J.YMSSP.2022.109621
14. Piezoelectric Discs, (n.d.). https://www.physikinstrumente.com/ (accessed March 1, 2022)
15. University of Salento AS.S.E. Lab. https://asselab.unisalento.it/en/, (n.d.)
16. Pico Technology. https://www.picotech.com/
17. Giurgiutiu, V.: Structural health monitoring (SHM) of aerospace composites. Polym. Compos. Aerosp. Ind., pp. 491–558 (2020). https://doi.org/10.1016/B978-0-08-102679-3.00017-4
18. Carrino, S., Maffezzoli, A., Scarselli, G.: Active SHM for composite pipes using piezoelectric sensors. Mater Today Proc. **34**, 1–9 (2019). https://doi.org/10.1016/J.MATPR.2019.12.048

19. Schindler, P.M., May, R.G., Claus, R.O., Shaw, J.K.: Location of impacts on composite panels by embedded fiber optic sensors and neural network processing. In: Smart Structures and Materials 1995: Smart Sensing, Processing, and Instrumentation, vol. 2444, pp. 481–489 (1995). https://doi.org/10.1117/12.207698
20. Paluszek, M., Thomas, S.: MATLAB Machine Learning Recipes: A Problem-Solution Approach, 2nd edn, pp. 1–347 (2019). https://doi.org/10.1007/978-1-4842-3916-2/COVER
21. Glorot, X., Bengio, Y.: Understanding the difficulty of training deep feedforward neural networks, (n.d.). http://www.iro.umontreal. (accessed November 15, 2022)
22. Kayri, M.: Predictive abilities of Bayesian regularization and Levenberg–Marquardt algorithms in artificial neural networks: a comparative empirical study on social data. Math. Comput. Appl. **21**, 20–21 (2016). https://doi.org/10.3390/MCA21020020
23. Møller, M.F.: A scaled conjugate gradient algorithm for fast supervised learning. Neural Netw. **6**, 525–533 (1993). https://doi.org/10.1016/S0893-6080(05)80056-5
24. de Luca, A., Perfetto, D., Lamanna, G., Aversano, A., Caputo, F.: Numerical investigation on guided waves dispersion and scattering phenomena in stiffened panels. Materials 2022, **15**, 74 (2021). https://doi.org/10.3390/MA15010074

Image-Based Transient Detection Algorithm for Gravitational-Wave Optical Transient Observer (GOTO) Sky Survey

Terry Cortez[1](✉), Tossapon Boongoen[1], Natthakan Iam-On[1], Khwunta Kirimasthong[1], and James Mullaney[2]

[1] Center of Excellence in AI and Emerging Technologies, School of Information Technology, Mae Fah Luang University, Chiang Rai 57100, Thailand
6151501002@lamduan.mfu.ac.th, {itschool,itschool,khwunta.kir}@mfu.ac.th
[2] Department of Physics and Astronomy, University of Sheffield, Sheffield S3 7RH, UK
j.mullaney@sheffield.ac.uk

Abstract. The paper proposes an alternative method to detect transients for an optical telescope using a deep learning algorithm. While the previous studies followed the conventional method, a classification, focusing on the manually extracted features of transients, the alternative method does the detection focusing on imagery instead. The algorithm is based on a famous UNET model which can do both classification and segmentation at the same time. Initial setups are used to test the capability of the model. In the same way, some data is fused with noise to see the model limitation. The result is a map showing where the objects are located, identified by binary class numbers. Both results either with or without noise are provided. This includes a comparison between different batch size setups as one of the key parameters for deep learning. As a preliminary to further studies, a list of possible parameters is also given.

Keywords: Sky survey · Transient detection · Convolutional neural network · UNET

1 Introduction

Being a computation-based counterpart of the advanced Laser Interferometer Gravitational-wave Observatory (aLIGO) which has an innate capability to detect gravitational-wave related astronomical events, given the name "transients" by astronomers, using a combination of scientific devices [2], the Gravitational-wave Optical Transient Observer (GOTO) relies on a data science approach, utilizing a large amount of data given by a telescope. The data is collected and necessarily analyzed in real-time to capture these events which are happening temporarily. However, it has not had an efficient method to detect transients due to limited resources, namely a storage space, and distinct data

N. Naik et al. (Eds.): UKCI 2023, AISC 1453, pp. 459–470, 2024.
https://doi.org/10.1007/978-3-031-47508-5_36

properties such as its sources being rather single-channel 64-bit float images. A strategy is to extract features from each interesting object in the images and let a machine learning algorithm decide whether it is transient or not [15,16]. Thus, a collaboration between two fields of study, astronomy and computer science, existed.

Although a conventional method suggests images, as raw data, to be preprocessed to retrieve several useful features for future analyses, namely a Transient Detection (TD), it comprises too many steps opposing the facts that a sky is persistently changing and transients are expected to be found in real-time. With feature-based TDs using classification algorithms, such as k-NN, tree, and forest, a huge chunk of storage space is required to contain features extracted from the raw data. Then, they are fed into an algorithm to proceed, consuming time depending on the number of resources available. At the same time, the features are considered not exactly the primary data but a mere product of an aforementioned preprocessing algorithm or a telescope pipeline whose results can be adjusted based on its initial parameters, affecting the quality of classification. Therefore, this research is going to bypass those two circumstances by using an image-based approach called a Convolutional Neural Network (CNN).

The contents of this paper are as follows: II. Related Work showing previous studies including CNNs in object classifications and detections, III. Methodology showing model, data preparation, parameters, and evaluation method used for the proposed algorithm, IV. Results including a comparison between different model setups, and V. Summary. In addition, the research is based on the same project under the GOTO direction and can be considered as the latter phase of a study.

2 Related Work

Usually, single-type-object detections can be considered as binary classifications of big data in which algorithms' potential is depending on factors including model setups and data characteristics. In a GOTO study, however, an artifact, bogus object, or background, which is one of the two classes, shares a large proportion against another class, which is a real source or transient event, causing the whole dataset to be imbalanced. This results in algorithms biasing toward the majority, ignoring and misunderstanding some of the transients. Studies in the former phase coped with data imbalance by implementing a variety of statistical methods, including undersampling and SMOTE technique, to manipulate several classes for the algorithms' training step [12,15,16]. One of them also suggested that the extracted features need to be filtered for the models' effectiveness. In other words, some features highly correspond to the classifications and vice versa [15,16]. While the former phase kept on with a conventional method, this study focuses on the raw data itself in the form of images retrieved from optical telescopes. This proposes an alternative algorithm that is expected to shorten the processes in TD, suitable for real-time detection. Unlike the conventional one, it does not require additional storage space to collect in-process data.

The concept of using deep learning algorithms or CNNs comes from object identifications (OIs) and change detections (CDs), commonly used in geography, military, and city surveys [3–5,7,9,10,14,18–21]. The uses of CNNs can also be found in industrial [8] and biomedical [6,11,22] fields. With the help of pixel-wise classifications or image-based algorithms that comprise multiple layers, different features are automatically extracted and analyzed, end to end. To put simply, low-level features, for example, lines, edges, and small objects are usually analyzed in the former CNNs' layers while high-level features such as large objects are usually analyzed in the latter CNNs' layers [14]. CNNs can be used in multiple ways depending on the purpose of the studies. For example, OIs train models to know different objects, and feed them a single input to get a feature map labeling those objects [6,8,11,20–22]. The objects which model interest or ignore rely on the training data given. On the other hand, CDs train models to find differences between two images (t1 and t2) at different times and feed them a pair of inputs to get feature map labeling changes [3–5,7,9,10,14,18,19]. The true changes, such as farmland, and environmental changes, such as clouds, can also be classified or ignored, relying on the training data given. They can be powerful to detect changes even in temporal images from slightly different viewpoints [5]. Moreover, with a multiple-layer structure, CNNs are quite flexible. The configurable layers decide how the networks' nodes communicate. CNNs can be combined [3–5,9,18], sharing their information inside while another algorithm lets two models perform dual operations [19]. Temporal images can also be interpreted as sequences, which are compatible with R-CNNs [3]. These configuration examples also make CNNs able to withstand smaller training sets and give them the ability to segment the objects precisely.

Similar to CDs, the raw data in TDs is sky imagery, instead of an earth or city image. Each night batches of sky images are captured by GOTO and kept in containers called FITS files for further studies. The files contain a single image with telescope metadata. The FITSs are fed through a preprocessing pipeline, as in conventional methods, to extract features to be analyzed. Because this phase uses pixel data inside FITS files, this step is ignored. Other information in FITS files has no benefit during this phase. In addition, there are writings about the depreciation of a FITS container; however, it is not much related to the study [13,17]. The details of the telescope and pipeline can also be read in the previous work [15,16].

3 Methodology

3.1 Model

One of CNN architectures called UNET is selected for GOTO image-based approach due to its advantages. Not only does UNET detect or classify objects but it also does segmentation. Moreover, it requires only a small amount of data during a training phase for its accuracy [11]. The latter benefits the experiments in which the training data is insufficient. In this case, an augmentation process is absent throughout this research to observe UNET's aforementioned ability.

As the name suggests, the architecture comprises two main parts that are an encoding process or convolution on the left side of U and a decoding process or deconvolution on the right side of U. To simplify, convolution layers do a pixel-wise analysis while deconvolution layers generate a feature map. The deeper layers are, the higher details are processed.

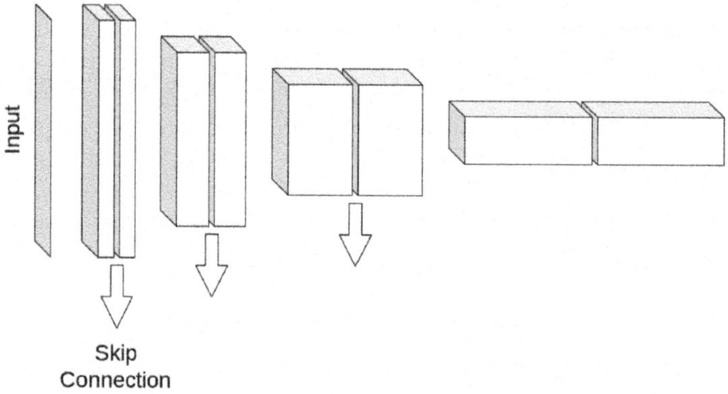

Fig. 1. A convolution section of UNET.

The convolution section contains a filtration system in which input images are analyzed pixel by pixel. In each depth, an image is convoluted followed by batch normalization. A single dropout layer is activated to prevent an overfit before the convolution and normalization are repeated. To sum up, each depth includes two convolution layers, resulting in a rescaled image with an additional dimension containing analyzed details. These depth-by-depth results are also retained for the deconvolution section. Between the depths, max pooling is implemented to resize the image by half, for higher details to be analyzed. After the last depth, input images are converted into tiny images with a very long dimension created by convolution layers, and then the deconvolution process begins.

The deconvolution section contains either transposed convolution layers or upsampling with convolution layers, to recreate output images with the same size as inputs, in which pixel values are replaced by class numbers. The details in the additional dimension are used to reconstruct the output, removing them from each layer of deconvolution. To improve classification and segmentation, the retained results from each depth in the last section are used in this part, called skipped connections. This connection concatenates that residue with an image from transposed convolution at the corresponding depth, followed by double convolution layers with no dropout to blend them. After the last layer, a sigmoid function is activated to convert its values into classes. However, the class numbers still range from zero to one as they require another threshold to separate them into binary classes, as a final output [5,7,11,14,19].

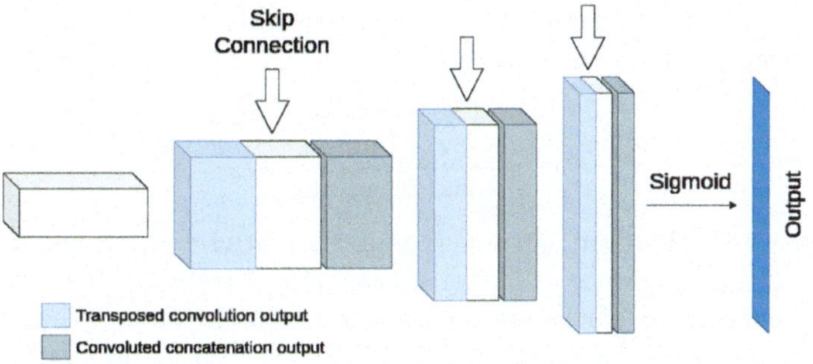

Fig. 2. A deconvolution section of UNET.

3.2 Data Preparation

Because transients are rare events, it is difficult to find them in images for experimentation. Thus, simulated transients are generated to be consistent with features collected from a conventional method. Several transients are injected into other sky images where transients are absent. This creates two images, one which contains no transient and vice versa, representing temporal images t1 and t2 respectively. The injection also creates binary maps to be used as ground truths for a training phase. Besides, the number of injections is limited as it is required to create a class imbalance situation, similar to feature-based experiments.

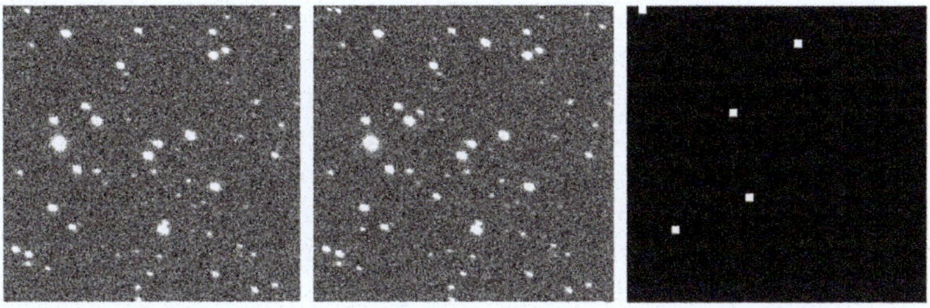

Fig. 3. Image pair (left–center) and ground truth (right).

The ground truths are similar to the feature map. They contain only zero and one. The sky images, however, contain many possible pixel values, which represent brightness or magnitude, as they are 64-bit float images. Thus, the pixel values should be normalized for ease of analysis. A proper range is between zero and one, where zero means no magnitude and one means maximum magnitude. As there is no absolute maximum and minimum magnitude for any object found, `PercentileInterval()` from the Python package called an Astropy is used to find the proper max–min range across a dataset [1]. Its parameter is set to

97. After the range is capped, normalized values can be calculated using the following formula.

$$x' = \frac{x - \min(x)}{\max(x) - \min(x)}$$

To create a single input, an image pair is created by stacking both temporal images into a single two-channel image. For a training phase, both image pairs and their ground truths are required to teach the model what needs to be detected. On the other hand, a testing phase, as well as, in practice required only image pairs for the algorithm to work. The idea is that the first image, t1, is a reference one where no transient exists, and the second image, t2, is a newly captured one which needs to be scanned for transients. Wherever an image, t2, is taken, there is always a reference image, t1, to create an input pair. Similar to CD, the pair is fed to the model, which is trained beforehand, and a feature map is given as a result. Note that, creating differences, on the other hand, can also be used at this step as an alternative method.

Unlike CDs, an OI approach might be achieved by ignoring a reference image, t1 both in training and in practice that results in an algorithm trying to classify every object in an image instead of finding a change. However, this can worsen the result because, from an optical viewpoint, transients share some similarities, such as brightness and shape, with other persistent objects. There is a chance for the algorithm to mistake stars and galaxies for transients. In CDs, the chance is very low since the persistent objects are the same in both images. In this case, the true obstacle is noise both from the telescope and the environment.

To examine the capability of the algorithm, images containing multiple levels of noise are used in a testing phase. To put it simply, one dataset is used to retest the algorithm again and again. This dataset is, on the other hand, fused with three levels of noise. The noise comes in the form of arrays of Gaussian distributions, having the mean of one with different standard deviations, with the same size as dataset images. By fusing, sky images are multiplied by those arrays, resulting in slight to moderate changes in pixel values. Images in the pairs are also modified separately, so the noise in each pair is not the same. This represents situations where environment and error affect capturing of sky images at different times. The three levels are as follows: (i) zero noise, (ii) gaussian distribution with 0.2 standard deviation, and (iii) gaussian distribution with 0.4 standard deviation. This, however, does not apply to the training phase to test the robustness of the algorithm fed with basic data without noise.

To sum up, one image pair has a shape of (256,256,2) while its ground truth has a shape of (256,256,1). The number of inputs in the training phase is 1,920 pairs, whose ten percent is retained for validation. The number of inputs in the testing phase is 484 pairs. During the evaluation, an instance is considered as a one-pixel value. Therefore, the total number of instances is 31,719,424 for the testing phase. According to the ground truths, 92,980 of them are transients whereas 31,626,444 are background. Note that the shape of transients in ground truths is 7x7 pixels, regardless of the shape of real transients in image pairs.

Fig. 4. Noise levels: No noise (left), SD = 0.2 (center), and SD = 0.4 (right).

This also means that one instance is just a part of one transient, and a square cluster of 49 instances should represent one transient.

3.3 Configuration

An epoch number is set to 30 while a training process can be stopped early when improvements in a validation loss are lower than 0.01 for ten consecutive epochs. After the process is stopped, the best weight, in which the validation loss is the lowest, is selected for the testing phase. Other parameters, which can affect the algorithm, and their values are as follow:

- Initial dimension size created by the first convolution layer: 64
- Dimension size multiplier in each convolution layer: 2
- Convolution/Deconvolution depth: 4
- Dropout rate: 0.05
- Deconvolution method (Upsampling/Transposed): Transposed
- Convolution layers' Activation function: Rectified linear unit (ReLU)
- Feature map threshold: 0.5
- Loss function: Binary cross entropy
- Optimizer: Adam
- Batch size: 8, 16, and 32

It is said that focal loss [19] and dice coefficient loss [10] functions are better for imbalanced classes. This paper, nonetheless, sticks to the default one, a binary cross-entropy loss function, to see whether the imbalance in GOTO's data can hinder the algorithm.

The experiment is conducted on Google Colab using the potential of its A100 GPUs. With these processors, the model runs faster, from days to minutes. One of the parameters affecting running speed is the batch size. When the batch size is large, the model runs faster. However, different sizes lead to different results during a process of trial and error. In this case, three sizes are also chosen for the experiment.

3.4 Evaluation

Three metrics, precision, recall, and F1 score, are selected for evaluation purposes. Both of them range between 0 and 100, where 100 is perfectly accurate. Precision represents the correctness of transients detected while recall represents the number of transients detected. According to the GOTO's purpose to find transient events as many as possible, recall is preferable. Thus, it is better to consider the two metrics separately. Anyway, the F1 score is also provided, showing a mean combination of those two metrics.

4 Results

After the best weights are selected, they are used to predict the testing datasets. In these circumstances, there are three models, both trained in the absence of noise with different batch-size setups. For predictions of the data without noise, all of them give good precisions, which are higher than 90. Their recalls are, however, varying. Only the model trained with the medium batch size gives a recall rate higher than 90 while the other models give recall rates close to 50. These show the outcomes of a change in batch size where batches larger are too rough and smaller are too fine.

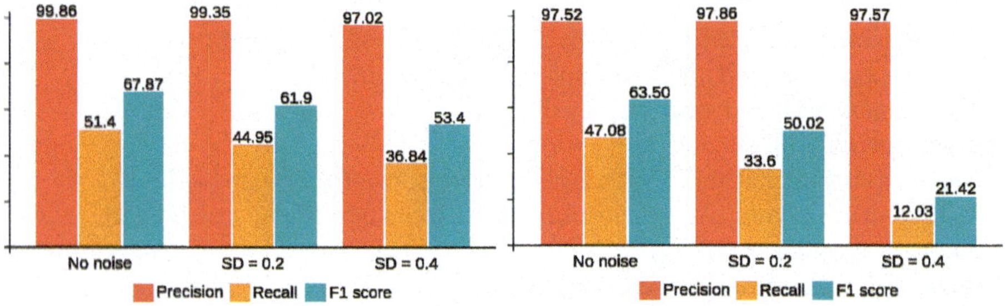

Fig. 5. Evaluation metrics for the models trained with batch sizes of 8 (left) and 32 (right).

In a situation where noise has roles, the model trained with a batch size of sixteen does not meet an expectation, especially with a noise-fused dataset with 0.4 standard deviations, resulting in a precision ratio of 6.77. Surprisingly, it is quite easy for human eyes to find transients in images fused with noise even if the noise levels in this study are high as the appearance of transients is still clear. A sudden drop in precision, on the other hand, maybe a sign of a lack of noise information. In other words, the models try to capture every change including the ones from noise even though these are not the focused objects as shown in Fig. 8e. These effects are gradually lifted when the batch sizes change as rough-tuned models ignore the noise and fine-tuned models reject small differences.

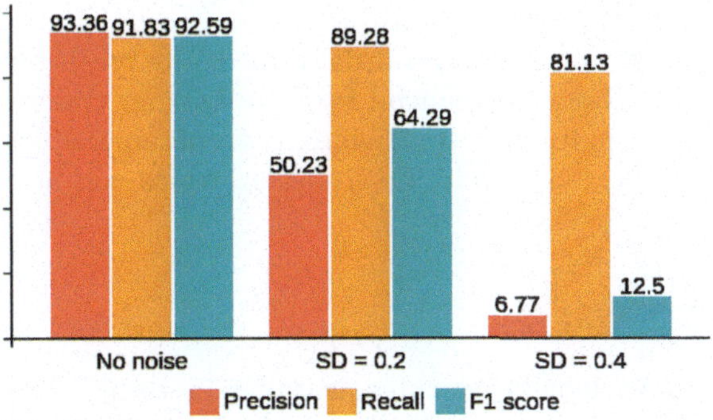

Fig. 6. Evaluation metrics for the models trained with a batch size of 16.

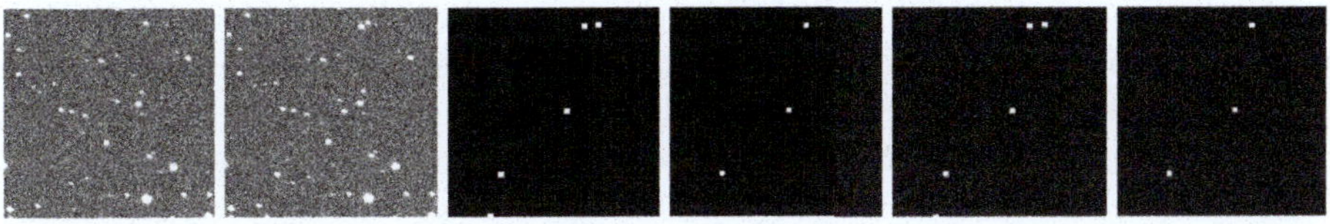

Fig. 7. Predictions where noise is absent. **a** first image. **b** second image. **c** ground truth. **d** prediction (batch sizes = 8). **e** prediction (batch sizes = 16). **f** prediction (batch sizes = 32).

As expected in the three models, the recall ratios are reasonably expected to drop when noise levels increase. In this case, the sixteenth batch size model gives the best rates. Its recall is higher than 80, regardless of noise levels. This can be a result of the model trying to capture all the changes as mentioned before. The other two model setups, however, give lower recall ratios, especially for the model trained with a batch size of 32, having a recall rate of 12.03 where the highest level of noise exists. This indicates that large batch sizes are not sufficient for this kind of detection even though they can make the training speed increase.

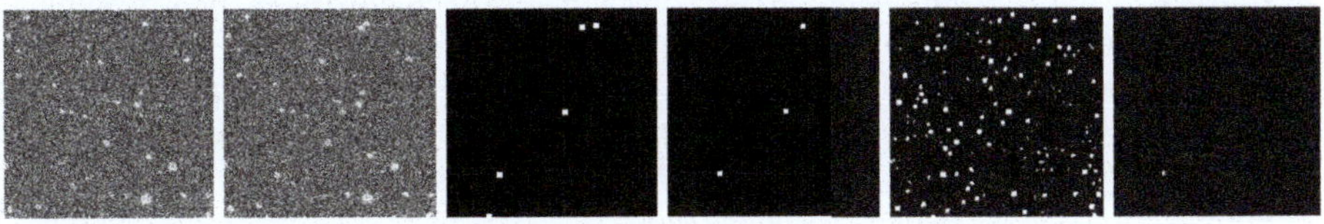

Fig. 8. Predictions where the noise level is highest. **a** first image. **b** second image. **c** ground truth. **d** prediction (batch sizes = 8). **e** prediction (batch sizes = 16). **f** prediction (batch sizes = 32).

Normally, TD mainly focuses on recall ratios rather than precision ratios for the reason that no object should be left behind as a false negative. This indicates that the second model trained with a sixteenth batch size is acceptable for future study as it can comprehend data due to its recall ability, although it lacks knowledge of noise. This can be fixed by making training datasets more general by adding noise-fused data. Besides, both parameter changes and model improvement are expected to help the situation, which will be conducted in the next study.

5 Summary

Sky survey seems to benefit machine learning in terms of data sizes which are enormous. Novel studies usually deal with the problem through statistical approaches and feature selection, using data derived from sky images. Alternatively, this study presents an approach from another perspective by learning from the images themselves. The UNET deep learning model is proposed to present how the system works. At the same time, little noise is added to measure the performance and robustness of the model.

The outcome of the UNET model can be seen through this paper although there are more complex setups of this model as shown in the past studies [6,8,10,11,20–22]. However, all of the models found in those papers belong to identification instead of detection. With this paper, it is shown that the model is also able to do the change detections. Interestingly, it is shown that the model can detect transients even though a small sign of underfit or overfit appears in each run. Thus, the work requires the inspection of parameter setups and training datasets for more robustness.

Acknowledgments. This work is partly funded by Mae Fah Luang University and also funded by Newton STFC–NARIT: Using astronomy surveys to train Thai researchers in Big Data analysis.

References

1. Astropy: PercentileInterval. https://docs.astropy.org/en/stable/api/astropy.visualization.PercentileInterval.html. [Accessed 15-Jul-2023]
2. Caltech: LIGO Technology. https://www.ligo.caltech.edu/page/ligo-technology. [Accessed 03-Jun-2023]
3. Cao, J., You, Y., Ning, Y., Zhou, W.: Change detection network of nearshore ships for multi-temporal optical remote sensing images. In: IGARSS 2020 - 2020 IEEE International Geoscience and Remote Sensing Symposium. IEEE (2020). https://doi.org/10.1109/igarss39084.2020.9323388
4. Daudt, R.C., Saux, B.L., Boulch, A.: Fully convolutional siamese networks for change detection. In: 2018 25th IEEE International Conference on Image Processing (ICIP). arXiv (2018). https://doi.org/10.48550/ARXIV.1810.08462. URL https://arxiv.org/abs/1810.08462
5. Guo, E., Fu, X., Zhu, J., Deng, M., Liu, Y., Zhu, Q., Li, H.: Learning to measure changes: Fully convolutional siamese metric networks for scene change detection. IEEE Transactions on Multimedia pp. 1–10 (2018). https://doi.org/10.48550/ARXIV.1810.09111. URL https://arxiv.org/abs/1810.09111

6. Hu, H., Zheng, Y., Zhou, Q., Xiao, J., Chen, S., Guan, Q.: MC-unet: multi-scale convolution unet for bladder cancer cell segmentation in phase-contrast microscopy images. In: 2019 IEEE International Conference on Bioinformatics and Biomedicine (BIBM). IEEE (2019). https://doi.org/10.1109/bibm47256.2019.8983121
7. Lan, L., Wu, D., Chi, M.: Multi-temporal change detection based on deep semantic segmentation networks. In: 2019 10th International Workshop on the Analysis of Multitemporal Remote Sensing Images (MultiTemp). IEEE (2019). https://doi.org/10.1109/multi-temp.2019.8866913
8. Lin, D., Cheng, Y., Li, Y., Prasad, S., Guo, A.: MLSA-UNet: End-to-end multi-level spatial attention guided UNet for industrial defect segmentation. In: 2022 IEEE International Conference on Image Processing (ICIP). IEEE (2022). https://doi.org/10.1109/icip46576.2022.9897416
9. Liu, T., Li, Y., Xu, L.: Dual-channel convolutional neural network for change detection of multitemporal SAR images. In: 2016 International Conference on Orange Technologies (ICOT). IEEE (2016). https://doi.org/10.1109/icot.2016.8278979
10. Peng, D., Zhang, Y., Guan, H.: End-to-end change detection for high resolution satellite images using improved UNet++. Remote Sens. **11**(11), 1382 (2019). https://doi.org/10.3390/rs11111382
11. Ronneberger, O., Fischer, P., Brox, T.: U-net: convolutional networks for biomedical image segmentation. In: 2015 Medical Image Computing and Computer-Assisted Intervention (MICCAI). arXiv (2015). https://doi.org/10.48550/ARXIV.1505.04597. URL https://arxiv.org/abs/1505.04597
12. Sangjan, T., Boongoen, T., Iam-on, N., Mullaney, J.: Classification of astronomical objects using light curve profile. In: 2019 IEEE Eurasia Conference on IOT, Communication and Engineering (ECICE). IEEE (2019). https://doi.org/10.1109/ecice47484.2019.8942673
13. Scroggins, M., Boscoe, B.M.: Once FITS, always FITS? astronomical infrastructure in transition. IEEE Ann. Hist. Comput. **42**(2), 42–54 (2020). https://doi.org/10.1109/mahc.2020.2986745
14. Slimani, Y., Hedjam, R.: Which cnn layer for which change? a CNN adaptation approach for change detection in remote sensing data. In: 2020 Mediterranean and Middle-East Geoscience and Remote Sensing Symposium (M2GARSS). IEEE (2020). https://doi.org/10.1109/m2garss47143.2020.9105168
15. Tabacolde, A.B., Boongoen, T., Iam-On, N., Mullaney, J., Sawangwit, U., Ulaczyk, K.: Transient detection modeling as imbalance data classification. In: 2018 1st IEEE International Conference on Knowledge Innovation and Invention (ICKII). IEEE (2018). https://doi.org/10.1109/ickii.2018.8569123
16. Tabacolde, A.B., Boongoen, T., Iam-On, N., Mullaney, J., Sawangwit, U., Ulaczyk, K.: Transient detection modelling for gravitational-wave optical transient observer (GOTO) sky survey. In: Proceedings of the 2018 10th International Conference on Machine Learning and Computing. ACM (2018). https://doi.org/10.1145/3195106.3195153
17. Thomas, B., Jenness, T., Economou, F., Greenfield, P., Hirst, P., Berry, D.S., Bray, E., Gray, N., Muna, D., Turner, J., de Val-Borro, M., Santander-Vela, J., Shupe, D., Good, J., Berriman, G.B., Kitaeff, S., Fay, J., Laurino, O., Alexov, A., Landry, W., Masters, J., Brazier, A., Schaaf, R., Edwards, K., Redman, R.O., Marsh, T.R., Streicher, O., Norris, P., Pascual, S., Davie, M., Droettboom, M., Robitaille, T., Campana, R., Hagen, A., Hartogh, P., Klaes, D., Craig, M.W., Homeier, D.: Learning from fits: Limitations in use in modern astronomical research. Astronomy and Computing **12**, 133–145 (2015). https://doi.org/10.48550/ARXIV.1502.00996. URL https://arxiv.org/abs/1502.00996

18. Varghese, A., Gubbi, J., Ramaswamy, A., Purushothaman, B.: ChangeNet: A deep learning architecture for visual change detection. In: 2018 European Conference on Computer Vision (ECCV). Springer International Publishing (2018). https://doi.org/10.1007/978-3-030-11012-3_10
19. Vinholi, J.G., Silva, D., Machado, R., Pettersson, M.I.: CNN-based change detection algorithm for wavelength-resolution SAR images. IEEE Geosci. Remote Sens. Lett. **19**, 1–5 (2022). https://doi.org/10.1109/lgrs.2020.3027382
20. Wang, Y., Gu, L., Jiang, T., Gao, F.: MDE-UNet: a multitask deformable UNet combined enhancement network for farmland boundary segmentation. IEEE Geosci. Remote Sens. Lett. **20**, 1–5 (2023). https://doi.org/10.1109/lgrs.2023.3252048
21. Zhang, D., Gade, M., Zhang, J.: SOF-UNet: SAR and optical fusion unet for land cover classification. In: IGARSS 2022 - 2022 IEEE International Geoscience and Remote Sensing Symposium. IEEE (2022). https://doi.org/10.1109/igarss46834.2022.9884504
22. Zyuzin, V., Chumarnaya, T.: Comparison of unet architectures for segmentation of the left ventricle endocardial border on two-dimensional ultrasound images. In: 2019 Ural Symposium on Biomedical Engineering, Radioelectronics and Information Technology (USBEREIT). IEEE (2019). https://doi.org/10.1109/usbereit.2019.8736616

Investigation of Efficient Approaches and Applications for Image Classification Through Deep Learning

Shruti Khandelwal and Shaligram Prajapat(✉)

International Institute of Professional Studies, Devi Ahilya Vishwavidyalya, Indore, M.P, India
shaligram.prajapat@iips.edu.in

Abstract. Deep learning has achieved significant success in image classification tasks. This study explores and compares efficient approaches for image classification using Convolutional Neural Networks (CNNs) and Recurrent Neural Networks (RNNs) within the context of deep learning. The primary goal is to enhance our understanding of the effectiveness of these architectures for image classification tasks. The research concludes with a comparative analysis of different methods employed, highlighting their strengths, weaknesses, and potential applications in blended learning and Decision Support Systems for the Indian Penal Code (IPC).

Keywords: Deep Learning · Image Classification · Convolutional Neural Networks (CNN) · Recurrent Neural Networks (RNN) · Benchmark Datasets · Hyperparameter Optimization

1 Introduction

Neural networks are models that are based on the principle of connectivism. They use interconnected nodes, called neurons in multiple layers to deduce the solution of a given problem. They are named so because they resemble the functioning and connectivity of neurons in the human brain. These networks consist of many layers. These layers are of three types: input layer, hidden layer, and output layer. The layers responsible for the main computation are the hidden layers.

In the context of image classification, neural networks offer the most efficient means for data extraction. It is the main contributor to the field of computer vision, along with reinforced machine learning algorithms. They are used for feature detection, image segmentation, and object recognition. Although several machine learning algorithms can also be used for computer vision, to tackle the problem of overfitting, deep learning algorithms offer the most optimized solution.

In recent years, deep learning has emerged as a transformative technology with remarkable achievements in various fields, particularly computer vision. In the domain of image classification, deep learning methods, such as Convolutional Neural Networks (CNNs) and Recurrent Neural Networks (RNNs), have demonstrated unparalleled success. Image classification, the task of categorizing images into predefined classes, is a

N. Naik et al. (Eds.): UKCI 2023, AISC 1453, pp. 471–487, 2024.
https://doi.org/10.1007/978-3-031-47508-5_37

fundamental problem in computer vision with numerous practical applications, ranging from medical diagnosis to autonomous vehicles.

The success of deep learning methods in image classification can be attributed to their ability to automatically learn intricate patterns and representations from vast amounts of data. CNNs, in particular, have shown exceptional performance by leveraging convolutional layers to capture local features in images and pooling layers to extract higher-level representations. On the other hand, RNNs excel at handling sequential data, making them suitable for tasks involving time series or sequential information.

Against the backdrop of these advancements, this study aims to investigate and juxtapose the efficiency of two deep learning methods in image classification tasks. Realizing the potency of CNNs and RNNs, this research dives into calibrating the limits and merits of these models when applied across diverse datasets. This study also explores the nuances of enhanced architecture, through intricate modifications, encompassing dropout layers, batch normalization, and diverse activation functions.

2 Literature Review

In recent years, deep learning techniques have achieved remarkable success in various domains, with image classification being one of the most prominent applications. Several researchers have explored and proposed different deep learning methods to enhance the efficiency of image classification tasks.

Liu et al. [7] investigated the effectiveness of transfer learning in image classification by leveraging pre-trained convolutional neural networks (CNNs). They demonstrated that fine-tuning CNN architectures, such as VGG16 and ResNet, on specific target datasets significantly improved classification accuracy and reduced training time. This approach allowed the models to benefit from the knowledge learned from vast source datasets, such as ImageNet, and adapt it to the target domain.

Zhang et al. [8] focused on exploring attention mechanisms in image classification tasks. They proposed a novel attention-guided CNN architecture that dynamically weighted different image regions based on their relevance to the classification decision. By attending to salient features, the attention-guided CNN achieved superior accuracy and better interpretability compared to standard CNN models.

Besides CNNs, recurrent neural networks (RNNs) have also been investigated for image classification tasks. Tan et al. [7] proposed a hybrid model that combined CNNs with bidirectional LSTM (Long Short-Term Memory) layers to capture both spatial and sequential information from images. This hybrid architecture demonstrated improved performance in tasks where temporal dependencies within images played a crucial role, such as action recognition and video classification.

In the pursuit of efficient deep learning models, Li et al. [10] introduced a method to reduce the computational complexity of CNNs by applying network pruning. By removing redundant connections and filters, the pruned CNN achieved a considerable reduction in inference time without significant loss in accuracy. This work provided insights into optimizing deep learning models for deployment in resource-constrained environments.

These works collectively highlight the diverse range of approaches and innovations in the field of deep learning for image classification. Transfer learning, data augmentation, attention mechanisms, hybrid architectures, and network pruning represent just a few of the strategies that researchers have explored to enhance the efficiency and performance of image classification models.

3 Background

Convolutional neural networks (CNNs) and recurrent neural networks (RNNs) are two types of deep learning algorithms that are commonly used for image classification.

CNNs are specifically designed for processing data that has a grid-like structure, such as images. They do this by using convolutional layers, which can learn spatial patterns in the data. This makes CNNs very effective for tasks such as image classification, object detection, and segmentation. Each layer of the convolutional neural network performs a very specific function, which can be seen in Table 1.

Table 1. Key features of CNN

Key components	Characteristics
Convolutional layers	• The core building blocks of CNNs are the convolutional layers • These layers apply convolutional operations to the input image using learnable filters or kernels • Convolution involves sliding the filters over the image to extract local features, such as edges, textures, and patterns
Pooling layers	• Pooling layers are used to downsample the spatial dimensions of the feature maps generated by the convolutional layers • Max pooling and average pooling are common techniques used in CNNs to reduce the size of the feature maps while retaining the most relevant information
Activation functions	• Activation functions introduce non-linearity to the network, allowing it to learn complex patterns • Common activation functions used in CNNs include ReLU (Rectified Linear Unit), sigmoid, and tanh
Fully connected layers	• The fully connected layers connect every neuron from the previous layer to every neuron in the next layer • These layers perform the final classification based on the high-level features learned from the previous layers

RNNs are designed for processing data that comes in sequences, such as text or speech. They do this by using recurrent connections, which allow them to remember information from previous inputs. This makes RNNs very effective for tasks such as natural language processing, machine translation, and speech recognition. Each layer of a Recurrent Neural Network performs a very specific function, which can be seen in Table 2.

Table 2. Key features of RNN

Key components	Characteristics
Recurrent connections	• The fundamental characteristic of RNNs is the presence of recurrent connections, which enable the network to use information from previous time steps while processing the current input • This recurrence allows RNNs to capture temporal dependencies within sequences
Hidden state	• The hidden state of an RNN serves as its memory, retaining information about the previous time steps • The hidden state is updated at each time step and influences the output for the current time step
Activation functions	• RNNs use activation functions, such as tanh or ReLU, to introduce non-linearity and process the input and hidden state

3.1 Activation Functions

Activation functions, also known as transfer functions, are used to determine the output of the network. They are used to bring non-linearity to the network, allowing the network to learn complex patterns in the data. They decide whether a neuron can contribute to the next layer. A neural network will almost always have the same activation function in all the hidden layers. The activation function should be *differentiable* so that the parameters of a network are covered during backpropagation. *Differentiability* means the change in the y-axis with respect to the x-axis. *Monotonic functions* are entirely non-increasing or non-decreasing. Activation functions can be linear and nonlinear (Fig. 1).

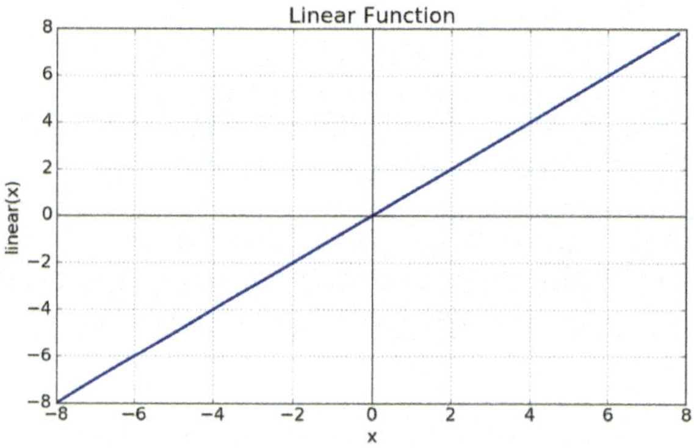

Fig. 1. Graph depicting linear function

3.2 Linear Activation Function

It is proportional to the input. It is represented as a straight line on a graph.

$$\begin{aligned} \text{Equation}: \ f(x) &= x \\ \text{Range}: \ (-\infty, \infty) \end{aligned} \tag{1}$$

The equation suggests that the gradient has no relationship with x. This also means that, during backpropagation, the adjustments made to weights and biases are not dependent on x.

3.3 Nonlinear Activation Function

Nonlinear functions make it easy for the model to generalize or adapt to a variety of data and to differentiate between the output. Non-linear activations are of many types. Some of them are examined in this study. Table 3 depicts the different activation functions, their graphs, ranges, and their major utility. Table 4 mentions the range of activation functions, as well as major criteria, such as non-linearity, monotony, and differentiability of functions. Table 4 also examines whether a function faces the vanishing gradient and dying ReLU problems. The *vanishing gradient problem* arises when, during the training of deep neural networks, the gradients that are used to update the network become extremely small or "vanish" as they are back-propagated from the output layers to the earlier layers. The *dying ReLU problem* arises when a ReLU function always outputs the same value, i.e., 0, on any input value. This condition is known as the dead state of ReLU neurons. In this state, it is difficult to recover because the gradient of 0 is 0. This becomes a problem when most of the input ranges are negative, or the derivative of the ReLU function is 0.

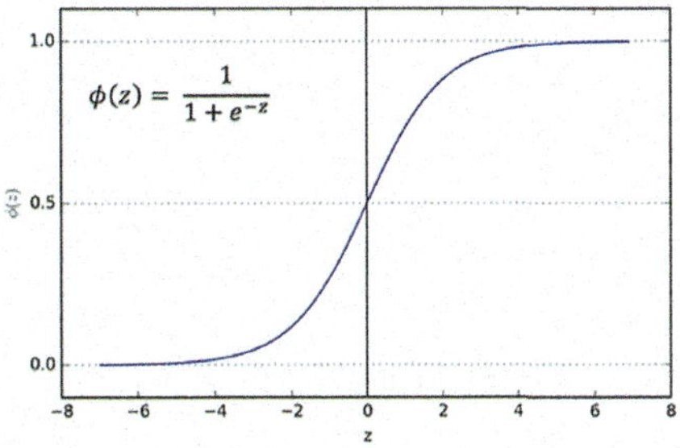

Fig. 2. Sigmoid function

4 Proposed Methodology

Convolutional neural networks(CNN) are specifically designed to process grid-like data, such as images. Recurrent Neural Networks, while primarily used to process sequential data, can also be used in image classification tasks involving sequential information, such as video classification. These models, when applied when different activation functions, produce results that vary in accuracy and efficiency.

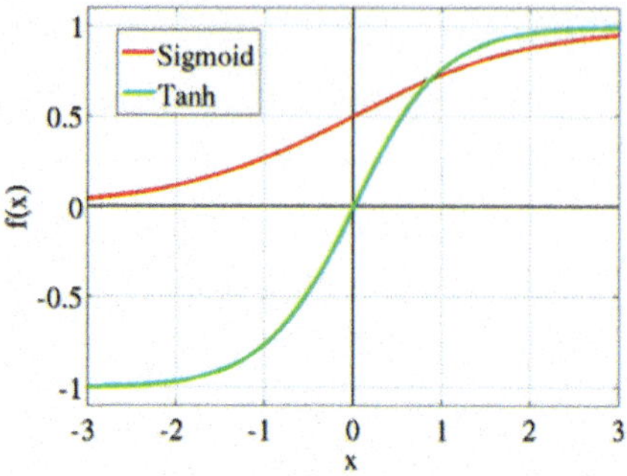

Fig. 3. Tanh function

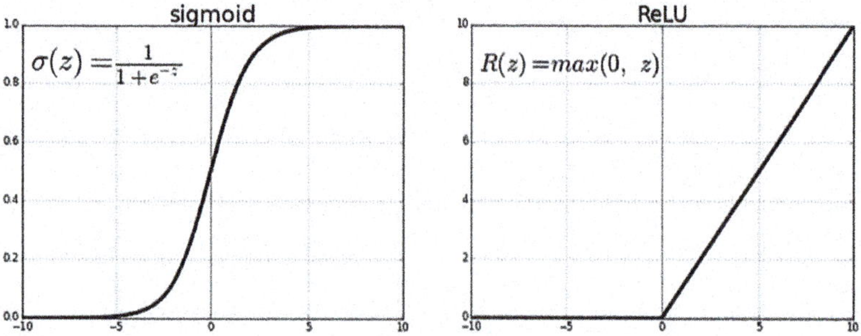

Fig. 4. ReLU function

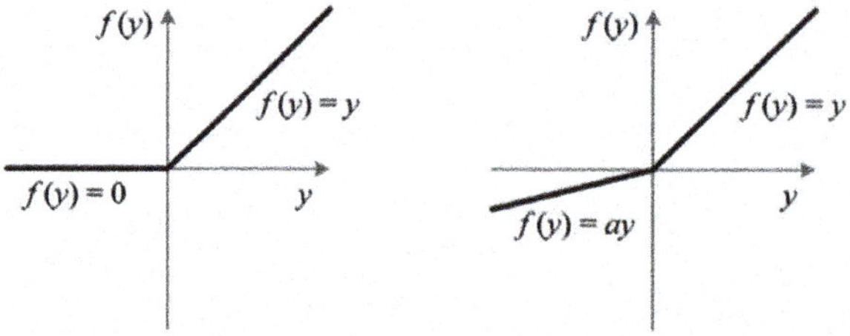

Fig. 5. Leaky ReLU function

As mentioned in Table 5, the two deep learning models examined in this study are compared to find out their benefits on various features. *Specialization* refers to the specific domain or type of data for which a neural network is designed. *Feature extraction* involves identifying and highlighting significant patterns or characteristics within data. *Spatial hierarchies* pertain to the arrangement of features or patterns in data with varying levels of abstraction. *Global information* encompasses the overall context or broader patterns present in the data. *Local information* refers to specific details or features within a smaller segment of the data. *Parallel processing* involves performing multiple computations simultaneously. *Transfer learning* is a machine learning technique where knowledge gained from training a model on one task is applied to a related task. *Image preprocessing*

Table 3. Non-linear activation functions

Non linear activation functions			
See Fig. 2 Range: (0,1)	See Fig. 3 Range: $(-1, 1)$	See Fig. 4 Range: $(0, \infty)$	See Fig. 5 Range: $(-\infty, \infty)$
$f(x) = 1/(1 + e^{-x})$ (2)	$f(x) = e^x - e^{-x}/(e^x + e^{-x})$ (3)	$f(x) = ½ + e^{-x}$ (4)	$f(x) = max(0,x)$ (5)
• It is specially used in models where the probability has to be calculated • Since the probability of anything exists only between 0 and 1, the sigmoid function offers the optimum function	• Since the tanh function has a similar range to the probability distribution of real-valued data • Therefore, it is often used in recurrent neural networks for natural language processing and speech recognition tasks	• Since the ReLU function outputs 0 for negative inputs. This means that the ReLU function cannot be used to represent probabilities that are less than 0 • The ReLU function can suffer from the dying ReLU problem, which can prevent the neural network from learning	• The leaky ReLU function is a good compromise between the ReLU function and the sigmoid function • The leaky ReLU function does not suffer from the dying ReLU problem, as it outputs a small positive value for negative inputs

Table 4. Activation functions and their Criteria

Activation Function	Range	Non-linearity	Monotonic	Differentiable	Vanishing Gradient Problem[a]	Dying ReLU Problem[b]
Sigmoid function	(0, 1)	Yes	Yes	Yes	Yes	Yes
Tanh function	(-1, 1)	Yes	Yes	Yes	Yes	Yes
ReLU function	$(0, \infty)$	Yes	Yes	Yes	No	Yes
Leaky ReLU function	$(-\infty, \infty)$	Yes	Yes	Yes	No	No

[a] Gradients of the sigmoid function can become very small, which can make it difficult for the neural network to learn

[b] The ReLU function can become stuck at 0 for some inputs, which can prevent the neural network from learning

involves preparing raw image data for analysis by applying various transformations, enhancements, or filters.

Table 5. Model and benefit of chosen algorithms for proposed problem domain

Aspect	Convolutional neural network (CNN)	Recurrent neural network (RNN)
Specialization	CNNs are specifically designed for image analysis	RNNs are more suited for sequential data and sequences
Feature extraction	Excellent at extracting hierarchical features like edges, textures, and shapes from images	Less focused on detailed pixel-level features
Spatial hierarchy	Can capture patterns at various spatial scales by using convolutional and pooling layers	May not inherently grasp spatial hierarchies
Global and local information	Balance both global context and local features due to convolutional layers and pooling operations	Might be more biased toward capturing temporal dependencies in sequences
Parallel processing	CNNs process image regions in parallel, leveraging shared weights for faster computation	Limited parallel processing due to sequential nature
Transfer learning	Pre-trained CNN models on large datasets can be fine-tuned for specific image classification tasks	Transfer learning may require adapting models differently for image classification tasks
Image processing	Robust to simple transformations like rotations, translations, and scaling	May require additional preprocessing for image data

As mentioned in Table 6, the models CNNs and RNN are compared on the basis of complexity, accuracy, and interpretability. *Complexity* refers to how intricate or involved the structure and operations of a neural network are. It encompasses the number of layers, connections, and parameters within the network architecture. *Accuracy* refers to how well a neural network can provide correct predictions or classifications for given input data. *Interpretability* pertains to the ease of understanding and explaining the inner workings of a neural network.

Table 6. Models and their comparison based on complexity, accuracy and interpretability.

Model	Complexity	Accuracy	Interpretability
Convolutional neural network (CNN)	Moderate to high complexity due to multiple layers, filters, and connections between layers	Excellent for image and grid-like data, excelling in feature extraction. Can achieve high accuracy in visual tasks	Moderate interpretability. CNNs can learn complex features, making it challenging to interpret every aspect
Recurrent neural network (RNN)	Moderate complexity in terms of layers and connections, can become complex in sequences	Good for sequence data, language modeling, and time series analysis	Moderate interpretability, although understanding inner workings might be difficult due to its sequential nature

4.1 Data Set Description

MNIST. The MNIST database[1] is a widely recognized benchmark dataset in the field of computer vision and machine learning. It comprises a collection of handwritten digits, each represented as a grayscale image. The dataset is commonly used for image classification tasks and serves as an initial step for evaluating the performance of various algorithms and models.

The MNIST dataset consists of two primary components: a training set and a test set. Each set contains grayscale images of handwritten digits ranging from 0 to 9. The training set contains 60,000 images, while the test set contains 10,000 images.

Each image in the MNIST dataset has a resolution of 28x28 pixels, resulting in a total of 784 pixels per image. The grayscale intensity of each pixel ranges from 0 (black) to 255 (white). The images are centered and normalized, contributing to consistency and ease of preprocessing.

CIFAR-20. The CIFAR-10 dataset[2] is a well-known benchmark in the field of computer vision and machine learning, specifically designed to evaluate the performance of image classification algorithms and models. It encompasses a diverse collection of colored images belonging to ten distinct classes.

The CIFAR-10 dataset is composed of 60,000 colored images, equally distributed across ten classes. These classes are: airplane, automobile, bird, cat, deer, dog, frog, horse, ship, and truck. The dataset is divided into two main subsets: a training set containing 50,000 images and a test set comprising 10,000 images.

Each image in the CIFAR-10 dataset has a resolution of 32x32 pixels. Unlike the MNIST dataset, CIFAR-10 images are colored, with three color channels (red, green, and blue) for each pixel. This results in a total of 3x32x32 = 3072 features per image.

4.2 Experimental Setup

The Datasets have been considered and were prepared using 10 cross-fold validation and percentage split. Data gets transformed to train and test data. The datasets have been divided into three subcategories. The sub-categories are: training dataset, validation dataset, and testing dataset.

A training dataset is used to train the data. A validation dataset is used to fine-tune the hyperparameters to produce accurate results. A testing dataset is used to test the data after the model has been trained. The dataset has been split into: 40% training dataset, 10% validation dataset, and 50% test dataset.

The models are iteratively trained and validated on these datasets. This method is known as cross-validation and is used to avoid the problem of overfitting.

4.3 Measures of Evaluation

Accuracy. It represents the number of correctly classified data instances over the total number of data instances.

$$Accuracy = \frac{TN + TP}{TN + TP + FN + FP} \tag{6}$$

Precision. It is defined as the ratio of correctly classified positive instances (True Positive) to the total number of classified positive instances (either correctly or incorrectly).

$$Precision = \frac{TP}{TP + FP} \tag{7}$$

Recall. It is calculated as the ratio between the number of Positive samples correctly classified as Positive to the total number of Positive samples. The recall measures the model's ability to detect positive samples. The higher the recall, the more positive samples detected.

$$Recall = \frac{TP}{TP + FN} \tag{8}$$

F1 Score. It is defined as the harmonic mean of precision and recall.

$$F1Score = 2 \times \frac{(Precision * Recall)}{(Precision + recall)} \tag{9}$$

4.4 Result Analysis

In the first experiment, a Convolutional Neural Network is paired with different activation functions, namely linear, sigmoid, tanh, and ReLU functions individually on the datasets. Table 7 depicts the various results obtained as a result of pairing a CNN with different activation functions. Figure 6 highlights the disparities between the different activation functions.

Table 7. Result Analysis of CNN paired with different activation functions.

Activation function	Precision	Recall	F1 Score	Accuracy
Linear	0.63	0.62	0.59	0.6226
Sigmoid	0.56	0.56	0.56	0.5574
TanH	0.65	0.65	0.65	0.6487
ReLU	0.50	0.49	0.47	0.6981

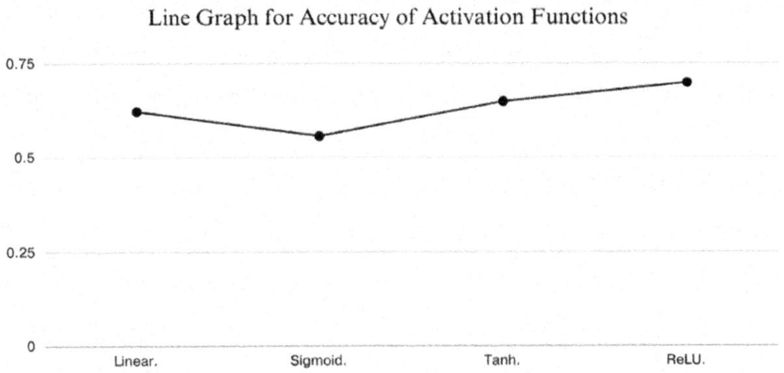

Fig. 6. Line Graph comparing accuracy of different activation functions with CNN model

In the second experiment, a Recurrent Neural Network is paired with different activation functions, namely linear, sigmoid, tanh, and ReLU functions individually on the 10 datasets. Table 8 depicts the various results obtained as a result of pairing an RNN with different activation functions. Figure 7 highlights the disparities between the different activation functions.

Table 8. Result Analysis of RNN paired with different activation functions.

Activation function	Precision	Recall	F1 Score	Accuracy
Linear	0.44	0.45	0.44	0.4469
Sigmoid	0.35	0.34	0.32	0.3416
TanH	0.40	0.39	0.37	0.3912
ReLU	0.40	0.40	0.39	0.4038

However, when comparing the accuracy of CNN and RNN models, the results show that the CNN model, when paired with any activation function, results in far superior results, when RNN is paired with the same function. This disparity is apparent in Figure 8.

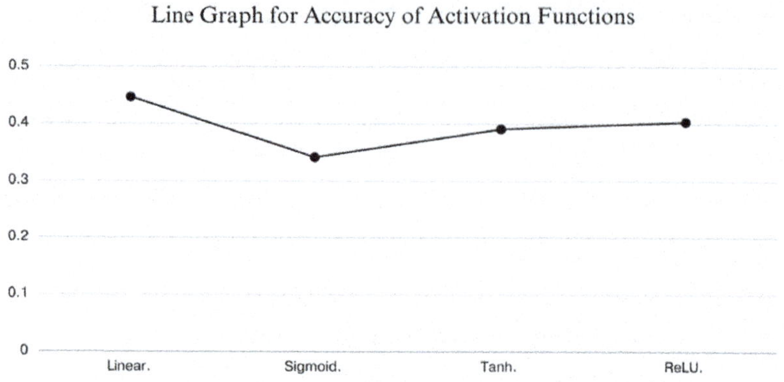

Fig. 7. Line Graph comparing accuracy of different activation functions with RNN model

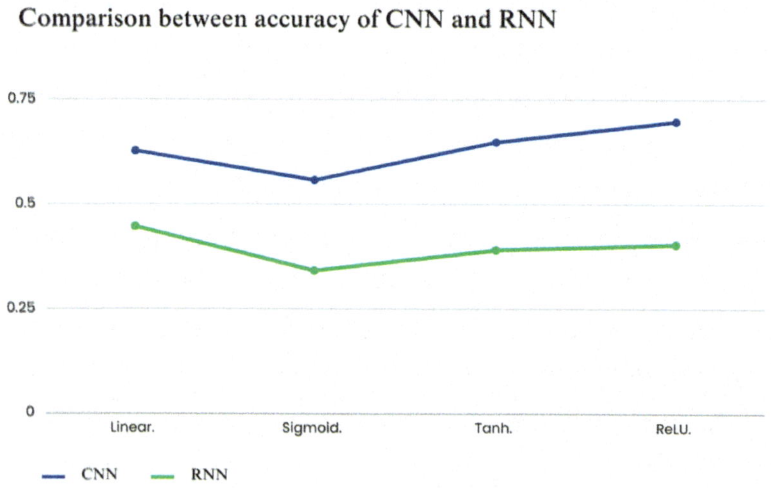

Fig. 8. Line Graph comparing accuracy of different activation functions with CNN, RNN models applications and future scope

5 Future Scope: Blended Learning and IPC Document Applications

In this section, we present the following two application domains where the proposed work can find versatility and usability.

(1) Blended Learning and e-Learning (4-Quadrants) [12]
(2) Decision Support System (DSS) for Indian Penal Code Section (IPC) [13] and [14]

The chosen applications are inherently tied to the investigation's core premise: enhancing efficiency in information classification. Whether applied to visual content in blended learning or textual content in legal analysis, the research underpins both domains with strategies to accurately categorize information. By exploring the potential of deep learning methods in image classification, we extend the reach of our investigation, fostering more effective and informed decision-making in diverse contexts.

Table 9 represents various benefits of Investigation of Efficient Approaches for Image Classification through Deep Learning in various domains.

Table 9. Possible scope towards blended learning and quadrants of MOOCs

Application	Remark	Benefits for MOOC
Improved image classification	Deep learning techniques have shown exceptional performance in image classification tasks	By investigating efficient approaches, the accuracy and efficiency of image classification models can be significantly improved
Enhanced learning analytics	Deep learning models generate vast amounts of data during the training and evaluation process	Analysis of training and evaluation data can provide valuable insights into learners' behavior, performance, and engagement patterns It can be used for personalized learning experiences
Automation and scalability	Efficient deep learning models can automate the image classification process, reducing the need for manual intervention	It makes it easier to scale image analysis tasks across a large number of learners and educational content
Real-time feedback	With faster and more efficient image classification algorithms, learners can receive real-time feedback on their performance,	It allows adjustment of learning strategies and improvement of understanding of the material
Personalized learning	By leveraging learning analytics from image classification, educators can tailor learning experiences to individual learners' needs, preferences, and learning styles	Educators can customize learning experiences to individual learners' needs, preferences, and learning styles

(1) **Applications for Blended Learning and e-Learning** from Investigation of Efficient Approaches for Image Classification through Deep Learning has the potential to revolutionize learning analytics and improve various aspects of blended mode education, making it more personalized, scalable, and effective for learners and educators alike.

By enhancing image classification accuracy, the study introduces a potential to revolutionize learning analytics in these educational modalities. The accurate categorization of visual data, made possible by the given research, can enable tailored content recommendations, adaptive assessment systems, and personalized learning pathways. This aligns with the paradigm of blended learning, where the integration of online and offline learning experiences is optimized to cater to individual learner needs. Our research offers the tools to create more engaging and efficient e-learning experiences by precisely analyzing and classifying visual materials, thereby enhancing learner engagement and knowledge retention. Following Table 10 depicts the application.

Table 10. Scopes for learning analytics and blended mode quadrants

Quadrants	Proposed application scope
E-tutorials	In e-tutorials, efficient image classification can be used to automatically assess learners' understanding of the material. ***By analyzing learners' responses to questions, educators can identify areas of difficulty and adjust the tutorials accordingly***
E-contents	Deep learning-based image classification can enhance e-contents ***by automatically tagging and categorizing images, making it easier for learners to search and access relevant content. It can also help identify suitable multimedia resources for different learning objectives***
Assessment	Image classification can play a crucial role in assessments by automatically ***grading and evaluating image-based questions***. This automation can save time for educators and provide immediate feedback to learners
Discussion forums	Deep learning-based image classification can be utilized in discussion forums to analyze and categorize images shared by learners. ***This can help identify trends, topics of interest, or potential issues that need further discussion***

(2) **Applications for Decision Support System (DSS) for Indian Penal Code (IPC)** from the Investigation of Efficient Approaches for Image Classification through Deep Learning can be found in various domains. Table 11 shows some potential applications for Decision Support System (DSS) for IPC:

The investigative nature of this study, focused on discerning optimal strategies within deep learning, complements the goal of the DSS to provide nuanced insights to legal professionals. This study's findings can contribute to the accurate identification of pertinent legal content from textual documents, a process akin to image classification. This resonates with the essence of decision support systems, where efficient and accurate data analysis underpins informed decisions. By showcasing efficient methods in the realm of image classification, the research indirectly strengthens the foundation of the DSS, enabling more effective legal analysis and decision-making.

Table 11. Possible Application, versatility and impact of DSS for IPC using deep learning image classification techniques

Application domain	Possible versatility and impact of DSS for IPC section
Criminal investigation and surveillance	DSS for IPC can be used in criminal investigation and surveillance systems to analyze and classify images captured from crime scenes, CCTV cameras, or other sources. ***It can assist law enforcement agencies in identifying suspects, vehicles, or objects related to criminal activities***
Evidence management	DSS can help in managing and categorizing visual evidence collected during criminal investigations. ***By automatically classifying images based on their relevance to the case, DSS can streamline the evidence management process and facilitate faster decision-making***
Facial recognition	DSS can be applied to facial recognition tasks, aiding in the identification of individuals involved in criminal activities or missing persons cases. ***It can be used to compare facial features against a database of known criminals or missing persons***
Document analysis	DSS can be extended to analyze and classify scanned documents related to legal cases. ***It can help in categorizing documents, extracting relevant information, and identifying key pieces of evidence***
Crime pattern recognition	DSS can assist law enforcement agencies in recognizing patterns and trends in criminal activities based on images and visual data. ***By identifying recurring patterns, investigators can better allocate resources and prevent future crimes***
Forensic analysis	DSS can support forensic experts in analyzing and interpreting forensic images related to crime scenes, accidents, or other incidents. ***It can aid in identifying critical details and reconstructing events***
Traffic violation detection	DSS can be used to detect traffic violations through image analysis of surveillance cameras or traffic sensors. ***It can help in identifying vehicles involved in speeding, red-light violations, or other traffic offenses***
Missing persons and amber alerts	DSS can contribute to identifying missing persons by analyzing and matching images of ***missing individuals against public databases or social media platforms***

(*continued*)

Table 11. (*continued*)

Application domain	Possible versatility and impact of DSS for IPC section
Border security	DSS can be employed in border security to ***analyze images and detect unauthorized border crossings, smuggling activities, or illegal immigration***
Cybercrime investigation	DSS can be extended to analyze digital images and multimedia content related to cybercrime cases. I*t can assist in detecting and categorizing images involved in cyberbullying, child exploitation, or other online crimes*
Criminal profiling	DSS can aid in criminal profiling by analyzing and classifying images related to the modus operandi of criminals. ***It can help investigators build profiles and anticipate future criminal behavior***
Witness identification	DSS can be utilized to assist witnesses in identifying suspects from a pool of images or to ***corroborate eyewitness testimonies***

6 Conclusion

This paper explored and compared efficient approaches for image classification using Convolutional Neural Networks (CNNs) and Recurrent Neural Networks (RNNs) within the context of deep learning methods. The investigation revealed the strengths and weaknesses of these architectures for image classification tasks on various datasets. For CNNs, the study found that they excel at capturing spatial features and achieving competitive results on datasets like MNIST and CIFAR-10. The use of different activation functions, such as linear, sigmoid, tanh, and ReLU, had varying impacts on the model's accuracy and efficiency. Tanh activation function performed the best among them, achieving a higher F1 score and accuracy. Whereas, RNNs showcased their strength in handling sequential data, particularly in video-related tasks. However, compared to CNNs, RNNs generally showed lower accuracy and F1 scores on the tested datasets. The incorporation of techniques like dropout layers, batch normalization, and activation functions demonstrated the potential for further improving model efficiency.

Further, the study discussed the potential applications of these efficient image classification models in the domains of blended learning and e-Learning, as well as in Decision Support Systems (DSS) for Indian Penal Code (IPC). In blended learning, the models can enhance image-based content, provide real-time feedback, and enable personalized learning experiences. For DSS in IPC, the models can assist in criminal investigation, evidence management, facial recognition, and other crime-related tasks. Future research in this domain could focus on exploring hybrid models that combine the advantages of CNNs and RNNs to address specific challenges in image classification and other computer vision tasks. Additionally, investigating the interpretability of these complex deep learning models could lead to improved transparency and trust in their decision-making processes, making them more suitable for real-world applications.

References

1. LeCun, Y., Cortes, C., & Burges, C.J.C.: The MNIST handwritten digit database (1998). Retrieved from http://yann.lecun.com/exdb/mnist/
2. Krizhevsky, A.: Learning multiple layers of features from tiny images. CIFAR-10 dataset (2009). Retrieved from https://www.cs.toronto.edu/~kriz/cifar.html
3. Ciresan, D. C., Meier, U., Gambardella, L.M., Schmidhuber, J.: Convolutional neural network committees for handwritten character classification. In: 2011 International Conference on Document Analysis And Recognition (pp. 1135–1139). IEEE (2011, September)
4. Ciresan, D. C., Meier, U., Masci, J., Gambardella, L.M., Schmidhuber, J.: Flexible, high performance convolutional neural networks for image classification. In: the Twenty-Second International Joint Conference on Artificial Intelligence (2011, June)
5. Simard, P.Y., Steinkraus, D., Platt, J.C.: Best practices for convolutional neural networks applied to visual document analysis. In *Icdar*, vol. 3, no. 2003 (2003)
6. Jarrett, K., Kavukcuoglu, K., Ranzato, M.A., LeCun, Y.: What is the best multi-stage architecture for object recognition? In: 2009 IEEE 12th International Conference on Computer Vision, pp. 2146–2153. IEEE (2009, September)
7. Han, T., Liu, C., Yang, W., Jiang, D.: Learning transferable features in deep convolutional neural networks for diagnosing unseen machine conditions. ISA Trans. **93**, 341–353 (2019)
8. Zhang, C., et al.: A deep learning image data augmentation method for single tumor segmentation. Front. Oncol. **12**, 782988 (2022)
9. Heaton, J., Goodfellow, I., Bengio, Y., Courville, A.: Deep Learning. The MIT Press (2016), 800 pp, ISBN: 0262035618. Genet. Program. Evolvable Mach., **19**(1–2), 305–307 (2018)
10. Prof. Li, J.J., Yeung, S.: Class Lecture, Topic: “Convolutional Neural Network for Visual Recognition.” CS231n, Stanford University, Stanford. Apr. 2019
11. Sagar Sharma. (2017, Sept. 6) *Activation Functions in Neural Networks* [Online].Available: https://towardsdatascience.com/activation-functions-neural-networks-1cbd9f8d91d6
12. Engati. (2021). *Vanishing Gradient Problem* [Online]. Available: https://www.engati.com/glossary/vanishing-gradient-problem#toc-what-is-the-importance-of-gradient-descent-in-training-neural-networks
13. Azhar, S.: What is the dying ReLU problem? (2023) [Online]. Available: https://www.educative.io/answers/what-is-the-dying-relu-problem
14. Prajapat, S., Agrawal, A., Anita, M.C.: Towards realization of assessment module using learning management system for integrated course. In: 2019 6th International Conference on Computing for Sustainable Global Development (INDIACom), pp. 1179–1183. IEEE (2019, March)
15. Harode, B., Prajapat, S., Bhurre, S.: Text processor for IPC prediction. In: Recent Innovations in Computing: Proceedings of ICRIC 2021, vol. 1, pp. 107–117. Springer Singapore, Singapore (2022)
16. Srivastav, A., Prajapat, S.: Text similarity algorithms to determine Indian penal code sections for offense reports. IAES Int. J. Artif. Intell. **11**(1), 34 (2022)

Healhcare Informatics

Deep Learning Based Lightweight Model for Brain Tumor Classification and Segmentation

Ifrah Andleeb[1], B. Zahid Hussain[1], Salik Ansari[1], Mohammad Samar Ansari[2(✉)], Nadia Kanwal[3], and Asra Aslam[4]

[1] Z. H. College of Engineering and Technology, Aligarh Muslim University, Aligarh, India
[2] University of Chester, Chester, United Kingdom
m.ansari@chester.ac.uk
[3] Keel University, Keele, United Kingdom
[4] Faculty of Medicine and Health, University of Leeds, Leeds, United Kingdom

Abstract. This paper presents two lightweight deep learning models for efficient detection and segmentation of brain tumors from MRI scans. A custom-made Convolutional Neural Network (CNN) is designed for identification of four different classes of brain tumors *viz.* Meningioma, Glioma, Pituitary brain tumor and normal (no tumor). Furthermore, another tailor-made lightweight model is presented for the segmentation of the tumor from the Magnetic Resonance Imaging (MRI) scans. The output of the segmentation model is the 'mask' depicting the tumor region. The overall performance in terms of detection accuracy, and segmentation accuracy, for the two models is found to be approximately 95% for both the cases individually. The proposed models are worthy additions to the existing literature on brain tumor classification and segmentation models due to their low-parameter count which make the models amenable for deployment on resource-constrained edge hardware.

Keywords: Brain tumor detection · Brain tumor segmentation · Convolution neural network · Deep learning · Lightweight models

1 Introduction

Brain tumors are among the most dangerous cancers on the planet. The carcinogenesis of glial cells in the spinal cord and brain causes glioma, the most common primary brain tumor. Glioma has multiple histological and malignancy categories, and glioblastoma patients have an average survival duration of less than 14 months following diagnosis [1]. Meningioma is a class of primary brain tumors that accounts for more than 30% of all brain tumors. A pituitary tumor occurs due to the growth of a mass of cancer cells in the pituitary gland. Magnetic Resonance Imaging (MRI), a prominent non-invasive method that provides

N. Naik et al. (Eds.): UKCI 2023, AISC 1453, pp. 491–503, 2024.
https://doi.org/10.1007/978-3-031-47508-5_38

a vast and diverse number of tissue contrasts in each imaging modality, is commonly utilized by medical professionals to identify brain tumors [2].

These various MRI techniques provide various forms of tissue contrast pictures, which provide essential structural information and allow for the identification and segmentation of tumours and subregions [3]. T1-weighted MRI (T1), T2-weighted MRI (T2), T1-weighted MRI with gadolinium contrast enhancement (T1-Gd), and Fluid Attenuated Inversion Recovery (FLAIR) are the four main MRI modalities used for glioma diagnosis. Although it varies by device, around 150 slices of 2D pictures are created during MRI capture to reflect the 3D brain volume. Furthermore, when the requisite standard modalities' slices are joined for diagnosis, the data becomes quite crowded and convoluted. T1 scans are often used to detect healthy tissues, whereas T2 images are used to define the edema area, which provides a strong signal on the image. The strong signal of the accumulated contrast agent (gadolinium ions) in the active cell zone of the tumor tissue in T1-Gd images clearly distinguishes the tumor boundary. Because necrotic cells do not interact with the contrast agent, they may be seen in a low-intensity section of the tumor core, allowing them to be clearly distinguished from the active cell region in the same sequence. The signal of water molecules is decreased in FLAIR pictures, which aids in identifying the edema area from the Cerebrospinal Fluid (CSF) Manual segmentation and analysis of structural MRI images of brain tumors, on the other hand, is a difficult and time-consuming process that can only be performed by experienced neuroradiologists [4,5]. Therefore, an automatic and robust brain tumor segmentation will have a significant impact on brain tumor diagnosis and treatment. It is critical to segment the tumor before beginning any treatment in order to protect healthy tissues while harming and eliminating malignant cells. It can also help with the early detection and treatment of neurological illnesses including Alzheimer's disease, schizophrenia, and dementia. To make therapeutic progress more successful and relevant, radiologists can use an automated approach for lesion segmentation to help them give critical information regarding the volume, localisation, and shape of tumors (including augmenting tumor core areas and full tumor regions). Size, bias field (undesirable artifact owing to faulty image capture), position, and shape [6] are some of the variations between the tumor and its normal adjacent tissue (NAT) that impede the efficacy of segmentation in medical imaging analysis. Brain tumor segmentation entails identifying, defining, and separating active cells, necrotic cores, and edema from normal brain tissues such as Gray Matter (GM), White Matter (WM), and CSF. This endeavor now entails manual annotation and segmentation of a huge number of multi-modal MRI images in clinical practice. However, because human segmentation takes a long time, the development of strong automatic segmentation algorithms that enable efficient and objective segmentation has been a major research topic in recent years. Deep learning algorithms currently have great segmentation performance, making them suitable contenders for this job.

Designing a potent network and training technique for CNN-based segmentation is a major challenge. 3D CNNs can make use of 3D characteristics, but they demand a lot of memory, which may restrict the size, depth, or number of features that the CNNs can process. However, using a 2D CNN in a slice-by-slice approach, ignoring 3D information offers a minimal memory demand.

The proposed DL model consists of two parts, one for the detection and classification of various types of brain tumors from four classes, namely, meningioma, glioma, pituitary brain tumor, and normal scan. The second model is used to create a binary mask of the affected parts of the brain for subsequent segmentation. Both models are trained on distinct datasets. The images from the dataset used to train the classification model are later passed through the segmentation model and are utilized to evaluate it.

Contribution: Two lightweight 2D-CNN models for classifying and segmenting malignant brain tumor tissues with the following features: (i) high accuracy; and (ii) a lightweight architecture amenable for medical imaging purposes unlike pre-existing models which are bulky, computationally extensive and have an unsuitable architecture for medical imaging or brain tumor segmentation.

The contents of the paper are organized as follows. Section II contains a summary of pertinent, related research of the work. In Section III, the proposed approach and methodology are described. Section IV evaluates the proposed system performance and results. Finally, Section V contains concluding remarks.

2 Related Works

Several models that try to find accurate and efficient boundary curves of brain tumors in medical images have been proposed. Deep learning methods extract crucial features automatically. These approaches have yielded outstanding results in various application domains, especially brain tumor segmentation [7].

Muhammad et al. in their study [8] bring forth a comprehensive literature of brain tumor classification methods using deep learning. Multiple deep learning models are compared and discussed on the basis of their segmentation processes, features, dataset, classifier and the total classes used. Transfer learning models like VGG-19, VGG-16, AlexNet, U-Net, Inception V3, ResNet34 along with custom Convolutional Neural Network (CNN) models were considered. The study investigates the respective CNNs by conducting experiments with and without data augmentation techniques and using two datasets i.e., [9] as "Data set 1" which consists of a multi-grade brain tumor dataset and [10] as "Data set 2" which includes the brain tumor dataset. Considering the experimental results of different CNN models it can be observed that VGGNet shows an exceptional accuracy of 94% over the selected brain tumor open-access dataset with image augmentation as well as obtained an accuracy of 93% over Multi-grade brain tumor dataset with image augmentation techniques.

In [11], Sobhaninia et al. researched about the distinct angles of brain MR images such as sagittal, coronal and axial and hence applied different convolutional neural networks for tumor segmentation. LinkNet Network which consists of an encoder and decoder is used. They utilized 2100 images for training the network, 20% of which are considered as validation and the rest is used for the test purpose. As the axial view contains lesser details, it obtained the lowest Dice score of 0.71 whereas sagittal view images acquired the best Dice score of 0.79. The results indicate that the LinkNet networks which are trained separately by different angles are finer and better than one LinkNet network which is trained by the whole dataset without segregation.

Sharif et al. [12] in their research work presented a four step procedure which comprises tumor preprocessing, tumor segmentation, feature extraction and classification. Among all the publicly available datasets, the proposed method is tested on Harvard dataset and on another private dataset which is collected from Nishtar Hospital Multan, Pakistan. The region of interest is extracted in the first step of preprocessing using manual skull stripping and by removing noise effects by Gaussian filter. The second step involves brain tumor segmentation which is implemented by binomial mean, variance and standard deviation. To filter the effective and more related features for best classification results in the third step nine geometric and four gray level co-occurrence matrix (GLCM) features are extracted for the classification of healthy and tumor images. In the final step, the Support Vector Machine (SVM) of linear kernel function is utilized for the classification of selected features. For segmentation results, a tanimoto coefficient (TC) of 0.926 and a DOI of 0.918 is obtained for the private dataset and a TC of 0.935 and a DOI of 0.961 is obtained for the Harvard dataset. Moreover, classification results are presented in different steps such as fusion of feature vectors and best features selection method. It is absolutely clear that the proposed selection method performs better as compared to fused features.

3 Proposed Methodology

We chose CNN for this use-case because it can be successfully employed in DL models to obtain human-like accuracy and precision. The dataset used, the suggested operational framework, the proposed network design, and evaluation metrics are all described in the following sections. Each section is divided into two parts, respectively for (i) Brain Tumor Classification (MRI) [13]; and (ii) Multimodal Brain Tumor Image Segmentation Benchmark (BRaTS 2021) [2,14,15].

3.1 Dataset

For the Brain Tumor Classification (MRI) The Brain Tumor Classification using MRI images is an open-access benchmark dataset[1]. The proposed DL model is trained and evaluated using this dataset, which contains 3264 images.

For the Multimodal Brain Tumor Image Segmentation Benchmark (BRaTS 2021) The training, validation, and testing data are based on a large number of multi-institutional regular clinically obtained multi-parametric MRI (mpMRI) scans of glioma with pathologically confirmed diagnosis and accessible MGMT

[1] https://www.kaggle.com/datasets/sartajbhuvaji/brain-tumor-classification-mri.

promoter methylation status. All BraTS[2] mpMRI scans are available as NIfTI files (.nii.gz), and they describe native (T1), post-contrast T1-weighted (T1Gd), T2-weighted (T2), T2 Fluid Attenuated Inversion Recovery (T2-FLAIR) volumes, and they were acquired using different clinical protocols and scanners from multiple data contributing institutions. All of the imaging datasets were manually annotated by experts using the same annotation technique. The GD-enhancing tumor (ET-label 4), the peritumoral edematous/invaded tissue (ED-label 2), and the necrotic tumor core (NCR-label 1) are all annotated.

3.2 Dataset Handling

For the Brain Tumor Classification For training the proposed model, the dataset includes 826 randomised MRI pictures of Glioma cases, 822 MRI images of meningioma cases, 827 MRI images of no tumor cases, and 395 MRI images of pituitary tumor. We scaled the photos to the required acceptable size of (250 × 250) in order to limit the classification accuracy reduction caused by deformations.

For the Multimodal Brain Tumor Image Segmentation Benchmark (BRaTS 2021) Since, pre-processing the raw images is the initial stage in every data-driven investigation, the following steps were performed. First, all the images are downsized to 240 × 240 pixels in order to feed them into the proposed model. Each subject has four scans with annotated masks in the dataset. The network is given all of the photos after categorizing them into ET, TC, and TC classes. The dataset consists of shuffled 2635 MRI images for training, and 465 MRI images for testing the proposed segmentation model.

3.3 Evaluation Metrics

For Tumor Detection Model The categorical cross-entropy loss function calculates an example's loss by adding the following values:

$$Loss = -\sum_{i=1}^{N} y_i - \log \hat{y}_i \tag{1}$$

where $\hat{y}_i$ is the output's i-th scalar value, and N is the output size. The equivalent goal value is y_i, and the number of scalar values in the output is output size. This loss is a useful indicator of how easily two discrete probability distributions can be distinguished from one another. In this case, y_i represents the likelihood that event i will occur, and the total of all y_i equals 1, implying that only one event will occur. When the distributions become closer to one other, the negative sign ensures that the loss gets lower.

[2] Available on: https://braintumorsegmentation.org.

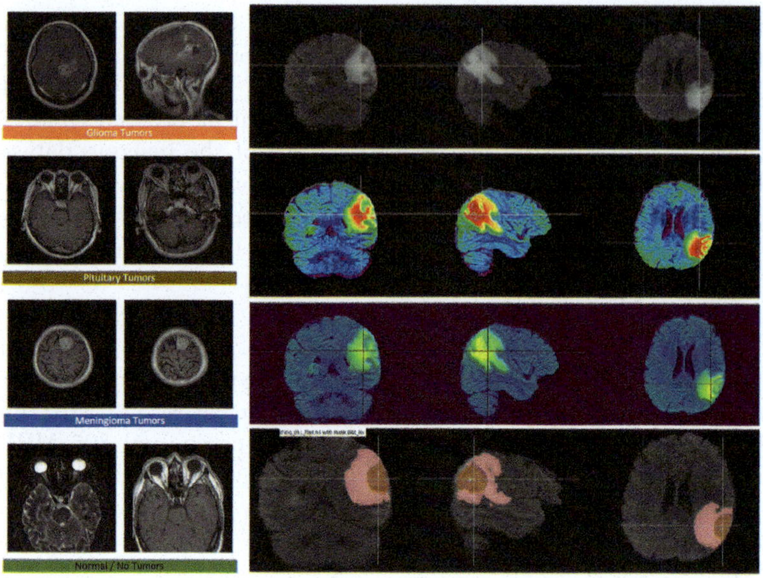

Fig. 1. **a** Samples from the MRI dataset; **b** Samples from the Multimodal Brain Tumor Image Segmentation Benchmark (BRaTS 2021) dataset

For Segmentation Model For evaluating the segmentation model, we use Binary Focal Dice Loss [16]. Focal loss [17] is a modified form of binary cross entropy in which the loss for confidently properly classified labels is scaled down, causing the network to focus more on wrong and low confidence labels rather than boosting its confidence in the labels that are already accurate. Focal Loss was created as a way to deal with very unbalanced datasets where positive instances were uncommon. Dice coefficient (2) is a common metric for pixel segmentation that can also be modified to act as a loss function.

$$D = \frac{2\sum_i^N p_i g_i}{\sum_i^N p_i^2 + \sum_i^N g_i^2} \tag{2}$$

where p_i and g_i represent pairs of matching prediction and ground truth pixel values, respectively. Values of p_i and g_i in a boundary detection scenario are either 0 or 1, indicating the pixel is a boundary (value of 1) or not (value of 0). Because the sum advances only when p_i and g_i match, the denominator is the sum of all boundary pixels from both prediction and ground truth, while the numerator is the sum of successfully predicted boundary pixels.

3.4 Experimental Design

Tumor Detection Model The procedure of fine-tuning different CNNs like VGG16, Resnet, GoogleNet, SqueezeNet, and MobileNetV2 followed in the prior

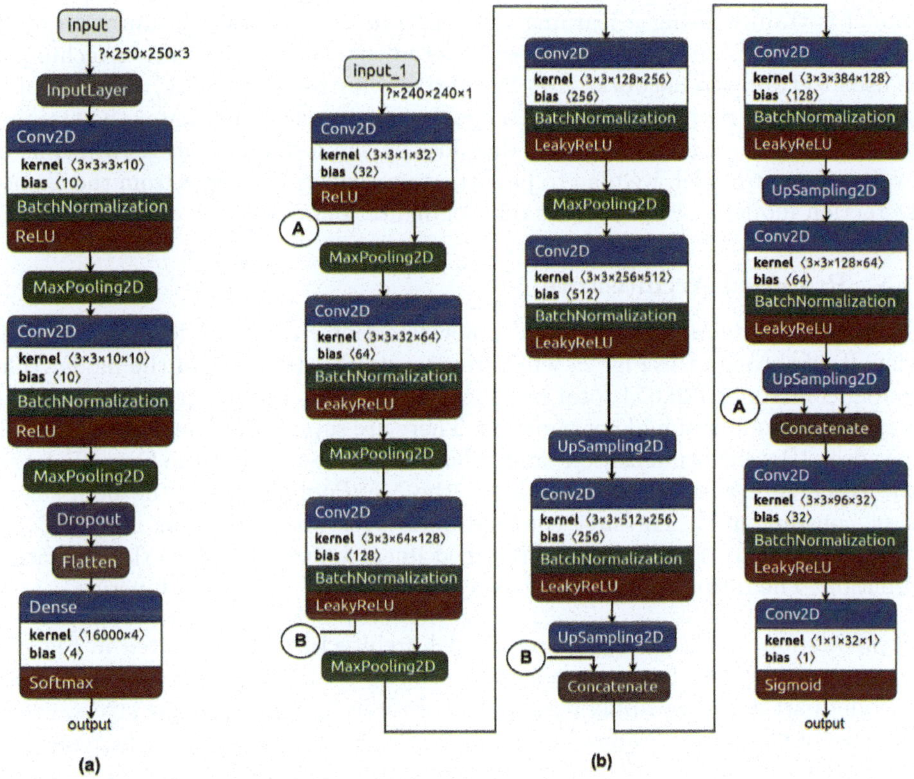

Fig. 2. Architecture of **a** Classification model; **b** Segmentation model

deep learning-based brain tumor diagnosis systems such as those in [7,9–12]. One significant disadvantage of fine-tuning such big CNNs is that the final on-disk size and number of layers of the trained model is too enormous, preventing these trained models from running smoothly at a suitable frame rate on low-cost hardware, as the end goal of all of these approaches is implementation for real-world applications. The emphasis on making the model lightweight cannot be highlighted enough. The dearth of lightweight models for classifying brain tumor presents a perfect ground for this research work.

Segmentation Model We suggest conducting the segmentation slice by slice from the axial perspective since the brains in the BRATS dataset lack resolution in the third dimension. As a result, our model analyses each 2D axial picture (slice) in a sequential manner, with each pixel linked with distinct image modalities, such as T1, T2, T1C, and FLAIR. Our technique predicts the class of a pixel by processing the MM patch centered on that pixel, similar to other CNN-based segmentation models. The next stage is to train the model to extract the multi-class tumor segments. Due to the complexity of the data, samples were handled in batches of 32. The training dataset is split 80–20 between train and test. The model is trained using a learning rate of 0.0001 and a training duration of 50 epochs. Adam[3], an adaptive first-order gradient-based optimization technique, is used to train the networks [19]. We also employ early stopping, which implies that if there is no progress after 10 epochs on the validation data, the training process will be ended. We utilize dice loss as the loss function. As a means of evaluating our model, we plan to pass the tumor positive images from the Tumor Detection model as validation to our Segmentation model.

3.5 Proposed Architecture

Tumor Detection Model The brain tumor detection custom model architecture is visualized through the Fig. 2a which represents the data flow of the model. The model consists of 140,654 total parameters including 140,194 trainable parameters and 460 non-trainable parameters. There are several convolutional layers followed by 'BatchNormalization' and 'MaxPool2D' layers. For the Conv2D layer, the kernel size is set to be 3 while for the MaxPooling layers, the pool size is kept constant at 3. The first convolutional layer takes an input image of size ($250 \times 250 \times 3$). Each layer has rectified linear units (ReLU) as the activation function. The output layer is set to have a 'Softmax' activation function.

Segmentation Model The proposed model architecture is elaborated in Fig. 2b, which gives the data flow representation of the model. The model comprises of an encoder part where we apply regular convolutions and max pooling layers, where the size of the image gradually reduces while the depth gradually increases. In the decoder part, we apply transposed convolutions along with regular convolutions. In the decoder, the size of the image gradually increases and the depth gradually decreases. The model consists of a total of nine convolutional layers with the first convolutional layer having a kernel size of 3 while the padding set to 'same', followed by four 'MaxPooling2D' layers along with four 'UpSampling' layers. There is a Batch Normalization layer after every convolutional layer. To provide a moderately higher flexibility, the rectified linear units (Relu) activation function is used. The filter size is set to 32 for the first convolutional layer and is successively increased upto 512 till the fifth convolutional layer and is then decreased upto 32 till the ninth convolutional layer. Moreover, to get better precise locations, the first convolutional layer is concatenated with the last UpSampling layer. The output layer is set to have a 'Sigmoid' activation function. Table 1 presents a summary of all the important details such as dataset name, loss function, batch size, etc. of the classification and segmentation models.

4 Results

Tumor Detection Model The model was trained over the training samples from all classes using the MRI dataset. The trained model was then put to the test

[3] Adam adds momentum and bias-correction to the conventional RMSprop optimizer which tends to help the Adam algorithm to marginally outperform RMSprop towards the close of the optimization process (when the gradients tend to get meager). [18].

Table 1. Details of the classification and segmentation models

Parameter	Classification Model	Segmentation Model
Dataset	Brain Tumor Public (MRI)	BRaTS 2021
Loss Parameter	Categorical Cross Entropy	Binary Focal Dice Loss
Batch Size	32	32
Epochs	50	Mini batches of 20
Input Image Size	250 × 250	240 × 240
Train-Test Split	80:20	80:20

on the remaining scans (test set) in each of the four classes (which were held back during training). The model has a precision of 92.81 percent and a total accuracy of 96.56 percent. Plots of the training and testing accuracy and loss for the detection model are presented in Fig. 3. Some sample images for correctly and incorrectly predicted classes are included in Fig. 4.

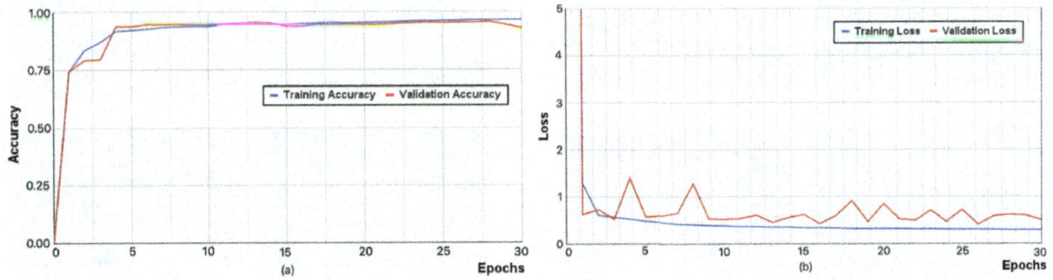

Fig. 3. a Accuracy and **b** Loss plots for the proposed lightweight brain tumor detection model

Table 2. Comparison of the proposed model with related works

Model	Accuracy	Sensitivity
[20]	85.69	86.26
[10]	81	91.28
[21]	88.41	94.58
Proposed Model	98.56	99.92

As a means of evaluating the model further, the images from the Brain Tumor Classification MRI dataset [13] were passed to the segmentation model to predict the segmentation mask of the classified brain tumor class. Figure 5 depicts the predicted segmentation mask of such images.

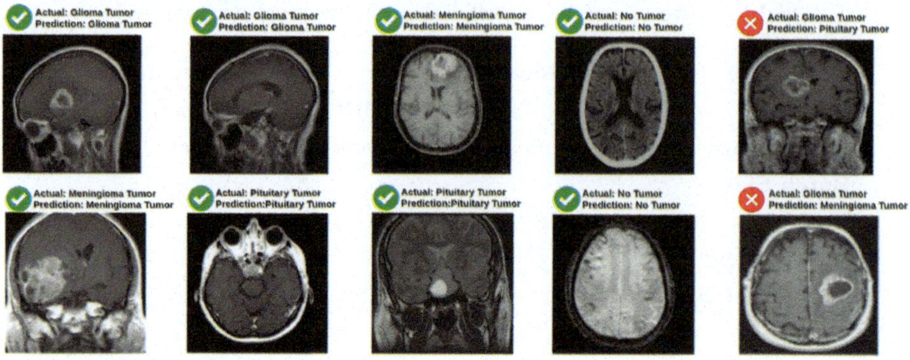

Fig. 4. Examples of some correct and incorrect predictions from proposed model

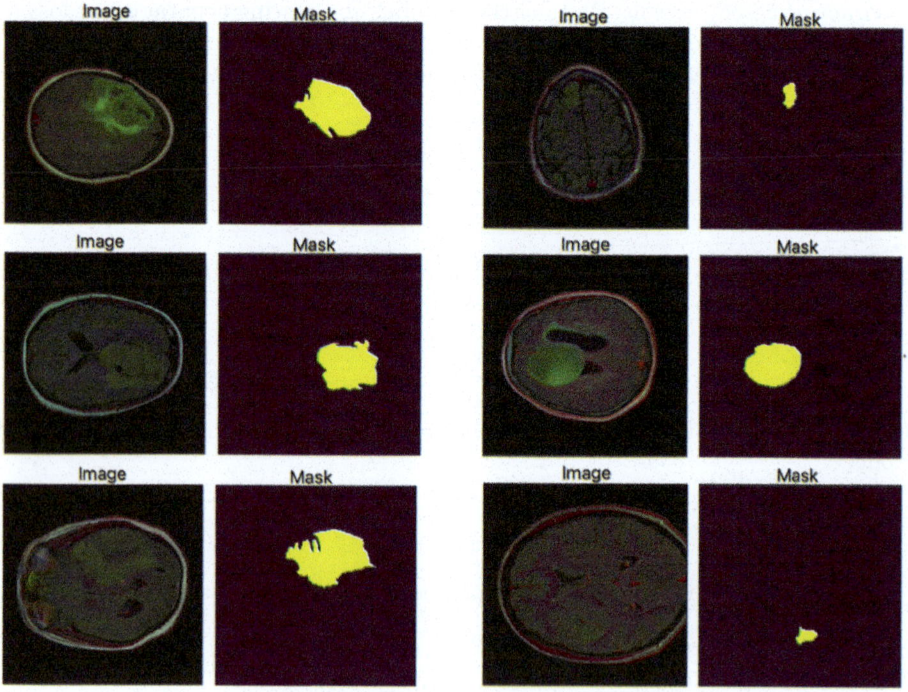

Fig. 5. Brain MRI images along with the generated masks from the proposed brain tumor segmentation model

Segmentation Model The segmentation of the brain tumour without any human intervention in completely automatic cause of action is demonstrated. A completely automated segmentation method, in general, combines artificial intelligence with previous information. Because brain tumour pictures typically have unclear and uneven borders with discontinuities and partial-volume effects, semi-

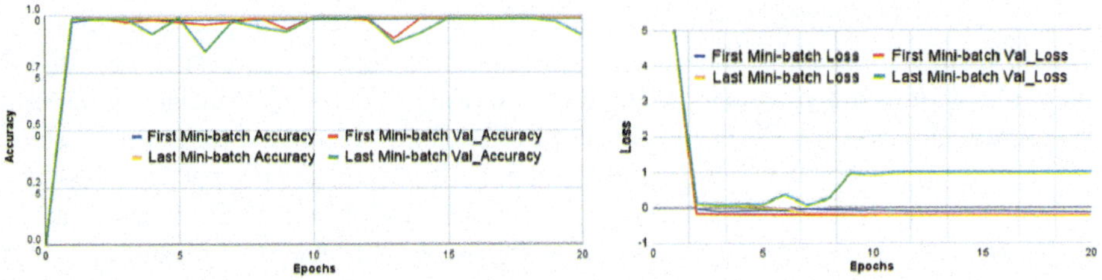

Fig. 6. Accuracy and loss plots for the segmentation model on BRATS dataset for the first mini-batch and the last mini-batch.

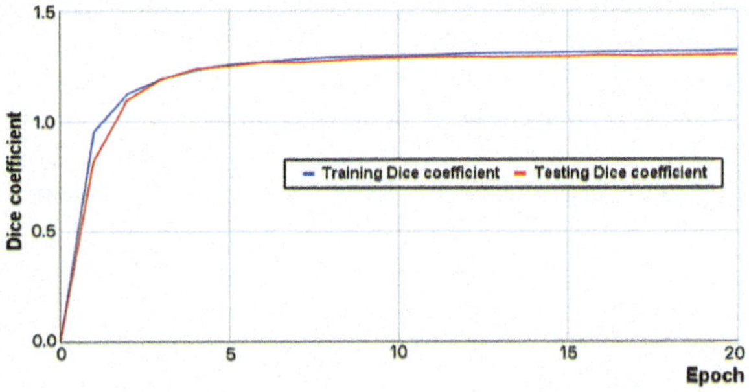

Fig. 7. Last mini-batch Dice Coefficient plot for the segmentation model on BRATS dataset

automated and completely automatic segmentation of tumour brain images presents significant obstacles. The segmentation model was trained on BRATS dataset which consists of a total of 2635 MRI images and 465 MRI images for testing. The segmentation model was trained on mini-batches of 20 images. On each mini-batch the accuracy metric comes out to be an average of 95%, the highest achieved accuracy was 99.11% with loss of −0.1711 and dice coefficient of 1.3285. For the evaluation to determine over– or under–segmentation of tumor sub-regions, sensitivity comes out to be 99.92 and 98.79% respectively, precision comes out to be 92.94%. Although automated segmentation produces fairly accurate results in most circumstances, in tough cases with low contrast and/or a limited tumour region, the automatic result may be sub-optimal, necessitating manual expert assessment and rectification. Plots of the training and testing accuracy and loss for the detection model are presented in Fig. 6. Plot for the Dice coefficient is presented in Fig. 7. Comparisons with some related works are presented in Table 2 which go on to highlight the superior performance of the proposed model as compared with the counterparts.

5 Conclusion

A lightweight deep-learning model using CNNs has been presented in this work for the detection, classification and segmentation of different classes of brain tumors, namely, Meningioma, Glioma, Pituitary brain tumor and normal (no tumor) from MR images. The idea is to provide a light architecture specifically for medical imaging purposes with high accuracy. The work consists of two different models, one for detection and classification, the other for segmentation. Segmentation model has been assessed using the training dataset for classification model and has shown great accuracy. The model can be further expanded in the future to include augmentation and possibly achieve even higher accuracy.

References

1. Van Meir, E.G., Hadjipanayis, C.G., Norden, A.D., Shu, H.K., Wen, P.Y., Olson, J.J.: Exciting new advances in neuro-oncology: the avenue to a cure for malignant glioma. CA: Cancer J. Clin. **60**(3), 166–193 (2010)
2. Bakas, S., Akbari, H., Sotiras, A., Bilello, M., Rozycki, M., Kirby, J.S., Freymann, J.B., Farahani, K., Davatzikos, C.: Advancing the cancer genome atlas glioma MRI collections with expert segmentation labels and radiomic features. Sci. Data **4**(1), 1–13 (2017)
3. Drevelegas, A.: Imaging of Brain Tumors with Histological Correlations. Springer (2002)
4. Sun, L., Zhang, S., Chen, H., Luo, L.: Brain tumor segmentation and survival prediction using multimodal MRI scans with deep learning. Front. Neurosci. **13** (2019)
5. Garali, I., Adel, M., Bourennane, S., Guedj, E.: Classification of positron emission tomography brain images using first and second derivative features. In: 2016 6th European Workshop on Visual Information Processing (EUVIP), pp. 1–5. IEEE (2016)
6. Bakas, S., Akbari, H., Sotiras, A., Bilello, M., Rozycki, M., Kirby, J., Freymann, J., Farahani, K., Davatzikos, C.: Segmentation labels and radiomic features for the pre-operative scans of the TCGA-LGG collection. Cancer Imaging Arch. **286** (2017)
7. Ganesh, P.S., Kumar, T.S., Kumar, M., Kumar, M.S.R.: Brain tumor detection and classification using image processing techniques. Brain **4**(3), 32–38 (2021)
8. Muhammad, K., Khan, S., Ser, J.D., de Albuquerque, V.H.C.: Deep learning for multigrade brain tumor classification in smart healthcare systems: a prospective survey. IEEE Trans. Neural Netw. Learn. Syst. **32**, 507–522 (2021)
9. Sajjad, M., Khan, S., Muhammad, K., Wu, W., Ullah, A., Baik, S.: Multi-grade brain tumor classification using deep CNN with extensive data augmentation. J. Comput. Sci. **30** (2018)
10. Cheng, J., Huang, W., Cao, S., Yang, R., Yang, W., Yun, Z., Feng, Q.: Enhanced performance of brain tumor classification via tumor region augmentation and partition. PLoS ONE **10** (2015)
11. Sobhaninia, Z., Rezaei, S., Noroozi, A., Ahmadi, M., Zarrabi, H., Karimi, N., Emami, A., Samavi, S.: Brain tumor segmentation using deep learning by type specific sorting of images. CoRR abs/1809.07786 (2018). arxiv:1809.07786
12. Sharif, M., Tanvir, U., Munir, E., Khan, M., Mussarat, Y.: Brain tumor segmentation and classification by improved binomial thresholding and multi-features selection. J. Ambient Intell. Humaniz. Comput. (2018)
13. Bhuvaji, S., Kadam, A., Bhumkar, P., Dedge, S., Kanchan, S.: Brain tumor classification (MRI) (2020). https://www.kaggle.com/dsv/1183165

14. Menze, B.H., Jakab, A., Bauer, S., Kalpathy-Cramer, J., Farahani, K., Kirby, J., Burren, Y., Porz, N., Slotboom, J., Wiest, R., et al.: The multimodal brain tumor image segmentation benchmark (brats). IEEE Trans. Med. Imaging **34**(10), 1993–2024 (2014)
15. Baid, U., et al.: The RSNA-ASNR-MICCAI brats 2021 benchmark on brain tumor segmentation and radiogenomic classification. CoRR abs/2107.02314 (2021). arxiv:2107.02314
16. Yeung, M., Sala, E., Schönlieb, C.B., Rundo, L.: Unified focal loss: generalising dice and cross entropy-based losses to handle class imbalanced medical image segmentation. Comput. Med. Imaging Graph. **95**, 102026 (2022). www.sciencedirect.com/science/article/pii/S0895611121001750
17. Lin, T., Goyal, P., Girshick, R.B., He, K., Dollár, P.: Focal loss for dense object detection. CoRR abs/1708.02002 (2017). arxiv:1708.02002
18. Ruder, S.: An overview of gradient descent optimization algorithms. arXiv:1609.04747 (2016)
19. Bock, S., Weiß, M.: A proof of local convergence for the adam optimizer. In: 2019 International Joint Conference on Neural Networks (IJCNN), pp. 1–8 (2019)
20. Zia, R., Akhtar, P., Aziz, A.: A new rectangular window based image cropping method for generalization of brain neoplasm classification systems. Int. J. Imaging Syst. Technol. **28**, 153–162 (2018)
21. Sajjad, M., Khan, S., Muhammad, K., Wu, W., Ullah, A., Baik, S.W.: Multi-grade brain tumor classification using deep CNN with extensive data augmentation. J. Comput. Sci. **30**, 174–182 (2019). www.sciencedirect.com/science/article/pii/S1877750318307385

Clinical Outcome Prediction Pipeline for Ischemic Stroke Patients Using Radiomics Features and Machine Learning

Meryem Şahin Erdoğan[1], Esra Sümer[1], Federico Villagra[2], Esin Öztürk Işık[1], Otar Akanyeti[3](✉), and Hale Saybaşılı[1]

[1] Institute of Biomedical Engineering, Boğaziçi University, 34684 Istanbul, Turkey
[2] Department of Life Sciences, Aberystwyth University, Ceredigion SY23 3DA, UK
[3] Department of Computer Science, Aberystwyth University, Ceredigion SY23 3DB, UK
otal@aber.ac.uk

Abstract. Ischemic stroke is a debilitating brain injury affecting millions of people, causing long-term disabilities. Immediately after stroke, it is not easy to predict the extent of the injury and its long-term effects, yet outcome prediction is desired to inform clinical decision-making processes. Apparent diffusion coefficient (ADC) maps, calculated from diffusion-weighted imaging, are widely used in clinics to diagnose and monitor ischemic stroke. Radiomics analysis is an emerging feature extraction method providing many quantitative imaging indicators from the ADC maps. Here, we have utilized these features to predict the clinical outcome of 43 ischemic stroke patients. We divided the clinical outcome into two groups (good and poor outcomes) based on the patients' modified Rankin Scale scores and trained a binary classifier to predict the correct outcome group. We compared various machine learning classifiers and feature selection and pre-processing techniques to create a parsimonious mRS score prediction pipeline. Our results showed that the best-performing classifier was a multi-layer perceptron classifier which used three radiomics features to achieve a classification accuracy of 0.94. This is a marked improvement compared to our previous results, where the classification accuracy was around 0.7 and matches the performance of previous studies reported in the literature. In the clinics, our pipeline can help doctors and stroke patients plan recovery and rehabilitation processes.

Keywords: Ischemic stroke · Apparent diffusion coefficient · Radiomics · Machine learning

1 Introduction

Stroke affects millions of people annually [1]. Many stroke survivors experience cognitive and motor impairments for the rest of their lives, affecting their independence and quality of life. The range of impairments, such as speech problems, detriment in learning and memory, and gait disturbances [2] depends on the ischemic brain region, lesion size, location, and previous health condition of the patient, and this heterogeneity makes clinical outcome prediction challenging.

N. Naik et al. (Eds.): UKCI 2023, AISC 1453, pp. 504–515, 2024.
https://doi.org/10.1007/978-3-031-47508-5_39

Modified Rankin Scale (mRS) is an ordinal scale that ranges from 0 (no symptoms at all) to 6 (death) and measures the patient's dependence on external help [3, 4]. It is rated by physicians, and a good outcome is commonly defined based on low mRS scores. Treatment or intervention where a patient goes from being bedridden (mRS score of 5) to being able to walk without assistance (mRS score of 3 or lower) is highly clinically relevant. The ability to predict long-term outcome can improve post-stroke planning for clinicians, patients, and their loved ones. It can also enable better resource allocation and design of individualized rehabilitation programs [5].

Magnetic resonance imaging (MRI) is commonly used to diagnose stroke and monitor changes in the brain during recovery. One MRI modality is the apparent diffusion coefficient (ADC) map which is derived from diffusion-weighted imaging (DWI) as a measure of the movement of water molecules within the brain. ADC maps can be analyzed to extract radiomics features, providing objective, reproducible, and comprehensive brain injury biomarkers [6]. Previous studies demonstrated the utility of the radiomics features in detecting lung cancer [7], and breast cancer [8], and predicting outcomes for Parkinson's disease [9] and Crohn's disease [10].

In our previous research [11], we evaluated the feasibility of using radiomics features to predict mRS scores in ischemic stroke patients. We proposed a data processing pipeline with three main modules: (1) feature extraction, (2) feature selection, and (3) outcome prediction. During feature extraction, we extracted 1132 radiomics features from the intensity-normalized ADC maps. During feature selection, first, we calculated Pearson's correlation to remove highly correlated features (various threshold values were tested). Then, we used Gini index to rank the remaining features and select the top-ranked features. Finally, during outcome prediction, we used these features to train a binary classifier (various machine learning classifiers were tested) to differentiate between low versus high mRS outcomes. We evaluated the performance of the proposed pipeline using the open-access Ischemic Stroke Lesion Segmentation (ISLES) 2017 dataset [12], and reported moderate mean outcome prediction accuracy of around 0.7.

In this paper, we aim to increase the outcome prediction accuracy by improving feature extraction and selection modules. In particular, (i) we expanded the initial radi-omics feature set by extracting additional filters, (ii) we evaluated the impact of normalizing the ADC maps before radiomics feature extraction, (iii) we evaluated the impact of including variance thresholding to remove low variable features during feature selection, and (iv) we tested a new feature selection method, least absolute shrinkage and selection operator (LASSO), which was previously shown to be effective in determining the key radiomics features for identification of salvageable brain tissue (penumbra) and clinical outcomes following acute stroke patients [13–15].

After these improvements, we ask the following questions: (i) can mRS score prediction accuracy be improved?, (ii) if so, what are the key radiomics features in predicting mRS score?, (iii) does classifier choice influence which features are deemed important? and (iv) can mRS prediction accuracy be further improved if radiomics features are combined with clinical information?

2 Related Work

DWI and ADC maps are commonly used in clinics to evaluate stroke. Many researchers attempted to automate this process. One of the utilized approaches in those studies is to extract quantifiable features from the DWI and ADC maps. Radiomics analysis is a tool that can provide over a thousand features based on the shape, size, texture, and intensity of the delineated region. These features can be utilized to understand the properties of abnormal (e.g., lesion or tumor) and healthy tissue [16, 17].

Zhou et al. analyzed DWI and ADC maps of 522 patients with ischemic stroke. They proposed a logistic regression classifier using 14 radiomics features and reported the Area Under the Curve (AUC) of 0.784 [18]. Quan et al. studied 190 ischemic stroke patients. They also proposed a logistic regression classifier using six radiomics features and reported a prediction accuracy of 0.700 and an AUC of 0.825. The radiomics features were extracted from both ADC maps and fluid-attenuated inversion recovery images, which allowed the detection of superficial brain lesions [14].

In [11], we proposed an mRS score prediction pipeline using radiomics features extracted from ADC maps without intensity normalization. To identify key features, we used a combination of Pearson's correlation (0.70, 0.75, 0.80, 0.85, 0.90, 0.95 cutoffs) and Gini index feature selection methods. We then used the selected features to train various machine learning classifiers, which achieved prediction accuracy up to 0.73. In this study, we evaluated whether this accuracy can be improved by tuning feature selection and pre-processing methods. The details are given in the next section.

3 Methodology

3.1 Dataset and Tools

The ISLES 2017 dataset is an open-access dataset with MR images and basic clinical information collected from 43 adult patients [12]. Patients' data were included in the dataset if they had an ischemic stroke with visible lesions without any endovascular therapy. The MRI data included perfusion-weighted imaging (PWI) images and acute ADC maps. The dataset also included manual segmentations of chronic lesions. These manual segmentations were based on 90-day T2 images instead of the acute lesions as seen in the provided ADC images. The 90-day T2 images were not provided in the dataset. An example ADC map with the segmented lesion is shown in Fig. 1. The basic clinical information provided included the mRS score, time to mRS (TTM), time since stroke (TSS), time to treatment (TTT), and thrombolysis in cerebral infarction (TICI). The time measurements were relative to stroke onset, and TICI indicates the success of thrombolysis.

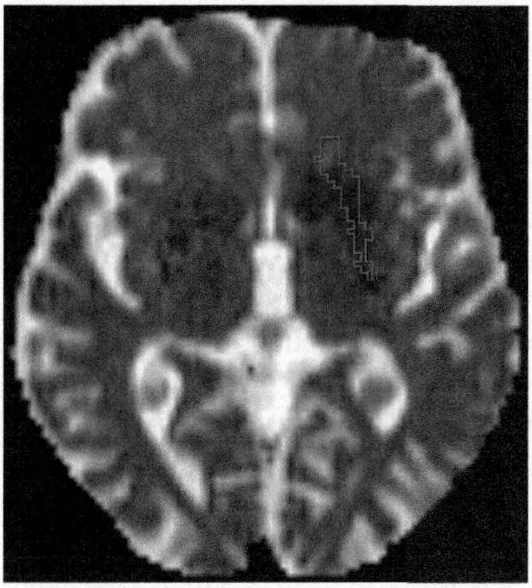

Fig. 1. A 2D slice of an acute axial ADC map of an acute ischemic stroke patient from ISLES 2017 dataset with 90-day lesion segmentation (magenta) superimposed. This patient had an mRS score of 1 out of 6 and a successful thrombolysis.

3.2 Data Preparation

The mRS scores were divided into two groups: good outcome (i.e., mRS < 2, n $= 22$, patients were independent) and poor outcome (i.e., mRS ≥ 2, n $= 22$, patients depended on external help). A similar dichotomous grouping was also applied in previous studies [14, 19, 20], with an mRS score of 2 or 3 chosen as the cutoff point.

3.3 Feature Extraction and Selection

The Pyradiomics toolbox was employed to extract radiomics features. [21]. All ADC maps were resampled to $2 \times 2 \times 2$ mm resolution. From the Pyradiomics toolbox, the following features were calculated: first-order statistics, shape-based (3D), shape-based (2D), gray level co-occurrence matrix, gray level run length matrix, gray level size zone matrix, neighboring gray-tone difference matrix, gray level dependence matrix features. All these features, except for shape, are calculated from original or derived images, which are obtained by applying LoG, wavelet, exponential, gradient, lbp3D, logarithm, square, and square root filters.

Prior to radiomics feature extraction, intensity normalization can be performed on ADC maps, but there is no clear evidence of whether the intensity normalization is beneficial (whether it leads to more robust features) given that ADC maps reflect absolute world values [14, 18, 22]. To evaluate the impact of intensity normalization, radiomics features were extracted from ADC maps with or without intensity normalization.

After radiomics feature extraction, all features were standard score (z-score) normalized. The initial radiomics feature set had 1820 features for each patient. To identify the most informative radiomics features, a three-step feature selection method was applied: (1) removing features with low variance, (2) removing highly correlated features, and (3) selecting features that are ranked by Gini or LASSO feature selection methods. The Gini index method ranks features based on the impurity of the feature, whereas the LASSO uses regression to minimize the cost function, and relevant features end up with non-zero coefficients.

To establish an optimal feature selection for this classification problem, we tried to select features using different parameters. First, we have applied variance thresholding. To focus on features that vary among patients, we excluded features with a lower variance than a threshold. This threshold was selected to be either 0.01 or 0.001. Secondly, we applied a Pearson's correlation coefficient threshold (0.90), and we excluded collinear features with higher pairwise correlation and higher column-wise average absolute correlation. Thirdly and lastly, we have utilized either Gini feature selection or LASSO) methods.

In the Gini index feature selection, the top 30 features were kept, and in LASSO, all the relevant features were kept. We employed 50 train-test splits for classification with each feature selection approach and ended up with eight different sets of features as a result of eight different approaches to feature selection. Each set of features was sorted among themselves with respect to the selection frequency out of 50 train-test splits.

Previously [11], we have shown that, generally, classification accuracy starts decreasing when the number of features exceeds 20–25. Here, after the list of most selected features was established, a forward bottom-up selection technique was utilized to obtain the most descriptive subset. We increased the number of features one by one to include the first 25 most frequently selected features (starting with the highest-ranked feature and incrementally adding the second, third, and so on until the accuracy plateaus). Out of these classifications, we reported the best accuracies achieved with the smallest possible number of radiomics feature subsets.

3.4 Machine Learning Classification and Performance Evaluation

Our feature selection module generated eight different feature sets. Each set was used for the binary classification of mRS scores using supervised learning. Sixteen different machine learning classifiers were used from Python's scikit-learn library [23]. These classifiers were: Adaptive Boosting Classifier (AdaBoost), Bagging Classifier (Bagging), Decision Tree Classifier (DecisionTree), Dummy Classifier (Dummy), Extra Trees Classifier (ExtraTrees), Gaussian Naïve Bayes Classifier (GaussianNB), Gradient Boosting Classifier (GradientBoost), k-Nearest Neighbor Classifier (kNN), Linear Support Vector Machine Classifier (LinearSVM), Logistic Regression (LogisticRegr), Multi-Layer Perceptron Classifier (MLP), Nu-Support Vector Machine Classifier (NuSVM), Random Forest Classifier (RandomForest), Stochastic Gradient Descent Classifier (SGD), Support Vector Machine Classifier (SVM), Extreme Gradient Boosting Classifier (XGB).

All classifiers were trained with default parameters except for the MLP classifier. With the MLP classifier, we selected the tolerance value to be 0.001 instead of the default 0.0001 to cut back on computation time. Stratified train-test splits were used with 80%

training and 20% testing data. The process was repeated 50 times with different seed values, and the performance of each classifier is reported as mean ± standard deviation of accuracy, sensitivity, specificity, and the area under the curve (AUC). The accuracy was calculated as the number of correctly classified instances divided by the total number of instances. Sensitivity was used to measure classifiers' ability to detect bad outcomes correctly. Specificity was used to evaluate the classifiers' ability to identify patients with good outcome correctly. AUC was calculated from the Receiver Operating Characteristic curve, and it took both sensitivity and specificity into account. Lastly, we compared the means of each feature set using Student's t-test to evaluate whether ADC map intensity normalization, variance thresholding, and LASSO feature selection had a statistically significant impact on improving classification accuracy.

4 Results and Discussion

4.1 Three Radiomics Features Were Enough to Accurately Predict the mRS Score

Table 1 lists the most prominent radiomics feature set out of 50 runs. These features were selected when using ADC map intensity normalization, variance thresholding of 0.001, and the LASSO feature selection algorithm.

Table 1. List of the most prominent radiomics features after intensity normalization, 0.001 variance thresholding, and LASSO feature selection. The frequency column shows how many times these features were deemed important during classifier training, and the highest possible frequency was 50.

Radiomics features	Frequency
logarithm_glcm_InverseVariance	32
wavelet-LHH_firstorder_Kurtosis	27
wavelet-HHH_glszm_SmallAreaHighGrayLevelEmphasis	24
wavelet-HLH_glszm_SmallAreaLowGrayLevelEmphasis	22
wavelet-HHL_glszm_SmallAreaEmphasis	17
lbp-3D-k_glszm_LowGrayLevelZoneEmphasis	17
exponential_gldm_DependenceEntropy	16
wavelet-HHH_glszm_SmallAreaEmphasis	15
wavelet-LLL_gldm_LargeDependenceHighGrayLevelEmphasis	13
wavelet-LLH_firstorder_Median	12
wavelet-LLH_glcm_ClusterShade	9
log-sigma-4–0-mm-3D_firstorder_Maximum	9
wavelet-LLL_firstorder_Minimum	8

(continued)

Table 1. (*continued*)

Radiomics features	Frequency
wavelet-HHH_gldm_LargeDependenceLowGrayLevelEmphasis	7
wavelet-HHL_firstorder_Maximum	6
wavelet-LHH_firstorder_Skewness	6
lbp-3D-m1_glszm_GrayLevelVariance	5
logarithm_glcm_Contrast	5
original_glcm_MaximumProbability	5
squareroot_glcm_Id	5
wavelet-HHL_glszm_SizeZoneNonUniformityNormalized	5
wavelet-HHL_glszm_SmallAreaLowGrayLevelEmphasis	4
wavelet-LHH_gldm_LargeDependenceLowGrayLevelEmphasis	4
gradient_glszm_SmallAreaLowGrayLevelEmphasis	4
wavelet-HHH_glszm_SmallAreaLowGrayLevelEmphasis	4

For selected features described in Table 1 (and the other seven feature sets, not shown here), 16 classifiers were tested. For each feature set, the best-performing classifier is presented in Table 2 (using the Gini feature selection method) and Table 3 (using the LASSO feature selection method).

Table 2. Classification accuracies (mean ± standard deviation) of best-performing classifiers when the Gini feature selection method was used.

Gini feature selection				
	Without intensity normalization		With intensity normalization	
Variance threshold	0.01	0.001	0.01	0.001
Best model	LogisticRegr	SVM	SVM	RandomForest
Accuracy	0.771 ± 0.141	0.773 ± 0.131	0.796 ± 0.111	0.780 ± 0.109
No of features	6	25	3	12

The best mRS score prediction accuracy was 0.949 ± 0.068, achieved by the MLP classifier using three radiomics features:

1. logarithm_glcm_InverseVariance
2. wavelet-LHH_firstorder_Kurtosis
3. wavelet-HHH_glszm_SmallAreaHighGrayLevelEmphasis

We next evaluated the performance of all the other classifiers using these three features only (Table 4). The performance of the logistic regression and linear support vector machine classifiers was not far-off from the performance of the MLP classifier.

Table 3. Classification accuracies (mean ± standard deviation) of best-performing classifiers when the LASSO feature selection method was used.

LASSO Feature Selection				
	Without Intensity Normalization		With Intensity Normalization	
Variance threshold	0.01	0.001	0.01	0.001
Best model	LogisticRegr	MLP	MLP	MLP
Accuracy	0.862 ± 0.086	0.900 ± 0.106	0.902 ± 0.104	0.949 ± 0.068
No of features	12	12	11	3

Table 4. Classification performance using three radiomics features (logarithm_glcm_InverseVariance, wavelet-LHH_firstorder_Kurtosis, and wavelet-HHH_glszm_SmallAreaHighGrayLevelEmphasis).

	Accuracy	Sensitivity	Specificity	AUC
MLP	0.949 ± 0.068	0.945 ± 0.105	0.952 ± 0.103	0.948 ± 0.068
LogisticRegr	0.942 ± 0.068	0.930 ± 0.113	0.952 ± 0.103	0.941 ± 0.068
LinearSVM	0.933 ± 0.078	0.910 ± 0.141	0.952 ± 0.103	0.931 ± 0.080
NuSVM	0.911 ± 0.095	0.905 ± 0.151	0.916 ± 0.128	0.910 ± 0.096
SVM	0.911 ± 0.095	0.905 ± 0.151	0.916 ± 0.128	0.910 ± 0.096
kNN	0.893 ± 0.095	0.895 ± 0.152	0.892 ± 0.152	0.893 ± 0.094
GaussianNB	0.884 ± 0.098	0.900 ± 0.143	0.872 ± 0.165	0.886 ± 0.095
SGD	0.873 ± 0.117	0.855 ± 0.190	0.888 ± 0.167	0.871 ± 0.118
ExtraTree	0.833 ± 0.115	0.825 ± 0.197	0.840 ± 0.185	0.832 ± 0.115
Bagging	0.811 ± 0.119	0.775 ± 0.227	0.840 ± 0.171	0.807 ± 0.123
GradientBoost	0.804 ± 0.122	0.770 ± 0.219	0.832 ± 0.187	0.801 ± 0.124
AdaBoost	0.796 ± 0.130	0.800 ± 0.196	0.792 ± 0.221	0.796 ± 0.126
RandomForest	0.796 ± 0.124	0.780 ± 0.187	0.808 ± 0.225	0.794 ± 0.119
DecisionTree	0.773 ± 0.145	0.775 ± 0.238	0.772 ± 0.206	0.773 ± 0.147
XGB	0.738 ± 0.134	0.710 ± 0.222	0.760 ± 0.229	0.735 ± 0.132
Dummy	0.556 ± 0.000	0.000 ± 0.000	1.000 ± 0.000	0.000

4.2 LASSO Feature Selection Improves Outcome Prediction Accuracy by 12%

Our results show that the ADC map intensity normalization and variance thresholding did not affect the mean prediction accuracy, whereas the LASSO feature selection method was significantly better than the Gini index feature selection method by 12% (Fig. 2).

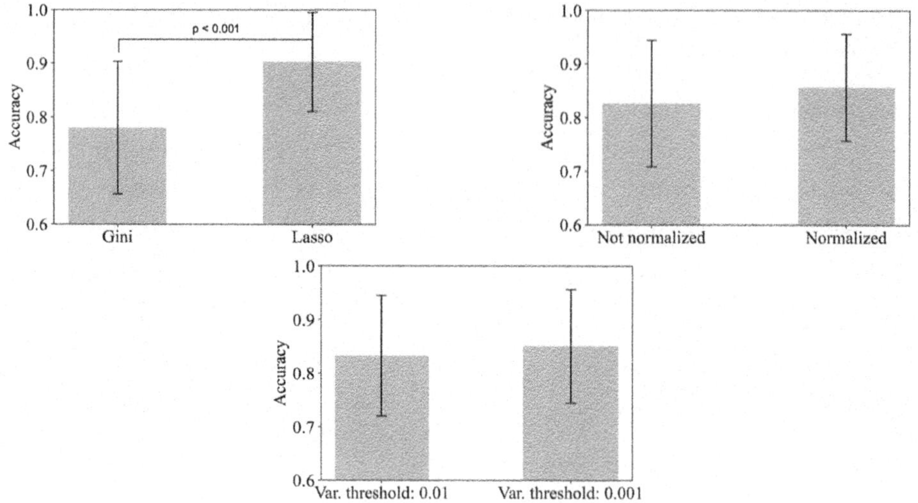

Fig. 2. Comparison of the effects of various feature extraction and selection parameters on prediction accuracy.

4.3 Covariate Clinical Information Does not Improve Prediction Performance

The ISLES 2017 dataset had basic clinical information (i.e., TTM, TICI, TSS, and TTT). When this information was combined with the three radiomics features identified in the previous section, the performance of the classifiers dropped by 8% (Table 5).

5 Conclusions and Future Work

We utilized radiomics features extracted from ADC maps to predict the clinical outcomes of patients with ischemic stroke. Our main goal was to reduce our radiomics features to the most descriptive and informative subset and improve upon our previous study [11]. To obtain the smallest and the most descriptive radiomics subset, we have tested various parameters for normalization, variance thresholding, and feature selection algorithms. For the ISLES 2017 dataset, we established that the LASSO feature selection enhances prediction accuracy by 12% compared to the Gini index feature selection. We also suggest using ADC map normalization prior to radiomics feature extraction and a variance threshold of 0.001 during feature selection. Even though the effects of ADC map normalization and variance threshold were not statistically significant, we still observed a slight improvement. Our results show that the MLP classifier can predict binary mRS scores with a very good accuracy (95%) compared to the literature, with classification accuracies varying from 70–75% [14, 20, 24]. In the future, we hope to test these feature extraction and selection methods for acute stroke lesions since predicting clinical outcome at an acute stage can inform clinical decision-making. We are also working on developing an AI algorithm for automatic stroke lesion segmentation and characterization from MRI scans [25], and a smartphone-based health monitoring technology to assess gait and balance [26, 27]. We have also started looking at potential biomarkers for

Table 5. Classification performance when seven features (three radiomics features and four clinical covariates were used.

	Accuracy	Sensitivity	Specificity	AUC
MLP	0.867 ± 0.114	0.855 ± 0.183	0.876 ± 0.161	0.865 ± 0.115
LogisticRegr	0.867 ± 0.108	0.845 ± 0.188	0.884 ± 0.140	0.864 ± 0.111
LinearSVM	0.867 ± 0.100	0.850 ± 0.160	0.880 ± 0.151	0.865 ± 0.100
NuSVM	0.853 ± 0.113	0.835 ± 0.186	0.868 ± 0.174	0.851 ± 0.113
SVM	0.836 ± 0.111	0.825 ± 0.191	0.844 ± 0.177	0.834 ± 0.111
SGD	0.833 ± 0.133	0.830 ± 0.198	0.836 ± 0.197	0.833 ± 0.132
ExtraTree	0.822 ± 0.123	0.795 ± 0.206	0.844 ± 0.168	0.819 ± 0.125
kNN	0.822 ± 0.112	0.735 ± 0.205	0.892 ± 0.135	0.813 ± 0.117
GradientBoost	0.787 ± 0.118	0.770 ± 0.225	0.800 ± 0.185	0.785 ± 0.121
Bagging	0.784 ± 0.135	0.750 ± 0.258	0.812 ± 0.187	0.781 ± 0.140
GaussianNB	0.782 ± 0.171	0.695 ± 0.259	0.852 ± 0.205	0.774 ± 0.174
RandomForest	0.776 ± 0.128	0.705 ± 0.235	0.832 ± 0.178	0.768 ± 0.132
DecisionTree	0.756 ± 0.147	0.750 ± 0.237	0.760 ± 0.185	0.755 ± 0.150
AdaBoost	0.753 ± 0.128	0.725 ± 0.238	0.776 ± 0.205	0.750 ± 0.129
XGB	0.742 ± 0.136	0.730 ± 0.262	0.752 ± 0.223	0.741 ± 0.137
Dummy	0.556 ± 0.000	0.000 ± 0.000	1.000 ± 0.000	0.000

multi-modal data fusing and analysis that can enhance clinical decision-making [28]. These additional neural and behavioral biomarkers may complement radiomics features to improve mRS prediction.

Acknowledgments. This work was funded by Aberystwyth University Centre for International Development Research at Aberystwyth (CIDRA) (grant number H1053–28, "Automatic Assessment of Gait Impairments in Stroke using Artificial Intelligence, Wearable Technology and Neuroimaging"); Welsh Government (Ser Cymru Cofund Fellowship, grant No. 663830-AU167, to OA, and Ser Cymru Tackling Covid-19, grant No. 009 to FVP and OA); Health and Care Research Wales (PhD Studentship to FVP and OA); Public Health Wales - Stroke Implementation Group (Stroke Research, Innovation, and Education fund to FVP and OA); European Commission (H2020-MSCA-RISE-2019, grant No. 873178, to OA); Turkey Council of Higher Education (YÖK) and Scientific and Technological Research Council of Turkey (TÜBİTAK) (PhD scholarship to MSE and ES).

References

1. Feigin, V.L., et al.: World stroke organization (WSO): global stroke fact sheet 2022. Int. J. Stroke **17**, 18–29 (2022). https://doi.org/10.1177/17474930211065917
2. Hankey, G.J.: Stroke. Lancet. **389**, 641–654 (2017). https://doi.org/10.1016/S0140-6736(16)30962-X
3. Rankin, J.: Cerebral vascular accidents in patients over the age of 60: II. Prognosis. Scott. Med. J. **2**, 200–215 (1957). https://doi.org/10.1177/003693305700200504
4. van Swieten, J.C., Koudstaal, P.J., Visser, M.C., Schouten, H.J.A., van Gijn, J.: Interobserver agreement for the assessment of handicap in stroke patients. Stroke. **19**, (1988). https://doi.org/10.1161/01.STR.19.5.604
5. Zhang, M.Y., Mlynash, M., Sainani, K.L., Albers, G.W.: Lansberg, MG: Ordinal Prediction Model of 90-Day Modified Rankin Scale in Ischemic Stroke. Front. Neurol. **12**, 727171 (2021). https://doi.org/10.3389/fneur.2021.727171
6. Schwier, M., et al.: Repeatability of multiparametric prostate MRI radiomics features. Sci. Rep. **9**, 9441 (2019). https://doi.org/10.1038/s41598-019-45766-z
7. Binczyk, F., Prazuch, W., Bozek, P., Polanska, J.: Radiomics and artificial intelligence in lung cancer screening. Transl. Lung Cancer Res. 10, 1186–1199 (2021). https://doi.org/10.21037/tlcr-20-708
8. Pesapane, F., et al.: How Radiomics Can Improve Breast Cancer Diagnosis and Treatment. J. Clin. Med. **12**, 1372 (2023). https://doi.org/10.3390/jcm12041372
9. Salmanpour, M.R., Shamsaei, M., Hajianfar, G., Soltanian-Zadeh, H., Rahmim, A.: Longitudinal clustering analysis and prediction of Parkinson's disease progression using radiomics and hybrid machine learning. Quant. Imaging Med. Surg. 12, 906–919 (2022). https://doi.org/10.21037/qims-21-425
10. Alyami, A.S.: The Role of Radiomics in Fibrosis Crohn's Disease: A Review. Diagnostics. **13**, 1623 (2023). https://doi.org/10.3390/diagnostics13091623
11. Şahin Erdoğan, M., Sümer, E., Villa, F., Öztürk Işık, E., Akanyeti, O., Saybaşılı, H.: Predicting Modified Rankin Scale Scores of Ischemic Stroke Patients Using Radiomics Features and Machine Learning. Presented at the UK Workshop on Computational Intelligence (to appear) September (2022)
12. Winzeck, S., Hakim, A., McKinley, R., Pinto, JAADSR, Alves, V., Silva, C., Pisov, M., Krivov, E., Belyaev, M., Monteiro, M., Oliveira, A., Choi, Y., Paik, M.C., Kwon, Y., Lee, H., Kim, B.J., Won, J.-H., Islam, M., Ren, H., Robben, D., Suetens, P., Gong, E., Niu, Y., Xu, J., Pauly, J.M., Lucas, C., Heinrich, M.P., Rivera, L.C., Castillo, L.S., Daza, L.A., Beers, A.L., Arbelaezs, P., Maier, O., Chang, K., Brown, J.M., Kalpathy-Cramer, J., Zaharchuk, G., Wiest, R., Reyes, M.: ISLES 2016 and 2017-Benchmarking Ischemic Stroke Lesion Outcome Prediction Based on Multispectral MRI. Front. Neurol. 9, 679 (2018). https://doi.org/10.3389/fneur.2018.00679
13. Zhang, R., Zhu, L., Zhu, Z., Ge, Y., Zhang, Z., Wang, T.: Apparent diffusion coefficient map based radiomics model in identifying the ischemic penumbra in acute ischemic stroke. Ann. Palliat. Med. 9, 2684–2692 (2020). https://doi.org/10.21037/apm-20-1142
14. Quan, G., et al.: FLAIR and ADC Image-Based Radiomics Features as Predictive Biomarkers of Unfavorable Outcome in Patients With Acute Ischemic Stroke. Front. Neurosci. **15**, 730879 (2021). https://doi.org/10.3389/fnins.2021.730879
15. Tang, T., et al.: Penumbra-based radiomics signature as prognostic biomarkers for thrombolysis of acute ischemic stroke patients: a multicenter cohort study. J. Neurol. **267**, 1454–1463 (2020). https://doi.org/10.1007/s00415-020-09713-7
16. Van Timmeren, J.E., Cester, D., Tanadini-Lang, S., Alkadhi, H., Baessler, B.: Radiomics in medical imaging—"how-to" guide and critical reflection. Insights Imaging **11**, 91 (2020). https://doi.org/10.1186/s13244-020-00887-2

17. Gillies, R.J., Kinahan, P.E., Hricak, H.: Radiomics: Images Are More than Pictures. They Are Data. Radiology. **278**, 563–577 (2016). https://doi.org/10.1148/radiol.2015151169
18. Zhou, Y., et al.: Feasibility of a Clinical-Radiomics Model to Predict the Outcomes of Acute Ischemic Stroke. Korean J. Radiol. **23**, 811 (2022). https://doi.org/10.3348/kjr.2022.0160
19. Ganesh, A., Luengo-Fernandez, R., Wharton, R.M., Rothwell, P.M.: Ordinal vs dichotomous analyses of modified Rankin Scale, 5-year outcome, and cost of stroke. Neurology **91**, e1951–e1960 (2018). https://doi.org/10.1212/WNL.0000000000006554
20. Wang, H., et al.: A Clinical-Radiomics Nomogram for Functional Outcome Predictions in Ischemic Stroke. Neurol. Ther. **10**, 819–832 (2021). https://doi.org/10.1007/s40120-021-00263-2
21. van Griethuysen, J.J.M., et al.: Aerts, HJWL: Computational radiomics system to decode the radiographic phenotype. Cancer Res. **77**, e104–e107 (2017). https://doi.org/10.1158/0008-5472.CAN-17-0339
22. Zhang, Y., et al.: MRI whole-lesion texture analysis on ADC maps for the prognostic assessment of ischemic stroke. BMC Med. Imaging **22**, 115 (2022). https://doi.org/10.1186/s12880-022-00845-y
23. Pedregosa, F., Varoquaux, G., Gramfort, A., Michel, V., Thirion, B., Grisel, O., Blondel, M., Prettenhofer, P., Weiss, R., Dubourg, V., Vanderplas, J., Passos, A., Cournapeau, D.: Scikit-learn: Machine Learning in Python. Mach. Learn. PYTHON
24. Tolhuisen, M.L., Hoving, J.W., Koopman, M.S., Kappelhof, M., van Voorst, H., Bruggeman, A.E., Demchuck, A.M., Dippel, D.W.J., Emmer, B.J., Bracard, S., Guillemin, F., van Oostenbrugge, R.J., Mitchell, P.J., van Zwam, W.H., Hill, M.D., Roos, Y.B.W.E.M., Jovin, T.G., Berkhemer, O.A., Campbell, B.C.V., Saver, J., White, P., Muir, K.W., Goyal, M., Marquering, H.A., Majoie, C.B., Caan, M.W.A.: Outcome Prediction Based on Automatically Extracted Infarct Core Image Features in Patients with Acute Ischemic Stroke. Diagnostics. 12, 1786 (2022). https://doi.org/10.3390/diagnostics12081786
25. Paudel, B., Zwiggelaar, R., Akanyeti, O.: Snapshot Ensemble on brain MRI segmentation. Presented at the UK Workshop on Computational Intelligence (to appear) September (2022)
26. Bunker, M.T., Sher, A., Akpokodje, V., Villagra, F., Parthaláin, N.M., Akanyeti, O.: Towards Fuzzy Context-Aware Automatic Gait Assessments in Free-Living Environments. In: Jansen, T., Jensen, R., Mac Parthaláin, N., Lin, C.-M. (eds.) Advances in Computational Intelligence Systems, pp. 463–474. Springer International Publishing, Cham (2022)
27. Sher, A., Langford, D., Villagra, F., Akanyeti, O.: Automatic Scoring of Chair Sit to Stand Test using a Smartphone. Presented at the UK Workshop on Computational Intelligence (to appear) September (2022)
28. Sahin Erdogan, M., Sümer Arpak, E., Kaya Keles, C.S., Villagra, F., Öztürk Isik, E., Afsar, N., Yucesoy, C.A., Mur, L.A.J., Akanyeti, O., Saybasili, H.: Biochemical, biomechanical, and imaging biomarkers of ischemic stroke: time for integrative thinking. Eur. J. Neurosci. (in print)

Binary Classification of Medical Images by Symbolic Regression

Ezekiel Allison(✉)

Aston University, Birmingham B4 7ET, UK
z21210@protonmail.com

Abstract. This experimental study investigates the application of symbolic regression to classification of medical diagnostic images, and the effectiveness of using alternative fitness functions for symbolic regression in order to account for the class imbalance that is often found in medical datasets. Three fitness functions are tested: one naïve function to act as a control, and two with different approaches to preventing performance bias. The performances of these fitness functions are compared and analysed, and the performance of symbolic regression for classification of medical images is evaluated. Analysis of these results shows such approaches to be insufficient to prevent performance bias for symbolic regression. However, symbolic regression is shown to be a promising machine learning algorithm for applications in medical diagnostic imaging nonetheless.

Keywords: Evolutionary Algorithm · Genetic Programming · Fitness Function

1 Introduction

Automated binary image classification can assist medical professionals in screening patients for diseases to ensure timely diagnosis and treatment [1–3]. The current dominant method for binary image classification is Convolutional Neural Networks (CNNs) [4, 5]. However, methods in the family of Genetic Programming (GP), such as symbolic regression, have some advantages over CNNs [6]. Most salient to medicine is the higher interpretability of the resulting programs when compared to CNNs. Interpretability is an important attribute of clinical application of Machine Learning (ML) [7]. Furthermore, with recent legislation restricting the use of black-box models to make decisions, interpretability and explainability have become important aspects for AI-based systems for automated decision-making [8]. Additionally, GP has the ability to learn and generalise well from relatively small datasets, which is highly relevant to the medical domain [9], an advantage over the CNN [6]. It is for these reasons, that the use of GP for binary image classification in a medical context is explored in this paper.

The image classification problem tackled in this paper is classification of chest radiographs by pneumonia diagnosis. The analysis of applying symbolic regression to screen for pneumonia, in addition to its direct application, also provides insights into the practicality of applying symbolic regression to the other medical diagnostic imaging problems. GP in general has seen little attention in medical diagnostic imaging, compared to more

N. Naik et al. (Eds.): UKCI 2023, AISC 1453, pp. 516–527, 2024.
https://doi.org/10.1007/978-3-031-47508-5_40

commonly used medical ML methods, such as Support Vector Machines (SVMs) or CNNs [10]. This makes the use of GP for medical diagnostic imaging an important area of exploratory research.

In this paper, the performances of three symbolic regressors with different fitness functions are assessed when learning from an imbalanced dataset. The objective of this paper is to determine whether symbolic regression with an appropriate fitness function is effective when learning from an imbalanced dataset. The questions investigated in this paper are:

1. Is weighting error costs according to class imbalance an effective way to account for class imbalance?
2. Is using a measure sensitive to class imbalance an effective way to account for class imbalance?
3. Do programs produced by symbolic regression generalise well?
4. Are programs produced by symbolic regression interpretable?

2 Literature Review

2.1 Machine Learning for Medical Image Classification

When comparing the performance of ML models, it is important to use relevant statistics. Shakya [11] analyses the performance of these models by calculating the accuracy, sensitivity, specificity, positive predictive value, and negative predictive value of each model operating over the testing data; Shakya acknowledges the importance of specificity to the classification problem being investigated, but only considers the accuracy as a differentiating factor between the performance of the tested algorithms. Conversely, Bhowan, Johnston, and Zhang [12] heavily criticise the use of accuracy as a measure of performance for classifiers, instead favouring Area Under receiving-operator-characteristic Curve (AUC); accuracy as a measure, is vulnerable to performance bias from class imbalance: accuracy on the majority class is valued over accuracy on the minority class by reason of prevalence. Further supporting this, other works in ML for image classification discuss the AUC when analysing the performance of a classifier [9, 13–16]. AUC is an appropriate and important measure of performance for classifiers.

GP has seen success in the particular classification problem of medical diagnostic imaging: for example, Bhardwaj et al. [17] achieve high accuracy in applying GP to medical diagnostic imaging of breast cancer from radiographs—a similar application to that investigated in this paper. However, Bhardwaj et al. do not consider sensitivity and specificity in their conclusion. In contrast, Choi and Choi [14] highlight the importance of both sensitivity and specificity in medical diagnostic imaging and centre the paper around the trade-off to be made. Choi and Choi also give an overview of how varying the discrimination threshold will tune the sensitivity/specificity trade-off without needing to retrain the model. Ryan et al. [16] plot the mean AUC of a population of programs against the generations, providing insight of AUC as a useful metric for performance as it increases with the fitness of the population. The increase of AUC as a GP model increases in other fitness functions shows AUC to be a balanced fitness function with regard to false positive and true positive rates.

2.2 Imbalanced Datasets

The performance on a ML model is underpinned by the quality of the dataset used for training [18]. Available datasets for medical diagnostic imaging are often imbalanced [19]. Bhowan, Johnston, and Zhang [12] describe imbalanced datasets as datasets in which one class makes up a minority of cases. The authors highlight the performance bias that using an imbalanced dataset can cause: high accuracy on the majority class, but low accuracy on the minority class. Bhowan, Johnston, and Zhang explain the cause of this performance bias as being the inappropriate use of conventional training measures such as accuracy or error rate. Bhowan, Johnston, and Zhang discuss two types of approach to this problem commonly found in literature: external and internal. External approaches seek to modify the dataset to reduce the class imbalance, by oversampling the minority class [20], undersampling the majority class [21], or through training multiple models on balanced subsets of the dataset [22]. Internal approaches, on the other hand, seek to adjust the learning algorithm to account for the class imbalance by either assigning different costs of misclassification to each class [23], or by using cost/objective functions that are not susceptible to performance bias from class imbalance [24]. Bhowan, Johnston, and Zhang conclude from the literature that AUC makes a better measure than accuracy for classifier performance when learning from an imbalanced dataset, such as datasets in the medical domain [19].

2.3 Feature Generation

The goal of feature construction is twofold: to transform raw features into a set of features that include highly discriminating features; and to reduce the dimensionality of the input [25]. However, the dimensionality of the constructed feature set may still be too high for ML to be effective. Depeursinge et al. [26] describe how some ML techniques are sensitive to the curse of dimensionality with regards to high dimensionality input vectors. Ortiz-Toro et al. [27] do not use feature selection in their work and achieve AUCs of 99% with all tested ML techniques. Ortiz-Toro et al. argue that textural features constructed from chest radiographs offer good performance for pneumonia detection. This shows textural features to be highly discriminating features for diagnosis of pneumonia from chest radiographs, without the need for feature selection. However, Ortiz-Toro et al. do not state the dimensionality of the input vector used in their work. Both Ortiz-Toro et al. and Tuncer et al. [28] use the radiomic features to great success, demonstrating their effectiveness as a discriminating feature for pneumonia from chest radiographs, with Tuncer et al. demonstrating the effectiveness of the Local Binary Pattern features (LBP) in particular.

2.4 Conclusion

GP algorithms are suitable for binary classification of medical images. However, GP is relatively understudied when compared to SVMs and CNNs, despite GP's advantage in generalisability and when learning from small datasets and, more specifically, symbolic regression's advantage in interpretability. The datasets available to train such classifiers are heavily imbalanced and of limited size, by nature of medicine. Although GP can suffer from performance bias with imbalanced datasets, certain fitness functions have been shown to generally be more suitable in such cases, such as fitness functions that weight the costs of misclassification of different classes, or use metrics such as AUC. Radiomic features, such as LBP, are effective textural features for classification of radiographs by pathology; additionally, if a small enough feature vector is generated, feature selection may be avoided.

3 Algorithms

3.1 Sigmoid Error

The sigmoid error fitness function uses the sigmoid function:

$$s(x) = \frac{2}{1 + e^{-x}} - 1 \tag{1}$$

to reduce the domain of the decision value to the range (-1, + 1). The error for a prediction is the difference between the resulting value and the true value of the diagnosis. The error of the GP model is the mean absolute error from predictions of the entire training dataset:

$$ff_{error}(f) = \frac{1}{N}\sum_{i=1}^{N}\left|y_i - (\frac{2}{1 + e^{-f(x_i)+\beta}} - 1)\right| \tag{2}$$

where f is the classifier, N is the size of the training dataset, y_i is the value of the true class of case i, x_i is the input feature of case i, and β is the threshold of the classifier. The GP model minimises this error. The prediction given by the program is -1 if the decision value is less than zero, and + 1 if not less than zero. This fitness function makes no attempt to account for an imbalanced dataset.

3.2 Weighted Sigmoid Error

The weighted sigmoid error fitness function is similar to the sigmoid error fitness function defined above. However, before calculating the mean absolute error, errors are weighted inversely proportional to the prevalence of the true class in the dataset. This weight for each class, w_j, is calculated by the function:

$$w_j = \frac{N}{2n_j} \tag{3}$$

where N is the total number of cases in the training dataset, and n_j is the number of cases of class j. The full fitness function is:

$$ff_{weighted}(f) = \frac{1}{N}\sum_{i=1}^{N} w_j\left|y_i - (\frac{2}{1 + e^{-f(x_i)+\beta}} - 1)\right| \tag{4}$$

where f is the classifier, N is the size of the training dataset, w_j is the weight for the true class of case i, y_i is the value of the true class of case i, x_i is the input feature of case i, and β is the threshold of the classifier. The GP model minimises this error. With this weighting, this fitness function accounts for imbalanced datasets by magnifying errors classifying the minority class.

3.3 Area Under Curve Score (AUC)

The AUC score fitness function evaluates the fitness of a binary classifier by calculating the true positive rate and false positive rate at various discrimination thresholds to produce a Receiving Operator Characteristic (ROC) curve. The area under this curve is the fitness of the classifier. The AUC is equal to the probability of a randomly chosen positive case being ranked higher than a randomly chosen negative case [29]; and AUC takes class imbalance into account [12]. The AUC of a classifier can be approximated by the function:

$$ff_{AUC}(f) = \frac{\sum_{x_+ \in X_+} \sum_{x_- \in X_-} \begin{cases} 1 \text{ if } f(x_-) < f(x_+) \\ 0 \text{ otherwise} \end{cases}}{|X_+| \cdot |X_-|} \tag{5}$$

where f is the classifier, X_+ is the set of positive input features, and X_- is the set of negative input features. The GP model maximises this score.

4 Methodology

The radiograph dataset selected for this paper is sourced from Large Dataset of Labeled Optical Coherence Tomography (OCT) and Chest X-Ray Images [30]. The class imbalance in this dataset is 74.22% majority class in training data, and 62.50% majority class in testing data.

As described by De Jong [31], a hyperparameter set well-suited for one fitness landscape may be inappropriate for another fitness landscape. To prevent one fitness function having an unfair advantage due to using a common hyperparameter set that is well-suited for one fitness landscape, but not the others, Latin hypercube sampling is used to generate 10 hyperparameter sets that cover a uniform range of possible values. After 50 generations, only the best performing hyperparameter set for each fitness function is continued to be trained. The performances of the fitness functions are compared after 100 generations of training.

5 Results

After the best performing hyperparameter sets for each fitness function are determined, and the corresponding symbolic regressors run through the 100th generation of training, the performances of the fitness functions are compared and analysed. It should be noted that, with an imbalanced binary dataset where the positive class is the majority class and the negative class is the minority class, such as the dataset used in this paper; sensitivity is equivalent to majority class accuracy and specificity equivalent to minority class accuracy. To reiterate the expectations and objectives of this study: ff_{error} is a baseline that does not account for class imbalance, and $ff_{weighted}$ and ff_{AUC} are expected to have greater specificity; however, ff_{error} is expected to have greater sensitivity, than either $ff_{weighted}$ or ff_{AUC}, as the performance bias of ff_{error} will favour majority class accuracy over minority class accuracy; the good generalisation of symbolic regression should lead to final models having performance on testing data that is not significantly worse than on training data; and finally, the resulting final programs will have high interpretability.

On the training data, the only performance metrics with a statistically significant difference are: the sensitivity when using ff_{error} is greater than that when using $ff_{weighted}$ ($p = 0.02238$). This is expected, as $ff_{weighted}$ prioritises accuracy on the minority class – specificity – at the cost of accuracy of the majority class – sensitivity. Such prioritisation comes at the cost of accuracy on the majority class, impacting overall accuracy. However, $ff_{weighted}$ did not show higher specificity than ff_{error} as expected.

On the testing data, the only performance metrics with a statistically significant difference is the specificity when using ff_{error}, and when using either of the other fitness functions. The difference is wholly contradictory to the expected result: the symbolic regressor using ff_{error} has greater specificity than either of the other fitness functions ($p < 0.0001$).

By all metrics but sensitivity, the performance of the symbolic regressor, across all fitness functions, is statistically significantly worse on testing data than on training data ($p < 0.0001$). This indicates a failure of all the tested fitness functions to generalise to unseen inputs.

In the context of a binary class dataset in which the majority class is positive, and the minority class is negative, sensitivity is equivalent to majority class accuracy and specificity is equivalent to minority class accuracy. All fitness functions, in training and in testing, have lower mean minority class accuracy than mean majority class accuracy ($p < 0.0001$). While expected for ff_{error}, this was unexpected for $ff_{weighted}$ and ff_{AUC} and indicates a failure of all of the tested fitness functions to account for class imbalance and mitigate performance bias.

The average ROC curves for the three fitness functions operating on training data, after 50 generations and after 100 generations, are shown in Fig. 1. Figure 2 similarly shows the average ROC curves on testing data.

6 Discussion

The questions that were investigated in this study are:

1. Is weighting error costs according to class imbalance an effective way to account for class imbalance?
2. Is using a measure sensitive to class imbalance and effective way to account for class imbalance?
3. Do programs produced by symbolic regression generalise well?
4. Are programs produced by symbolic regression interpretable?

Each of these questions are addressed in the following discussion. The effectiveness of $ff_{weighted}$ and ff_{AUC} at mitigating performance bias is evaluated by comparing their majority/minority class accuracies when compared to ff_{error} and the results of similar papers. The generalisation of the symbolic regressor using all three fitness functions is evaluated by comparing their performance on training data and on testing data. Finally, the interpretability of the best program produced is investigated. After the discussion of results, the methodology employed is critiqued to make recommendations for how to carry out further investigation into ML on imbalanced datasets.

The fitness function that resulted in the greatest minority class accuracy was ff_{error}. This contradicts the expected result: that the fitness function that does not take into

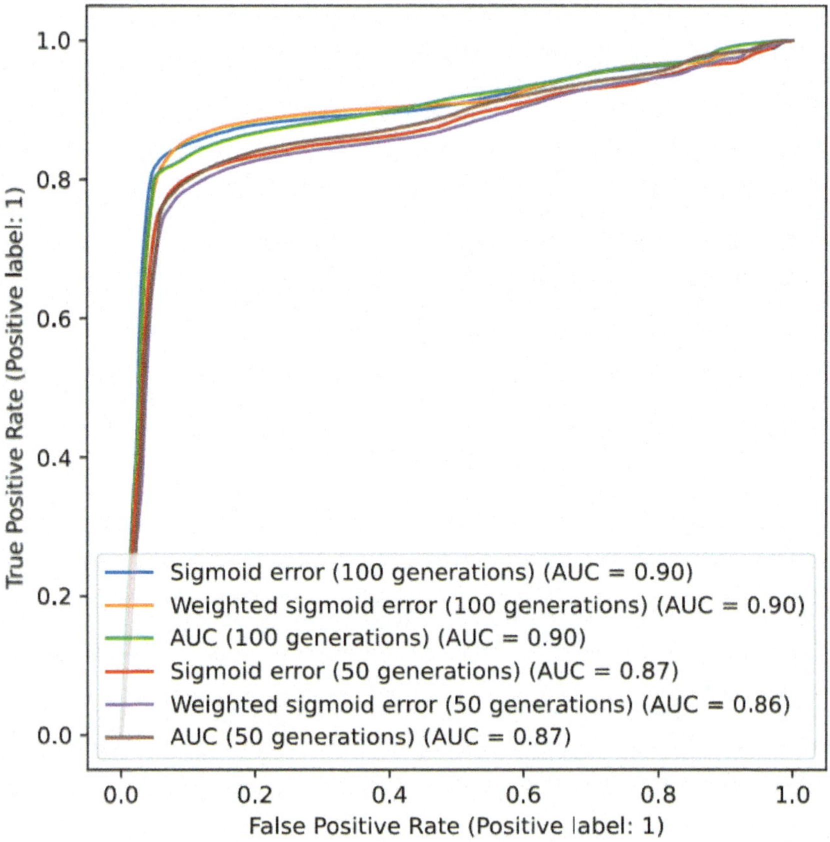

Fig. 1. The average ROC curves for the three fitness functions operating on training data, after 50 generations and after 100 generations

account class imbalance will have lesser specificity than the fitness functions that do [12]—ff_{error} makes no attempt to account for class imbalance. $ff_{weighted}$ is shown to have even worse performance on testing data than the unweighted variant. However, this supports Bhowan, Johnston, and Zhang's [12] conclusion that weighted fitness functions are only effective internal approaches to account for class imbalance in some applications, even within the field of computer vision. ff_{AUC} resulting in a lower specificity than ff_{error} comes as more of a surprise, as it contradicts Choi and Choi's [14] claim that using AUC as a fitness function will lead to few false positives.

Across all three of the fitness functions, the symbolic regressor has lesser performance than the ML techniques used by Ortiz-Toro et. al [27] to perform the same task, trained on the same dataset. A primary difference between the ML pipeline tested here, and those tested by Ortiz-Toro et al., is the range and number of features generated from the training data. Ortiz-Toro et al. generate several larger sets of textural features, with radiomic features producing far better performing models than those here. LBP is a radiomic feature, however it is not one Ortiz-Toro et al. elect to use in their work. The

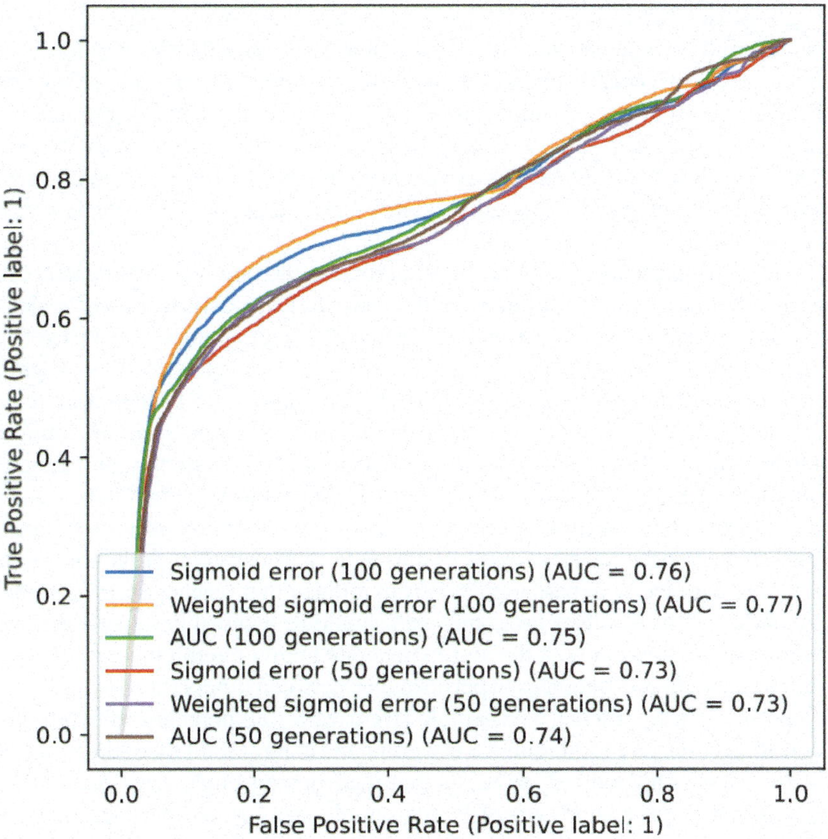

Fig. 2. The average ROC curves for the three fitness functions operating on testing data, after 50 generations and after 100 generations

difference in the performance of the models produced in this paper, and those produced by Ortiz-Toro et al. indicate that LBP alone may not be an appropriate feature when not using feature selection. Further work to investigate the performance of the proposed GP model when using different radiomic features would grant insight into this matter.

Analysis of the ROC curves in Fig. 1 shows that GP is a viable algorithm for medical diagnostic imagery, as it performs better than a random classifier. However, a statistically significantly difference across all three of the fitness functions, is the minority class accuracy being lower than the majority class accuracy. This is a failure of any of the proposed fitness functions to adequately account for class imbalance. A potential area for further work is to investigate the difference in performance of symbolic regression using the proposed $ff_{weighted}$ fitness function, with a range of more extreme weighting functions, in order to determine whether the weights used were insufficiently different, or whether weighting misclassification costs – with any weights –, is wholly insufficient to account for class imbalance in this application.

By all metrics but sensitivity, the performance of the symbolic regressor is statistically significantly worse on testing data than on training data. This failure to generalise is surprising, given generalisation is a strength of symbolic regression. However, the parameter sets used in the symbolic regressors that were run to 100 generations, all have low values of the parsimony coefficient. This reduces the pressure to evolve smaller programs at a slight cost to the fitness function, leading to overly complex programs that overfit the training data, instead of simpler programs consisting of fewer, better generalising rules.

Of all the programs produced in this experiment, the one with the greatest AUC is analysed and interpreted. The program tree is fairly large, consisting of 313 nodes. This is likely a result of the small parsimony coefficient hyperparameter used (0.0001300). The variable importance of the features used as hyperparameters for this program are analysed. Of the features 33 features used by the program, only 9 have a variable importance greater than 0.01, with the most contributing feature having a variable importance of 0.03644. The most important features correspond to the central regions of the radiographs, where the lungs are located. This reveals that GP is capable of extracting common, fixed regions of interest during training. However, this interpretability of this program is not as great as desired, as a result of the large and complex programs produced. Sampling the parsimony hyperparameter from a range that reaches so low harmed the interpretability of the resulting programs, and should be avoided in future work seeking to use symbolic regression with the aim to generate highly interpretable programs.

Additionally, most differences in performance between fitness functions are revealed to be statistically insignificant. The sample size used for all populations in this study is 30—considered the bare minimum for assuming the central limit theorem and enabling the use of t-tests. Andrade [32] suggests that inadequate sample sizes can lead to statistically insignificant results, where larger sample sizes would produce more useful results. The cause for the small sample size in this study is the limited computing power available and the large number of populations to sample from. The hyperoptimisation by sampling from the space of possible hyperparameters lead to a multiplicative increase in the computing required for this study. Although the "no free lunch" theorem supports this decision—"any two algorithms are equivalent when their performance is averaged across all possible problems" [33, p. 1]—good performance would still be possible using hyperparameters known to perform well at an adjacent problem. For example, Burks and Punch [34] specify the relevant hyperparameters used in their GP model for tuberculosis screening from chest radiographs. When comparing the performance of GP models in exploratory work, the use of hyperparameters found in existing literature will greatly reduce the computational costs, and allow more rigorous hypothesis testing, when compared to performing hyperoptimisation of the models used.

7 Conclusion

As found in the literature review: symbolic regression is a relatively understudied algorithm in the application of medical diagnostic imaging, despite its highly relevant strengths in generalisation and interpretability [6, 9]; imbalanced datasets are a fact of machine learning in the medical domain, and can lead to performance bias if not prevented by modification of the dataset or machine learning algorithm used [9, 12, 19]; textural features, such as LBP, are effective discriminating features for classifying radiographs by the presence of a pathology [27, 28]; and computationally expensive feature selection is not necessary with feature generation that produces sufficiently small feature vectors [25, 27]. This informed the research carried out in this paper: comparing the performance of three fitness functions for symbolic regression on an imbalanced dataset for the task of classifying chest radiographs by the presence of pneumonia. One fitness function acted as a control, making no attempt to account for class imbalance; the other two fitness functions took a different approach to preventing performance bias. The machine learning pipeline consisted of preprocessing the radiographs to reduce noise, LBP feature generation, hyperoptimisation of the symbolic regressors, and training for 100 generations.

The research process included selecting a high quality dataset, researching and selecting appropriate preprocessing and feature generation techniques, optimising hyperparameters, training and evaluating symbolic regressors, and implementing statistical tests to compare the final performance of the resulting programs. The dataset used [30] is a high quality dataset and suitable example of an imbalanced dataset. The feature generation used LBP, which although an effective radiomic feature, is not suitable for the binning used to reduce the feature space to a workable dimensionality. The decision to optimise hyperparameters greatly increased the computationally requirements for the study, limiting the number of iterations each model could be run for, and the sample sizes used in the study. This consequence significantly limited the discussion of results, as most differences measured were not statistically significant, partly due to the small sample size. Additionally, the opportunity to log low-level performance data during training was missed. However, symbolic regression still proved a promising algorithm for classifying medical images: Fig. 1; and demonstrated the ability to extract fixed regions of interest during training. Further research into applying symbolic regression to medical diagnostic imaging tasks could demonstrate its strengths over more popular algorithms such as CNN or SVM.

Although GP is a competent algorithm for medical diagnostic imagery – as shown by the resulting ROC curves – it is vulnerable to performance bias when trained on imbalanced datasets. Internal approaches, such as using an appropriate fitness function, are not guaranteed to prevent this performance imbalance; the use of external approaches such as data balancing is recommended instead.

Acknowledgements. I would like to thank my supervisor, Professor Aniko Ekárt, for sparking my interest in genetic programming, and providing invaluable guidance and support throughout this study.

References

1. Chowdhury, M., et al.: Can AI help in screening viral and COVID-19 pneumonia? IEEE Access **8**, 132665–132676 (2020). https://doi.org/10.1109/access.2020.3010287
2. Grzybowski, A., et al.: Artificial Intelligence for Diabetic retinopathy screening: A Review. Eye **34**(3), 451–460 (2020). https://doi.org/10.1038/s41433-019-0566-0

3. McKinney, S., et al.: International evaluation of an AI system for breast cancer screening. Nature **577**(7788), 89–94 (2020). https://doi.org/10.1038/s41586-019-1799-6
4. Kido, S., Hirano, Y., Hashimoto, N.: Detection and classification of lung abnormalities by use of convolutional neural network (CNN) and regions with CNN features (R-CNN). In: International Workshop on Advanced Image Technology (IWAIT) 2018, Chiang Mai, Thailand, pp. 1–4. IEEE (2018). https://doi.org/10.1109/iwait.2018.8369798
5. Chan, H., Hadjiiski, L., Samala, R.: Computer-aided diagnosis in the era of deep learning. Med. Phys. **47**(5), e218–e227 (2020). https://doi.org/10.1002/mp.13764
6. Evans, B., Al-Sahaf, H., Xue, B., Zhang, M.: Evolutionary deep learning: a genetic programming approach to image classification. In: IEEE Congress on Evolutionary Computation (CEC) 2018, Rio de Janeiro, Brazil, pp. 1–6. IEEE (2018). doi:https://doi.org/10.1109/cec.2018.8477933
7. Teng, Q., Liu, Z., Song, Y., Han, K., Lu, Y.: A survey on the interpretability of deep learning in medical diagnosis. Multimedia Syst. **28**(6), 2335–2355 (2022). https://doi.org/10.1007/s00530-022-00960-4
8. Ferreira, L., Guimaraes, F., Silva, R.: Applying genetic programming to improve interpretability in machine learning models. In: IEEE Congress on Evolutionary Computation (CEC) 2020, Glasgow, UK, pp. 1–8. IEEE (2020). https://doi.org/10.1109/cec48606.2020.9185620
9. Esteva, A., et al.: Dermatologist-level classification of skin cancer with Deep Neural Networks. Nature **542**(7639), 115–118 (2017). https://doi.org/10.1038/nature21056
10. Jiang, F., et al.: Artificial Intelligence in healthcare: past, present and future. Stroke Vasc. Neurol. **2**(4), 230–243 (2017). https://doi.org/10.1136/svn-2017-000101
11. Shakya, S.: Analysis of artificial intelligence based image classification techniques. J. Innov. Image Process. **2**(1), 44–54 (2020). https://doi.org/10.36548/jiip.2020.1.005
12. Bhowan, U., Johnston, M., Zhang, M.: Developing new fitness functions in genetic programming for classification with unbalanced data. IEEE Trans. Syst., Man, Cybern., Part B (Cybern.) **42**(2), 406–421 (2011). https://doi.org/10.1109/tsmcb.2011.2167144
13. Gulshan, V., et al.: Development and validation of a deep learning algorithm for detection of diabetic retinopathy in retinal fundus photographs. JAMA **316**(22), 2402–2410 (2016). https://doi.org/10.1001/jama.2016.17216
14. Choi, W.-J., Choi, T.-S.: Computer-aided detection of pulmonary nodules using genetic programming. In: IEEE International Conference on Image Processing 2010, Hong Kong, China, pp. 4353–4356, IEEE (2010). doi:https://doi.org/10.1109/icip.2010.5652369
15. Zhang, K., et al.: Clinically applicable AI system for accurate diagnosis, quantitative measurements, and prognosis of covid-19 pneumonia using computed tomography. Cell **181**(6), 1423–1433 (2020). https://doi.org/10.1016/j.cell.2020.04.045
16. Ryan, C., Krawiec, K., O'Reilly, U.-M., Fitzgerald, J., Medernach, D.: Building a stage 1 computer aided detector for breast cancer using genetic programming. In: Nicolau, M., et al. (eds.) Genetic Programming. EuroGP 2014. Lecture Notes in Computer Science, vol 8599, pp. 162–173. https://doi.org/10.1007/978-3-662-44303-3_14
17. Bhardwaj, H., Sakalle, A., Tiwari, A., Verma, M., Bhardwaj, A.: Breast cancer diagnosis using simultaneous feature selection and classification: A genetic programming approach. In: IEEE Symposium Series on Computational Intelligence (SSCI) 2018, Bangalore, India, pp. 2186–2192, IEEE (2018). https://doi.org/10.1109/ssci.2018.8628935
18. Aroyo, L., Lease, M., Paritosh, P., Schaekermann, M.: Data excellence for AI. Interactions **29**(2), 66–69 (2022). https://doi.org/10.1145/3517337
19. Rahman, M., Davis, D.: Addressing the class imbalance problem in medical datasets. International Journal of Machine Learning and Computing, 224–228 (2013). doi:https://doi.org/10.7763/ijmlc.2013.v3.307

20. Barandela, R., Sánchez, J., García, V., Rangel, E.: Strategies for learning in class imbalance problems. Pattern Recognition 36(3), 849–851 (2003). doi:https://doi.org/10.1016/s0031-3203(02)00257-1
21. Kubat, M., Matwin, S.: Addressing the curse of imbalanced training sets: one-sided selection. In: Fisher, D. (ed.) International conference on machine learning 1997, Nashville, TN, USA, vol. 97, no. 1, pp. 179–186. Morgan Kaufmann Publishers Inc. (1997)
22. Breiman, L.: Bagging predictors. Mach. Learn. **1**, 123–140 (1996)
23. Holmes, J.: Differential negative reinforcement improves classifier system learning rate in two-class problems with unequal base rates. Genet. Program. **1**, 635–642 (1998)
24. Bradley, A.: The use of the area under the ROC curve in the evaluation of machine learning algorithms. Pattern Recogn. **30**(7), 1145–1159 (1997). https://doi.org/10.1016/s0031-3203(96)00142-2
25. Al-Sahaf, H., Song, A., Neshatian, K., Zhang, M.: Two-tier genetic programming: towards raw pixel-based image classification. Expert Syst. Appl. **39**(16), 12291–12301 (2012). https://doi.org/10.1016/j.eswa.2012.02.123
26. Depeursinge, A., et al.: Comparative performance analysis of state-of-the-art classification algorithms applied to lung tissue categorization. J. Digit. Imaging **23**(1), 18–30 (2008). https://doi.org/10.1007/s10278-008-9158-4
27. Ortiz-Toro, C., García-Pedrero, A., Lillo-Saavedra, M., Gonzalo-Martín, M.: Automatic detection of pneumonia in chest X-ray images using textural features. Comput. Biol. Med. **145**, 105466 (2022). https://doi.org/10.1016/j.compbiomed.2022.105466
28. Tuncer, T., Ozyurt, F., Dogan, S., Subasi, A.: A novel Covid-19 and pneumonia classification method based on F-transform. Chemom. Intell. Lab. Syst. **210**, 104256 (2021). https://doi.org/10.1016/j.chemolab.2021.104256
29. Fawcett, T.: An introduction to ROC analysis. Pattern Recogn. Lett. **27**(8), 861–874 (2006). https://doi.org/10.1016/j.patrec.2005.10.010
30. Kermany, D., Goldbaum, M., Cai, W., et al.: Identifying Medical Diagnoses and Treatable Diseases by Image-Based Deep Learning. Cell **172**(5), 1122–1131 (2018). https://doi.org/10.1016/j.cell.2018.02.010
31. De Jong, K.: Parameter setting in EAS: A 30 Year perspective. Parameter Setting in Evolutionary Algorithms **1**, 1–18 (2007). https://doi.org/10.1007/978-3-540-69432-8_1
32. Andrade, C.: Sample size and its importance in research. Indian J. Psychol. Med. **42**(1), 102–103 (2020). https://doi.org/10.4103/ijpsym.ijpsym_504_19
33. Wolpert, D., Macready, W.: Coevolutionary Free Lunches. IEEE Trans. Evol. Comput. **9**(6), 721–735 (2005). https://doi.org/10.1109/tevc.2005.856205
34. Burks, A., Punch, W.: Genetic programming for tuberculosis screening from raw X-ray images. In: Aguirre, H. (ed.) Genetic and Evolutionary Computation Conference 2018, Kyoto, Japan, pp. 1214–1221. Association for Computing Machinery (2018). https://doi.org/10.1145/3205455.3205461

Tumour Detection and Segmentation in MRI Scans of the Gut Area

Olatunji Azeez and Raluca Lefticaru(✉)

Department of Computer Science, University of Bradford, Bradford, UK
{o.y.o.azeez,r.lefticaru}@bradford.ac.uk

Abstract. A tumour is a dangerous ailment formed when abnormal cells group together to form unwanted tissues. Over the past years, many have suffered greatly and some even died as a result of late detection of tumour tissues growing in them. In this paper, a means of automatically detecting and mapping out tumour-infected regions in gut MRI scans of patients is proposed with the aid of artificial neural networks and computational algorithms. The dataset used in this paper comprises 38,496 MRI 16-bit greyscale scan slices of the gut area of various patients. Each scan represents a slice of the gut area of a patient, a single patient may have multiple scans of various slices of their gut area. Here, the means of improving model robustness via data augmentation is devised, alongside a suitable metric function for the estimation of loss and accuracy as well as gradient computation in model training were discussed. The methodology proposed yields a model that achieved an average accuracy of 89–90% on inference data.

Keywords: Magnetic resonance imaging · Artificial neural networks · Convolution neural networks · Augmentation · Segmentation

1 Introduction

Magnetic Resonance Imaging (MRI) is without a doubt one of the wonders of contemporary medicine. It is an amazing imaging tool that can display dynamic physiologic changes in real time and provides superb soft tissue contrast with great spatial resolution [13]. Medical images, such as MRI scans, are critical tools used by doctors to diagnose and treat patients. However, despite careful examination, there is a significant risk of errors being made. These errors can occur due to the visual interpretation of the images or the cognitive evaluation of the data. In some cases, these errors have had a significant impact on patients' lives [10]. Even if the number of incorrect diagnoses is not particularly high, the consequences are severe when compared to the accurate interpretation. As fresh technologies have developed to enhance imaging capturing techniques and as previous data has been used to diagnose images with increasing accuracy, this disparity has improved over time [13].

N. Naik et al. (Eds.): UKCI 2023, AISC 1453, pp. 528–541, 2024.
https://doi.org/10.1007/978-3-031-47508-5_41

The issue of analysing MRI scans within a short time frame has been somewhat problematic upon its discovery. It is worth noting however that the discovery of MRI and the development of MRI systems have brought about a more efficient way of the diagnosis of internal organ issues and failures, compared to previous methods. The development of MRI for use in medical research has significantly improved diagnosis, especially with the reduction of exposure to potentially hazardous ionising radiation [7]. In order to investigate the fundamental mechanisms of both healthy and sick biological functions, it is both an incredibly flexible diagnostic tool and a data-rich research tool [2].

Another prominent issue is during radiation therapy where patients diagnosed with tumours are treated daily by applying some amount of X-ray or photon therapy radiation on affected areas. These affected areas are manually mapped out by the radiation oncologist prior to the radiotherapy. Manually mapping out affected regions can be time consuming and inefficient, coupled with the fact that due to human error and differences between affected and unaffected areas in the scan that cannot be visually distinguished, the oncologist may be prone to unintentionally applying some dose of radiation where there is no tumour, thus exposing the patients' organs to unnecessary radiation which may cause the damage of relevant human cells and tissues.

Therefore, numerous Artificial Intelligence (AI) techniques have been applied to MRI data, in order to assist with image segmentation and reconstruction, disease classification, for example tumour detection [4–6,11,14].

The technique proposed in this paper for segmenting tumours in MRI scans utilizes a neural network architecture that includes 5 residual connections between the encoder and decoder heads, as well as sub-residual connections in both. This architecture draws inspiration from the variational auto-encoder and surpasses the segmentation techniques from past papers.

Moreover, this paper introduces a unique data augmentation technique in which various data transformation modules are stacked in a transformation pipeline, and the order of transforms in the pipeline is shuffled for each function call. This technique ensures that the model is trained to be robust to altered and slightly corrupted data. One of the major challenges encountered during the implementation of the model architecture was ensuring its computational efficiency without compromising on performance.

The paper is structured as follows: Sect. 2 introduces the fundamental techniques used for medical imaging and machine learning, then related works for tumour detection are presented in Sect. 3. Section 4 illustrates the new proposed approach. Section 5 presents the experiments and evaluation performed to validate the proposed model. The paper then draws the conclusions in Sect. 6.

2 Background

2.1 Magnetic Resonance Imagery

MRI is a medical imaging technique used for visualising internal organs and tissues as well as detecting foreign entities and ailments like tumours and cancer.

The MRI works by using magnetic fields alongside computer-generated radio waves to generate detailed images of the internal organs and tissues. The MRI machine generates static magnetic waves that interfere with the protons of the body temporarily. The protons in the human body spin, thus creating a very small magnetic charge, the MRI machine produces a magnetic field that causes the protons to align with its direction of flow, and as this process goes on, a supply of radio frequency (RF) signal is turned on and the RF signals flow in a tangential direction to the direction of flow of the magnetic field; these RF signals tend to push these protons against the flow direction of the magnetic field till they are turned off; when turned off, the protons revert back to the original direction of the flow of the magnetic field and do so with an outburst of faint electromagnetic energy. This energy is captured by the MRI and is used to differentiate tissues based on how quickly the protons of a tissue release the energy. The MRI machines are tube-like magnet enclosures and the patients are told to lay inside while the MRI technologist operates the machine from the outside. The MRI machines generate high-resolution and detailed cross-sectional images of organs and tissues in the body, thus making diagnosis of internal ailments seamless. With its exceptional soft tissue contrast, high spatial resolution, tomographic 3D presentation, and ability to show dynamic physiologic changes, it is an amazing imaging tool [13].

Medical imaging is vital for healthcare, covering prevention, early detection, diagnosis, therapy, and follow-up [3]. Before MRIs, X-rays and CT scans were used for diagnosis.

2.2 Overview on Data Modelling with Machine Learning

Data modelling extracts hidden information by establishing relationships between features, finding applications in various domains. Machine learning algorithms, like ANNs (Artificial Neural Networks), are employed for data modelling. ANNs perform deep learning, extracting high-level features from unstructured and structured data. They are suitable for big data but prone to overfitting, mitigated by techniques like parameter dropout, weight decay, and data augmentation. Artificial networks, represented by a network of nodes, are widely used [8] and have gained attention in healthcare, including the field of MRI, in order to assist healthcare professionals [11].

2.3 Localisation, Detection and Segmentation

Object localisation belongs to the field of computer vision; the aim is to locate a target within an image or video and map out the location. It can take various forms including localization within a bounding box; this can be achieved by making the model predict bounding box coordinates of the object(s) in the given image sample, usually this type of localisation is accompanied with object classification to make up object detection. In object detection, the coordinates of the bounding boxes are predicted alongside the probability class score(s) of the object(s) in each image.

The form of localisation that is of interest in this research is *image segmentation*. One of the most influential publications prior to the use of Machine Learning (ML) in medical imaging segmentation today was the multispectral tissue classification report by Vannier et al. in 1985 [15], which used statistical pattern recognition techniques (and satellite image processing software from NASA). Unlike in the previous type of localisation that was briefly discussed, in image segmentation, each pixel of the image is assigned a class corresponding to the object they each belong to; this leads to a much better fine grain mapping out of objects in each image. Object segmentation is of two types, *instance segmentation* and *semantic segmentation*. In instance segmentation, multiple objects of the same class are treated like distinct individual instances, whereas in semantic segmentation multiple objects of the same class are treated as one entity. In this paper, object segmentation technique is used for the detection of tumours in MRI scans of patients.

3 Related Works

In recent years, various methodologies and techniques have been employed to detect tumours in medical images. Brindha et al. [4] proposed a methodology that utilizes Convolutional Neural Networks (CNNs), a type of ANN specialised in spatial data analysis, to analyse and predict tumour presence in MRI brain scans. Their technique achieved a validation accuracy, F1-score, and precision of 94, 88.4 and 89%, respectively.

Husssain et al. [9] employed a multimodal feature extraction strategy and machine learning for brain tumour detection, achieving high accuracy with decision trees and SVM (Support Vector Machine) classifiers.

In [14] Senan et al. proposed an early diagnosis of brain tumours using a hybrid approach combining deep learning and machine learning. Their method involved a combination of a CNN, a softmax activation layer, and an SVM. The CNN served as trainable feature extractors, the softmax layer computed probability scores, and the SVM acted as the final classifier.

Chen et al. [5] utilized AI and CNNs to automatically detect metastatic carcinoma clusters and atypical cancer clusters on bone marrow aspiration smears, achieving good sensitivity, specificity, and accuracy on the validation set. The model achieved a sensitivity score of 56.6%, a specificity score of 91.3%, and an accuracy of 82.2% on the validation set.

Some surveys analysed *deep learning* in radiology or medical imaging, with a focus on MRI [11,12] while others on MRI brain tumour *segmentation* [6]. However, checking the relevant literature surveys [12], it is notable the large amount of works focusing on brain tumours, then on lungs or breast MRIs, compared with gut/ abdomen area, that are less explored by the AI community.

4 Methodology

A detailed description of this project methodology is presented also in the Master's dissertation [1] and summarised as follows:

1. A relevant dataset for this project is identified.
2. The dataset is loaded and arranged for exploration, visualisation and analyses, to discern visual and statistical cues from various samples.
3. The data transformation pipeline is implemented by chaining together various data augmentation techniques, to improve model robustness for cases with slightly corrupt image samples.
4. The data pipeline is implemented for quick data sample and mask retrieval, augmentation, resizing and normalisation, as well as batch retrieval of data samples with the implemented data loader.
5. The neural network architecture of the proposed model is implemented here after. The architecture comprises two separate neural networks that act as an encoder and a decoder.
6. The loss function for measuring model performance is implemented, the loss is based on three metric functions, namely, the binary cross entropy loss, the Hausdorff distance and the Dice coefficient.
7. Furthermore, data splicing is performed; the dataset is spliced into training, augmented and non-augmented testing sets, and the various splices are readied for model training and testing process.
8. The model is trained on a mix of pure and augmented training set and tested for each cycle after training on the augmented testing set similar to the training set.
9. The model is further tested on a subset of data without augmentation, prior to the final metric evaluation of the model.
10. The model performance is graphically visualised and analysed, the model predictions are also visualised and compared with the corresponding target segments.

5 Experiments

5.1 Dataset Description

The dataset[1] utilised was provided by the University of Wisconsin Madison on the kaggle platform. It comprises 38,496 MRI 16-bit greyscale scan slices of the guts of various patients, each scan represents a slice of the gut area of a patient in each day, a single patient may have multiple scans of various slices of their gut area. The dataset also comes with a csv metadata for data lookup and referencing. The csv file contains three columns which are id, class and segmentation; each ID represents a given slice image and is expressed in the format case_day_slice; each case slice has three distinct classes large bowel, small

[1] https://www.kaggle.com/competitions/uw-madison-gi-tract-image-segmentation/data.

bowel and stomach, each class of each slice has a corresponding RLE of the mask segment; thus, in the csv file, each ID is repeated three times for each of the different classes. The RLE stands for Running Length Encoding and is a means of compressing data like text and binary image mask into a sequence of numbers. In RLE, each number at an even index represents the starting point index with the succeeding number being the length of the mask pixel in a flattened image matrix.

5.2 Implementation

It is worth noting that the project's source code is publicly available[2]. This allows the project to be transparent and allows users to browse the code freely, with detailed comments to support the reproducibility of experiments.

Exploratory data analysis was carried out on the MRI dataset, prior to transformation, to extract statistical information and for data visualisation:

1. Metadata loading and visualisation.
2. Conversion of data frame from long to wide.
3. Slice info extraction via string manipulation on slice paths.
4. RLE visualisation.
5. RLE to mask conversion.
6. Image and mask visualisation.
7. Count plot visualisation of sample slices with and without masks

Data transformation and augmentation represented the next step in our approach. Training neural networks is challenging due to overfitting. Regularization techniques like parameter dropout and L1/L2 regularization address this. Data augmentation generates additional samples to enhance model robustness.

The image and mask undergo similar geometric transformations during object segmentation using the torch vision library in PyTorch. The following *transformation techniques* were used as part of the pipeline: random resized crop; random rotation; random flip; random Gaussian blur; random noise addition.

The *data pipeline* consists of the Dataset and DataLoader classes for preparing and delivering data to the neural network model. Sample images are stored in a container to minimize I/O overhead. The Dataset class retrieves metadata, applies transformations, and resizes images and masks before normalisation.

The data loader class retrieves data batches from the dataset class. It takes the dataset instance, thread workers, batch size, and shuffle value as arguments. The data loader is a generator, returning image mask batches in sequence. Concurrency depends on thread workers but doesn't exceed CPU thread count. The shuffle value argument is a boolean value and when set to true, the algorithm shuffles the data when the iterator is created in the data loader.

[2] https://www.kaggle.com/code/ollatunji/tumor-detection-in-mri-scans-v2.

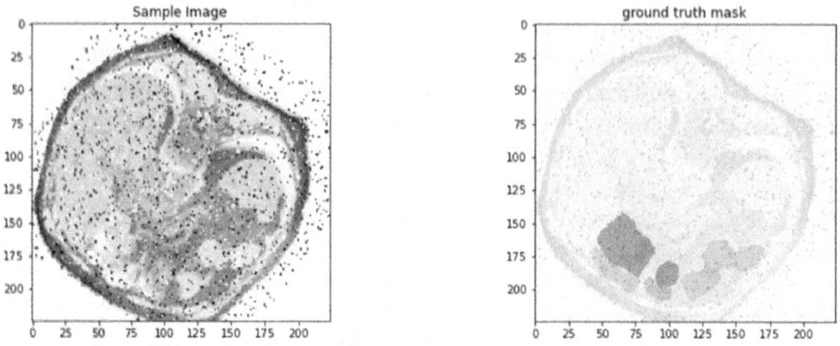

Fig. 1. Resulting sample images from the dataset class

5.3 Data Modelling and Inference

The architecture of the neural network model, the loss functions used, the optimization algorithm employed, and the learning rate scheduler utilised, all constitute a training pipeline.

The **model architecture** in this implementation is an ANN architecture, specifically an FCN (Fully Convolutional Network). The architecture comprises an encoder and a decoder and in total has 24,476,547 trainable parameters.

The encoder is a ResNet-34 architecture without the pooling and dense layers. The ResNet architecture is a neural network architecture with residual connections between the various layers and the ResNet-34 is a ResNet architecture with 34 layers. Residual connection also known as skip connection entails that feature maps of prior layers before the last layer to a given hidden layer are added or concatenated in some manner to the input of that layer; the reason behind this is so that relevant latent and high-level information of original input is not lost as the network depth increases. The ResNet architecture originally comprises a single convolutional layer followed by a batch normalisation layer, a ReLU activation layer, a max pooling layer, 4 grouped layers, an average pooling layer and finally a dense layer. A single grouped layer comprises a varying number of residual blocks depending on the ResNet architecture. In ResNet-34, the 4 grouped layers consist of 3, 4, 6 and 3 blocks respectively, with each residual block consisting of two convolutional and batch normalisation layers and a ReLU activation layer. The encoder of the model architecture takes a batch of sample images as input, and down samples them into batches of latent feature maps at each layer transformation step. The encoder has no dense and pooling layers but rather an extra convolutional layer prior to the four grouped layers, the encoder returns 5 of these latent feature maps of varying dimensions; the first feature map is the output of the first convolutional layer before the first grouped layer, the second to fifth latent feature maps are outputs of the first to fourth grouped layers respectively. Table 1 shows the sizes of the encoder input and the latent feature maps.

Both input and feature maps are 4D matrices. The first dimension represents the batch size, second is the number of channels, and third and fourth are

Table 1. Encoder input and feature map sizes

Encoder	Size (N = batch size)
Input	N, 1, 224, 224
Latent f-map 1	N, 64, 112, 112
Latent f-map 2	N, 64, 56, 56
Latent f-map 3	N, 128, 28, 28
Latent f-map 4	N, 128, 14, 14
Latent f-map 4	N, 512, 7, 7

height and width. The decoder consists of 5 blocks with a 2D transpose layer, instance normalization, convolutional layer, another instance normalization, and ReLU activation. Residual connections are established between the encoder and decoder. The output undergoes sigmoid activation (Table 2).

Table 2. Decoder blocks' input output structure table

Decoder block	First input	Input latent feature map	Decoder block output	Output size
1	Latent f-map 5	Latent f-map 4	Output 1	N, 256, 14, 14
2	Output 1	Latent f-map 3	Output 2	N, 128, 28, 28
3	Output 2	Latent f-map 2	Output 3	N, 64, 56, 56
4	Output 3	Latent f-map 1	Output 4	N, 64, 112, 112
5	Output 4	None	Output 5	N, 3, 224, 224

Metric functions are mathematical functions used to evaluate and assess the performance of a machine learning model. They provide a measure of how well or poorly a trained model is performing. In this context, there are two main categories of metric functions: loss functions and accuracy functions.

The **loss function**, also known as the criterion or cost function, is used to penalise the model for incorrect predictions during training. It quantitatively and qualitatively measures the difference between the model's predictions and the corresponding targets. During neural network training, the loss function is utilized to compute gradients for updating the model's parameters. In this particular implementation, the loss function consists of three components: the probability distribution loss function, the region-based loss function, and the boundary-based loss function. The three losses used are the BCE loss (L_{bce}), Hausdorff distance metric (L_{hd}), and Dice loss (L_{dc}). The weights assigned to the Hausdorff and Dice losses in this implementation are both set to 0.5 (w_1 and w_2). These losses collectively form the three components of the overall loss function.

The BCE loss, which stands for Binary Cross Entropy loss, is a probability distribution loss function that measures the discrepancy between the predicted probability scores and the class labels. As the name suggests, class labels can only be either 0 or 1. Therefore, the BCE loss is particularly effective for binary classification problems and other classification problems where class categories are not mutually exclusive. To prevent the loss function from assuming an insignificant value when the model does not converge, 1 is added to the BCE loss before multiplying it with other losses.

The **Dice loss** is a region-based loss function that assesses the similarity between the intersection and union ratios of two mask segments. It is especially useful for measuring similarity between areas in problems like object detection, localization, and segmentation. The *Dice loss function* can be mathematically expressed as:

$$L_{dc} = 1 - (2 * \frac{Y \cap P}{Y + P}) : \ 0 \leq L_{dc} \leq 1 \tag{1}$$

where y and p are targets and predicted mask segments. In this loss function, all pixel probability values of the prediction are rounded to whole numbers for comparison.

The **Hausdorff distance** measures the maximum distance between boundary points. It aligns prediction and target mask segments. An **accuracy function** quantifies prediction closeness to targets, complementing the loss function's minimization. In this implementation, the *accuracy function* is determined by:

$$A(P, Y) = (w_1(1 - L_{hd}) + w_2(1 - L_{dc}) : \ 0 \leq A(P, Y) \leq 1 \tag{2}$$

The **optimization function** used in this implementation is the ADAM (Adaptive Moment Estimation) optimizer. This is a gradient descent optimization algorithm that combines RMS (Root Mean Square) propagation and momentum gradient descent together.

The *learning rate decay function* gradually reduces the learning rate as training cycles increase. It prevents overshooting of the global minimum and improves model convergence. The implementation uses the *Cosine Annealing Warm Restart* technique for adaptive learning rates and transfer learning. This is mathematically expressed as:

$$\alpha_t = \alpha_{min} + \frac{1}{2}(\alpha_{max} - \alpha_{min})(1 + cos(\frac{T_{cur}}{T_i}\pi)) \tag{3}$$

where α_t is the next learning rate, α_{max} and α_{min} are the maximum and minimum learning rates respectively, T_{cur} is the current epoch in an i-th run and is the maximum number of steps in an i-th run. When $T_{cur} = T_i$, the cosine operation will be -1 and hence $\alpha_t = \alpha_{min}$. In this implementation, there are only two i-th runs and T_i is set to 20 with an exponential increasing factor of 2 for each i-th run, such that if $i = m; m \in \{1, 2, 3, 4, 5\}$, then $T_i = v; v \in \{20, 40, 80, 130, 320\}$, respectively.

5.4 Evaluation

First Notebook Execution Session. The model underwent 60 epochs of training, and its performance was evaluated using the Dice coefficient, Hausdorff distance metric, and overall loss. The plots in Fig. 2 depict model performance over 60 epochs.

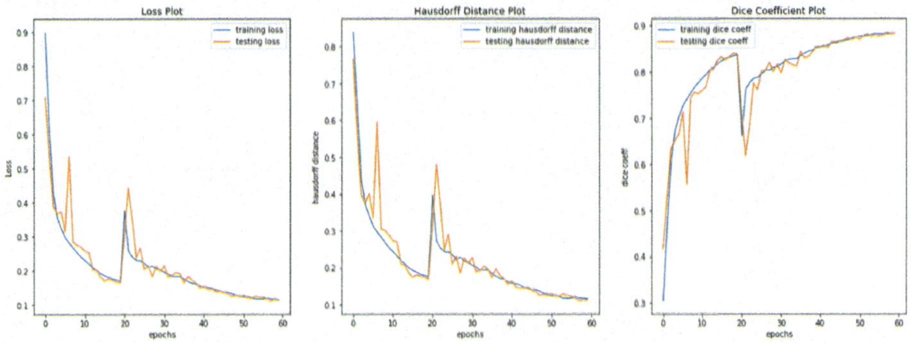

Fig. 2. Dice coefficient, Hausdorff distance and overall loss plot (First Notebook)

The plots show that as the model loss decreases, the Hausdorff distance decreases and the Dice coefficient increases. The model loss rapidly decreases from 0.8890 to 0.1711 over 20 epochs as the learning rate decreases. The model accuracy is computed for each epoch, and the best accuracy is saved. The model is tested on a non-augmented testing set (Table 3).

Table 3. First notebook metric values of best training and testing accuracies

	Best accuracy (%)	Loss	Hausdorff coefficient	Dice coefficient	Epoch number
Training set	88.41	0.1168	0.1167	88.49	59
Augmented testing set	88.72	0.1135	0.1117	88.62	67

Model Predictions 1. A few samples of model predictions on the non-augmented testing set were plotted for visualisation and comparison with the corresponding targets as shown in Fig. 3.

From the sample predictions, it is seen that the model predicts no mask for sample images with no tumour present, as it predicts almost accurately, the masks for samples where tumour is present (Table 4).

Table 4. First notebook model performance on non-augmented testing set

	Accuracy (%)	Loss	Hausdorff	Dice coefficient (%)
Non-augmented testing set	92.15	0.0789	0.0830	92.60

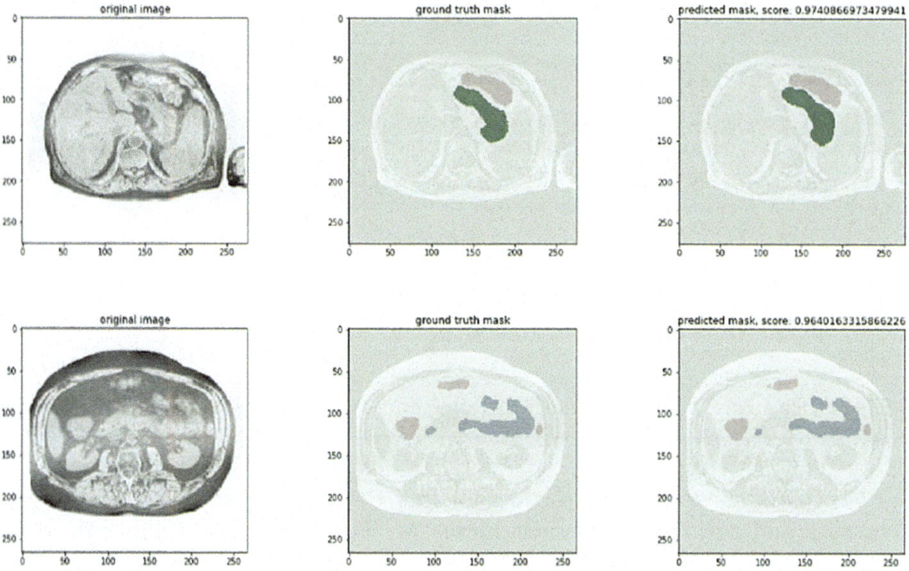

Fig. 3. First notebook model results on non-augmented testing data

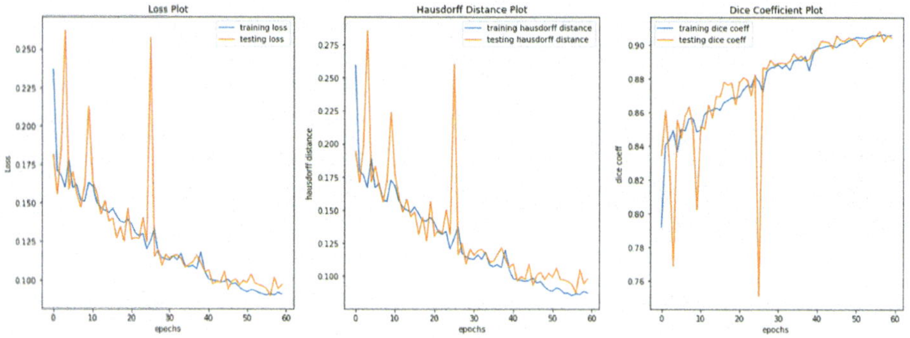

Fig. 4. Dice coefficient, Hausdorff distance and overall loss plot (second notebook)

Second Notebook Execution Session. In the second notebook run the model weights were initialised with the trained weights from the first phase and the model underwent another 60 epochs of training and testing like the first execution. The evaluation of the model on the metrics are shown and discussed. Furthermore, the model predictions are shown and compared to target masks. The model performance was measured with the Dice coefficient, the Hausdorff distance and the overall loss metric. The model performance on the metric functions is shown in Fig. 4.

From the plots, it can be observed that the model loss varies greatly with the Hausdorff distance and the Dice coefficient metric, just like in the metric evaluation of the first notebook execution. At the first epoch (epoch 0), the model had an accuracy value of 76.62 and 82.03% on the training and augmented testing set, the relatively high accuracy is to be expected at the first epoch, given that the model initial weights are already trained weights from the first notebook execution. In this notebook execution session, the learning rate was not restarted, in other words, the learning rate decreased from its maximum to its minimum value in the entire training span.

The model was able to attain a maximum accuracy of 91.05% at epoch 56 was attained on the augmented testing set, with a corresponding training accuracy of 90.95%. A maximum training accuracy 91.02% was attained at epoch 57, with a corresponding accuracy of 89.88% on the augmented testing set. Tables 5 and 6 show details pertaining to the maximum training and testing accuracy.

Table 5. Second notebook metric values of best training and testing accuracies

	Best accuracy(%)	Loss	Hausdorff	Dice coefficient	Epoch number
Training set	91.02	0.0903	0.0862	90.66	57
Augmented Testing set	91.05	0.0901	0.0876	90.86	56

Table 6. Second notebook model performance on non-augmented testing set

	Accuracy (%)	Loss	Hausdorff	Dice coefficient (%)
Non-augmented testing set	93.12	0.0682	0.0713	93.54

Model Predictions 2. Similar to the first notebook run, few samples of the model predictions on the non-augmented testing set were plotted for visualisation and comparison with the corresponding targets as shown in Fig. 5.

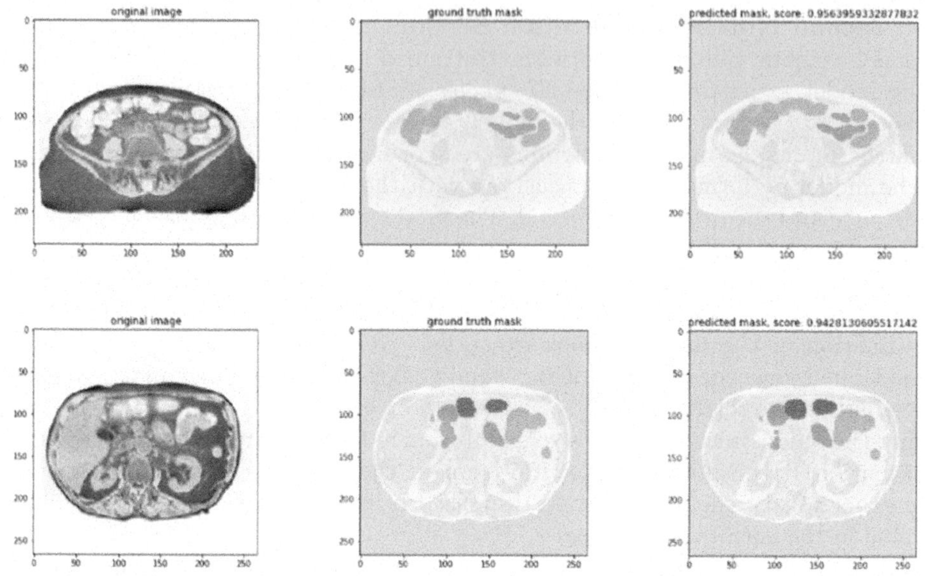

Fig. 5. Second notebook model results on non-augmented testing data

6 Conclusions

The implemented fully convolutional network model has made significant contributions in tumour detection and segmentation in gut area MRI scans. It outperformed previous works, demonstrating higher accuracy and Dice coefficient scores. This model enables better context understanding, leading to more accurate results. It is applicable in real-life scenarios for tumour detection and segmentation. The slight difference between the second and first notebook executions suggests limited improvement with further training. Modifying the model architecture, like using deeper ResNet or ensemble networks, may enhance convergence and accuracy.

References

1. Azeez, O.: Tumour detection and segmentation in MRI scans of the gut area. Master's thesis, University of Bradford (2023)
2. van Beek, E.J., Kuhl, C., Anzai, Y., Desmond, P., Ehman, R.L., Gong, Q., Gold, G., Gulani, V., Hall-Craggs, M., Leiner, T., Lim, C.C.T., Pipe, J.G., Reeder, S., Reinhold, C., Smits, M., Sodickson, D.K., Tempany, C., Vargas, H.A., Wang, M.: Value of MRI in medicine: more than just another test? J. Magn. Reson. Imaging : JMRI **49**(7), e14 (2019)
3. Bercovich, E., Javitt, M.C.: Medical imaging: from Roentgen to the digital revolution, and beyond. Med. J. **9**(4), 2076–9172 (2018)

4. Brindha, P.G., Kavinraj, M., Manivasakam, P., Prasanth, P.: Brain tumor detection from MRI images using deep learning techniques. In: IOP Conference Series: Materials Science and Engineering, vol. 1055(1) (2021)
5. Chen, P., Chen Xu, R., Chen, N., Zhang, L., Zhang, L., Zhu, J., Pan, B., Wang, B., Guo, W.: Detection of metastatic tumor cells in the bone marrow aspirate smears by artificial intelligence (AI)-based Morphogo system. Front. Oncol. **11**, 742,395 (2021)
6. Gordillo, N., Montseny, E., Sobrevilla, P.: State of the art survey on MRI brain tumor segmentation. Magn. Reson. Imaging **31**(8), 1426–1438 (2013)
7. Grover, V.P., Tognarelli, J.M., Crossey, M.M., Cox, I.J., Taylor-Robinson, S.D., McPhail, M.J.: Magnetic resonance imaging: Principles and techniques: lessons for clinicians. J. Clin. Exp. Hepatol. **5**(3), 246–255 (2015)
8. Heaton, J.: Deep learning. Genet. Program Evolvable Mach. **19**(1), 1573–7632 (2018)
9. Hussain, L., Saeed, S., Awan, I.A., Idris, A., Nadeem, M.S.A., Chaudhry, Q.U.A.: Detecting brain tumor using machines learning techniques based on different features extracting strategies. Curr. Med. Imaging **15**(6), 595–606 (2019)
10. Krupinski, E.A.: Current perspectives in medical image perception. Attent. Percept. Psychophys. **72**(5), 1205–1217 (2010)
11. Lundervold, A.S., Lundervold, A.: An overview of deep learning in medical imaging focusing on MRI. Z. Med. Phys. **29**(2), 102–127 (2019)
12. Mazurowski, M.A., Buda, M., Saha, A., Bashir, M.R.: Deep learning in radiology: an overview of the concepts and a survey of the state of the art with focus on MRI. J. Magn. Reson. Imaging **49**(4), 939–954 (2019)
13. Plewes, D.B., Kucharczyk, W.: Physics of MRI: a primer. J. Magn. Reson. Imaging **35**(5), 1038–1054 (2012)
14. Senan, E.M., Jadhav, M.E., Rassem, T.H., Aljaloud, A.S., Mohammed, B.A., Al-Mekhlafi, Z.G.: Early diagnosis of brain tumour MRI images using hybrid techniques between deep and machine learning. Comput. Math. Methods Med. **2022**, 17 (2022)
15. Vannier, M.W., Butterfield, R.L., Jordan, D., Murphy, W.A., Levitt, R.G., Gado, M.: Multispectral analysis of magnetic resonance images. Radiology **154**(1), 221–224 (1985)

AI Applications

Artificial Intelligence (AI) Applications in Chemistry

Ishita Naik[1], Dishita Naik[2], and Nitin Naik[3](✉)

[1] King Edward VI Five Ways School, Birmingham, United Kingdom
17inaik@fws.kevibham.org
[2] Birmingham City University, Birmingham, United Kingdom
dishita.naik@mail.bcu.ac.uk
[3] School of Computer Science and Digital Technologies, Aston University, Birmingham, United Kingdom
n.naik1@aston.ac.uk

Abstract. The production of chemicals is a complex task due to the highly nonlinear behaviour of chemical processes; therefore, traditional approaches may not be very effective in developing or predicting such processes and their outcomes at optimal level. Consequently, it has always been challenging to find ways to improve efficiency and productivity while reducing the time and cost. Artificial Intelligence (AI) techniques are becoming valuable in chemistry due to several reasons such as easy to learn and use, simple implementation, easy designing, effectiveness, generality, robustness, and flexibility. AI is comprised of several techniques within it, such as artificial neural networks, evolutionary algorithms and fuzzy logic. AI techniques have been widely used in various areas of chemistry including molecule design, molecular property prediction, retrosynthesis, reaction outcome prediction and reaction conditions prediction. Therefore, this paper investigates AI applications in the aforementioned areas, wherein it explains about each aforementioned area with a suitable example, limitations of traditional techniques, and types of AI techniques which are utilised within those areas.

Keywords: Artificial intelligence · AI · Machine learning · Deep learning · De Novo design · QSAR · QSPR · Molecule design · Molecular property prediction · Retrosynthesis · Reaction outcome prediction · Reaction conditions prediction

1 Introduction

The integration of artificial intelligence (AI) techniques in chemistry has enabled chemists and chemical engineers to explore, develop or predict new chemical processes and their outcomes at optimal level. A chemistry-focused AI approach facilitates chemists and chemical engineers to develop an intelligent system, which is trained on the structures and properties of molecules to perform the desired chemical task or process. There are numerous chemical tasks and

N. Naik et al. (Eds.): UKCI 2023, AISC 1453, pp. 545–557, 2024.
https://doi.org/10.1007/978-3-031-47508-5_42

processes, wherein datasets often feature complicated input and output relationships. Using a chemistry-focused AI approach, chemists and chemical engineers can extract valuable insights from a large amount of data that would be difficult or impossible to analyse using traditional approaches. That helps to reduce the effort, time and cost required to design and perform chemical experiments. These AI techniques can be effectively used in several areas of chemistry such as: molecule design, molecular property prediction, retrosynthesis, reaction outcome prediction and reaction conditions prediction. Therefore, the aim of this paper is to simplify the aforementioned areas of chemistry with a suitable and consistent example based on Acetaminophen (Paracetamol) molecule. It presents generalised AI techniques in each area of chemistry for easy understanding of AI applications, and avoiding any complexity and confusion due to a large number of existing AI techniques in each area.

2 Applications of AI in Chemistry

This section investigates AI applications in the five main areas of chemistry: molecule design, molecular property prediction, retrosynthesis, reaction outcome prediction and reaction conditions prediction as shown in Fig. 1. It explains about each aforementioned area with a relevant example of Acetaminophen (Paracetamol) molecule, limitations of traditional techniques, and what AI techniques are useful within those areas as well as their strengths and limitations.

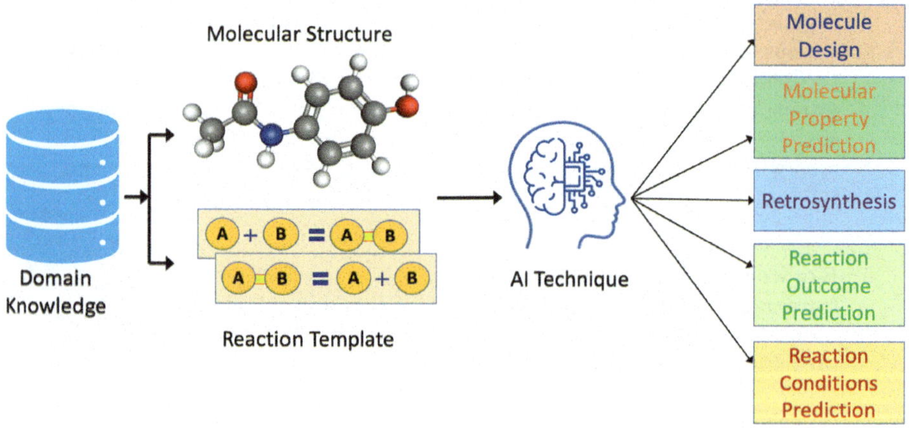

Fig. 1. AI applications in chemistry

2.1 Applications of AI in Molecule Design

2.1.1 Molecule Design

A molecule is a group of two or more atoms that form the smallest identifiable unit wherein a pure material may be split while preserving its content and chemical characteristics. A molecular formula is used to represent a molecule using its chemical symbols to indicate the types of atoms followed by subscripts to show the number of each type of atom in the molecule. However, a subscript is used only when more than one atom of a given type is present. A molecule design is a complex scientific process performed by chemists to combine two or more components to generate a new biological agent. Chemists utilise their domain knowledge and creativity to propose new molecular structures by introducing and removing functional groups, groupings of atoms and bonds, with specified qualities. Later they are tested either virtually (in silico) or experimentally in the relevant applications.

The molecule design process involves the construction of each alternative molecular structure for testing, which is a time-consuming operation even if researchers only intend to tweak a single carbon atom, researchers must select among thousands of different possibilities to build this structure. Chemists generally perform each modification step manually despite using such techniques that predict ideal desired attributes. This manual design work has several limitations, such as natural bias in human creativity, that means chemists may prefer certain functional groups or molecular patterns (consciously or unconsciously), and may exclude molecules that they find abnormal, unsuitable or unfavourable. Moreover, if a large number of new molecules are needed for virtual screening, then the human brain cannot design such large sets of molecules in a reasonable time. Finally, this manual design process can take a long time at each stage and yet not produce molecules with the appropriate characteristics.

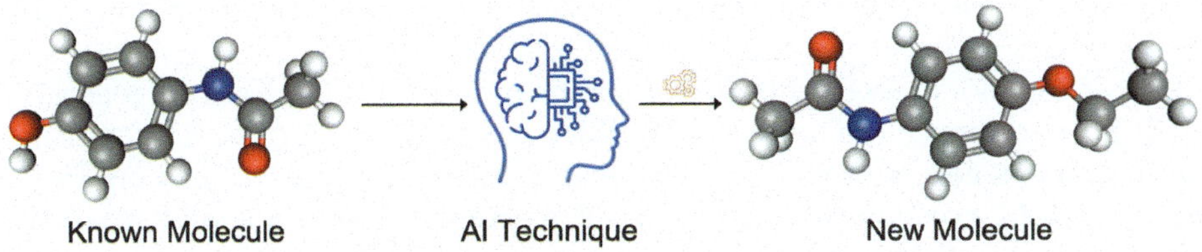

Fig. 2. Molecular design using AI technique

2.1.2 AI Based De Novo Design

Computational de novo molecular design is the process of generating new chemical structures that optimally satisfy a desired molecular profile. Generally in drug discovery, molecular profiling is used to produce a desirable biological response, while maintaining acceptable pharmacokinetic properties [8]. There are several different de novo methodologies, largely differing in the process of chemical structure generation such as whether molecular design is modelled on an atom-based,

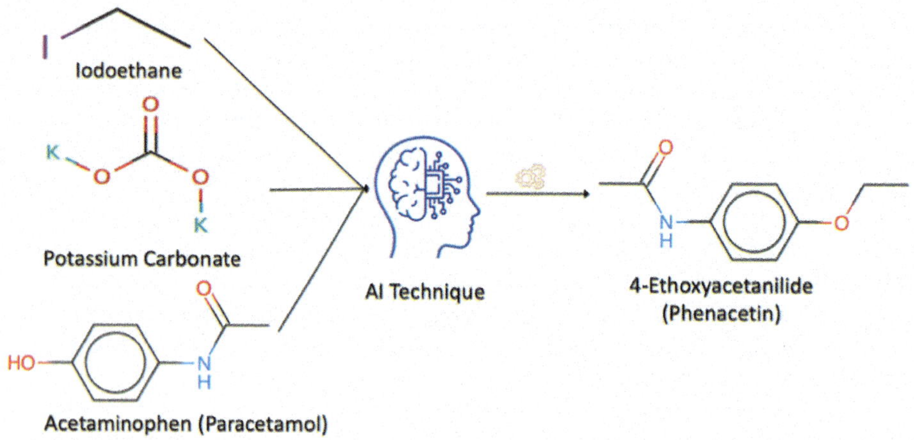

Fig. 3. An illustrative example of designing phenacetin using paracetamol

fragment-based, or reaction-based paradigm, and the type of scoring technique employed.

Several employed AI techniques for de novo design can be classified into two categories: gradient-free molecular optimization and gradient-based molecular optimization. Gradient-free or metaheuristic AI techniques for de novo design utilise population-based stochastic optimisation procedures to navigate chemical space, such as evolutionary algorithms or swarm intelligence [8]. Gradient-based AI techniques are often pretrained on large datasets of existing molecular structures and learn to navigate arbitrary property surfaces toward optimal solutions, such as, variational autoencoders (VAEs), generative adversarial networks (GANs), and recurrent neural networks (RNNs) [8].

Most commonly used AI techniques in de novo molecular design is generative artificial intelligence (AI). Generative AI refers to a category of AI algorithms that generates new outputs based on the data they have been trained on. Unlike traditional AI systems that are designed to recognize patterns and make predictions, Generative AI creates new content in the form of text, images, audio, and synthetic data. Generative AI uses a type of deep learning called generative adversarial networks (GANs) to create new content. A GAN consists of two neural networks: a generator that creates new data and a discriminator that evaluates the data. The generator and discriminator work together, with the generator improving its outputs based on the feedback it receives from the discriminator until it generates content that is indistinguishable from real data. The trained generative models allow the user to sample molecules from the learnt chemical space and, when coupled with an optimization process, such as Bayesian optimization (BO) or reinforcement learning (RL), it can efficiently identify desirable molecular profiles.

Figure 2 shows how an AI technique can design a new molecule from the known molecule; and Fig. 3 illustrates an example of designing the Phenacetin molecule using the known Paracetamol molecule.

2.2 Applications of AI in Molecular Property Prediction

2.2.1 Molecular Property Prediction

Molecules can be formed by assembling atoms in several different ways, such as by combining the same atoms in different quantities or the same atoms may combine in the same ratios but in a different physical structure. The physical structure of a molecule determines its properties. Molecular properties can be classified in several categories such as chemical properties, physical properties, structural properties and rheological properties.

Chemical scientists have been predicting the chemical properties of molecules manually, which is a very costly and time-consuming process. These techniques are usually based on physical laws or empirical relationships relating the structure of molecules to their properties through a set of chosen descriptors. However, in many cases, the physical laws that determine the molecular properties are not exactly known, or empirical relationships would be too complicated to establish due to strong non-linearities or correlations between parameters.

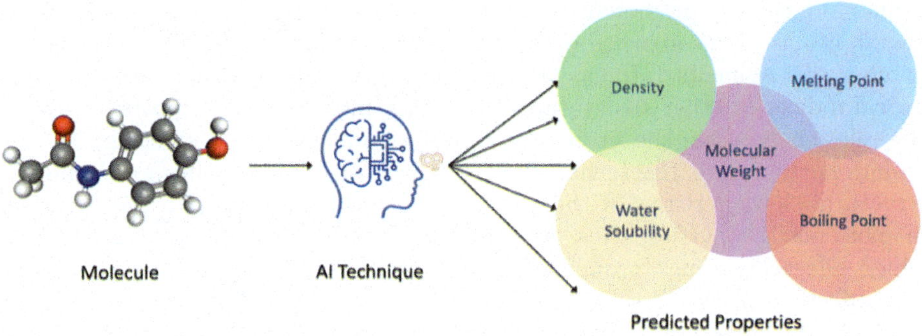

Fig. 4. Molecular property prediction using AI technique

2.2.2 AI Based QSAR/QSPR

Quantitative Structure-Activity Relationships (QSAR) and Quantitative Structure-Property Relationships (QSPR) are one of the oldest and most popular techniques for molecular property prediction and molecule design. In the last several decades, these QSAR/QSPR techniques have evolved from the simple regression/classification analysis of a small set of similar compounds to more sophisticated AI techniques that can extract the chemical, physical, and biological functions embedded in a massive data set of complex molecular structures. Screening of vast chemical spaces at high screening rates is what makes

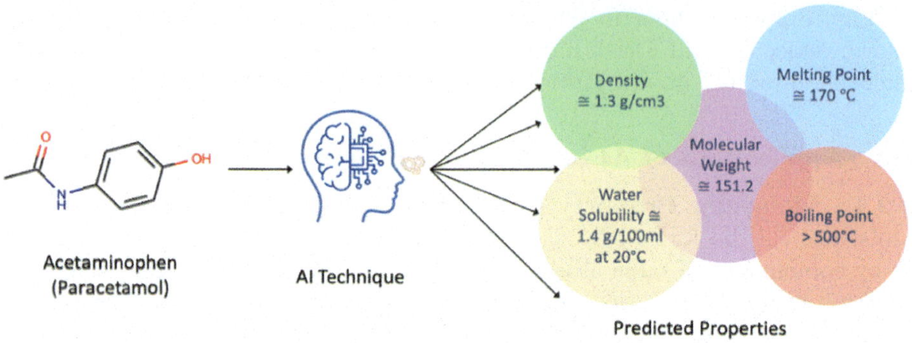

Fig. 5. An illustrative example of predicting properties of paracetamol (values are based on World Health Organisation (WHO) and International Labour Organization (ILO) guidelines [3])

QSAR/QSPR an affordable and highly productive technique provided important issues such as the applicability range and model validation are properly taken into account.

The core concept of a QSAR/QSPR technique is that the structure of a molecule determines its properties; meaning that the biological activity of a compound can be predicted by its molecular structure. These QSAR/QSPR techniques use some structural characteristics or molecular descriptors, such as number of fragments (atoms, bonds or group of atoms in a molecule), topological indices or other structural information, molecular descriptors, to express the relation between the property of interest and molecular structure of compound. Furthermore, this technique can identify and describe the most important structural characteristics of the compounds that are relevant to the variations in molecular properties, thus, affording an insight into the structural factors which affect molecular properties. Additionally, QSAR/QSPR techniques are very effective in virtual screening in drug discovery, which reduces the number of candidate compounds that need to be experimentally tested, thus reducing development costs and speeding up the drug discovery process. Currently, the most commonly used AI techniques in QSAR/QSPR are variations on random forest (RF) [11], support vector machine (SVM) [5], and deep neural network (DNN) [6].

Figure 4 shows an AI technique can predict various properties of a molecule; and Fig. 5 illustrates an example of predicting the desired properties of a Paracetamol molecule.

2.3 Applications of AI in Retrosynthesis

2.3.1 Retrosynthesis When chemical scientists or researchers have identified a promising molecule, they want to synthesize it by performing retrosynthesis. Retrosynthesis is a process of breaking down a molecule recursively into its components in order to find a set of commercially available components in that molecule [1]. This retrosynthesis process is mostly performed manually by chemists, which is time-consuming and indeterminate.

2.3.2 AI Based Retrosynthesis Existing AI-based retrosynthesis techniques can be broadly divided into two categories: template-based techniques and template-free techniques.

A template-based technique combines the reaction templates with machine learning algorithms for the prediction of chemical reactions and retrosynthesis routes. Template-based techniques predict reactant sets using mainly reaction templates. A reaction template encodes atom connectivity changes during a chemical reaction and can be used to transform a product molecule to reactants [9]. Reaction templates are normally extracted from reaction databases or hand-coded by chemists. These template-based AI techniques usually view the problem as a classification task in which the templates are modelled as distinct classes [9]. For retrosynthesis prediction, each template is applied to a product to find a match, i.e., a subgraph isomorphism. If a suitable subgraph isomorphism is found, a product is transformed to a reactant set depending on that template. This process continues until chemically suitable pathways are identified [13].

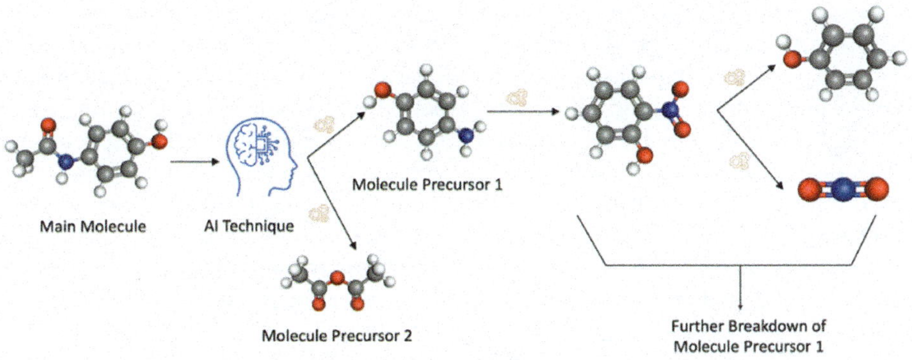

Fig. 6. Retrosynthesis using AI technique

These template-based techniques can also be categorised based on how reaction templates are created, i.e., a manual encoding technique or an automated extraction technique. Synthia (formerly Chematica) is one of the most popular AI tools for retrosynthesis, which is an expert-encoded and template-based retrosynthetic analysis tool. This is a commercial program developed by Grzybowski and their team. This retrosynthesis tool utilises a manually collected knowledge database comprising around 10,000 hand-encoded reaction transformation rules [12]. Some other examples of template-based tools are ReactionPredictor and AiZynthFinder.

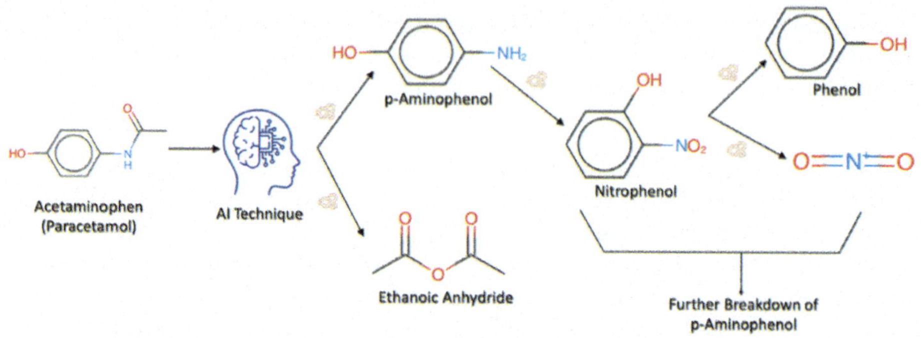

Fig. 7. An illustrative example of performing retrosynthesis of paracetamol

A template-free technique does not require to construct an external reaction template database, and it directly transforms products into potential reactants. Template-free techniques can be further subdivided according to the molecular representation protocol into graph-based techniques, and sequence-based techniques. A graph-based technique has a natural representation of human interpretable molecular graphs, and it is based on the concept that the graph topology of precursor molecules is largely unaltered during a chemical reaction. In this category, G2Gs, RetroXpert and GraphRETRO techniques are very effective [10]. A sequence-based technique treats molecules as one-dimensional token sequence, and formulates the problem of reaction pathway planning as a language translation problem using a string representation of molecules (e.g., SELFIES, SMILES, INCHI, and DEEPSMILES - all aim to map a string of tokens to a molecular graph). In this category, the deep learning approaches such as LSTM based encoder-decoder frameworks, and transformer-based techniques are very effective [10].

Figure 6 shows an AI technique can perform retrosynthesis of a molecule; and Fig. 7 illustrates an example of performing retrosynthesis of Paracetamol molecule.

2.4 Applications of AI in Reaction Outcome Prediction

2.4.1 Reaction Outcome Prediction Determining the main product of a chemical reaction while the input reactants and certain conditions are specified is an important part of the reaction prediction problem in chemistry. Chemical reactions are driven by a complex physical interaction of electronic and structural attributes of the reactants at specified reaction conditions. Therefore, chemists rely on their knowledge and experience to predict the outcome of reactions. The true measure of success for any synthesis process is whether the predicted outcome matches what is observed experimentally.

Traditional techniques require significant expertise and experience in the certain area of chemistry, which allows those expert chemists to predict the outcome of reactions that have not yet been conducted. For example, physical simulations of transition states using various quantum mechanical and other approximation techniques can be used to predict reactions and its outcomes. However, evaluation of all possible candidates experimentally is virtually impracticable as it requires significant effort, time, expertise, and money. Therefore, it is essential to have an effective approach to predict reaction outcomes without performing extensive experiments in the laboratory [7].

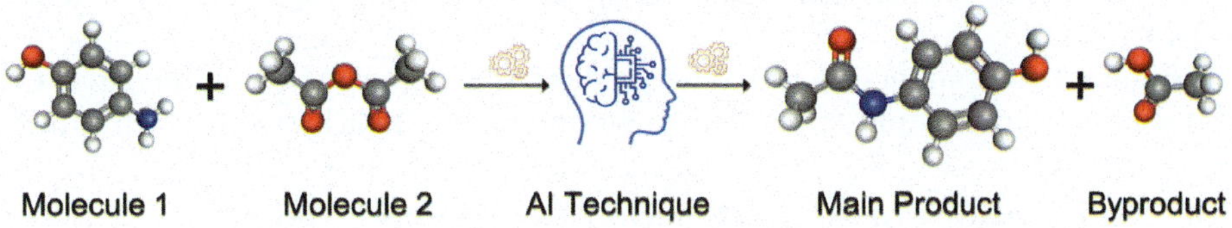

Fig. 8. Reaction outcome prediction using AI technique

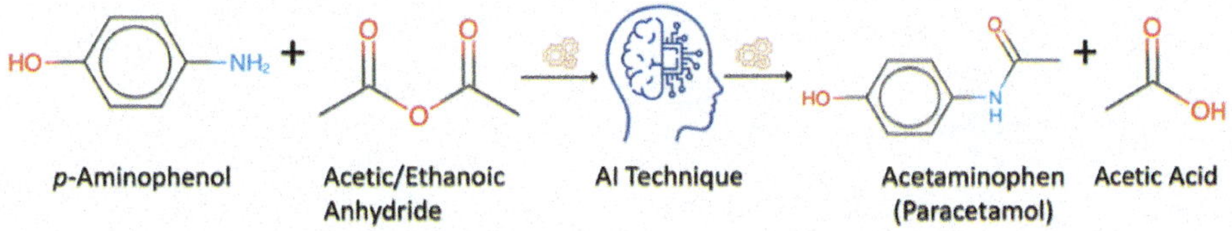

Fig. 9. An illustrative example of predicting an outcome in synthesis of paracetamol

2.4.2 AI Based Reaction Outcome Prediction AI-based techniques used for the prediction of chemical reaction outcomes can be classified into two categories: data-driven AI techniques and hybrid AI techniques.

A data-driven AI technique uses complex, end-to-end machine learning architectures that encode the reactions, discover transformations, and predict the outcomes with little to or no explicit incorporation of prior chemistry knowledge [6]. The data-driven technique is normally a two-stage process mostly based on neural networks, wherein, the first stage generates candidates by modelling interactions between molecular orbitals, and then the second stage identifies the most productive reaction by the ranking of candidates [6]. These data-driven AI techniques can be categorised into several categories such as deep neural networks, sequence-to-sequence techniques, graph convolutional neural networks, and recurrent neural networks [6]. Ranking the most productive reaction can be performed by using any scoring technique such as Weisfeiler-Lehman Networks-based technique for scoring candidate molecules by modelling high-order interactions between changes occurring in a molecule [4].

A hybrid AI technique combines the traditional reaction templates (submolecular patterns that encode changes in atom connectivity) with machine learning techniques for the prediction of chemical reaction outcomes. These hybrid techniques incorporate the reaction templates to complement data-driven techniques, and utilise several graph-based machine learning approaches to predict outcomes or products. These graphs are normally used to predict reaction types followed by the application of reaction templates to predict outcomes or products. These hybrid AI techniques are based on several types of graph-based techniques such as molecule-reaction graphs, and graph convolutional neural network.

Figure 8 shows reaction outcome prediction using an AI technique; and Fig. 9 illustrates an example of predicting an outcome in synthesis of Paracetamol.

2.5 Applications of AI in Reaction Conditions Prediction

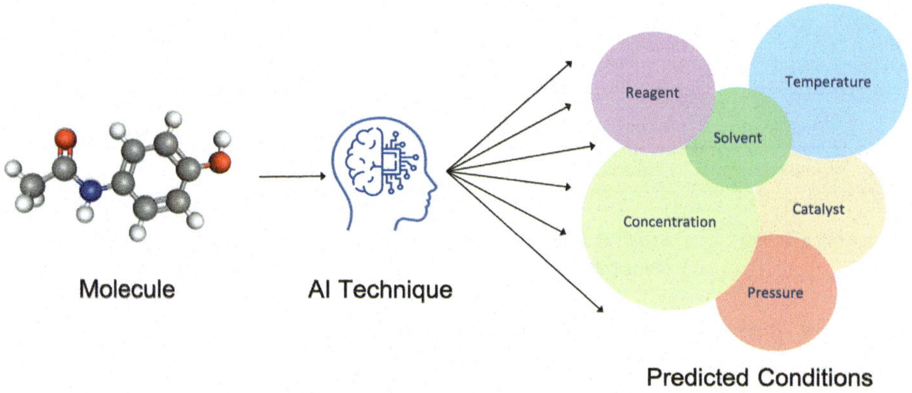

Fig. 10. Reaction conditions prediction using AI technique

2.5.1 Reaction Conditions Prediction When chemists identify a reaction which they want to run, they still have to find optimal reaction conditions such as solvent(s), catalyst, reagent(s), temperature, concentrations, reaction time, purification method. However, finding the optimal reaction condition that produces a high product yield may require thousands of experiments, as it requires control over several parameters. In many cases, small changes in reaction conditions can lead to significantly different reaction outcomes. Therefore, recent research on reaction outcome prediction has started to include reaction conditions to improve the accuracy and specificity of predictions [2].

Chemists use experience, intuition, domain knowledge and empirical optimization techniques to identify the best reaction conditions. However, this approach has several limitations and challenges. Their recommended reaction conditions may be biased based on the chemists' preference and familiarity with

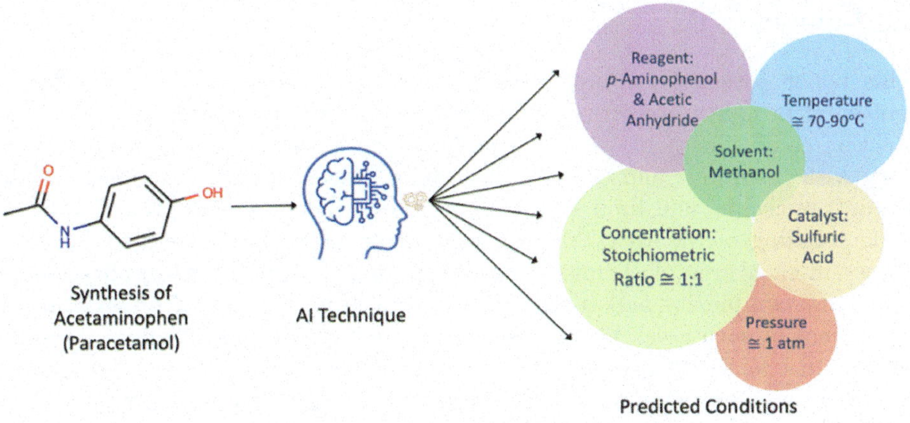

Fig. 11. An Illustrative example of predicting reaction conditions in synthesis of paracetamol (conditions are based on World Health Organisation (WHO) and International Labour Organization (ILO) guidelines [3])

certain types of reactions; the heuristic rules might not be all-encompassing or too abstract to narrow down to specific chemicals, and conditions of a precedent reaction may not be applicable to the new reactions even if the reactants are structurally similar [2].

2.5.2 AI Based Reaction Conditions Prediction Similar to the AI-based reaction outcome prediction, AI-based techniques used for the prediction of chemical reaction conditions can be classified into two categories: data-driven AI techniques and hybrid AI techniques.

A data-driven AI technique has been trained with databases of millions of organic reaction conditions to predict the optimal reaction condition with little or no explicit incorporation of prior chemistry knowledge. The success of a data-driven technique is dependent on the large dataset, and if the data is limited then it may be affected by low accuracy, and chemically invalid or implausible predictions. These data-driven techniques can be categorised into several categories such as deep neural networks, sequence-to-sequence techniques, graph neural networks, and small-sample techniques [14].

A hybrid AI technique combines the domain knowledge (chemical facts, rules, and logical structures) with machine learning techniques which are mostly data-driven for the prediction of chemical reaction conditions. These hybrid techniques incorporate the domain knowledge to complement data-driven techniques. These hybrid techniques can be categorised into several categories such as chemistry-rich representations, chemistry-informed search, the expertise of human chemists, and chemistry-driven constraints [14].

Figure 10 shows reaction conditions prediction using an AI technique; and Fig. 11 illustrates an example of predicting reaction conditions in synthesis of Paracetamol.

3 Conclusion

This paper presented several AI applications in chemistry: molecule design, molecular property prediction, retrosynthesis, reaction outcome prediction and reaction conditions prediction. It simplified the aforementioned areas of chemistry with a suitable and consistent example based on Acetaminophen (Paracetamol) molecule, and limitations of traditional techniques to provide a better understanding of the area. Subsequently, it presented the generalised AI techniques in each area of chemistry for easy understanding of AI applications to avoid any complexity and confusion due to a large number of pre-existing AI techniques in each area. This investigation provided an insight about AI applications in chemistry and its great success. However, there are some areas of chemistry wherein AI applications require more improvements. The progress of AI in chemistry will rely heavily on scaling up the training data to fit evolving needs and ensuring the accuracy of data. Additionally, it also requires expert knowledge and verification to ensure the best fit scenario. In the future, it would be beneficial to explore more AI applications in those areas of chemistry which were not investigated in this paper. Additionally, it is also important to evaluate some specific AI techniques and how they work in the specific area of chemistry to find their strengths and limitations.

References

1. Chauhan, N.S.: How artificial intelligence is used in chemical industry (2022), www.theaidream.com/post/how-artificial-intelligence-is-used-in-chemical-industry
2. Gao, H., Struble, T., Coley, C., Wang, Y., Green, W., Jensen, K.: Using machine learning to predict suitable conditions for organic reactions. ACS Cent. Sci. **4**(11), 1465–1476 (2018)
3. Inchem.org: Paracetamol: Physical & Chemical Information (2023), www.inchem.org/documents/icsc/icsc/eics1330.htm
4. Jin, W., Coley, C., Barzilay, R., Jaakkola, T.: Predicting organic reaction outcomes with Weisfeiler-Lehman network. Adv. Neural Inf. Process. Syst. **30** (2017)
5. Liu, H., Zhang, R., Yao, X., Liu, M., Hu, Z., Fan, B.T.: Qsar study of ethyl 2-[(3-methyl-2, 5-dioxo (3-pyrrolinyl)) amino]-4-(trifluoromethyl) pyrimidine-5-carboxylate: an inhibitor of ap-1 and nf-κb mediated gene expression based on support vector machines. J. Chem. Inf. Comput. Sci. **43**(4), 1288–1296 (2003)
6. Ma, J., Sheridan, R.P., Liaw, A., Dahl, G.E., Svetnik, V.: Deep neural nets as a method for quantitative structure-activity relationships. J. Chem. Inf. Model. **55**(2), 263–274 (2015)
7. Mann, V., Venkatasubramanian, V.: Predicting chemical reaction outcomes: a grammar ontology based transformer framework. AIChE J. **67**(3), 17190 (2021)
8. Meyers, J., Fabian, B., Brown, N.: De novo molecular design and generative models. Drug Discovery Today **26**(11), 2707–2715 (2021)
9. Seidl, P., Renz, P., Dyubankova, N., Neves, P., Verhoeven, J., Wegner, J.K., Segler, M., Hochreiter, S., Klambauer, G.: Improving few-and zero-shot reaction template prediction using modern hopfield networks. J. Chem. Inf. Model. **62**(9), 2111–2120 (2022)

10. Sun, R., Dai, H., Li, L., Kearnes, S., Dai, B.: Towards understanding retrosynthesis by energy-based models. Adv. Neural. Inf. Process. Syst. **34**, 10186–10194 (2021)
11. Svetnik, V., Liaw, A., Tong, C., Culberson, J.C., Sheridan, R.P., Feuston, B.P.: Random forest: a classification and regression tool for compound classification and qsar modeling. J. Chem. Inf. Comput. Sci. **43**(6), 1947–1958 (2003)
12. Synthiaonline: SYNTHIA Retrosynthesis Software (2023), www.synthiaonline.com/
13. Ucak, U.V., Ashyrmamatov, I., Ko, J., Lee, J.: Retrosynthetic reaction pathway prediction through neural machine translation of atomic environments. Nat. Commun. **13**(1), 1186 (2022)
14. Venkatasubramanian, V., Mann, V.: Artificial intelligence in reaction prediction and chemical synthesis. Curr. Opin. Chem. Eng. **36**, 100749 (2022)

Demystifying the Working, Types, Benefits and Limitations of Chatbots

Ishita Naik[1], Dishita Naik[2], and Nitin Naik[3](✉)

[1] King Edward VI Five Ways School, Birmingham, United Kingdom
17inaik@fws.kevibham.org

[2] Birmingham City University, Birmingham, United Kingdom
dishita.naik@mail.bcu.ac.uk

[3] School of Computer Science and Digital Technologies, Aston University, Birmingham, United Kingdom
n.naik1@aston.ac.uk

Abstract. A chatbot is one of the most innovative tools for interaction between humans and machines. It simulates human behaviour and interacts with a user in natural language to provide required information, services, thus enhancing the user experience. An increasing number of organisations are utilising chatbots to offer better services to users in real time to improve user satisfaction, streamline the user journey, and provide user-centric support. However, this requires building specific chatbots for specific services, which leads to variational designs of chatbots based on their application areas.This paper aims to demystify fundamentals of a chatbot including its working, types, benefits and limitations. It will mainly focus on two main categories of chatbots rule-based and AI-based chatbots, and explain all the aforementioned facts as well their comparative analysis to simplify the chatbot topic in general.

Keywords: Artificial intelligence · AI · Rule-based chatbot · AI-based chatbot

1 Introduction

Innovation in technology has offered a faster, easier and real time solution in all areas of human life. A chatbot is such a kind of technological innovation that can directly interact with humans to offer various kinds of support services. A user can ask a question and the chatbot responds with the relevant information based on its knowledge base. Chatbots can be advantageous in many aspects of the user experience, for example, providing desired information and services to the user, engaging users in a business and recommending suitable products or solutions to them. Additionally, several chatbots can learn from previous conversations with a user to offer personalised experience. Additionally, chatbots are easy to develop, implement and customise depending on the specific requirements. Due to the abundant information available on the chatbot regarding its working, types,

N. Naik et al. (Eds.): UKCI 2023, AISC 1453, pp. 558–566, 2024.
https://doi.org/10.1007/978-3-031-47508-5_43

benefits and limitations, it is often difficult to understand precise information regarding the different types of chatbots. Therefore, this paper aims to demystify fundamentals of a chatbot including its working, types, benefits and limitations. It will mainly focus on two main categories of chatbots, rule-based chatbots and AI-based chatbots, and explain all the aforementioned facts as well their comparative analysis to simplify the chatbot topic in general.

This paper consists of the following sections: Section 2 explains what is a chatbot and its types: Rule-Based and AI-Based Chatbots; Section 3 explains the working of both Rule-Based and AI-Based Chatbots; Section 4 presents a comparative analysis of Rule-Based and AI-Based Chatbots; Section 5 presents various benefits of Chatbots; Section 6 presents various limitations of Chatbots; Section 7 presents the summary of the paper and future work.

2 Chatbot and Its Types

2.1 What is a Chatbot?

A chatbot is a computer program that interacts with a human through text or speech and answers questions based on its knowledge base. A chatbot is generally embedded on an organisational platform to provide automated, affordable, effortless services to users without requiring human intervention. Therefore, it is also referred to as a virtual assistant or virtual agent. There are several popular chatbots such as Apple's Siri, Google Assistant, Amazon Alexa, Google Bard and OpenAI's ChatGPT.

2.2 Types of Chatbot

Mainly chatbots can be classified into two categories: Rule-Based Chatbots and AI-Based Chatbots.

2.2.1 Rule-Based Chatbot

A rule-based chatbot is based on a set of predefined rules and patterns that controls the communication with users, and responds to a specific query by drawing the necessary information from its rule base. These rules and patterns are normally based on keywords and phrases, and the chatbot will respond based on the best match it can find in its rule base. If the rule base is more dense and complete, the chatbot can answer more questions effectively. It generally provides users a choice of options or buttons to obtain a response for a specific query. A rule-based chatbot can be useful for simple tasks or providing information to users; however, it may not be able to handle more complex or open-ended conversations. This chatbot has limited customisation capabilities, however, it is easy, swift and affordable to develop and implement.

2.2.2 AI-Based Chatbot

An AI-based chatbot utilises AI techniques to facilitate natural language conversations between humans and AI systems. It is generally based on conversational AI that enables machines to understand, process and respond to human language. Conversational AI leverages AI techniques such as Natural Language Processing (NLP), Natural Language Understanding (NLU) and Natural Language Generation (NLG) [1]. An AI-based chatbot can learn from previous conversations, allowing it to adapt its answers to new situations, environments and patterns. This self-improvement is performed through repeated conversions with users and obtains more data in order to expand the existing knowledge base as well as enhance the relevancy and accuracy of their responses. It also has the ability to understand a query outside of its pre-programmed instructions and provide an appropriate response based on existing data.

3 Working of a Chatbot

Generally, chatbots are designed to deal with users queries and provide an appropriate response to meet their requirements. However, the working of any chatbot is dependent on the type of chatbot and its specific context.

3.1 Working of a Rule-Based Chatbot

A rule-based chatbot allows a user to start interaction (or ask a question) by clicking on buttons or using predefined options to obtain an answer to a specific query. A rule-based chatbot uses a set of predefined rules and patterns to understand user query and provides a rule-based response (see Fig. 1). The rule base contains mostly *if–then* rules and acts as the domain knowledge for the rule-based chatbot. There is an inference engine, also known as the semantic reasoner which is responsible for interpretation of the rules and taking action accordingly. The inference engine generally works in three steps: match, conflict-resolution, and act.

3.2 Working of an AI-Based Chatbot

Modern AI chatbots are also leveraged with several AI techniques, in particular, Natural Language Processing (NLP) to recognise users' intent from the context of their input and generate correct responses. These chatbots are trained using a large volume of user data, and can operate independently using NLP and other AI techniques. NLP is a branch of AI that bridges human and machine language to enable more natural human-to-computer communication. It enables computers to understand human language in both written and verbal forms.

NLP alongside with its sub-fields NLU and NLG provides chatbots a greater conversational capability (see Fig. 2). It enables computers to process and analyse vast amounts of natural language data, understand the meaning of it and

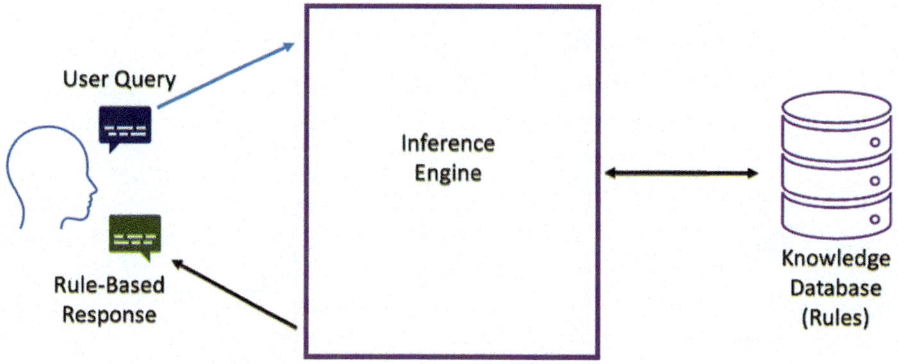

Fig. 1. Working of a Rule-Based Chatbot

communicate with humans. The more the chatbot interacts, the greater the volume of data accumulated. It updates its programming with the growing data to deliver targeted answers. NLP enables chatbots to convert users' text and speech into machine-readable data to be understood by a machine. NLU is a subfield of NLP which focuses on understanding the meaning of human speech by recognising patterns in unstructured speech input. NLG is another subfield of NLP which generates a human language text response based on its knowledge base. A knowledge base is a library of information that the chatbot relies on to fetch the data used to respond to users. Different knowledge bases are utilised for different application areas.

4 Comparative Analysis of Rule-Based Chatbot and AI-Based Chatbot

As earlier discussed, chatbots can be classified into two main categories based on their type of internal logic and knowledge base: rule-based chatbot and AI-based chatbot. There are several differences between them, and some of the main differences are explained in Table 1 [6].

5 Benefits of Chatbots

Chatbots offer several benefits in various application areas, here are some of the main benefits [2,4,5].

5.1 Availability

Unlike humans, chatbots are available 24/7 and respond to users instantaneously. Therefore, chatbots as virtual assistants help organisations and businesses to handle a large volume of inquiries and support requests from the users or customers round the clock. In case, if the chatbot is not able to resolve the issue, it

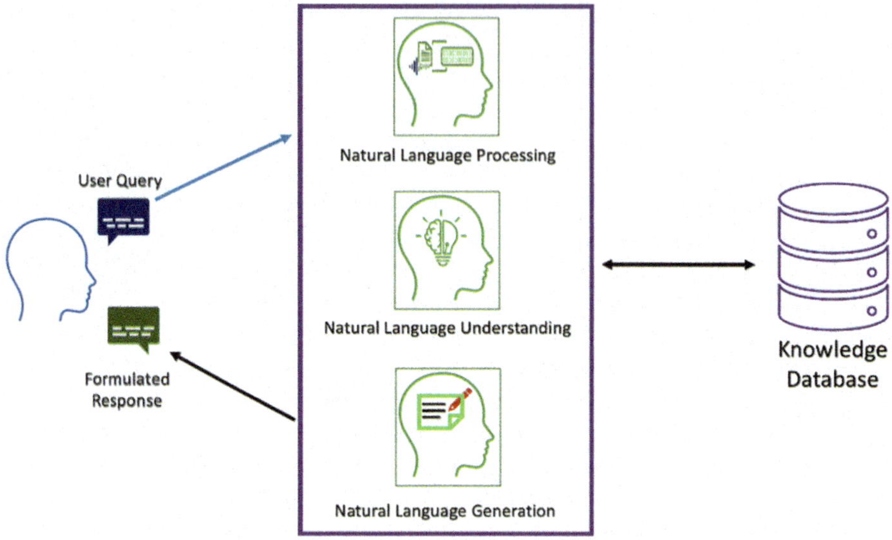

Fig. 2. Working of an AI-Based Chatbot

Table 1. Comparative Analysis of Rule-Based Chatbot and AI-Based Chatbot

Criteria	AI-Based Chatbot	Rule-Based Chatbot
Underlying Mechanism	It normally relies on Natural Language Processing (NLP) to extract meaning from user input and provide a response.	It relies on predefined rules, keyword matching, and regular expressions to identify patterns in user input and provide a response.
Intent and Context	It has the ability to understand the context, intent, and sentiment behind user queries.	It lacks the ability to understand context, intents, and sentiment beyond its pre-programmed conversational flow.
Adaptability	It is more flexible and adaptable as it can learn and improve over time from the collected data and user interactions.	It has limited adaptability due to the predefined rules and responses set by developers.
Scope	It can handle variations in language, synonyms, misspellings, and language nuances, providing more accurate and contextually relevant responses.	It is limited to how much variations it can handle as it requires developers to manually define rules for all possible user inputs, which is a challenging task.
Personalised Response	It can provide personalised responses based on user preferences, user profiles and historical data.	It lacks the ability to provide personalised responses beyond its pre-programmed conversational flow.
Training	It requires a large amount of data for training underlying AI models and its iterative improvements.	It requires a limited amount of training data and number of rules depending on the functionality and problem domain.
Computational Resources	It requires significantly greater computational resources for development and deployment.	It requires fewer computational resources for development and deployment.

can still gather information from users, assess the urgency, and send the query to the relevant staff or department to be resolved as soon as possible.

5.2 Adaptability

Modern chatbots, in particular AI-based chatbots, are more flexible and adaptable as they can quickly review data and make decisions based on their analysis. They can learn and improve over time from the collected data and user interactions. However, some chatbots, in particular rule-based, are less adaptable due to the predefined rules.

5.3 Affordability

Chatbots offer the most cost-effective support services when compared to the same services offered by humans. Therefore, the total cost of employing and managing chatbots is low compared with employing additional human members of staff. Additionally, rule-based chatbots are more cost-effective as they do not require extensive resources for development and implementation.

5.4 Customer/User Engagement

With advancements in machine learning and natural language understanding, chatbots are now more capable than ever to interact with users in a human-like manner, catering to their needs effectively and efficiently. Chatbots can improve customer engagement by making their experience more interactive through conversation. They provide the real-time personalised response to users based on their previous interactions, profiles, and historical data. Therefore, chatbots deliver clear and concise responses to users that avoid any irrelevant information for users.

5.5 Multilingual

Chatbots can perform conversation and reply in multiple languages, and particularly in users' own language, enhancing the user experience for seamless questioning.This is particularly crucial for global service providers, operating in different geographical market and want to deliver a high-quality experience to their users. Therefore, chatbots can provide support, resources, and personalised buying experiences in languages most suited to user. By offering support in the customers own languages, services, sale deals, and brands, can be made more empathetic and human like.

5.6 Efficiency

Chatbots allow organisations and businesses to automate their support services at a large scale. They can take over a lot of tedious and repetitive tasks that are currently performed by customer support staff which can be very dull and draining for them. They can automate such routine and repetitive tasks, relieving employees to focus on other complex and imperative task, thus increasing efficiency and productivity of an organisation and business.

5.7 Scalability

Scalability refers to the ability of a chatbot to handle an increasing number of users and queries without affecting their performance or functionality. Generally, chatbots are scalable given the availability of resources, whilst also being able to handle multiple users and queries simultaneously without any decline in performance. This is particularly useful for organisations and businesses that have a high volume of user queries.

6 Limitations of Chatbots

Chatbots have several limitations depending on the type of chatbots, and some of them are explained here [3].

6.1 Lack of Practical AI

Chatbots are very good at mimicking humans and providing solutions for simple and repetitive tasks; however, they may lack the practical intelligence to comprehend complex problems and situations. Therefore, they may struggle to deal with very complex issues that require human intervention. Several chatbots, in particular rule-based chatbots, may lack the contextual understanding, unable to reason and adapt to new situations. Even though AI-based chatbots have the contextual understanding that allows them to learn and improve, they are still limited by their capabilities being dependent on the available data and underlying AI techniques employed.

6.2 Lack of Customer Perspective

The design and goal of any chatbot should be focused on users. Nonetheless, there are several chatbots, which are designed to implement user specific tasks but without any regard to user experience (UX). This lack of user centrality in the process of chatbot designing often focuses on technological aspects and does not add any value to users. In many cases, UX failures can lead to the failure of chatbots. Therefore, the design and goal of any chatbot should be based on UX rules and require UX designers to test to discover the flaws of the chatbot.

6.3 Lack of Emotions

Rule-based chatbots generally lack emotional understanding. Though AI-based chatbots employ AI techniques including NLP to understand the context and intent of users, they are not yet able to fully replicate genuine emotional intelligence and empathetic understanding of humans. The emotional understanding is very crucial in human communication, allowing individuals to understand and respond appropriately to emotions, concerns, and personal circumstances. In several scenarios, where a user query requires some emotional support or sensitivity, this lack of emotional understanding may be a significant disadvantage for chatbots and can affect its ability to provide a meaningful and satisfying user experience.

6.4 Chatbots Are Often Repetitive

Chatbots are generally very repetitive and only deliver a set number of responses. This is specifically applicable to the rule-based chatbots due to the limited set of rules and responses. It would be impossible to program them with all of the infinite variations of human language so that they can fully engage in authentic conversation. In general, this is very boring for users and customers, and in some cases, this could lead to an uncomfortable scenario where a chatbot repeatedly sends error messages or annoying responses to users. Consequently, users may feel disenchanted with the chatbots and eventually with the business or brand.

6.5 Lack of Platform Independency

Most chatbots are often designed to work on specific platforms, that can affect their reach to a wider audience and provide support to all users. Even the same organisation and businesses have to develop multiple chatbots for their platforms such as websites, apps, and messengers. Developing chatbots for multiple platforms can be time-consuming in addition to protecting data. Additionally, it requires the significant resources whilst having to develop separate intelleginent chatbots for multiple platforms can be costly. These requirements may affect small organisations and businesses, as a result of this they may settle with basic chatbots that only respond to a limited set of queries. Therefore, multi-platform chatbots would reduce expenses and provide a streamlined experience across every platform to users.

6.6 Lack of Extensibility and Connectivity

Chatbots may be limited by their lack of extensibility and connectivity. It may not be able to connect with the existing IT infrastructure and other systems or APIs, which can limit its functionality and effectiveness. Chatbots are generally implemented as a part of a broad IT infrastructure of organisations or businesses, and it is important that they can be successfully integrated with the workflows, communication channels and existing technologies. This can limit the ability of the chatbot to provide a complete solution to users.

7 Conclusion

This paper explained the fundamentals of a chatbot including its working, types, benefits, and limitations. It focused on two main categories of chatbots, rule-based and AI-based chatbots, and explained all the aforementioned facts as well as their comparative analysis to simplify the chatbot topic in general. A rule-based chatbot is based on a set of pre-defined rules and patterns and answers the user query based on its rule base. Whereas an AI-based chatbot utilises AI techniques to facilitate natural language conversations between humans and AI systems. Both have their benefits and limitations and can be used in different types of applications. This study aimed to explain the basic concepts of a chatbot; however, it is a vast and diverse topic that requires further study to explore different designs of chatbots in different application areas alongside their effectiveness.

References

1. Boost.ai: What is conversational AI, anyway? (2023). www.boost.ai/knowledge/what-is-conversational-ai
2. Cm.com: What is an AI chatbot? (2023). www.cm.com/en-gb/glossary/what-is-ai-chatbot/
3. iAdvize.com: What is an AI chatbot? (2023). www.iadvize.com/en/blog/ai-and-chatbot
4. Jassova, B.: What are the top benefits of chatbots in business (2022). www.landbot.io/blog/benefits-of-chatbots
5. Rachel, N.: 10 benefits of chatbots you need to know now (2022). www.eleviant.com/insights/blog/benefits-of-chatbots/
6. Ventoniemi, J.: What is an AI chatbot? here's what you need to know (2023). www.giosg.com/blog/what-is-ai-chatbot

A Comparative Analysis of GPT-3 and BERT Models for Text-based Emotion Recognition: Performance, Efficiency, and Robustness

Enguerrand Boitel(✉), Alaa Mohasseb, and Ella Haig

University of Portsmouth, Portsmouth, UK
{enguerrand.boitel,alaa.mohasseb,ella.haig}@port.ac.uk

Abstract. This paper presents a comparative analysis between the GPT model and a BERT model for emotion recognition. While GPT has shown remarkable performance across various tasks, the question arises as to whether its results can be compared to models specifically trained for particular tasks. The paper explores the presentations of GPT, including its models (Davinci, Curie, Babbage and Ada) as well as the fine-tuning approach. Similarly, the paper discusses the presentation of BERT, specifically DeBERTa v3, and its fine-tuning methodology. A comparison is then drawn between the two models, considering the general vs. task-specific nature of AI. Furthermore, the paper investigates the reasons behind GPT's exceptional performance and examines why the obtained results for emotion recognition may not be significantly better than those of task-specific models. Ultimately, the conclusion reflects on the better accuracy often exhibited by BERT, but highlights the potential for future advancements, such as the new GPT-4, that may surpass existing models and offer unparalleled versatility in answering various questions.

Keywords: GPT model · BERT model · Emotion recognition · General vs. task-specific AI

1 Introduction

Deep learning has revolutionised artificial intelligence (AI) by enabling computers to learn from complex data [1]. Two widely recognised models, GPT and BERT, have demonstrated impressive performance in natural language processing tasks.[1]

This research paper focuses on comparing these models specifically for emotion recognition. Emotion recognition is vital for applications such as sentiment analysis and affective computing. Deep learning models like GPT-3.5 and

[1] URL: https://www.ibm.com/blog/how-bert-and-gpt-models-change-the-game-for-nlp/.

N. Naik et al. (Eds.): UKCI 2023, AISC 1453, pp. 567–579, 2024.
https://doi.org/10.1007/978-3-031-47508-5_44

BERT offer promising solutions for addressing this challenge. In this paper, the strengths, limitations, and suitability of GPT and BERT for emotion recognition are analysed. The architectures, training methods, and fine-tuning approaches of these models are discussed. By comparing their performance in terms of accuracy, efficiency, and robustness, valuable insights are provided for researchers and practitioners in deep learning and AI.

In this paper, the architectures, training methods, and fine-tuning approaches of GPT and BERT models are explored within the context of emotion recognition. The performance of these models is evaluated based on accuracy, efficiency, and robustness, with the goal of offering valuable insights to researchers and practitioners in the field of deep learning.

Understanding the performance of GPT and BERT in emotion recognition is crucial for real-world applications. Findings will assist in making informed decisions when selecting and implementing these models. Additionally, this research contributes to advancements in emotion recognition and the development of intelligent systems capable of understanding and responding to human emotions. Overall, this paper aims to provide insights into the strengths and weaknesses of Davinci and BERT models for emotion recognition, benefiting researchers and practitioners in the field of deep learning and AI.

This paper begins with an exploration of related works in the field of emotion recognition, focusing on approaches that utilise GPT and BERT models. This is followed by a section dedicated to the BAUM data-set, which is the data source of this research. Furthermore, the GPT model is discussed, with specific emphasis on the Davinci variant and the results obtained from its implementation. The BERT model is then introduced, with a focus on the DeBERTa architecture, the process of fine-tuning, and the results achieved. Finally, a comprehensive comparison and discussion between GPT3 and BERT in the context of emotion recognition is presented.

2 Related Works

This section presents an overview of the related works in the field of emotion recognition, focusing on GPT-based and BERT-based approaches. It explores how previous methods have been replaced by models such as GPT, which have shown significant advancements in capturing the complexities of human emotions. The section discusses various studies that have utilised GPT and BERT for text-based emotion recognition tasks, highlighting their capabilities in detecting and understanding emotions in textual data.

2.1 GPT & BERT

Emotion recognition plays a vital role in natural language processing, enabling various applications such as sentiment analysis, chatbots, and virtual assistants to better understand and respond to human emotions. In recent years, the emergence of advanced machine learning models has significantly impacted the field of emotion recognition. Two prominent models, GPT (Generative Pre-trained Transformer) and BERT (Bidirectional Encoder Representations from Transformers), have gained considerable attention and success in natural language processing tasks. GPT, based on the transformer architecture, is a generative model that has demonstrated impressive language generation capabilities. On the other hand, BERT, a pre-trained transformer model, has shown exceptional performance in various language understanding tasks. Both models have been widely applied in sentiment analysis, text classification, and other emotion recognition tasks. Therefore, in this study, we aim to conduct a comparative analysis between GPT and BERT models in the context of emotion recognition. By exploring their respective strengths, weaknesses, and performance, we can gain insights into their applicability and effectiveness in capturing emotional nuances in text data.

2.2 Emotion Recognition from Text

Authors in [2] suggested employing a deep learning model at the SemEval 2018 Task 1 Competition [3]. They designed a multi-layer self-attention system incorporated into a Bidirectional Long-term Short-Term Memory (Bi-LSTM). According to the experimental findings, random initialisation outperformed this transfer learning approach. Moreover, a supervised logistic regression method was revealed in [4] to identify emotions in text. Sentences with emotions were supplied into a logistic regression model and emotion labels throughout the training procedure. In their testing process, the trained classifier was only fed texts with hidden emotion labels. Precision, recall, and F1-score were used to assess the effectiveness of their model. For the feelings studied (joy, fear, sadness, guilt and shame), they obtained an F1 score between 0.57 and 0.76. Over 144,000 tweets were used in [5] to construct an emotion recognition model using CNN. The emotion of "happiness" was identified with the highest accuracy rate (73%), whereas "anger" had the lowest precision level (37%). Also, to automatically forecast the authors' personality sentiments, authors in [6] derived contextualised word embedding from text data using the BERT pre-trained model and the bagged-SVM classifier. Comparing their performance to baseline methods, they obtained an increase of 1%.

2.3 GPT-based Emotion Recognition Approaches

Emotion recognition has been a significant research area in the field of natural language processing and affective computing [7]. Traditional approaches to emotion recognition relied on manual feature engineering and rule-based systems, which often struggled to capture the complex nuances of human emotions [8]. However, with the advent of deep learning and language models, such as GPT-3 and BERT, there have been notable advancements in the field [9].

Several studies have explored the use of GPT for text-based emotion recognition tasks. For instance, authors in [10] highlighted the advanced language processing capabilities of GPT-4 as an ideal tool for chatbot emotion recognition. By analysing the user's text-based communication, GPT-4 can detect the emotional tone of the message and respond accordingly.

Another study [11] discussed how AIs like ChatGPT are already imagining what feelings they will end up with. The author explored the idea that artificial intelligence might develop feelings and questioned whether we have been thinking about AI emotions all wrong.

These studies collectively highlight the potential of GPT in text-based emotion recognition tasks. They demonstrate the capability of GPT to capture contextual information and nuances related to emotions in textual data.

2.4 BERT-based Emotion Recognition Approaches

BERT has also shown promising results in text-based emotion recognition tasks. For instance, an uncased BERT ("bert-large-uncased"[2]) was found [12] that has an accuracy of 0.89 while the cased version has an accuracy of 0.90, showing slightly higher performance.

Authors in [13] achieved a state-of-the-art accuracy of 0.91 on emotion recognition in tweets using an ensemble model combining BERT and SVM.

Furthermore, researchers who focused GPT efficiency [14] found that a BERT-based emotion classifier outperformed pre-existing models and produced results with 98% accuracy on English sentences.

Also, a novel paradigm for emotion recognition in conversation (ERC) was proposed [15] by exploring contextual information and dialogue structure information in the fine-tuning step of BERT. Their model, BERT-ERC, achieved substantial improvement on four datasets.

These studies collectively demonstrate how BERT has the potential to be effective in text-based emotion recognition tasks by capturing contextual information and subtle nuances related to emotions in textual data.

3 Data: the BAUM Data-set

The BAUM dataset [16] was used in this research. It is composed of 1184 videos from 31 actors, which resulted in more than 3 hours of audio-visual recordings. The video clips were created by taking frontal and half-profile views of the subjects with stereo and mono cameras, respectively. A series of images and a short video clip were first shown to the actors, which have been selected to inspire different emotions and mental states. Secondly, they expressed their thoughts and feelings regarding the pictures and videos they have just viewed.

The recordings were labelled with the six fundamental emotions according to Ekman [17] (anger, disgust, fear, happiness, sadness and surprise), as well as six emotional states (boredom, bothered, concentrating, contempt, thinking and unsure). The "neutral" expression was finally added. The distribution of each expression, after normalisation, is presented in Table 1.

[2] https://huggingface.co/bert-large-uncased.

Table 1. Distribution of emotion classes in the BAUM dataset

Emotion	Number of recordings	Percentage (%)
Anger	95	8
Happiness	273	23
Sadness	237	20
Fear	47	4
Disgust	154	13
Surprise	59	5
Neutral	319	27
Total	1184	100

It is worth noting that this study is part of a larger research focused on multimodal emotion recognition, incorporating text, audio, images, and motion sentiment analysis. By utilising the subtitles provided in the BAUM dataset, emotions within the text domain can be extracted, contributing to the broader investigation of multimodal emotion understanding. However, the presence of associated non-verbal cues may introduce limitations and potential biases when performing text-based analysis. Nevertheless, through careful consideration of the dataset's strengths and limitations, the objective is to explore the text-based aspect of emotion recognition within a multimodal context and offer valuable insights into the emotional content conveyed in the text domain. Finally, the video clips of the BAUM's dataset have been categorically annotated by five labelers, further ensuring the accuracy and reliability of the dataset for subsequent analysis.

4 GPT Model

4.1 Davinci

The GPT-3.5 model is a fine-tuned version of the GPT-3 model [18]. It was developed in January 2022 and has 3 variants each with 1.3B, 6B and 175B parameters. The main feature of GPT-3.5 was to eliminate toxic output to a certain extent. Davinci is based on GPT-3.5 but works within specific policies of human values and has only 1.3 billion parameters fewer than the previous version by 100X. The Davinci variant of the GPT-3.5 model has 175 billion parameters, making it one of the largest and most powerful language models in existence [19]. It has been designed to generate high-quality natural language responses to a wide range of tasks, including language translation, summarising, question-answering, and more.

4.2 Fine-Tuning

GPT-3, an advanced language model, was trained using a massive 175 billion parameters. Known as "Generative Pre-trained Transformer 3", this model excels in learning from contextual information. During training, GPT-3 analysed a vast amount of unlabelled text data to grasp language patterns. It was then fine-tuned for various NLP tasks, such as emotion recognition, text completion, and question answering. To evaluate Davinci's performance, it underwent rigorous testing on different NLP datasets, including well-known benchmarks and novel tasks. The evaluation covered three conditions: few-shot learning, one-shot learning, and zero-shot learning. GPT showed impressive adaptability across these conditions, achieving exceptional results [20].

Table 2. GPT variants

Model	Parameters	Pricing ($/1K tokens)
Ada	2.7 billion	0.0004
Babbage	6.7 billion	0.0005
Curie	13 billion	0.0020
Davinci	175 billion	0.0200

While GPT-3 displayed considerable strengths, it also faced challenges in tasks like natural language inference and specific reading comprehension datasets. This research aims to contribute to the understanding of few-shot learning in language models and identify areas for improvement.

Davinci, Curie, Babbage, and Ada are different models of the GPT series that are distinguished by their capability and cost. Davinci is the most capable and expensive model, while Ada is the least capable but faster and cheaper. Each model family has a series of models that are further distinguished by capability. These capabilities are typically identified by names, and the alphabetical order of these names generally signifies the relative capability and cost of that model within a given model family. Particularly, GPT-3 variants indicate relative capability and cost. Table 2 presents OpenAI's variants models.[3]

4.3 Results

Figure 1 shows the summary chart for the fine-tuning of the Davinci (GPT-3.5) model.

The fine-tuning process achieved a good validation accuracy of 82.8%. This indicates that the model was able to accurately classify the input data with a high degree of precision. The corresponding loss for the fine-tuned model was 12%, indicating that the model performed relatively well in minimising the errors during training. Other attempts were able to achieve better loss results, but the overall accuracy was worse (Fig. 1).

Figure 2 presents the confusion matrix of the fine-tuned Davinci model. The confusion matrix provides insights into the performance of the model by showing

[3] URL: https://openai.com/pricing.

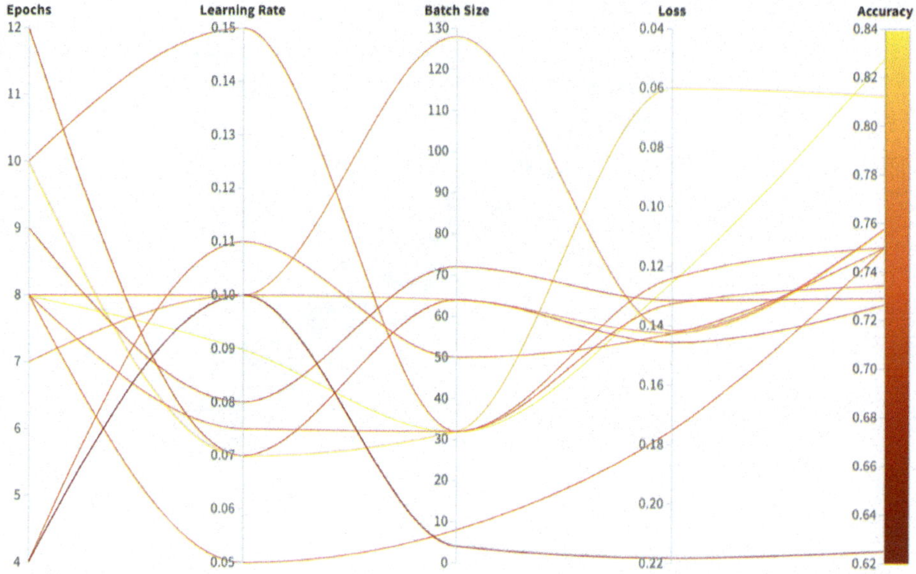

Fig. 1. Fine-tuning summary chart–davinci

the distribution of predicted labels compared to the true labels of the validation dataset. The model achieved high accuracy in classifying the majority of the samples, especially "Neutral" and "Sadness" with a 90%-accuracy. Also, some emotions might be under-predicted (e.g. the "Disgust" emotion).

5 BERT Model

BERT [21] is a language model that excels in various NLP tasks and is trained on a vast quantity of text data. The BERT model is dependent on the context of the sentence and learns contextual information from input texts. The BERT technique defeats word-based representations (Word2Vec [22]) in many contexts, by utilising the sub-word representation.

5.1 DeBERTa

More specifically, the well-known DeBERTa (Decoding-enhanced BERT with disentangled Attention) [23] significantly improves the efficiency of BERT by representing each word with two vectors that separately code its position and content. Also, the output softmax layer is replaced with an enhanced mask decoder in order to anticipate the masked tokens for the model pre-trained. This model was the first to surpass human performance on SuperGLUE.[4]

[4] URL: https://www.microsoft.com/en-us/research/blog/microsoft-deberta-surpasses-human-performance-on-the-superglue-benchmark/.

TRUE CLASS LABELS \ PREDICTED CLASS LABELS	Angry	Disgust	Fear	Happy	Neutral	Sad	Surprise
Angry	87%	0%	0%	2%	6%	4%	2%
Disgust	4%	88%	0%	4%	3%	1%	3%
Fear	0%	4%	83%	4%	0%	4%	4%
Happy	1%	0%	1%	85%	14%	0%	0%
Neutral	1%	1%	1%	4%	90%	2%	1%
Sad	4%	2%	1%	0%	3%	90%	0%
Surprise	0%	0%	9%	6%	0%	0%	84%

Fig. 2. Confusion matrix davinci (GPT-3.5)

Authors in [23] present that existing approaches to relative position encoding use a separate embedding matrix to compute the relative position bias in computing attention weights. However, it is argued that the position-to-content term is also important since the attention weight of a word pair depends not only on their contents but on their relative positions, which can only be fully modelled using both the content-to-position and position-to-content terms. In this implementation, both content-to-position and position-to-content terms are used, and the position-to-position term is removed. DeBERTa architecture follows Equation 1, for two tokens at positions i, j in a sequence and two vectors $\{H_i\}$ and $\{P_{i|j}\}$ (from [23]):

$$\begin{aligned} A_{i,j} &= \{H_i, P_{i|j}\} \times \{H_j, P_{j|i}\}^T \\ &= H_i H_j^T + H_i P_{j|i}^T + P_{i|j} H_j^T + P_{i|j} P_{j|i}^T \end{aligned} \tag{1}$$

In the approach proposed in this paper, the DeBERTa-V3-Large pre-trained model was used, as presented in Fig. 3. This specific model was trained by [23] using unlabelled data from English Wikipedia (12GB), BooksCorpus (6GB), and Reddit content (38GB). This model is composed of 24-layered transformer blocks where each block contains 1024 hidden outputs and has 304 million parameters in total. Additionally, randomly initialised fully connected layers were developed with 768×4 parameters along with softmax and fine-tuning for emotion recognition on the BAUM dataset using a categorical cross-entropy loss as the objective function, an Adam optimiser with a learning rate of 1×10^{-5}, and a batch size of 16. Finally, the PyTorch framework for the BERT model training was used.

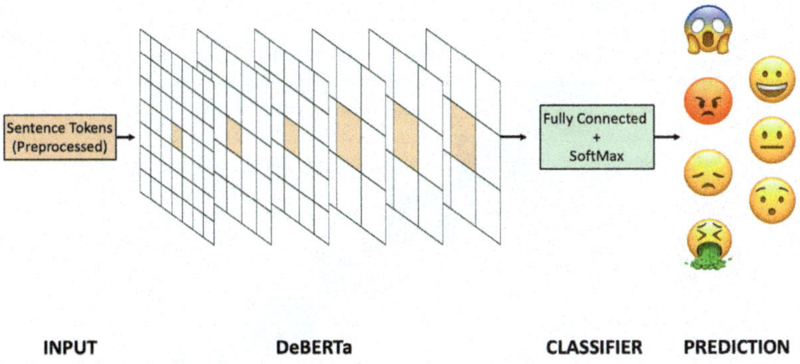

Fig. 3. DeBERTa framework for text-based emotion recognition

5.2 Fine-Tuning

In the fine-tuning process for emotion recognition using DeBERTa, several key parameters have been utilised to optimise the model's performance. The "deberta-v3-large" was used to represent the pre-trained DeBERTa model used as the base architecture. The learning rate parameter, set to 1e-5, enables efficient weight updates during training. The "batch size" parameter, set to 16, determines the number of training instances processed together in each iteration, balancing computational efficiency and model convergence. Lastly, 10 epochs were used to fine-tune the BERT model.

5.3 Results

Regarding BERT, Fig. 4 showcase the predicted emotions compared to actual true labels. An impressive 95%-validation-accuracy was obtained on the BAUM dataset.

6 Comparison & Discussion: GPT3 vs BERT on Emotion Recognition

When comparing GPT and BERT for emotion recognition, several key points emerge. Firstly, in terms of processing speed and efficiency, BERT takes approximately 10-20 seconds to process a single sentence, while GPT demonstrates a significantly faster processing rate, able to process 60 sentences per minute.

Another aspect to consider is the consistency of the models in emotion recognition. In this regard, Davinci exhibits higher consistency compared to BERT. The standard deviation of emotion recognition accuracy for GPT is 3%, while BERT shows a higher standard deviation of 6%. This indicates that GPT3 provides more stable and reliable predictions across different inputs, enhancing its performance in emotion recognition tasks.

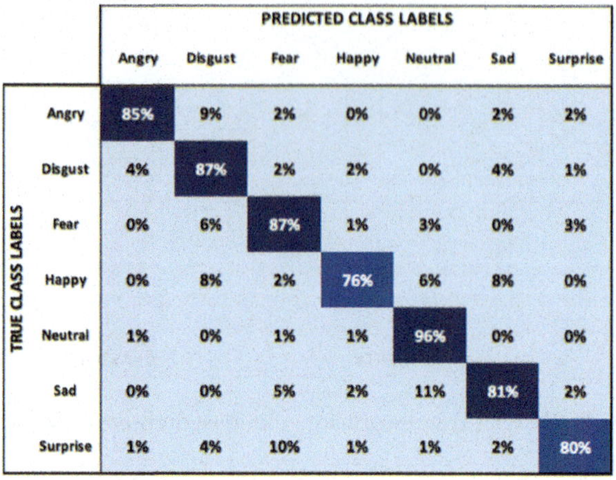

Fig. 4. Confusion matrix BERT

Furthermore, when examining specific emotion categories, some interesting observations can be made. The dataset used for emotion recognition has imbalanced data, particularly in the categories of Happiness and Neutral emotions. BERT tends to struggle with these imbalances, whereas GPT3 performs better in recognising emotions within imbalanced datasets. Additionally, the emotions of Surprise and Fear are found to be very similar in terms of their predicted labels by both models.

Moreover, the relationship between positive sentiments and other emotions is worth noting. Negative sentiments are consistently linked with other negative emotions, such as Anger and Sadness. This suggests that Davinci captures the interplay between different emotions and their sentiments, providing a more nuanced understanding of emotional expressions.

Finally, GPT3 shows better performance in recognising more complex emotions, such as Surprise, which can exhibit both positive and negative characteristics. This ability to capture the complexity of emotions adds to the overall effectiveness of GPT in emotion recognition tasks.

The findings of this comparison highlight the strengths of BERT in emotion recognition tasks. BERT exhibits slower processing speed but higher consistency, and robust performance in handling imbalanced data and recognising complex emotions. These specific characteristics make BERT a valuable choice for applications that require efficient and accurate emotion recognition from textual data.

Also, according to [24], the GPT-3.5 model achieves an 85.5%-accuracy in "Commonsense reasoning around everyday events", and this result further confirms the high accuracy of 82.4% that was achieved through the fine-tuning process.

Furthermore, one important aspect to consider is the potential bias that can arise from the training data. For example, if the training data primarily consists of expressions and emotions from certain demographics, such as Western cultures, it may not accurately capture the nuances of emotional expression from other cultures, such as Asian cultures, where individuals may exhibit less overt expressiveness. This can lead to biased predictions and inaccurate interpretations of emotions. Indeed, the BAUM's dataset for fine-tuning was created by Turkish individuals. To ensure fairness and mitigate potential biases, future research should strive to include diverse and representative training data that encompasses a wide range of cultural backgrounds, allowing for more accurate and culturally sensitive emotion recognition.

Additionally, the quality and characteristics of the datasets used for fine-tuning are also very important in the performance of the models. Factors such as dataset size and data quality can significantly impact the accuracy and robustness of the fine-tuned models in emotion recognition. Future studies could investigate the effects of dataset variations, the inclusion of multi-modal data (e.g., text, images, audio), and the importance of domain-specific datasets to further enhance the performance and adaptability of the models.

7 Conclusion

Both GPT and BERT have shown impressive performance in text-based emotion recognition tasks. Davinci has demonstrated exceptional capabilities across various NLP tasks, highlighting its proficiency in capturing contextual information and nuances associated with emotions. On the other hand, BERT has also achieved high accuracy in emotion recognition tasks and has been employed in state-of-the-art models.

Based on the analysis, it is evident that while GPT has exhibited remarkable performance in diverse tasks, its results in emotion recognition may not significantly surpass those of task-specific models such as BERT. The broad nature of GPT could potentially limit its ability to excel in specific tasks compared to models specifically tailored for those tasks. BERT, being a task-specific model, has demonstrated better accuracy in emotion recognition tasks. However, it is important to acknowledge that both Davinci and BERT have their respective strengths and limitations.

The outstanding performance of GPT across various tasks highlights its versatility and potential for future advancements. The forthcoming release of GPT-4, with its anticipated improvements, holds promise for GPT models to exceed existing models and provide unparalleled versatility in addressing a wide range of questions.

To conclude, while BERT currently exhibits superior accuracy in emotion recognition, the progress of GPT models like GPT-4 might possess the potential to surpass existing models and further enhance the capabilities of general AI models in various NLP tasks, including emotion recognition. The field of emotion recognition will continue to evolve, necessitating careful consideration of the strengths and limitations of different models when selecting and implementing them in real-world applications.

References

1. LeCun, Y., Bengio, Y., Hinton, G.: Deep learning. Nature **521**(7553), 436–444 (2015)
2. Baziotis, C., Athanasiou, N., Chronopoulou, A., Kolovou, A., Paraskevopoulos, G., Ellinas, N., Narayanan, S., Potamianos, A.: Ntua-slp at semeval-2018 task 1: Predicting affective content in tweets with deep attentive rnns and transfer learning. ArXiv Preprint arXiv:1804.06658 (2018)
3. Mohammad, S.M., Bravo-Marquez, F., Salameh, M., Kiritchenko, S.: Semeval-2018 task 1: affect in tweets. In: Proceedings of International Workshop on Semantic Evaluation (SemEval-2018). New Orleans, LA, USA (2018)
4. Alotaibi, F.M.: Classifying text-based emotions using logistic regression. In: VAWKUM Trans Comput Sci (2019)
5. Park, S.-H., Bae, B.-C., Cheong, Y.-G.: Emotion recognition from text stories using an emotion embedding model. In: 2020 IEEE International Conference on Big Data and Smart Computing (BigComp), pp. 579–583. IEEE (2020)
6. Kazameini, A., Fatehi, S., Mehta, Y., Eetemadi, S., Cambria, E.: Personality trait detection using bagged svm over bert word embedding ensembles. ArXiv preprint arXiv:2010.01309 (2020)
7. Kratzwald, B., Ilić, S., Kraus, M., Feuerriegel, S., Prendinger, H.: Deep learning for affective computing: Text-based emotion recognition in decision support. Decis. Support Syst. **115**, 24–35 (2018)
8. Nandwani, P., Verma, R.: A review on sentiment analysis and emotion detection from text. Soc. Netw. Anal. Min. **11**(1), 81 (2021)
9. Acheampong, F.A., Nunoo-Mensah, H., Chen, W.: Transformer models for text-based emotion detection: a review of bert-based approaches. Artif. Intell. Rev. **54**(4), 5789–5829 (2021)
10. Walsworth, A.: The role of gpt-4 in chatbot emotion recognition. In: Adam Walsworth-Streaming Media Engineer (2023)
11. Gorvett, Z.: The ai emotions dreamed up by chatgpt. www.bbc.com/future/article/20230224-the-ai-emotions-dreamed-up-by-chatgpt (2023)
12. Chiorrini, A., Diamantini, C., Mircoli, A., Potena, D.: Emotion and sentiment analysis of tweets using bert. In: EDBT/ICDT Workshops, vol. 3 (2021)
13. Albu, I.-A., Spînu, S.: Emotion detection from tweets using a bert and svm ensemble model (2022)
14. Kumar, A., Jain, A.K.: Emotion detection in psychological texts by fine-tuning bert using emotion-cause pair extraction. Int. J. Speech Technol. **25**(3), 727–743 (2022)
15. Qin, X., Wu, Z., Zhang, T., Li, Y., Luan, J., Wang, B., Wang, L., Cui, J.: Bert-erc: Fine-tuning bert is enough for emotion recognition in conversation. Proceed. AAAI Conf. Artif. Intell. **37**, 13492–13500 (2023)
16. Zhalehpour, S., Onder, O., Akhtar, Z., Erdem, C.: Baum-1: A spontaneous audio-visual face database of affective and mental states. In: IEEE Transactions on Affective Computing, vol. 4, pp. 1–1 (2016)
17. Ekman, P.: Strong evidence for universals in facial expressions: a reply to russell's mistaken critique. In: American Psychological Association (1994)
18. IQ, O.: Gpt-3.5 model architecture. www.iq.opengenus.org/gpt-3-5-model/ (2022)
19. GetGenie.: Davinci vs turbo-a detailed comparison than you need. www.getgenie.ai/davinci-vs-turbo/ (2023)

20. Goyal, T., Li, J.J., Durrett, G.: News summarization and evaluation in the era of gpt-3. ArXiv preprint arXiv:2209.12356 (2022)
21. Devlin, J., Chang, M.-W., Lee, K., Toutanova, K.: Bert: Pre-training of deep bidirectional transformers for language understanding. ArXiv preprint arXiv:1810.04805 (2018)
22. Liu, H.: Sentiment analysis of citations using word2vec, CoRR arXiv:1704.00177 (2017)
23. He, P., Liu, X., Gao, J., Chen, W.: Deberta: Decoding-enhanced bert with disentangled attention. ArXiv preprint arXiv:2006.03654 (2020)
24. OpenAI, Gpt-4 technical report. arxiv.org/abs/2303.08774 (2023)

A Human-friendly Verbal Communication Platform for Multi-Robot Systems: Design and Principles

Christopher Carr[1], Peng Wang[2(✉)], and Shengling Wang[3]

[1] Loughborough University, Loughborough LE11 3TU, UK
ttcc4@lunet.lboro.ac.uk
[2] Manchester Metropolitan University, Manchester M15 6BH, UK
p.wang@mmu.ac.uk
[3] University of Sheffield, Sheffield S10 2TN, UK
swang119@sheffield.ac.uk

Abstract. Aiming at the first steps toward de-coupling the application of multi-robot systems from reliance on network infrastructure, this paper proposes a human-friendly verbal communication platform for multi-robot systems, following the deliberately designed principles of being adaptable, transparent, and secure. The platform is network independent and is subsequently capable of functioning in network-deprived environments–from underwater scenarios to planet exploration. A series of experiments were conducted to demonstrate the platform's capability in multi-robot systems communication and task co-ordination, showcasing its potential in infrastructure-free applications. To benefit the community, we have made the code open source at https://github.com/jynxmagic/MSc_AI_project GitHub.

Keywords: Human-robot interaction · Verbal communication · Heterogeneous agents · Multi-robots

1 Introduction

The continuous development of robotics-based technologies along with the significant advancements achieved by Artificial Intelligence (AI) techniques such as Natural Language Processing (NLP) and Computer Vision (CV) have driven the massive deployment of robots in sectors like manufacturing, warehousing, and healthcare. Recently, being still in its nascent stage, multi-robot systems (MRS) have attracted attention as they excel a single robot in terms of robustness and efficiency in task execution, and can deliver tasks that are impossible for a single robot [1]. Despite the advantages, MRS cannot be simply considered as the generalisation of techniques that are applied to a single robot [2], which further stimulates the research and applications of MRS [1].

Ever since the introduction of MRS in the 1980s [1], an array of topics have been investigated ranging from biological inspired MRS techniques, communication, architectures, task allocation, localisation, mapping, exploration, object

N. Naik et al. (Eds.): UKCI 2023, AISC 1453, pp. 580–594, 2024.
https://doi.org/10.1007/978-3-031-47508-5_45

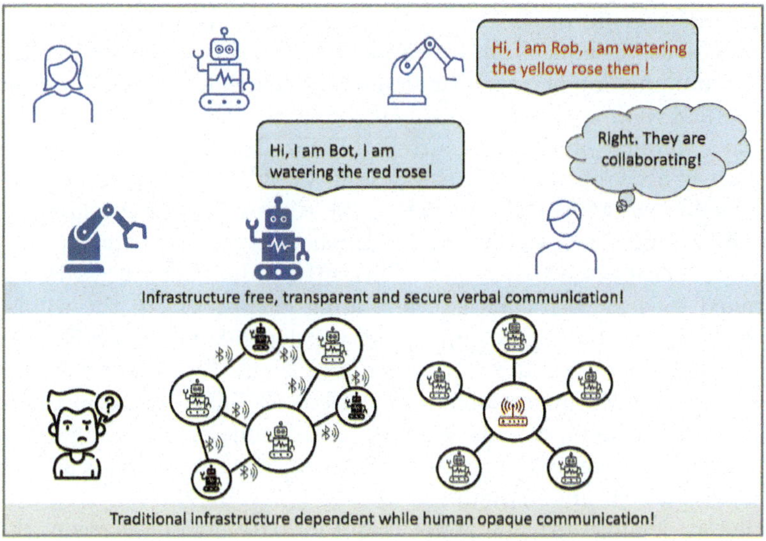

Fig. 1. Comparison of the proposed verbal communication platform (top) against traditional Bluetooth based decentralised MRS (bottom middle) and Wi-Fi based centralised MRS (bottom right) which rely on infrastructure and the communication is opaque to humans. The proposed verbal communication platform is, in contrast, infrastructure-free, transparent, and secure to humans.

transport & manipulation, motion coordination, to re-configurable robots [3]. A few taxonomy paradigms have been proposed aiming at categorising the topics [2,4]. Roughly, there are three features that are widely used for MRS classification: **(1)** Types of robots. MRS can be broadly broken down into two types, i.e., homogeneous and heterogeneous. In a homogeneous MRS, all robots are of equivalent physical structure and follow similar objectives, whereas robots in a heterogeneous MRS can have differing characteristics and objectives [5]. **(2)** Control architectures. The control architecture is the 'brain' of the MRS, and it is crucial for task co-ordination to deliver collaborative and cooperative jobs. There are mainly three control architectures, i.e., decentralised control (reactive control), centralised control (deliberative control), and hybrid architectures, which is a combination of centralised and decentralised controls. The hybrid control is mostly used as it works in a hierarchical way: in the low level, decentralised control is used to co-ordinate (percept each other then act based on perceptions) between robots directly while in the high level, centralised control is used to provide the MRS with rich sensor perception and the full environment information for coordination and collaboration. **(3)** Communication mechanisms [4]. Communication can be broken down into what is known as 'explicit' and 'implicit' communications. Implicit communication refers to communication which is discovered via the state-change of an environment. For example, if a robot moves a box and another robot discovers the box had been moved, the result is implicit communication. Explicit communication occurs when robots communicate with each other for specific reasons solely with the intent of communication and no environmental state changes [3]. Researchers have discovered that in many cases, explicit communication of even a small amount can lead to drastic improvements within the MRS [6].

Whether implicit or explicit, current MRS relies on certain 'infrastructure'. While there is no doubt that Wireless Fidelity (Wi-Fi) network has become the *de facto* facility for MRS communication [7], some constraints remain: 1) Wi-Fi routers operate on the 2.4 GHz band, and the distance the signals can reach is affected by blockage and bypass, etc. Currently, redundant Access Points (APs) are installed to deal with such issues, adding extra costs to the whole system and making the system more infrastructure dependant. 2) Wi-Fi signals can only reach centimetres underwater, which makes any Wi-Fi network based MRS solutions ineligible to underwater applications. 3) The communication within MRS is opaque to humans, which makes it difficult, or at least requires extra efforts to monitor issues such as ethical biases in the communication. Please refer to Fig. 1 for a visual understanding.

Aiming at dealing with the above issues in MRS communication, this paper proposes a human-friendly verbal communication platform (HVCP). The HVCP refers to the scenario where a robot speaks verbally to another robot (or human/agent), which speaks back through microphone or speakers. Figure 1 shows a comparison of the HVCP and traditional Bluetooth and Wi-Fi based communication frameworks. The HVCP will enable MRS to communicate verbally and network-infrastructure free, henceforth possessing the potential of being used in challenging tasks such as planet/underwater explorations. The contributions of this paper include 1) Robot-to-Robot (R2R) communication carried out verbally, i.e., robots are verbally talking to each other directly for coordination and collaboration, etc. Compared to Wi-Fi signals, verbal signals can travel as far as 180 metres in air and 2,030 metres underwater, which enables MRS to work in a larger area. 2) No network infrastructure is needed for communication. As robots are talking verbally, the network that was traditionally meant to connect robots becomes unnecessary. This drastically expands the applications of MRS to scenarios where network is not applicable. 3) The communication is transparent to humans, i.e., humans can monitor the communication at any time and anywhere to avoid biases in communications that would lead to discrimination or biases in task execution stage.

The remainder of the paper is organised as follows. Some related works are introduced in Sect. 2. Section 3 elaborates on the principles and design of the human-friendly verbal communication platform. Implementation, results, and analysis are provided in Sect. 4, and the paper is concluded in Sect. 5.

2 Related Works

Verbal communication is continuously attracting attention from both industry and academia, which as Marin, et al. claims is driven by the desire of developing and achieving human-like interaction between humans and robots [8]. The benefits of verbal communication between humans and robots includes but is not limited to robot therapy [9], educational robotics [10], and assistive robotics [11]. Compared to non-verbal communication solutions, verbal communication possesses the potential of bridging the knowledge gap between techniques and non-expert humans, and breaking the ethical barriers to eventually make robots a part of the human society [12]. In literature, the communication interface through which a human communicates and delivers/extracts data to a robot can either be statically programmed or learned through an AI model, e.g., chat-GPT has been used in robotics very recently [13]. Both of these have their own advantages, with AI models being much more adaptive to circumstances. Among all the techniques related, text-to-speech and speech-to-text are at the heart of verbal communication. Consequently, the challenges in these two areas undermine the capacity of verbal communication. For instance, the capture quality of speech is affected by both the distance from the speaker to the receiver and the ambient noise level [8,14]. When multiple speakers communicate to one or more receiver robots simultaneously, there is the risk of intentions not delivered to the correct destinations due to the lack of speaker identification [8,15]. Even if there is speaker identification mechanisms, the changing of voice along aging process could still hinder the performance of the verbal communication systems [8,16].

It is surprising that while human-to-robot (H2R) verbal communication has been so popular, R2R verbal communication has not been as researched. To the best of the authors' knowledge, there is no published solutions for R2R verbal communication. The most popular R2R communication is mostly achieved by Wi-Fi [17], Bluetooth [18], or Ultra-wideband (UWB) [19] techniques, which are efficient but not human friendly. There is also work that aims at enabling MRS to develop their own language to achieve communication. The issue still is the developed language cannot be understood by humans [20]. One would argue that R2R verbal communication can be naturally solved by mature H2R solutions. The authors agree that there is the possibility, but believe an independently designed human-friendly verbal communication platform following the principles of adaptability, transparency, and security will massively benefit the MRS community. This will be discussed in the next section.

3 Design Principles and Architecture

Inspired by the H2R achievements in literature, our vision in designing and implementing the R2R verbal communication platform is such that it is infrastructure (cable or APs) and robot type (heterogeneous or homogeneous) independent, which makes it an adaptable tool for general coordination and collaboration purposes. The long-term ambition to eventually bring MRS based on R2R verbal communication platform into human society also requires it to be secure and transparent to humans. We will detail each of the three aspects as follows.

3.1 Adaptability

The development of a verbal communication platform through which the robot communicates requires for the ability of robots to not only encode the information which it is attempting to transfer, but also for the receiving robot to be able to decode that information into useful data. Therefore, there is the need of co-evolution of encoders and decoders [21].

Encoders and decoders for human language of high quality are freely available to most people. The co-evolution of decoders and encoders would be offloaded onto a completely separate area of research, mainly text-to-speech and speech-to-text, which are hot spots in NLP and successful application such as DALL-E are able to locate objects according to natural language commands. With the development of text-to-speech and speech-to-text technologies, one can foresee that robots will be able to communicate heterogeneously, to any other different types of robot, using a verbal communication platform. They would all be using the same definitions and protocols for communication (language), thus allowing a team of many different types of robots and software to communicate for coordination and collaboration, and the usage of verbal signals for communication will massively reduce the dependency of infrastructures and make MRS adaptable.

3.2 Transparency

There is no doubt that a verbal communication based MRS can be stand-alone and robot-only, if required. However, the anticipation for robots being -as-a-service is prevailing that it will inevitably play its part in human society. For instance, Iqbal and Riek performed some research on the dynamics of a multi-robot, multi-human heterogeneous team [22]. They discovered that adding a robot to a human team can alter its dynamics drastically. In their work, they based most of what could be done based on what the human team could see the robot team was doing. If the robots within the experiments were communicating verbally, the human team would have been much more apt at adapting to the robot teams' behaviour. Verbal communication in MRS gives the added advantage of transparency to human operators around them. It will be much easier for such teams (humans and robots) to exist together and perform more complex tasks. The added level of transparency not only allows humans to understand what the robot team is doing, but also intervene if the robot team is doing something wrong. This is extremely important to the ethics of AI [23]. Ethical issues which are present in current AI research also present themselves in robotics-based applications of AI. The issue of discrimination in robotics can be partly solved by using verbal communication. For instance, the verbal communication platform will enable humans, being operators or just normal humans, will be able to hear inter-process communication as it happens, and are much more likely to spot discrimination as it happens, and subsequent actions can be taken to deal with the situation.

Another advantage of such transparency is the ability to 'plug-and-play' human operators into an MRS. Humans, at any intelligence level, can easily join in to the MRS and begin understanding or giving commands to the robots within the team. As long as the robots are communicating verbally, it makes no difference to add a human operator at any point to join in on the task. This is extremely advantageous. Whilst robot teams may not need to take elongated breaks, humans do. There will be periods of over 12 hours where no human operators could be in the mix and the robots are still working (after people finish work). Or, if shift work is in place, another group of human operators would need to come and replace the existing human operators within the team. It would be extremely simple for these humans to add themselves to the robot team and understand what the robots are currently doing.

3.3 Cybersecurity

Robots, being traditionally connected to a network with open ports to enable communication, are presenting themselves to multiple digital attacks [24]. An MRS functioning as such exposes itself to hackers and can lead to malicious attacks–consequences could be unbearable if a human is involved in the task. The interests of businesses and people alike need to be protected against hackers by developing mechanisms to stop hackers from gaining unauthorised access to an MRS [25].

As mentioned earlier, communicating verbally makes an MRS less infrastructure dependent, and almost free of traditional networks. This makes cyber-attacks on the robots very difficult, as hackers will have to connect to the robots manually (i.e. 'physically connect' to the robot). This can be easily tackled by simple security measures and reported by humans that are collaborating with the robots. A solution such as this still leaves issues whereby human operators are not present. In that case, verbal communication between robots could be deliberately encrypted, with each robot containing its own access key to decrypt the communications. Hackers would then have no way of gaining access to the communications, without studying the communications at length. Overall, communication through the sound channel (verbally), rather than a network channel, makes the MRS almost impervious to an attack.

3.4 Architecture

Bearing the three principles in mind, the architecture of the proposed verbal communication platform is designed to be as generalised and adaptable as possible. The Open Systems Interconnection (OSI) model is used as a basis for the platform to ensure heterogeneity and usability. In this paper, each OSI model layer as shown in Fig. 2 is mapped to its functionality in the proposed platform. The OSI physical layer corresponds to the *robot hardware* layer. The Data Link Layer and Network Layer in the OSI model are replaced by verbal signals, indicating the independence of our platform on infrastructure. The Network Layer, however, may need adjustments for long distance verbal communication, or scenarios where there are noises and echoes. The Transportation Layer in the proposed platform is implemented by speakers to either record speech or read out loud text information for communication. The Session Layer, the Presentation layer, and the Application Layer are meant to maintain the conversations among robots, perform text-to-speech and speech-to-text translation, and integrate the platform with ROS to make it usable to the ROS driven robotics community, respectively.

Given the general architecture, a complete communication between any two robots for any task will include three steps, i.e., 1) greeting, 2) task definition, and 3) roles & responsibilities coordination. The initial greeting is a template: 'Hi, I am X'. This initial greeting serves the purpose of sharing a unique identifier to other robots, so that other robots may contact that robot specifically. After greeting, each robot has the unique identifier of all other robots within its vicinity. Also, a unique identifier can be used to contact specific groups of robots, or all robots. For example: 'Team A', 'Team B', and 'All'.

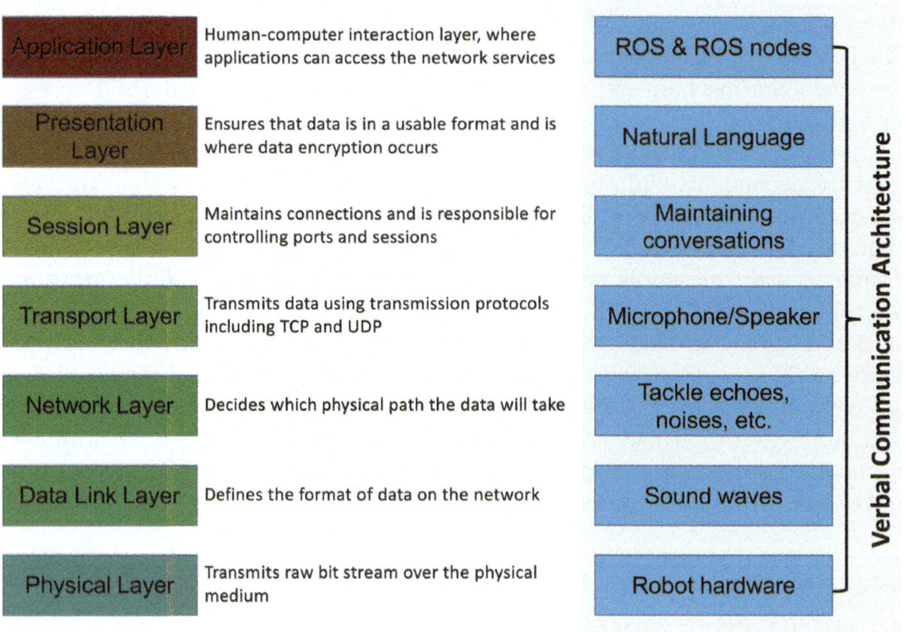

Fig. 2. Architecture of the verbal communication platform and its correspondence with the OSI model.

The next stage within communication is task definitions. The robots can simply share names of the tasks they are completing. This helps to identify robots which are collaborating on a task: 'Hi all, I am X, member of Team A, completing task: water the red rose'. These communications can be broken down into separate statements, though it should be noted that all robots are required to specify their name during communications as

'*Hi*, [robot identifier], *I am X*, [statement]'.

This allows other robots to understand which robot the communication is coming from and what they are doing. It is worth noting that in our implementation, we use a task repository to maintain tasks and the statement of each task is broken down to a series of actions such as *moving forward* to deliver the task.

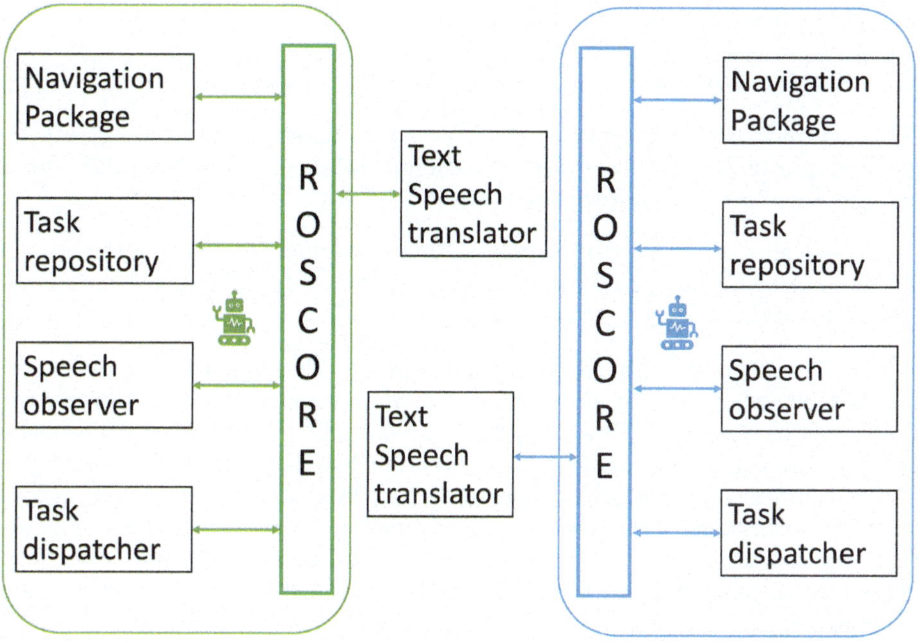

Fig. 3. Illustration of two robots running in decentralised mode for coordination and modules that are involved to achieve coordination.

Robots, along with defining the task, must also define their responsibilities within a task. This mostly relates to collaboratively delivering a task between robots. Assuming two robots are completing the task ‘water plants’, they need to distribute the sub-tasks between them so that each will water some different plants. To achieve efficient coordination and collaboration, robots are sharing task repositories. It defines specifically what the task is and what the responsibilities of the robots are within that task. For the ‘water plants’ task, the statement looks as follows: ‘Hi, [team plant], I am X, I will complete task: 1’. Other robots should also assign themselves to responsibilities within the task: ‘Hello [team plant], I am Y, I will complete task: 2’. Repeating in this fashion, the robots will distribute the tasks between themselves until no tasks remain. Again, it is very important to note that each of the robots knows exactly what the other robots are doing so as to not assign themselves to the same task.

The implementation of the verbal communication platform for two robots is shown in Fig. 3. As can be seen, the two robots are running their own roscore for verbal communication, which indicates that the proposed platform is decentralised (N.B. traditional MRS based on Robot Operating System (ROS) usually uses only one roscore). The robots use a task repository to manage tasks and update the repository upon task completion. To benefit the community, the code is opensourced at GitHub.

4 Results and Analysis

4.1 Setting Up

An array of experiments and simulations were conducted to demonstrate the performance of the proposed verbal communication platform for MRS. To be specific, there are mainly two types:

1. Assessments of the verbal communication performance of the platform.
2. Assessments of the verbal communication platform for MRS coordination and collaboration.

Two Turtlebot Burger robots with Raspberry Pi 3B were used for experiments. Unbuntu 20.04 LTS and ROS were used for implementation. Any speakers that adapt to Raspberry Pi will work for this case. We also believe as the emerge of large language models, they will be used in cases like ours in the future. The Picovoice library is adopted for both speech-to-text and text-to-speech generation. The advantage of Picovoice is that it allows for offline translation of speech and text in a bi-direcitonal manner. Note that all the systems and software are chosen this way so that the verbal communication platform does not rely on any communication infrastructure, regardless of being wired or wireless.

4.1.1 Effectiveness of Wake Words Wake words are used following the work pattern of Amazon Alexa and Apple Siri, etc. to invoke robots to collaborate in this paper. One critical issue is the effective distance in which the robots could be invoked and accept and execute commands conveyed by verbal signals. Two robots installed with speakers (also work as receivers) were set to speak to each other, using the wake words. The distance between the two robots was increased at a step of 10 cm, and at each distance 20 tests were carried out. The wake words success rate at each distance was calculated, as shown in Fig. 4. Note all the results were obtained at a 20 dB ambient noise level. One can see a sharp decrease of success rate after a certain distance, which is 120 cm in this case. For convenience, we call this distance the *pivotal distance*. The existence of the pivotal distance is due to the degrading strength of verbal signals, which makes it more challenging for Picovoice to translate verbal signals into texts correctly.

4.1.2 Multi-Robot Coordination Based-on the Verbal Communication Platform To demonstrate that the verbal communication platform can be used by an MRS for coordination, two robots installed with the verbal communication platform were used to carry out the experiments. Two distinct sets of experiments were carried out based on the operational modes of the Multi-Robot System (MRS). The first mode, referred to as the “Coordinating mode,”

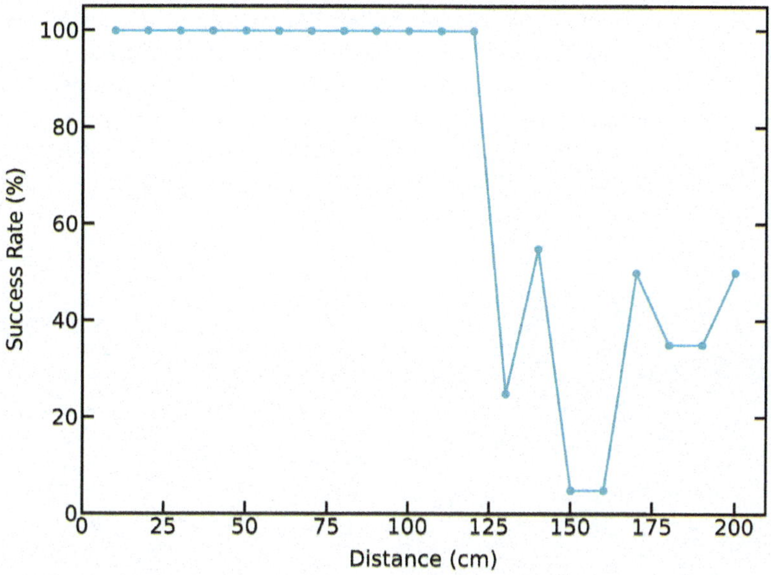

Fig. 4. Success rate of wake-word detection at different distances. The ambient noise level is 20 dB.

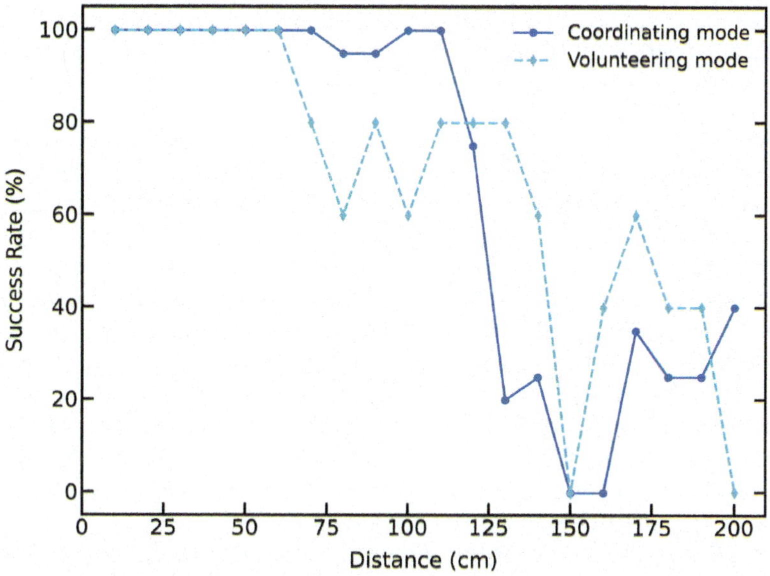

Fig. 5. Success rate of the verbal communication platform in different working modes. The ambient noise level is 20 dB.

involved a designated robot (referred to as the coordinator) issuing commands to another robot to execute specific tasks. For example, a coordinator robot might communicate as follows: '*Greetings, Rob. I am Bot. Please water the yellow rose.*'. In contrast, the second mode, known as the "Volunteering mode," entailed robots engaging in mutual communication to collaborate and accomplish a given task. As an illustration, if the task at hand was 'watering plants,' robot Rob could voluntarily offer to water the yellow rose, while robot Bot might proceed to water the red rose.

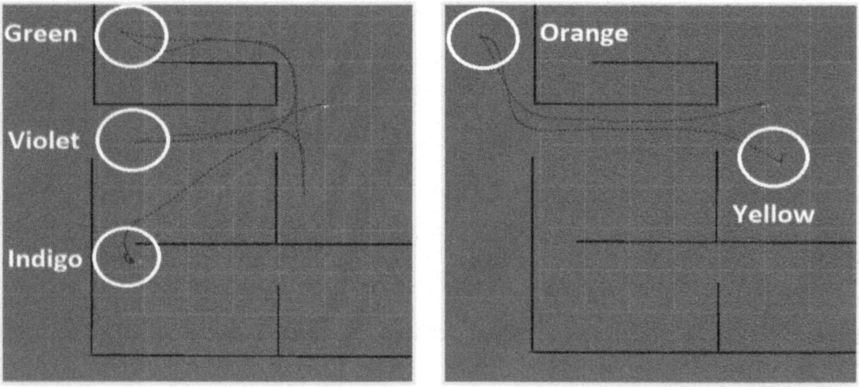

(a) Robot *Rob*'s Task Execution Trails (b) Robot *Bot*'s Task Execution Trails

Fig. 6. Simulation of two robots, *Rob* and *Bot*, collaborating to water plants. The plants are highlighted with a white circle. The execution trails of the robots are demonstrated to showcase how the tasks are achieved separately.

Figure 5 shows the success rate of the two experiments. To be specific, in coordinating mode, the leading robot asks the other robot to perform a task at various distances, with 20 times at each distance. In volunteering mode, the two robots verbally communicate first to allocate tasks and then execute the tasks. One can see the significant decline of success rate after the pivotal distance in both cases. This is because the wake words are difficult for the robots to recognise over the pivotal distance. The success rate before hitting the pivotal distance is, however, quite high for both cases. It is worth mentioning that for the volunteering mode, the success rate starts dropping at the distance of 70 cm. This is due to the fact that after one robot, such as Rob, volunteers to do a task, it starts moving straight away from the other robot, and the increased distance eventually undermines the success rate.

Figure 6 shows the simulated collaboration of two robots, i.e., *Rob* and *Bot*, using the proposed verbal communication platform for the '*water plants*' task. In the task, there are five plants to water, i.e., plant *Green*, *Violet*, *Indigo*, *Orange*, and *Yellow*. They are registered in the task repository. After initial greetings, *Rob* and *Bot* will randomly pick a task for themselves from the task repository, and execute the tasks. After the successful completion of the current task, either robot will pick the next task and update the task repository. The process carries on until no tasks left in the task repository. As shown in Fig. 6, *Rob* has watered three plants and *Bot* has watered the remaining two.

4.2 Ablation

As text-to-speech and speech-to-text carried out by Picovoice is a critical factor in determining the performance of the proposed verbal communication platform, we further investigated how it impacts the verbal communication platform. It is worth noting that Picovoice is not opensource, so all our results are generated by calling Picovoice API.

4.2.1 Text to Speech In our simulations and experiments, we noticed that most text-to-speech translations were correct, but there were still chances to get confusing results. In the English language, some words are spelt the same but have multiple pronunciations, known as homophones. For instance, the word *read* can be pronounced 'r-eed', as in 'comprehend the meaning of symbols', or 'r-ehd'–past tense form of 'r-eed'. These two words are both spelt the same, though with a different meaning and pronunciation. In every case during text to speech, the word is encoded as 'r-eed', regardless of context used. This can introduce some major errors into the translation process, which is discussed some more in the speech to text section below. Table 1 gives more examples of English words and their various methods of communication, along with how Picovoice pronounces the word.

4.2.2 Speech to Text The homophones issues also occur in the speech to text stage. Again, using the pronunciation of 'read' as an example. There are two possible pronunciations for this word. As listed in Table 1, Picovoice uses the word as pronounced 'r-eed'. In translation of the word 'red' as in color, the speech to text module translated the word into 'read'. The following text to speech module will use the 'r-eed' pronunciation and eventually deliver 'red' as in color to 'read' as in 'comprehend the meaning of symbols'.

In this paper, the issue was mitigated by using alternatives of such words. For instance, when the robot failed to translate a word correctly, the phrase which triggered instead of the correct phrase was recorded. Table 2 shows the different phrases which triggered according to some specific keywords–the objectives. For example, a frequent error by the translation was from 'indigo' to 'in the go'. Should 'in the go' be listed as an alternative phrase to 'indigo', it would be possible to improve the robots performance drastically. With many such small improvements as this, it is possible to improve the perfect translation performance up to a distance of 354 cm with ambient noise level at 20 dB.

Table 1. Possible and used pronunciations for homophones

Word	Possible Pronunciations	Used Pronunciation
read	r-ehd, r-eed	r-eed
bass	b-ass, b-ase	b-ase
row	r-oh, r-ow	r-oh
bow	b-ow, b-oh	b-oh
tear	t-e-rh, t-ee-r	t-e-rh
live	l-iv, l-eye-ve	l-eye-ve
wind	w-in-d, w-eye-nd	w-in-d
wound	w-oo-nd, w-ow-nd	w-oo-nd
close	cl-oh-s, cl-oh-ze	cl-oh-s
excuse	eh-cs-oos, eh-cs-use	eh-cs-oos
polish	p-oh-l-i-sh, p-o-l-i-sh	p-o-l-i-sh
lead	l-ed, l-eed	l-eed
dove	d-oh-ve, d-uh-ve	d-uh-ve
graduate	g-rad-u-ate, g-rad-u-at	g-rad-u-at

Table 2. Incorrect translations of objectives in complex communications

Word	Translation	Count
Indigo	in the go	3
Indigo	him to go	4
Indigo	do it again	3
Indigo	you to know	1
Yellow	you know	2
Orange	a range	1

5 Conclusion

A verbal communication platform is proposed in this paper following the principles of being adaptable to environments and robot types, and transparent and secure to humans. The platform enables a multi robot system to communicate verbally and therefore become less infrastructure dependent. This makes the platform promising in expanding the application of multi-robot systems into network lacking scenarios. A series of experiments were conducted to demonstrate the effectiveness of the verbal communication platform in terms of communication efficiency as well as its application in multi-robot coordination. Future works will deal with coordination in a larger number robots using the verbal communication platform.

Acknowledgment. This work was funded by 'Constructing accurate 3D human with deep learning in digital twin environments for safe human-robot collaboration' (Grant No. 914175).

References

1. Parker, L.E., Rus, D., Sukhatme, G.S.: Multiple mobile robot systems. In: Springer Handbook of Robotics, pp. 1335–1384. Springer (2016)
2. Verma, J.K., Ranga, V.: Multi-robot coordination analysis, taxonomy, challenges and future scope. J. Intell. Robot. Syst. **102**(1), 1–36 (2021)
3. Arai, T., Pagello, E., Parker, L.E., et al.: Advances in multi-robot systems. IEEE Trans. Robot. Autom. **18**(5), 655–661 (2002)
4. Ismail, Z.H., Sariff, N., Hurtado, E.: A survey and analysis of cooperative multi-agent robot systems: challenges and directions. In: Applications of Mobile Robots, pp. 8–14. IntechOpen (2018)
5. Quinn, M.: A comparison of approaches to the evolution of homogeneous multi-robot teams. In: Proceedings of the 2001 Congress on Evolutionary Computation (IEEE Cat. No. 01TH8546), vol. 1, pp. 128–135. IEEE (2001)
6. Balch, T., Arkin, R.C.: Communication in reactive multiagent robotic systems. Auton. Robot. **1**(1), 27–52 (1994)
7. Behera, S., Panigrahi, B., Rath, H.K., Pal, A.: Wireless characteristics study for indoor multi-robot communication system. In: Proceedings of the 1st Workshop on Complex Networked Systems for Smart Infrastructure, pp. 1–6 (2018)
8. Marin Vargas, A., Cominelli, L., Dell'Orletta, F., Scilingo, E.P.: Verbal communication in robotics: a study on salient terms, research fields and trends in the last decades based on a computational linguistic analysis. In: Frontiers in Computer Science, p. 63 (2021)
9. Melo, F.S., Sardinha, A., Belo, D., Couto, M., Faria, M., Farias, A., Gamboa, H., Jesus, C., Kinarullathil, M., Lima, P., et al.: Project inside: towards autonomous semi-unstructured human-robot social interaction in autism therapy. Artif. Intell. Med. **96**, 198–216 (2019)
10. Budiharto, W., Cahyani, A.D., Rumondor, P.C., Suhartono, D.: Edurobot: intelligent humanoid robot with natural interaction for education and entertainment. Proced. Comput. Sci. **116**, 564–570 (2017)
11. Zhou, B., Wu, K., Lv, P., Wang, J., Chen, G., Ji, B., Liu, S.: A new remote health-care system based on moving robot intended for the elderly at home. In: Journal of Healthcare Engineering, vol. 2018 (2018)
12. Kuipers, B.: Perspectives on ethics of ai. In: The Oxford Handbook of Ethics of AI, p. 421. Oxford University Press (2020)
13. Vemprala, S., Bonatti, R., Bucker, A., Kapoor, A.: Chatgpt for robotics: Design principles and model abilities. Microsoft Auton. Syst. Robot. Res **2**, 20 (2023)
14. Kumatani, K., Arakawa, T., Yamamoto, K., McDonough, J., Raj, B., Singh, R., Tashev, I.: Microphone array processing for distant speech recognition: Towards real-world deployment. In: Proceedings of The 2012 Asia Pacific Signal and Information Processing Association Annual Summit and Conference, pp. 1–10. IEEE (2012)
15. Goeldner, M., Herstatt, C., Tietze, F.: The emergence of care robotics-a patent and publication analysis. Technol. Forecast. Soc. Chang. **92**, 115–131 (2015)

16. Kennedy, J., Lemaignan, S., Montassier, C., Lavalade, P., Irfan, B., Papadopoulos, F., Senft, E., Belpaeme, T.: Child speech recognition in human-robot interaction: evaluations and recommendations. In: Proceedings of the 2017 ACM/IEEE International Conference on Human-Robot Interaction, pp. 82–90 (2017)
17. Chen, K.-C., Lin, S.-C., Hsiao, J.-H., Liu, C.-H., Molisch, A.F., Fettweis, G.P.: Wireless networked multirobot systems in smart factories. Proc. IEEE **109**(4), 468–494 (2020)
18. Mahmood, M., Hong, O.C.S., Sanhoury, I.M., Norrazi, M.A.M., Amin, S.H., et al.: Implementation of multi robot communication via bluetooth. In: FIRA RoboWorld Congress, pp. 441–453. Springer (2013)
19. Shule, W., Almansa, C.M., Queralta, J.P., Zou, Z., Westerlund, T.: Uwb-based localization for multi-uav systems and collaborative heterogeneous multi-robot systems. Proced. Comput. Sci. **175**, 357–364 (2020)
20. Mordatch, I., Abbeel, P.: Emergence of grounded compositional language in multi-agent populations. In: Proceedings of the AAAI Conference on Artificial Intelligence, vol. 32, 1st edn (2018)
21. Lipson, H.: Evolutionary robotics: emergence of communication. Curr. Biol. **17**(9), R330–R332 (2007)
22. Iqbal, T., Riek, L.D.: Coordination dynamics in multihuman multirobot teams. IEEE Robot. Autom. Lett. **2**(3), 1712–1717 (2017)
23. Hagendorff, T.: The ethics of ai ethics: An evaluation of guidelines. Mind. Mach. **30**(1), 99–120 (2020)
24. Carr, C., Wang, S., Wang, P., Han, L.: Attacking digital twins of robotic systems to compromise security and safety. ArXiv preprint arXiv:2211.09507 (2022)
25. Lacava, G., Marotta, A., Martinelli, F., Saracino, A., La Marra, A., Gil-Uriarte, E., Vilches, V.M.: Cybsersecurity issues in robotics. J. Wirel. Mob. Networks Ubiquitous Comput. Dependable Appl. 12(3), 1–28 (2021)

Exploring Community Detection Algorithms and Their Applications in Social Networks

Mukesh Sakle[1(✉)] and Shaligram Prajapat[2]

[1] Shri G.S. Institute of Technology and Science, Indore, India
mukeshsakle@gmail.com
[2] International Institute of Professional Studies, Devi Ahilya University, Indore, India

Abstract. Community detection is a critical process for locating cohesive groups within social networks. The segment of strongly interconnected vertices into distinct communities yields valuable insights for various applications such as pattern recognition, recommendation systems, and data analysis. Since learners of higher education system are also on social networks forming strong communities, these can overlap with weaker natural communities. Identifying hidden weak communities within dominant ones is a challenge.This paper analyses Community detection algorithm for the online leaning student group recommendations based on interests. This study conducts a comparative analysis of existing community detection algorithms, assessing their performance across different parameters and applications, including online learning environments.

Keywords: Social Networks · Community · Clustering · Modularity · Community detection algorithms

1 Introduction

Networks are everywhere. There are different types of networks, one of which is social networks. The high concentration of edges between some groups of vertices generates a denser network within the broad network, resulting in the formation of communities. Communities are groups of vertices which share some common properties. These communities can be best analyzed and represented in the form of graphs. These communities are not the representation of mere connection amongst the individuals. They are the source of valuable information which can be used in various networks (especially, social networks) benefitting both the individual and the organizations.

Over the past several decades, community detection has turned out to be a boon in the field of networks which allows for better analysis of the network to understand its capabilities. Earlier the use of community detection was based upon finding disjoint communities by performing partitioning on the nodes of the network. Later, it was realized that the interconnected communities are quite important too and thus various algorithms were designed to find out such overlapping communities. There exist some partitioning techniques as well which can very well discover the overlapping communities [5].

N. Naik et al. (Eds.): UKCI 2023, AISC 1453, pp. 595–611, 2024.
https://doi.org/10.1007/978-3-031-47508-5_46

As we know about the progress in the field of community detection, the real catch is the idea of hidden community which is getting popular amongst the researchers these days. In general the real world has vast networks of sparse communities like organizations or groups which are not as tightly connected as the denser ones like families and friends. In case the intersection of the members between denser and weak communities is considerably same then in such cases such communities are ignored [5].

1.1 Applications of Community Detection

1.1.1 Scope in General Applications

The hidden communities are of great help in many different applications including multiple fields and therefore, working upon finding these communities becomes important. For instance, in PPI (Protein-Protein Interaction) networks, the hidden community detection can help to identify the genes performing the same functions. Similarly in various forms the hidden community detection helps the biologists to figure out the unobvious groups which are of great importance. Evaluating these hidden communities from the sparse community structures of the network is more crucial but also suffers from some major problems in doing so [5].

1.1.2 Scope in Social Network Applications

Many social networking applications used these days have their own algorithm, through which they can fetch every user's details and partition them into various communities. Many social media websites are also aiding manually community creation features to engage with people having similar interests.

1.2 Motivation

Community detection is very useful in evaluating and analyzing the large complex networks. It uses different parameters like edge weights of the graph and therefore is applicable for analyzing the network structures [11]. The process of finding the connected groups or clusters in the network is called community detection. It is one of the most important steps in social network analysis. Finding these communities is essential in many applications especially when decisions depend on the group. For example, recommending a product to a group of interested people [1].

1.2.1 Techniques for Community Detection

Due to the rapid growth of users at the social networking platforms, the visualisation graphs representing these networks are getting really complicated to understand and analyze. On detecting communities in these networks we get the absolute analyzed result required and thus, it becomes easier to understand them. The realization of communities in social networks can be useful in many ways. Some of the applications of these communities are listed below [1].

1.2.2 Expansion of Communities in Social Media

As the number of social networking sites are increasing with time, the scope of these sites also has a proportional impact. The sites are expanding both in terms of scope and focus. Nowadays there are sites for enhancing social interaction like Facebook and also there are sites for sharing pictures like Flickr. Considering Twitter, one can analyze the community structure by analyzing the tweet-retweet and the followers-followee. General nature of the tweets can also be analyzed and community detection algorithms could be used then to evaluate the underlying communities[1].

1.2.3 Enhancing Recommender Systems with Community Detection

The principle of recommendation system is based on the use of data with users of similar likings to generate a recommendation. This is similar to finding such groups in a graph. Therefore, community detection can work wonders in enhancing the working of these recommender systems [1].

1.3 Literature Survey

Puran Bedi and Chhavi Sharma [1] review the various algorithms and techniques available for discovering communities in social networks and their applications. They evaluated different algorithms based on different research activities. The article talks about common databases and applications. With the development of the web and the emergence of SNS i.e. Social Networking Sites, it has been effortless for users to communicate with each other.. Graphs can be used to represent social networks connected by edge nodes. Each node represents an entity or unit, and each edge represents an interaction between them. Virtual groups or communities are formed when people with similar interests come together on social networks. This approach has numerous practical applications, such as identifying communities, discovering shared search spaces in collaborative networks, pinpointing similar users for marketing and recommendations, and detecting protein interaction networks in biological systems.

Boccaletti, Stefano, et al. [2] In this article summarizes the latest developments in the study of complex networks, as well as their applications in fields as diverse as linear science, biology, statistical mechanics, and medicine. In addition to biological and chemical systems, nervous systems, social interactions, and the Internet (World Wide Web) is an example of a highly complex but interconnected dynamic system.

Muhammad et al. [3] performed a literature review of algorithms used for community detection, encompassing both conventional and advanced algorithms designed for detecting overlapping communities. The paper covers a range of algorithms, including those that employ dimensionality reduction methods such as principal component analysis (PCA) and non-negative matrix factorization (NMF). Additionally, the authors discussed the possible use cases of community detection in different fields.

Khawaja, F.R., Sheng, J., Wang, B., and Memon, Y. [4] perform a comprehensive analysis of the concealed community organization with the help of a hidden community detection (HCD) algorithm. Additionally, the paper discusses previous research in this area, such as clustering, overlapping communities, and the HCNC (Hidden Community based on Neighbourhood Similarity Computation) approach.

He K., Li Y., Soundarajan S., and Hopcroft J.E. [5] introduced a novel concept called HICODE, which stands for Hidden Community Detection. They utilized various community detection algorithms and structure reducing methods such as Re70 moveEdge, ReduceEdge, RemoveEdge, and ReduceWeight to implement HICODE. The hidden communities detected by HICODE are subparts of dominant communities. The authors also provide an overview of previous research in the field of community detection, including disjoint communities, overlapping communities, and multi-view clustering.

In their work, Günce Keziban Orman and Vincent Labatut [6] present a thorough comparison of multiple algorithms for community detection. They provide an in-depth analysis of various community network properties, including transitivity, modularity, community, and degree. The authors evaluate the performance of Walktrap Algorithm, Fast Greedy Algorithm, Label Propagation Algorithm, Eigenvector Algorithm, and Spinglass Algorithm based on different metrics.

The paper [7] proposes an innovative algorithm for detecting overlapping communities in dynamic social networks that evolve and change over time. The algorithm utilizes a sliding window methodology and modularity optimization to identify communities in a given time window. It also employs a graph similarity measure to match communities across different windows and track their evolution over time. The algorithm proposed performs better than existing methods in terms of both accuracy and computational efficiency. The flexible window size approach is effective in capturing the temporal evolution of communities. This study contributes to the domain of dynamic community detection in social networks by introducing a valuable algorithm.

The paper [8] proposes a technique for detecting novel communities in social networks using community embedding. The method involves mapping nodes into a low-dimensional space using the deep autoencoder neural network and then clustering the nodes based on their embeddings. The method suggested demonstrates better performance than various existing community detection algorithms when tested on synthetic and real-world datasets. The method is scalable and can handle large networks with millions of nodes and edges. The paper's contribution to the field of community detection lies in its application of deep learning techniques for accomplishing the task of detecting communities in social networks.

The paper [9] presents a novel community detection algorithm that uses multi-task learning in attributed graphs. The algorithm simultaneously performs multiple tasks, including node classification and community detection, to enhance the precision of community detection. It considers both the structural properties of the graph and the node attributes for community detection. The method suggested demonstrates better performance than various existing community detection algorithms in terms of accuracy and efficiency. This paper contributes to the field of community detection by providing a new approach that effectively incorporates node attributes into the community detection process.

The paper [10] proposes a multi-layered approach to detect hidden communities in social networks. It applies graph embedding to represent the network structure and extracts features from multiple layers of the embedding. The approach then utilizes a clustering algorithm to group the nodes into communities based on their feature similarities. The method is evaluated on various real-world social networks, and the results show that it demonstrates better performance than various existing community detection algorithms in terms of accuracy and stability. The proposed approach provides a promising solution for detection of hidden communities in social networks and has practical implications for various applications such as marketing and social network analysis.

2 Community Detection Algorithms

The algorithms of community detection are used to find more cohesive and dense communities. Community detection methods are basically classified under two subcategories i.e. Divisive and Agglomerative approach. Talking about the former, we remove edges one by one from a given graph. In the latter, we keep on adding the edges to the graphs which have only a single node. We move from a more important edge to a less important edge while adding them. Some techniques are listed below.

2.1 Girvan Newman Algorithm

This Algorithm runs by eliminating edges from the original given network to find out the desired communities. Girvan-Newman's algorithm does not work in order to detect the most important edges, rather its centre of attention is on those edges which are most probably to be "in between" communities. More the value of edge-betweenness, the more likely it will be to get removed first.

2.2 Greedy Modularity Maximization Algorithm

This is a modularity-based algorithm and the approach used is hierarchical agglomerative approach [6]. The algorithm works in a manner that starting with every node present in its own community it joins other communities iteratively until the modularity calculated is maximum.

2.3 The Louvain Community Detection Algorithm

This algorithm is also a Modularity based algorithm. Iteratively change in modularity is observed by removing an edge from a community and connecting it with neighbouring communities. It follows a hierarchical agglomerative approach with the same aim as above of achieving maximum Modularity.

2.4 Label Propagation Algorithm

This algorithm uses the assumption that an edge connected between two nodes depicts similarity between them.We start by assigning an initial label to all the nodes. Following a repetitive approach every node keeps on taking the label which is more extensively present in the surrounding. When there's no change in the labels of the nodes the process stops. The final label obtained at the last is the resulting label of the communities [6].

Girvan-Newman Algorithm

The Girvan-Newman algorithm is a well-known technique for clustering communities. It operates by iteratively breaking down the network into smaller sub-networks and creating distinct communities, accomplished by removing edges with the highest number of shortest paths between nodes.

The notion of identifying the betweenness centrality of edges was first introduced by Michelle Girvan and Mark Newman. Betweenness centrality is a measure of the number of shortest paths that pass through a specific edge. Edges that link different communities tend to have high betweenness centrality. By removing the edges with the highest betweenness centrality, the community structure of the network becomes more refined, making it easier to detect and identify communities [14]. The algorithm consists of four main steps:

1. The betweenness centrality of each edge is computed through a calculation process.To accomplish this, no action is necessary. Choose a node and apply a breadth-first search (BFS) algorithm to calculate the number of shortest paths from the selected node to all other nodes. These counts are then allocated as scores to each individual node. The credit assigned to an edge is calculated based on the leaf nodes, using the subsequent equation:

   Credit of edge = (1 + (Σedge credits to the node)) * (score of destination node/score of starting node).

   Compute the edge credits for all edges present in the graph.

   Continue iterating from the first step until no nodes remain unprocessed.

   The edge betweenness of each edge is equivalent to half of the sum of all edge credits.
2. Eliminate the edge with the greatest betweenness centrality.
3. Compute the modularity Q for the divided communities.

   The concept of modularity is employed to measure the strength of communities by comparing the fraction of internal edges to the fraction of edges that would be present in a randomly connected network. The fundamental idea is that a community should exhibit more links between its members than what would occur in a chance gathering of nodes. A Q value of 0 indicates that the proportion of internal edges within communities.

   is no different than what would be expected in a random network, while a Q value of 1 implies community structure having the highest possible strength.

$$Q \propto \sum_{s \in S} \left[(\#\text{edges inside group } s) - (\text{expected } \#\text{edges inside group } s) \right] \quad (1)$$

$$Q(G, S) = 1/2w * \sum_{s \in S} \sum_{i \in s} \sum_{j \in s} \left(W_{ij} - \left(d_i d_j \right) / 2w \right) \quad (2)$$

   $W_{ij} = 1$ if $i \rightarrow j$, 0 else.

   Normalizing cost: $-1 < Q < 1$.
4. If Q is between 0.3 and 0.7, repeat the process starting from step 1.

The function takes a NetworkX graph G as input and returns an iterator that produces sets of nodes in G as tuples. Each node set represents a community, while each tuple denotes the communities at a specific step in the algorithm.

Community Visualization:

See Figs. 1 and 2.

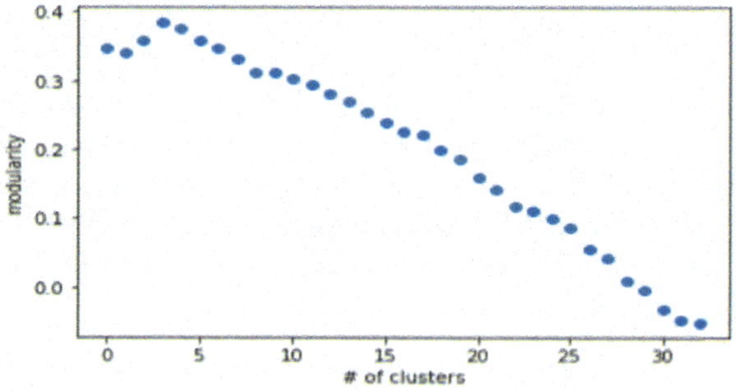

Fig. 1. Graph of modularity variation

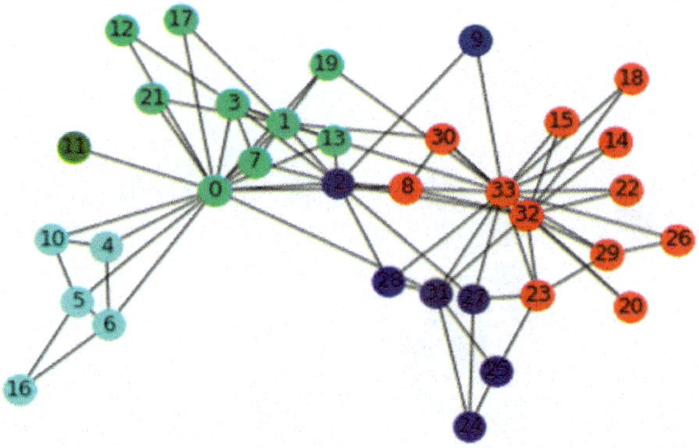

Fig. 2. Different communities (represented by different colors) detected using Girvan-Newman Algorithm

2.5 Greedy Modularity Algorithm

The Greedy Modularity Algorithm is an agglomerative hierarchical clustering technique that employs Clauset-Newman-Moore greedy modularity maximization to determine the community partition that yields the highest modularity score[1]. The algorithm initiates with each node being considered as a community, and subsequently, it merges the pair of nodes that generates the maximum modularity score.

The steps involved in the algorithm can be outlined as follows:

1. In the beginning, the algorithm creates |X| communities, where each individual node is considered to belong to its unique community.

2. Then, the algorithm proceeds by iteratively merging pairs of communities that maximize the resulting modularity. The modularity change after merging communities Ci and Cj can be calculated using the equation:

$$\Delta Q_{Ci,Cj} = [[|E_{Ci,Cj}| / |E| - (|E_{Ci}| / |E|) * (|E_{Cj}|/|E|)] \quad (3)$$

   where $|E_{Ci,Cj}|$ is the no. of edges between communities Ci and Cj, $|E_{Ci}|$ is the total degrees of nodes in community Ci, and $\Delta Q_{Ci,Cj}$ can be calculated in constant time. The algorithm continues until all nodes are part of a single community after (|X|-1) steps of joining.
3. The algorithm ends once all nodes in the network have been combined into a single community, a process that requires (|X| - 1) successive merging steps.
4. This results in a total of |X| partitions, with the initial partition defined by the first step and next partitions resulting from each following (|X| - 1) joining steps. The output of the algorithm is the partition with the highest modularity value, which approximates the maximum modularity.

The method takes in a NetworkX graph G as its input parameter and returns a list of frozensets of nodes, each corresponding to a community. The list is sorted in descending order of length, with the largest communities appearing first.

Community Visualization:

See Fig. 3.

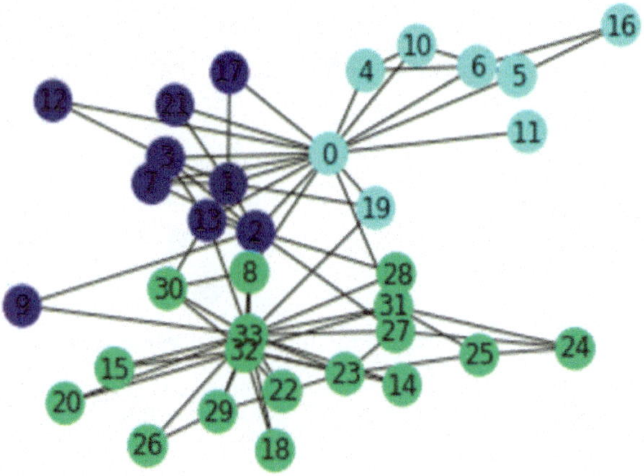

Fig. 3. Different communities (represented by different colors) detected using Greedy Modularity Algorithm

2.6 Louvain Algorithm

One commonly used approach for detecting communities is the Louvain algorithm, with the objective of maximizing the modularity score of every community. This hierarchical clustering method joins communities recursively into one node, and then performs modularity clustering [1].

The Louvain algorithm consists of two main steps: Modularity Optimization and Community Aggregation. The process is repeated iteratively until no additional modularity enhancements can be made.. In the Modularity Optimization step, the algorithm maximizes the modularity score for individual communities. This is achieved by computing the change in modularity when a node is moved from its current community to another, and selecting the move that results in the largest increase in modularity. In the Community Aggregation step, the communities that share the same parent node are merged into a single node, and the process repeats until all nodes belong to a single community.

$$\begin{aligned} M &= 1/2w * \Sigma_{i,j}\left(W_{ij} - p_{ij}\right)\delta(C_i, C_j) \\ M &= 1/2w * \Sigma_{i,j}\left(W_{ij} - (d_i d_j)/2w\right)\delta(C_i, C_j) \end{aligned} \tag{4}$$

The algorithm uses various notations, such as Wij to represent the weight of the edge connecting nodes i and j, di to represent the degree of node i, Ci to represent the respective community, and δ-function δ(u, v) to indicate if u = v or not.

Modularity Optimization:

In the initial step, the algorithm initially assigns each node to its own community. Subsequently, it attempts to maximize the increase in modularity obtained by relocating every node to its neighboring communities. In case relocating to any other community does not result in a higher modularity score, the node is retained in its current community[11].

The first stage iterates until no further modularity improvement is possible through node movement.

$$\begin{aligned} \Delta M = &\left\{\left[\left(\Sigma_{in} + 2d_{i,in}\right)/2w\right] - \left[(\Sigma_{tot} + d_i)2w\ \right]^2\right\} \\ &- \left\{\Sigma_{in}/2w - (\Sigma_{tot}/2w)^2 - (d_i/2w)^2\right\} \end{aligned} \tag{5}$$

One approach to enhance the efficiency of the algorithm is to simplify Eq. (3.3.2) by computing ΔM w instead of the entire expression, where Σ_{in} represents the total weight of the links within community C, Σ_{tot} is the sum of the weights of all connections to nodes within the community C, di denotes the total weight of all links incident to node i, di, represents the total weight of links from node i to nodes within C, and w is the total weight of all edges in the graph.

$$\Delta M = d_{i,in}/w - 2\Sigma_{tot} d_i/(2w)^2 \Leftrightarrow \Delta Mw = d_{i,in} - \Sigma_{tot} d_i/2w \tag{6}$$

Although the values of di, in and Σtot have to be computed for each community being tested, di/(2m) pertains only to the node under analysis. Therefore, this expression only needs to be recomputed when a new node is examined in the Modularity Optimization process.

Community Aggregation: The algorithm's second step involves creating a new network Wherein the communities detected during the first stage become the nodes. The links among the new nodes are assigned weights based on the total weight of links between nodes belonging to the relevant communities.. After this phase, the first step can be reapplied to form larger communities with higher modularity.

These two steps are repeated until no further modularity gain can be achieved.

The method takes a NetworkX graph G as input and decomposes it into communities. The output is a dictionary representing the partition, where each community is assigned a number from 0 to the total number of communities.

Community Visualization:

See Fig. 4.

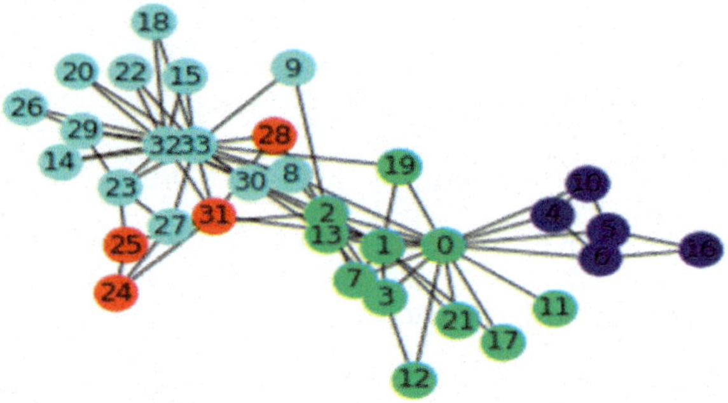

Fig. 4. Different communities (represented by different colors) detected using Louvain Algorithm

2.7 Label Propagation Algorithm

This is a type of machine learning algorithm that works semi-supervised, meaning it can assign labels to unlabeled data to classify observations in a dataset. Its main purpose is to label unclassified data to aid in classification.[1]. In the context of a community network, this algorithm modifies the assigned label of every node based on the labels of the communities to which its connected nodes belong.

Principle of label propagation Algorithm is as follows:

1. Every individual node within the network. is assigned a label representing its initial community membership.
2. The labels of the nodes indicate the distinct communities to which they belong.
3. During each iteration, an individual node modifies its community membership on the basis of the community labels of its neighbors until nodes with the same label belong to the same community.
4. The updated community membership of each node is determined by the community label that the maximum number of its neighbors belongs to.
5. Over time, Nodes that have a high degree of interconnectivity tend to converge to a common label community.

The algorithm takes an undirected graph G as input and employs a semi-synchronous label propagation technique to detect communities within G, leveraging the combined benefits of synchronous and asynchronous models. The resulting output is a dict_values object that groups nodes into their respective communities.

Community Visualization:

See Fig. 5.

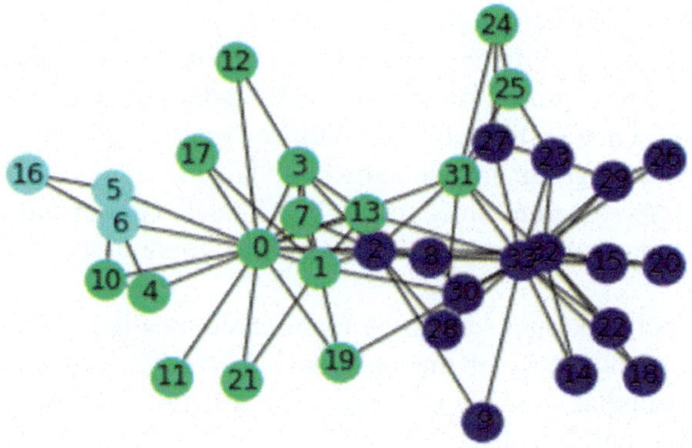

Fig. 5. Different communities (represented by different colors) detected using Label Propagation Algorithm

3 Experimental Setup

In this study, the representation of communities is in the form of a graph. A graph is a non-linear data structure comprising of vertices, also known as nodes, and edges that link any two nodes in the graph. To detect communities, four different methods were used, and the analysis was carried out on Zachary's karate club dataset.

3.1 Libraries Required

We employed the following libraries in our project.

Pandas: It is a Python package that provides tools for manipulating databases.

NetworkX: A software tool written in Python that enables the generation, handling, and investigation of intricate networks.

Community: This class has the objective of computing and measuring the community structure, and its functions are not directly imported into the main networkx namespace. To access these functions, one can import the networkx.algorithms.community module and then use the functions as attributes of the community object.

Matplotlib:This Python library is designed to generate visualizations that can be static, animated, or interactive and we have employed its pyplot submodule which helps in creating, editing figures.

3.2 Dataset

Zachary Karate Club dataset served as our project's dataset. Wayne Zachary (1977) gathered information from members of a university karate club.. Using this dataset and a network conflict resolution model, he was able to provide a rationale for the division of the group that occurred due to disagreements among the members. The karate club network comprises 34 members. Zachary documented the connections between pairs of club members who had interactions outside of the club. In terms of a network, Each member of the club can be depicted as a node, and each connection between two members can be represented as an edge. In the course of his research, a conflict emerged between the club's administrator and the instructor, which resulted in the club splitting into two separate factions. One faction formed a new club under the instructor, while the other faction either found a new instructor or discontinued practicing karate. Zachary was able to accurately predict the group memberships of all but one member of the club following the split based on his research. Since its use by Michelle Girvan and Mark Newman in 2002, the network has become a popular example of community structure in networks [12].

The 78-edge network data set can be represented as a list of integer pairs, where each integer corresponds to a club member, and a pair indicates that two members have interacted. It is a popular example of community structure and we have employed the same data set for community detection [13].

4 Results

4.1 Parameters Used

Table 1 shows the list of parameters used by each method to obtained results:

Table 1. Parameter used in algorithms

Algorithm	Parameters
Girvan-Newman	With the flexibility to choose the number of communities result are obtained. With 34 nodes so we can have the results of 4 communities or 6 communities • Edge-betweenness-centrality • Modularity
Greedy modularity	Main aim here is to maximise modularity • Modularity • Weight
Louvain	• Modularity • Weight
Label propagation	• Modularity • Number of edges • Number of nodes • Weight

4.2 Graph of Communities Obtained.

Graphs in Table 2 of distinct communities obtained in each method along with the specification of what all nodes comprises a community present in the graph and modularity of each algorithm:

Table 2. Community visualization

Algorithm	Community visualization
Girvan-Newman	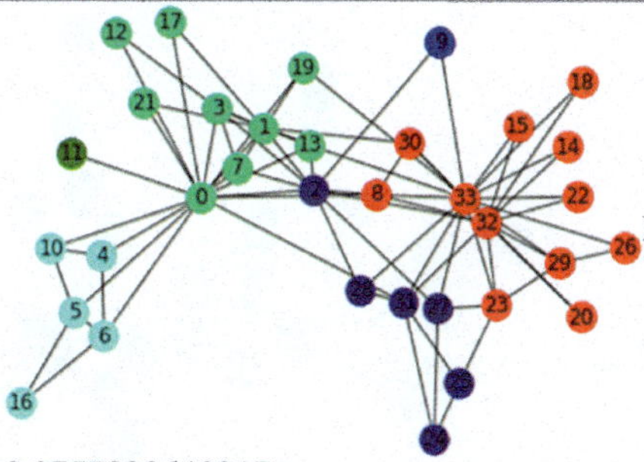 Modularity = 0.3757800640917 Total communities obtained: 5
Greedy modularity	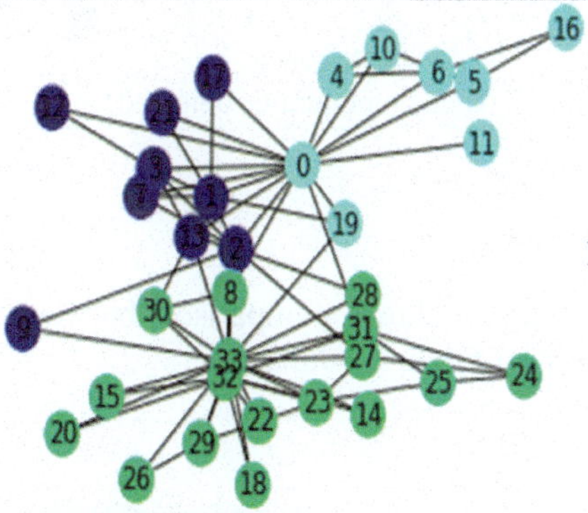 Modularity = 0.4109649369389 Total communities obtained: 3
Louvain	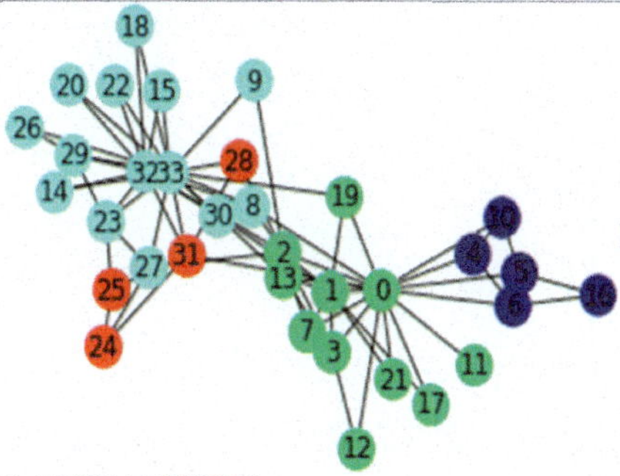 Modularity = 0.4438541256723 Total communities obtained: 4

(*continued*)

Table 2. (*continued*)

Algorithm	Community visualization
Label propagation	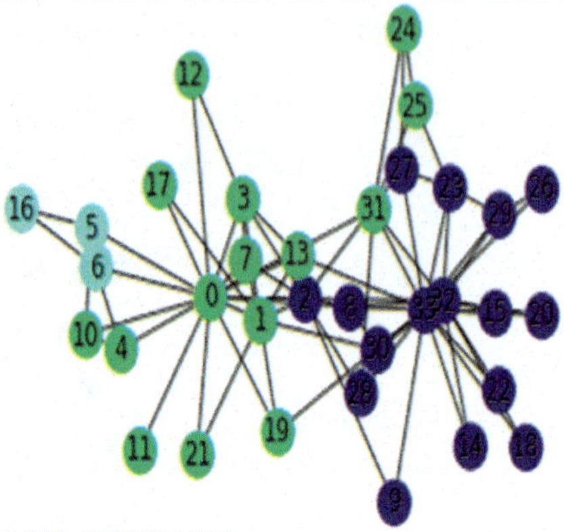 Modularity = 0.3094863289668 Total communities obtained: 3

4.3 Comparison Graph

Following is the graph that shows comparison between the different methods we implemented on the basis of their modularity:

See Fig. 6.

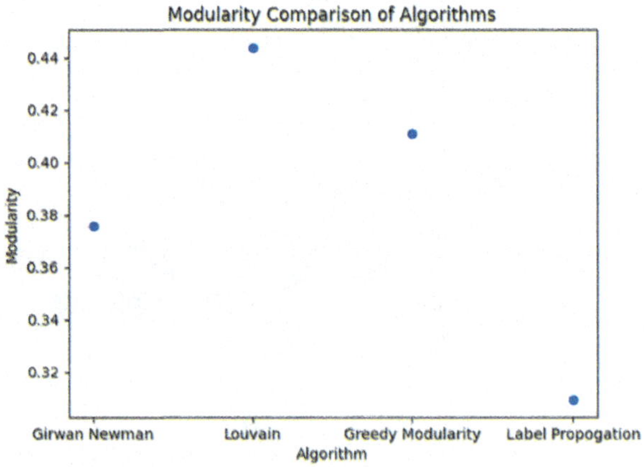

Fig. 6. Comparison of algorithms

5 Community Detection Algorithm for Online Learning Environments

Following section describes possibilities of usage of community detection algorithms for online learning environments.

1. To identify learners who are struggling with the same concepts. This information can be used to provide targeted support to learners who need it most.
2. To identify learners who are interested in the same topics. This information can be used to connect learners with each other and to create online communities where learners can discuss and learn from each other.
3. To identify learners who are at the same level of expertise. This information can be used to group learners together for collaborative learning activities.
4. To identify learners who are highly engaged and active. This information can be used to identify learners who are likely to benefit from additional support or resources.

From Tables 3 and 4, it may be inferred that community detection algorithms as a tool for improving the learning experience in online learning environments. By identifying communities of learners with similar interests and needs, we can provide targeted support, create opportunities for collaboration, and improve the overall learning experience.

Table 3. Overview of community detection algorithm for online learning

Approach	Algorithm	Benefits	Challenges
Divisive	Girvan-Newman	Finds overlapping communities of learners with similar interests	Can be computationally expensive for large networks
Agglomerative	Louvain	Finds non-overlapping communities of learners with similar interests	Can be less accurate than divisive methods
Modularity-based	Greedy Modularity Maximization	Finds communities of learners with high modularity	Can be sensitive to the initial partitioning of the network
Label propagation	Label Propagation	Simple to implement and scalable to large networks	Can be less accurate than other methods

Table 4. Illustration of community detection algorithm application in online learning

Algorithm	Application Scope	Example
Girvan-Newman algorithm	To identify overlapping communities of learners with similar interests For example, the algorithm	It can be used to identify learners who are interested in the same course, the same topic, or the same learning style
Louvain algorithm	To identify non-overlapping communities of learners with similar interests	It can be used to identify learners who are enrolled in the same course, who have the same learning goals, or who have the same level of expertise
Greedy modularity maximization algorithm	To identify communities of learners with high modularity. Modularity is a measure of how well the edges within a community are connected to each other compared to the edges between communities To identify communities of learners with high modularity	It can be helpful for identifying communities of learners who are highly engaged and active
Label propagation algorithm	To identify communities of learners with similar interests The algorithm may start by assigning each learner to a random community, and then iteratively assigns each learner to the community with the most learners that have the same label. This process continues until the labels of the learners no longer change	It can be a simple and effective way to identify communities of learners with similar interests, but it can be less accurate than other methods

6 Conclusion

Community detection algorithms can be a useful tool for improving the learning experience for students in India. By identifying communities of students with similar interests and needs, we can provide targeted support, create opportunities for collaboration, and improve the overall learning experience. Together with these benefits following 4 challenges that are common in Indian online learning environments (1) Community detection algorithms can be used to identify smaller groups of students within large classes making it easier for instructors to provide personalized support to students and to create a more engaging learning environment. (2) From Indian perspective, which is a country with a rich cultural diversity. Community detection algorithms can be used to identify students from similar cultural backgrounds. This can help students to feel more comfortable and supported in their learning environment. (3) Not all students in India have access to computers or the internet. Community detection algorithms can be used to identify students who do not have access to technology. (4) This information can be used to provide those students with the resources they need to participate in online learning.

References

1. Bedi, P., Sharma, C.: Community detection in social networks. Wiley Interdiscip. Rev.: Data Min. Knowl. Discov. **6**(3), 115–135 (2016). https://doi.org/10.1002/widm.1178
2. Boccaletti, S., et al.: Complex networks: structure and dynamics. Phys. Rep. **424**(4–5), 175–308 (2006). https://doi.org/10.1016/j.physrep.2005.10.009
3. Javed, M.A., Younis, M.S., Latif, S., Qadir, J., Baig, A.: Community detection in networks: a multidisciplinary review. J. Netw. Comput. Appl. **108**, 87–111 (2018). https://doi.org/10.1016/j.jnca.2018.02.011
4. Khawaja, F.R., Sheng, J., Wang, B., Memon, Y.: Uncovering hidden community structure in multi-layer networks. Appl. Sci. **11**(6), 2857 (2021). https://doi.org/10.3390/app11062857
5. He, K., Li, Y., Soundarajan, S., Hopcroft, J.E.: Hidden Community Detection in Social Networks. Inf. Sci. **425**, 128–147 (2018). https://doi.org/10.1016/j.ins.2017.10.014
6. Orman, G.K., Labatut, V.: A Comparison of Community Detection Algorithms on Artificial Networks. In International Conference on Discovery Science (pp. 242–256). Springer (2009). https://doi.org/10.1007/978-3-642-04747-3_20
7. Huang, X., Li, B., Zhou, Y.: Dynamic overlapping community detection via modularity optimization with flexible window size. Inf. Sci. **547**, 357–371 (2021). https://doi.org/10.1016/j.ins.2020.08.068
8. Lyu, C., Shi, Y., Sun, L.: A community detection method for social network based on community embedding. IEEE Access **9**, 43811–43822 (2021). https://doi.org/10.1109/ACCESS.2021.3067203
9. Yu, L., Chen, Y., Li, Y., Wang, X., Wu, Y.: Community detection in attributed graphs via multi-task learning. Knowl.-Based Syst. **238**, 107472 (2022). https://doi.org/10.1016/j.knosys.2021.107472
10. Pourhabibi, T., Ong, K.L., Boo, Y.L., Kam, B.H.: Detecting covert communities in multi-layer networks: a network embedding approach. Futur. Gener. Comput. Syst. **124**, 467–479 (2021). https://doi.org/10.1016/j.future.2021.06.027
11. Chen, M., Kuzmin, K., Szymanski, B.: Community detection via maximization of modularity and its variants. IEEE Transactions on Computational Social Systems **1**, 46–65 (2014). https://doi.org/10.1109/TCSS.2014.2307458
12. Zachary, W.: An information flow model for conflict and fission in small groups. J. Anthropol. Res. **33**, 452–473 (1977)
13. UCI Network Data Repository. (n.d.). Zachary's karate club network dataset. Retrieved from http://konect.cc/networks/ucidata-zachary/
14. Analytics Vidhya. (2020). Community Detection Algorithms: A Comprehensive Overview. Retrieved from https://www.analyticsvidhya.com/blog/2020/04/community-detection-graphs-networks

Probability Approximation Based Link Prediction Method for Online Social Network

Praveen Kumar Bhanodia[1](✉), Aditya Khamparia[3], Shaligram Prajapat[2], Babita Pandey[3], and Kamal Kumar Sethi[1]

[1] Computer Science Engineering, Acropolis Institute of Technology & Research, Indore, India
kumarpkb2@gmail.com

[2] Computer Science, IIPS DAVV, Indore, India

[3] Department of Computer Science Engineering, Baba Bhimrao Ambedkar Central University, Lucknow, India

Abstract. Social media is significantly contributing to information sharing between humans. Social networking sites Facebook, Twitter and LinkedIn are popular in connecting people across the world. For understanding and studying the behavior, social networks are represented through graphs referred as sociographs in which nodes are users and links between nodes are relationships. Social networks are dynamic as millions of new nodes are added to the networks making it large and complex to study and analyze. Predicting links in social network is a known computation problem wherein future link between the nodes is to be predicted. As to predict the links similarity between nodes deduced by calculating the similarity measure such that higher similarity measure value inferences existence of new link between nodes. In this work we are focused to detect and recognize the future links between the nodes by exploiting the node neighbourhood property. The proposed computation techniques compute the probability contribution measure for having link between the nodes.

Keywords: Link prediction · Probability · Nodes · Links · Dynamic social network

1 Introduction

1.1 A Subsection Sample

These days, data and information exchange between individuals has converged on online social networks due to the rapid development of internet technology. There are many formal and popular online social networking sites, such as Facebook, LinkedIn, Twitter, Flickr, and Google+. Almost every individual is connected to each other through these network systems, whether professionally or personally. With millions of users, these social networks are highly dense and complex, and the flow of information across them is very fast. The prediction of future links between non-connected nodes will make it more efficient. The link prediction problem is a known computation problem in the social network. The link prediction problem is of two types: the first is the prediction

N. Naik et al. (Eds.): UKCI 2023, AISC 1453, pp. 612–624, 2024.
https://doi.org/10.1007/978-3-031-47508-5_47

of missing links [1], and the second is the prediction of future links between the nodes. Missing link prediction may be limited to the analysis of previous and static networks only, whereas future link prediction involves the analysis of evolving social networks on a timeline. Therefore, link prediction uses the network data at different time instants, t1, to identify possible future links, validating them with the network data at the next time instant, t2, such that t1 $<$ t2. In this work, we focus on future link prediction problems in growing online social networks. For identification and prediction of future links, node properties have been used.

The study of the evolution of networks has revealed insights across various domains like chemistry, biology, physics, and sociology [3–5]. Analysis of these large online social networks has influenced the technology of evolving social networks and changed our perspective on life. The virtual model of society has provided benefits by enhancing the work efficiency of individuals and creating new opportunities. The other way around, it has badly affected the natural phenomena of human societies like fake news spreading, antisocial connections, virus propagation, etc. [6, 7]. Therefore, studying and analyzing the structural behavior of large and complex online social networks along with their properties is of utmost importance.

In this paper, we have proposed an algorithm that measures the similarity between the nodes across the network; the proposed approach approximates the probability contribution using the mutual neighbourhood information of the node pair to predict a potential link between them. The probability contribution computed is directly proportional to the number of mutual neighbours of the selected node pair. Suppose the pair of nodes have more mutual neighbours' nodes in the network, then it is assumed that the probability of having a potential link between them will be high. The proposed algorithm has been tested over real-world network data sets. It is observed that the method significantly contributes in predicting future links in a large social network.

In online social network link prediction techniques, usually approximate similarity between nodes by using the properties of the nodes. In [2], Liben-Nowell and Kleinberg introduced techniques that exploit the structural topology of the social network. According to them, ranks were assigned to the node pairs based on the similarity measure computed, and the probability of having a link between the nodes will be high for the higher-order ranks of the node pairs. Hasan et al. [20] have further extended this and exploited the topological features for link prediction in social networks. Adamic and Adar proposed a method that computes similarity between the nodes by taking mutual neighbours into account, referred to as the Adamic-Adar (AA) Index [8]. Exploiting the hierarchy of the networks, Newman has also proposed an algorithm [9]. Zhou et al. have proposed two similarity indices, namely Resource Allocation and Local Path [10].

Community detection is another issue in the study of large social networks. While working on link prediction problems, mostly local information has been exploited. There are some nodes in the graph that are part of community structure, so addressing the issue of community detection in the large social network may also give significant information about the nature of the network. In [11], Girvan and Newman introduced a modularity-based community detection algorithm exploiting the topological structural properties of the network. Newman also explored collaboration networks specifically for biology and physics for solving link prediction problems [15]. In collaboration networks, two authors are supposed to be linked if they have at least one publication together. Here, perhaps the number of publications will increase the likelihood of link prediction between the node pair.

The existing proposed methodologies dominate in simplicity and genericity, and the methods exploit the local entity attributes. Contrarily, these methods are usually intricate and include excessive parameters. Further online social network evolution research has increased after the link prediction methods proposed by Barbasi and Albert [21], The work proposed by Kleinberg in [22], and some scientific reports in [23]. Kleinberg [22] demonstrated and exploited the properties of social networks, such as the small world phenomenon and the power law distributions [21], during the evolution of the links of the network. The relative comparison in between says that the link prediction model exploits the entire structural properties of the network, whereas the evolutionary model considers local properties of the network for the approximation of missing links. In this work, we emphasized the prediction of future links between the selected node pairs. The exponential growth of online social networks like Facebook and Twitter has posed various challenges in solving the link prediction problem dynamically. Therefore, existing techniques and methods seem effective and efficient when implemented on today's large and complex online social networks.

We have organized the paper as follows: Sect. 2 illustrates the related work done in link prediction. Section 3 describes the proposed method for link prediction, followed by Sect. 4, which discusses the experimental analysis performed. Section 5 and Sect. 6 are the results of the discussion and conclusion, respectively.

2 Link Prediction

Formally the problem of link prediction can be defined as; In a sociography = (V,E), where each edge e = {u,v} belongs to E shows relationship between u and v at an instant of time t. Besides the nodes and links in the social network, local node and link attributes can also be obtained from the networks structure analysis. Thus each node u has list of node attributes A = {a1,a2,a3,a4,.....an}. The similarity of two nodes across the network is calculated with the help of these attributes from the list.

In a link prediction model, between a time stint; [t0,t0'] and [t1,t1'] such that t1 > t0' a subgraph G'{u'v'} is extracted. In this subgraph 'z' is the feature vector between node pair(u'v') having no relationship link between them at [t0,t0']. The label y = [0,1] belongs to the link to be predicted between node pair (u'v') based on the computed feature vector z of the node pair. if y = 0, represents no link between the nodes; y = 1 represents link between the nodes.

2.1 Link Prediction Techniques

Link prediction techniques can be categorized as 1. Similarity based techniques, 2. Maximum likelihood Algorithms and 3. Probabilistic models. According to Lu et al. the second and third category techniques are quite complex in terms of time for large complex networks [12]. These days online social networks are large, complex, and dynamic in nature. The similarity-based techniques can further divide in to local based (node attribute based) and global based (generally paths are used).

Common neighbours (CN)
It quantifies the similarity feature by measuring the total number of mutual nodes between the selected node pair u and v [15].

$$\mathrm{CN}_{xy} = |\Gamma(x) \cap \Gamma(y)| \tag{1}$$

Jaccard Coefficient
It computes the similarity by considering the mutual neighbours of the selected node pair between which link is to be predicted [13]. The formulae used to measure are.

$$\mathrm{JC}_{\mathrm{xy}} = \frac{|\Gamma(\mathrm{x}) \cap \Gamma(\mathrm{y})|}{|\Gamma(\mathrm{x}) \cup \Gamma(\mathrm{y})|} \tag{2}$$

Adamic Adar
It computes the similarity by characterizing the neighbours which are overlapping between the nodes pair. Assign more weights to node pairs having a smaller number of mutual neighbour nodes [14].

$$\mathrm{AA}_{\mathrm{xy}} = \sum_{\mathrm{z}\in(\mathrm{x})\cap\Gamma(\mathrm{y})}^{\infty} \frac{1}{\log|\mathrm{N}(\mathrm{z})|} \tag{3}$$

Average path weight
Measures the sum of minimum weights on the directed path between the selected node pair divided by total number of possible paths between the node pair such that the selected path length l <= 3; it is to optimize the size of the network and complexity of the method.

$$Avg_{Path(u,v)} = \frac{1}{n.(n-1)} \tag{4}$$

Preferential Attachment
The approach considers the product of node degrees of both the nodes between which the link is to be predicted. The measure obtained is directly proportional to the probability of having a link between them [15].

$$PA_{(x,y)} = \Gamma(X) \times \Gamma(Y) \tag{5}$$

Resource Allocation
In order to measure it considers the amount of resource equally shared amongst the peer neighbours that are connected with the individual nodes [16].

$$RA = \sum_{z\in(x)\cap(y)}^{\infty} \frac{1}{N(z)} \tag{6}$$

Salton Index
Measures the number of common neighbours relative to their geometric mean [13]

$$Sore.Ind(u, v) = \frac{|\Gamma(u) \cap \Gamma(v)|}{\sqrt{k_u k_v}} \tag{7}$$

Sorenson Index
Measure the number of common neighbours relative to their arithmetic mean. This index is similar to J, however J counts the number of unique nodes in the shared neighbourhood. The index was previously used to establish equal amplitude groups in plant sociology based on the similarity of species [17].

$$Sore.Ind(u, v) = \frac{|\Gamma(u) \cap \Gamma(v)|}{k_u + k_v} \tag{8}$$

3 Methods

Thorough study of online social network evolution has inferred that the growth of the social network often displays certain patterns to understand. The attributes of nodes and links contribute significantly in forecasting the estimation of new links between the nodes across the network. The proposed model of link prediction is demonstrated in Fig. 1 In our method we will exploit the node neighbourhood properties to compute the probability contribution in prediction of link between the nodes.

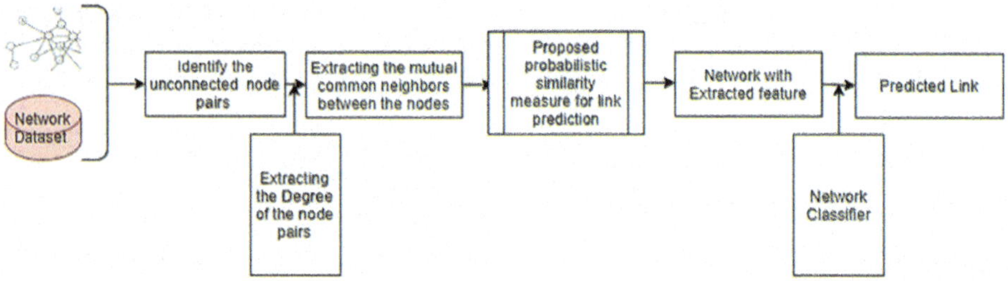

Fig. 1. Process to predict future links between the node pairs in a large and complex online social network

Probabilistic Similarity Measure Algorithm
According to the method, assume that the probability contribution of having a link between node pair is computed as.

Let '*p*' be the probability of two nodes getting connected when having one common neighbour node. Then the probability of not getting connected in the future with the network mathematically will be computed as:

$$1 - p \tag{9}$$

As we are approximating the probability contribution for having a link between node pairs where there are n number of common nodes, therefore, for 'n' number of common nodes, the above mathematical equation may be written as:

$$(1-p)^n \tag{10}$$

Thus, the probability of having a link between the node pairs with n number of neighbour nodes is determined by subtracting the equation no. 10 from 1, so the equation of probability similarity measure for prediction of a link between the node pairs with n number of neighbor nodes across the network can be written as follows:

$$Prob_{sim}.(u,v) = 1-(1-p)^n \tag{11}$$

where 'n' is the number of common neighbour nodes between the node pair u and v in an online social network.

Algorithm:
x, y: Nodes in the network
p: probability similarity
1. *Input nodes u and v between which link is to be predicted, such that there is no link in between.*
2. *For each node pair*
I. Compute the degree of node x and node y
3. *For each node pair where node x! = y*
II. Compute 'n', the common neighbour nodes between nodes x-y in the network.
*III. Compute: prob_sim=1- (1-p)*n*
4. *Initialize the value for p such that it should lie between the range 1 to 0.*
5. *If prob_sim> threshold value*
IV. Predict link between x and y

4 Methods

The proposed algorithm has extracted the node features used for further approximation of the potential link between the node pair across the entire online social network. the proposed algorithm is implemented using python to test and verify the process of link prediction between the node pairs in an online social network. typically, a thousand-node undirected social network has been created, such that the network will dynamically evolve based on random similarity calculated by exploiting the node degrees of the unconnected node pairs. to further classify the possibility of a future link network classifier built using supervised machine learning techniques.

4.1 Result and Discussion

Proposed method is tested on real time online social network dataset downloaded from https://snap.stanford.edu/data/. [24].

Deezer
This network is developed for the users of the European region; it demonstrates the mutual relationship between the citizens. The node feature is basically fetched based on likes given by users to the artists. The network graph can be exploited for binary node classification where the gender of the user is to be predicted and the feature is derived from the respective names of the user nodes [18].

Facebook
The network set used is a page-by-page graph of Facebook pages that have been validated. Facebook pages are nodes, whereas reciprocated likes between sites are connections. The site descriptions that the page owners produced to explain the site's purpose are used to extract node characteristics. This graph was compiled in November 2017 using the Facebook Graph API and was limited to pages from four Facebook-defined categories. Politicians, governmental bodies, television shows, and businesses fall within these categories. Multi-class node classification for the four site types is the job associated with this dataset [19].

GitHub
In June 2019, a significant social network of GitHub developers was gathered using the public API. Edges are mutual follower relationships between nodes and edges are developers who have starred at least 10 repositories. The location, starred repositories, employer, and e-mail address are used to extract the vertex features. The graph task is binary node classification, which requires determining if the GitHub user is a web or machine learning developer. This target characteristic was developed from each user's job title [19].

Twitch
Twitch user-user networks of gamers who broadcast in a certain language are the datasets utilised for node classifications and transfer learning. The users are the nodes, and the ties between them are mutual friendships. The games played and loved, as well as location and streaming habits, are used to extract vertex characteristics. Transfer learning across networks is achievable because datasets have the same set of node attributes. These social networks were gathered in May of this year. Binary node classification is a supervised problem connected to these networks in which one must predict if a streamer uses explicit language [19]. Table 1 summarizes the statistics from the detailed datasets.

Table 1. Online social network dataset description and statistics

Dataset Properties	Twitch Social Network	Facebook social network	Github Social Network	Deezer Social Network
Nodes	4385	22470	37700	28281
Edges	37304	171002	289003	92752
Density	.004	0.001	0.001	0.0001

The selected online social networks are equally dense except twitch which is slightly denser as compared to others.

4.2 Experimental Setup

The proposed technique is implemented using Python. Initially A hundred social networks have been created with random edges between the nodes just to understand the evolution of the network with respect to time. The local properties of the network are processed repetitively, leading to the calculation of the similarity measure between the node pairs. The probabilistic link prediction technique is applied to four different online social networks for further evaluation on the scale of accuracy, precision, and recall. The obtained network dataset is then applied with machine learning classification techniques. Weka 3.9 is used for further classification of legitimate links predicted between the node pairs. Weka is a term coined by the University of Waikato for an efficient and effective tool for machine learning and data mining functions like classification, pattern extraction, and visualization.

The observed experimental developments in the network have been demonstrated in Figs. 2(a) and (b). Figure 2(a) illustrates the social network obtained at time 't, and Fig. 2(b) is the illustration of the subsequent network, which is dynamically generated by calculating the probability similarity measure between unconnected node pairs u and v. It is visible from the relative comparison that new links have been formed between nodes across the entire network.

The similarity between the nodes is identified by using the probability similarity algorithm described in the above section. The obtained probability similarity measures are tabulated in Table 2. The threshold value for the probability contribution '*p*' is set in the range of 0 to 1. The higher the probability contribution, the higher the probability of having a link between the node pair.

Experimental analysis was carried out for evaluation of proposed technique with state-of-the-art link prediction techniques. Classifier built using machine learning methods trained with the social networking data set along with extracted features from node properties. The link prediction observations obtained by applying the algorithm on Twitch, Facebook and Deezer online social networks are tabulated below. For better understanding of the proposed algorithm, it has been applied all the three different datasets. Table 3 shows the comparative analysis of proposed similarity algorithm Twitch online social network. Similarly, Table 4, Table 5 and Table 6 are for Facebook, Deezer and GitHub online social networks respectively.

The performance of the probability-based algorithm while processing the Twitch online social network supersedes with over 98% precision compared to conventional link prediction techniques (Admic-Adar, Jaccard, Preferential attachment, and Resource allocation). Though the feature exploited for computation of similarity measures is similar for all the techniques, it is the mutual neighbor node. The sensitivity is also reasonably high compared to JC and PA; with this, it is deduced that the true positivity rate is quite significant. The high value of the F measure justifies the level of precision and TPR.

In the same fashion, the proposed algorithm exhibits similar characteristics while processing the Facebook social network dataset (refer to Table 4). The accuracy here is also high, with a precision of 98.5% and a recall of 98%. The high trade-off between

Table 2. Probability similarity obtained between the node pairs for future link prediction.

Node u	Node v	Prob_similarity
4	56	0
4	57	0
4	58	0
4	59	0.01
4	60	0
4	61	0.01
4	62	0
4	63	0.01
4	64	0.01
4	65	0.01
4	66	0
4	67	0.01

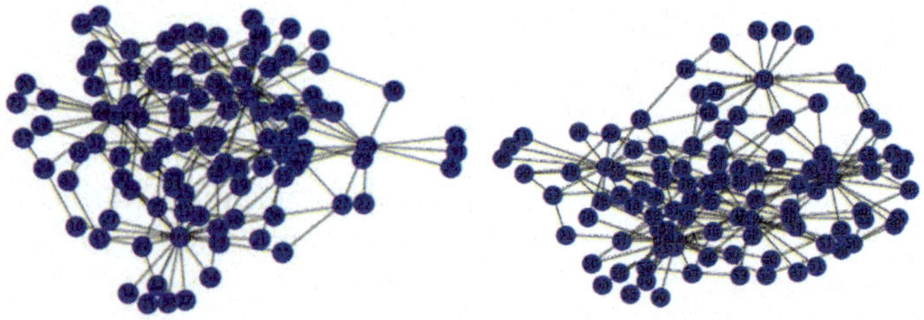

Fig. 2. **a** Social network generated at time 't'; **b** the network obtained at time 't + 1' after applying probability-based technique

Table 3. Performance analysis of algorithm on Twitch online social network

Method	Precision	Recall	F-measure
AA	0.975	0.985	0.985
JC	0.969	0.979	0.979
PA	0.954	0.951	0.948
RA	0.98	0.981	0.981
Proposed Algorithm	0.982	0.981	0.982

precision and sensitivity means the model is predicting genuine links correctly and efficiently.

Table 4. Performance analysis of algorithm on Facebook online social network

Method	Precision	Recall	F-measure
AA	0.956	0.953	0.95
JC	1	1	1
PA	0.959	0.957	0.953
RA	0.976	0.975	0.974
Proposed Algorithm	0.985	0.98	0.981

Table 5. Performance analysis of algorithm on Deezer online social network

Method	Precision	Recall	F-measure
AA	0.943	0.938	0.935
JC	1	1	1
PA	0.982	0.982	0.982
RA	0.953	0.95	0.949
Proposed algorithm	0.989	0.989	0.987

On the Deezer network (Table 5), the algorithm generates better results than on Twitch, Facebook, and Github networks. On this network dataset, the precision of link prediction is close to 99% (exactly 98.9%); the true positivity rate is also very high (98.9%), which shows that the prediction classification of genuine links between nodes is significantly high. The tradeoff between precision and sensitivity is also very high, as shown by the F-measure (98.7%).

Table 6. Performance analysis of algorithm on Github online social network

Method	Precision	Recall	F-measure
AA	0.946	0.944	0.945
JC	1	1	1
PA	0.98	0.977	0.978
RA	0.979	0.979	0.979
Proposed Algorithm	0.985	0.982	0.98

The proposed probability-based technique has significant results on Github social network for prediction of future links between nodes. It is 98.5% precise compared to other existing techniques. Sensitivity representing true positive rate and F-measure representing precision and recall tradeoff are 98.2% and 98% respectively (refer Table 6).

According to the results obtained it is clearly visible that the probabilistic similarity technique is contributing good in comparison with the other state of the art link prediction techniques. The results obtained for Jaccard coefficient it is quite unrealistic to obtain nearly 100% result, this may be due to the size of the dataset and the features extracted. The most of the similarity feature obtained for JC are similar for most of the node pairs. Therefore, in this respect we can ignore it.

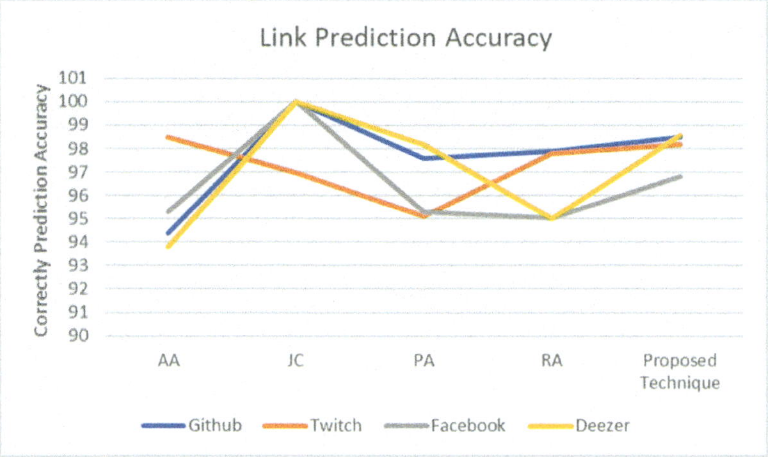

Fig. 3. Relative comparison of correctly classified link prediction accuracy across Facebook, Twitch, Deezer and GitHub networks.

5 Conclusion

The relative comparison of correctly predicted the future links between the node pairs in the social network has shown that the proposed probabilistic method is significantly contributing in addressing the link prediction problem. Figure 3 demonstrates the accuracy analysis of the proposed method with the state-of-the-art link prediction techniques across the selected online social networks: Facebook, Twitch, Deezer and GitHub.

The main objective of this work was an effort in the direction of proposing a new similarity measure which uses the probability contribution for identification and prediction of links between the node pairs in evolution of online social network. This paper answers a certain question: Do the probability contribution similarity measure contributes significantly to solving the link prediction problem? How does this link prediction technique perform in comparison with classical techniques? Use of similarity metrics for approximation of future association are the majorly accepted techniques in addressing the link prediction problem. The proposed probabilistic similarity measure considers the mutual neighbour node properties of node pair in a social network.

As link prediction is a binary classification problem where possibility of a legitimate link between nodes the nodes is being obtained, therefore supervised machine learning framework is used to classify whether an interaction is there or not. The classifier will be learning the probability similarity feature extracted to understand the significance of the probabilistic similarity measure. Standard online social network datasets have been used for the experimental analysis of the technique. It is evaluated using the standard performance parameter confusion metrics. The obtained results have shown that the proposed probabilistic similarity technique is significantly contributing in prediction of future links and has performed better. Moreover, the technique is performing better in evolution or online social network dynamically i.e. with respect to time. In future the amalgamation of similarity measures technique and machine learning techniques can be explored extensively for link prediction in social networks.

References

1. Narang, K., Kristina, L., Ponnurangam, K.: Network flows and the link prediction problem. In: Proceedings of the 7th Workshop on Social Network Mining and Analysis (p. 3). ACM (2013)
2. Liben-Nowell, D., Kleinberg, J.: The link prediction problem for social networks. In: Twelfth International Conference on Information and Knowledge Management ACM (2003)
3. Ozkanlar, A., Clark, A.E.: Chem networks: a complex network analysis tool for chemical systems. J. Comput. Chem. **35**(6), 495–505 (2014)
4. Jeong, H., et al.: The large-scale organization of metabolic networks. Nature **407**(6804), 651–654 (2000)
5. Rapoport, A., Horvath, W.J.: A study of a large sociogram. Behav. Sci. **6**(4), 279–291 (1961)
6. Serazzi, G., Zanero, S.: Computer virus propagation models. In Tutorials of the 11th IEEE/ACM International Symposium on Modeling Analysis and Simulation of Computer and Telecommunications Systems (MASCOTS) (pp. 26–50) 03 2965 (2010)
7. Albert, R.I., Albert, Nakarado, G. L.: Structural vulnerability of the North American power grid. Phys. Rev. E Stat. Nonlinear & Soft Matter, Phys. **69**(2), 292–313 (2004)
8. Adamic, L.A., Adar, E.: Friends and neighbours on the Web. Foreign Aff. **25**(3), 211–230 (2003)
9. Clauset, A., Moore, C., and Newman, M.E.: Hierarchical structure and the prediction of missing links in networks. Nature **453**(17191), 98–101 (2008)
10. Zhou, T., Lo, L., Zhang, Y.C.: Predicting missing links via local information. Phys. Condens. Matter **71**(4), 623–630 (2009)
11. Newman, M.E., Girvan, M.: Finding and evaluating community structure in networks. Phys. Rev. E: Stat., Nonlin, Soft Matter Phys. **69**(2 Pt 2), 026113 (2004)
12. Lu, Z., Savas, B., Tang, W., Dhillon, I.: Supervised link prediction using multiple sources. In: IEEE 10th International Conference on Data Mining (ICDM) (pp. 923–928). IEEE (2010)
13. Salton, G., McGill, M.: Introduction to modern information retrieval (1986)
14. Krebs, V.E.: Mapping networks of terrorist cells. Connections **24**(3), 43–52 (2002)
15. Newman, M.E.J.: Clustering and preferential attachment in growing networks. Phys. Rev. E **64**, 025102 (2001)
16. Zhou, T., Lu, T., Zhang, Y.: Predicting missing links via local information. Eur. Phys. J. B-Condens. Matter Complex Syst. **71**, 623–630 (2009)
17. Sørensen, T.: A method of establishing groups of equal amplitude in plant sociology based on similarity of species and its application to analyses of the vegetation on danish commons. Biol. Skr. **5**, 1–34 (1948)
18. Rozemberczki, B., Sarkar, R.: Characteristic functions on graphs: birds of a feather, from statistical descriptors to parametric models. CIKM (2020) https://arxiv.org/abs/2005.07959

19. Rozemberczki, B., Allen, C., and Sarkar, R.: Multi-scale Attributed Node Embedding. arXiv (2019)
20. Hasan, M. A., Chaoji, V., Salem, S., Zaki M.: Link prediction using supervised learning. In: Proc. of SDM 06 Workshop on Link Analysis, Counterterrorism and Security (2006)
21. Barabasi, A.-L., Albert, R.: Emergence of scaling in random networks. Science **286** (5439), 509–512 (1999). https://doi.org/10.1126/science.286.5439.509, http://science.sciencemag.org/content/286/5439/509, arXiv:http://science.sciencemag.org/content/286/5439/509.full.pdf
22. Kleinberg J.M.: Navigation in a small world. Nature **406**(6798), 845 (2000)
23. Zhang, Q.M., Xu X.-K,. Zhu, Y.-X, Zhou, T.: Measuring multiple evolution mechanisms of complex networks. Sci. Rep. 5, 10350 (2015). https://doi.org/10.1038/srep10350arXiv:1410.3519
24. https://snap.stanford.edu/data/

Predicting the Popularity of YouTube Videos: A Data-Driven Approach

Alaa Aljamea[1,2](✉) and Xiao-Jun Zeng[1]

[1] Department of Computer Science, University of Manchester, Manchester M13 9PL, UK
x.zeng@manchester.ac.uk

[2] Department of Computer Science, CCSIT, King Faisal University, AlAhsa 31982, South Africa
alaa.aljamea@postgrad.manchester.ac.uk

Abstract. This paper presents a data-driven approach to predict the popularity of YouTube videos. The existing conducted studies focus on identifying whether the video is trending currently or after a month of publishing. Where this study aims to leverage a comprehensive dataset of YouTube videos, analyze various features and employ machine-learning techniques to forecast next-hour video popularity accurately. By considering factors such as video metadata and engagement metrics, we develop predictive models that can assist content creators and marketers in understanding the likelihood of a video's success on the platform. The results demonstrate the efficacy of our approach and provide practical insights into predicting YouTube video popularity.

Keywords: YouTube · Popularity prediction · Content analysis · Engagement metrics · Feature selection · Regression model · Time series analysis

1 Introduction

The widespread popularity of YouTube as a platform for sharing and consuming video content has sparked significant interest in understanding the factors that contribute to the success and popularity of videos [7,12]. By predicting video popularity, content creators, marketers, and platform administrators can gain valuable insights and make informed decisions to optimize video visibility and maximize audience reach [23]. Predicting the popularity of YouTube videos is a challenging task due to the complex interplay of various factors such as video content, metadata, and user engagement [28]. A couple of studies conducting to help in this challenge.

The conducted study [25] proposes the utilization of Support Vector Regression with Gaussian Radial Basis Functions (SVR with RBF) to predict the popularity of online video content, specifically measured in terms of the number of user views. This research focuses on examining visual and social features, as well

N. Naik et al. (Eds.): UKCI 2023, AISC 1453, pp. 625–639, 2024.
https://doi.org/10.1007/978-3-031-47508-5_48

as video view numbers by individual users. However, this study miss includes the video topic and the sentiment of the social interactions. Study [5] presents an approach for early-stage prediction of YouTube music video popularity by utilizing data available at the time of video upload. The method involves employing a supervised learning technique and an ensemble model to build a customized prediction model specifically designed for video popularity. The study demonstrates its strength in accurately predicting views for new videos. However, its limitation lies in the inability to forecast video popularity over continuous time periods. This may restrict the practical value of the study for content creators and marketers in determining the viability of ongoing video content, as well as its potential for targeted marketing purposes. To further enhance video popularity prediction, the researchers suggest incorporating sentiment analysis of text features and conducting the study across different YouTube video categories. The study [26], utilizes 13 YouTube features and employs a multivariate linear model to predict the video view number after a duration of 30 days from the video's collection. This study tries to enhance the view popularity prediction by removing the outliers values. Study [21] conducted research on 11 YouTube video features to predict video popularity using the XGBoost algorithm. In the next study the study [21] researchers aim to incorporate sound features and geographic location in their future studies to enhance prediction accuracy. Additionally, they plan to conduct predictions on a classified dataset On the other hand, study [11] analyzed YouTube videos from seven countries, considering 14 video features. Feature selection algorithms were applied, and videos were classified into binary categories as either trendy or not. The study suggested utilizing metadata and tags features to identify video trends. Additionally, it proposed enhancing predictions by determining the correlation between the view feature and other popular YouTube video features that were collected.

Despite the existing body of research on predicting YouTube video popularity, there is still a need for advancements in this field. Previous studies have often focused on a limited number of YouTube video features, lacking a comprehensive analysis that considers the interplay of various features and provides a broader understanding of the most influential factors affecting video popularity. Furthermore, there is a need for further investigation into the temporal dynamics and trends of video popularity, especially concerning next-hour view prediction. Where existing study determines whether a video will continue being trendy in the coming month's time or not.

In this paper, we propose a data-driven approach to predict the popularity of YouTube videos. Our study aims to explore the relationships between video features and their influence on viewer engagement, leveraging a large dataset of YouTube videos. By analyzing and extracting insights from this dataset, we seek to develop predictive models that can accurately forecast the popularity of videos in the next hour. To accomplish our goals, we employ advanced machine learning techniques such as feature selection, regression modelling, and ensemble methods. We utilize a comprehensive set of video features including metadata, social engagement indicators, temporal dynamics to capture the multifaceted nature of video popularity on YouTube, video age science video published and video comments polarity and score.

The contributions of this paper lie in both the theoretical and practical domains. From a theoretical perspective, we extend the understanding of the factors that contribute to the popularity of YouTube videos in the next-hour and shed light on their interdependencies. From a practical standpoint, our proposed data-driven approach provides insights for content creators, marketers, and platform administrators to enhance the visibility and reach of their videos, thereby increasing the chances of attracting a larger audience. We evaluate the effectiveness of our proposed approach using a large-scale dataset and compare the performance of our predictive models against existing methods. The experimental results demonstrate the superiority of our approach in accurately predicting the popularity of YouTube videos in the next-hour time, underscoring its potential as a valuable tool for video content optimization.

The rest of the paper is organized as follows: Sect. 2 presents the proposed next-hour video forecasting model, Sect. 3 reports the experiment results and finally, Sect. 4 expresses the conclusions.

2 Proposed Methodology

This research offers an AI-based approach to predicting a YouTube video's popularity across various video categories in the UK. The overall operation of the suggested methodology is shown in Fig. 1.

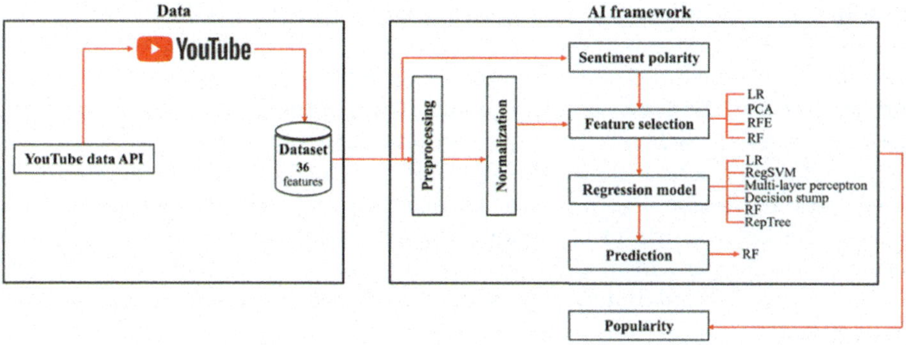

Fig. 1. The overall framework of the proposed methodology.

2.1 The Data

This study utilizes the YouTube Application Programming Interface (API[1]) to gather a dataset comprising approximately 140 videos from the top 10 most

[1] https://developers.google.com/youtube/v3/docs.

popular videos in 14 distinct YouTube video categories. Additionally, the dataset includes the top 10 related videos associated with each of the aforementioned categories, obtained through YouTube recommendations. The selected categories encompass Film & Animation, Autos & Vehicles, Music, Pets & Animals, Sports, Travel & Events, Gaming, People & Blogs, Comedy, Entertainment, News & Politics, How-To & Style, Education, and Science & Technology. Consequently, the collection process yields a total of 1,400 related videos.

Data collection took place in March 2023, with hourly intervals to ensure a representative sample. This resulted in a comprehensive dataset comprising a total of 653,800 videos. The collected video meta-data encompasses various aspects, including video-specific details such as the video ID, category ID, category name, publication date, title, description, channel title, channel ID, channel view count, channel subscriber count, tags, video language, audio language, duration, dimension, definition, caption availability, transcript source, transcript language, transcript language code, transcript generation details, available transcript translation languages, transcript text, and duration, projection type, region restriction details (allowed and blocked regions), rating and comment disablement status.

Furthermore, the dataset incorporates video engagement metrics, including the number of views, likes, dislikes, and comments received. Notably, these data points are collected on an hourly basis for each video and its associated related videos.

2.2 AI Framework

In addition to the data collected from the YouTube Data API, several derived features have been computed for analysis in this study. One of these features is the video age, which represents the time elapsed in seconds between the video's publication and the data collection point. The sentiment polarity of comments has also been assessed, categorizing them into positive, negative, or neutral sentiments. Furthermore, a sentiment score for the video has been generated using the logistic regression algorithm and incorporated into the dataset.

The dataset consists of 36 distinct features, which may contain redundant, outlier, or irrelevant features. To enhance the experimental results, a preprocessing step has been employed to convert the dataset into a machine-readable format using standard normalization techniques. Subsequently, unsupervised machine learning algorithms have been applied to eliminate redundancy and irrelevant features value that is approaching zero, while the interquartile range algorithm has been utilized to identify and remove outliers.

In order to investigate the correlation between the number of views and other significant video features and eliminate the least important feature for views number prediction, linear and non-linear feature selection algorithms have been employed. Furthermore, different regression algorithms have been utilized to identify a model for view number prediction. These regression models include Linear Regression, Support Vector Machine for Regression (RegSVR), Multilayer Perceptron, Decision Stump, Random Forest, and Reduced Error Pruning Tree (RepTree). The performance of each model has been evaluated using the Mean Absolute Percentage Error (MAPE) and Root Mean Squared Error (RMSE) metric. Subsequently, the best-performing regression model has been selected to predict the number of views for the next hour.

2.2.1 Data Preprocessing In order to prepare the dataset for the experiment and to enhance the experimental results, the dataset needs to be converted into a format that can be effectively processed by machine learning algorithms. Standard normalization techniques were applied to ensure consistent scaling of the data. Hence, within the dataset, the features exhibit varying ranges of values, with certain measures in the hundreds, such as the like/dislike count, while others reach tens of thousands, such as the view count [11]. To mitigate potential prediction inaccuracies stemming from this discrepancy, a data normalization technique is employed to standardize the data within a range of 0 to 1. This normalization process is achieved through the utilization of the Min-Max method, which scales the feature values accordingly. By applying this normalization technique, the disparate scales of the features are harmonized, facilitating fair and meaningful comparisons and analysis.

$$f(x) = (\ x - min)/(max - min) \tag{1}$$

Subsequently, unsupervised machine learning algorithms were utilized to remove redundant, small features close to zero, focusing on retaining only the most informative ones for the specific task. This improves computational efficiency. Additionally, the interquartile range algorithm was employed to detect and eliminate any outliers present in the data. By addressing outliers, we aimed to enhance the robustness and reliability of the subsequent analysis. Through these preprocessing steps, the dataset was appropriately transformed, irrelevant features were eliminated, and outliers were handled, establishing a solid foundation for more accurate and reliable analysis in the following stages of the experiment.

2.2.2 Comment Sentiment Polarity and Score Following the comprehensive collection of video comments, encompassing a substantial number of up to 8000 comments for the most popular videos, it is evidenced that highly popular videos tend to attract a greater volume of comments compared to others [14]. In order to ascertain the sentiment polarity and corresponding scores of these comments, a logistic regression model is employed. Prior to applying the model, a thorough cleaning process is conducted to eliminate irrelevant comments, advertisements, and other forms of noise. Subsequently, the model is utilized to discern the sentiment of the comments, categorizing them as positive, negative, or neutral, while also assigning a score ranging between 0 and 1.

2.2.3 Feature Selection Techniques Feature selection plays a crucial role in machine learning by controlling the number of features to mitigate over-fitting and enhance the predictive capability of the model on the training dataset [16]. The dataset comprises 36 features, necessitating the need for feature selection to identify important video features that will facilitate predicting the number of video views in the coming hour. The importance of feature selection lies in the fact that a model trained on a reduced set of features demonstrates improved robustness and reproducibility compared to the model built on the original, larger feature set [16]. In the process of feature selection, the focus is primarily on identifying relevant features or identifying correlations among features. Features that do not contribute valuable information are considered irrelevant, while redundant features are those that do not offer additional information beyond what is already captured by the selected features [18]. Noise, which refers to features that lack association or correlation with the target class variables, introduces bias and hampers predictive performance. Therefore, managing noise becomes crucial to enhance prediction accuracy, and this can be achieved through dimensionality reduction [16]. Dimensionality reduction can be accomplished through feature extraction or feature selection techniques [8]. However, different feature selection methods may produce different valuable features. Consequently, it is advisable to employ multiple feature selection techniques and make decisions based on a majority vote, as recommended by [10]. In this study, the following feature selection techniques have been utilized to address these considerations:

1. Linear Regression (LR): Linear regression is a basic algorithm that assumes a linear relationship between the target variable and the predictors [3]. It estimates the coefficients of the linear equation that best fits the data, enabling prediction based on the learned model [13].
2. Mutual Information (MI): It measures the statistical dependence between features and the target variable [6]. It quantifies the information that a feature provides about the target variable, considering both linear and non-linear relationships. By evaluating each feature independently, MI calculates the mutual information score for each feature, ranking them based on their relevance to the target variable. Features with higher MI scores are considered more informative and selected for further analysis, while features with lower scores are deemed less relevant.
3. Correlation Matrix (CM): It identifies relevant features by examining their correlation with the target variable and each other [15]. It constructs a correlation matrix to quantify the linear relationship between feature pairs. Features highly correlated with the target variable are considered influential, while high inter-correlations may indicate redundancy. By analyzing the correlation matrix, this technique selects features with strong positive or negative correlations to the target variable, indicating their potential relevance for prediction.
4. Probability Value (p-value): It assesses the statistical significance of the relationship between features and the target variable [24]. It calculates the probability of obtaining the observed data under the assumption of no relationship. Statistical tests, such as t-tests or ANOVA, are used to calculate the p-value.

A significance level, such as 0.05, is set to determine if the relationship is statistically significant. Features with low p-values are considered relevant and selected for further analysis, while features with high p-values are deemed less influential. This technique eliminates irrelevant features, leading to a more focused and accurate model.

5. Principal Component Analysis (PCA): It uses to reduce the dimensionality of a dataset while retaining important information. It transforms the original features into a set of uncorrelated variables known as principal components [11]. PCA identifies directions in the feature space that exhibit the highest variance, capturing the maximum amount of information. These directions are represented by the principal components, with the first component explaining the most variance, followed by the second component, and so on. During PCA feature selection, the original features undergo a linear transformation to create the principal components. The number of components to retain depends on the desired level of dimensionality reduction. By selecting a subset of components that explain a significant portion of the variance, the dataset's dimensionality can be effectively reduced.
6. Recursive Feature Elimination (RFE): according to [4] RFE iteratively removes less important features to identify the most relevant ones. It starts by training a model on the entire feature set and assigning importance scores to each feature. Features with the lowest scores are pruned from the dataset. The model is then retrained using the reduced feature set, and the process continues until the desired number of features or stopping criterion is reached. Cross-validation is often used to evaluate model performance during each iteration. RFE provides a ranking of feature importance, allowing researchers to focus on the subset of features that contribute the most to the predictive power of the model.
7. Random Forest (RF): it identifies the most significant features in a dataset. According to [3] RF utilizes multiple decision trees and combines their predictions to achieve accurate results. During the feature selection process, RF assesses the contribution of each feature to the model's performance by measuring the decrease in accuracy when a specific feature is randomly permuted or removed from the dataset [27].
   The importance of each feature is determined by calculating the average decrease in model accuracy across all decision trees in the random forest [27]. Features that result in a significant decrease in accuracy when removed or permuted are considered more important, as they provide valuable information for prediction. By analyzing the feature importance scores obtained from the RF model, researchers can identify the most relevant features for prediction. Features with high scores are considered more influential, while those with low scores may have less informative value and can be excluded from further analysis.

2.2.4 Regression Model Regression models are commonly used in machine learning to predict the relationships between variables [19]. These algorithms are particularly effective when the target variable is continuous or numeric [1]. In this study, we employed different regression algorithms to investigate the strength of the relationship between the target variable, namely the views number, and other important selected features. The best-performing regression algorithm from the list below would aid in predicting the popularity of the video in the subsequent hour.

1. Support Vector Machine for Regression (RegSVR): aims to find a hyperplane that maximizes the margin between data points and predicts the target variable based on its distance from the hyperplane [9]. RegSVR could handle nonlinear relationships and capture complex patterns in the data. By employing a nonlinear kernel function, RegSVR can effectively model the intricate interactions between the target variable and the selected features, potentially leading to accurate predictions of video popularity.
2. Multi-layer Perceptron (MLP): a type of artificial neural network with multiple layers of interconnected nodes (neurons) [17]. It can learn complex patterns and non-linear relationships between variables, making it suitable for regression tasks [17]. Moreover, MLP can capture temporal dependencies and capture the intricate dynamics of video popularity over time.
3. Decision Stump: considers only one node at a time. It selects a single feature and a threshold to split the data, making predictions based on the outcome of this split [2]. As it is a single level decision tree, it provides a baseline for comparison. Moreover, its simplicity and interpretability allow for a straightforward examination of the relationships between the target variable and the selected features, aiding in understanding the impact of individual features on video popularity.
4. Reduced Error Pruning Tree (RepTree): a decision tree algorithm that uses reduced error pruning to remove less informative branches, reduce overfitting and improve the model's ability to generalize [22]. This makes RepTree robustness in handling noisy and uncertain data. This is crucial for accurate time series forecasting, particularly when dealing with potential noise and outliers in the data.
5. Random Forest (RF): ensembles learning method that constructs multiple decision trees and combines their predictions [27]. RF can effectively handle complex relationships and capture nonlinear interactions between the target variable and the selected features. By leveraging the aggregation of predictions from multiple trees, RF enhances prediction accuracy and reduces the risk of overfitting [27].

Each regression algorithm was trained on the dataset and evaluated based on its performance using the accuracy measures. By comparing the accuracy measures scores of different algorithms, we can determine the most effective model for predicting the next hour's view number.

2.2.5 Prediction The chosen best regression model with the lowest MAPE would be utilized to predict the video view number in the next hour, providing insights into future viewership trends.

3 Experiments and Results

3.1 YouTube Data API v3

The YouTube Data API library offers a means to incorporate API-based access to YouTube videos, providing various functionalities for working with video content [11]. By utilizing this library, can retrieve video-specific information such as the title, description, and video thumbnail, as well as perform actions like a playlist and channel creation, among other attributes. To enable access to this library from a web application, it is necessary to create a YouTube Data API project and obtain an API key. This API key needs to be included in the web application's API requests when accessing the desired YouTube services.

The YouTube data API v3 library is employed for extracting data from a given video URL. When making a request to access the YouTube Data API, the video ID is transmitted along with the API key. Subsequently, the resulting API response needs to be parsed to extract the desired video data, such as the view count, like count, and description, among other relevant information.

3.2 Feature Selection and Importance Value

In order to identify the most relevant YouTube features and eliminate less informative ones, a series of feature selection algorithms were applied. Table 1 shows the applied method with its accuracy measures. Tables 2, 3, 4, 5, 6, 7 and 8 show the methods' importance features.

Table 1. Different feature selection algorithms.

Accuracy measure	LR	MI	CM	P-value	PCA	RFE	RF
MAPE (%)	1.58	1.57	1.57	1.57	1.64	1.56	0.005
RMSE	0.033	0.033	0.033	0.033	0.034	0.032	0.001

Initially, linear regression was employed for this purpose. However, the limitations of linear regression, such as its inability to consider non-linearity and its poor percentage error of 1.58%, led to reduced reliability of the features selected through this approach. To overcome these limitations and enhance the feature analysis, RFE and RF were subsequently employed as alternative feature selection methods.

To achieve this, RFE and RF algorithms were implemented. Performance comparison revealed that RF outperformed RFE, yielding a remarkably lower MAPE of 0.0052% compared to RFE's 1.56% MAPE. This notable disparity in performance underscored the superiority of the RF algorithm in identifying salient features.

The significance weights assigned to the features by the RF algorithm, which exhibit correlations with the "view" number, are presented in Table 8. These weights play a pivotal role in predicting video forecasting and will be employed in subsequent analyses, contributing to the overall understanding and predictive capabilities of our model.

Table 2. LR importance features with their weights.

Feature	Importance value
Likes	0.9253
Comments count	0.02086
Channel subscriber	0.01821
Channel view	0.01282
Comment sentiment score	0.00847
Video audio language	0.00536
Video age	0.00574
Category ID	0.00267
Video language	0.00055
Region restriction blocked	0.00002

Table 3. MI importance features with their weights.

Feature	Importance value
Likes	0.9253
Comments count	0.02086
Channel subscriber	0.01821
Channel view	0.01282
Comment sentiment score	0.00847
Video audio language	0.00536
Video age	0.00574
Category ID	0.00267
Video language	0.00055
Region restriction blocked	0.00002

Table 4. CM importance features with their weights.

Feature	Importance value
Likes	0.9253
Comments count	0.02086
Channel subscriber	0.01821
Channel view	0.01282
Comment sentiment score	0.00847
Video audio language	0.00536
Video age	0.00574
Category ID	0.00267
Video language	0.00055
Region restriction blocked	0.00002

Table 5. P-value importance features with their weights.

Feature	Importance value
Likes	0.9253
Comments count	0.02086
Channel subscriber	0.01821
Channel view	0.01282
Comment sentiment score	0.00847
Video audio language	0.00536
Video age	0.00574
Category ID	0.00267
Video language	0.00055
Region restriction blocked	0.00002

3.3 Regression Model

In order to predict the view count of videos in the next hour, various regression models were assessed using the dataset consisting of the most important features. To address the complexity and size of the dataset, various dataset clustering techniques were employed to enhance the regression results. By partitioning the dataset into smaller clusters that exhibit distinct behaviours and distributions, improved predictions were achieved. The evaluation encompassed two scenarios: non-clustered data and clustered data.

For non-clustered data, the regression models were trained and tested using the entire dataset. In contrast, for clustered data, different approaches were considered based on the type of clustering performed. This involved analyzing the dataset from three perspectives: input-only clustering, output-only clustering, and input-output clustering. The input-only clustering involved clustering the

Table 6. PCA importance features with their weights.

Feature	Importance value
Likes	0.9253
Comments count	0.02086
Channel subscriber	0.01821
Channel view	0.01282
Comment sentiment score	0.00847
Video audio language	0.00536
Video age	0.00574
Category ID	0.00267
Video language	0.00055
Region restriction blocked	0.00002

Table 7. RFE importance features with their weights.

Feature	Importance value
Likes	0.9253
Comments count	0.02086
Channel subscriber	0.01821
Channel view	0.01282
Comment sentiment score	0.00847
Video audio language	0.00536
Video age	0.00574
Category ID	0.00267
Video language	0.00055
Region restriction blocked	0.00002

Table 8. RF importance features with their weights.

Feature	Importance value
Likes	0.9253
Comments count	0.02086
Channel subscriber	0.01821
Channel view	0.01282
Comment sentiment score	0.00847
Video audio language	0.00536
Video age	0.00574
Category ID	0.00267
Video language	0.00055
Region restriction blocked	0.00002

data based on all attributes except for the views attribute. On the other hand, output-only clustering focused solely on clustering based on the views number. Furthermore, input-output clustering considered clustering based on both input attributes and the views attribute.

Table 9, 10, 11 and 12 present the accuracy measures for each of the applied algorithms.

Based on the observed results, the random forest regression model demonstrated remarkable performance, particularly in the input clustering scenario. Therefore, it would be emerged as the promising choice for predicting video popularity in the next hour.

Table 9. No clustering data regression model.

Model	MAPE	RMSE
Linear Regression	14.02%	0.02
RegSVM Puk Kernel	4.9%	0.072
Multi-Layer Perceptron	9.5%	0.14
Decision Stump	1.09%	0.0029
Random Forest	1.06%	0.0006
RepTree	1.07%	0.0009

Table 10. Input clustering data regression model.

Model	MAPE	RMSE
Linear Regression	9.29%	0.05
RegSVM Puk Kernel	5.23%	0.06
Multi-Layer Perceptron	6.77%	0.05
Decision Stump	1.8402%	0.0047
Random Forest	0.084%	0.0008
RepTree	0.65452%	0.0024

Table 11. Output clustering data regression model.

Model	MAPE	RMSE
Linear Regression	5.15%	0.2
RegSVM Puk Kernel	5.25%	0.004
Multi-Layer Perceptron	2.5%	0.003
Decision Stump	1.09%	0.0027
Random Forest	1.01%	0.0007
RepTree	1.06%	0.001

Table 12. Input & Output clustering data regression model.

Model	MAPE	RMSE
Linear Regression	9.29%	0.05
RegSVM Puk Kernel	5.23%	0.06
Multi-Layer Perceptron	6.77	0.05
Decision Stump	1.8402%	0.0047
Random Forest	0.0841%	0.0008
RepTree	0.6545%	0.0024

3.4 Prediction

As clearly noticed from Tables 9, 10, 11 and 12, the random forest provides optimal prediction accuracy compared to the other discussed algorithms. Therefore, the random forest technique is utilized for time series forecasting to predict the prospective popularity of a video within the subsequent hour. The used dataset comprises information about video updates at each hour over a collection period of 480 hours. The dataset is divided into 80% for training and 20% for testing purposes. Notably, the achieved MAPE performance was 0.41%. The disparity between the anticipated and predicted outcomes of video next-hour forecasting is visualized in Fig. 2.

3.5 Comparison with Existing Methods

In order to evaluate the effectiveness of the proposed model, a comparison was conducted with other existing models, as illustrated in Table 13. The performance of each model was assessed using appropriate evaluation metrics.

The results clearly indicate that the proposed model outperforms the other models in terms of performance. These findings validate the efficacy of the proposed model and demonstrate its potential for providing more accurate predictions compared to existing methods.

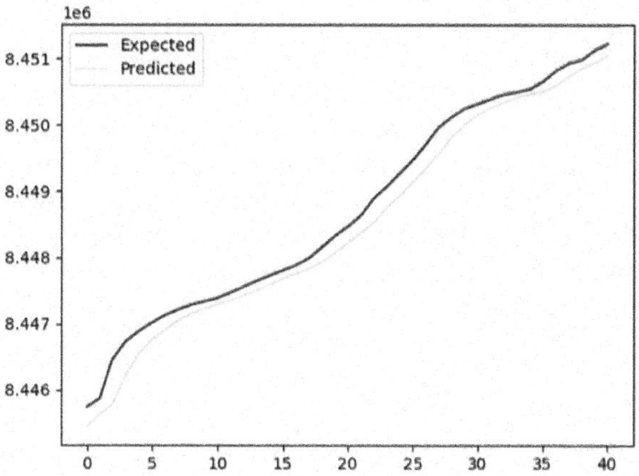

Fig. 2. Random forest next hour video view popularity.

Table 13. Comparison of the proposed method with other related approaches.

Study	MAPE
Proposed approach	0.41%
[25]	0.98%
[21]	0.7%

4 Conclusion

In this study, an analysis was conducted on YouTube video features, specifically focusing on a dataset of videos from the UK. Feature selection techniques were applied to determine the importance of each video feature and to explore the relationship between the "view" feature and other YouTube features. Both linear and non-linear feature selection methods were employed to identify the significant features and their impact on view prediction.

The results obtained from the best feature selection method revealed the most important features for enhancing the visibility of YouTube videos. Furthermore, regression models were employed to predict the next hour's view count, with the aim of identifying the most effective model in terms of performance. The proposed approach was compared to existing works, demonstrating substantial improvements in terms of the MAPE.

Overall, the proposed work provides valuable insights into the significance of different video features in determining the popularity of YouTube videos. The regression models and feature selection techniques employed in this study offer promising results and pave the way for further advancements in predicting and optimizing video viewership on the platform.

Future work for this study may involve extending the application of the proposed model to different regions, beyond the UK, with an expanded set of

YouTube features. Furthermore, incorporating additional data sources, such as social media trends, could offer valuable insights into the broader social dynamics influencing video popularity. Additionally, exploring advanced time series modelling techniques, including deep learning models like RNNs or LSTMs, would enhance the ability to capture complex patterns and dynamics in video popularity. Moreover, developing real-time prediction models that provide instantaneous forecasts of video view numbers, utilizing streaming analytics and online learning techniques, would enable continuous updating and refinement of the prediction model based on incoming data.

References

1. Alghanmi, N., Zeng, X.: A hybrid regression model for mixed numerical and categorical data. In: Advances in Computational Intelligence Systems: Contributions Presented at the 19th UK Workshop on Computational Intelligence, September 4–6, 2019, pp. 369–376. Portsmouth, UK 19 (2020)
2. Andriushchenko, M., Hein, M.: Provably robust boosted decision stumps and trees against adversarial attacks. In: Advances in Neural Information Processing Systems, p. 32 (2019)
3. Chakraborty, D., Elhegazy, H., Elzarka, H., Gutierrez, L.: A novel construction cost prediction model using hybrid natural and light gradient boosting. Adv. Eng. Inform. **46**, 101201 (2020)
4. Chen, X., Jeong, J.: Enhanced recursive feature elimination. In: Sixth International Conference on Machine Learning and Applications (ICMLA 2007), pp. 429–435 (2007)
5. Chen, Y., Chang, C.: Early prediction of the future popularity of uploaded videos. Expert Syst. Appl. **133**, 59–74 (2019)
6. Dave, V., Vakharia, V., Singh, S.: Ball bearing fault diagnosis using mutual information and Walsh-Hadamard transform. In: Reliability, Safety and Hazard Assessment for Risk-Based Technologies: Proceedings of ICRESH, vol. 2019, pp. 607–616 (2020)
7. Davidson, J., Liebald, B., Liu, J., Nandy, P., Van Vleet, T., Gargi, U., Gupta, S., He, Y., Lambert, M., Livingston, B.: The YouTube video recommendation system. In: Proceedings of the Fourth ACM Conference on Recommender Systems, pp. 293–296 (2010)
8. Drotár, P., Gazda, J., Smékal, Z.: An experimental comparison of feature selection methods on two-class biomedical datasets. Comput. Biol. Med. **66**, 1–10 (2015)
9. Guo, Z., Bai, G.: Application of least squares support vector machine for regression to reliability analysis. Chin. J. Aeronaut. **22**, 160–166 (2009)
10. Halim, Z., Atif, M., Rashid, A., Edwin, C.: Profiling players using real-world datasets: clustering the data and correlating the results with the big-five personality traits. IEEE Trans. Affect. Comput. **10**, 568–584 (2017)
11. Halim, Z., Hussain, S., Ali, R.: Identifying content unaware features influencing popularity of videos on youtube: A study based on seven regions. Expert Syst. Appl. **206**, 117836 (2022)
12. Hong, S., Kim, H.: A comparative study of video recommender systems in big data era. In: 2016 Eighth International Conference on Ubiquitous and Future Networks (ICUFN), pp. 125–127 (2016)

13. Hundi, P., Shahsavari, R.: Comparative studies among machine learning models for performance estimation and health monitoring of thermal power plants. Appl. Energy **265**, 114775 (2020)
14. Hussain, M., Tokdemir, S., Al-Khateeb, S., Bandeli, K., Agarwal, N.: Understanding digital ethnography: socio-computational analysis of trending YouTube videos. In: The Eight International Conference on Social Media Technologies, Communication, and Informatics (2018)
15. Jiang, B.: Covariance selection by thresholding the sample correlation matrix. Stat. Prob. Lett. **83**, 2492–2498 (2013)
16. Khaire, U., Dhanalakshmi, R.: Stability of feature selection algorithm: A review. J. King Saud Univ.-Comput. Inf. Sci. 34, 1060–1073 (2022)
17. Khalil Alsmadi, M., Omar, K., Noah, S., Almarashdah, I.: Performance comparison of multi-layer perceptron (Back Propagation, Delta Rule and Perceptron) algorithms in neural networks. In: 2009 IEEE International Advance Computing Conference, pp. 296–299 (2009)
18. Kumar, V., Minz, S.: Feature selection: a literature review. SmartCR. **4**, 211–229 (2014)
19. Maulud, D., Abdulazeez, A.: A review on linear regression comprehensive in machine learning. J. Appl. Sci. Technol. Trends **1**, 140–147 (2020)
20. Meseguer-Martinez, A., Ros-Galvez, A., Rosa-Garcia, A.: Linking YouTube and university rankings: Research performance as predictor of online video impact. Telemat. Inf. **43**, 101264 (2019)
21. Nisa, M., Mahmood, D., Ahmed, G., Khan, S., Mohammed, M., Damaševičius, R.: Optimizing prediction of YouTube video popularity using XGBoost. Electronics **10**, 2962 (2021)
22. Pham, B., Jaafari, A., Nguyen-Thoi, T., Van Phong, T., Nguyen, H., Satyam, N., Masroor, M., Rehman, S., Sajjad, H., Sahana, M.: Ensemble machine learning models based on reduced error pruning tree for prediction of rainfall-induced landslides. Int. J. Digit. Earth **14**, 575–596 (2021)
23. Pinto, H., Almeida, J., Gonçalves, M.: Using early view patterns to predict the popularity of youtube videos. In: Proceedings of the Sixth ACM International Conference on Web Search and Data Mining, pp. 365–374 (2013)
24. Tang, J., Alelyani, S., Liu, H.: Feature selection for classification: A review. In: Data Classification: Algorithms and Applications, pp. 37 (2014)
25. Trzciński, T., Rokita, P.: Predicting popularity of online videos using support vector regression. IEEE Trans. Multimedia **19**, 2561–2570 (2017)
26. Wongsuparatkul, E., Sinthupinyo, S.: View count of online videos prediction using clustering view count patterns with multivariate linear model. In: Proceedings of the 8th International Conference on Computer and Communications Management, pp. 123–129 (2020)
27. Wu, L., Xiao, Y., Ghosh, M., Zhou, Q., Hao, Q.: Machine learning prediction for bandgaps of inorganic materials. ES Mater. Manufact. **9**, 34–39 (2020)
28. Wu, S., Rizoiu, M., Xie, L.: Beyond views: Measuring and predicting engagement in online videos. In: Proceedings of the International AAAI Conference on Web and Social Media, p. 12 (2018)

Analyzing and Comparing Clustering Algorithms for Student Academic Data

Shraddha Bhurre[1], Sunny Raikwar[2(✉)], Shaligram Prajapat[1], and Deepika Pathak[2]

[1] International Institute of Professional Studies, DAVV, Indore, India
shaligram.prajapat@iips.edu.in
[2] Dr. APJ Kalam University, Indore, India
falseraikwar.sunny23@gmail.com

Abstract. The online learning data could be analyzed in order to improve the teaching-learning process. The Performance of students depends on several factors, that could be managed to improve their academic growth. It is difficult to acquire a complete picture of the state of the student's performance and concurrently uncover critical information from their patchy performance when using the typical student grouping based on average scores. Clustering algorithms can be used to identify patterns and clusters in learners' data in e-learning scenarios. There are several clustering algorithms that can be used to do this task. In this paper, k-means clustering, hierarchical clustering, and the farthest first clustering algorithm will be used to analyze students' academic data and form clusters based on 4 major categories: demographic features, academic background, parent participation, and behavioral features. The dataset used in this study is taken from Kaggle, which is originally collected from LMS (Learning Management system) Kalboard-360 and it consists of 480 student records with 16 features. The main aim of this study is to identify the clusters with given features and the best-performing clustering algorithm which gives the highest correctly classified instances. Results show that the highest correctly clustered instances are given by the k-mean clustering algorithm.

Keywords: Clustering Algorithms · K-Means Clustering · Hierarchical Clustering · Farthest First · Machine Learning · Online Learning

1 Introduction

Numerous things could prevent students from achieving and maintaining a high performance that represents their overall achievement in the classroom throughout their time at the university. By focusing on these elements while creating strategies, faculty members may help students learn more effectively and perform better in a class by keeping track of how they are progressing. Performance evaluation is thus one of the foundations for tracking the development of student performance in higher education institutions. Based on this important issue, it is now difficult to divide pupils into different categories according to their performance. With traditional student grouping based on average

N. Naik et al. (Eds.): UKCI 2023, AISC 1453, pp. 640–651, 2024.
https://doi.org/10.1007/978-3-031-47508-5_49

scores, it is challenging to get a full picture of the state of the student's performance and simultaneously uncover crucial information from their sporadic performance. It is feasible to identify the essential qualities of student performance and maybe use those characteristics for future prediction with the aid of data mining techniques like clustering algorithms. Clustering has numerous uses. It is frequently used as a standalone data mining tool to examine each cluster's properties and narrow in on a selected group of clusters for additional investigation. E. Amrieh et al. in a study [1] describe the effect of behavioral features on students' academic performance using several data mining techniques such as Artificial Neural Networks, Naïve Bayesian, and Decision Tree classifiers. The same dataset will be used in this study with three clustering methods: K-mean Clustering, Hierarchical Clustering, and Farthest First Clustering, to evaluate clusters and compare these algorithms. Correctly Clustered Instances will be evaluated in order to show the best clustering method. The detailed Experiment and result will be discussed in a separate section.

2 Literature Review

Data mining, learning analytics, and machine learning are making effective changes in educational data. Several studies have presented their work on Clustering as well as Classification techniques to analyze student academic data based on features like demographic, social information, geographical, academic, behavioral, etc. This paper mainly focuses on clustering algorithms (k-means clustering, hierarchical clustering, and the farthest first clustering). Following are some studies that work to analyze student academic data, in order to predict their performance, analyze affecting features, compare more efficient methods, etc.

Based on socio-demographic information about the students and their performance during the first (i.e., preparatory) year of the study program, a classifier that uses a decision tree algorithm is employed to forecast student success at the program's conclusion. Critical courses that can be utilized as performance markers for students are also derived using the classifier. To identify the characteristics of students who have dropped out of the degree program, on the other hand, a clustering technique is used. The findings highlight the significant potential for data mining tools to offer insights to students, academic advisers, and university administration, which can serve to enhance learning and raise the standard of educational processes [2].

Moreover, research is done on the LMS (Learning Management System) Assessment module through a variety of features and parameters that influence online learner performance [3].

C. Anuradha et al. describe in a study [4] that, Among the many data mining techniques, clustering helps to classify the student in a well-defined cluster to find the behavior and learning style of students. Although their primary objective is to describe how clustering algorithms classify the students in a well-defined cluster which enables academicians to predict student performance.

In order to divide the students into several clusters according to their learning capacities, Kausar et al. developed a clustering strategy [5]. They discovered that clustering via quick search and discovering density peaks using heat diffusion produces more reliable findings.

In a study [6], Sharmila and M. Kumar work on three clustering algorithms: EM algorithm, farthest first algorithm, and k-means algorithm. They describe that different clustering algorithms will be used according to the behavior of the data. According to their study, Farthest first algorithm is suitable for the large dataset but it creates a non-uniform cluster.

A. Dutt et al. in a study [9] give a detailed review of various data mining approaches like clustering, classification, and prediction algorithms applied in educational data mining. They are mostly used to evaluate student performance, predict dropout students, identify significant variables, analyze log data, etc.

3 Clustering Algorithms

Grouping of unlabeled data is known as Clustering. It is a task of segregating data into groups or clusters in a way that data in the same groups are more alike to other data in the same group and different from the data in another group. In this paper, k-means clustering, hierarchical clustering, and the farthest first clustering algorithm will be used to analyze students' academic data.

3.1 K-Mean Clustering

K-mean is an unsupervised learning algorithm, based on how close a data point is to the cluster centroid, the algorithm repeatedly allocates the data points to one of the K clusters. There are 2 main tasks that perform by the k-mean cluster. The first is to find the most suitable k value. The second is to assign data points to the nearest cluster. The main goal of this algorithm is to minimize the sum of distances between the data points and their corresponding clusters [7].

The algorithm has the following steps:

Input:

1. Learner performance data (e.g., test scores, completion rates, interaction patterns).
2. Number of clusters, K.

Output:

1. Cluster assignments for each learner.
2. Cluster centroids.

Procedure:

1. Specify the number of clusters, K.
2. Randomly select K data points from the learner performance data as initial centroids.
3. Repeat until convergence:
   a. *Assign each learner to the nearest cluster centroid:*
      i. For each learner, calculate the distance to each centroid using an appropriate distance metric (e.g., Euclidean distance).

ii. Assign the learner to the cluster with the closest centroid.

b. *Update the cluster centroids:*

i. For each cluster, compute the mean of the learner performance attributes within the cluster.

ii Update the centroid of each cluster with the computed mean.

c. *Check for convergence*:

i. If the cluster assignments and centroids do not change significantly (below a predefined threshold), exit the loop.

ii. Output the final cluster assignments and centroids.

4. Perform performance assessment:
   i. Analyze the characteristics and patterns of learners within each cluster.
   ii. Evaluate the performance metrics of learners in each cluster
   iii. Compare and interpret the performance differences among clusters.
   iv. Identify clusters with similar performance patterns.

End.

3.2 Hierarchical Clustering

This method divides the dataset into clusters to produce a structure that resembles a tree and is known as a dendrogram. By removing the appropriate amount of the tree, it is possible to choose the observations or any number of clusters. It uses a bottom-up agglomerative strategy. When one progresses up the hierarchy, pairs of clusters are joined, starting analyses from a finite number of clusters. An agglomerative algorithm selects two closet groups to merge after each iteration. This distance is the shortest one between any two elements of a group that may be connected by a single link in single-link clustering [4].

Input:

1. Learner performance data (e.g., test scores, completion rates, interaction patterns)
2. Distance metric (e.g., Euclidean distance, cosine similarity)
3. Number of desired clusters, K (optional)

Output:

1. Cluster assignments for each learner

Procedure:

1. Initialize the algorithm:
   i. Specify the distance metric to measure the dissimilarity between learners' performance data. Common distance metrics include Euclidean distance and cosine similarity.
   ii. Specify the number of desired clusters, K (optional). If not specified, the algorithm will create a hierarchical structure representing all possible clustering.
2. Calculate the dissimilarity matrix:

   i. Compute the pairwise dissimilarities between learners' performance data using the specified distance metric.
   ii. The dissimilarity matrix will have dimensions NxN, where N is the number of learners.
3. Build the dendrogram:
   i. Initialize the dendrogram with N individual clusters, where each learner is initially assigned to its own cluster.
   ii. While the number of clusters is greater than K (or until a stopping criterion is met):
   iii. Identify the two closest clusters based on the dissimilarity matrix.
   iv. Merge the two closest clusters into a single cluster.
   v. Update the dissimilarity matrix to reflect the new distances between the merged cluster and the remaining clusters.
   vi. The resulting dendrogram represents the hierarchical structure of clusters.
4. Determine the optimal number of clusters:
   i. If the number of desired clusters, K, is not specified, you can use a criterion (e.g., the elbow method, silhouette coefficient) to determine the optimal number of clusters based on the dendrogram.
   ii. Cut the dendrogram:
   iii. Based on the desired number of clusters (K) or the optimal number of clusters determined in the previous step:
   iv. Output the final cluster assignments.

End.

3.3 Farthest First

In the Farthest First traversal, there are k cluster centers, where the first cluster center point is selected randomly and each successive cluster point is selected as far as possible. The left-out center will be determined by choosing the farthest point from the already-determined center and the rest are added to the group of points whose center is closet.

The remaining points are added to the cluster whose center is closest after choosing the point that is the furthest from the set of previously picked centers to decide each remaining center [6].

The farthest First algorithm has the following steps:

Input:

1. Learner performance data (e.g., test scores, completion rates, interaction patterns)
2. Number of desired clusters, K

Output:

1. Cluster assignments for each learner

Procedure:

1. Initialize the algorithm:

i. Specify the number of desired clusters, K.
ii. Randomly select the first cluster center point from the learner performance data.
iii. Initialize cluster centers:
iv. Create an empty list to store the cluster centers.
v. Add the first cluster center to the list.

2. Assign learners to clusters:
   i. For each learner, calculate the distance to each cluster center using an appropriate distance metric (e.g., Euclidean distance, cosine similarity).
   ii. Assign the learner to the cluster with the maximum distance, ensuring that no two learners are assigned to the same cluster initially.
3. Update Cluster Centers:
   i. For each cluster, calculate the new cluster center by computing the mean or centroid of the learners' performance data within the cluster.
4. Repeat steps 3 and 4 until K clusters are formed:
   While the number of Cluster is less than K:
   i. Calculate the distance between each learner and the existing cluster centers.
   ii. Select the learner with the maximum minimum distance to any existing cluster center.
   iii. Add the selected learner as a new cluster center.
   iv. Update the cluster assignments for all learners based on the new cluster centers.
   v. Recalculate the cluster centers.
5. Output final cluster assignment.

End

4 Experiment

4.1 Data Description

In this study, an educational dataset was taken from Kaggle, which is originally collected from a learning management system (LMS) called Kalboard 360. This dataset consists of 480 student records with 16 features. The features are classified into four major categories: (1) Demographic features (2) Academic background features (3) Behavioral features (4) Parent Participation. A sample description of the attributes of the data is given in the Table 1.

4.2 Experiment Setup and Confusion Matrix

The dataset has 16 features (Gender, Nationality, Place of birth, Educational Stages, Grade Levels, Section ID, Topic, Semester, Parent Responsible, Raise Hand, Visited Resources, Viewing Announcements, Discussion Groups, Parent Answering Survey, Parent School Satisfaction, Student Absent Days) and 480 instances. First, we assign the cluster value for each algorithm 3, because we have 3 classes defined in the dataset (L (0 >= 69), M (70 >= 89), H (90 >= 100)). In "classes to cluster evaluation" first we ignore the class attribute and after training assign it to the cluster formed.

Table 1. Attributes description

Feature category	Attribute	Description	Values
Demographical Feature	Gender	Student's gender	{M,F}
	Nationality	Student's nationality	List of nationality
	Place of birth	Student's Place of birth	List of places
Academic Background Features	Educational Stages	Educational level student belongs	{'Lowerlevel',' MiddleSchool',' HighSchool'}
	Grade Levels	Grade student belongs	'G-01', 'G-02', 'G-03', 'G-04', 'G-05', 'G-06', 'G-07', 'G-08', 'G-09', 'G-10', 'G-11', 'G-12
	Section ID	Classroom student belongs	{A,B,C}
	Topic	Course topic	List of topics
	Semester	School year semester	'First',' Second'
	Student Absent Days	The number of absence days for each student	above -7, Under-7
Parent Participation	Parent Responsible	Parent responsible for student	'mom',' father'
	Parent Answering Survey	Parents answered the surveys which are provided by school or not	Yes, No
	Parent School Satisfaction	The Degree of parent satisfaction from school	Yes, No
Behavioral feature	Raise Hand	Student raises his/her hand in the classroom	0–100
	Visited Resources	Student visits a course content	0–100
	Viewing Announcements	Checking new announcements	0–100
	Discussion Groups	Students participate in discussion groups	0–100

Confusion matrices (Table 2) represent counts from predicted and actual values. The output "TN" stands for True Negative which shows the number of negative examples classified accurately. Similarly, "TP" stands for True Positive which indicates the number of positive examples classified accurately. The term "FP" shows a False Positive value, i.e., the number of actual negative examples classified as positive; and "FN" means

Table 2. Confusion matrix

Total count		Predicted value	
		Negative	Positive
Actual value	Negative	TN (True Negative)	FP (False Positive)
	Positive	FN (False Negative)	TP (True Positive)

a False Negative value which is the number of actual positive examples classified as negative.

Confusion Matrix Obtain through "Class to Cluster" for K-Means Clustering

K-means update cluster centroids until a local minimum is discovered. It is a centroid-based algorithm, where each cluster is associated with a centroid. The main aim of this algorithm is to minimize the sum of distances between the data point and their corresponding clusters [8]. The number of clusters will be 3, hence it will generate a 3*3 confusion matrix after "class to cluster evaluation".

For, Table 3.

Cluster 0 – H (High Level).

Cluster 1 – M (Middle Level).

Cluster 2 – L (Low Level).

Table 3. Confusion matrix for K-means clustering

Cluster 0	Cluster 1	Cluster 2	
75	73	63	**M**
9	14	104	**L**
86	46	10	**H**

From Table 3 CCI (Correctly Clustered Instances), can find out for correctly clustered students into 3 classes M, L, H. It can be find out by following formula.

CCI (Correctly Clustered Instances)

The sum of TP (True Positive) and TN (True Negative) is CCI.

$$\text{CCI} = \text{TP} + \text{TN} \tag{1}$$

ICI (Incorrectly Clustered Instances)

The sum of FP (False Positive) and FN (False Negative) is ICI.

$$\text{ICI} = \text{FP} + \text{FN} \tag{2}$$

CCI for k-means will be: $86 + 104 + 73 = 263$.

ICI for k-means will be: $75 + 63 + 9 + 14 + 46 + 10 = 217$.

As a result in Table 6. K-mean has an accuracy of 54.79% for 263 correctly classified instances out of 480 total instances.

Confusion Matrix Obtain through "Class to Cluster" for Hierarchical Clustering

The hierarchical clustering procedure ends when the number of clusters is equal to k. The results of an agglomerative hierarchical clustering are typically presented as a hierarchical structure called a dendrogram [9]. The number of clusters will be 3 for this algorithm as well. The Confusion matrix generated after "class to cluster evaluation" with a 3*3 matrix is given in Table 4.

For, Table 4.

Cluster 0 – M (Middle Level).

Cluster 1 – H (High Level).

Cluster 2 – No Class.

Table 4. Confusion matrix for hierarchical clustering

Cluster 0	Cluster 1	Cluster 2	
207	4	0	**M**
126	1	0	**L**
138	3	1	**H**

CCI (Correctly Clustered Instance) for hierarchical clustering confusion matrix will be:

$$\text{CCI} = 210\ (\text{Out of 480 instances})$$

That is 43.75% Accuracy by Hierarchical Clustering.

Confusion Matrix Obtain through "Class to Cluster" for Farthest First Clustering

Farthest First approach chooses the centroid and assigns the data like in k-means, however, it uses maximum distance rather than the minimum distance. Every cluster center is placed in turn at the location that is the furthest away from the available cluster center. In this algorithm, the cluster value will be 3. The Confusion matrix generated after "class to cluster evaluation" with a 3*3 matrix is given in Table 5.

For, Table 5.

Cluster 0 – H (High Level).

Cluster 1 – M (Middle Level).

Cluster 2 – L (Low Level).

Table 5. Confusion matrix for farthest first clustering

Cluster 0	Cluster 1	Cluster 2	
71	**85**	55	**M**
27	10	**90**	**L**
71	66	5	**H**

CCI (Correctly Clustered Instance) for Farthest First clustering confusion matrix will be:

$$\text{CCI} = 85 + 90 + 71 = 246(\text{Out of 480 instances})$$

That is 51.25% Accuracy by Farthest First Clustering.

4.3 Result Discussion

In order to analyze clusters formed by all three algorithms (k-means clustering, hierarchical clustering, farthest first algorithm), the "classes to cluster evaluation" mode will be used. In which it first ignores the class attribute and generates the clustering. Then during the test phase, it assigns classes to the clusters, based on the majority value of the class attribute within each cluster.

The students were classified into 3 numeric intervals (Classes: L, M, H) based on their total grades/marks:

L:0 >= 69.

M:70 >= 89.

H:90 >= 100.

Here, the value for k will be 3 for each of the algorithms. The completion time of algorithms is also given in Table 6.

Table 6. Performance analysis of all three algorithms

Algorithms	Completion time	Clustered Instances for k = 3			Correctly clustered instances
		0	1	2	
K-means	0 s	35%	28%	37%	54.79%
Hierarchical	0.39 s	98%	2%	0%	43.75%
Farthest First	0.01 s	35%	34%	31%	51.25%

K-means clustering makes 3 clusters (0,1,2), each for 170, 133, and 177 instances respectively for 3 classes (H (High), M (Medium), L (Low)), as given in Table 6.

Here all 16 features were considered. From Fig. 1, it is clear that the K-means algorithm gives the highest correctly clustered instances.

Hierarchical Clustering makes 3 clusters (0,1,2), each for 471, 8 and 1 instance respectively for 3 classes (M (Medium), H (High), "No Class").

Farthest First Clustering makes 3 cluster (0,1,2) each for 169, 161 and 150 instances respectively for 3 classes (H (High), M (Medium), L (Low)).

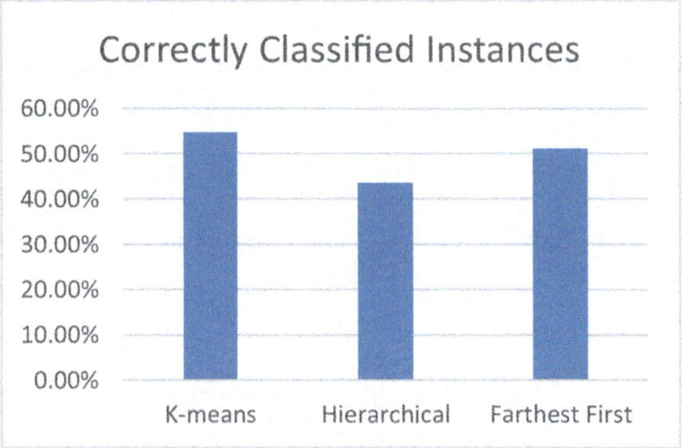

Fig. 1. Correctly Classified Instances by all three algorithms

5 Conclusion

This study mainly focuses on three algorithms (Simple K-means, Hierarchical algorithm, and Farthest First). All three were applied on 480 instances dataset of student academic performance, with 16 features. The algorithms were implemented in "Class to cluster" modes. From the above tables, we conclude that k-means has a maximum number of correctly clustered instances among all three. Out of 480 instances, 263 instances were correctly clustered by it. In the future, which features among these 4 major categories (demographic features, academic background, parent participation, and behavioral features), will affect student performance most could find out. Also, the appropriate cluster based on these categories could be evaluated.

References

1. Amrieh, E., Hamtini, T., Aljarah, I.: Preprocessing and analyzing educational data set using X-API for improving student's performance. In: 2015 IEEE Jordan Conference on Applied Electrical Engineering and Computing Technologies (AEECT) (2015)
2. Alsheddy, A., Habib, M.: On the application of data mining algorithms for predicting student performance: a case study. Int. J. Comput. Sci. Net. Secur. **17**(10), 189–197 (2017)
3. Prajapat, S., Agrawal, A., MC, A.: Towards realisation of assessment module using learning management system for integrated course. In: 2019 6th International Conference on Computing for Sustainable Global Development (INDIACom) (pp. 1179–1183) (2019)

4. Anuradha, C., Velmurugan, T., Anandavally, R.: Clustering algorithm in educational data mining: a review. Int. J. Power Control. Commun. (IJPCSC) **7**, 47–52 (2015)
5. Kausar, S., Huahu, X., Hussain, I., Wenhao, Z., Zahid, M.: Integration of data mining clustering approach in the personalized e-learning system. IEEE ACCESS **6**, 72724–72734 (2018)
6. Sharmila, Kumar, M.: An optimized farthest first clustering algorithm. In: 2013 Nirma University International Conference on Engineering (NUiCONE) (pp. 1–5) (2013)
7. Parveen, Z., Alphones, A., Naz, S.: Extending the student's performance via k-means and blended learning (IJEACS). Int. J. Eng. Appl. Comput. Sci. **2**(4) (2017)
8. Triayudi, A., Iksal, Haerani, R.: Data Mining K-Means Algorithm for Performance Analysis. J. Phys.: Conf. Ser. (2022)
9. Dutt, A., Aghabozrgi, S., Ismail, M., Mahroeian, H.: Clustering algorithms applied in educational data mining. Int. J. Inf. Electron. Eng. **5**(2) (2015)
10. Chandran, P., Shaikh, S.: Educational data mining: predicting student's performance using clustering. Int. J. Manag., IT Eng. **8**(6) (2018)

Investigation of Decision Support System for Indian Penal Code Section Using Similarity Algorithm and Fuzzy Logic

Ambrish Srivastav(✉) and Shaligram Prajapat

International Institute of Professional Studies, Devi Ahilya Vishwavidyalya, Indore, India
a.srivastav30@gmail.com, shaligram.prajapat@iips.edu.in

Abstract. This research study focuses on the development and investigation of a Decision Support System (DSS) for Indian Penal Code (IPC) Sections using a combination of similarity algorithm and fuzzy logic. The DSS aims to assist law enforcement agencies and legal professionals in effectively analyzing and interpreting the applicability of IPC sections to specific legal cases. The proposed system leverages a similarity algorithm to compare the textual content of crime reports and provisions of the IPC sections. By calculating the similarity score, the system can identify the most relevant IPC sections which are likely to be applicable to a given case. In addition, fuzzy logic techniques are employed to provide help in decision-making to determine the applicable IPC sections for crime reports.

Keywords: Fuzzy Logic · Similarity Algorithm · Jaccard Algorithm · Count Vector · TF-IDF · BERT · Decision support systems

1 Introduction

Decision support systems (DSS) are computer-based systems that support organizational decision-making processes. They use data, models, and analytical tools to provide information and insights to organizations and users, which help them, make better decisions. DSS can be used in various domains, including the judiciary for supporting decision-making related to a crime and its associated punishment.

In the context of the Indian judiciary System, the crime committed is assessed on the basis of the Indian Penal Code (IPC) and other alternatives mentioned in Table 1. IPC is the official criminal code of India and is available in a semi-structured format that covers all various offenses and punishments [1]. As shown below I have created a graphical representation displaying AndOR graphs for both general IPC document that is divided in 23 chapters and specified IPC Chapter 16 with list of related IPC sections. Alongside displaying the Structure of a Standard IPC Section (Exemplified by Section 309). AndOR Graph representation is chosen here to support applicability of one or more than one chapter or section for a specific case (as and when required).

Figures 1, 2, and 3, show the entire structure of the IPC document. Conventionally the determination of IPC Sections is based on the experience and reference manual. The conventional methods for manually determining the IPC are displayed in Table 1.

N. Naik et al. (Eds.): UKCI 2023, AISC 1453, pp. 669-670, 2024.
https://doi.org/10.1007/978-3-031-47508-5_50

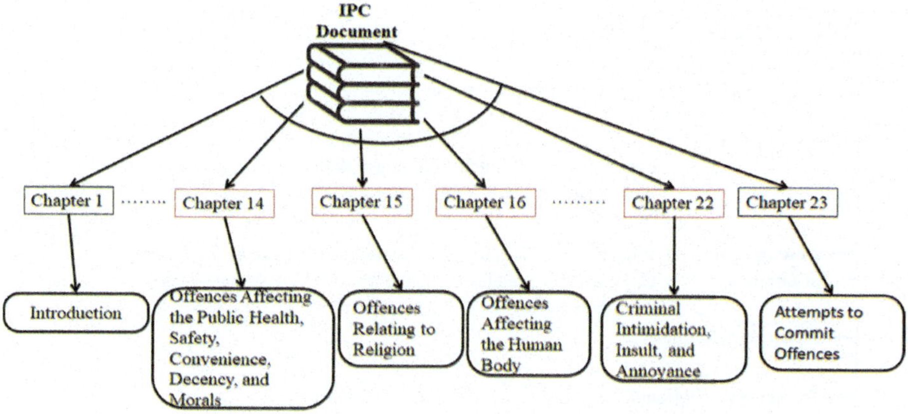

Fig. 1. AndOR graph representation of IPC Document

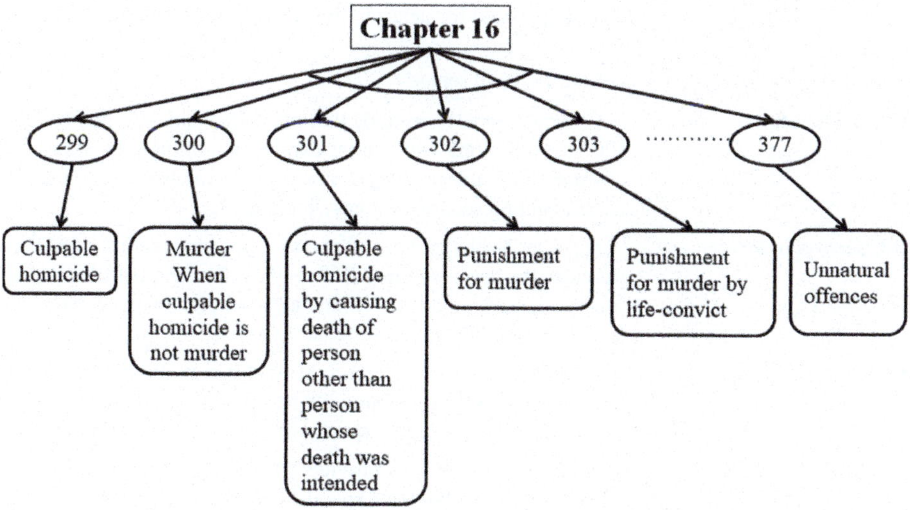

Fig. 2. AndOr Representation of a specific IPC Chapter (Chapter 16)

From Table 1 it is clear that the conventional process of determining appropriate IPC sections involves manually searching (Including the past experience of users) through the IPC text or referring to physical reference guides (such as booklets, pocket guides, or section indexes). The manual process of determining appropriate IPC sections is time-consuming, confusing, and may lack comprehensive features and functionalities.

Online/Offline platforms with DSS may offer advantages such as efficiency, accessibility, enhanced searchability, and improved user experience. Thus DSS for IPC empowers the facility of quick and accurate determination of relevant IPC sections. The task of such a decision support system is to identify the applicable IPC sections for the crime reported which is a cognitive process and is performed by legal experts in this area based

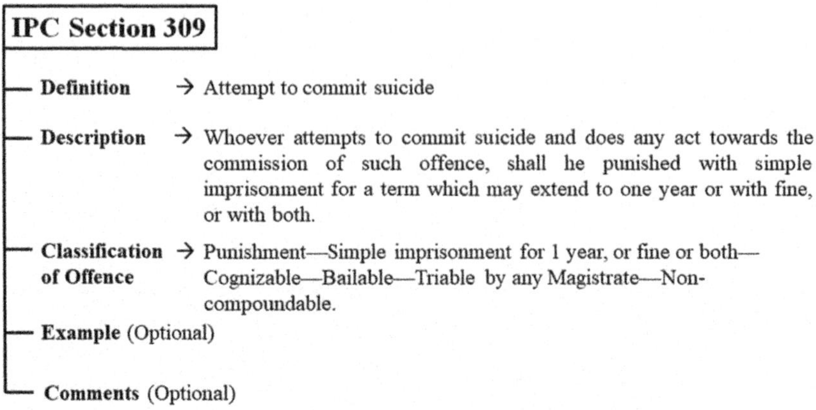

Fig. 3. Structure of a specific IPC Section (Sect. 309)

Table 1. The traditional approach for determining IPC section manually

Traditional approach	Description and remark
IPC quick reference guide	A Compact booklet or guide that provides a quick reference to IPC sections along with a brief explanation of the offenses covered. It includes a categorized index of offenses and their corresponding sections for easy lookup
IPC section index	A reckoner book that lists offenses in alphabetical order and provides the corresponding IPC sections for each offense. Helps users to quickly find the relevant section based on the offense or legal concept they are looking for
IPC pocket guide	A portable and handy reference book that contains an organized listing of IPC sections along with its brief explanations. Designed to be carried on the go for quick and easy section determination
Online IPC reckoners	Various online platforms provide IPC reckoner tools or databases where users can search for specific offenses or keywords to determine the relevant IPC sections. e.g www.indiankanoon.org etc. These online reckoners offer user-friendly interfaces and search functionalities for quick and convenient section determination

on their experience. Figure 4 illustrates the various steps involved in developing a DSS to determine the relevant Indian Penal Code (IPC) sections for reported crimes. Here are the following steps to develop such a system.

Step 1: Develop a knowledge base for the IPC decision to support DSS. Each IPC section defines an offense with a number of conditions; therefore a knowledge base has been developed with a unique row section number for each possible condition to increase the accuracy of the system.

Note: - A knowledge base was developed for research purposes, incorporating four chapters (14, 15, 16, and 22) of the IPC document. Within this knowledge base, unique identification and definition of possible conditions for each section were established.

Step 2: Compute the similarity between the crime report and every section of the IPC document is calculated.

Step 3: Use Fuzzy logic to map the result of similarity to find a list of applicable IPC sections with a degree of accuracy.

Step 4: Provide the list of the most applicable IPC sections obtained for the crime report.

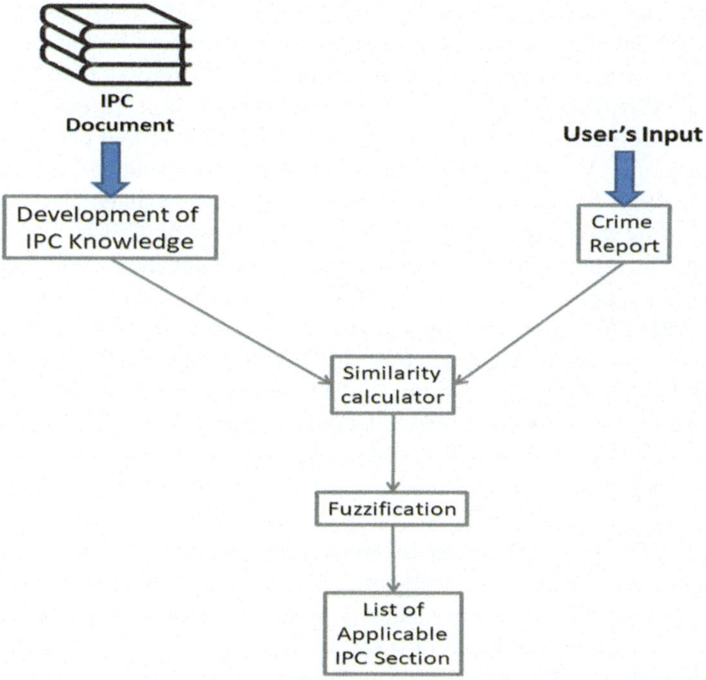

Fig. 4: Steps of Developing a Decision Support System to Identify the IPC Sections for the Committed Crime in the Crime Report

The subsequent sections describe a literature review (Sect. 2) on various types of decision support systems and fuzzy logic-based systems for exploring the candidate systems. Section 3 presents lexical and semantic similarity algorithms based on the inference of Sect. 3. Section 4, describes a fuzzy-based implementation as a result of a similarity algorithm. Section 5 presents the result of the modeled DSS using fuzzy logic. Section 6 concludes the research work and presents some future work.

2 Literature Review

Methods for comparing the text similarity of a particular offence report to the relevant Indian Penal Code (IPC) sections in order to determine the suitable IPC sections. Using different text similarity techniques, such as cosine similarity and Jaccard similarity, the crime report texts can be compared with the IPC sections. [2]. To determine IPC codes using machine learning (ML) and natural language processing (NLP) algorithms, the authors suggested a methodology for selecting an appropriate IPC segment based on the input text supplied by contrasting the analysis of bag of words (BOW), Word2Vec, TF-IDF, and BERT [3]. Some methods were developed to calculate semantic similarity using fuzzy formal concept analysis by using page counts and snippets. Authors grouped the various lexical patterns to find semantic relations between two words accurately [4].

A keyword-based question-answering (QA) system was built for Indian court texts in addition to providing answers to questions about crime. According to responses kept in the corpus, this QAS chooses keywords and question kinds from the query and response [5, 6]. Some information retrieval systems (IRS) are made to find pertinent responses to legal questions. A system used natural language processing to process the user's question [7]. With legal judgment prediction (LJP) task misjudgment problems can be resolved using the Criminal Element Extraction Network (CEEN) technology. The reinforcement learning (RL) framework serves as the foundation for CEEN. The term frequency-inverse document frequency (TF-IDF) similarity technique and the similarly worded fact descriptions of different perpetrators are issues that the authors have highlighted in earlier work [8]. To forecast the results of murder-related cases, a model was created. The model was developed using Delhi District Court rulings and machine learning classification methods. The authors identified and gathered crucial legal aspects for 86 murder-related instances and discovered encouraging outcomes [9].

A respiratory disease retrieval model retrieves information based on the concept of fuzzy logic from vague information and uncertain on-text documents for supporting diagnosis [11]. The fuzzy Euclidean distance clustering algorithm approach categorizes documents into more meaningful categories. The authors showed mathematically how texts can be clustered [12]. To automatically produce questions based on Bloom's cognitive levels, an architecture based on a Fuzzy Restricted Boltzmann machine coupled with Gaussian Markov-based softmax was designed. A key component of the system is the disambiguation of related action verbs and phrases [13].

3 Similarity Calculation Algorithm

Word embedding techniques are very useful for converting text into numbers for natural language processing-based applications. These numbers are represented as vectors for text, and they help solve various problems in natural language processing (NLP).These vectors are also used to calculate the similarity between two documents. Count Vector, TF-IDF, Word2vec, Doc2vec, and Sentence Transform are the most popular word embedding techniques. The similarity between the two test documents can be estimated in two ways:

1. Lexical similarity
2. Semantic similarity

Lexical similarity algorithms find similarities between two documents based on common words present in them and semantic similarity is based on the meaning of the words present in the document. Jaccard similarity algorithm, Count Vector, and TF-IDF are examples of lexical similarity and Word2vec, Doc2vec, and Sentence Transform using BERT are examples of semantic similarity. Since the information on offences in the IPC sections and crime reports is different. Consequently, the result of lexical similarity and semantic similarity is not very useful individually. Furthermore, many non-applicable IPC sections also produce noisy results by holding the value of a similarity score. Some crime report samples were taken to calculate lexical and semantic similarity, for example:

> "I was returning to my home on a bicycle when I met "X", who was drunk, on the way. He abused and attacked me for some old thing, which caused serious injuries to my eyes and nose, and on the way he also threatened to kill me."

IPC Sects. 320B and 506 are relevant based on the offense details in this report owing to grievous hurt and the threat of causing death or grievous hurt. But the similarity algorithms (Jaccard similarity algorithm, CountVector, TF-IDF, word2vec, doc2vec, and BERT) generate so many different sections as a result on the basis of calculation similarity in its own manner. This result is not very useful for decision-making purposes because the generated sections are not consistent across different algorithms and may not provide a clear indication of the appropriate sections to apply in a given case. Table-2 showed the top five sections as a result of different similarity algorithms.

Table 2. Top five results of similarity algorithms

Jaccard	CountVector	TF-IDF	word2vec	doc2vec	BERT
279	279	279	370	318	320H
376C	283	283	273	505A	282
280	376C	376C	374	505B	304B
283	280	320B	367	505D	305
304B	304B	505B	342	353	364

In order to make the result useful fuzzy logic-based estimates may be investigated as, a different approach (To calculate the similarity between these two documents). By finding all the words and their synonyms in the keyword list of each section in the crime report, the similarity of the section with the hit ratio of the keyword list is calculated and the most suitable IPC sections are selected by applying fuzzy logic.

4 Implementation of Similarity Calculation and Fuzzification

The definition and description of crimes are added as the knowledge base with a set of keywords. These keywords' frequency in the crime report demonstrates the applicability of the corresponding IPC section. We have first derived fuzzy rules to calculate the degree of membership and four steps have been identified to develop a decision support system to identify the IPC sections for the committed crime in the crime report.

Algorithm: Developing Decision Support System to Identify the IPC Sections for the committed offense in the crime report shown in Figs. 5 and 6.

Figure 7 shows the membership degree of appropriate IPC sections. This value lies from 0 to 1. If the value is '0' then the section is not applicable and if the value is '1' then the section is applicable. If the value lies between '0' and '1', it shows the degree of involvement. Based on the degree of involvement, an appropriate IPC section can be identified for the committed crime in the crime report.

5 Analytical Framework for Similarity of Crime Report & IPC Document

Jaccard similarity algorithm finds similarity between two documents on the basis of the ratio of intersection and union of words present in documents. The similarity ratio in this algorithm lies between 0 and 1. Here 'n' is the total number of IPC Section and 'k' is individual Section

$$\text{Jaccard Similarity }(k) = \sum_{k=1}^{n} \frac{\text{Document 1 (Crime Report)} \cap \text{Document 2 (IPC Section } k)}{\text{Document 2 (Crime Report)} \cup \text{Document 2 (IPC Section } k)} \tag{1}$$

Count Vector (CountVectorizer) is a word embedding technique used to transform a given text into a vector on the basis of the counting of tokens in text. TF-IDF stands for term frequency-inverse document frequency. Like CountVectorizer, TfidfVectorizer takes term frequency and inverse document frequency as parameters to decrease the weight of the terms appearing commonly in all the sentences. Formulas of calculating TF-IDF stepwise are:

$$\text{TF}(t, d) = \frac{\text{count of } t \text{ in } d}{\text{number of words in } d} \quad \text{//Term Frequency} \tag{2}$$

$$\text{DF(t)} = \text{Occurrence of in documents} \quad \text{//Document Frequency} \tag{3}$$

$$\text{IDF}(t, d) = \log \frac{N}{df(t)} \quad \text{//Inverse Document Frequency} \tag{4}$$

$$\text{TF - IDF}(t, d) = \text{TF}(t, d) * \text{IDF}(t, d) \tag{5}$$

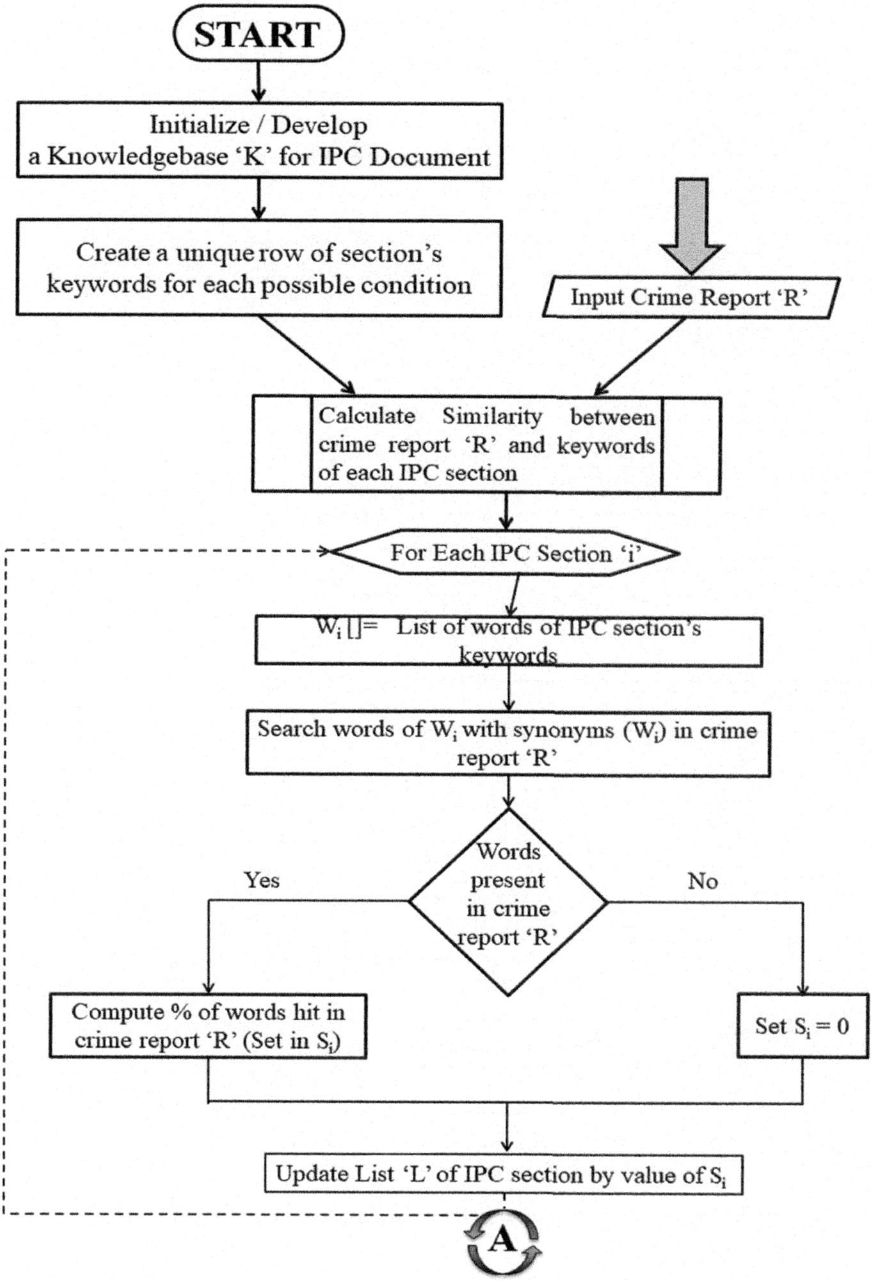

Fig. 5. First Segment of the IPC Similarity Algorithm Flowchart

These algorithms calculate the lexical similarity between IPC sections and crime report. If common words are available in both documents (crime report and IPC section)

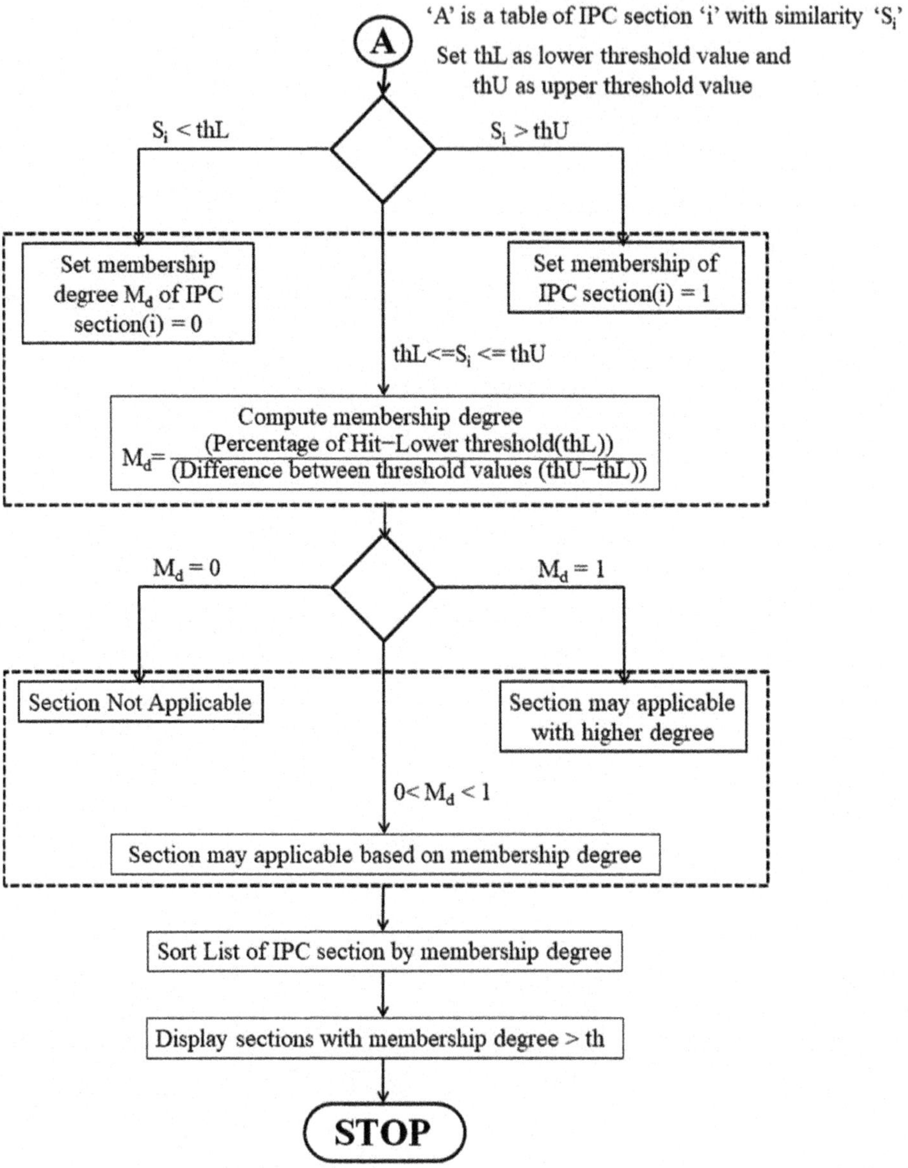

Fig. 6. Second Segment of the IPC Similarity Algorithm Flowchart

then only these algorithms calculate similarity. Word2vec, doc2vec, and sentence transformation using BERT are word embedding techniques to calculate semantic similarity. These word embedding technique find vectors of two documents in different manners. To calculate the similarity value, the cosine similarity between the crime reports and

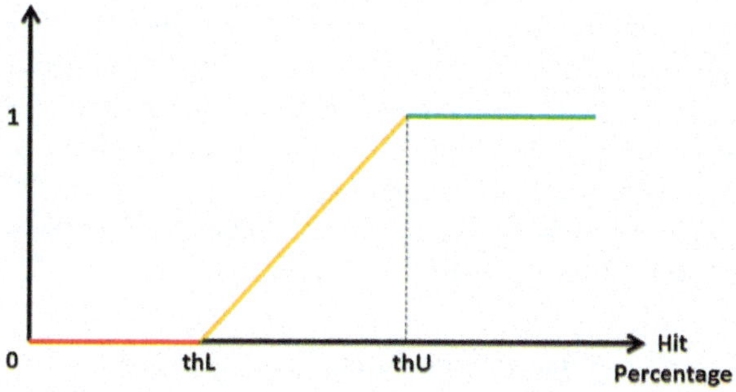

Fig. 7. Membership degree for appropriate IPC Section

each of the IPC sections is obtained. The formula for calculation of cosine similarity is-

$$\text{Similarity}(d1, d2) = \frac{d1.d2}{||d1||X||d2||} = \frac{\sum_{i=1}^{n} d1_i X d2_i}{\sqrt{\sum_{i=1}^{n} d1_i^2} X \sqrt{\sum_{i=1}^{n} d2_i^2}} \quad (6)$$

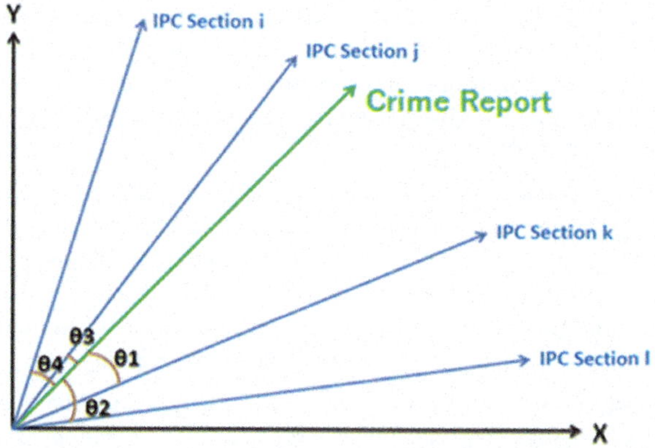

Fig. 8. Cosine similarity between crime report and IPC sections

Figure 8 shows the working of cosine similarity. It calculates the angle between two vectors that represent two different documents. The values range of cosine_similarity lies between -1 and 1, where -1 is perfectly dissimilar and 1 is perfectly similar. Here one vector is for the crime report and another vector for the IPC section keywords (IPC Section i, j, k, l, etc. represent different sections of the IPC document).

Words of the keywords list of many IPC sections are also present in other sections due to which many unrelated sections are also shown in extracting similarity between keywords and the crime report. And many times the keywords of the applicable IPC section are not exactly available in the crime report. This inaccuracy of the result can be minimized by applying fuzzy logic. By fuzzy logic, the fuzzy value of the overlaid sections is set to '0' or near to '0' by which it is filtered. But still, some unrelated sections are shown in the result.

To test the algorithm, the similarity scores between keywords from the text of five different crime reports and IPC knowledge sections were obtained as a sample. As a result, applicable IPC sections were present in all the crime reports. Along with these, some related sections were also shown with similarity scores. Keywords were found in 251 rows to identify 118 sections of chapters 14, 15, 16, and 22 in the IPC knowledgebase. Table 2 shows samples of input crime reports and Table 3 shows the summary result of all input crime report with fuzzy result.

Table 3. Sample of input crime report

Sample id	Sample content of Crime Report (A: Complaint registrant)
Input-1	I, A, was returning to my home on a bicycle when I met X, who was drunk, on the way. He abused and attacked me for some old thing, which caused serious injuries to my eyes and nose, and on the way he also threatened to kill me
Input-2	I A, want to request that on 15/01/2023 my neighbor X came to my house in an inebriated condition and started asking for a money loan. When I refused to give the money, he abused and threatened to beat me and left my house. Please take appropriate action
Input-3	I A, want to bring to your notice that today on 14/05/2022 my son has not returned home from school. We tried to find him everywhere but couldn't find him. In the evening I got a call from an unknown number that they have kidnapped my son and demanded 2 lack rupees from me. Has threatened to kill him for not giving the money or telling the police
Input-4	I A, would like to bring to your notice that I was admitted to X hospital on 25/06/2022 due to a worsening of my health. On the midnight of 27/06/2022 while I was sleeping I felt someone touching me. I saw a ward boy on his arm. Before I could say anything, he covered my mouth with his hands. I was not in a position to oppose him. He rape me in the same room and told me not to tell anyone. If I did not do so, I and my family were threatened with death
Input-5	I A, want to bring to your notice that I was married on 15/05/2021 to X, a resident of Vijaygarh, after marriage, when May became pregnant, my husband did not want to increase the family at that time, I explained to him a lot but he did not agree. Other people in my in-law's house also supported her. Those people fraudulently performed an abortion in a private hospital against my wishes in a state of unconsciousness, after which I became very mentally disturbed. I want to punish the people of my in-laws for their acts

Table 4 shows the top five Jaccard similarity, count vector, and TF-IDF results for the input crime reports, along with the actually applicable section. As seen in the result, there is no guarantee that these similarity algorithms will give the actual result in the top. Table 5 shows the result of similarity calculation with fuzzification. In this method, the actual result shows with higher fuzzy value.

The majority of the non-implemented IPC sections received a similarity value of 0, some received a score between 0 and 0.5 and 0.5 to 1, and very few received a score of 1, as shown in the results of Table 6. It is surprising that the relevant IPC sections received

Table 4. Top 5 result of jaccard similarity, count vector and TF-IDF Crime report/ applicable section

Crime Report with applicable section	Jaccard Similarity		Count Vector		TF-IDF	
	Section	Similarity	Section	Similarity	Section	Similarity
Input 1 320B 506	279	0.1176471	279	0.4743416	279	0.4628439
	279	0.1176471	283	0.4082483	283	0.3456661
	376C	0.0909091	376C	0.2041241	376C	0.2734777
	280	0.0833333	280	0.1767767	280	0.1768739
	283	0.0625	304B	0.1767767	304B	0.1569108
Input 2 506	Nil	Nil	Nil	Nil	Nil	Nil
	Nil	Nil	Nil	Nil	Nil	Nil
	Nil	Nil	Nil	Nil	Nil	Nil
	Nil	Nil	Nil	Nil	Nil	Nil
	Nil	Nil	Nil	Nil	Nil	Nil
Input 3 363 506	376C	0.0714286	376C	0.5773503	376C	0.6470112
	Nil	Nil	Nil	Nil	Nil	Nil
	Nil	Nil	Nil	Nil	Nil	Nil
	Nil	Nil	Nil	Nil	Nil	Nil
	Nil	Nil	Nil	Nil	Nil	Nil
Input 4 376D 506	376D	0.1052632	376D	0.5033415	376D	0.5033415
	375C	0.0952381	375C	0.3674911	375A	0.3756034
	375A	0.0555556	375A	0.3756034	375F	0.3756034
	375F	0.0555556	375F	0.3756034	376	0.3756034
	376	0.0555556	376	0.3756034	375C	0.3674911
Input 5 312	294	0.0588235	294	0.2886751	366	0.3026952
	376D	0.0588235	376D	0.2357023	376D	0.2887407
	304A	0.0555556	304A	0.2357023	304B	0.2462941

(*continued*)

a score of 1 in many cases and between 0.5 and 1 in others. *By harnessing the power of fuzzy logic, a sophisticated filtering process is implemented with the aim of effectively removing irrelevant sections. This meticulous filtration plays a pivotal role in elevating the quality and accuracy* of the decision-making process to a significant degree.

Table 4. (*continued*)

Crime Report with applicable section	Jaccard Similarity		Count Vector		TF-IDF	
	Section	Similarity	Section	Similarity	Section	Similarity
	315	0.0555556	315	0.2357023	304B	0.2411311
	366	0.0555556	366	0.2041241	294	0.2367957

Table 5. Result of similarity calculation with fuzzification

Input 1		Input 2		Input 3		Input 4		Input 5	
Section	Fuzzy Value	Section	Fuzzy Value	Section	Fuzzy Value	Section	Fuzzy Value	Section	Fuzzy Value
320G	1	294	0.5	294	1	506	1	312	1
320H	1	295A	0.5	278	1	351	1	376C	0.916667
321	1	506	0.5	508	1	352	1		
322	1	509	0.5	363	1	355	1		
323	1	320H	0.5	304A	0.916667	375A	1		
325	1	278	0.5	309	0.916667	375F	1		
326	1	370	0.5	315	0.916667	376	1		
351	1	503	0.5	364	0.916667	364	0.916667		
352	1					376C	0.916667		
506	1					375E	0.916667		
						294A	0.916667		
						377	0.916667		
						376D	0.916667		

Table 6. Observation Table of Similarity Calculation and Fuzzification

<table>
<tr><th>Sample</th><th>Applicable Section</th><th>Fuzzy Result</th><th>Applicability Ratio</th><th>Result</th></tr>
<tr><td>Input 1</td><td>320B
506</td><td><table><tr><th>Fuzzy Value</th><th>Section Count</th></tr><tr><td>0</td><td>180</td></tr><tr><td>0 to 0.5</td><td>49</td></tr><tr><td>0.5 to 1</td><td>12</td></tr><tr><td>1</td><td>10</td></tr></table></td><td>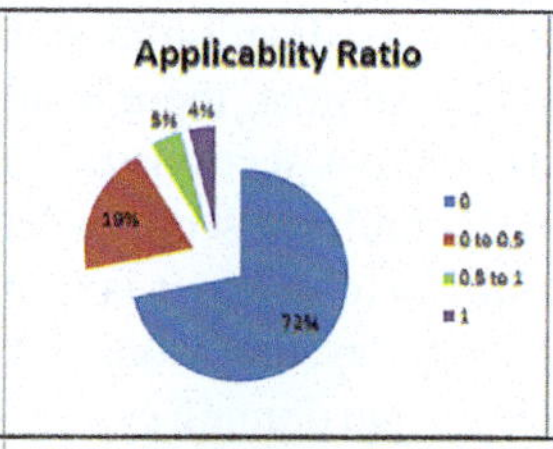</td><td>Section found</td></tr>
<tr><td>Input 2</td><td>506</td><td><table><tr><th>Fuzzy Value</th><th>Section Count</th></tr><tr><td>0</td><td>221</td></tr><tr><td>0 to 0.5</td><td>27</td></tr><tr><td>0.5 to 1</td><td>2</td></tr><tr><td>1</td><td>1</td></tr></table></td><td>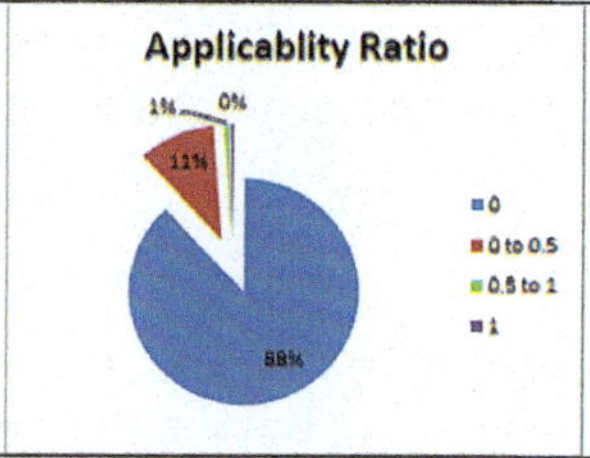</td><td>Section found</td></tr>
<tr><td>Input 3</td><td>363
506</td><td><table><tr><th>Fuzzy Value</th><th>Section Count</th></tr><tr><td>0</td><td>202</td></tr><tr><td>0 to 0.5</td><td>36</td></tr><tr><td>0.5 to 1</td><td>9</td></tr><tr><td>1</td><td>4</td></tr></table></td><td>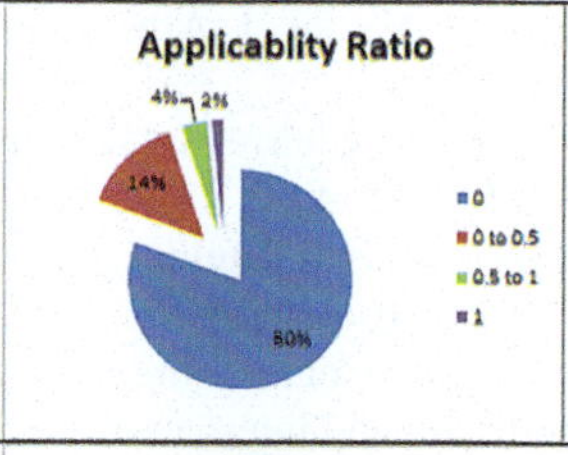</td><td>Section found</td></tr>
<tr><td>Input 4</td><td>376D
506</td><td><table><tr><th>Fuzzy Value</th><th>Section Count</th></tr><tr><td>0</td><td>213</td></tr><tr><td>0 to 0.5</td><td>25</td></tr><tr><td>0.5 to 1</td><td>6</td></tr><tr><td>1</td><td>7</td></tr></table></td><td>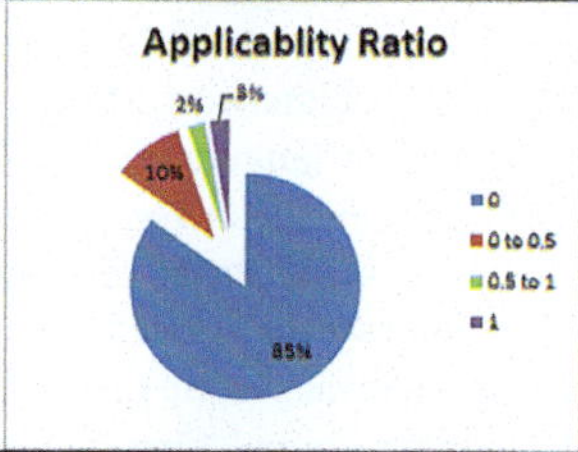</td><td>Section found</td></tr>
<tr><td>Input 5</td><td>312</td><td><table><tr><th>Fuzzy Value</th><th>Section Count</th></tr><tr><td>0</td><td>237</td></tr><tr><td>0 to 0.5</td><td>12</td></tr><tr><td>0.5 to 1</td><td>1</td></tr><tr><td>1</td><td>1</td></tr></table></td><td>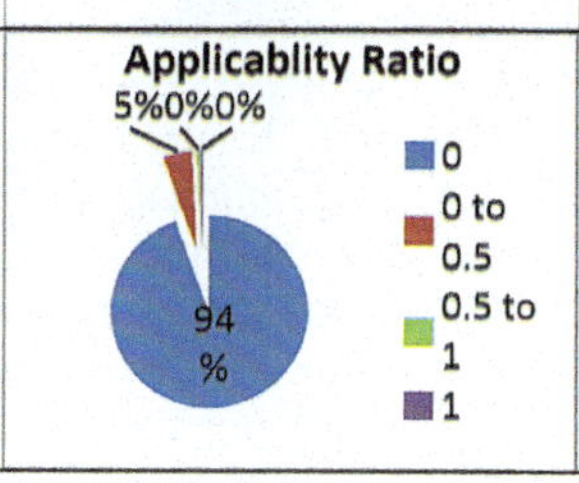</td><td>Section found</td></tr>
</table>

6 Conclusion

DSS for IPC Section using similarity algorithms and fuzzy logic has shown great promise in enhancing the efficiency and effectiveness of legal decision-making processes. By leveraging advanced technologies such as machine learning and fuzzy logic, this research has explored innovative approaches for analyzing crime reports and recommending appropriate IPC sections to legal professionals. As choosing the appropriate IPC sections requires experience-based cognitive insights. Sections of a new crime report can be acquired using classification models if a labeled data collection of crime reports is available. But at the moment, there is no such data set. Word embedding can be used to extract similarities across sections of a report, but this makes decision-making challenging because even unrelated sections have some degree of similarity, whereas linked sections have less of it. However, it is important to acknowledge that the development and deployment of such a system would require careful consideration of various factors, including data quality, legal domain expertise, and ethical considerations surrounding the automation of legal processes. Furthermore, the DSS should be continuously updated and refined to ensure its accuracy, reliability, and alignment with evolving legal frameworks.

References

1. The Indian Penal Code, 1860, India, URL: https://www.ncib.in/pdf/indian-penal-code.pdf
2. Srivastav, A., Prajapat, S.: Text similarity algorithms to determine Indian penal code sections for offense report. IAES Int. J. Artif. Intell. **11**(1), 34–40 (2022). https://doi.org/10.11591/ijai.v11.i1.pp34-40
3. Harode, B., Prajapat, S., Bhurre, S.: Text processor for IPC prediction. In: Singh, P.K., Singh, Y., Kolekar, M.H., Kar, A.K., Gonçalves, P.J.S. (eds) Recent Innovations in Computing. Lecture Notes in Electrical Engineering, vol. 832. Springer, Singapore (2022). https://doi.org/10.1007/978-981-16-8248-3_9
4. Selvi, K., Suresh, R.M.: Context similarity measure using Fuzzy Formal Concept Analysis. In: Proceedings of the Second International Conference on Computational Science, Engineering and Information Technology - CCSEIT '12 (2012). https://doi.org/10.1145/2393216.2393287
5. Putra S.J., Naf'an, M.Z., Gunawan, M.N.: Improving the scoring process of question answering system in Indonesian language using Fuzzy Logic," 2018 International Conference on Information and Communication Technology for the Muslim World (ICT4M), Kuala Lumpur, Malaysia, 2018, pp. 239–242.https://doi.org/10.1109/ICT4M.2018.00051
6. Muangon, W., Muangprathub, J., Saelee, J., Soonklang, T, Pongpinigpinyo, S., Sitdhisanguan, K.: An information retrieval system on thailand tourism community websites. In: Proceedings of the 2018 10th International Conference on Information Management and Engineering (pp. 101–105). https://doi.org/10.1145/3285957.3285968
7. Ayeldeen H., Hassanien A. E. and Fahmy A. A., "Fuzzy clustering and categorization of text documents," *13th International Conference on Hybrid Intelligent Systems (HIS 2013)*, Gammarth, 2013, pp. 262–266. https://doi.org/10.1109/HIS.2013.6920493
8. Sairaj, R.T., Balasundaram, S.R.: Improving the Cognitive Levels of Automatic Generated Questions using Neuro-Fuzzy Approach in e-Assessment. In: 2020 IEEE 5th International Conference on Computing Communication and Automation (ICCCA), Greater Noida, India, pp. 454–458 (2020). https://doi.org/10.1109/ICCCA49541.2020.9250716
9. Tirpude, S.C., Alvi, A.S.: Closed domain keyword based question answering system for legal documents of IPC sections & Indian laws. Int. J. Innov. Res. Comput. Commun. Eng. (2015)
10. Kamdi, R.P., Agrawal, A.: Keywords-based closed domain question answering system for Indian penal code sections and Indian amendment laws. Int. J. Intell. Syst. Appl. **7**, 57–67 (2015). https://doi.org/10.5815/ijisa.2015.12.06
11. Sangeetha, D., Kavyashri, R., Swetha, S., Vignesh, S.: Information retrieval system for laws. In: 2016 Eighth International Conference on Advanced Computing (ICoAC), Chennai, India, 2017, pp. 212–217. https://doi.org/10.1109/ICoAC.2017.7951772

12. Lyu, Y., et al.: Improving legal judgment prediction through reinforced criminal element extraction. Inf. Process. Manage. **59**(1), 102780 (2022). https://doi.org/10.1016/j.ipm.2021.102780
13. Shaikh, R.A., Sahu, T., Anand, V.: Predicting outcomes of legal cases based on legal factors using classifiers. Procedia Comput. Sci. **167**, 2393–2402 (2020). https://doi.org/10.1016/j.procs.2020.03.292

Author Index

N. Naik et al. (Eds.): UKCI 2023, AISC 1453, pp. 669-670, 2024.
https://doi.org/10.1007/978-3-031-47508-5

The manufacturer's authorised representative in the EU is Springer Nature Customer Service Centre GmbH, Europaplatz 3, 69115 Heidelberg, Germany. If you have any concerns regarding our products, please contact ProductSafety@springernature.com

Printed and bound by CPI Group (UK) Ltd, Croydon, CR0 4YY

01/10/2026

02225860-0019